CW01573013

HANDBOOK OF RESEARCH METHODS IN COMPLEXITY SCIENCE

Handbook of Research Methods in Complexity Science

Theory and Applications

Edited by

Eve Mitleton-Kelly

Director, Complexity Research Group, London School of Economics and Political Science, UK

Alexandros Paraskevas

Chair in Hospitality Management, University of West London, UK

Christopher Day

Senior Research Associate, Complexity Research Group, London School of Economics and Political Science, UK

Edward Elgar
PUBLISHING

Cheltenham, UK • Northampton, MA, USA

Published by
Edward Elgar Publishing Limited
The Lypiatts
15 Lansdown Road
Cheltenham
Glos GL50 2JA
UK

Edward Elgar Publishing, Inc.
William Pratt House
9 Dewey Court
Northampton
Massachusetts 01060
USA

A catalogue record for this book
is available from the British Library

Library of Congress Control Number: 2017947082

This book is available electronically in the **Elgar**online
Social and Political Science subject collection
DOI 10.4337/9781785364426

ISBN 978 1 78536 441 9 (cased)
ISBN 978 1 78536 442 6 (eBook)

Typeset by Servis Filmsetting Ltd, Stockport, Cheshire
Printed and bound by CPI Group (UK) Ltd, Croydon, CR0 4YY

Contents

Contributors

Peter Allen is Emeritus Professor at Cranfield University. He developed and ran the Complex Systems Research Centre in the School of Management at Cranfield University. He is still actively involved in ongoing research. He has a PhD in theoretical physics and from 1970 to 1987 worked with Professor Ilya Prigogine at the Université Libre de Bruxelles on research that led on to the development of complexity science. He is an Editor in Chief of *Emergence: Complexity and Organization*. He has written and edited several books in the field of complexity and socio-economic modelling and published well over 200 articles in a range of fields including ecology, social science, urban and regional science, economics, systems theory, and physics. In 2011 he co-edited the *Sage Handbook on Complexity and Management* and in 2015 co-authored *Embracing Complexity* (Oxford University Press) with Jean Boulton and Cliff Bowman. He has been a consultant to multiple governmental and non-governmental organizations, the United Nations University, the European Commission and the Asian Development Bank. He has managed a number of large European and UK research contracts.

Pierpaolo Andriani is Professor in Complexity and Innovation Management at Kedge Business School, France. He got his BA/MSc in Physics. He started his career as scientist in industrial R&D in Italy where he was project manager for various European research and development projects in the laser industry. Then he moved to academia and received his PhD in the social sciences from Durham University, UK. His research interests are focused on the impact of complexity and evolutionary theory on innovation. His research has been published in journals such as *Organization Science*, *Journal of International Business Studies*, *Research Policy*, *Long Range Planning* and *Complexity*.

Santo Banerjee was a senior research associate in the Department of Mathematics, Politecnico di Torino, Italy from 2009 to 2011. Currently he is working as an Associate Professor (UDQ9), in the Institute for Mathematical Research (INSPEM), University Putra Malaysia, Malaysia. He is also a founder member of Malaysia–Italy Centre of Excellence in Mathematical Science, UPM, Malaysia. His research is mainly concerned with nonlinear dynamics, chaos, complexity and secure communication. He has 15 books in the field of nonlinear dynamics and complex systems, and has also published more than 90 international research articles. Currently he is involved in many international and industry projects on signal processing, biomedical measures and devices, network and cryptography. He is an editor of *EPJ plus* (Springer) and other journals with regular and special issues. For more detail, please visit https://sites.google.com/site/santobanhome/.

Yaneer Bar-Yam is Founding President of the New England Complex Systems Institute. He has contributed to founding the field of complex systems science, introducing fundamental mathematical rigor, real world application, and educational programs for new concepts and insights of this field. In developing new mathematical methods and in their application, he has published on a wide range of scientific and real world problems ranging from cell biology to the global financial crisis. He is the author of two books:

a textbook *Dynamics of Complex Systems* (1997, Perseus Books), and a popular book *Making Things Work* (2005, Knowledge Press), describing the use of complex systems science for solving problems in military, healthcare, education, systems engineering, international development, and ethnic violence. Professor Bar-Yam received his SB and PhD in physics from MIT in 1978 and 1984 respectively. His recent work quantitatively analyses the origins and impacts of market crashes, social unrest, ethnic violence, military conflict and pandemics.

Patrick Beautement, formerly Research Director of The abaci Partnership LLP, has developed and implemented many practical solutions for taking advantage of the opportunities offered by real-world complexity. In this he has used a wide range of novel approaches, methods and modelling and simulation techniques – adapting them appropriately to the contexts of interest. Patrick has a successful track record of partnering with commerce, government and world-class researchers in multidisciplinary contexts and in developing toolsets which direct exploit insights from complexity research. Recent work has involved: carrying out a systematic analysis of techniques for decision-making in complex environments; developing and employing a framework for assessing the 'real-world-readiness' of agent-based modelling; examining the practicality of using causal and influence networks for exploring complex multi-agency / multi-stakeholder intervention situations; and implementing an adaptation tool for use by communities mitigating climate change in Africa.

Chase R. Booth graduated from Grinnell College in May 2016 with a degree in Classics, and he won the prestigious Thomas J. Watson Fellowship award for 2016. He is interested in classical art and literary representations of sex, gender, and ethnicity as well as the politics and identities of selfhood in classical antiquity.

Jane Bromley is a lecturer on the Digital and Technology Solutions degree apprenticeship programme at the Open University. Her PhD from Imperial College London was on human vision and her postdoctoral work at Bell Labs on neural networks for visual tasks. She now works with the Complex Systems and Natural Language Processing groups at the Open University and has a particular interest in the application of Deep Learning to understand human perception. She is currently applying hypernetworks to complex systems data from UK police forces. This research is funded by the Higher Education Funding Council for England.

Hannah L. Brown graduated from Grinnell College in May 2016 with a degree in Gender, Women's and Sexuality Studies. At Grinnell, she has worked on a variety of research opportunities in interdisciplinary studies, such as complexity science and digital humanities. She aims for a career in Policy Analysis.

Julian Burton is Director of Delta7 Change Ltd. As a consultant, facilitator and visual artist, he is passionate about improving the quality of dialogue in organizations, where he believes most change happens. Visuals can be a powerful way to ground the meaning of words in everyday experience, supporting people to make better sense of change through having better conversations that can build the relationships needed for organizations to be more human, effective and successful. Julian was inspired to develop Delta7 in 1997 after first learning about complexity theory at the NECSI conference in 1997 and then through being an artist in residence at the LSE Complexity Research Group from 1998–2004.

Delta7 has since grown into a change consultancy that has partnered with many organizations such as Coutts Bank, BAE Systems, Rolls Royce, NHS, BT, MOD and Network Rail who wanted a highly innovative approach to employee engagement, culture change and leadership development. Their visual dialogue process helps to bridge the gap between strategy and implementation by creating space for employees and leaders to come together to make sense of change, connect strategy to action and take ownership of the parts each has to play in successful implementation. Prior to 1997 Julian had a career as a medical illustrator, landscape painter and web designer. His nine-year-old son has described his work best: "my dad makes pictures of problems so people can talk about them".

Giuseppe Carignani is an engineer and a professor at Technical High School Malignani (Udine, Italy, EU). He holds a Master's degree in civil engineering and a PhD in industrial and information engineering. He is presently (2017) a PhD candidate for a second PhD in management at the University of Udine.

He had an eclectic career as a consultant in diverse areas of engineering and management, including industrial and civil design and prototyping, software development, New Product Development, supply chain management, innovation management, and quality management.

His research interests are focused on the evolutionary theories of technological innovation and their implications for management and organizations, the practice of web-enabled open innovation, and teaching technological creativity. His research is published in peer-reviewed edited books, proceedings and journals, including the *Academy of Management Proceedings*, the *International Journal of Entrepreneurship and Innovation Management*, and *Research Policy*.

Brian Castellani is Professor of Sociology and Head of the Complexity in Health and Infrastructure Group, Kent State University, USA (cch.ashtabula.kent.edu). His work focuses on the development of computational and complexity science methods and their application to health, healthcare and infrastructure. Topics include allostatic load, community health and resilience, grid reliability, health trajectories and well-being across time, medical professionalism and addiction. He is specifically known for his development of the SACS Toolkit, a case-based, mixed-methods, computationally-grounded platform for modelling complex systems. For more information, see www.personal.kent.edu/~bcastel3/brian%20castellani.html.

G. Christopher Crawford is an Assistant Professor of Strategy and Entrepreneurship at Ohio University in Athens, Ohio. After working in industry for two decades and founding his own consulting company, Dr Crawford received his PhD in entrepreneurship from the University of Louisville in 2013. His dissertation, "Causes of Extreme Outcomes in Entrepreneurship," received a $20,000 Kauffman Dissertation Fellowship Award. Dr Crawford's recent work has been published in the *Journal of Business Venturing*, *Small Business Economics*, and *Journal of Management Inquiry*. His research interests include: new venture strategy, creation, and growth; complexity science and agent-based simulation modelling applications to entrepreneurship; and understanding the mechanisms that generate power law distributions and outliers in social systems. Dr Crawford won the 2015 National Federation of Independent Business Award at the Babson College Entrepreneurship Research Conference and was the keynote speaker at the 2016 Strategic Management Society Conference extension in Bolzano, Italy, with a presentation titled "The Emergence of Outliers in Social Systems: A Rockstar Theory." His $205,000 National Science Foundation grant will use the maximum-likelihood estimation methods outlined in this book, along with agent-based modelling, to build and test a universal theory about the emergence of new order (for example, outlier ventures, breakthrough technology, disruptive innovations, or extreme events).

Christopher Day currently works at the United Nations in New York to support peace-keeping field missions and political field missions. His previous assignments include work in Uganda and Afghanistan in collaboration with the United Nations, the German Gesellschaft fuer Internationale Zusammenarbeit and the German Centre for International Migration and Development. Prior to entering the field of peace keeping and international development, he worked in the private sector as a managing editor, communications expert and consultant. Christopher holds an Executive Master's degree in consulting and coaching for change – a programme jointly run by HEC Paris and Saïd Business School, University of Oxford. His areas of interest are social innovation, leadership and organizational change. He is Senior Research Associate in the Complexity Research Group at the London School of Economics and Political Science, UK.

Carl J. Dister is a Principal Systems Engineer at ReliabilityFirst with 30 years of systems engineering experience. Mr Dister's experience is in the design and analysis of highly reliable electromechanical devices and complex systems. Mr Dister is a graduate of Cleveland State University with a Bachelors of electrical engineering and University of Wisconsin-Madison with a Masters of electrical engineering and is a Certified Systems Engineering Professional (CSEP) with INCOSE.

Robin Durie is Senior Lecturer in Politics at the University of Exeter. His early research focused on the phenomenology of time. He edited and co-translated an edition of Bergson's notorious book on special relativity theory, including the debate in Paris between Bergson and Einstein. He then edited a collection of essays by theoretical physicists and philosophers looking at fundamental problems of time and temporality. This formed the basis for his work on complexity theory, where his research has focused on fundamental ontological questions concerning the relational dynamics of complex systems. His work has subsequently contributed to projects ranging from transformative community regeneration, to creating a culture of co-creative publicly engaged research, to the evolution of culture within societies of robots.

Elizabeth G. Eason graduated from Grinnell College in May 2017 with a degree in Mathematics and Statistics. She worked on a variety of mathematical-based research projects centred around topics such as gender and epistemology. She now works as a Pricing Analyst on the Research & Development team at Nationwide Mutual Insurance Company in Des Moines, Iowa, USA.

Krista M. English MBA PhD(c), is the Co-Principal Investigator of the Complexity Science Lab at the School of Population and Public Health in the Faculty of Medicine at the University of British Columbia and the Associate Director of the WHO Collaborating Centre for Complexity Science for Health Systems. Her interests focus on using complexity science to understand how research knowledge is translated for health system decision-making. Ms English has worked at the British Columbia Centre for Disease Control since 2005 and most recently as Associate Director of the Centre's Division of Mathematical Modeling. She is an applied researcher whose work in health systems and policy design has supported the advancement of the discipline through innovative programme design and strategic collaborations. Her research interests more generally relate to public health policy design, health promotion, knowledge translation, ethics, equity, social determinants of health, global health, and topics at the intersection of complex systems, health systems and decision-making.

Joyce Fortune is Emeritus Professor of Technology Management at the Open University, UK. Her main teaching and research interest is the use of systems thinking to promote learning from failure and to predict failures, particularly in the area of project management. She has published three books and numerous papers on systems failures. She has also undertaken research in other topics such as technology adoption in healthcare and taught quality management for more than two decades. She is currently working on a piece of research funded by the Higher Education Funding Council for England that is using systems thinking and complexity science to look at the policing of organized crime.

Michael Gabbay is a Senior Principal Physicist in the Applied Physics Laboratory at the University of Washington. His current research involves the development and application of mathematical models and computational simulations of network dynamics, focusing on social and political systems. Dr Gabbay's work seeks to advance both basic and policy-relevant research. He has conducted empirical research using political rhetoric, human subjects experiments, and analyst input and has applied his models and methods towards understanding and anticipating the behaviour of real-world militant networks and government leadership groups. His publications on political networks have appeared in both academic and policy-oriented venues. Dr Gabbay has also conducted research in the dynamics of nonlinear oscillators, nonequilibrium pattern formation, and signal processing. He received his PhD in physics from the University of Chicago and a BA in physics from Cornell University.

Jeffrey A. Goldstein, PhD, is a Full Professor at the Willumstad School at Adelphi University, New York, USA. Professor Goldstein is well-known as a pioneer in the application of the sciences of complex systems to social and organizational dynamics. He also does research and publishes in the areas of the philosophy of science and the philosophy of mathematics. Since 2004, he has been an editor-in-chief of the only journal fully devoted to these themes, *Emergence: Complexity and Organization*. He is also one of three trustees of the empirical research, complexity-based journal *Nonlinear Dynamics, Psychology, and Life Sciences*. Professor Goldstein is the author/editor of 17 books, and hundreds of book chapters, scholarly monographs, and journal papers. Professor Goldstein has been invited to lecture and hold seminars/courses/workshops at major universities and institutions throughout the world. He is also a consultant to many public and private organizations in the US and abroad.

James K. Hazy is Full Professor at Adelphi University in New York where for ten years he has been teaching entrepreneurship, strategy, and leadership. He has published over 50 articles in such venues as *The Leadership Quarterly*, *Leadership*, and *Nature's Frontiers in Psychology* winning numerous awards including two *Academy of Management Best Paper Awards* and the *Bender Body-of-Work Award*. He co-wrote the Amazon Top-100 Management Science book: *Complexity and the Nexus of Leadership* (2010, Palgrave Macmillan) and co-edited two other volumes.

For 25 years before joining the academy, Dr Hazy held leadership positions in industry including EVP for an Ernst & Young business unit and Financial VP at AT&T. He was also a former Board member and treasurer for Goodwill Industries International. He has earned a doctorate with "distinguished honors" from George Washington University, an MBA "with distinction" from The Wharton School, and a BS in mathematics from Haverford College.

Kate Hopkinson is Director of Inner Skills Services Ltd, a company she set up in 1995, which offers management and organization development consultancy drawing on her innovative methodology, *Landscape of the Mind*. She has worked with a wide range of household names, from blue chip commercial companies, to the public sector and not-for-profit organizations. Kate's academic background is in psychology. She also holds a Master's degree in management learning. Her publications include two papers in *The British Journal of Psychiatry*, as well as articles in management and business journals. From 2000 to 2015 she was a senior research associate with the Complexity Research Group at London School of Economics, where she contributed to projects and workshops, and gave seminars.

Kate's film on *Landscape of the Mind, An Introduction to Inner Complexity*, is available via her website (www.innerskills.co.uk). Kate's expertise centres around leadership, especially navigating successfully in uncertainty and constant turbulent change; and reaping the Divergence Dividend – which involves recognizing and utilizing significant hidden assets. A member of the British Psychological Society, she has been a designated expert on cognitive science for the European Union. Kate is currently taking Inner Skills to the next level, setting up hubs – centres of excellence in applying *Landscape of the Mind* – in the commercial, public, and not-for-profit sectors. She will be training consultants to work via these hubs as *Landscape of the Mind* Facilitators and Practitioners, in their respective fields.

Nathaniel Hupert, MD, MPH, is a practicing clinician and public health researcher focusing on healthcare processes and emergency response logistics. He is an Associate Professor of Healthcare Policy and Research and of Medicine at Weill Cornell Medicine, Cornell University. Using a variety of methods including Process Mining and Discrete Event Simulation, his research seeks to improve the effectiveness of care delivery in both conventional and crisis settings. He led the development of two US healthcare planning documents, the "Community Guide for Public Health Preparedness" (2004) and the "Guidebook for Hospital Preparedness Exercises" (2010). Dr Hupert was the founding Director of the Preparedness Modeling Unit at the US Centers for Disease Control and Prevention (CDC) (2008–2010). Currently, in addition to his academic position, he is Senior Advisor for both the CDC's Division of Preparedness and Emerging Infections and the US Department of Health and Human Services (DHHS) National Healthcare Preparedness Program.

Emily S. Ihara is an Associate Professor of Social Work at George Mason University. Her research focuses on the social determinants of health inequities across the life course, particularly for older adults, racial and ethnic minorities, and other underserved populations. She is actively engaged in implementing intervention studies for individuals with Alzheimer's disease and related dementias, using agent-based modelling to address the complexity of caregiving, and assessing the role of health inequities for dialysis patients. She uses multiple quantitative and qualitative methods in her work, and has gained a solid foundation in agent-based modelling through her selected participation in the Institute of Systems Science and Health, sponsored by the Office of Behavioral and Social Science Research (OBSSR) and the National Institutes of Health (NIH). Dr Ihara is a graduate of UC Berkeley (AB), UCLA (MSW), and The Heller School for Social Policy and Management at Brandeis University (PhD, MA).

Damian G. Kelty-Stephen received his PhD in psychology from the University of Connecticut-Storrs and continued his work as postdoctoral fellow at Harvard University's Wyss Institute for Biologically Inspired Engineering. He is currently an Assistant Professor in Psychology at Grinnell College where he had formerly served for four years as visiting faculty.

William G. Kennedy is a Research Assistant Professor with the Krasnow Institute for Advanced Study at George Mason University working in the areas of cognitive science and computational social science. He has earned a BS from the US Naval Academy (majoring in mathematics), a MS from the Naval Post Graduate School (in computer science), and a PhD from George Mason University (in artificial intelligence). He is also a retired Navy Captain (30 years of service in submarines and the Naval Reserve) and a retired civil servant (Nuclear Regulatory Commission and Department of Energy). He joined the Mason faculty in 2008 after a postdoctoral fellowship at the Naval Research Laboratory in cognitive robotics and human–robot interaction. His research interests focus on advancing computational cognitive modelling of human interactions with other intelligent agents and on integrating cognitive science and cognitive modelling into computational social science.

Sohee Kim received her PhD in public administration from the School of Public Affairs at the Pennsylvania State University at Harrisburg. Her research interests are in policy networks, governance networks, health policy implementation, and partnerships and networks in healthcare. She co-authored an article (2013 *Administration & Society*) about the urban governance network in Center City, Philadelphia and an article (2011 *NITTE Management Review*) about the effectiveness of disaster management networks in Gujarat, India.

Lesley Kuhn is an Adjunct Fellow with the School of Business at Western Sydney University, Australia, following her retirement after 27 years as an academic. Dr Kuhn's applied philosophical work concerns exploration of the contours of an epistemology that privileges curiosity, kindness and humility over foundational, fundamental and totalitarian habits of thought. With degrees in music, education, environmental science and philosophy and her doctoral work focusing on the nature of epistemology and belief, Dr Kuhn's transdisciplinary antecedence informs her research into how the complexity sciences might be drawn upon to improve understanding of the lived human condition.

Dr Kuhn has, over the past 20 years, been engaging complexity informed habits of thought in philosophical and social inquiry. Along with writing *Adventures in Complexity for Organisations Near the Edge of Chaos* (2009, Triarchy Press), Dr Kuhn has edited numerous volumes, authored more than 40 book chapters and published papers, and led more than 30 research projects.

Henrik Jeldtoft Jensen is a Professor of Mathematical Physics and Leader of the Centre for Complexity Science at Imperial College London. He works on the statistical mechanics of complex systems. He has worked on the dynamical properties of condensed matter systems and developed the Tangled Nature model of evolving ecosystems, which is currently used to develop the Tangled Finance approach. His two books on complexity science *Self-organized Criticality* (1998, Cambridge University Press) and *Stochastic Dynamics of Complex Systems* (with Paolo Sibani, 2013, ICP) has attracted very broad interest. Jensen has more recently worked on brain dynamics and structure by analysing fMRI and EEG data and he is involved in a project with the Guildhall School of Music and Drama concerning quantitative analysis of EEG time series from improvising or non-improvising classical musicians.

Jeffrey Johnson is Professor of Complexity Science and Design in the Faculty of Mathematics, Computing and Technology at the Open University. His research includes applications in social and economic systems, business, artificial intelligence and robotics, machine vision and neural systems, and city planning and large road traffic systems. His research into social systems is situated in policy – which he sees as designing, planning and managing the future. He has worked in mathematics, geography, design, and engineering departments and believes the same mathematical theory of multilevel systems is required for social, economic, physical and engineering systems. His current work includes developing an operational theory of narratives that can underpin computational policy support systems. He is a chartered mathematician and a chartered engineer, Executive Committee member of the Complex Systems Society, and a Board member of the UNESCO UniTwin Complex Systems Digital Campus that federates scientific research and teaching worldwide.

Benyamin Lichtenstein PhD, is Associate Professor of Entrepreneurship and Management at UMass Boston. His research specialty is the study of emergence – the creation and re-creation of new ventures, organizations, and collaborations – which he explores through complexity science. He has published four books and more than 50 articles and chapters on this theme, as well as on sustainability, entrepreneurship, and organizational change. Benyamin is a core faculty member in the Organizations and Social Change PhD Track at the College of Management; is Academic Director of the Entrepreneurship Center at UMass Boston, and is a Research Fellow at the Center for Sustainable Enterprise (SERC).

Craig Lundy is a Senior Lecturer in Social Theory at Nottingham Trent University. Prior to commencing at NTU Craig held a three-year Research Fellowship at the Institute for Social Transformation Research, University of Wollongong (Australia). He was previously employed in various teaching and research positions at the University of Exeter, Royal Holloway (University of London) and Middlesex University. The majority of Craig's research is concerned with processes of transformation – an interest that he has pursued through cross-disciplinary projects that explore and make use of developments in complexity studies, socio-political theory and 19th/20th century European philosophy. He is the author of *History and Becoming: Deleuze's Philosophy of Creativity* (2012), *Deleuze's Bergsonism* (forthcoming) and co-editor with Daniela Voss of *At the Edges of Thought: Deleuze and Post-Kantian Philosophy* (2015), all published by Edinburgh University Press.

Bill McKelvey (PhD, MIT, 1967) is Professor Emeritus: Strategic Organizing, Complexity Science, and Econophysics at the UCLA Anderson School of Management. His book, *Organizational Systematics* (1982, University of California Press) remains the definitive treatment of organizational taxonomy and evolution. He chaired the Building Committee that produced the $110,000,000 Anderson Complex. He has directed over 170 field study teams on six-month projects concerned with strategic and organizational improvements to client firms; co-edited *Variations in Organization Science* (with J. Baum, 1999, Sage), a Special Issue of *Emergence* (with S. Maguire, 1999), and a Special Issue of *Journal of Information Technology* (with J. Merali, 2006); edited a Special Issue of *International Journal of Complexity and Management* (2013); co-editor of *Sage Handbook of Complexity and Management* (2011, Sage); editor of a Routledge Major Works Series: *Complexity Concepts* (2012, Routledge; five-volumes, 2447 pages). He has published 95 articles and chapters since 1996 applying complexity-, power-law-, and econophysics-related topics to organization science and management.

Eve Mitleton-Kelly was Founder and Director of the Complexity Research Group, London School of Economics and Political Science, London, UK (1994–2014); is Senior Fellow in LSE IDEAS, Centre for the Study of International Diplomacy, External Affairs and Strategy; Visiting Professor at the Open University; Fellow at Cambridge University's Department of Engineering; Member of the World Economic Forum's Global Agenda Council on Complex Systems (2012–2014).

EMK has had two careers. One in the British Civil Service, Department of Trade and Industry (1967–1983) and another as an academic at the London School of Economics (since 1988). At the DTI she was involved in the negotiation of European Directives on behalf of the UK Insurance Industry and in international affairs. Earlier in her career she was involved in the writing of the first report on Computers and Privacy in the 1970s.

She has been Policy Advisor to European and USA organizations, the European Commission, the UN, several UK government departments; Scientific Advisor to six government administrations in developing a framework of governance for government based on complexity theory. She has led and participated in 49 research and other projects. Has edited and co-authored five volumes and published extensively on the application of complexity science principles in organizations and institutions. EMK is an international speaker and teacher with over 600 presentations worldwide. She has developed a methodology based on complexity science to address difficult, complex problems by identifying the multidimensional problem space and co-creating an Enabling Environment. Researchers and practitioners around the world are being trained by EMK, to use the methodology to address organizational, societal and global problems.

Sam Mockett was introduced to complexity theory when he started working for change management consultancy Delta7 Change Ltd in August 2013. He previously worked in care for adults with learning difficulties before completing a postgraduate degree in philosophy and public policy at the LSE in 2012. His experience in the care sector exposed him to the difficulties of coping with a complex policy environment while attempting to ensure that a number of individuals with a diverse range of needs were provided for within a fractured systemic framework. While at Delta7 Sam learned this type of problem was not limited to care provision. He joined Delta7's sister company, Visual Meaning Ltd, in July 2015 and is still looking for a solution.

Göktuğ Morçöl is a Professor of Public Policy and Administration at Penn State Harrisburg. His research interests are complexity theory, metropolitan governance, business improvement districts, and research methodology. He has authored, edited, or co-edited the following books: *Challenges to Democratic Governance in Developing Countries* (2014, Springer), *A Complexity Theory for Public Policy* (2012, Routledge), *Complexity and Policy Analysis* (2008, ISCE Publishing), *Business Improvement Districts* (2008, Wiley), *Handbook of Decision Making* (2007, Taylor & Francis), *A New Mind for Policy Analysis* (2002, Praeger), *New Sciences for Public Administration and Policy* (2000, Chatelaine Press). His articles have appeared in *Public Administration Review, Administration and Society, Administrative Theory and Praxis, Policy Sciences, Public Administration Quarterly, Politics and Policy, International Journal of Public Administration, Journal of Urban Affairs, Emergence*, and others. He is an editor-in-chief of the journal *Complexity, Governance & Networks*. He served as the founding Chair of the Section on Complexity and Network Studies of the American Society for Public Administration.

Sayan Mukherjee received his MPhil and PhD degree in pure mathematics from Calcutta University (India). He has been working as an Assistant Professor at the Department of Mathematics at Sivanath Sastri College, Kolkata (India) since 2006. He is author and co-author of more than 20 scientific articles in international/national journals, books and proceedings of the conferences with reviewing committee. His research interests include complexity theory, complex networks theory and fractal and multi-fractal theory in nonlinear economical, biomedical and musical time series analysis.

Sanjay Kumar Palit received his PhD degree in pure mathematics from Calcutta University (India). He has been working as an Assistant Professor of Mathematics at Calcutta Institute of Engineering and Management, Kolkata (India) since 2004. He has also worked as Guest Faculty in the Department of Electronics and Instrumentation, Jadavpur University. He is author and co-author of more than 30 scientific articles in international/national journals, books and proceedings of the conferences with reviewing committee. His research interests include dynamical systems, nonlinear time series analysis, complexity analysis, complex networks, time and frequency domain analysis of biomedical and music signals, fractals and multi-fractals, phase space analysis through recurrence plot and high dimensional data analysis.

Alexandros Paraskevas (BSc, MSc, PhD, CertTHE) is Professor in Strategic Risk Management and Chair in Hospitality Management at London Geller College of Hospitality and Tourism, University of West London. Alexandros researches the governance and management of risks/crises both in an organizational and tourism destination context and he often uses complexity theory principles in his investigations. He has led numerous hotel industry projects in the areas of risk, crisis, disaster management and business continuity and authored several academic articles and book chapters on these topics.

A visiting scholar in Austria, Finland, Hong Kong, Mexico, Spain and Taiwan, Alexandros has worked with governments and tourism professional associations on safety and security issues and on crisis communications strategies. He has served as advisor of the International Hotel and Restaurant Association's (IH&RA) Global Council on Security, Safety and Crisis Management and is a member of ASIS (the American Society for Industrial Security professionals) and HEAT (ASIS' Council for Hospitality and Tourism). His work on risk management and business continuity with InterContinental Hotels Group (IHG) has been honoured by the Institute of Risk Management (IRM) with the 2013 Global Award for Best Partnership.

Babak Pourbohloul, PhD, is trained as a theoretical physicist (chaos theory and nonlinear dynamics) and has been active in the development and application of complex quantitative methods in public and global health systems policy design since 1999. He is an Associate Professor and Principal Investigator of the Complexity Science Lab, at the School of Population and Public Health, Faculty of Medicine, University of British Columbia. He was the founding Director of the Division of Mathematical Modeling, British Columbia Centre for Disease Control (2001–2016). He has been the Principal Investigator on several international modelling projects in the application of mathematical and computational tools to mitigate emerging infectious disease outbreaks and was designated as Director of the *World Health Organization Collaborating Centre for Complexity Science for Health Systems (CS4HS)*. He aims to develop and employ methods of complex systems analysis, through multidisciplinary collaborations, to optimize health policy design at the local, national and international levels.

Rajeev Rajaram is an Associate Professor of Mathematics and a member of the Complexity in Health and Infrastructure Group, Kent State University, USA (cch.ashtabula.kent.edu). His primary training is in control theory of partial differential equations and he is currently interested in applications of differential equations and ideas from statistical mechanics and thermodynamics to model and measure complexity. For more information, see www.kent.edu/math/profile/rajeev-rajaram.

Fatimah Abdul Razak has obtained her PhD from the Imperial College London in the area of complex systems and networks. Prior to that she did the Mathematical Tripos Part III at Cambridge. She is currently a Senior Lecturer at the School of Mathematical Sciences, Universiti Kebangsaan Malaysia (UKM). Her PhD research was on investigating information theoretic measures and its usage for determining directionality on complex systems. She continues to do research on the different information theoretic measures in relation to complex systems and collective behaviour. She is also currently doing research on synchronization of Malaysian fireflies as well as looking at networks related to Malaysian Twitter data and Malaysian meteorological data using persistent homology.

Kurt A. Richardson is an engineer, physicist and publisher. As an engineer he has built data-driven web-based applications and has designed microchips for many companies including DIRECTV, Panasonic, Thuraya, SES, Lockheed Martin, SLAC, General

Dynamics, and NASA. He was also a Senior Systems Analyst for NASA on their Gamma-Ray Large Area Space Telescope (GLAST, now Fermi). As a physicist he is an active researcher in fundamental complexity theory with a particular interest in the nature of boundaries, and the relationship between form and function in complex dynamical networks. He is currently leading a project to develop customized processing chips (graph processing units) to support dynamic ultra-fast graph-based analysis. As a publisher he owns and runs Emergent Publications which specializes in publishing academic works concerning complex systems thinking. Its flagship publication is the international journal *Emergence: Complexity & Organization*.

J. Rowan Scott is an Associate Clinical Professor of Psychiatry with the University of Alberta. He is a psychotherapy supervisor in the University of Alberta Psychiatric Residency Training Program and teaches in the area of family therapy, conscious studies and interpersonal neuroscience. Dr Scott has an active outpatient general psychiatry practice with an interest in psychopharmacology, individual, couples and family therapy. He completed his medical and psychiatric training at the University of Alberta, and a postgraduate Fellowship in the area of individual, couples and family psychotherapy as well as sex therapy at McMaster University. The study and application of system models and complexity theory in psychiatric and psychotherapy contexts is a long-standing interest. The relationship between consciousness and the limits of reductive natural science has become an important focus of ongoing work. The collaboration with Dr Yakov Shapiro is a much-valued personal relationship.

Yakov Shapiro is a Clinical Professor of Psychiatry and Psychotherapy supervisor at the University of Alberta. He obtained his MD with Honors in research from the University of Alberta in 1988; a certificate of training from the Menninger Foundation in Topeka, Kansas in 1991; and became Fellow of the Royal College of Physicians in 1993. He runs an outpatient practice specializing in integrated individual/group psychotherapy and psychopharmacology, and teaches postgraduate seminars in neuroscience of psychotherapy, evolutionary psychiatry, psychodynamic psychopharmacology, and dynamical systems approaches to psychiatric formulation and treatment. He has given courses and workshops for the Alberta, Canadian and American Psychiatric Associations, American Academy of Psychoanalysis and Dynamic Psychiatry, International Association of Relational Psychoanalysis and Psychotherapy, and Society for the Exploration of Psychotherapy Integration. He is a member of the APA Psychotherapy Caucus and CPA Research Network. He has been working on complexity applications to psychiatric diagnosis, treatment and training for the past 15 years.

Jonathan Stead (FRCGP) has recently retired after 30 years as a GP. He has held an academic post as a Senior Clinical Research Fellow with University of Exeter Medical School since 1980. His special interest is in prevention and treatment of long-term conditions, particularly diabetes. He has worked with Hazel Stuteley for over a decade and together with research Fellows from the Health Complexity Group, they developed the 'Connecting Communities' programme, known as C2, using insights from complexity science as the theoretical framework to understand community transformation.

Jonathan has always mixed clinical time with health management, working in the Strategic Health Authority, and more recently Clinical Commissioning Groups. He also works with the Royal College of General Practitioners to promote the C2 approach to community transformation, trying to encourage CCGs to commission this way of working to improve the health of those who are most disadvantaged. He is currently the national Co-Director of the C2 Learning Programme (drjwstead@gmail.com).

Hazel Stuteley: Following nurse training at Kings College Hospital London, Hazel was a practicing Health Visitor for 25 years. In 1995 she led the multi-award winning Beacon Project, reversing the decline of a deeply disadvantaged community in Falmouth. Following a year's secondment with the Department of Health in 2001, she accepted a Senior Research Fellowship at the University of Exeter Medical School and became a founder member of the Health Complexity Group, which uses insights from complexity science to understand and enable transformative community change.

Hazel is currently Director of the C2 Connecting Communities Programme at the Institute of Health Service Research, Exeter University. A recent *Health Service Journal* awards ceremony at Royal College of Physicians, listed Hazel within the top 100 most influential and innovative clinical NHS leaders in healthcare, England. An active Executive Board member of the NHS Alliance, Hazel was appointed OBE in 2001 for services to the community of Falmouth (h.stuteley@exeter.ac.uk).

Andrew Tait is Chief Technology Officer of Decision Mechanics. Decision Mechanics specializes in decision science and the design/development of related technology. With a background in artificial intelligence and operations research he is currently involved in helping organizations make use of machine learning to tease insights from their data lakes. Prior to founding Decision Mechanics, Andrew held a range of posts in business, government and academia. He blogs at decisionmechnanics.com and tweets in bursts at @decisionmech.

Catherine J. Tompkins joined the faculty at George Mason University in 2003. She is currently an Associate Professor in the Department of Social Work, coordinator for the gerontology programs and the Assistant Dean for Undergraduate Studies in the College of Health and Human Services. Dr Tompkins teaches courses in both the undergraduate and graduate social work programmes. Her research areas include: family caregiving, interventions for dementia care, kinship care, and emotional well-being and retirement. She is currently on the editorial board for three journals and serves on other local and national committees, including the Fairfax County Kinship Care Committee. Dr Tompkins is a John A. Hartford Faculty Scholar in Geriatric Social Work. She received both her MSW and PhD from the University of Maryland School of Social Work.

Liz Varga is Professor of Complex Infrastructure Systems and Director of the Complex Systems Research Centre at Cranfield University, UK. Professor Varga has expertise in transdisciplinary and interdependent infrastructure systems (energy, transport, water, waste and telecoms) via her research projects, engagements with practice and feedback from dissemination. The evolution of infrastructure systems and the connection to food and housing systems are a key focus, providing nexus challenges for policy and technology in increasingly stressed cities. She applies both qualitative and quantitative methods to implement agent-based models to explore potential futures within different scenarios and governance regimes, and to assess the impact of infrastructure systems on economies, society, the environment and more generally sustainability. In this context, she has been teaching mixed methods research to doctoral students since 2009, encompassing broad questions of research design, philosophy, and paradigm, as well as focused and targeted questions about methods and data integration, skills requirements, and critical evaluation of knowledge.

Xiaogeng Wan is a lecturer in the Department of Mathematics at Beijing University of Chemical Technology, China. She did her postdoc at the Mathematical Sciences Department at Tsinghua University, China and obtained her PhD in Mathematics from Imperial College London in January 2015 with a thesis titled "Time Series Causality Analysis and EEG Data Analysis on Music Improvisation". Her research interest is in time series causality analysis and data analysis protein structural classification and developing geometric methods on proteins.

Peter R. Wolenski is the "Russell B. Long" Professor of Mathematics at Louisiana State University (LSU) in Baton Rouge, Louisiana. He has published over 60 peer-reviewed research articles in a variety of mathematics journals, and collaborated on many other papers of an interdisciplinary nature. He co-authored the books *Nonsmooth Analysis and Control Theory* with F.H. Clarke, Yu. Ledyaev, and R.L. Stern (1998, Springer), and *Linear Mathematical Models in Chemical Engineering* with M. Hortso (2010, Wspc), and has co-edited two others.

Dr Wolenski received his PhD in mathematics from the University of Washington in 1988 under the supervision of R.T. Rockafellar. He held postdoctoral positions at Imperial College London, the International Institute for Advanced System Analysis (IIASA) (Vienna), and Centre de Recherches de Matematiques (Montreal) before beginning his tenure at LSU in 1990, where he has been happily ensconced to the present day.

Michael E. Wolf-Branigin is Chair and Professor of Social Work at George Mason University. His research focuses on complex adaptive systems and their application to social work practice in relation to his substantive areas including behavioural health and intellectual and developmental disabilities. He worked for two decades in the addictions and disabilities fields. He received his graduate diploma in economics from the University of Stockholm, an MSW from the University of Michigan, and a PhD in research and evaluation from Wayne State University. He has consulted for governmental and non-governmental organizations in the United States and abroad. For 18 years he has been an accreditation surveyor with CARF International. He is an Accredited Professional Statistician™ through the American Statistical Association, has written about 50 peer-reviewed articles and one book, and serves on the editorial boards and reviews for several academic journals.

Katrina Wyatt is Professor of Relational Health and part of the Child Health and Health Complexity Groups at the University of Exeter Medical School. Her research covers three principal areas; the development and evaluation of complex system-based programmes, supporting and enabling patients, service users and communities to be active partners in research projects and in the provision of methodological expertise for evaluations of public health programmes. Ongoing research projects include looking at how the conditions for transformational change are created such that the changes are sustained, creating a culture of publicly engaged research across the University of Exeter and the development and trialling of a novel childhood obesity prevention programme. The research is informed by complexity science and participatory methods are used in the design and delivery of the research such that service users, patients and carers are partners throughout the research process.

Reviewers

Professor **Peter Allen**, Cranfield University, UK
Dr **Helen Bevan**, National Health Service, UK
Dr **Richard A. Blythe**, University of Edinburgh, UK
Dr **Chris Crawford**, Ohio University, USA
Professor **Paul Davis**, Pardee RAND Graduate School, USA
Gene Downum, Department of Defense, USA
Professor **Lasse Gerrits**, University of Bamberg, Germany
Professor **Nigel Gilbert**, University of Surrey, UK
Professor **Dan Hanfling**, George Washington University, USA
Dr **Dimitris Kugiumtzis**, Aristotle University of Thessaloniki, Greece
Professor **Fiona Lettice**, University of East Anglia, UK
Dr **Terry Marks-Tarlow**, The Insight Center, Los Angeles, USA
Dr **David Marsay**, University College London, UK
Professor **Bill McKelvey**, UCLA Anderson School of Management, USA
Nigel Monaghan, Public Health Wales, UK
Professor **Alfonso Montuori**, California Institute of Integral Studies, USA
Professor **Frank Moulaert**, University of Leuven, Belgium
Dr **Aaron Pycroft**, University of Portsmouth, UK
Professor **Michael Small**, University of Western Australia, Australia
Professor **Daniel Solow**, Case Western Reserve University, USA
Rick Stern, NHS Alliance, UK
Professor **Raymond-Alain Thietart**, ESSEC Business School, France
Professor **Daniela Vaz**, Superior School of Health Sciences of Leiria, Portugal
Professor **Michael Ward**, Duke University, USA
Dr **Theo Zamenopoulos**, Open University, UK

Acknowledgements

We would like to thank our 52 authors, who contributed 26 brilliant chapters, each offering a different perspective on research methods underpinned by complexity science. They use approaches inspired by philosophy, art, psychology, a range of qualitative and quantitative methods, and multi-level networks. The depth and erudition of the scientific enquiry on which each chapter has been based is truly inspiring and we would like to thank them all for their distinctive and collective contribution.

The final high quality of the chapters is also due to our 25 reviewers, who offered their expertise in a very wide range of topics to review each chapter. Some chapters had only a few edits while others had to be almost re-written, but the end result was well worth the effort. Reviewing is one of those tasks which academics are expected to do in their 'spare' time, and to give each chapter a great deal of thought. We all benefit from each other's expertise and it never fails to amaze us how generous our fellow academics who have acted as reviewers (see List of Reviewers) were with the time they offered to help us in this Handbook.

The Handbook has six parts and each one covers different types of research methods. Early on, the co-editors agreed that between us we did not have the expertise to do justice to all the chapters and invited some of the authors to review the chapters in each part and provide an introduction highlighting the contribution to the Handbook of each chapter. We hope that the readers will find these introductions of help in choosing which chapter they need to study. We would like to thank those authors, who also contributed by introducing all the chapters in their section, as it makes the Handbook much more useful to the reader.

Finally we would like to thank our publishers Edward Elgar, who took the initiative and invited us to edit this Handbook and in particular, Alexandra O'Connell who answered our endless questions with patience and charm.

Our heartfelt thanks to everyone who contributed to this Handbook, which could become a milestone in research methods in complexity science.

Eve Mitleton-Kelly
Alexandros Paraskevas
Christopher Day
26 February 2017

Editors' introduction

Eve Mitleton-Kelly, Alexandros Paraskevas and Christopher Day

This Handbook has been in preparation since June 2013 when the first emails were exchanged with Edward Elgar, but the long journey has been worth it. What we have achieved is a truly comprehensive coverage of research methods in complexity science.

The Handbook is addressed to both academic researchers and practitioners and attempts to cover most research methods based on the sciences of complexity, and applying them to practical cases. Its six parts include 26 chapters, contributed by 52 authors and each part includes different types of research methods. They range from Research Philosophy (Part I) to Qualitative Techniques (Part II), Visual Methodologies (Part III), Modelling and Statistical Analysis of Empirical Data (Part IV), Multi-Level Networks (Part V) and Mixed Methods (Part VI). The research methods complement each other and most are presented in a way that other researchers and decision makers may be able to use.

Each chapter has been peer-reviewed, and although the process was a long one, it has provided us with a Handbook of much higher quality than would have been the case without the contribution of our reviewers.

The Handbook includes not just the usual qualitative and quantitative research methods, but also unusual ones based on art and psychology; also ambitious methods that explore multi-level networks. The case studies are truly astonishing in the changes achieved by using a complexity science approach. They range from the regeneration of communities to musical performance; from complex governance networks to psychotherapy; from gender dynamics to agent-based modelling and appropriate response to pandemics. They also explore important concepts such as exaptation, emergence, self-organization and co-evolution. As the author of Chapter 24 (referencing Edmondson and McManus, 2007) points out, the choice of a research method should be based on the kind of question being asked, rather than the method most familiar to the researcher. The rationale for this Handbook is based on that principle and it is hoped that both researchers and practitioners will be inspired to explore new research methods to address the complex challenges they are facing.

Each of the six parts has its own introduction which highlights the contribution of each chapter and we will therefore not be repeating that exercise in the Editors' Introduction. Instead we would like to offer an overview, by describing briefly what you should expect from the volume and how the different parts complement each other and offer a wide spectrum of approaches underpinned by complexity science. By studying the Editors' Introduction and the introductions which precede each part, the reader should be able to judge which chapters would be relevant to their needs. We would, however, also strongly recommend that readers explore other methods that just excite their interest.

Part I '*Complexity Science Research Philosophy*' offers some reflections on the contribution that complexity science has made to our understanding of physical, biological and social systems and how approaches have developed from cross-disciplinary to trans-disciplinary and unified efforts to address very complex problems. These problems are discussed in Chapter 1 *'Introduction to the Strategy and Methods of Complex Systems'* and include the study of biomolecular interaction, the workings of the mind, global socio-economic risks, pandemics and environmental disasters. Chapter 2 *'Complex Evolving Social Systems: Unending, Imperfect Learning'*, contrasts the 'hard science' approach in the natural sciences, where repeated experiments allow us to discover general laws and patterns and to make predictions about 'similar' situations elsewhere, with complexity science, which acknowledges that such experiments are not feasible in social co-evolving systems. Furthermore complexity reveals the limits to knowledge and the impossibility of guaranteed prediction not only in the social world, but also in the physical and biological one. In other words, complexity science has changed our way of seeing the world, in terms of whole systems within a co-evolving context, focusing on relationships of the interacting entities and their emergent patterns. This is in sharp contrast to a reductionist approach, which looks at increasingly smaller sub-divisions of individual entities. Complex systems, especially in the social realm, also include individual agents with reflection and intentionality capable of changing their behaviour on the basis of new knowledge. Finally, Chapter 3 *'Information Theoretic Measures of Causality: Music Performance as a Case Study'*, looks at causation and probabilistic approaches to causality measures, informed by studies of time series on the brain and finance. The application is an unusual one and the authors discuss the findings of the analysis of music performance, where causation amongst the musicians and audience was studied by using EEG time series.

Part II *'Case-Based Qualitative Techniques'* offers a very rich palette of qualitative methods applied to a wide range of successful cases. Chapter 4 *'Addressing Global Challenges: the EMK Complexity Methodology'* offers new material on a highly effective methodology, able to address apparently intractable organizational, societal, and global problems. The author provides detailed advice on how to use the approach and illustrates it with one case, specifically showing how the two parts of the methodology were used effectively in a very difficult context. The first part was the identification of the *multi-dimensional problem space* (where the dimensions may be social, cultural, political, economic, physical, financial, technical and other) and the *co-evolutionary dynamics* between the dimensions, which provide a starting point for decision-making. The second part of the methodology acknowledges that complex problems do not have single solutions, but need a broader *enabling environment*, capable of addressing the challenge over time as it changes and evolves.

Chapter 5 *'Complexity Informed Social Research: From Complexity Concepts to Creative Applications'*, describes and demonstrates a complexity informed qualitative social research approach, associated methods and techniques and concomitantly explains the translations and interpretations made in utilizing complexity to effectively comprehend human social experience and sense making. The perspective assumed in the chapter is that qualitative research, like complexity, takes a radically relational approach to interpreting interrelationships between sense makers, fragments of knowledge, cultures, histories, futures, aspirations and so on. It introduces *Vortical postmodern ethnography* as an inquiry approach, along with the narrative generating method of *coherent conversations* and the narrative analysis techniques of *fractal narrative analysis* and *attractor narrative analysis*. The approach is used in an indicative research project, on community enabling.

For decades, costly interventions have tried and failed to 'fix' Britain's most disadvantaged communities, where poverty, crime, unemployment and poor health are rife. Chapter 6 *'From Isolation to Transformation with C2'*, offers a very effective, not costly and practical seven-step approach, underpinned by complexity science, called 'C2' (Connecting Communities), which does offer a solution. The C2 intervention is delivered on site and in the classroom by a small team of experienced practitioners, working alongside local residents and service providers to enable them to implement the framework. The key to the effectiveness of the approach is the direct involvement of community leaders, academics and frontline primary healthcare professionals, who have helped transform many communities across the UK in the last decade, using this approach.

Chapter 7 *'Using Complexity Principles to Understand the Nature of Relations for Creating a Culture of Publically Engaged Research within Higher Education Institutes'*, discusses how to secure culture change within higher education institutes such that public engagement becomes part of how research is done within that institution. In 2008 RCUK funded six 'Beacons for Public Engagement'. Support was provided for each funded partner organization to create a culture of engaging with the public to inform the design and delivery of research. Using maximum variation sampling, seven case studies were selected. Analysing the findings from a complex systems perspective, the researchers conceptualized an 'engagement cycle' with three phases or elements: (a) creating the conditions; (b) co-creation of research; and (c) feedback loops to inform ongoing and future research. The chapter shows how it is possible to understand the dynamics of successful public engagement with research, using complexity theory, and what implications this has had for the methods used.

Part III *'Visual Methodologies'*, is unusual in that it discusses visual methodologies using art, psychology, dynamic network topologies and two tools used to examine governance networks. The work described in Chapter 8 *'The Art of Complexity: Using Visual Artefacts and Dialogue to Bridge the Gap between Strategic Plans and Local Actions in Organisations'* started fourteen years ago when an artist started working with the Complexity team at the London School of Economics. What emerged and evolved was the visual representation of organizational strategy combined with facilitated dialogue ('Visual Dialogue') as a complexity-inspired tool for culture change and organizational development. The approach has since been implemented in over 100 large organizations. The process creates spaces for employees and leaders to come together to make sense of what is happening, what needs to change, and what actions are required. It helps people connect strategy to action, learn about each other's roles and take ownership of the parts each has to play in successful implementation. The process creates the conditions from which learning, innovation and self-organization can emerge. As the process evolved the artist/researchers learned that its principal impact lay in improving the quality of conversations about change between leaders and front line employees, working to create *enabling environments* that fostered mutual sense making, rich connection and creativity.

Most social science research applying complexity principles has concentrated on complex systems in the external world. Chapter 9 *'Inner Complexity: Using* Landscape of the Mind *to Catalyse Change in Organisations'*, uses a tool based on psychology to look at the inner complexity of humans, by identifying their preference profiles. In other words how individuals and teams prefer to make decisions, learn, innovate and so on. At the same time that an artist started working with the Complexity Group at LSE so did a psychologist using the visual methodology called *Landscape of the Mind* (referred to as LoM) which is described in this chapter. Case studies and examples illustrate its practical applications and show how it can be used to catalyse change in organizations, with particular reference to the implications for leadership and innovation.

The tools and methods described in Chapters 4, 8 and 9 have often been used together to look at complex organizational problems as they provided a much deeper understanding of the problem space than a single approach could on its own. Together they used qualitative analysis based on interviews to identify the multidimensional problem space; art and visual dialogue to explore conversations about change; and psychology to look at preference profiles. The first two were qualitative, but LoM offered quantitative data on the change process. The combination of the three tools and methods was quite unique and provided an effective means not only of understanding the deep nature of the problem, but to also facilitate the co-creation of enabling environments to address the challenge. This Handbook is offering many more tools and methods, which could be used together to support and complement each other.

Chapter 10 '*On the Visualization of Dynamic Structure: Understanding the Distinction between Static and Dynamic Network Topology*', looks at the structure of complex networks. In the past decade there has been a significant increase in interest in complex network data, and specifically in biological and social networks. These types of networks are highly interconnected and large scale; there is therefore a need for tools that enable us to make sense of these structures. The chapter explores how, for a given network, there are a range of emergent dynamic structures that support the different behaviours exhibited by the network's various state space attractors. A selected Boolean Network was used to calculate a variety of structural and dynamic parameters, and to explore the various dynamic structures that are associated with it (using both space-time diagrams and power graphs), to consider the activities (Shannon entropy) associated with each of the network's nodes when in certain modes/attractors. This work contributes to the development of robust complexity-informed tools to support the wealth of tools associated with Network Theory, with particular emphasis on network dynamics. Chapter 11 '*Network Text Analysis and Social Network Analysis in Investigating Complex Governance Networks: Applications of AutoMap and ORA*', also examines networks; it discusses how network text analysis and social network analysis can be used to investigate complex *governance networks*. The importance of examining their complex structural properties and the roles actors play in them is explored and two tools are described and illustrated. They are AutoMap (network text analysis software) and ORA (dynamic network analysis software), which were used in the cases of an urban governance network and that of statewide policymaking processes. The protocol used for the applications of AutoMap and ORA is given in the appendix to the chapter.

Part IV '*Modelling and Statistical Analysis of Empirical Data*', moves fully into quantitative approaches and is the longest part of the volume with six chapters. Chapter 12 '*Using Maximum Likelihood Estimation Methods and Complexity Science Concepts to Research Power Law-Distributed Phenomena*', looks at extreme events that skew what we consider 'average' and explores the basic causes of skewed distributions, specifically power laws and also examines horizontal scalability processes. These processes are generated by scale-free mechanisms (that is, the same cause at multiple levels of analysis) that result in self-similar fractal structures within organizations. Bak's (1996) 'self-organized criticality' is used to explain how and why the addition of small inputs, once beyond some critical threshold, have the potential to cause extreme outcomes and co-evolutionary effects at multiple levels of analysis. Using three longitudinal datasets of entrepreneurial ventures at different states of emergence, the authors demonstrate a method to determine whether data are power law distributed and, subsequently, how critical thresholds can be calculated. The chapter offers a conceptual-empirical link for moving beyond loose qualitative metaphors to rigorous quantitative analysis so as to enhance the generalizability and utility of complexity science theories.

Chapter 13 '*Multifractal Signatures of Intersectionality: Nonlinear Dynamics Permits Quantitative Modeling of Hierarchical Patterns in Gender Dynamics at the Cultural Level*', focuses on gender studies, but moves away from qualitative approaches and offers an introduction to multifractal analysis; it suggests that cascade dynamics and multifractal analysis might provide the logical formalism and corresponding statistical framework to make intersectionality a quantitatively tractable model for gendered experience. The authors review recent cognitive science advances, where multifractal analysis laid bare key features of the cascades driving cognitive performance, and provide a demonstration of similar cascade structure in gender dynamics through multifractal analysis of web-traffic data for gender terms on Wikipedia. The chapter proposes that cascade formalisms and multifractal analysis may provide new avenues for gender studies that balance both logical formalisms and dynamic concepts.

To improve reliability and resilience in infrastructures, it is necessary to adopt a 'complex, smart territory' modelling strategy, acknowledging the importance of social complexity. Chapter 14 '*Modeling Social Complexity in Infrastructures: A Case-based Approach to Improving Reliability and Resiliency*', discusses a case study on a segment of the United States power grid with a simple goal: to create a first proof-of-concept, to show how thinking about infrastructures in 'complex systems' terms, especially in their social aspects, can prove beneficial. The case study uses the SACS Toolkit, which is part of the new approach to modelling complex systems, called case-based complexity. As a technique, the SACS Toolkit is a computationally grounded, case-comparative, mixed methods platform for modelling complex systems as sets of cases. The chapter provides an overview of the research process, and a summary of the novel insights the approach was able to achieve.

Chapter 15 '*Phase Transitions and Social Contagion as Enabling Mechanisms for Coordinated Action in Populations: A Mathematical Framework*', presents a general mathematical framework to study discontinuous change in human interaction dynamics. There are two complementary perspectives, the macro and the micro. Regarding the macro context, the authors propose that levels of ordered structure in complex human organizing can be represented by a category theoretic representation that reflects informational influence acting on individual agents from two sources. First, external to the population, resource and competitive conditions in the ecosystem provide generalized influence on the actions of individuals. Second, internal to the population, relational and cooperative conditions provide localized influence on individual action. These independent influences interact to change the set of interaction rules that are enacted locally. Regarding the micro context, contagion is given as the mechanism whereby a common organizing state is adopted across multiple agents. The chapter shows that the ordered structure, which emerges within a population, can be indexed as the number of active degrees of freedom embedded in local rules of interaction that are guiding groups of agents. The authors argue that like the natural sciences, research in social science should use category theoretic mathematical approaches to suggest deductive hypotheses that can be tested empirically with definitive results.

Chapter 16 '*Applying Complex Adaptive Systems to Agent-Based Models for Social Programme Evaluation*'. Measuring the effectiveness of social programmes is a significant challenge and forecasting the outcomes of these programmes can be difficult to justify to funders and other key stakeholders. Understanding the logic and assessing the impact behind the intervention can be difficult because commonly-used tools are based primarily on linear methods that assume that a set amount of input, throughput, and output will result in a set outcome. This chapter takes a complexity science approach and facilitates the use of agent-based modelling. It provides the requisite background for evaluators and researchers to frame their efforts as complex adaptive systems; a requisite in moving to the next step of developing agent-based models. The authors address both qualitative and quantitative aspects of complexity through two applications of agent-based modelling that consider related social policy issues.

Chapter 17 '*Complexity, the Bridging Science of Emerging Respiratory Outbreak Response*' looks at the incidence of emerging infectious diseases (EIDs) which is increasing. Despite the best efforts to prepare for such events, real-time management of emerging disease outbreaks is often marked by confusion and uncertainty. During these outbreaks, decision makers were challenged to make impactful decisions with little time and incomplete information. Health authorities typically approach such threats by focusing on the safety and effectiveness of individual-level interventions, such as vaccines and antivirals. This does not, however, detail how these countermeasures should be used to optimally benefit population health as a whole. Decisions around how to best use the limited supply of pharmaceuticals, targeted and/or social interventions in these situations require a unique combination of scientific fields, and integrating these fields in real time requires a bridging science. Mathematical modelling of complex systems represents that bridging science. Chapter 17 discusses the conceptual design and structure of mathematical models of communicable diseases, using transmission dynamics in the context of respiratory-borne pathogens within human populations. The authors then demonstrate the necessity of assembling appropriate expertise related to mathematical modelling, epidemiology, public health, virology, and clinical management to ensure valuable quantitative decision-support tools to assist policymakers at the time of crisis.

Part V '*Multi-Level Networks*' examines such networks and Chapter 18 '*Multilevel Systems and Policy*' defines what these networks mean for the implementation of computer models to investigate the multi-level consequences of policy. In order to inform policymaking part–whole aggregation and taxonomic aggregation are described as methods of representing multi-level structure, and it is shown how they are interleaved in the construction of vocabulary to describe multi-level systems. This enables complex nested structures to be represented as a kind of backcloth that supports patterns of aggregate and disaggregate numbers that describe the day-to-day traffic of people, resources and responsibility that are essential for systems to function.

Chapter 19 '*Complex Scenarios in Socio-economic Data: A Comprehensive Analytical*

Study', explores the analysis of socio-economic conditions of different countries, which reflect the country's social, economic, political, ideological, ethical, cultural and communicative habits, using new methods based on complexity science. To show the effectiveness of different nonlinear tools in analysing socio-economic data, the authors have implemented three nonlinear tools – $\tau-$ recurrence rate, Mean conditional recurrence (MCR), Complex networks (CN) to analyse country level GDP and population data, and successfully validated the derived results with the standard conclusions based on general theories of economics. $\tau-$ recurrence rate is used to show how two non-identical systems get synchronized through their phase spaces. MCR detects the driver and response system in synchronized states and CN reflects the overall scenarios of the complex systems by its various statistical measures. The datasets are collected from NASA's Earth Observing System Data and Information System and are downscaled projected based on Special Report on Emissions Scenarios (SRES).

Chapter 20 *'Employment of Tools and Models Appropriate to Complex, Real-world Situations'* advises practitioners how to judge which tools and methods are appropriate to use and when, where and how to apply them. The chapter provides a model of practice, and a framework for judging appropriateness of tools based on that model (with three examples of the framework in use). It also offers a critique of two sets of example tools: examining the applicability of autonomous agents and multi-agent systems to a range of situations; and explains how to employ multi-modal, multi-level influence networks to bring about ongoing change. Finally, the chapter presents a list of principles of practice, drawn from experience in the field, to be used to inform real-world decision- and policymaking.

Chapter 21 *'Leadership Network Structure and Influence Dynamics'*, describes a quantitative methodology for the analysis and modelling of leadership networks which leverages research in complex systems, in particular nonlinear dynamical systems theory and network science. A prototype software package, PORTEND, is introduced which implements the methodology using data from expert analysts in order to help assess policy and factional outcomes with respect to the internal dynamics of a system of political actors. The methodology includes structural analysis methods, such as algorithms for analyzing issue positions and community structure, and a simulation of nonlinear social influence dynamics. PORTEND's capabilities are illustrated for an application to Iran involving 15 leadership elites and seven issues. The factional structure of the Iranian leadership group is analysed first based on their issue positions, then with respect to the network of inter-actor influence relationships, and finally by a synthesis of the issue and network data. An application of the nonlinear social influence simulation to the nuclear issue is presented and its implications are discussed with respect to Iranian decision-making concerning the 2013–2015 nuclear negotiations.

Part VI *'Mixed Methods and Complex Analogies'* not only offers several examples of the use of mixed methods, but also explores some complexity principles in depth. Chapter 22 *'Complex Analogy and Modular Exaptation: Some Definitional Clarifications'*, discusses the *complex analogy* between biological evolution and technological innovation, focusing in particular on the novel construct *modular exaptation*. The chapter defines *exaptation*, which is a biological concept whose technological analogue is useful in innovation studies and explores its epistemological bases, arguing that the *etiological* concept of function – a biological tenet – is valid also in the technological domain. The complex analogy extends to biological and technological *functional modules*, providing the main building block on which *modular exaptation* can be founded. Establishing a complex analogy enables the description of the two *domains* via the same *relational structure*. In turn this allows the transferability of knowledge from the *base* domain to the *target* domain, and vice-versa. The complex analogy can therefore be considered a methodological tool for understanding complex systems in general and technological innovation in particular.

Chapter 23 '*Emergence and Radical Novelty: From Theory to Methods*' questions the strong association between self-organization and emergence and explores how the self-organization model and the methods of researching complex systems stemming from it are misleading as to an accurate account of emergence. It further questions the assumptions underlying the idea of self-organization and the insufficiency of the concept in supplying a cogent account of the radical novelty characterizing emergent phenomena. Radical novelty is what supports the needed explanatory gap of emergence between the antecedent and lower micro-level and the consequent and higher macro-level. Although not mysterious nor calling for some kind of supra-natural explication, the explanatory gap is what challenges and motivates a conception of emergence which remains true to the fundamental claims made about emergence.

Chapter 24 '*Applying the 15 Complexity Sciences: Methods for Studying Emergence in Organizations*' introduces the 15 sciences of complexity, each one being an effective method, shown by an ongoing stream of research that uses it. Each of these sciences will be described in terms of its empirical method, its primary research question(s), and the data needed for analysis. These sciences and even broader conceptualizations of complexity can be organized into three main paradigms or approaches: (a) Computational Agent-based Modeling, developed through Holland's complex adaptive systems, and scientists at the Santa Fe Institute; (b) Natural Sciences and Idiographic Analogies, in which the dynamics of order-creation in, for example, physics, thermodynamics and biology, is applied through a symmetrical analogy to human systems; (c) Narrative and Multi-Method Studies. These have perhaps the most to offer for seeing and enacting complexity in the social world. Likewise, multi-method longitudinal studies allow for a deep understanding of a system's dynamics, through repeated measures and interviews taken over time. The chapter presents a set of complexity methods and models that researchers can use to help identify the appropriate complexity methods to use to answer a specific research question.

Chapter 25 '*Mixed Methods Research: A Method for Complex Systems*' illustrates the need for multiple perspectives, philosophically and theoretically, and new stances to solving paradigmatic dilemmas. Frameworks are compared to assist in the alignment with research questions and research purpose as well as recognizing practical influences on research design choice. Numerous mixed methods research designs demonstrating the integration of mixed methods are considered, as are techniques for integrating the data between traditional methods and the benefits and challenges of mixed methods research. Overall, mixed methods research has critical mass but continues to evolve and become ever more relevant to address complex systems problems.

Chapter 26 '*Dynamical Systems Therapy (DST): Complex Adaptive Systems in Psychiatry and Psychotherapy*' as the last chapter offers an ambitious application of complexity science to psychiatric treatment. The nonlinear dynamical systems approach to psychiatric nosology allows practitioners to shift from linear categorical diagnostic and treatment algorithms to integrative process models, conceptualizing individual and group dynamics as *complex adaptive systems (CAS)* with emergent properties of subjective and cultural experience. Dynamical systems therapy (DST) represents a complexity derived treatment application that puts individual capacity for self-system coherence and flexible adaptation to changing environmental demands at the cornerstone of psychological health. Recurrent patterns of feeling, thinking and relating can be analyzed by using modified fitness diagrams (adaptive or *A-landscapes*), which integrate objective, subjective and intersubjective clinical data. They enable psychiatric practitioners to chart the patient's unique life trajectory through its malleable *attractor/repellor states* in health and psychopathology that uniquely informs both psychosocial and psychopharmacological treatments.

The 26 chapters cover research methods, which range from the philosophical to the visual; from the qualitative to the quantitative that includes modelling and statistical analysis, as well as multi-level networks. It ends with mixed methods approaches, but this last part also includes the exploration of the theoretical concepts of exaptation, self-organization and emergence as research methods; the Handbook is completed with an unusual application in psychobiology.

PART I

COMPLEXITY SCIENCE RESEARCH PHILOSOPHY

Professor Henrik Jeldtoft Jensen

Complex systems science considers situations where multiple components evolve together in tangled webs of interdependencies. And complexity science acknowledges that what at one level is considered to be the components may themselves, as one so to speak zooms in, be studied in their own right as collections of subcomponents, for example, the components of sociology are individual humans (or classes of humans), each composed from myriads of individual cells and organs. Typical fields of science considered by complexity theory include evolutionary ecology, neuroscience, sociology, economics and finance.

Is complexity an altogether new kind of science? Not really. Any particular 'thing', say a glass of water, is always part of the larger surrounding world and what we see as a glass of water at our everyday level of experiencing our surroundings, may be seen as an intricate organisation of zillions of subatomic particles into water molecules, which in turn organise themselves into the liquid water with its ripples and bubbles. The traditional hard sciences are used to hierarchical organisation of reality, but physics, say, can very much be seen as a science for which meaningful results can be derived using ideas such as closed system, equilibrium or steady state dynamics where the components of the system don't not undergo any fundament change.

Physics has had great success in approximating the world by building blocks treated as things and derived laws applicable to each level: subatomic, atomic, molecule, liquids, solids and so on. However, this description is, after all, an approximation. The different levels of reality are connected and the world does not consist of 'things' with their own once and for all given properties. Bar-Yam discusses these aspects in Chapter 1.

The philosopher Alfred North Whitehead argued that not 'things' but processes are the basic ontological constituents of the world. Of course, we often pretend it is otherwise and sometimes, for instance like in physics, we even get away with simplistic approaches that ignore the basic interconnectedness and co-evolutionary nature of the world.

But as we move to aspects of the surrounding world involving processes typically classified as biology, sociology, economics and so on, it becomes very problematic to insist on, for example, equilibrium or a closed system. When analysing and modelling complex systems, it is important from the onset to apply a methodology that allows for co-evolution of hierarchical structures. Building blocks with a fixed set of attributes can then not form the basis for our analysis and model building. A much more appropriate starting point consists of co-evolving processes. In Chapter 4 of this Handbook, Mitleton-Kelly discusses verbal models of co-evolution; such models can of course be developed further to go beyond the verbal descriptive level and allow for more quantitative approaches to co-evolution.[1]

The fact that the world really consists of co-evolving processes obviously complicates the research into the behaviour of complex systems significantly and perhaps also encourages the complexity scientist to be more humble and sober concerning what theory of complex systems can do. A complexity scientist is unlikely to think that it makes much sense to imagine that a 'theory of everything' exists, as some suggested by physicists. Such dreams seem unrealistic to a complexity scientist since complexity science is persistently confronted with the realisation that the hierarchical structure of our world leads to a staircase of emergent ontological levels. The collective cooperative processes at one level, say a tornado in the atmosphere are of course related to the individual air molecules, since the tornado is the flow pattern of air molecules. But a single air molecule doesn't possess any microscopic 'tornado property'. The tornado is entirely a collective process that only exists at the level of the cooperative motion of many air molecules. Hence, rather than looking for a model allowing for the prediction of the fluctuations of the stock market from the cells composing the traders' brains, a complexity scientist will be more modest and acknowledge that phenomenological theories are what makes best sense. And, importantly, when trying to understand the stock market complexity science will investigate if collective emergent behaviour of the involved agents may perhaps typically be more important than the idiosyncratic personalities of the individual agents.

How about prediction? Does the co-evolution, the interconnectedness, the contingency, the hierarchies and so on force us to give up predicting the behaviour or the future of complex systems? These questions are discussed by Allen in Chapter 2. No, we can still develop models that help us to understand the behaviour and even to forecast future events. But our quantitative theories will predict with a precision that is meaningful for the given situation. Of course, we don't expect predictions that deterministically determine when a given trader will submit a bid of a given value for a given stock. But we do think that it is possible to do much better than just storytelling. The experience from complexity sciences suggests that well-specified bounded quantitative predictions are possible. Think of the forecasting of the weather. The meteorologists are now able to attach likelihood estimates to the forecasting of the weather some days out in the future. These likelihoods are computed from an understanding of the lack of precision of the current state of the system and together with knowledge of the dynamics of the components involved. The consequences of the lack of precision of the initial state depend strongly on which dynamical regime the meteorological system is in at the moment of forecasting. Smooth dynamics will not suffer too much from inaccuracies in determining the initial configuration. This is because close by configurations will evolve more or less in the same manner and lead to more or less the same weather in a couple of days' time. But if the forecaster determines that the configuration of the atmosphere is deep in a regime of turbulence, or chaos, the inaccurate initial configuration makes it impossible to predict with certainty which of a range of possible scenarios for the next few days is actually going to happen. In this case the specific forecast will be assigned a much lower likelihood.

Sociology, or economics, is of course even more complicated and involved than meteorology and we are unlikely ever to be able to have models that predict with certainty trajectories of behaviour. But examples from, for example, finance, suggest that it is possible to determine which kind of dynamics that can be consistent with, say, the observation that crashes are recurrent phenomena.[2]

However, before we start thinking of forecasting we need to develop an understanding of how components are influencing each other. Network theory is well-suited as a way to describe sets of entities linked together in a web of interdependencies. To be able to capture the evolutionary aspects and the hierarchical emergent structures we'll typically need a flexible structure of dynamically evolving layered and/or nested networks. But the nodes and the edges, or links, of the relevant networks have to be identified before we can commence on a theoretical analysis of a specific network model of our given system. Pierson's correlation coefficient is often taken as a starting point for assigning a measure of interdependence between two entities. This approach can be applied whenever we have access to data relating one quantity to another. A famous example is how it was realised that smoking causes a specific type of lung cancer by studying the correlations between incidents of cancer and smoking habits. Correlations are symmetric and will therefore not give us any indication of causal direction. In the case of cancer, a causal direction is hardly needed, since clearly it seems unlikely that cancer makes people smoke, but quite plausible that smoke may stress the lung tissue and thereby induce cancer. But if we look at, for instance, financial time series it may not at all be clear which, say, stock price could be the driver and which could be the driven. In such cases, a directional interdependence measure is of great interest. At present, there is very active research effort in developing information theoretic causality measures (see Chapter 3 by Razak, Wan and Jensen) based on the ideas introduced a long time ago by Granger. Such approaches identify 'causal' directions by analysing the likelihood that we are able to better predict one time series if we have knowledge of another. These techniques are by now very powerful and can be used to establish directed network models of a given system. The nodes will be identified with the time series, so, for example, a node could be a given stock represented by the dataset of recorded prices for that stock. The directed edged are then obtained from the information theoretic analysis and allows one to assign a direction and a weight to an edge between two nodes, for example, two stocks.

Complexity science is complicated, but no more so than we have experienced during the last few decades a huge increase in the arsenal of tools and methodologies and approaches developed to quantify and model the co-evolutionary hierarchical dynamics of many types of complex systems. Some of these developments are discussed in this Handbook.

NOTES

1. See, for example, H.J. Jensen and E. Arcaute, Complexity, Collective Effects and Modelling of Ecosystems: Formation, Function and Stability. *Annals of the New York Academy of Science*, **1195**, E19–E26 (2010).
2. E. Viegas, M. Takayasu, W. Miura, K. Tamura, T. Ohnishi, H. Takayasu and H.J. Jensen, Ecosystems Perspective on Financial Networks: Diagnostic Tools. *Complexity*, **18**, 34 (2013).

1. Introduction to the strategy and methods of complex systems

Professor Yaneer Bar-Yam, New England Complex Systems Institute

1.1 OVERVIEW

The structure of scientific inquiry is being transformed by broad relevance of the strategies and methods of complex systems science for understanding physical, biological and social systems. Disciplinary and cross-disciplinary interactions are giving way to trans-disciplinary and unified efforts to address the relevance of large amounts of information to the description, understanding and control of complex systems. From the study of biomolecular interactions to the workings of the mind to global socio-economic risks, pandemics and environmental disasters, complexity has arisen as a unifying feature of challenges to understanding and action. In this arena, information, structure, function and action are entangled. New approaches that recognize the importance of collective patterns of behavior, the multiscale space of possibilities, and evolutionary or adaptive processes that select systems or behaviors that can be effective are central to advancing our understanding and capabilities.

Complex systems analyses range from detailed studies of specific systems, to studies of the mechanisms by which patterns of collective behaviors arise, to general studies of the principles of description and representation of complex systems. These studies enable us to understand and modify complex systems, design new ones for new capabilities or create contexts in which they self-organize to serve our needs without direct design or specification. The need for applications to biological, cognitive, social, information and other systems is apparent.

For example, biology has followed the approach of accumulating large bodies of information about the parts of biological systems, and looking for interpretations of system behavior in terms of these parts. Yet, it has become increasingly clear that biological systems and their health and disease conditions are better understood as emergent collective behaviors of spatially structured networks, so that dependencies rather than components are the essential property to be understood. The role of information in biological action and the relationships of structure and function are only beginning to be probed by those who are interested in biological systems designed by nature for their functional capabilities. Underlying these systems are a wealth of design principles in areas that include the biochemical networks (Gallagher and Appenzeller 1999; Service 1999; Normile 1999; Weng, Bhalla and Iyengar 1999; Hartwell et al. 1999), immune systems (Perelson and Wiegel 1999; Noest 2000; Cohen and Segel 2001; Pierre et al. 1997) and neural systems (Anderson and Rosenfeld 1988; Bishop 1995; Kandel, Schwartz and Jessell 2000), as well as animal behaviors such as the swimming mechanisms of fish (Triantafyllou and Triantafyllou 1995) and the gaits of animals (Golubitsky et al. 1999). These systems and architectures point to patterns of function that have a much higher robustness to failure and error and a higher adaptability than conventional human engineered systems.

Computers have made a transition from systems with tightly controlled inputs and outputs to networks that respond on demand as interactive information systems (Stein 1999). This has changed radically the nature of their design. The collective behaviors of these networked computer systems, including the Internet, limit their effectiveness. Whether these have to do with the dynamics of packet loss in Internet traffic (Paxson 1996), or cyberattacks (Kephart 1994; Forrest, Hofmeyr and Somayaji 1997; Kephart et al. 1997; Goldberg et al. 1998) that, at times, have incapacitated a large fraction of the Internet, these effects are not small. The solution to these problems is understanding collective behaviors and designing computer systems to be effective in environments with complex demands and to have a higher robustness to attack.

The human brain is often considered the paradigmatic complex system. The implications of this recognition are that cognitive function is distributed within the brain and mechanisms may vary from individual to individual. Complete explanations of cognitive function must themselves be highly complex. Major advances in cognitive science are currently slowed by a combination of efforts to explain cognitive function directly from the behavior of individual molecular and cellular components, and on the other hand by aggregating or averaging the cognitive mechanisms of different human beings. Still, diverse advances that are being made are pointing the way to improvements in education (National Institute of Mental Health n.d.), man–machine interfaces (Norman and Draper 1986; Nielsen 1993; Hutchins 1995) and retention of capabilities during aging (Stern and Carstensen 2000; Lawton 1981; Mandell and Schlesinger 1990; Davidson, Teicher and Bar-Yam 1997).

Recent global crises, including the global financial crisis, the global food crisis, social unrest including the Arab Spring, and the Ebola epidemic and other pandemics, have demonstrated that global connectivity leads to vulnerabilities due to the high rate of global travel, and the rapid propagation of economic and social influences (Lagi, Bertrand and Bar-Yam 2011a; Merchant 2014; Lagi et al. 2011b; Harmon et al. 2011; Harmon et al. 2010; Rutherford et al. 2014; Rauch and Bar-Yam 2006). Many of the key problems today have to do with 'indirect effects' of human activities that may have substantial destructive effects on the human condition. These include global warming and ecological deterioration due to overexploitation of resources. Effective approaches to these problems will require an understanding of both the environmental and socio-economic implications of both current actions, and of actions that are designed to alleviate these problems (NSF 2001). For example, the problem of global warming includes the effects of large-scale human activity interacting with both the linear and potentially nonlinear climactic responses. Despite the grave risks associated with global warming, a key factor impeding actions to alleviate it are fears of major impacts of such efforts on socio-economic systems. Better understanding of the potential effects of such interventions should enable considered actions to be taken.

Other diverse social system problems may be linked to increasing societal complexity in healthcare, the education systems and governance more generally. Current approaches continue to be dominated by large-scale strategies that are not effective in addressing complex problems. Even with the appearance of more holistic approaches to, for example, third world development (World Bank 1998; Wolfenson 1999), the basic concept of existing strategy remains weakly informed by complex systems insights. This gap is an opportunity for major contributions by the field of complex systems, both at the conceptual and technical levels. Further contributions can be made based upon research projects that emphasize the intrinsic complexity of these systems.

1.2   THE METHOD OF MULTISCALE ANALYSIS

The traditional approach of science to take things apart and assign the properties of the system to its parts has been quite successful, but the limits of this approach have become apparent in recent years. When properties of a system result from dependencies and relationships, but we assign them to their parts, major obstacles to understanding, design, regulation and control arise. Once the error of assignment is recognized, some of the obstacles can be overcome quickly, while others require rigorous inquiry. While many scientists think that the parts are universal but the way parts work together is specific to each system, it has become increasingly clear that how parts work together can also be studied in general and by doing so we gain insight into every kind of system that exists, including physical systems like the weather, as well as biological, social and engineered systems.

One of the central insights about complex systems is that the effect of dependencies among components cannot be fully represented by traditional mathematical and conceptual approaches based in calculus and statistics. A key to their limitation is that they are applicable only to systems in which there is a separation of behavior between the micro and macro scales. Microscale behaviors are averaged using statistics, and macroscale behaviors are treated mechanistically. Interactions among the parts that cause behaviors across scales violate this separation.

But many systems are not well described by separate micro and macro scales. Consider a flock of birds. If all the birds flew independently in different directions, we would need to describe each one separately. If they instead all went in the same direction, we could describe their average motion. However, if we are interested in their movement as a flock, describing each bird's motion would be too much information and describing the average would be too little information. Understanding complex behavior that is neither independent nor coherent behavior is best described across multiple scales. This requires knowing which information can be observed at a scale of interest.

Multiscale analysis (Bar-Yam 2016, 2004a, 2002) can be used to identify the complex relationships between the behavior of parts and the whole, across scales. In multiscale analysis, we represent the behavior of a system completely at a consistent scale, and are able to vary that scale. Quantifying this strategy has been done through a variety of mathematical techniques, but the most widely applicable approach is that of renormalization group and its generalization to multiscale information theory (Bar-Yam 2016). The overall complexity of a system, or the amount of information required to describe a system, can be analyzed as a function of scale. If the parts of a system are independent, then the whole system exhibits fine scale random behavior. If the parts are highly correlated, the system has large-scale coherent behavior. In a case where there are fully dependent components in groups, the number of elements of the group is the scale and the behavior of that group occurs at that scale. More generally, if the parts are interdependent, the system can perform complex behaviors that can be characterized to identify key properties as a function of the scale they occur at. Many of the real world systems we are interested in are interdependent and the analysis of the scale dependent behavior is a technical challenge that requires analysis of how the aggregation of components gives rise to the behavior at larger scales, and the independence of those components gives rise to behavior at finer scales.

While the mathematical implementation can be challenging, multiscale analysis is ultimately essential to the study of biological and social systems because it is impossible to represent all of the information about a system, and such a representation would not be useful as each instantiation of a system is different at the microscopic scale. Without the ability to generalize, we cannot anticipate the behavior of systems that we have not fully characterized (an impossible task), inform decisions about how to respond to new circumstances that arise in the world (that is, disease conditions or global crises), or design a system that we rely upon for such responses. Thus, characterizing the important information about a system is critical for both scientific knowledge that can be generalized across systems, and our ability to respond to real world circumstances. Case studies have been made but widespread application of this approach is necessary.

Additional background on the methodological approach and a set of diverse examples are provided elsewhere including application to evolutionary biology with relevance to ecology, biodiversity, pandemics, and lifespan, and in the context of social systems with relevance to ethnic violence, global food prices, and stock market panic (Bar-Yam 2016).

As one example, consider the application of multiscale analysis to the vulnerability of species to ecological catastrophes.

If we consider just the biodiversity itself and not the importance of scale, we can arrive at an incorrect conclusion. When considering the loss of biodiversity to a catastrophic event, extinctions are unlikely because they require the complete loss of all closely related types. A quantitative analysis implies that extinction of 95 percent of species would only eliminate 20 percent total diversity of the tree of life (Nee and May 1997). The reason is that random losses, even when high, are unlikely to remove all individuals belonging to a deep branch of the species tree even when it forms a small proportion of the population, thus preserving most of the diversity. However, to analyze the full effect we should consider not just the diversity, but the number of repetitions of specific genomes or of members of the same species, that is, the multiplicity-scale (Allen, Kon and Bar-Yam 2009). In contrast to the analysis of biodiversity, an analysis of multiplicity (Rauch and Bar-Yam 2004) suggests that the small immediate loss of species is followed by a much greater loss over time due to the vulnerability of small residual populations to extinction. The loss of a large fraction of a group of closely related species (or of closely related organisms) leaves the remainder of the group highly vulnerable to extinction.

Other examples show how the role of both scale and complexity are important for biological and social dynamics. The selection of biological traits, such as altruism, is strongly affected by the role of interactions in space that lead to collective behaviors manifest as patches of genetic and behavioral types. In social systems, ethnic violence is linked to the geographical size of ethnic groups as they are embedded/surrounded by other groups, market prices behavior can be better understood by modeling the collective effects of trend following by traders, and market panic can be understood by considering the co-movement of prices. In each case, characterizing the scale of behaviors provides insight into the essential dynamical properties of interest.

Understanding complex systems does not mean that we can predict their behavior exactly. It is not just about massive databases or massive simulations, even though these are important tools of research in complex systems. The main role of research in the study of complex systems is recognizing what we can and cannot say about complex systems given a certain level (or scale) of description, and how we can generalize across diverse types of complex systems. It is just as important to know what we can know, as to know. Thus, the concept of deterministic chaos appears to be a contradiction in terms: how can a deterministic system also be chaotic? It is possible because there is a rate at which the system behavior becomes dependent on finer and finer details (Cvitanovic 1989; Devaney 1992, 1989; Strogatz 1994; Ott 1993). Thus, how well we know a system at a particular time determines how well we can predict its behavior over time. Understanding complexity is neither about prediction or lack of predictability, but rather a quantitative knowledge of how well we can predict and, only within this constraint, what the prediction is.

1.3 MAJOR DIRECTIONS OF INQUIRY IN COMPLEX SYSTEMS

Complex systems science combines approaches that recognize the importance of patterns of behavior, the multiscale space of possibilities, and evolutionary or adaptive processes that select systems or behaviors that can be effective in a complex world (Bar-Yam 1997). Each of these is informed by multiscale analysis and its ability to describe behaviors at the largest scales.

1.3.1 Self-organization, Pattern Formation, and Design of Systems

Self-organization is the process by which elements interact to create spatiotemporal patterns of behavior that are not directly imposed by external forces. To be concrete, consider the patterns on animal skins, spontaneous traffic jams and heart beats. The robustness of self-organized systems is also a desired, and difficult to obtain, quality in conventional engineered systems. For biomedical applications, the promise is to understand processes like the development of the fertilized egg into a complex physiological organism, like a human being. In the context of the formation of complex systems through development or through evolution, elementary patterns are the building blocks of complex systems. This is diametrically opposed to considering parts as the building blocks of such systems.

Spontaneous (self-organizing) patterns arise through symmetry breaking in a system when there are multiple inequivalent static or dynamic attractors. In general, in such systems, a particular element of a system is affected by forces from more than one other element and this gives rise to 'frustration' as elements respond to aggregate forces that are not the same as each force separately. Frustration contributes to the existence of multiple attractors and therefore of pattern formation.

Pattern formation can be understood using simple rules of local interaction, and there are identifiable classes of rules (universality) that give rise to classes of patterns. These models can be refined for more detailed studies. A useful illustrative example of pattern-forming processes is local-activation long-range inhibition models. Local activation leads to similar behavior among nearby elements, while long range inhibition leads to breakpoints so that patches of a certain size arise. There can be many reasons for the local activation and long range inhibition. In chemical systems, the local activation can arise from slowly diffusing species that engage in self-reinforcing chemical reactions, while the long range inhibition arises from more rapidly diffusing species that are produced by the reaction but inhibit it and have their effect in a larger area around locations where the reaction takes place due to their rapid diffusion. Social system patterns can arise from within group mimicry. These models may be used to describe the complex patterns of animal skins, magnets, air flows in clouds, wind driven ocean waves, and swarm behaviors of insects and animals. Studies of spontaneous and persistent spatial pattern formation were initiated by Turing (Turing 1952) and the wide applicability of patterns has gained increasing interest in recent years (Bar-Yam 1997; Meinhardt 1994; Murray 1989; Nijhout 1992; Segel 1984; Ball 1999).

The use of multiscale analysis to characterize patterns that self-organize involves understanding the universality of these patterns in their macroscopic description, including how this description changes or responds in the presence of external forces, perturbations or changes in initial conditions (Bar-Yam 1997).

1.3.2   Description and Representation

The study of how we describe complex systems is itself an essential part of the study of such systems. A description is a map of the 'actual' system onto a mathematical, graphical or linguistic object. Shannon's information theory (Shannon 1948 [1963]) has taught us that the notion of description is linked to the space of possibilities. Thus, while description appears to be very concrete, any description must reflect not only what is observed but also an understanding of what might be possible. The 'space of possibilities' is an essential and deep concept about the behavior of complex systems. The space of possibilities is captured in the representation we use – the parameters and variables of its mathematical description.

Among the essential concepts relevant to the study of description is the role of universality and non-universality (Wilson 1983) as a key to the classification of systems and of their possible representations. In this context, rather than studying a single model of a system, effective studies are those that identify the class of models that can capture properties of a system or a group of systems. Related to this issue is the problem of testability of representations through the validation of the mapping of the system to the representation.

An important practical objective is to capture information and create representations that allow human- or computer-based inquiry into the properties of a system. The construction of human-usable representations must grapple with the finite complexity of a human being, and other human factors due to properties of our sensory and information processing systems.

The combination of multiscale analysis with the problem of description/representation gives rise to a theory of structure in which each piece of information is characterized as to its redundancy (Allen, Stacey and Bar-Yam 2014). The amount of information as a function of scale is the 'complexity profile' (Bar-Yam 2004a, 2002, 1997) which is the amount of information necessary to specify the system as a function of the scale of description. The complexity profile has been used to study a variety of questions ranging from the mathematical behavior of coupled variables to the effectiveness of social organizations, including the healthcare, education and military systems (Bar-Yam 2004b). In each case, the way a system is organized leads to the scale and complexity of its behaviors, which have to match the demands of its tasks for it to be effective. This is as true about military organizations as it is about healthcare and educational ones. For example, in healthcare, organizational structures that are effective for simple tasks such as providing flu shots and blood tests are different from organizational structures that are effective at diagnosis and treatment of diverse medical conditions. Absent an understanding of this distinction, efforts to reduce medical costs may mistakenly apply approaches to improvement that are appropriate to industrial (large-scale) processes to complex medical services instead of the ones that would benefit from them like flu shots and screening tests. Applications to education include recognizing the role of standardized testing as a large-scale strategy for evaluation, and the contrast to the complexity of student abilities and their eventual professional diversity. Military applications include the distinction in scale and complexity between conventional conflicts as compared to insurgencies and combating terrorism.

1.3.3   Evolutionary Dynamics

The formation of complex systems, and the structural/functional change of such systems, occurs through a process of adaptation, especially through evolution. Evolution (Darwin 1859 [1964]) is the adaptation of populations through intergenerational changes in the composition of the population (the individuals of which it is formed). Learning is a similar process of adaptation of a system through changes in its internal patterns, including, but not exclusively, the changes in its component parts.

Characterizing the mechanism and process of adaptation, both evolution and learning, is a central part of complex systems research (Holland 1992; Kauffman 1993; Goodwin 1994; Kauffman 1995; Holland 1995). This research generalizes the problem of biological evolution by recognizing the relevance of processes of incremental and competitive evaluation-based change to the formation of all complex systems. It is diametrically opposed to the notion of creation in engineering which typically assumes that new systems are invented without precursor. The reality of incremental changes in processes of creativity and design reflect the general applicability of evolutionary concepts to all complex systems.

Multiscale analysis and the multiscale characterization of biological and social complex systems inform our understanding of how evolution is responsible for the creation of structure. Rather than understanding evolution as a generic process based upon energy flows that counter equilibration by entropy increase, we must understand evolution as a process that results in multiple scales of patterns of structure from the microscopic to the macroscopic.

1.3.4   Choices and Anticipated Effects: Games and Agents

Game theory (von Neumann and Morgenstern 1944; Maynard Smith 1982; Fudenberg and Tirole 1991; Aumann and Hart 1992) explores the relationship between individual and collective action using models where there is a clear statement of consequences (individual payoffs), that depend on the actions of more than one individual. A paradigmatic game is the 'prisoner's dilemma.' Traditionally, game theory is based upon logical agents that make optimal decisions based upon full knowledge of the possible outcomes, though these assumptions can be usefully relaxed.

Underlying game theory is the study of the role of anticipated effects on actions and the paradoxes that arise because of contingent anticipation by multiple anticipating agents, leading to choices that are undetermined within the narrow definition of the game, and thus sensitive to additional properties of the system.

Game theory is relevant to fundamental studies of various aspects of collective behavior: altruism and selfishness, and cooperation and competition. It is relevant to our understanding of biological evolution, socio-economic systems and societies of electronic agents. At some point in increasing complexity of games and agents the models become agent-based models directed at understanding specific systems.

Multiscale analyses of game theory provide new insights into the relevance of game theory to collective social processes (Stacey, Gros and Bar-Yam 2011).

1.3.5 Generic Architectures

The concept of a network, describing the connectivity, accessibility or relatedness of components in a complex system, is widely recognized as important in understanding these systems. So much so, that many names of complex systems include the term 'network.' Among the systems that have been identified thus are: artificial and natural transportation networks (roads, railroads, waterways, airways) (Geographic Information Services 2001; Maritan et al. 1996; Banavar, Maritan and Rinaldo 1999; Dodds and Rothman 2000), social networks (Wasserman and Faust 1994), military forces (Alberts, Garstka and Stein 1999; Joint Vision 2010; Future Combat Systems; Bar-Yam 2001; National Defense University 1997; Priest 2001; Cares 2002), the Internet (Cheswick and Burch n.d.; Claffy, Monk and McRobb. 1999; Zegura, Calvert and Donahoo 1997), the World Wide Web (Lawrence and Giles 1999; Huberman et al. 1998; Huberman and Lukose 1997), biochemical networks (Service 1999; Normile 1999; Weng, Bhalla and Iyengar 1999; Hartwell et al. 1999), neural networks (Anderson and Rosenfeld 1988; Bishop 1995; Kandel, Schwartz and Jessell 2000), and food webs (Williams and Martinez 2000). Networks are anchored by topological information about nodes and links, with additional information that can include nodal locations and state variables, link distances, capacities and state variables, and possibly detailed local functional relationships involved in network behaviors.

Networks may be understood as universal properties in a multiscale analysis in which system properties require characterization of the network for description of its collective behavior (Bar-Yam 2016).

1.4 APPLICATIONS OF MULTISCALE ANALYSIS

The full richness of complex systems applications for multiscale analysis cannot be captured here. However, a few examples should provide a sense of the integral nature of complex systems science to advances in biomedicine, cognitive science, and social and global systems.

1.4.1 Biomedical Systems

Applications of complex systems methods in biomedical systems include the study of biochemical networks (gene regulatory networks, metabolic networks and so on) that reveal the functioning of cells and the possibilities of medical intervention (Service 1999; Normile 1999; Weng, Bhalla and Iyengar 1999; Hartwell et al. 1999), detailed studies of the mechanisms and function of specific biochemical systems (von Dassau et al. 2001), and high throughput data acquisition in genomics and proteomics (Strausberg and Austin 1999). The key to a broader perspective on such applications is recognizing that the large quantities of data that are currently being collected are being organized into databases that reflect the data acquisition process rather than the potential use of this information. The description of cellular and multicellular organisms must capture the spatiotemporal dynamics of the system as well as the biochemical network and its dynamics. More significantly, the multiscale analysis of this data will enable characterizing the collective properties of the system, including health and disease.

The challenge is to develop comparative multiscale descriptions, including the variety across organisms (for example, human beings) and the variety that exists across types of organisms. Ultimately, the purpose is to develop an understanding/description of the patterns of biological systems today as well as their evolution. The objective of understanding variety and evolution requires us to understand not just any particular biochemical system, but the space of possible systems, their general properties, their specific mechanisms, how these general properties carry across organisms and how they are modified for different contexts. Approaches that study large-scale biological structure and function as well as information flow are necessary. For healthcare in particular, abstracting the large-scale behavior from molecular interactions will lead to an effective knowledge resource about interventions.

1.4.2 Cognitive Systems

The problem of understanding the brain and mind can be understood quite generally through the role of relationships between patterns in the world and patterns of neuronal activity and synaptic change. While the physical/biological structure of the system is the brain, the properties of the patterns identify the psycho-functioning of the mind. The relationship of external and internal patterns is further augmented by relationships between multiple patterns that are possible within the brain. This complex nonlinear dynamic system has a great richness of valid statements that can be made about it, but identifying an integrated understanding of the brain/mind system cannot be captured by perspectives focusing on particular representations. Indeed, the potential contributions of the diverse approaches to studies of brain and mind have been limited by the difficulty in relating them to each other.

A key way to make progress is the adoption of a multiscale analysis that identifies the universality of representations, that is, relates different representations to each other as to what they actually represent. Since many kinds of representations represent the same things, such an effort would unify or help to distinguish the unique contributions of different approaches to neuroscience.

The multiscale approach can further contribute principles that are necessary for the understanding of practical issues in cognitive function, including teaching and learning, and the role of complexity in individual and societal function. An approach that recognizes the differences between individuals is needed.

1.4.3   Global Systems

In our increasingly complex, interdependent world, it is important to recognize how changes in one part of the world can have important effects in another. Complex systems science, using multiscale analysis to identify the largest scale effects, has the ability to describe dependencies and infer their policy implications. National and international policies should be informed by complex systems science to evaluate global consequences. For example, these methods can be used to trace the cause of the Arab Spring to market policies in the US. The wave of social unrest known as the Arab Spring was preceded by food riots, the result of spiking global food prices. In turn, the cause of the fluctuations in the food markets can be traced to commodities deregulation in the US, which allowed for rampant speculation, as well as ethanol fuel mandates which promoted the inefficient conversion of food into fuel (Lagi, Bertrand and Bar-Yam 2011a; Merchant 2014; Lagi et al. 2011b).

Similar policy decisions in the US precipitated the 2008 economic crisis, as well as other market crashes (Harmon et al. 2011; Harmon et al. 2010). Global interconnectedness also plays a role in the incidence of ethnic violence (Rutherford et al. 2014). Increasing long-distance travel is crucial for the modern global economy, but it also acts a vector for the transmission of pathogens (Rauch and Bar-Yam 2006) including a new strain of Ebola virus that spread internationally in 2014.

Among the key problems in studies of global systems is understanding the indirect effects of global human activity, which in many ways has reached the scale of the entire earth/biosphere. The possibility of human impact on global systems through overexploitation or other byproducts of industrial activity has become a growing socio-political concern. The cascading effects of societal problems are also a concern. Our effectiveness in addressing these questions will require a greater level of understanding and representations of indirect effects, effective interventions, and which aspects of a system can be understood or predicted based upon available information.

In general, the ability of humanity to address these global problems must rely upon the collective behavior of people around the world. Global action is now almost standard in responses to everything from local natural disasters to wars to environmental concerns. The high complexity of these problems implies that many individuals, who are diverse and yet coordinated, must be involved in addressing these problems.

1.5   CONCLUSIONS

The excitement in the study of complex systems arises not from a complete set of answers but rather from the appearance of a new set of questions. These questions differ from the conventional approaches and provide an opportunity for advances in understanding and in applications. Human civilization, across multiple scales from biological molecules to international economic systems, and its environmental context, are all complex. The most reliable prediction possible is that this complexity will continue to increase. The increasing complexity suggests that there will be a growing need for understanding of complex systems as a counterpoint to the increasing specialization of professions and professional knowledge. The insights of complex systems research and its methodologies, including multiscale analysis, may become pervasive in guiding research and policy decisions, across disciplines as diverse as biomedical, information, cognitive, and global systems.

REFERENCES

Alberts, D.S., J.J. Garstka and F.P. Stein (1999), 'Network centric warfare', *DoD C4ISR Cooperative Research Program*.

Allen, B., M. Kon and Y. Bar-Yam (2009), 'A new phylogenetic diversity measure generalizing the Shannon index and its application to phyllostomid bats', *The American Naturalist*, **174** (2), 236–243, http://doi.org/10.1086/600101.

Allen, B., B.C. Stacey and Y. Bar-Yam (2014), 'An information-theoretic formalism for multiscale structure in complex systems', *arXiv:1409.4708*, September 16.

Anderson, J.A. and E. Rosenfeld (eds) (1988), *Neurocomputing*, Cambridge, MA, USA: MIT Press.

Aumann, R.J. and S. Hart (eds) (1992), *Handbook of Game Theory with Economic Applications*, vol. 1–2, Amsterdam: North-Holland.

Ball, P. (1999), *The Self-made Tapestry: Pattern Formation in Nature*, Oxford, UK: Oxford University Press.

Banavar, J.R., A. Maritan and A. Rinaldo (1999), 'Size and form in efficient transportation networks', *Nature*, **399**, 130–132.

Bar-Yam, Y. (1997), *Dynamics of Complex Systems*, New York, NY, USA: Perseus.

Bar-Yam, Y. (2001), 'Multiscale representation phase I', *Final Report to Chief of Naval Operations Strategic Studies Group*.

Bar-Yam, Y. (2002), 'Complexity rising: From human beings to human civilization, a complexity profile', in *Encyclopedia of Life Support Systems*, Oxford, UK: EOLSS UNESCO Publishers.

Bar-Yam, Y. (2004a), 'Multiscale variety in complex systems', *Complexity*, **9** (4), 37–45.

Bar-Yam, Y. (2004b), *Making Things Work: Solving Complex Problems in a Complex World*, Reading, MA, USA: Knowledge Press.

Bar-Yam, Y. (2016), 'From big data to important information', *Complexity*, doi: 10.1002/cplx.21785.

Bishop, C.M. (1995), *Neural Networks for Pattern Recognition*, Oxford, UK: Oxford University Press.

Cares, J.R. (2002), *New Perspectives on Conventional Military Force and the War on Terrorism*, Washington, DC, USA: Sponsored by Joint Chiefs of Staff.

Cheswick, B. and H. Burch (n.d.), *Internet Mapping Project*, http://www.cs.bell-labs.com/who/ches/map/ (accessed July 25, 2017).

Claffy, K., T.E. Monk and D. McRobb (1999), 'Internet tomography', *Nature*, January 7.

Cohen, I. and L.A. Segel (eds) (2001), *Design Principles of the Immune System and Other Distributed Autonomous Systems*, Oxford, UK: Oxford University Press.

Cvitanovic, P. (ed.) (1989), *Universality in Chaos: A Reprint Selection*, 2nd edition, Bristol, UK: Adam Hilger.

Darwin, C. (1859 [1964]), *On the Origin of Species (By Means of Natural Selection)*, reprinted 1964, Cambridge, MA, USA: Harvard University Press.

Davidson, A., M.H. Teicher and Y. Bar-Yam (1997), 'The role of environmental complexity in the well-being of the elderly', *Complexity and Chaos in Nursing*, **3** (5).

Devaney, R.L. (1989), *Introduction to Chaotic Dynamical Systems*, 2nd edition, Reading, MA, USA: Addison-Wesley.

Devaney, R.L. (1992), *A First Course in Chaotic Dynamical Systems: Theory and Experiment*, Reading, MA, USA: Addison-Wesley.

Dodds, P.S. and D.H. Rothman (2000), 'Scaling, universality, and geomorphology,' *Annual Review of Earth and Planetary Sciences*, **28**, 571–610.

Forrest, S., S.A. Hofmeyr and A. Somayaji (1997), 'Computer immunology', *Communications of the ACM*, **40**, 88–96.

Fudenberg, D. and J. Tirole (1991), *Game Theory*, Cambridge, MA, USA: MIT Press.

Future Combat Systems, http://www.darpa.mil/fcs/ (accessed July 25, 2017).

Gallagher, R. and T. Appenzeller (1999), 'Beyond reductionism', *Science*, **284** (79).

Geographic Information Services (2001) *National Transportation Atlas, Transportation Networks*, http://www. bts.gov/gis/ (accessed July 25, 2017).

Goldberg, L.A., P.W. Goldberg, C.A. Phillips and G.B. Sorkin (1998), 'Constructing computer virus phylogenies', *Journal of Algorithms*, **26** (1), 188–208.

Golubitsky, M., I. Stewart, P.L. Buono and J.J. Collins (1999), 'Symmetry in locomotor central pattern generators and animal gaits', *Nature*, **401** (6754), 693–695.

Goodwin, B. (1994), *How the Leopard Changed its Spots: The Evolution of Complexity*, New York, NY, USA: Charles Scribner's Sons.

Harmon, D., B. Stacey, Y. Bar-Yam and Y. Bar-Yam (2010), 'Networks of economic market interdependence and systemic risk', *arXiv:1011.3707v2*, November 17.

Harmon, D., M. de Aguiar, D. Chinellato, D. Braha, I. Epstein and Y. Bar-Yam (2011), 'Predicting economic market crises using measures of collective panic', *arXiv:1102.2620v1*, February 13.

Hartwell, L.H., J.J. Hopfield, S. Leibler and A.W. Murray (1999), 'From molecular to modular cell biology', *Nature*, **402 supplement** (6761), C47–C52.

Holland, J.H. (1992), *Adaptation in Natural and Artificial Systems*, 2nd edition, Cambridge, MA, USA: MIT Press.

Holland, J.H. (1995), *Hidden Order: How Adaptation Builds Complexity*, Reading, MA, USA: Helix Books, Addison-Wesley.

Huberman, B.A. and R.M. Lukose (1997), 'Social dilemmas and internet congestion', *Science*, **277**, 535–538.

Huberman, B.A., P. Pirolli, J. Pitkow and R.M. Lukose (1998), 'Strong regularities in world wide web surfing', *Science*, **280**, 95–97.

Hutchins, E. (1995), *Cognition in the Wild*, Cambridge, MA, USA: MIT Press.

Joint Vision (2010), http://www.dtic.mil/jv2010/jv2010.pdf (accessed July 25, 2017).

Kandel, E.R., J.H. Schwartz and T.M. Jessell (eds) (2000), *Principles of Neural Science*, 4th edition, New York, NY, USA: McGraw-Hill.

Kauffman, S.A. (1993), *The Origins of Order: Self Organization and Selection in Evolution*, Oxford, UK and New York, NY, USA: Oxford University Press.

Kauffman, S.A. (1995), *At Home in the Universe*, Oxford, UK and New York, NY, USA: Oxford University Press.

Kephart, J.O. (1994), 'A biologically inspired immune system for computers', in R.A. Brooks and P. Maes (eds), *Artificial Life IV, Proceedings of the Fourth International Workshop on the Synthesis and Simulation of Living Systems*, Cambridge, MA, USA: MIT Press, pp. 130–139.

Kephart, J.O., G.B. Sorkin, D.M. Chess and S.R. White (1997), 'Fighting computer viruses: Biological metaphors offer insight into many aspects of computer viruses and can inspire defenses against them', *Scientific American*, November.

Lagi, M., K. Bertrand and Y. Bar-Yam (2011a), 'The food crises and political instability in North Africa and the Middle East', *arXiv:1108.2455*.

Lagi, M., Y. Bar-Yam, K. Bertrand and Y. Bar-Yam (2011b), 'The food crises: A quantitative model of food prices including speculators and ethanol conversion', *arXiv:1109.4859*, September 21.

Lawrence, S. and C.L. Giles (1999), 'Accessibility of information on the web', *Nature*, **400**, 107–109.

Lawton, M.P. (1981), *Environment and Aging*, Monterrey, CA, USA: Brooks/Cole.

Mandell, A.J. and M.F. Schlesinger (1990), 'Lost choices: Parallelism and topo entropy decrements in neurobiological aging', in S. Krasner (ed.), *The Ubiquity of Chaos*, Washington, DC, USA: American Association for the Advancement of Science, pp. 35–46.

Maritan, A., F. Colaiori, A. Flammini, M. Cieplak and J. Banavar (1996), 'Universality classes of optimal channel networks', *Science*, **272**, 984–986.

Maynard Smith, J. (1982), *Evolution and the Theory of Games*, Cambridge, MA, USA: Cambridge University Press.

Meinhardt, H. (1994), *The Algorithmic Beauty of Sea Shell Patterns*, New York, NY, USA: Springer-Verlag.

Merchant, B. (2014), 'The math that predicted the revolutions sweeping the globe right now', *Motherboard*, February 19.

Murray, J.D. (1989), *Mathematical Biology*, New York, NY, USA: Springer-Verlag.

National Defense University (1997), '1997 Strategic Assessment', *Institute for National Strategic Studies*, http://www.ndu.edu/inss/sa97/sa97exe.html (accessed July 25, 2017).

National Institute of Mental Health (n.d.), *Learning and the Brain*, http://www.edupr.com/brain4.html (accessed July 25, 2017).

Nee, S. and R.M. May (1997), 'Extinction and the loss of evolutionary history', *Science*, **278**, 692–694.

Nielsen, J. (1993), *Usability Engineering*, Boston, MA, USA: Academic Press.

Nijhout, H.F. (1992), *The Development and Evolution of Butterfly Wing Patterns*, Washington, DC, USA: Smithsonian Institution Press.

Noest, A.J. (2000), 'Designing lymphocyte functional structure for optimal signal detection: Voila, T cells', *Journal of Theoretical Biology*, **207** (2), 195–216.

Norman, D.A. and S. Draper (eds) (1986), *User Centered System Design: New Perspectives in Human–Computer Interaction*, Hillsdale, NJ, USA: Erlbaum.

Normile, D. (1999), 'Complex systems: Building working cells "in silico"', *Science*, **284** (80).

NSF (2001) *Biocomplexity in the Environment, NSF-02-010*, Washington, DC, USA: NSF.

Ott, E. (1993), *Chaos in Dynamical Systems*, Cambridge, MA, USA: Cambridge University Press.

Paxson, V. (1996), 'End-to-end routing behavior in the internet', paper presented at the ACM SIGCOMM '96 Conference on Communications, Architectures and Protocols, Stanford University, August.

Perelson, A.S. and F.W. Wiegel (1999), 'Some design principles for immune system recognition', *Complexity*, **4**, 29–37.

Pierre, D.M., D. Goldman, Y. Bar-Yam and A.S. Perelson (1997), 'Somatic evolution in the immune system: The need for germinal centers for efficient affinity maturation', *Journal of Theoretical Biology*, **186**, 159–171.

Priest, D. (2001), 'Special forces may play key role', *Washington Post*, September 14.

Rauch, E.M. and Y. Bar-Yam (2004), Theory predicts uneven distribution of genetic diversity within species, *Nature*, **431**, 449.

Rauch, E.M. and Y. Bar-Yam (2006), 'Long-range interaction and evolutionary stability in a predator-prey system', *Physical Review E*, **73**, 020903.

Rutherford, A., D. Harmon, J. Werfel, S. Bar-Yam, A.S. Gard-Murray, A. Gros and Y. Bar-Yam (2014), 'Good fences: The importance of setting boundaries for peaceful coexistence', *PLoS ONE*, **9** (5), e95660.

Segel, L.A. (1984), *Modeling Dynamic Phenomena in Molecular and Cellular Biology*, Cambridge, MA, USA: Cambridge University Press.

Service, R.F. (1999), 'Complex systems: Exploring the systems of life', *Science*, **284** (80).

Shannon, C.E. (1948 [1963]), 'A mathematical theory of communication', *Bell Systems Technical Journal*, reprinted in C.E. Shannon and W. Weaver (1963), *The Mathematical Theory of Communication*, Urbana, IL, USA: University of Illinois Press.

Stacey, B.C., A. Gros and Y. Bar-Yam (2011), 'Beyond the mean field in host-pathogen spatial ecology', *arXiv:1110.3845*, October 18.

Stein, L.A. (1999), 'Challenging the computational metaphor: Implications for how we think', *Cybernetics and Systems*, **30** (6), 473–507.

Stern, P.C. and L.L. Carstensen (eds) (2000), *The Aging Mind: Opportunities in Cognitive Research*, Washington, DC, USA: National Academies Press.

Strausberg, R.L. and M.J.F. Austin (1999), 'Functional genomics: Technological challenges and opportunities', *Physiological Genomics*, **1**, 25–32.

Strogatz, S.H. (1994), *Nonlinear Dynamics and Chaos. With Applications to Physics, Biology, Chemistry, and Engineering*, Reading, MA, USA: Addison-Wesley.

Triantafyllou, G.S. and M.S. Triantafyllou (1995), 'An efficient swimming machine', *Scientific American*, **272**, 64–70.

Turing, A. (1952), 'The chemical basis of morphogenesis', *Philosophical Transactions of the Royal Society of London B*, **237**, 37–72.

von Dassau, G., E. Meir, E.M. Munro and G.M .Odell (2001), 'The segment polarity is a robust developmental module', *Nature*, **406**, 188.

von Neumann, J. and O. Morgenstern (1944), *Theory of Games and Economic Behavior*, Princeton, NJ, USA: Princeton University Press.

Wasserman, S. and K. Faust (1994), *Social Network Analysis*, Cambridge, MA, USA: Cambridge University Press.

Weng, G., U.S. Bhalla and R. Iyengar (1999), 'Complexity in biological signaling systems', *Science*, **284** (92).

Williams, R.J. and N.D. Martinez (2000), 'Simple rules yield complex food webs', *Nature*, **404**, 180–183.

Wilson, K.G. (1983), 'The renormalization-group and critical phenomena', *Reviews of Modern Physics*, **55** (3), 583–600.

Wolfenson, J.D. (1999) *A Proposal for a Comprehensive Development Framework (A Discussion Draft)*, Washington, DC, USA: World Bank.

World Bank (1998), *Partnership for Development: Proposed Actions for the World Bank*, Washington, DC, USA: World Bank.

Zegura, E.W., K.L. Calvert and M.J. Donahoo (1997), 'A quantitative comparison of graph-based models for Internet topology', *IEEE/ACM Transactions on Networking*, **5**, 770–787.

2. Complex evolving social systems: unending, imperfect learning
Professor Peter Allen, Cranfield University

2.1 INTRODUCTION

In this chapter we will attempt to explore the basis on which the understanding of social systems could be made. If such understanding could be gained then clearly, it would give decision and policymaking a better chance of achieving any of its aims. Also, it might inform us as to what 'aims' are desirable, possible and achievable. Since the biological and human worlds are the result of ongoing evolutionary processes, then the question of what methodology might help us understand comes down to deciding if and how systems of evolutionary processes could possibly be understood. What exactly is the 'science' of social systems?

In the natural sciences, for many situations we can perform repeatable experiments repeatedly. This allows us to find general laws and patterns and make predictions about 'similar' situations elsewhere. This has led to the extraordinary bounty of science and technology. We can both find robust 'laws' by induction, and deduce specific behaviour from deduction. This is 'hard science'.

In social/biological and especially human systems, repeatable experiments are much more difficult to establish, because the people or organisms that inhabit a situation are really in co-evolution with each other and their environment, learning imperfectly from their individual experiences and changing over time. So, it is much harder to use induction to find general rules and patterns that seem to hold, and much more dubious to use such

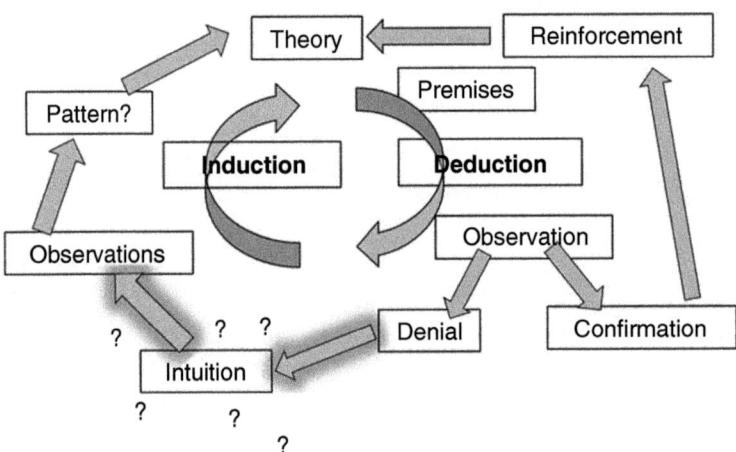

Figure 2.1   The scientific induction–deduction learning loop. The glowing arrows cannot be defined scientifically

rules or patterns to 'predict' the outcome of a particular case. This is because the essence of a 'repeatable experiment' is that you believe the internal constituents and their behaviours and responses are the same in the 'repeat' experiments and that they are operating in the same environment. But, as we know, particular local histories and emergent cultures modify the beliefs and responses of the elements and individuals in any particular social system, and so it can never be totally clear that the knowledge, values and actions will be the same as for any previous experiments. In everyday life we almost certainly learn pragmatically rules that work well enough for us to avoid immediate death, but more benign behaviours may leave trails that are difficult to distinguish in the noise of ordinary life (Allen 2014).

In his remarkable book *The Alchemy of Finance* George Soros (1987, pp. 304–305) points out that we should not expect the social sciences to be as 'hard' as physical science based on repeatable experiments:

> How could social science be protected against this interference? I propose a simple remedy: recognize a dichotomy between the natural and social sciences. This will ensure that social theories will be judged on their merits and not by a false analogy with natural science. I propose this as a convention for the protection of scientific method, not as a demotion or devaluation of social science. The convention sets no limits on what social science may be able to accomplish. On the contrary, by liberating social science from the slavish imitation of natural science and protecting it from being judged by the wrong standards, it should open up new vistas.

This is a very important point. But we can also now see that the development of complexity science itself already destroys the 'infallibility' of hard science, because open systems of interacting elements can be shown to exhibit multiple possible futures and to defy conclusive prediction. Such ideas applied to social systems will be the central theme of this chapter showing that not only is Soros right concerning social systems of thinking individuals, but what he says is right for complex systems of purely physical entities whose nonlinear interactions make them capable of self-organizing behaviour, with emergent characteristics and capabilities. Instead of aiming at complete prediction and control, as expected for the old 'hard' science, complex systems models can be built which allow different possible future trajectories, but can nevertheless allow us to reflect upon and formulate possible policies and interventions into human systems. Then, when a plan is chosen and put into action, the model can be used to compare the expected outcome with reality. Then unexpected deviations and new phenomena can be spotted as soon as possible and plans and models revised to explore a new range of possible futures (Allen 2013).

2.2   INTERPRETIVE FRAMEWORKS: EVOLUTION, DYNAMICS AND EQUILIBRIUM

In social science we seek an understanding and an explanation of a situation involving people. What are the characteristics and elements that make up the structure? Why did it become what it is? Should we act in some way? How might it change and what is a good idea, and for whom? We see that this implies a description of the emergence of the 'present' from the past, and possibly then an ability to reflect on possible futures. This certainly sounds like a sensible goal if we are to move beyond laissez-faire.

2.2.1   Understanding by Making Successive Assumptions/Simplifications

The surprise is that when we try to 'frame' problems and work out what is what, what is going on and why, we actually proceed by making successive simplifications/assumptions. We move from a totally undefined literary 'impression' of a social situation, towards the identification of descriptors, of characteristic variables, and of interactions and mechanisms and that make it what it is. Looking into its history shows what happened through time as new people, ideas and activities came in and shaped the organization, the land and city scape. In a fundamental diagram (Figure 2.2), we identify the vital steps that take us to interpretive frameworks that start off as literary and impressionistic, through an evolutionary tree of changing forms and actions, towards short term mechanical representations that may move towards possible equilibrium views of reality (Allen, Strathern and Baldwin 2007).

In Figure 2.2 as we move to the right from Reality (the vague 'cloud' on the left) we come to successively simpler, more understandable and less detailed representations of that reality. And although Reality, on the left, may evolve qualitatively over time, adding new variables and structures, the models on the right of Figure 2.2 cannot. They can only 'run' but not 'evolve', or are assumed to have run already and led to a stable situation. So if we monitor 'reality' against our models, then we shall be forced to create successive modified dynamical systems – which we shall only be able to do post-hoc.

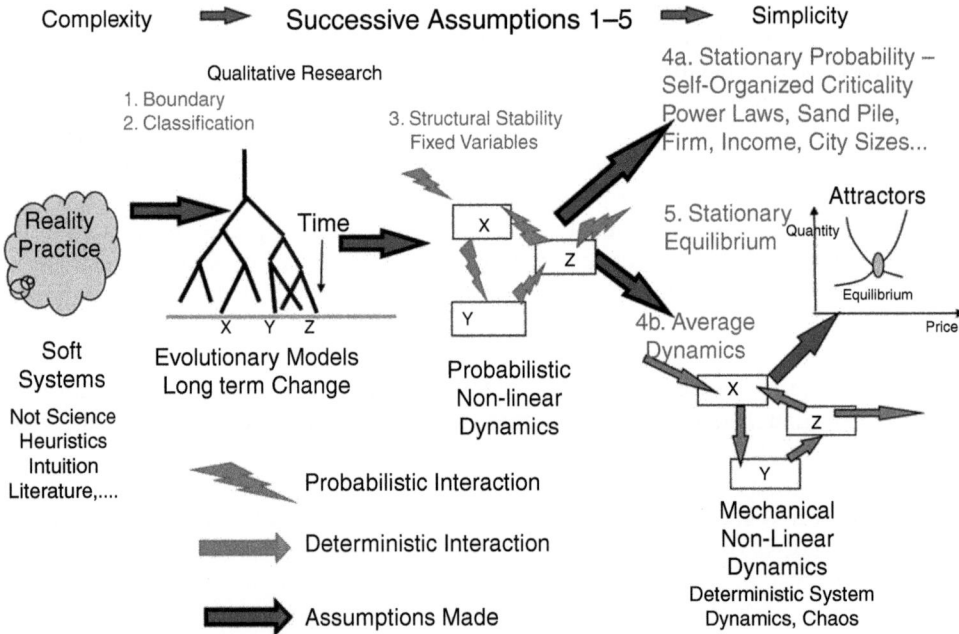

Figure 2.2   *Successive assumptions take us from incomprehensible 'reality', through evolutionary systems, to probabilistic nonlinear dynamics, then either to stationary solutions of this, or to average, 'system dynamics'. The final assumptions take us to the attractors of the dynamics*

In other words, in an evolving world, our representations will always be pictures of the past!

In science, understanding and prediction are achieved in practice by making models that express successive assumptions concerning the situation. This means that we exchange uncertainty about the system for uncertainty about the truth of our assumptions. If no assumptions are made then we are in the realm of narrative, where we are limited in our ability to learn or generalize or predict. When we make lots of assumptions we appear to have simple clear predictive models but are uncertain whether our assumptions are still true. Uncertainty is an irreducible fact.

Assumption 1: The boundary

Our first step (assumption) is to decide on the entity that we wish to study – what is 'inside' and what is outside, in the environment. We wish to try to understand the behaviour of what is 'inside' in the context of its environment. In fact it may not be clear where exactly the boundary should lie, and so we really proceed by choosing an 'experimental' boundary and seeing whether the model that results is useful. In fact this first assumption is actually tied up with the second, because the boundary can also be defined as being the decision to include or exclude a particular element or variable from the 'model' and leave it in the 'environment'. The elements within the 'system' will affect each other and so this is the core, qualitative specification of what will be studied – the descriptors and variables of the problem.

Assumption 2: Evolutionary complex models – changing taxonomy

The 'contents' of the system could be the different species in an ecosystem, different types of people, diverse economic activities, or different types of job within an organization. The functioning of the system will be about the flows of energy, materials, money or products that characterize the running of the system in time. However, while the constituents of the system might remain the same for some time, there will have been moments when new activities and types appeared and when some older ones disappeared. This is important because it could lead to changes as to what is 'in' the system and what things are in the environment. Complex systems can possess emergent behaviours that connect them to new factors in the environment. In this way, one of the important properties of a complex system is that it can itself change the boundary of the system. This means that the 'seeker after knowledge' must be sufficiently humble to admit that the initial choice of a boundary might need revision at some later time. The point is that 'learning' is an experiment that seeks a representation that is useful.

Social and economic systems such as markets have all evolved and changed over time as innovations, new technologies, new practices and markets have emerged. New types and activities emerge and others leave. Clearly, a model could simply be a description of what was present and when, but such statistical models offer no understanding. There is an absolutely critical question: How do systems get 'free' of the predictable behaviour that the underlying hard science appears to predict? After all every small piece of a system is subject to the hard science that is the repeatable experiments of physics and chemistry.

The wonderful example of origami shows us how (Figure 2.3). If I have sheet of paper then by folding it in particular sequences I can create a whole range of different 'objects' with emergent characteristics, features and capabilities! However, the laws of physics and

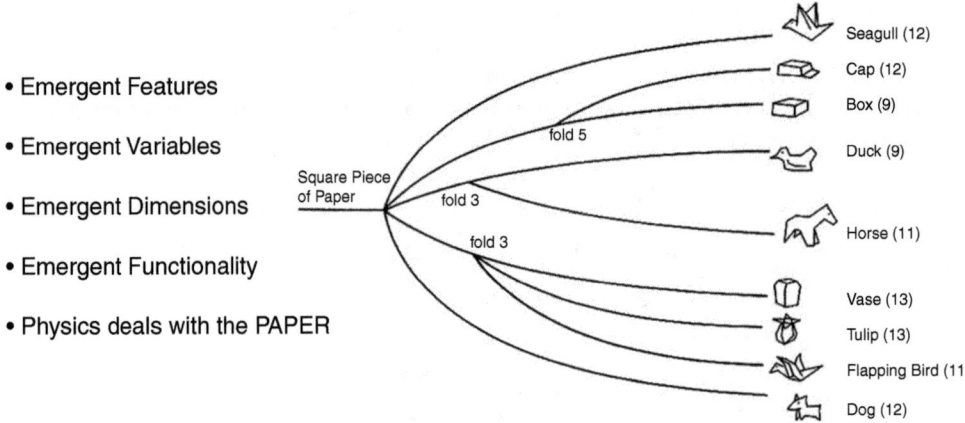

Figure 2.3   Origami illustrates the reality of emergence as symmetry breaking leads to new 'dimensions' of performance, and variables and selection change qualitatively

chemistry can only tell us about the colour, elasticity, weight and strength of the paper – but nothing about the emergent characteristics, features and capabilities of the objects that have been created. By folding the paper, we utilize the freedom of morphological creativity, to escape the boring determinism of the microscopic level. And if there is anything more morphologically creative than a sheet of paper it is a long string with stick bobbles in it – such as macromolecules like proteins and enzymes – whose creative accidental folding led eventually to living cells, and the many splendours of biology and indeed of humanity. This does not 'bring down' humanity to mere folding molecules but demonstrates that this freedom and creativity that emerges at the levels above that of physics and chemistry is virtually infinite.

Of course the laws of physics and chemistry are always obeyed, but they are channelled by emergent structures that create levels of organization and structure that go way beyond anything imaginable on the basis of the laws of 'hard science'. Folding leads to new forms, technologies, techniques and practices – can lead to entirely new emergent properties and dimensions of performance (Allen 1982).

At the mundane level of trying to understand some social situation, we see then that over time new types of individual can emerge, new variables and new levels of organization and structure with emergent functionalities and capabilities. This is the qualitative change that characterizes evolutionary processes. Molecules, elements and systems can adopt new morphologies that break some previous symmetry leading to emergent features, capabilities, and functionalities at the level above. Any particular mathematical model or representation of a system, developed at a given moment, will be in terms of the existing current taxonomy and capabilities. But in reality, 'reality and the model' will diverge over time! This means that the assumption of structural stability (systems with a fixed cast of players, a fixed taxonomy), cannot be taken for granted.

As the paper is folded features like heads, tails and wings emerge, as also do functionalities such as flapping wings, ability to hold water and so on. These features and functionalities allow us to give names to the emergent morphologies, and to potentially make more of the ones we like. This leads us to the longer term evolution of an origami (or a biological or human) world. While at lower levels different folds and changes are occurring, these will be retained or eliminated according to the emergent properties of their morphology. And the selection at this 'upper level' of morphology will not be able to suppress experimentation at the lower levels inside of individuals or groups. If the resulting morphology still 'works' as before then the interior details will be invisible to the outside world. This means that evolution at the lower level cannot be stopped and there will be diversity among individuals, possibility involving several steps before the new morphology, characteristics or capabilities reveal themselves. This can lead to both radical and marginal innovations hidden within populations and allowing both adaptation to changing circumstances and also genuine innovations. It is the continuing functionality of an existing morphology that allows experimentation and drift at the lower levels, which in turn makes qualitative evolution an inevitability. Instead of discussing 'a' or 'the' model of a social system, evolutionary and co-evolutionary changes lead us successively to a series of qualitatively different models – changed behaviours and new ideas.

So, these first two assumptions do not lead to any single clear picture of the system, but instead to an open, diverging series of possible 'models' that correspond to an evolving system with changing taxonomy. If we think of our assumptions as corresponding to 'constraints' on what the elements in the system can do, then we see that with only these first two assumptions the elements, and the organizations they are part of, are still free to change. Structural stability is not guaranteed when that much freedom is present in the system. Big change may well occur after perhaps a long period of protected 'exploration' at the lower, individual level below. This is an absolutely vital point. Novelties, potential innovations and new ideas need to be nurtured and protected within an organization until they are ready to face the outside world (Allen 1990, 1994).

Assumption 3: Dynamics of the current system
Despite the beauty of the ideas in the previous section, the real world tends only to pay for the promise of prediction. Not that social science should in any way be predicated on who will pay for it. This seems to depend on making further assumptions that take us further to the right of our fundamental diagram of Figure 2.1. The next assumption therefore is that our system is structurally stable – the variables, taxonomy, types of individual or agent do not change.

Take the variables we currently see in our system and see how they change over time. Hopefully, we may be able to make predictions about what will happen, or more interestingly, what will happen if I act on it in a particular way. We then 'calibrate' our model on the situation under study, and assume that no new surprise elements or variables can disturb the system. The changing values of the variables result from the rates of production, sales, growth, declines and movements in and out of the system. But the underlying rates of the different microscopic, local events will depend on each individual detailed event that actually occurs, and this kind of detail we do not find in the overall figures. We therefore replace the detail that we don't know with probabilities set around the average observed parameter values. We develop a dynamic model of the probabilities of different states for the variables of our system. These kinds of equations, known as the Master or Chapman–Kolmogorov equations (Allen 1988), govern the changing probability distribution of the different variables. The title 'Master' is used because they allow for the occurrence of *all possible sequences of events*, taking into account their relative probability, rather than simply assuming that only the most probable events occur. In other words instead of simply considering that only average events occur, this model allows all events to occur, according to their relative probability. Episodes of good and bad luck are thereby included and such models can explore the resilience of the system to possible, but improbable extreme conditions or accidents. For any single system this allows into our scientific understanding the vital notion of 'freedom', 'luck' and 'uncertainty' in its behaviour.

Although, a system that is initially not at the peak of probability will *more probably* move towards the peak, it can perfectly well move the other way; it just happens to be *less probable* that it will. It is precisely these fluctuations that probe the stability of the most probable state. Such probabilistic systems can 'tip' spontaneously from one type of solution/attractor to another. It also underlines the very important idea that the 'average' for a system should be calculated from the distribution of its *actual* possible behaviours – not that the distribution of its behaviour should be calculated by simply adding a Gaussian distribution around the average. The Gaussian is a distribution much loved of economists which expresses the spread of random shots around a target. In reality though, the actual distribution is given by the full probabilistic dynamics and can be calculated precisely. It will in general have a complicated mathematical form, and will be changing over time according to the probabilistic dynamics.

Assumption 4: Either a) or b)
In order to proceed further with any 'prediction' about what must happen, we have two possible paths forward:

a) **Assume the probability is stationary:**
   The assumption that the probability distribution has reached a stationary state leads to the idea of the nonlinear dynamics leading to 'self-organized criticality' and to the power law structure that often characterizes them (Bak 1996).

b) **Use average trajectory, assuming probability remains sharply peaked: system dynamics**
   Instead of a) we can look at the 'average' dynamics of the system and see where this leads. The full probabilistic dynamics is a rather daunting mathematical problem of the changing probability distributions over time. However, much of the uncertainty can be taken away as if by magic by changing slightly the question that is being asked of the model. The change is so slight that many people simple do not realize that the problem has been vastly simplified by the artifice of making this particular assumption.

In a quest for funds, it is often better to change the exact question from 'what can actually happen to this system?' to 'what will most probably happen?' This seemingly innocent change takes us to a much simpler problem, avoiding all the very heavy maths involved in probabilistic dynamics. But most people do not realize the magnitude of the difference between these two questions. It is the difference between a heavy set of probabilistic equations describing the spreading and mixing of possible system trajectories into the future, and a representation in which the system moves cleanly into the future along a single, narrow trajectory. This simple deterministic trajectory appears to provide predictions and do 'what if' experiments in support of decision-making or policymaking. It appears to tell us exactly what will happen in the future with or without whatever action we are considering taking. These 'system dynamic' models are immensely appealing to decision makers since they seem to provide prediction and certainty. They allow the calculation of the expected outcomes of different possible actions or policies and allow 'cost/benefit' calculations. They can also show the factors to which the real situation is potentially very sensitive or insensitive, and this can provide useful information. But systems dynamics models are deterministic; they still only allow for one solution or path from a particular starting point.

Nevertheless, their 'predictions' allow comparisons with reality and can reveal when the model is failing. Without this, we might not know that the system had changed! And so this provides a basis for a learning experience where the each model is constantly monitored against reality to see when something has changed.

Assumption 5: Solutions of the dynamical system
The final assumption that one can make to simplify a problem even further is to consider the possible long-term solutions of the dynamical equations – the 'attractors' of the dynamics. This means that instead of studying how the system will run, one looks simply at where it might 'run to'. Of course, nonlinear interactions can lead to different possible 'attractors' – equilibrium points, permanent cycles, or chaotic attractors. This could be useful information – at least for some time.

But, of course, over longer times, the system will evolve and the equations that worked previously will become untrue, and the possible attractors will also. In reality there is a trade-off between the utility and simplicity of predictions, and the strength of the assumptions that are required in order to make them. Of course, it is much easier to 'sell' a model that appears to make solid predictions. Because of this scientists often have had to underplay the real level of uncertainty and doubt about the possible consequences of interventions, actions, technologies and practices, allowing the seemingly solid business plans and policy consequences to be presented as persuasively as possible. In any case, usually people wish to hear clear statements that imply knowledge and certainty and find the actual uncertainty and risk much more disturbing. People will often prefer a lie that comforts to the uncomfortable truth.

In giving advice it is of critical importance to know how long the assumptions made in the calculations may hold. This is how long the actual complexity and uncertainty of 'reality' can be expected to remain hidden from view. Of course, believers in 'free markets' can get around this problem by simply stating that whatever occurs is by definition the best possible outcome. But if we wish to give advice to particular players within the system then we will need to develop models that can explore possible futures as well as possible.

At any given time, of course, we would not be able to 'value' different types of micro-diversity, since we would not know which would in fact be important for some future problem. A theory based on the evolutionary emergence of micro-diversity, and the way that evolution itself adjusts its range (The Theory of Evolutionary Drive) was developed some time ago (Allen and McGlade 1987) but has not been much commented upon, even by evolutionary economists. Instead of a complex system being successfully described by any fixed set of components and mechanisms we see that the system of components and mechanisms is not fixed, but is itself changing with the events that occur. As the system runs, so it is changed by its running.

2.3   MODELLING HUMAN SYSTEMS

Behaviours, practices, routines and technologies are invented, learned and transmitted over time between successive actors and firms, and we shall discuss how the principles of Evolutionary Drive can be used to understand them. The models described here are what are called Multi-Agent Models or Agent-based Models. These were developed first in the urban and fishing models described in, for example, Allen (1997) and Allen and McGlade (1986). The fishing models anticipated the more general results of March (1991) in which his ABM models showed the importance of 'Discovery and Exploitation'.

2.3.1   Emergent Market Structure

Since the 'invisible hand' of Adam Smith, the idea of self-organization has been present in economic thought (Veblen 1898). However, towards the end of the 19th century mainstream economics adopted ideas from equilibrium physics as the basis for understanding. This led us to neoclassical economics that was strong on very general and rigorous theorems concerning artificial systems, but rather weak on dealing with reality in practice. Today, with the arrival of computers able to 'run' systems instead of us having to solve them analytically, interest is burgeoning in complex systems simulations and modelling. Complex systems thinking offers us an integrative paradigm, in which we retain the fact of multiple subjectivities, differing perceptions and views, and indeed see this as part of the complexity, as a source of creative interaction and of innovation and change. Building the model is in itself extremely informative – since it shows us mechanisms and ideas that were not apparent.

For example, in building a model we quickly find that a decision rule for the agents in charge of firms to expand production when there are profits and decrease it when there are losses, will not allow firms to launch a new product, since every new product must start with an investment – a loss. However, in the real world, firms are created and new products and services are developed. Therefore the 'equation' governing the increase or decrease of production volume cannot be based on the actual profits made instantaneously. This point is discussed in Allen (2014).

The next idea that occurs is that an economic agent, running a firm, would use 'expected' profits to adjust their production volume. So, firms moving into a new market area must be doing so because, they think that on balance their investment cost will be more than balanced by future profits. However, if we try to put this in our model we find that it is actually impossible for an agent to calculate expected profits for different pricing strategies because he does not know the strategies of other firms. Profits in each firm will depend on the products and prices of other firms and none of them know what the others will do.

Of course we could use the sort of neoclassical economics idea which would say that if firms are present then they must be operating with a strategy that maximizes profits. And if no firms ever went bankrupt then we might have to accept such an idea – but in reality we know that many firms do go bankrupt and therefore cannot have been operating at an optimal strategy. An examination of the statistics concerning firm failures (Foster and Kaplan 2001; Ormerod 2005) shows us that whatever it is that entrepreneurs or firms believe, they are quite often completely wrong. The bankruptcies, failure rates and life expectancies of firms all attest to the fact that the beliefs of the founders, managers or investors are often not correct. In trying to build our model we are faced with the fact that firms cannot know what strategy will maximize profits. Markets are not populated by competing rational agents (*Homo economicus*), but instead are the arenas of possible learning for people driven by self-belief, and not scared of risk.

By participating, players may find strategies, products, and mark-ups that work. Schumpeter (1962) was correct. At a given moment a market contains interacting firms that have entered and have not yet gone bankrupt. Some firms are growing and others shrinking. But with the entry into the market of new firms and products will come innovations and innovative organizations, and so over time the 'bar' will be raised by successive 'generations' of firms. Instead of supposing 'magical entrepreneurs and consumers' with perfect information and knowledge, our model shows us how real agents may behave with knowledge limited to what is realistically possible. They cannot calculate strategies and behaviours that fulfil (magically) the assumptions of (touchingly naïve) neoclassical economists. Our model shows us the many possible market trajectories into the future. None of these correspond to a 'global' optimum (maximum profits and utility) and indeed there is no global agent to oversee the process. Each different trajectory is a possible future history of the system and will bring corresponding winners and losers, and particular patterns of strategies, imitations and routines.

The complex, evolutionary market model has been presented before (Allen, Strathern and Baldwin 2007).

Figure 2.4 shows us the model's structure. There can be any number of interacting firms, but in the examples we employ there will be up to 18 present at any given moment. The internal structure of each firm is represented in the illustration labelled 'firm 1'. Production has fixed and variable costs, that depend on the quality of the product. It also needs sales staff to sell the stock to potential customers. On the right of the figure, there are three different types of potential customer, and we have chosen here to distinguish between three groups differing in their price sensitivity. The point is that potential customers are sensitive to the price/quality of the different products on offer, and so will be attracted differentially to the different firms competing in the marketplace, thus creating the 'selection mechanism'. Profits from sales allow increased production and pay off any debts. In this way, our model provides an evolutionary theatre within which competing and complementary strategies are generated, tested and retained or rejected.

The firm tries to finance its growth and avoid going near its credit limit. If it exceeds its credit limit then it is declared bankrupt and closed down. The evolutionary model then replaces the failed firm with a new one, with new credit and a new strategy of price and quality. Again this firm either survives or fails. The model assumes that managers want to expand to capture their potential markets, but are forced to cut production if sales fall. So, they can make a loss for some time, providing that it is within their credit limit, but they much prefer to make a profit, and so attempt to increase sales, and match production to this.

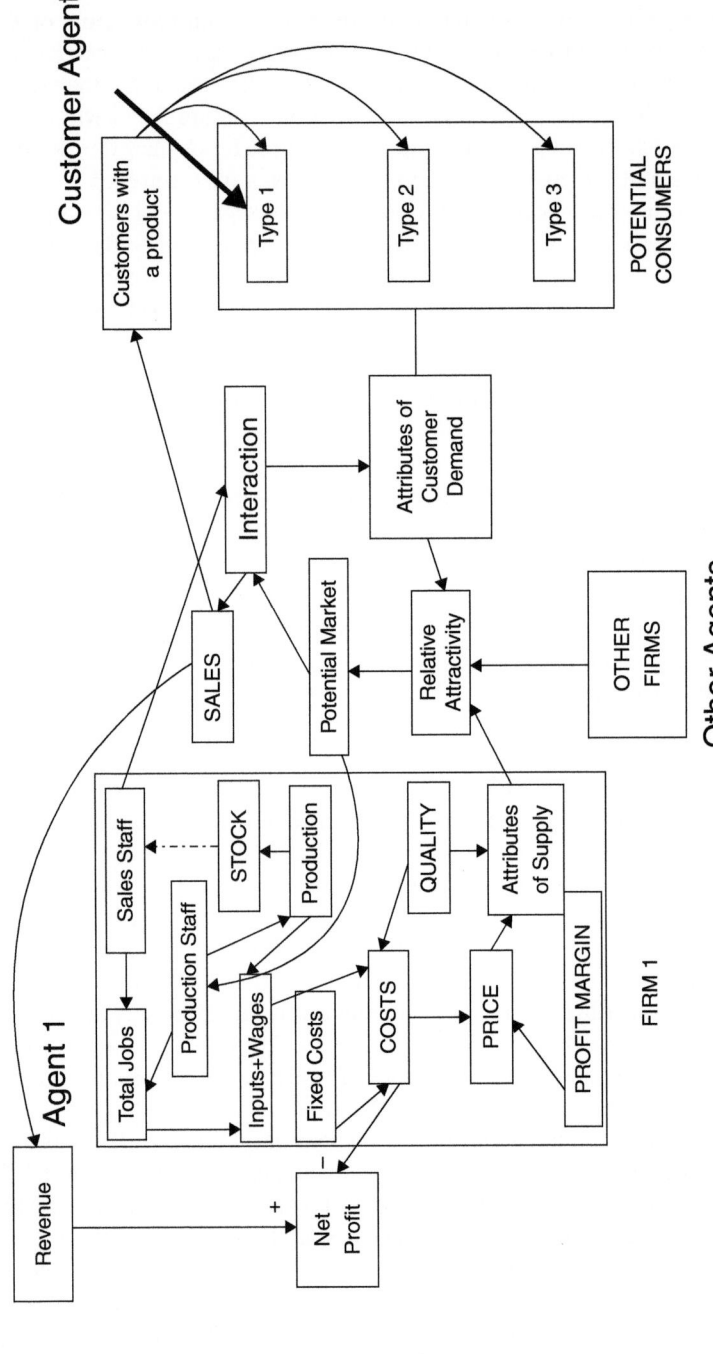

Figure 2.4 On the left we show one firm/agent in detail. It hires staff and facilities in order to make products of a certain quality and tries to sell them a price that leads to profit. The potential customers on the right choose to buy from the various products on the market, each with a given quality and price. Customers have different sensitivities to price. There are economies of scale of production

Before 1982, market models assumed simply that markets would move almost instantaneously towards the optimal outcomes (the neoclassical approach) – maximum profits for firms and maximum utility for consumers. But this idea was changed when Nelson and Winter (1982) developed an evolutionary model of markets where selection operated on changes in firms processes or technology. This was a great step forward but although the supply side was viewed dynamically the demand side was inferred in an abstract way. The model we present here is agent-based with both supply and demand agents creating an evolutionary process of market development. Here, we model explicitly both supply and demand, though in a relatively simple manner, and the 'uncertainty' resides in the impossibility of the firm agents knowing beforehand what the real pay-off will be for a given price/quality strategy.

2.3.2   Exploring the Three Meta-strategies

Running the model tells us that it is not the exact fixed strategy of a firm that matters. It is how successfully it can be changed if it isn't working! Learning what works for you is what matters, though the 'success' of any particular behaviour will always be temporary.

The question that we want to investigate is how firms change their strategies (quality, mark up, publicity, research and so on) over time and in the light of experience. We now consider the results of running a model with 18 competing firms with three different ways (meta-strategies) to CHANGE what they are doing. They can be:

- **Learners** – test the profits that would arise from small changes in quality or price, and move their production in the direction of increasing profits. The individual firms learn by experiment.
- **Imitators** – move their product strategy towards that of whichever firm is currently making most profits.
- **Darwinists** – adopt a strategy 'intuitively' and then stick with it. The individuals do not 'learn' by experiment but the 'species' or 'meta-strategy' might through differential survival.

In the simulations here we study the interaction and outcomes of six learning firms, six imitating firms and six Darwinists. Any firm that goes bankrupt is replaced by another with the same meta-strategy but starting from a different initial position of price and quality. We can then run our model repeatedly and see how well the three meta-strategies (Learning, Imitation, Darwinist) perform.

If we repeat the simulations for ten different random sequences then we find the overall results of Figure 2.5.

The 'average' message seems clear. Learning by experiment is the best meta-strategy. Using intuition and individual belief (Darwinist) is good, and imitating winners is the least successful meta-strategy. Imitators seem to arrive late to a strategy, and then suffer the competition of both the original user and that of other imitators. But the spread of the trajectories in Figure 2.6 shows that we cannot guarantee than any meta-strategy will always win. There is some overlap of outcomes and so in a particular case we can probably never say with absolute certainty that the 'learning' strategy will 'definitely' be better than the others, only that it will 'most probably' be better than the others. If a manager

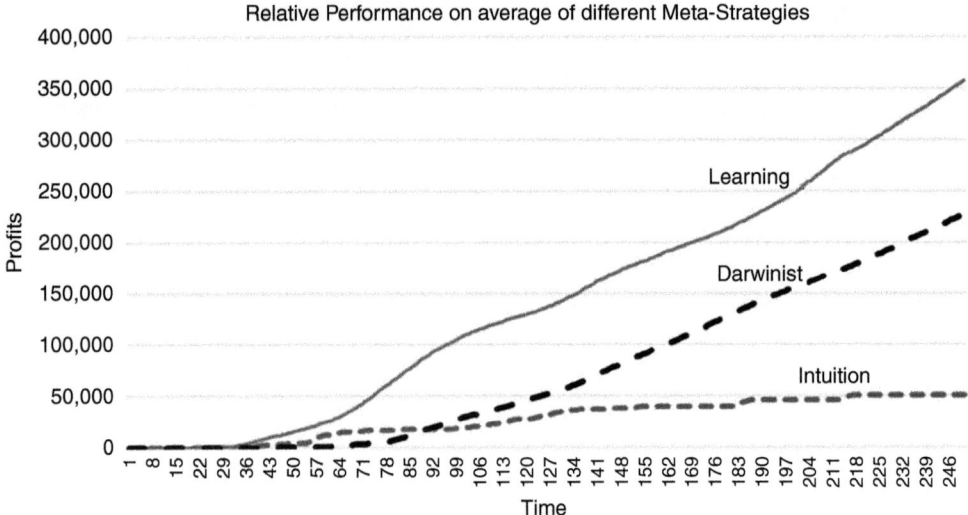

Figure 2.5   For 10 different random series, the average performance of the learning, imitator and Darwinist meta-strategies are clearly quite different

owned the simulation model, then it would still not guarantee they would definitely win, but only increase the probability of winning.

There are clear 'limits to knowledge' (Allen, Strathern and Baldwin 2007), in that future trajectories and strategies of other firms cannot be known, and therefore there cannot be a corresponding 'perfect' strategy. This puts a real limit on any predictive 'horizon' which may be until the next firm changes its strategy, or if it is pursuing a meta-strategy (of learning or imitating for example) how this will change over time. This example is limited to a discussion of innovations concerning the quality and price of a product or service, but the model can equally well look into more ambitious innovations of technology or research. The model shows us that the basic process of 'micro-variation' and differential amplification of the emergent behaviours is the most successful process in generating a successful market structure, and is good both for the individual players and for the whole market, as well as its customers (Foster and Metcalfe 2001; Metcalfe and Foster 2004).

Human agents will be reflecting and experimenting with their various approaches, technologies and ideas in their different roles as producers, suppliers and consumers. Clearly, the strength and success of these innovative flows will depend on the 'regimes' operating in the organizations concerned. So, there are important local conditions that affect this. First, there is the possibility of individuals being able to think new thoughts, and here the richness of the local cultural and technological environment will feed new ideas. Second, there has to be freedom, and mechanisms, by which new ideas, techniques and technologies can be tried out and tested. These are underlying conditions of endogenous externalities that really will be important in allowing evolutionary change to occur. Again any model that has fixed products, costs, prices and consumer preferences will rapidly be overtaken by the changing reality created by the evolving firms. Clearly, the simplistic ideas of neoclassical economics are quite ludicrous.

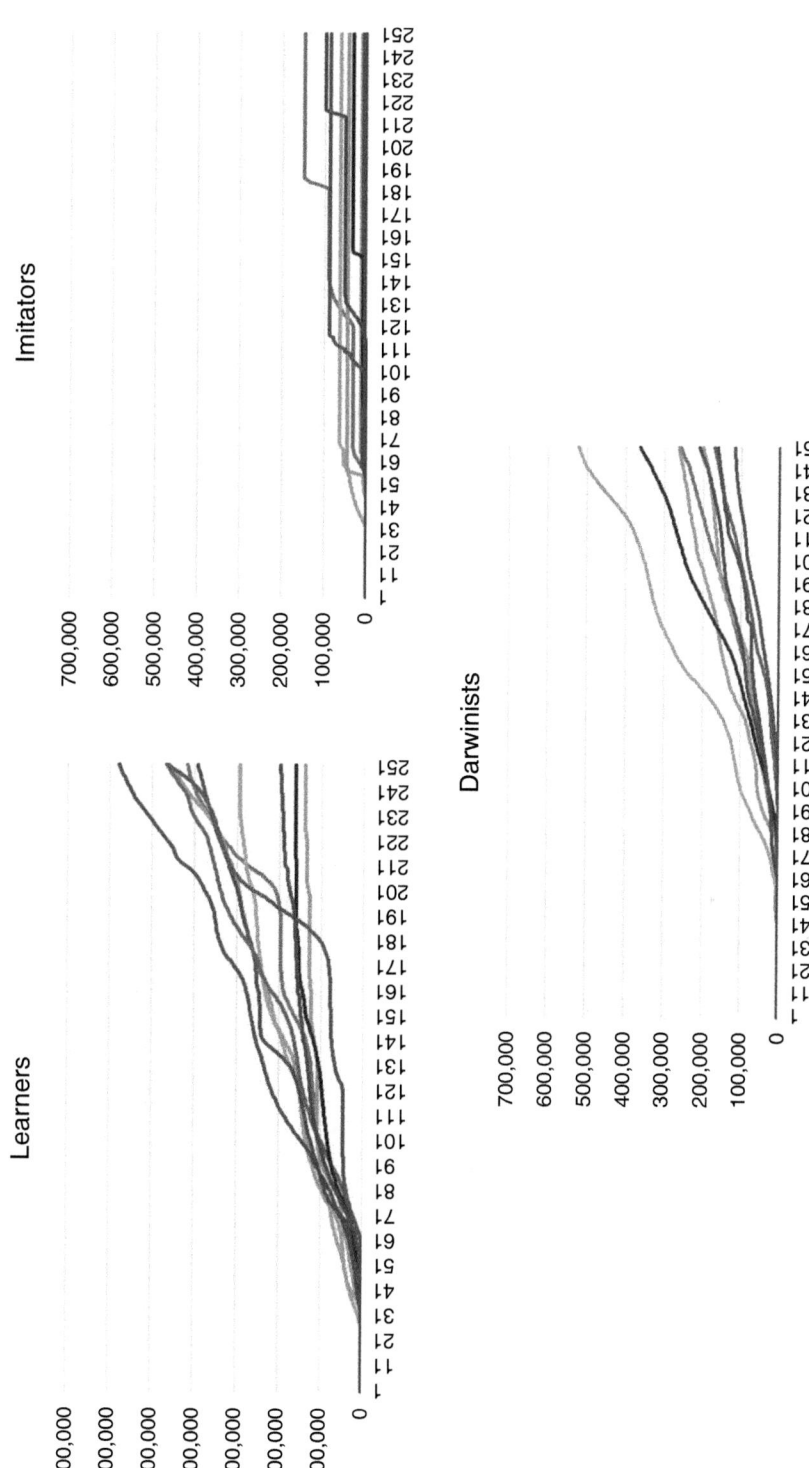

Figure 2.6   The trajectories of each computer run for the different meta-strategies. It shows that learners perform better than the others on average, but this result is probabilistic

Schumpeter had the genius to see that markets were not mechanical systems of robotic producers and consumers giving rise to equilibrium markets of homogeneous goods with maximal profits for producers and maximum utility for consumers. He saw that what mattered was what came into the system and what went out. His phrase for this was 'Creative Destruction', and this the same view as that coming out of complexity science. Both Darwin and Schumpeter saw and spoke a truth, which has taken a long time for science to come to terms with.

2.3.3   Organizational Evolution

In discussing how firms change their performances through product and process innovation, we can refer briefly to an example that has been published before (Allen, Strathern and Baldwin 2006; Allen and McGlade 1987). Changing patterns of practices and routines are studied using the ideas of Evolutionary Drive. For a particular industrial/business sector we find a 'cladistic diagram' (a diagram showing evolutionary history) showing the succession of new practices and innovative ideas within a particular economic activity. This idea looks at organizational change in terms of the emergence of particular 'bundles' of practices and techniques with performances that allow survival in the market. The ideas come from McKelvey (1982, 1994), McCarthy (1995) and McCarthy et al. (1997).

For the automobile sector the observed bundle of possible 'practices', our 'dictionary', allows us to identify 16 distinct organizational forms – 16 bundles of practice that actually exist:

- Ancient craft system; Standardized craft system; Modern craft system;
- Neocraft system; Flexible manufacturing; Toyota production;
- Lean producers; Agile producers; Just in time;
- Intensive mass producers; European mass producers;
- Modern mass producers; Pseudo lean producers; Fordist mass producers; and
- Large-scale producers; Skilled large-scale producers.

Cladistic theory calculates backwards the most probable evolutionary sequence of events. The key idea in the work presented here is to use a survey of manufacturers that explores their estimates of the ***pair-wise interactions*** between the practices. In this way we can 'predict' the synergetic 'bundles' of working practices and understand and make retrospective sense of the evolution of the automobile industry.

The evolutionary simulation model examines how the random introduction of new practices and innovations is affected by the changing 'receptivity' reflecting the overall effects of the positive or negative pairwise interactions. As a result of the particular sequence of attempted additions particular bundles of practices and techniques emerge, structural attractors, which correspond to different organizational forms. The model can generate the history of a particular 'firm' which launches new practices randomly, and grows where there is synergy between the practices. Figure 2.7 shows us one possible history of a firm. The particular choices of practices introduced and their timing allows us to assess how their performance evolved over time, and also assess whether they would have been eliminated by other firms.

Overall performance of each firm is a function of the synergy of the practices that

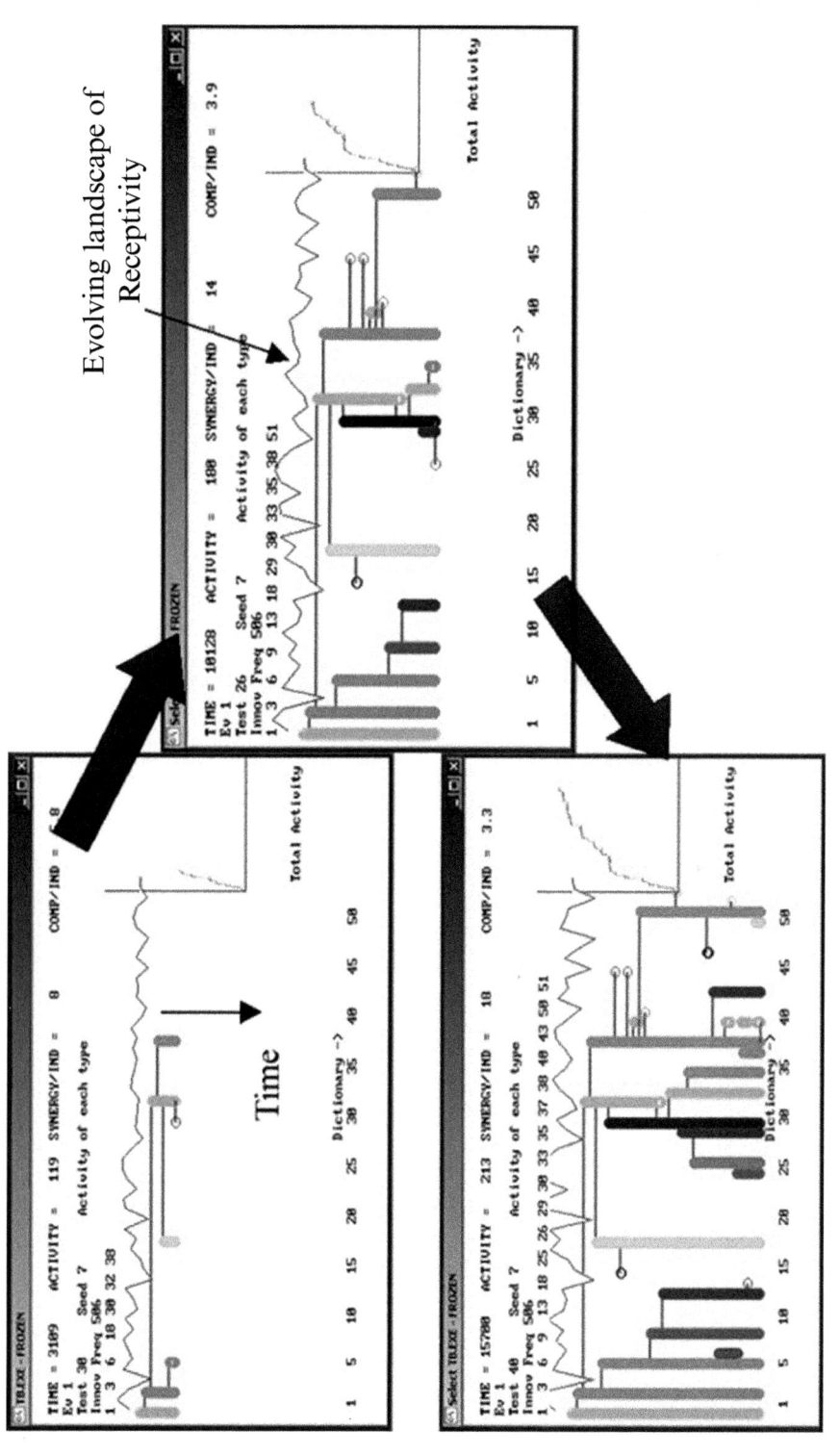

Figure 2.7   Successive moments (t=3000, 10000 and 15000) in the evolution of a particular firm as new practices are added randomly. The evolutionary tree of the organization emerges over time

33

are tried successfully in the context of the other evolving firms. The particular emergent attributes and capabilities of the organization result from the particular combination of practices that constitute it. Different simulations lead to different organizational structures. The actual emergent capabilities and qualities that would be desirable for any particular firm cannot be predicted in advance, since the performance of any particular organization will depend on that of the others with which it is co-evolving. Firms, markets and life are about an ongoing, imperfect learning process that both creates and requires uncertainty.

2.3.4   Emergent Supply Chain Performance

These ideas were applied to the study of the aerospace supply chain (Rose-Anderssen et al. 2008). The aerospace supply chain actually needs a series of different capabilities if it is to succeed. They are:

1)   Quality.
2)   Cost Efficiency.
3)   Reliable Delivery.
4)   Innovation and Technology.
5)   Vision.

The stage in the life cycle of the product, or the market situation, determines what mix of these is required as the platform or product moves from design and conception, through initial prototyping and production to an eventual lean production phase. In addition 27 key characteristics or practices were identified that could characterize supply chain relationships. A questionnaire was formulated to enquire into the opinion of important individuals within these key aerospace supply chains in order to understand better the underlying beliefs that affect the decisions concerning the structure of supply chains. The questionnaire considered the intrinsic improvement of a given practice and the possible interaction between pairs of practices. We summarize this in Table 2.1 and Figure 2.8.

The important point about the evolution of systems is that they concern both the elements inside a system, that constitute its identity, and also the external environment in which they are attempting to perform and the requirements that are perceived for successful performance.

Our model then explores the random launching of practices under different performance selection criteria corresponding to our five basic performance qualities. The practices retained are on the whole synergetic (Rose-Anderssen et al. 2008). We can look at the patterns of synergy – the Structural Attractors – that have been selected (Figure 2.9).

Knowledge of the effects of interaction can therefore be of considerable advantage in creating a successful supply chain. Even if the pair interaction terms are considered to be 50 times smaller than the direct effects of a practice there are still synergy effects of up to 75 per cent. This shows the importance of considering the systemic, collective effects of any organization or supply chain. It is another example of the 'ontology of connection' and not that of 'isolation'.

Table 2.1   The scores for the separate practices

Characteristics	Success criteria factors				
Rate of characteristic to success factor criteria High (9), None (0)	Product quality	Cost efficiency	Delivery precision	Techn./ innovation	Vision for the future
1. Outsourcing competitive advantage	9	9	7	9	9
2. Outsourcing what is easily imitated	8	8	9	0	0
3. High level of collaborative relationship	5	5	3	8	9
4. Arm's-length relationships	0	0	0	0	0
5. Long-term relationship	8	7	8	7	7
6. Formal partnership	9	2	6	8	8
7. Subcontracting whole systems and sections	9	8	9	8	8
8. Flexibility of operations	8	6	7	5	7
9. Risk-sharing	6	3	4	9	9
10. Sharing knowledge	9	7	8	8	6
11. Offsets as part of sales contract	7	0	2	0	5
12. Culture of continuous improvement	8	7	8	7	6
13. Ability to handle cutural differences	5	5	5	7	6
14. High level of dominance over supplier	8	8	8	1	2
15. High level of planning and control	7	5	8	1	3
16. Easy dialogue with supplier	0	0	0	6	8
17. IT system integration	0	1	6	1	8
18. High levels of integration of chain	8	6	7	2	8
19. Responsive to market change	7	7	7	7	7
20. Transparent organisation	0	0	0	0	0
21. TQM procedures	9	7	6	5	6
22. Just-in-time delivery	5	2	9	6	6
23. Lean practice	7	7	7	7	7
24. Explorative learning practices	8	3	3	6	8
25. Investment in training	9	8	7	4	8
26. Supplier development	8	8	8	7	7
27. Monitoring supplier	7	7	7	2	1

2.4   REFLECTIONS ON SOCIAL SCIENCE METHODOLOGY

Returning to the issues involved in trying to understand situations involving social systems. We see that evolution is the most general framework for our understanding. Within this long-term view however, under the assumptions noted above, we can represent the behaviour of the system as the mechanical dynamics of its variables, and if we ask where this dynamics drives the system, we can discuss the possible long-term outcomes, the 'attractors', of the system. The assumptions, if true, lead us to simpler representations of the situation under study. This would seem to correspond to a clearer understanding of what is going on. But over longer times, the mechanisms and variables in question change as values, technology and human activities change.

Explanation of a situation does indeed imply the historical course of events at successive

Characteristics

Strongly synergetic (+5), indifferent effects (0), strongly conflicting (−5)

Column headers (diagonal):
1. Outsourcing competitive advantage
2. Outsourcing what is easily imitated
3. High level of collaborative relationship
4. Arm's length relationship
5. Long-term relationship
6. Formal partnership
7. Subcontracting whole systems and sections
8. Flexibility of operations
9. Risk-sharing
10. Sharing knowledge
11. Offsets as part of sales contract
12. Culture of continuous improvement
13. Ability to handle cultural differences
14. High level of dominance over supplier
15. High level of planning and control
16. Easy dialogue with supplier
17. IT system integration
18. High levels of integration in chain
19. Responsive to market change
20. Transparent organisation
21. TQM procedures
22. Just-in-time delivery
23. Lean practices
24. Explorative learning practices
25. Investment in training
26. Supplier develop

Characteristic	1	2	3	4	5	6	7	8	9	10	11	12	13	14	15	16	17	18	19	20	21	22	23	24	25	26
2. Outsourcing what is easily imitated	−5																									
3. High level of collaborative relationship	3	−4																								
4. Arm's-length relationship	−5	4	−5																							
5. Long-term relationship	4	−5	5	−5																						
6. Formal partnership	4	0	−4	0	4																					
7. Subcontracting whole systems and sections	1	2	0	−4	4	5																				
8. Flexibility of operations	4	0	3	0	3	4	5																			
9. Risk-sharing	5	4	−5	−5	4	5	5	4																		
10. Sharing knowledge	5	5	−3	4	4	4	5	2	5																	
11. Offsets as part of sales contract	−5	5	−3	4	0	−4	−4	3	−3	5																
12. Culture of continuous improvement	3	−3	3	−2	4	4	4	4	3	3	1															
13. Ability to handle cultural differences	3	−2	4	−3	3	3	3	4	4	3	3	5	0													
14. High level of dominance over supplier	−5	5	−5	4	−4	−4	−5	−3	−3	−2	5	−4	0													
15. High level of planning and control	4	−4	0	0	−3	−4	−4	4	3	2	−4	−3	2	5												
16. Easy dialogue with supplier	4	0	5	−3	4	4	5	4	3	5	0	5	3	−4	5											
17. IT system integration	4	0	4	−5	4	4	5	3	4	4	−3	4	4	−2	−3	0										
18. High levels of integration in chain	0	3	−4	4	5	5	4	4	5	4	0	2	2	−3	−4	3	4									
19. Responsive to market change	2	0	5	−3	3	4	4	5	3	0	−4	0	0	−4	2	4	2	4								
20. Transparent organisation	0	0	2	2	0	2	0	0	0	4	0	0	4	0	0	0	0	−1	4							
21. TQM procedures	4	2	0	3	4	3	4	2	4	1	3	4	−3	3	−3	4	3	0	0	2						
22. Just-in-time delivery	3	3	−3	5	4	5	3	4	3	2	4	0	−2	0	−3	3	1	3	4	2	4					
23. Lean practices	4	−4	4	0	5	5	4	4	3	4	−2	5	−3	−2	−4	4	5	0	1	0	4	5				
24. Explorative learning practices	−4	−3	5	−4	4	4	4	4	4	−3	3	3	2	−4	−3	0	4	3	0	4	0	5	5			
25. Investment in training	−3	−3	1	0	3	4	2	0	4	4	2	0	−4	−3	0	0	4	3	4	4	−1	4	4	4		
26. Supplier development	−3	5	3	−4	0	−3	−5	−3	5	0	4	−4	0	5	2	4	3	3	3	4	4	3	4	4	5	
27. Monitoring suppliers	−3	0	−4	4	−3	−3	−2	−2	4	4	0	4	−1	4	1	−3	−2	3	−3	0	0	−1	−1	−1	0	2

Figure 2.8   The pair interactions as estimated from the interviews

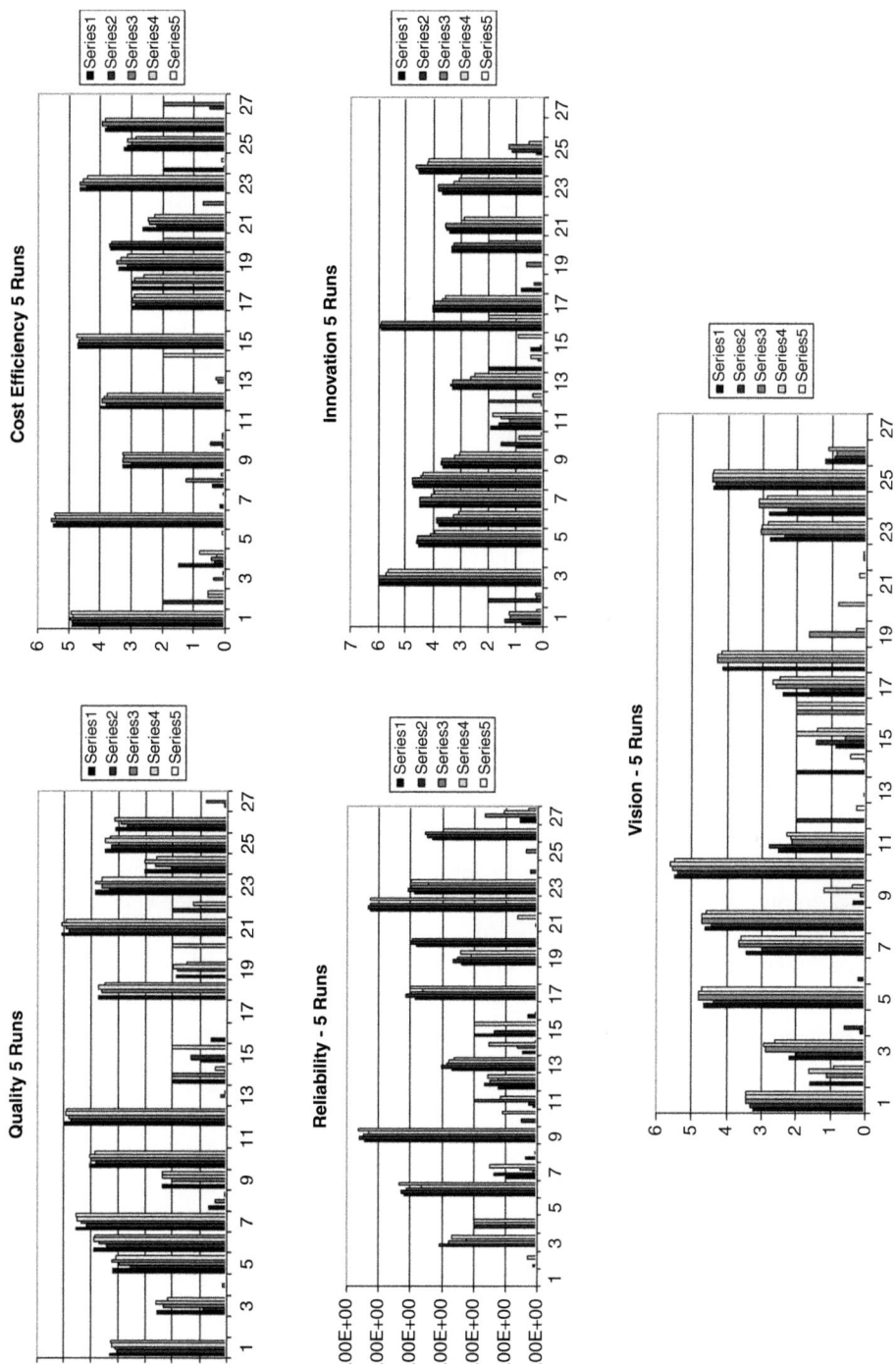

Figure 2.9   These are the Structural Attractors (synergetic bundles) in the space of possible practices that correspond to the five different dimensions of performance sought

moments and localities. The characteristics of our societies, cities and organizations really result from the accumulation of successive changes over time. At the centre of many large cities today one will find churches, cathedrals, palaces and great houses, which provide a physical homage to past 'power structures' and views of what was important. Our current values would not erect them at such key locations today, the site would be much too expensive, but they were built in different times when different values and powers decided on urban priorities. We live among the remnants of our earlier beliefs, and sometimes we build around our past excesses, and sometimes we demolish them. This is the 'historical' explanation of our surroundings.

The chapter has so far presented some examples of how one can create models of economic and organizational systems that show the importance of complexity and evolution. These help us to better imagine the rich possibilities that may be present, and take into account the degree of uncertainty that may really characterize our pathway into the future. We will find that many elements that are present do not necessarily 'make sense' within any overall teleology, having no obvious role or function. We will not necessarily understand 'better' by analysing more data. Social evolution evolves and changes, and may not in fact 'make sense', except when historians post-rationalize.

In trying to deal with the world we develop an 'interpretive framework' with which we attempt to navigate reality and to understand opportunities and dangers. This is really a set of beliefs about the entities that make up reality and the connections that exist between them. We see that this is really a qualitative 'model' and perhaps, sometimes, this can even be transformed into a quantitative mathematical model. But over time, our interpretive frameworks are constantly tested by our observations and experiences. Our beliefs provide us with expectations concerning the probable consequences of our actions and when these are confirmed then we tend to reinforce our beliefs. When our expectations are denied however, we must face the fact that our current interpretive framework – set of beliefs – is inadequate. But there is no scientific method to tell us how to modify our views. Why is it not working? Are there new types or behaviours present? Or are their interconnections incorrect? Importantly however, there is no scientific, unique way to change our beliefs. In reality, we simply have to experiment with modified views and try to see whether the new system seems to work 'better' than the old.

In addition, when our expectations are denied, we can only draw on our beliefs and experiences to decide 'how to change our ideas'. This may suggest to us which of our beliefs are most likely mistaken, whose ideas or comments we should trust and listen to, and who's we should discard. Of course, some people may be happy to take on new ideas every day, while others may choose never to modify their beliefs, feeling that the increasing evidence of inadequacy is merely a test of their faith. Figure 2.10 was originally drawn so as to represent 'the honest scientist' seeking the truth (a reflection of the author's naivety in approaching the problem). But it was pointed out to me (thanks to Professor Jane McKenzie) that in reality people are much more complex than that. Although some people may learn, others will simply find reasons to ignore or reject any evidence that is contrary to their beliefs or may detract from their own status or prestige. So micro-diversity encompasses not only the different interpretive frameworks people may have, but also how willing and equipped people are to change their beliefs and understanding and to adapt to what is happening. The real picture is more like that of Figure 2.11.

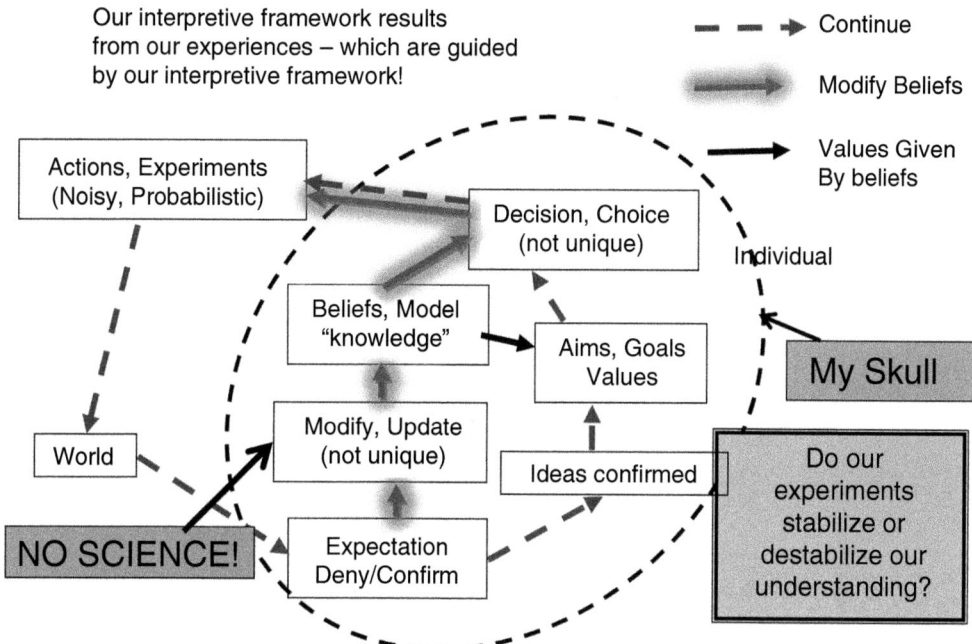

Figure 2.10   Experiences confirm or deny the expectations we have that are based on our interpretive framework

Complexity therefore suggests a 'messy cognitive evolution' in which some people change their beliefs and models, generating different behaviours and responses, which lead to differential success. A multi-agent model must reflect the fact that it is both an evolution in 'reality' and also in the interpretive frameworks of the participants. This allows some beliefs and interpretive frameworks to evolve with the real world, as they are tested and either retained or dropped according to their apparent success. It is therefore clearly very important for an organization that wants to 'learn', to have employees that are willing to participate honestly in the learning process. This implies first, that individuals are diverse and that the local organizational ambiance encourages open exchanges and discussions. It probably requires continual disagreement and rivalry among staff, but which recognizes the overall good of the organization. We should realize that actions and events are really 'experiments' that test our understanding of how things work. Clearly, given the lack of any clear scientific method on how to change one's own beliefs, many may simply adopt the views of their preferred group, and simply mimic their responses without, necessarily, understanding the basis of these. Culture may play an important role here where hierarchical cultures may fail to learn from 'bottom-up' experiences, and individualistic cultures may allow a spread of different lessons to be learnt by different groups. This may explain the importance of 'social networks', and the 'wisdom' or 'idiocy' of crowds.

After a crisis such as that experienced in 2007, the old model had failed, and nobody knew what should be changed. Since it is impossible to know what the 'correct view' really is, there is really a social process that occurs in which various new 'models' gradually

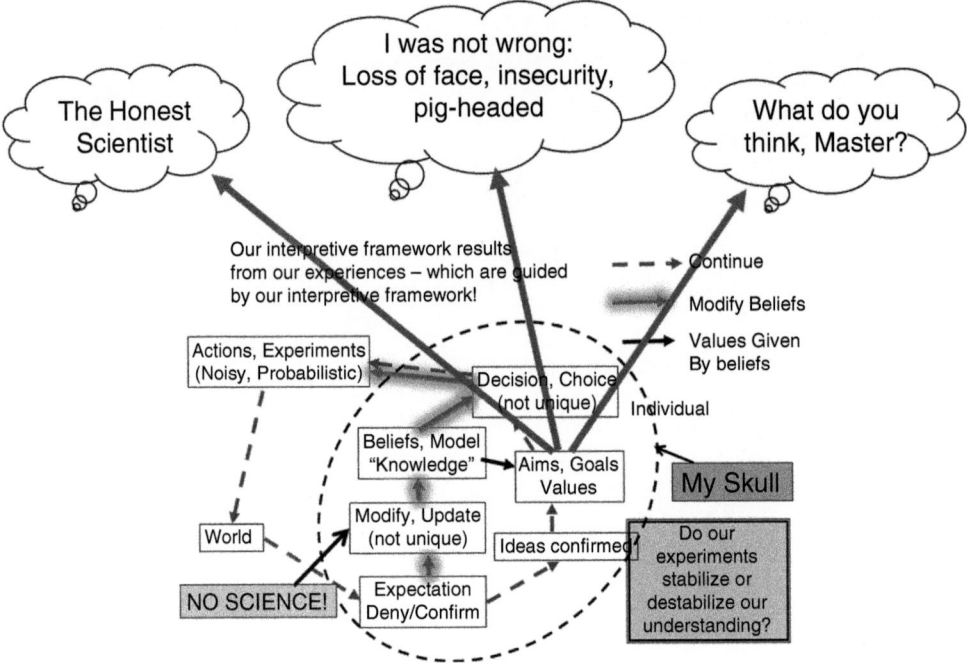

Figure 2.11   The more realistic view of the learning process has some 'honest scientists', but also people intent on defending their own status, or just asking what the new view should be

emerge and people adopt the new view about the economic/banking system that seems more attractive to them.

The point is though, that the new views are not necessarily any nearer the 'truth' of the complex system than the previous view before the crisis. The result is that often two or more 'schools', as in Figure 2.12 will open up with different views on the 'new' reality. In the UK there is still disagreement as to the amount of 'austerity' or 'growth' that would have been best for recovery. The complexity of the economic and financial system defeats the ability to prove convincingly that any particular view is completely true or false. What we have is a vast and worrying pragmatic process by which beliefs and behaviours are tested, resulting in successive crises, but very imperfect 'learning'. After all these are not repeatable experiments since circumstances and behaviours change.

Evolution is driven by the noise/local freedom and micro-diversity to which it leads – meaning that not only are current average 'types' explained and shaped by past evolution but so also are the micro-diversity and exploratory mechanisms around these. The need to remain adaptive leads to the retention of noise and micro-diversity generating mechanisms. This means that aggregate descriptions will always only be short-term descriptions of reality, though useful perhaps for operational improvements. Micro-diversity and freedom, though not profitable in the short term, are the only

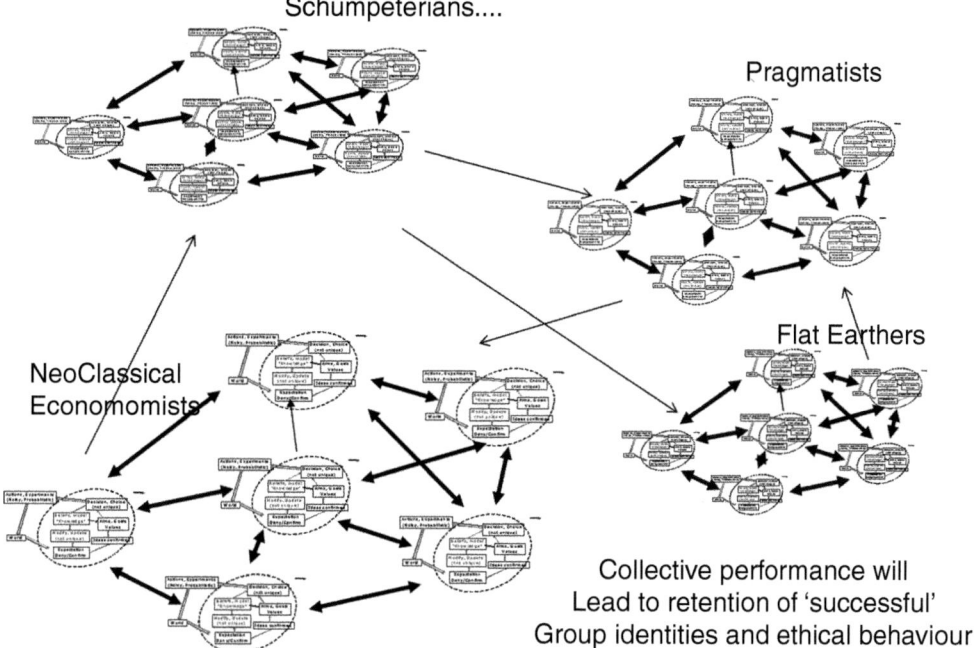

Figure 2.12   Following a crisis when previous views failed, various new 'schools' may emerge such as, for example 'growth therapy' or 'austerity therapy', which cannot be proven or disproven until perhaps the next crisis

guarantee of survival into the longer term. Living systems create a world of connected, co-evolving, multi-level structures, at times temporally self-consistent and at other times inconsistent.

An important part of all this is the recognition of reflexivity within social systems (Allen 2014; Soros 1987). Often we initially look at a situation in terms of available government data. But this does not tell us what interactions and mechanisms may link those variables, nor what those affected think or how they may respond. A first model of a situation may simply take government statistics or available data to get a preliminary view. This resembles the famous situation of 'Plato's Cave' where we only perceive the cave wall shadows of what is happening in the real world outside the cave. Our 'model' does not tell us what is really happening and why. In other words we try to make a model by looking at whatever data exists about the system. But, a better way forward is to look at the system as an evolving 'multi-agent' interaction in which the goals and choices open to the different agents present can be put into the model allowing it to explore forwards in time under various possible actions or scenarios. This is an excellent step forward, and will usually involve getting from the active agents some information about their motives, means and objectives. However, if the model is run forward in time, and the outcomes shared with the participating agents, knowledge of the 'predicted' behaviour of the system may cause some agents to change their behaviour. Figure 2.13 shows this 'evolution of the model'. This 'reflexivity' would invalidate the model and make the model part of the

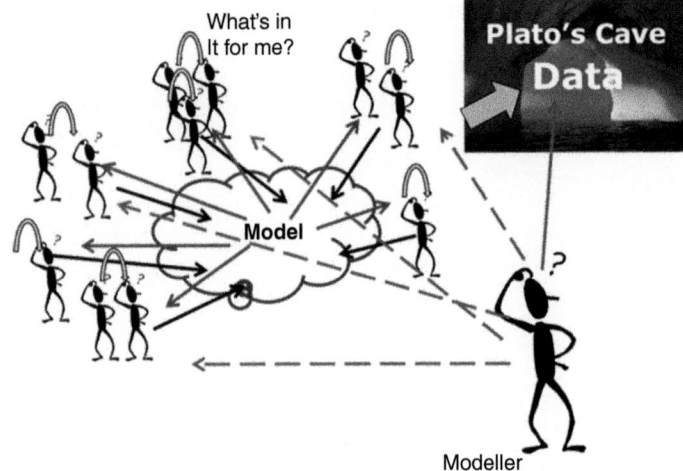

Figure 2.13   *Reflexivity exists because as a situation is 'modelled' in a social system, different agents may change their behaviour as a result of knowing the outcome – thus invalidating the model and changing the outcome*

complex system. One could conceive of models that included changes in behaviour of agents as a result of the model predictions, but clearly this would introduce new levels of complexity. Still, even today with the publication of data and statistics, people do change their behaviour as a result of their extrapolations of these, and so the reflexivity of complex systems is just an extension of this. It is better to take it all into account than to revert to simple hope.

This new understanding of the world might appear to say that the complexity of the world is such that we cannot find a firm base for any actions or policies, and so we should perhaps just pursue our own self-interest and let the world look after itself. In many ways this would resemble behaviour resulting from belief of the 'invisible hand' of neoclassical economics. But this would be a false interpretation of what we now know of complexity. We know that we must operate in a multi-level, ethically and politically heterogeneous world where both wonderful and terrible things happen. The models discussed here provide us with a better view than before of some possible futures, and allow us to get some idea of the likely consequences and responses to our actions and choices. We could even imagine 'Machiavellian' versions of complexity models that contained several layers of expected responses and countermoves on the part of the multiple agents interacting. Although one could include 'responses' to the behaviour of other agents in the behaviour of each agent, for this to be effective would require that the behaviour of agents was transparent and declared and not covert and silent. The idea that one could capture agents' responses as they occur, and counter them instantaneously would imply that the only 'model' of reality was 'reality' itself. Hard scientists find it hard to accept that there can be no perfect model of a multi-agent complex system. But the view put forward here is to accept that making a model is better than not making one, but any model can only provide limited information about the actual future. Multi-agent models of the kind we are suggesting are really useful to reflect on the system and its possible futures, and will always require updating as reality deviates from the model. Complexity tells us that our understanding of the system may be good for the short term, reasonable for the medium but will inevitably be inadequate for the long. It also tells us that sometimes there can be very sudden major changes – such as the 'financial crisis' of 2007/8. This means that policies should always consider resilience as well as efficiency or cost, as the one thing we do now know is that systems that are highly optimized for a single criteria such as profits or costs will crash at some point. Creative destruction data tells us that most firms fail quickly, some enjoy a period of growth, but all eventually crash (Foster and Kaplan 2001).

As mentioned in the introduction George Soros (1987, 2009) reinforces the point that there is a problem if social science is asked to be as 'hard' as physical science. Hard science is the part of natural science that is based on problems for which we can make repeatable experiments. But, already for complex systems in natural science this is not the case, and certainly for social and even biological situations this is not the case either. We would therefore agree with George Soros and would add to what he is saying. Social science must not be judged by the measures used for hard natural science and neither must complex systems both in the natural and social domains. Any system that is the result of a long evolutionary process, involving multiple scales and levels of interaction and organization will behave in ways that are less predictable than that for 'repeatable experiments' of simple physical situations. The dichotomy occurs not simply between natural and social science but between simple, hard natural science and complex systems science – that which includes evolutionary processes and structural developments that give rise to emergent characteristics and capabilities. The origami picture of Figure 2.3 is one of the most important here in that it shows the emergent 'freedom' of objects above the molecular level. Physics can only give us hard facts about the paper! But life is about what emerges from folded structures (such as proteins and other biomolecules) and populations of such multi-level structures.

In reality the real shift occurs when one studies open and not closed systems. Open physical systems display self-organization and emergent structures – and some of these become capable of reproduction. This leads eventually, through evolutionary processes, to living organisms and social systems, and to multiple levels of organized structures, leading to cognitive evolution and systematic studies which can be called science.

From the cases discussed in this chapter we see that there always need to be 'sub-optimal' redundancies, seemingly pointless micro-diversity and freedom if long-term survival is to occur. Not only is evolution 'messy' but this lack of clarity is the secret of its success. There are multiple understandings, values, goals and behaviours that co-habit a complex system at any moment, and these change with the nature of the elements in interaction as well as with their changing interpretive frameworks of what is going on. There is no end to history, no equilibrium and no simple recipes for success. But, how could there be? If social science could give us predictions or tell us what to do then we wouldn't be free. But that very creativity and freedom mean that social science will inevitably be an unending, imperfect learning process.

REFERENCES

Allen, P.M. 1982, Evolution, modelling and design in a complex world, *Environment and Planning B*, 9, 95–111.

Allen, P.M. 1988, Dynamic models of evolving systems, *System Dynamics Review*, 4 (1–2), 109–130.

Allen, P.M. 1990, Why the future is not what it was, *Futures*, 22, 554–570.

Allen, P.M. 1994, Coherence, chaos and evolution in the social context, *Futures*, 26 (6), 583–597.

Allen, P.M. 1997, *Cities and Regions as Self-Organizing Systems: Models of Complexity*, London: Taylor & Francis.

Allen, P.M. 2013, Complexity, uncertainty and innovation, *Economics of Innovation and New Technology*, Papers in Honour of Stan Metcalfe, 22 (7), 702–725.

Allen, P.M. 2014, Evolution: complexity, uncertainty and innovation, Schumpeter Conference Papers 2012, *Journal of Evolutionary Economics*, doi: 10.1007/s00191-014-0340-1, February.

Allen, P.M. and McGlade, J.M. 1986, Dynamics of discovery and exploitation: The case of the Scotian Shelf Fisheries, *Canadian Journal of Fisheries and Aquatic Sciences*, 43 (6), 1187–1200.

Allen, P.M. and McGlade J.M. 1987, Evolutionary drive: The effect of microscopic diversity, error making and noise, *Foundation of Physics*, 17 (7, July), 723–728.

Allen, P.M., Strathern, M. and Baldwin, J.S. 2006. Evolutionary drive: New understanding of change in socio-economic systems, *Emergence, Complexity and Organization*, 8 (2), 2–19.

Allen, P.M., Strathern, M. and Baldwin, J.S. 2007, Complexity and the limits of learning, *Journal of Evolutionary Economics*, 17, 401–431.

Bak, P. 1996, *How Nature Works: The Science of Self-Organized Criticality*, New York: Copernicus.

Foster, R. and Kaplan, S. (2001), *Creative Destruction*, New York: Doubleday.

Foster, J. and Metcalfe, J.S. (eds) 2001, *Frontiers of Evolutionary Economics, Competition, Self-Organization and Innovation Policy*, Cheltenham, UK and Northampton, MA, USA: Edward Elgar Publishing.

March, J.G. 1991, Exploration and exploitation in organizational learning, *Organization Science*, 2 (1), 71–87.

McCarthy, I. 1995, Manufacturing classifications: Lessons from organisational systematics and biological taxonomy, *Journal of Manufacturing and Technology Management – Integrated Manufacturing Systems*, 6 (6), 37–49.

McCarthy, I., Leseure, M., Ridgeway, K. and Fieller, N. 1997, Building a manufacturing cladogram, *International Journal of Technology Management*, 13 (3), 2269–2296.

McKelvey, B. 1982, *Organizational Systematics*, Oakland, CA: University of California Press.

McKelvey, B. 1994, Evolution and organizational science, in Baum. J. and Singh. J (eds), *Evolutionary Dynamics of Organizations*, Oxford: Oxford University Press, pp. 314–326.

Metcalfe, J.S. and Foster, J. (eds) 2004, *Evolution and Economic Complexity*, Cheltenham, UK and Northampton, MA, USA: Edward Elgar Publishing.

Nelson, R. and Winter, S. 1982, *An Evolutionary Theory of Economic Change*, Cambridge, MA: Harvard University Press.

Ormerod, P. 2005, *Why Most Things Fail*, London: Faber and Faber.

Rose-Anderssen, C., Ridgway, K., Baldwin, J., Allen, P., Varga, L. and Strathern, M. 2008, Aerospace supply chains as evolutionary networks of activities, *Creativity and Innovation Management*, 17 (4), 304–318.

Schumpeter, J. 1962, *Capitalism, Socialism and Democracy* (3rd edition), New York: Harper Torchbooks.

Soros, G. 1987, *The Alchemy of Finance*, London: John Wiley & Sons.

Soros, G. 2009, General theory of reflexivity, *Financial Times*, 26 October.

Veblen, T. 1898, Why is economics not an evolutionary science, *The Quarterly Journal of Economics*, 12 (1898).

3. Information theoretic measures of causality: music performance as a case study

Fatimah Abdul Razak, Universiti Kebangsaan Malaysia, Xiaogeng Wan, Tsinghua University and Professor Henrik Jeldtoft Jensen, Imperial College London

3.1 INTRODUCTION

Big data in the form of multiple times series are now frequently available for complex systems and this brings about the question of how best to identify 'causal' interdependencies between such data streams. Looking at it from a network point of view, the question is how to form directed networks when directions are not clearly indicated in the dataset. In networks literature the references to 'causality' take many guises. The terms directionality, information transfer and sometimes even independence can possibly refer to some sort of 'causal' interdependencies.

For a very long time scientists have used correlation measures such as the Pearson correlation coefficient as an indicator of some sort of interdependence between two observed quantities. However, correlation coefficients are symmetric. If A is correlated to B then B is correlated with the same strength to A. Measures of causality obviously should be asymmetric. Either A causes B or B is the cause for A – but not both ways simultaneously. Recent information theoretic approach within the Wiener–Granger framework of 'causality', have been applied to time series from complex systems.

Here we first review the information theoretic and probabilistic approach to 'causality' measures in the Wiener–Granger framework. Moreover, we explain the difference between direct and indirect measures. Finally we discuss the findings of an analysis of music performance where causation amongst the musicians and audience was studied by use of EEG time series. We show how one of the most recent causality measures (called MIME) is able to identify differences in the way information flow between the different brain regions within the same player or listener or between brain regions in different persons.

3.2 WIENER–GRANGER FRAMEWORK OF CAUSALITY

Our definition of 'causality' is based on how well a variable helps the prediction of another variable within the Wiener–Granger framework. Granger (while crediting Wiener for the idea) outlined two things about 'causality' in his Nobel lecture (Hlavackova-Schindler et al. 2007). The first is that the cause must come before the effect (the arrow of time) and second, that the cause should contain unique information about the effect that cannot be found in any other variable (Granger 1988). Therefore, the term causality in the chapter will refer to causality in the Wiener–Granger sense. To emphasize that we use the term causality in this restricted sense we'll sometimes place quotation marks around the word.

The idea in this framework goes as follows: Time series $A(t)$ 'causes' variable $B(t)$ if the ability to predict $B(t)$ is improved by incorporating information about $A(t)$ in the prediction of $B(t)$. In other words, say we want to test whether $A(t)$ causes $B(t)$. The first step would be to predict the current value of $B(t)$ using the historical values of B such as $B(t-1), B(t-2)$ and so on. The second step is to do another prediction the using historical values of A such as $A(t-1), A(t-2), \ldots$ in addition to using the historical values of B to predict the current value of $B(t)$. The third step would be to compare the former to the latter. If the second prediction is judged to be better than the first one, then one can conclude that time series $A(t)$ 'causes' time series $B(t)$ in the Wiener–Granger sense. Of course there isn't really certainty that an increase in the predictability is due to a direct mechanistic causal relation between A and B. But on the other hand if A does in fact cause B, we would certainly be able to predict B's behaviour better if we know A.

First introduced by Nobel Prize winning Clive Granger in the context of linear autoregressive model, G-causality is the most commonly used 'causality' indicator (Bressler and Seth 2011). However, the very fact that we need to linearly regress the data to obtain the prediction means that we will lose a lot of nonlinear information. The usual argument of the proponents of G-causality is that linear approximation works well on large-scale interactions. To Granger's credit (Granger 1980, 1988) he has always been clear that G-causality is not absolute 'causality' and he himself acknowledges that the optimal predictor may very well be nonlinear. By definition, the emergence and criticality expected in a complex system is a nonlinear phenomenon, thus it has been pointed out by many (Bressler and Seth 2011; Lungarella et al. 2007) that in complex systems, using G-causality may not be suitable.

3.3   COMPLEX SYSTEMS SCIENCE, BIG DATA AND THE NEED FOR MEASURES OF INTERDEPENDENCE

A complex system can roughly be defined as, a system composed of a large number of interacting components that give rise to emergent hierarchical structures where the components typically change with time (Jensen 2009). The study of complex systems has flourished in the advent of big data where we seek to learn and understand how systems work from huge datasets. More often than not, the 'causal connections' in the datasets are not clearly indicated even though from our knowledge of the originating system, there must exist directions and 'causality' for communication purposes.

Take the brain for example. A simple way to view the brain as a complex system is to imagine the neurons as individual components and the brain as a whole that emerges from cooperation of the neurons. Neuroscientists agree that different parts of the brain controls certain specific functions. However, these controlling areas need to communicate with other areas especially when reacting to external inputs. The communication links exist mostly as a series of action potentials forming a connection. Therefore identifying the 'causal' directed connections are very important for understanding of the inner workings of the brain.

Unfortunately, in the case of the brain the most common datasets available are in the form of EEG and fMRI where there is no clear direction of 'causality'. The key word here is directed, having a starting point and an end point. The study of complex systems came about due to the need to understand how these components interact. Understanding the direction of interaction is crucial to understanding how these components interact.

Moreover, when there are many components, there is also the notion of indirect causality to take into account. Take lightning and thunder for example (Granger 1980); we now know that the reason we usually observe lightning before thunder is because light is much faster than sound. We also know the fact that lightning and thunder are both essentially the same event manifested at different times and caused by the same electrical discharge. Let *A* be thunder, *B* the lightning and *C* the electrical discharge. If we only look at *A* and *B* we will mistakenly say that *B* causes *A*, that is, lightning causes thunder since thunder always comes after lightning. However if we include *C* then we will be able to infer that *C* (the electrical discharge) is the real cause of *A* and *B* as well as the fact that the very existence of a relationship between *A* and *B* depends completely on *C*.

The actual causal link in the case above is the following. *C* influences *A* directly and *C* directly influences *B*, but that there is no direct causal relation between *A* and *B*. Direct measures of causality would correctly determine that non-direct causal link exists from *B* to *A*, but indirect measures can be misleading and falsely detect a causal link from *B* to *A*. Therefore not only does one need a measure that can detect directionality (causal connection), but ideally a measure that can also uncover indirect causal connections. Measures have been developed that with a certain success can distinguish direct from indirect causal dependencies.

3.4   INFORMATION THEORETIC CAUSALITY MEASURES

Very many causality measures have been proposed and at present a systematic understanding of applicability, strengths and weaknesses of these measures doesn't exist. For a comparison of a number of measures on a range of time series we refer to Wan (2014). In practical terms it seems that at present one will have to simply test measures that are likely to be able to handle the specific challenges of a given situation. We illustrate the ideas behind some of these measures by discussing a few we have studied theoretically. Concerning their effectiveness, our experience is that when applying measures of causality to data streams, it is difficult a priori to know which is the most appropriate. When we analysed EEG data streams of performing musicians and audience we found MIME to deliver the most reliable indications of the flow of 'causation'. This is our reason for describing MIME below.

Generalized forms of G-causality include nonparametric approaches intended to be model free. Most of these generalizations are rooted in information theory and related to entropy-based Mutual Information value. Mutual Information (Cover and Thomas 1999) is defined as:

$$I(X,Y) = E\left[log\frac{P(X,Y)}{P(X)P(Y)}\right] = \sum_x \sum_y p_{XY}(x,y) log\frac{p_{XY}(x,y)}{p_X(x)p_Y(y)} \qquad (3.1)$$

where *X* and *Y* are random variables (time series) with discrete probability distributions $p_X(x), p_Y(y)$, as well as joint distribution $p_{XY}(x,y)$. $E[*]$ is the expectation of *. Taking into account conditional variables, the conditional Mutual Information (Cover and Thomas 1999; Schreiber 2000) is defined as:

$$I(X,Y|Z) = \sum_x \sum_y \sum_z p_{XYZ}(x,y,z) \log \frac{p_{XY|Z}(x,y|z)}{p_{X|Z}(x|z)p_{Y|Z}(y|z)} \qquad (3.2)$$

where Z is similarly another random variable (time series). Many variants of conditional Mutual Information are used for causality detection. The Transfer Entropy (TE) is one of the more frequently used of such measures. If we replace Y in equation (3.2) with the historical values of Y say Y^h and Z with the historical values of X say X^h, then we get the TE (Schreiber 2000) of Y to X:

$$T_{\{Y \to X\}} = I(X, Y^h | X^h) = E\left[\log \frac{P(X, Y^h | X^h)}{P(X|X^h)P(Y^h|X^h)}\right] = E\left[\log \frac{P(X|Y^h, X^h)}{P(X|X^h)}\right] \qquad (3.3)$$

TE can be compared to G-Causality by revisiting the three steps for ascertaining 'causality' in the Wiener–Granger sense. Say we have time series (variables) $A(t)$ and $B(t)$. We wish to test whether $A(t)$ causes $B(t)$. If one was to do this with TE, the value of $T_{\{A \to B\}}$ needs to be obtained using equation (3.3.3). The first step would be to predict the current value of B using the historical values of B, $B^h = \{B(t-1), B(t-2) \ldots\}$. This will be the process of obtaining the value of transition probability $P(B|B^h)$ in the denominator. The second step would be another prediction where the historical values of A and B are both used to predict the current value of B. For TE this requires the estimation of $P(B|A^h, B^h)$ in the numerator of equation (3.3.3).

In order to compute the the the TE of observed time signals, one obviously needs to estimate a number of probability densities. If the time signals consist of continuous variables one will need to bin the data to obtain histograms. The resulting TE depends on the details of the binning in a non-straightforward way. This somewhat unfortunate difficulty was studied in some detail in Abdul Razak and Jensen (2014); see also Schreiber (2000).

Consider two time series A and B. To determine whether a causation – and its direction – exists between A and B, one compares the predictions for $A \to B$ to $B \to A$. If the first prediction is judged to be better than the last (in the Transfer Entropy case if $T_{\{A \to B\}}$ is large), one can say that A Granger causes B. In the case of TE one utilizes the expected log ratio between the two probability distributions to compare the predictions.

Let us briefly address the relationship between Granger causations and TE causation estimates. TE can be considered a generalization of G-causality (Hlavackova-Schindler et al. 2007) which is valid for nonlinear interdependencies. In the case of linearly related and Gaussian distributed signals TE can be proven to be identical to G-causality.

So which measure of causal interdependence should one use in a concrete situation? As mentioned above a general answer doesn't seem to exist at the moment. Using EEG to try to access the communication, interaction or information flow between performing musicians and their audience lead us to conclude that the recently suggested measure Mutual Information from Mixed Embedding (MIME) (Vlachos and Kugiumtzis 2010) was particularly useful for the high frequency data stream measuring the presumably strongly nonlinear electromagnetic fields produced by the neuronal activity of the brains. MIME was for this kind of data remarkably better than, for example, TE at detecting differences in the 'causal flow' for different modes of performance, essentially improvisation in contrast to non-improvisation.

Now we briefly describe MIME, details can be found in Vlachos and Kugiumtzis (2010) and algorithmic codes are available from Kugiumtzis's website http://users.auth. gr/dkugiu/. MIME is a time domain information-based nonlinear causality measure that can be defined for any stationary processes. Let X and Y be two stationary vector processes, $V_F = (X_{n+1}, X_{n+2}, ..., X_{n+T_X})$ be the future vector of X with time horizon T_X and $B = \{X_n, X_{n-1}, ..., X_{n-L_X}, Y_n, ..., Y_{n-L_Y}\}$ be a uniform state-space embedding vector (Vlachos and Kugiumtzis 2010; Barnard, Aldrich and Gerber 2001; Boccaletti et al. 2002) containing the history of both X and Y with the maximum time lags L_X and L_Y, which are taken as input to a progressive scheme (Barnard, Aldrich and Gerber 2001; Boccaletti et al. 2002; Garcia and Almeida 2005). The progressive scheme starts with V_F, B and an initial selected embedding vector $b_0 = \phi$. At each iterative cycle s, the scheme uses kNN method (k-th nearest neighbour estimator) (Vlachos and Kugiumtzis 2010; Faes, Porta and Nollo 2008; Tsimpiris, Vlachos and Kugiumtzis 2012; Vejmelka and Palus 2008) to estimate the conditional mutual information rates between V_F and elements in $B \backslash b_{s-1}$. The progressive scheme seeks the element in $B \backslash b_{s-1}$ that satisfies the maximum criterion: I : $\max_{* \in B \backslash b_{s-1}} \{I(V_F;*| b_{s-1})\}$. The element that satisfies the maximum criterion is next moved from B to b_{s-1} to obtain $b_s = [b_{s-1}, *]$ for the s-th iterative cycle. The progressive scheme stops at an s+1-th iterative cycle when the stopping criterion (Schreiber 2000) $I(V_F;b_s)/I(V_F;b_{s+1}) > A$ is satisfied and one has b_s as the final selected non-uniform state-space embedding vector. Here, $A \in (0,1)$ is a significance threshold near 1, and $A=0.95$ is the empirical optimum choice (Vlachos and Kugiumtzis 2010). Higher A will allow more cycles to be iterated so that more components will be included in the selected embedding vector even if these only contribute very little. A lower value of A reduces the number of iterative cycles and can prevent a sufficient number of components to be included in the selected embedding vector (Vlachos and Kugiumtzis 2010; Kugiumtzis 2013). The empirical A=0.95 was found by Vlachos and Kugiumtzis (2010) to stick a good balance as this value of the threshold allowing for an effective size of the embedding vector and prevents false positives. Finally, the $MIME_{Y \to X}$ is estimated by the ratio between the conditional mutual information rates:

$$MIME_{Y \to X} = 1 - \frac{I(V_F;b_s^X)}{I(V_F;b_s)} = \frac{I(V_F;b_s^Y|b_s^X)}{I(V_F;b_s)}, \tag{3.4}$$

where b_s^X and b_s^Y are the X and Y components of b_s (Vlachos and Kugiumtzis 2010). The MIME ratio measures the amount of information that Y contains about X in the final selected embedding vector (Barnard, Aldrich and Gerber 2001; Boccaletti et al. 2002; Garcia and Almeida 2005), and is normalized by the total mutual information rate to scale the causality between 0 and 1. MIME indicates an existence of causality from $Y \to X$ if $MIME_{Y \to X} > 0$, otherwise no causality is detected from Y to X. Generally. $MIME_{Y \to X}$ is a non-negative value around 0, no matter how small the $MIME_{Y \to X}$ is, it indicates an existence of causality. Higher value of $MIME_{Y \to X}$ indicates stronger causal dependence of X on Y, because b_s^Y occupies a larger portion in b_s, when there is no causal dependence of X on Y, $b_s^Y = \phi$, the numerator vanishes (Vlachos and Kugiumtzis 2010).

MIME is an indirect causality measure whose direct version was introduced by Kugiumtzis (2013) as the *partial* MIME or PMIME for short. PMIME is defined on multivariate stationary processes, which uses the same progressive scheme and criteria as MIME does, except in a multivariate manner (Kugiumtzis 2013). Consider a system of K variables $X = [X_1,\dots,X_K]$, the multivariate uniform state-space embedding vector considers the lagged values of all variables in the system (Kugiumtzis 2013): $B = \{X_{1,n}, X_{1,n-1},\dots,X_{1,n-L_1},\dots,X_{K,n}, X_{K,n-1},\dots, X_{K,n-Lk}\}$, L_i is the maximum time lag of each X_i, $i = 1,2,\dots,K$. The multivariate uniform state-space embedding vector (Barnard, Aldrich and Gerber 2001; Boccaletti et al. 2002; Garcia and Almeida 2005) is taken as input to the progressive scheme, which is followed by the same maximum and stopping criteria as MIME does. $A=0.97$ is the empirical optimum choice for the stopping criterion for PMIME (Kugiumtzis 2013), which is slightly higher than that of MIME ($A=0.95$) (Vlachos and Kugiumtzis 2010), the higher value be because the partialization of the conditional mutual information rates on the remaining (confounding) variables makes the maximum criterion more rigid, the stopping criterion needs to be increased in order to obtain more iteration cycles, which will allow enough relevant components to be included (Faes, Porta and Nollo 2008; Kugiumtzis 2013; Anderson 2004). When the stopping criterion is met at iterative cycle number s+1, the non-uniform state-space embedding vector b_s is selected as the final embedding vector for PMIME. The PMIME causality for $X_j \to X_i$ is given by (Vejmelka and Palus 2008):

$$PMIME_{X_j \to X_i|Z} = \frac{I(V_F; b_s^j | b_s^i, b_s^z)}{I(V_F; b_s)} \tag{3.5}$$

where $Z = X \backslash \{X_i, X_j\}$ represents the remaining variables in X excluding X_i and X_j, b_s^i, b_s^j, and b_s^z are the X_i, X_j, and z components of b_s, respectively (Vejmelka and Palus 2008).

3.5   MIME ANALYSIS OF EEG OF MUSICIANS AND AUDIENCE

Two separate experiments were performed: one in which a pianist's solo performed (Wan 2014) and another experiment with a trio (Wan, Crüts and Jensen 2014; Doland et al. 2013). Musicians and audience were connected to EEG recorders. The musicians performed different types of music using different modes of play: composed music (music performed according to scores) and improvisation (instantaneous creative performance of music), and the musicians used different playing modes: strict mode (mechanical rendition of music) and 'letting-go' mode (music performance with free emotional expression) (Doland et al. 2013). We used causality measures to identify the neural differences between different namely the music types and different modes of performance. Synchronized EEG data was recorded from cortical large brain regions (frontal, central, temporal, parietal and occipital regions) (Wan 2014; Wan, Crüts and Jensen 2014; Doland et al. 2013) of both musicians and listeners.

As described in Wan, Crüts and Jensen (2014) we analysed the EEG recordings by three different measures; namely the frequency-based Partial Directed Coherence (PDC) (Baccala and Sameshima 2001) together with the time dominated measures TE and MIME. As mentioned above, the reason for choosing exactly these three measures are not entirely rigorous but was to some degree a matter of trial and error. It is natural when dealing with EEG data to apply a frequency-based method since much EEG analysis refer to the different frequency bands of brain activity. However, we found that PDC returns unreliable causal inference by which we mean that PDC returned causal relations which clear didn't make sense. We assume that one reason may be that the data is strongly non-linear, another that the causal relations may not be hosted in particular frequency bands, but spread out across all frequencies. More detailed analysis of this point is desirable and interesting.

After having experienced this shortcoming of a frequency-based approach we studied the causal relations by use of the two time domain-based measures TE and MIME to analyse the causalities (Wan 2014; Wan, Crüts and Jensen 2014). TE is computation-ally fairly simple to apply. It has the shortcoming of being an indirect measure whereas MIME is in principle able to identify direct causal relations. For our datasets we found that TE didn't convincingly resolve the direction of the causal relation, for example, we would expect the flow to be predominately from musicians to listeners and not the other way around (Wan 2014).

Among the measures we investigated, only MIME presented clear and robust causality detections, which we could use to construct both intra-brain and cross-brain networks (Wan 2014; Wan, Crüts and Jensen 2014). The results of the analysis are discussed in detail in Wan (2014) and Wan, Crüts and Jensen (2014). Here we summarized some of the salient features. In the intra-brain networks, the behavioural differences between playing composed music and improvisation were observed in the difference of distribu-tion of network links. We found that improvisation tends to trigger more widely dis-tributed networks than composed music, this is in agreement with the results reported in Berkowitz (2010). When comparing 'strict' or 'letting-go' mode we found that the intra-brain networks for 'strict' mode tend to have directed links that are opposite to the directed links in the networks of 'letting-go' mode (Wan 2014; Wan, Crüts and Jensen 2014). Not surprisingly, in both modes the importance of frontal regions is clear. Interestingly, the analysis suggests that the neural information flow to the left and right frontal regions reverse directions when composed music is changed to improvisation and strict mode is changed to 'let-go' mode. More detailed and systematic experiments are needed to verify the generality of these results and to determine its significance in terms of cognitive processing.

One important use of 'causality measures' is to uncover the network of leader–follower relationships in complex systems. As a very explicit example of such a study we mention now our analysis of information flows amongst the musicians and between the musicians and the listeners (Wan 2014; Wan, Crüts and Jensen 2014). As to be expected, the flow is predominately from the musicians towards the audience. This is consistent with a reading experiment done to assess the validity of MIME's predictions, where we found that the information flow suggested by the MIME analysis was from reader to listener. In the case of the preforming musicians MIME sometimes suggest a flow from the audience to the musicians. That the audience influences the improvising musician is exactly the idea of improvisation in a chamber music setting. A much more systematic study of this feedback flow to check validity and generality will be of great interest.

One of our experiments consisted of a trio (harp, viola and flute) performing in dif-ferent modes of play. Our analysis suggested that the flow direction between the three musicians kept changing during the performance – which obviously is what one will expect to happen. However, it did appear that overall the harpist was typical; the source of the information flow during improvisation with the flautist and viola play often in a kind of ping-pong relationship with each other. That the harp would act as a kind of lead was confirmed by the flautist who mentioned that although the trio didn't agree any arrangement of roles before their improvisation, he thought in hindsight that the harp might very well by its ten tones often suggest a harmonic structure the two other players then responded to.

3.6   DISCUSSIONS AND CONCLUSIONS

We repeat that at the moment we do not know of some rigorous a priori procedure to determine which causality measure one should apply in a given situation. From model studies (Wan 2014; Abdul Razak and Jensen 2014) and from applications to EEG (Wan 2014; Wan, Crüts and Jensen 2014) analysis, we have found that nonlinear measures such as TE, MIME and PMIME are suitable for use on both linear and nonlinear data analyses. They possess several strengths, for example they are model-free (Schreiber 2000; Faes, Porta and Nollo 2008; Kaminski and Blinowska 1991) and the nonlinear measures seem more applicable than linear measures because real world time series such as EEG and financial time series are most often nonlinear (Wan 2014; Abdul Razak and Jensen 201). In addition, one needs to consider the issue concerning direct and indirect 'causality'. For the measures introduced here, PMIME (Kaminski and Blinowska 1991) are direct measures, while TE (Schreiber 2000), MIME (Faes, Porta and Nollo 2008) are indirect measures.

Another important factor is the computation speed. Computation speed are influenced by certain parameters such as the embedding dimension (time lag) and the number of variables under consideration (Wan 2014). The computation speed of indirect measures such as TE and MIME are determined by the time lag rather than the number of variables (Wan 2014). However, the computation of direct measures such as PMIME, are influenced by both the time lag and the number of variables, with computation speed decreasing polynomially with the number of variables (Wan 2014).

Some forms of TE suffer from several computational costs when the embedding dimension is large (Abdul Razak and Jensen 2014; Anderson 2014; Wan 2014). In contrast, MIME and PMIME are more suited for experimental data analysis since large enough embedding dimensions can be obtained with efficient computations (Wan 2014). In particular, MIME is faster for small systems (Faes, Porta and Nollo 2008; Anderson 2004) while PMIME is faster for large systems (Kaminski and Blinowska 1991; Anderson 2004).

The notorious question of 'causality' has long been debated. The definition of it is yet to be agreed upon and can lead to intense philosophical debate. Here we have defined 'causality' as being related to predictability as formulated within the Wiener–Granger framework. A few types of causality measures were presented with discussions of their strengths and weaknesses. For time series from a range of different types of systems (models as well as real data, see Wan 2014) we found MIME (Vlchos and Kugiumtzis 2010) to identify correct (for models) or sensible (EEG data, for example) causal network structures.

REFERENCES

Abdul Razak, F. and H.J. Jensen 2014, Quantifying 'causality' in complex systems: Understanding TE. *PLoS ONE*, 9(6): e99462.

Anderson, H.M. 2004, Choosing lag length in nonlinear dynamic models. In R. Becker and S. Hurn (eds), *Contemporary Issues in Economics and Econometrics*. Cheltenham, UK and Northampton, MA, USA: Edward Elgar Publishing, 126–204.

Baccala, L.A. and K. Sameshima 2001, Partial directed coherence: A new concept in neural structure determination. *Biological Cybernetics*, 84, 463–474.

Barnard, J.P., C. Aldrich and M. Gerber 2001, Embedding of multidimensional time-dependent observations. *Physical Review*, E 64, 046201.

Berkowitz, A.L. 2010, *The Improvising Mind*. Oxford: Oxford University Press.

Boccaletti, S., D.L. Valladares, L.M. Pecora, H.P. Geffert and T. Carroll 2002, Reconstructing embedding spaces of coupled dynamical systems from multivariate data. *Physical Review*, E 65, 035204.

Bressler, S.L. and A.K. Seth 2011, Wiener–Granger causality: A well-established methodolgy. *NeuroImage*, 58: 323–329.

Cover, T. and J. Thomas 1999, *Elements of Information Theory*. New York: Wiley.

Doland, D., S. Sloboda, H.J. Jensen, B. Crütz and E. Feygelson 2013, The improvisatory approach to classical music performance: An empirical investigation into its characteristics and impact. *Music Performance Research*, 6, 1–38.

Faes, L., A. Porta and G. Nollo 2008, Mutual nonlinear prediction as a tool to evaluate coupling strength and directionality in bivariate time series: Comparison among different strategies based on k nearest neighbors. *Physical Review*, E 78, 026201.

Garcia, S.P. and J.S. Almeida 2005, Multivariate phase space reconstruction by nearest neighbor embedding with different time delays. *Physical Review*, E 72, 027205.

Granger, C.W.J. 1980, Testing for causality: A personal viewpoint. *Journal of Economic Dynamics and Control*, 2: 329–352.

Granger, C.W.J. 1988, Some recent developments in a concept of causality. *Journal of Econometrics*, 39: 199–211.

Hlavackova-Schindler, K. et al. 2007, Causality detection based on information-theoretic approaches in time series analysis. *PhysicsReport*, 441: 1–46.

Jensen, H.J. 2009, Probability and statistics in complex systems, introduction to. In R. Meyers (ed.), *Encyclopedia of Complexity and Systems Science*. Berlin: Springer, 7024–7025.

Kaminski, M. and K.J. Blinowska 1991, A new method of the description of the information flow in the brain structures. *Biological Cybernetics*, 65, 203–210.

Kugiumtzis, D. 2013, Direct coupling information measure from non-uniform embedding. *Physical Review*, E 87, 062918.

Lungarella, M. et al. 2007, Methods for quantifying the causal structure of bivariate time series. *Journal of Bifurcation Chaos*, 12: 903–921.

Schreiber, T. 2000, Measuring information transfer. *Physical Review Lettters*, 85: 461–464.

Takahashi, D.Y., L.A. Baccala and K. Sameshima 2010, Frequency domain connectivity: An information theoretic perspective. The 32nd Annual International Conference of the IEEE EMBS, Buenos Aires, Argentina.

Tsimpiris, A., I. Vlachos and D. Kugiumtzis 2012, Nearest neighbor estimate of conditional mutual information in feature selection. *Expert Systems with Applications*, 39, 12697–12708.

Vejmelka, M. and M. Palus 2008, Inferring the directionality of coupling with conditional mutual information. *Physical Review*, E 77, 026214.

Vlachos, I. and D. Kugiumtzis 2010, Nonuniform state-space reconstruction and coupling detection. *Physical Review*, E 82, 016207.

Wan, X. 2014, Time series causality analysis and EEG data analysis on music improvisation. PhD Thesis, Imperial College London. Available as open access via spiral.imperial.ac.uk.

Wan, X., B. Crüts and H.J. Jensen 2014, The causal inference of cortical neural networks during music improvisations. *PLoS ONE* 9(12), e112776.

PART II

CASE-BASED QUALITATIVE TECHNIQUES

Professor Peter Allen

This section tries to address the underlying question of how far the new ideas of 'complexity' and 'complex systems' might actually improve understanding and outcomes in the social sphere compared to that of well-intentioned common sense? First, it should be admitted that common sense is not necessarily densely scattered in our world, and so the question this section addresses is about the approach and the process of how a method would help us apply 'complex systems thinking' to social situations, with a view to doing better than before.

One of the main thinkers behind the discovery of 'complex systems' was Ilya Prigogine who used the title *From Being to Becoming* for one of his books, to capture an important difference that complexity and open systems thinking signalled. Instead of considering an understanding of the world in terms of either an equilibrium (possibly expressing some optimum), or as mechanical systems simply churning forward with given interactions, he saw that complexity was really an evolutionary vision. Complex systems, like people are not 'passive'. They are both creative and exploratory themselves, if permitted, and may be embedded in creative and exploratory surroundings as well.

Although the fundamental laws of physics, are always true, they do not provide an explanation of what goes on in the living and social world. Complex systems are multi-level systems with networks, organizations and structures and over time they evolve as some parts wither and others grow, and entirely new aspects can emerge. So, in origami we see that a piece of rectangular paper, when folded, can exhibit emergent characteristics and functions as a result of symmetry breaking in its morphology – which the mechanics of the paper cannot possibly 'explain'. The origami object interacts with its environment entirely differently from a sheet of paper. It can be a bird, a vase, a hat, a box, and all sorts of things. The laws of physics can only tell us about the elasticity and density of the paper, but nothing about the forms and functions that can emerge. The paper and the laws of physics are simply the materials on which the forces of complexity can create all kinds of amazing things. From macro-molecules, this includes living matter and people. And people can also get 'locked in' to existing situations which seem be unfair and very bleak, and not know how to change things.

They are what they are now, but what could they become?

This is where the ideas coming from complexity provide a framework for reflection of what could be – becoming instead of just being. Remarkably, the chapters in this section present real cases in which the ideas coming from complexity science shapes their interventions, and leads to some amazingly successful outcomes.

This Handbook is focused particularly on Research Methods in Complexity Science, and this section is very much devoted to the methods that have been created and used in real cases in social science. They share the idea of setting up a method that can be used in different possible situations and circumstances; in order to create an enabling environment for joint reflection on who is who, what are the issues and what might be done to improve things. In the chapter by Eve Mitleton-Kelly this:

> Enabling Environment has to be co-created by the people involved in the actual case and those guiding the process. An Enabling Environment is not a 'free for all'. It sets very clear aims and objectives, both in the short and longer term. It also provides an enabling infrastructure with suitable training, support, facilities and so on for local leaders to enable them to respond appropriately and in a timely manner. To work well, the Enabling Environment needs very good communications.

The chapter describes the EMK complexity methodology in sufficient detail to enable researchers to use it and illustrates how part of the approach was used in a case study. The method identifies the multiple dimensions in the problem space and their co-evolutionary dynamics (how the different dimensions and issues interact and change each other in the process). The methodology recognizes that there is no single, simple 'solution' to a complex problem, but that the problem can be addressed effectively by setting up an enabling environment, which directly addresses the co-evolving clusters and evolves with the problem space.

Another important point that is raised in this first chapter is that of different cultural perspectives. In the Indonesian case described by EMK, it is completely unclear what would be a 'good' outcome for the many different players. The underlying issue is deforestation, and we might think that it is simply a question of reducing deforestation. But locally there may be very complex relationships and loyalties that come into play, and complex social relations and loyalties will affect what decisions may seem 'good' for different participants. The customs of the culture may appear to us to resemble nepotism and corruption, but all we can do is to provide an 'enabling arena' of discussion, so that over time perhaps they may choose to modify such practices. The chapter points out that it is important to be aware of these differences in culture and to have the courage to articulate the distinction between what was culturally acceptable and what amounted to corruption, when money changed hands. But this is really all part of the 'becoming' and there is no clear evidence of how long it may take. We can perhaps only contribute to the discussion of different aspects, hoping that some of the more glaring problems can be partially reduced, and perhaps changed more in the future.

The chapter contributed by Lesley Kuhn describes some creative applications of a complexity informed social research. A complexity perspective suggests that the only viable long-term strategy for the development of a complex social system, such as the Mount Druitt community, an example used in the chapter, is for the system to learn about itself. Informing key or relevant people is a powerful way of nudging complex social systems towards self-managed improvement.

A complexity informed approach that corresponds to a depiction of life as self-organizing, dynamic and emergent is well suited to inquiry into human experience as it brings a sense of openness to appreciating and studying human experience as it emerges. The outcome from complexity informed social research is that the researcher, the situation and the theory are adaptively engaged and simultaneously improved. The chapter introduces some fairly new ideas (for me) such as: vortical postmodern ethnography, which can be seen as having five main parts. First it focuses on narratives, discourse and behaviour as these unfold during the research. Second, it relies on qualitative data in the form of formal and reflective narratives pertaining to the research. Third, it takes a holistic perspective that sees the research as being focused on one level, but implicating issues at higher and lower levels. Fourth, expects new ideas, insights, hypotheses and questions to arise from the vorticity generated by the research process. Fifth, analysing data for identification of themes that have a strong resonance for participants and as treating data generation as a further source of vorticity that leads to new cycles of emergence. Also, as an ex-physicist I find the invention of phrase space as some kind of social version of the phase space of physics, rather interesting. They might represent the space of the dimensions of the whole problem.

The chapter then illustrates the application of these ideas to a community in New South Wales, Australia, in Mount Druitt, an area that is culturally and linguistically diverse, with a large Indigenous population and low median income. The program was begun in response to citizen requests for assistance in developing leaders and involved participants who were: engaged in the community in some way (paid or voluntary); recommended by a community organization, government agency or religious organization; between 15 and 90 years of age; committed to Mount Druitt. Once again the intervention seems to have had extremely beneficial effects on the community, and demonstrates the potential importance of these 'complexity' related approaches.

In the chapter by Hazel Stutely and Jonathan Stead we find an approach which is also inspired by complexity thinking. The Connecting Communities (C2) project has developed its own seven-step application of complexity science to underpin a delivery framework. C2 offers an intervention, delivered on site by a small team of experienced practitioners, working alongside local residents and service providers to enable them to implement the framework. "It placed great value on widespread connectivity, the creation of new relationships and dialogue based on trust. Conversations, humility and respect, I now realised, contributed hugely to the creation of that all-important enabling environment, which released the resourcefulness of this community to become self-organising and achieve such significant and dramatic outcomes." These ideas had been tried out initially in the Beacon project set in a severely disadvantaged social housing estate (pop. 6,000) in Falmouth. The results were in my view staggering: overall crime rate down 50 per cent; unemployment down 71 per cent; improvement to 1,000 homes; educational attainment up 100 per cent; child protection rates down 42 per cent; post-natal depression down 70 per cent; childhood asthma down 50 per cent; teenage pregnancy zero! Other projects have followed. REACH (Redruth Enabling Active Community Health) 2004–2006, is an example of close collaboration between a community and the emergency services. Outcomes included: 210 patients treated between 2004 and 2006 on site; 30 per cent drop in incidence of under-age problem drinking and an 18 per cent reduction in emergency call outs.

Since then, the approach has been used in a series of further projects: OPERATION GOODNIGHT: with the police and residents, in Redruth in 2008. More recently the

Greenfingers project was run in partnership with Duchy Agricultural College. It was highly successful and has transformed not only estate gardens and green spaces but the lives of its participants who have accessed further education in greater numbers. The TR14ers Camborne: Cornwall 2005–2008: It was founded in response to significant police concerns about rising levels of antisocial behaviour (ASB) and health inequalities affecting the youth of Camborne. The majority of young people that attend the TR14ers live on remote social housing estates with little social or play facilities and their families are often troubled by a raft of health and socio-economic issues. The project has led to a large drop in antisocial behaviour, and to much improved health and educational outcomes.

The improvements in the lives of underprivileged communities that these projects have attained are truly remarkable.

In the next chapter, Robin Durie, Craig Lundy and Katrina Wyatt use complexity principles to understand the nature of relations for creating a culture of publicly engaged research within Higher Education Institutes. The chapter presents an approach to social systems which is based on the complex systems approach, and the behaviour of complex adaptive systems. That is, the behaviour of the parts of complex systems results from a co-evolution with the other parts. Similarly, the behaviour of the whole system is affected by its co-evolution with the environment of which it forms a part. Thus, complex adaptive systems are continually responding and adapting to changes in their environment, just as the environment itself changes and adapts to changes amongst its elements. The ongoing behaviour of the whole system, as well as of its parts, remains to a greater or lesser degree unpredictable; such novel, unpredictable, behaviours – often a consequence of the 'self-organization' of the system – are thus said to be 'emergent'.

The chapter discusses the reasons why it makes sense to use a complexity approach to public engagement in research. It offers the potential for comparing phenomena such as networks, sustainability and resilience in biological systems with similar phenomena in social systems, and thereby opens the possibility of transferable co-learning about the causes of such phenomena. It provides a broader framework for ideas and confirmation than just the actual examples studied. In particular, it looks at notions such as co-evolution which seem appropriate for the way that relations between communities and researchers may develop. The principle of emergence offers a potentially insightful means of capturing the tendency for community engagement to yield novel and unexpected research outcomes. Most of all, the potential for phenomena such as co-evolution, self-organization and emergence stems, according to complexity theory, from the distinctively nonlinear nature of the dynamic relations between agents, and this focus on dynamic relational processes appears to offer a particularly effective means for making sense of the relations that underpin research projects involving community–university partnerships.

In this section on methodology of interventions the chapter then offers a three level method of analysis. An initial analysis records interviews and focus groups, which are then transcribed and checked. A number of cross-cutting themes are identified, and these are presented during 'negotiated feedback' sessions. The resulting discussions lead to the refinement of the cross-cutting themes. Then there is a complexity analysis. This is a secondary analysis of the emergent cross-cutting themes from the perspective of complexity. Our provisional assumption here is that the cross-cutting themes should exemplify principles pertaining to complex systems. The third phase is the critical-reflective analysis. This is a rigorous interrogation of the complexity themes involving the lead researcher and Robin Durie and Katrina Wyatt. The purpose of this final iteration of analysis was to critically reflect on the appropriateness of the use of complexity theory as an interpretative framework for the primary data, and at the same time to reflect on whether this bringing together of the first two levels of analysis poses questions of complexity theory itself, and of its applicability to social systems in general, and systems of community–university research partnerships in particular.

There is an important issue that the chapter also highlights. It is not about the research carried out, but is about the relations between the people being studied and the researchers doing it. In the projects here, there had been a commitment to 'stay the distance' from the outset, and this helped greatly in creating the kind of relationships that lead to successful engagement. It is an important part of an 'enabling environment' for these projects. Community participants were concerned about researchers just coming in to 'do' the project, tick their boxes and then leave, without having a real commitment to the outcome. Fortunately, many academic partners recognized this issue as important in gaining the trust of communities. This was seen to be an important step in the creation of meaningful community–academic relations. This raises the problem of the short-term nature of research funding in this kind of domain.

The research projects sought to identify patterns of enabling behaviours for successful community engagement in research. It also sought to undertake a complexity informed case study approach to try and identify patterns of dynamic processes by which successful engagement occurs. A multi-phased framework was developed aimed at understanding the distinctive dynamics involved in the processes constituting the engagement cycle. The perspective of complexity theory suggests that the focus for community engagement with academic research should be on the dynamic relations that constitute the processes of engagement, and, in particular, on how such relations might emerge, or be co-created, within engagement processes.

This short introduction to Part II of the Handbook focuses on some the chapters that describe research methods being developed to look at social systems using the perspective of complexity thinking. The different chapters included provide clear descriptions of methods based on complexity thinking, and provide an optimistic view of what has been happening, and how the approaches should be further developed. Clearly, the way that complex systems evolve, co-evolve and change qualitatively over time is of great relevance and importance for our understanding of social systems. These applications move us away from the 'laissez faire' approach to neighbourhoods and society, and show us that we can help people to help themselves improve their communities, their health and their lives by using the evolutionary model offered by complex systems thinking.

REFERENCE

Prigogine, I. (1980), *From Being to Becoming: Time and Complexity in the Physical Sciences*, New York: W.H. Freeman & Company.

4. Addressing global challenges: the EMK complexity methodology

Professor Eve Mitleton-Kelly, London School of Economics and Political Science and Cambridge University

Is it possible to effectively address complex problems such as pandemics, deforestation and gender inequality, when there are multiple and often conflicting interests, as well as multiple interacting causalities, within a constantly changing and complex environment? When Tolstoy analysed the process and causes of war, in *War and Peace* (Tolstoy [1869] 2007) he identified an endless list of contributing factors, which made it impossible to understand why a battle actually took place in the way it had unfolded. The concatenation of all the contributing factors, some of which contradicted each other, provided a picture with no linear causality, and no overall coherent meaning. The events and the intentions appeared too random to help explain the complex interactions that led to a battle. Many policy makers today make the same assumption! What they perceive is a plethora of issues and they feel paralysed by the enormity of the task. Understanding the characteristics of complex organisations and institutions and the challenges they face, from a complexity science perspective, however, can change that approach and can enable policy makers to effectively address very difficult and complex challenges.

The key is to identify not only the multiple interacting issues and causalities, but their *co-evolutionary dynamics*. The chapter provides detailed advice on how to use the approach and real life examples of its application, specifically showing how the two parts of the methodology were used effectively in a very difficult context. The first part was the identification of the *multidimensional problem space* (where the dimensions may be social, cultural, political, economic, physical, financial, technical and other) and the *co-evolutionary dynamics* between the dimensions, which provide a starting point for decision-making. The second part of the methodology acknowledged that complex problems do not have single solutions, but need a broader *enabling environment*, capable of addressing the challenge over time as it changes and evolves. This thoroughly tested methodology is able to address apparently intractable organisational, societal, and global problems.

4.1 THE UN CASE

It will be easier to start with an example before describing the method employed, as it will bring the theory to life. The example is a recent project for the UN (Mitleton-Kelly 2015a). The author was asked to look at gender inequalities in ocean communities and other coastal waters, in the context of decision-making and the involvement of women in that process. These communities were primarily dependent on fishing and the assumption was that the men were the fishermen, they were physically stronger and could control a boat in turbulent seas; as a consequence they were the main income earners and hence the primary decision makers. The UN policy makers formulated their policies based on that assumption.

The research however, found that this model was incomplete and did not represent the actual state of affairs in different communities around the world. It also did not represent an accurate picture of the involvement of women in decision-making in those very communities, where men were the main decision makers, as this only applied to the fishing industry. By contrast, in the home and in the community, women often were the main decision makers. They looked after the family and helped to educate their children and were often the ones who took the lead in setting up new initiatives in the community. Furthermore, although the men did do the fishing, it was the women who prepared and sold the fish, and repaired and maintained the fishing nets. The assumption, therefore, that because the men did the fishing and took the decisions in the fishing industry, was an accurate representation of decision-making in the community, was erroneous.

The project looked at 10 different communities around the world and found that the model of 'man as fisherman and main decision maker' did apply to some communities, but not all and one of the main insights from the research was that *gender asymmetries were context dependent and influenced* by history, culture, religion and the economic structure of society. Gender, therefore, needed to be looked at within specific contexts such as the home, the village, the co-operative, or at policy level. Furthermore, gender could also be seen as a power relationship between men and women, influenced by that context. The inevitable conclusion was that a single model could not be used to develop global policy, as it would be irrelevant in many communities, yet that is how policy may often be formulated. The argument against taking a context dependent approach is that there would be too many issues to be taken into account. That is correct, but is based again on an erroneous assumption that policy necessarily provides 'a solution' to a particular problem.

4.2   THE TWO FUNDAMENTALS

From a complexity science perspective, a single 'solution' is likely to address very specific conditions in the problem space. When those conditions change, as they invariably will over time, the 'solution' however optimal at the outset, becomes ineffective, irrelevant or inappropriate. If however, we start with the assumption that the problem space itself is going to change and evolve over time then the obvious approach would be a 'solution' that also changes and evolves over time. The challenge here is to align the appropriateness and timeliness of the solution-evolution to the change in the problem space. This is not easy and most organisations and institutions often fail to ensure timely alignment, and the response usually lags behind the change in the problem space. Part of the reason for this lag in time is that policy rests in the hands of the few; for example, the CEO and Executive Team or their equivalent in the public sector.

An alternative approach would be to co-create an *Enabling Environment* that enables those directly involved, to take action. This is one of the two fundamentals in effectively addressing complex social problems. This approach however, assumes *distributed leadership*, and gives power to local leaders. Most policy makers would find this threatening and would argue that it would lead to chaos. But that is not necessarily the case. An *Enabling*

Environment is not a 'free for all'. It sets very clear aims and objectives, both in the short and longer term. It also provides an *enabling infrastructure* with suitable training, support, facilities and so on for local leaders to enable them to respond appropriately and in a timely manner. To work well, the *Enabling Environment* needs very good communications. Local initiatives cannot exist in isolation, they need to be assessed in terms of their contribution to a greater whole, and their progress reported to the management team as well as to other colleagues. Again this effort to communicate may appear cumbersome, but it need not be. Short but frequent get-togethers over coffee could help to inform a few colleagues, who can cascade any news or developments to their own colleagues. A major advantage of sharing the success of a local initiative is that others in the organisation become inspired and wish to also try out their own local initiatives. Also a member of the management team could be invited to these short meetings and report back to the whole management team. The assumption is that local initiatives are guided, but not totally constrained, by a clear overarching strategy. This again is a stumbling block as many organisations do not achieve such clarity and the role of the senior management team needs to be to develop that strategy and to communicate it as clearly as possible. That team should also be constantly aware of changes in the internal and external environments, to enable it to make any necessary adjustments to the strategy.

But before attempting to address a problem or challenge, the nature of that challenge needs to be identified, with its underlying multiple, interacting and co-evolving causalities. This is the first fundamental.

The above exposition has, however, jumped too far ahead and we now need to take a few steps back and to look at the methods and tools that would make such an approach feasible.

Over a 20-year period the Complexity Research Group at the London School of Economics, has worked with organisations in the private, public and voluntary sectors; and with policy makers in the UK and at least six other government administrations, to address complex problems. In the process it has developed a methodology based on complexity science, which is able to effectively address what may appear to be intractable problems, in both single organisations and global institutions. It has been tested in a range of environments that included SMEs (small and medium-sized enterprises/businesses), global businesses, NGOs (non-governmental organisations), charities and government departments.

The approach will be referred to as the *EMK Complexity Methodology*, which includes a variety of tools and methods. Many have been described elsewhere (Mitleton-Kelly 2003; Mitleton-Kelly and Puszczynski 2006; Mitleton-Kelly 2011) but what will be given in this chapter are some of the core methods, tools and concepts.

To address any complex problem effectively, two fundamentals are needed:

1. Identification of the *multiple dimensions in the problem space* with its underlying, interacting and co-evolving* causalities; and
2. Co-creation of an *Enabling Environment*.
    * *Co-evolution is defined as reciprocal influence which changes the behaviour of the interacting entities.*

Let us explore the process in each of the above.

4.3   THE MULTIDIMENSIONAL PROBLEM SPACE

A complex problem has multiple dimensions in the sense that it manifests itself in a cultural, social, economic, physical, technical (and other) context. Unfortunately, most often, a complex problem is seen simplistically in a single dimension; "it is just a technical problem, if only we had the correct IS system, all these issues will be resolved"; or "it is just a financial issue, with the right financial strategy we will solve it"; or "it is primarily a matter of restructuring the organisation" and so on.

No complex problem can be addressed effectively by focusing on a single dimension. Why? Because complex behaviour arises through the interaction of multiple related elements and organisational issues are very rarely confined within a single dimension. An information system does not involve only technology (technical dimension); it impacts, influences and changes human interaction (social/political/organisational dimensions), physical space (physical dimension), the budget (financial dimension), and the culture of the organisation (cultural dimension). While restructuring an organisation is likely to impact all of the above dimensions and many more, as it is also likely to affect external relationships, when staff are moved or changed in a major operation.

It is therefore essential to analyse the problem space in depth to identify the multiple dimensions that are likely to be impacted. The process is intensive, but with an experienced facilitator it can be completed in a couple of days involving a core of key staff (10 minimum to 30 maximum). However, a list of dimensions is not enough; in addition the key issues within each dimension need to be explored. The exercise itself is simple enough and is described below.

However, before that process starts there has to be a fair amount of preliminary work.

4.4   PREPARATION

In another project, working with a government agency in Indonesia, to enable it to address its major challenge of deforestation (Mitleton-Kelly 2015b) a set of interviews were conducted by two local facilitators in the local language and the findings were translated for the benefit of the lead researcher (Mitleton-Kelly) who also trained the local facilitators on the method of conducting and analysing interviews. A set of in-depth interviews are essential as a preliminary to analysis workshops, which identify the multidimensional problem space, as well as to identifying themes, dilemmas and underlying assumptions. These analysis workshops with organisational representatives will be described later in the chapter.

To return to the interviews, ideally they should be done in person, but if that is not possible because of distance or geographic spread, they can be undertaken by Skype or telephone. They should be recorded with the express permission of the interviewee, as the analysis will be based on full transcripts not notes. Notes are already edited and filtered by the brain of the note taker and the 'voice' of the interviewee is lost.

Furthermore, after conducting a few interviews, they tend to merge in the interviewer's memory; in addition, attention cannot be consistent for 90 minutes and some comments may not have registered. The interviewer's memory and notes are therefore totally inadequate for a full in-depth analysis.

Ninety minutes was mentioned above and this is the ideal length for an interview. It takes time to get started with all the preliminaries and to gain the trust of the interviewee. That trust or rapport is essential. If the interviewee feels that the interviewer is simply going through the motions, and is not genuinely interested in what is being discussed, then the material will be superficial and of little value. While if a rapport is established, on the basis of mutual trust and deep interest in what the interviewee can offer, then the interviewee will reflect more deeply and will answer more truthfully. This kind of interview is meant to help the interviewee reflect on the topic under discussion. It is not adversarial and should never be aggressive. If the rapport is genuine, then very difficult questions can be asked, provided that they are asked in the right way. For example I usually preface such awkward questions by saying that my next question is quite provocative and that it is up to the interviewee if he/she wishes to answer it. Pressure is never exerted. I have not had a single interviewee not respond to these 'provocative' questions with openness and thoughtfulness.

Almost one hour into the interview, the really difficult topics can be broached. It is important to also note that the interviews are based on a few topics to be reflected upon, a maximum of eight, not endless questions. Also the experienced interviewer does not need to keep to the list of topics in strict order. They should be known well enough that an opening in the 'conversation' can be followed by one of the topics on the list, without making an obvious shift in the topic under discussion.

The interviews provide first-hand data and start the process of understanding the problem in depth. That understanding is then deepened and enhanced during the analysis process. In-depth analysis, based on and coupled with first-hand in-depth interviewing, is essential in understanding the multidimensional problem space. Superficial interviewing and a quick analysis are highly unlikely to provide the depth of understanding needed to effectively address a complex problem. In effect, a superficial process is not only a complete waste of time, it may also do harm by focusing on symptoms instead of the genuine underlying and interacting causalities. Unfortunately many organisations are in such a hurry that they prefer a quick process and then wonder why they cannot address a complex problem.

Complex problems are nonlinear (*a* does not necessarily cause *x*, except superficially, when in reality it takes a, b, c, d, e, f and so on interacting and in the process changing each other, to bring about '*x*') and using simplistic linear methods to address them, simply does not work, as these methods very often simply identify symptoms not deep, underlying causalities. Such initially quick and cheap methods may end up costing the organisation a great deal in time, money and even reputation.

The findings from the analysis can then be used either at a Reflect Back Workshop to report back on the problem space or as preparation for the Problem Space Analysis and Enabling Environment Workshop with organisational representatives, which will be described below.

The Reflect Back Workshop offers the key findings from the interview analysis, as clusters of related themes, dilemmas, questions and underlying assumptions. The presentation describes the problem space and highlights some of the multiple, inter-related contributing causalities. It also provides some suggestions or recommendations on how to address the problem by describing the key conditions necessary for setting up an Enabling Environment. The researchers should not fall into the trap of offering a 'solution' to the organisation's problem. They should provide the evidence and an explanation as to 'why' that problem exists; they may also offer some suggestions of how to go about addressing it, but the job of creating the Enabling Environment to address the problem is that of the organisation itself. The researchers may participate in the process of co-creating the Enabling Environment, but they cannot set it up. They do not have the organisational or moral authority to do so. In addition it is essential that the organisation 'owns' the problem and that 'problem owners' take responsibility to address it. Providing a 'solution' will simply not work. The recommendations will not be implemented in the correct way, that is, in the way that is part of the culture, while at the same time changing the thinking and practices of that organisational culture.

Because of the depth of interviewing and analysis, the number of interviews does not need to be extensive. As a preliminary exercise for a Problem Space Workshop, then six to eight may be quite enough, provided that they cover a cross section of the organisation. If the end point is a Reflect Back Workshop, to report back findings in depth, then many more would be needed. As a guide, 12 may be enough when dealing with a single business; we have done up to 22 within a single organisation with several businesses, as each business had to be adequately covered, and 44 on an international project that covered three countries.

The interviews, analysis and Reflect Back Workshop described above are led by researchers, external to the organisation. There is, however, another process which involves organisational representatives directly in the analysis and the identification of the Enabling Environment. This process helps the participants to identify not only the multiple dimensions in the problem space, but also the co-evolving clusters, which provide the deep insights into what needs to change to effectively address a complex problem.

4.5   PROBLEM SPACE AND ENABLING ENVIRONMENT WORKSHOP: THE PROCESS

The preparatory analysis ensures that the researcher/facilitator knows enough about the issues to guide the workshop process and to ensure that all critical issues are included. It also provides the workshop participants with a good starting point. However, it is important to also allow for the unexpected and the emergent in such a workshop. The element of surprise when the participants recognise a new pattern or fresh insight is an essential part of the process and the researcher should enable these surprises and not constrain them by over-controlling the process.

Thorough preparation is necessary and whenever the research partners have refused to allow time for preliminary interviews, the workshop analysis has not been as comprehensive and the outcomes have not been as thorough. There is also an obvious lack of familiarity with the issues, which the participants do notice, irrespective of how thorough the briefing may appear to have been. Actually interviewing some of the participants beforehand changes the dynamic of the workshop and gives the facilitator greater credibility.

The process:

1.   A set of flip chart paper is pinned/blu-tacked around the walls, headed with the different dimensions, for example, Social, Cultural, Political, Physical, Economic/

Financial, Technical and so on. Some blank sheets are also made available for the participants to add dimensions.

2. The process is explained to the participants, who are first asked to discuss the problem/challenge and identify some of the dimensions, in small groups. If preliminary interviews have been conducted, then a short presentation of the key issues identified by the interview analysis is given and the participants are then asked to build on that analysis for their discussion and to add any other issues they wish. The presentation does not attempt to separate the issues into their different dimensions, but a couple of examples may be given. It is the task of the participants to identify the dimensions and to disaggregate the challenge.

3. The participants then move around the room and write the issues under each dimension, on their own, without further consultation with others.

4. The issues and sub-issues are listed by individual participants. The exercise of separating issues by dimension is difficult and counter-intuitive. Participants begin to experience a significant discomfort when assigning issues to a single dimension. This is quite deliberate and reinforces the idea that complex issues cannot be allocated neatly into single categories.

5. The next stage takes advantage of that difficulty and the participants are asked to note linkages. This can be done by a simple notation. For example all Cultural issues are named and numbered C1, C2, C3 and so on, on the left hand side of the flip chart. On the right hand side are noted the links. So that if C3 is related to P5 (Physical Dimension, issue 5), and to T7 (Technical Dimension, issue 7), and to E11 (Economic Dimension, issue 11), then these are noted on the right hand side of the relevant issue and both issues are marked. So that C3 will show that it is related to P5; and P5 will also show that it is related to C3 and so on. If the exercise is done correctly, then the key issues will show up quite distinctly, as richly connected clusters.

6. The beauty of this exercise is its physicality, with the active involvement of all participants moving around the room, making notes on the flip chart paper. It is also highly visual. The richly connected clusters are easily identified.

7. Rich clusters are then identified and discussed in terms of how they are related and how a change in one issue might impact other issues. The questioning at this stage has to be very thorough with the question 'why?' drilling down into the multiple, interconnected and underlying causalities. The objective is not just to identify links or connections, as anything can potentially be connected to everything else, but to identify *reciprocal influence which changes the behaviour of the interacting entities*. The phrase in italics defines co-evolution; hence the richly connected and reciprocally influencing issues are co-evolving clusters. In other words, if an intervention were to be made in one issue within a single dimension, what would be its impact or consequence in other key issues and dimensions? Reciprocally, what effect would that impact have on the initiators of the change? It is not enough to identify influence in one direction; that would be quite straightforward and show a link, but would not provide any deep insight. It is reciprocal influence which is the key concept. Identification of co-evolutionary clusters of issues and dimensions which is the distinctive characteristic of the *EMK Complexity Methodology* and what provides it with its deep insights.

8. The co-evolving clusters are discussed and the team focuses on those that appear to

be critical, in the sense that if they are not addressed the problem cannot be resolved. Interestingly, the critical clusters will invariably touch upon many of the other issues not directly included in a particular cluster.

9. The critical clusters of dimensions and issues are fairly comprehensive descriptors of the multidimensional problem space. They need to be listed separately as they will provide the starting point for the following day.

10. Someone needs to make extensive notes of the discussion, as the bullet points on the flip charts cannot capture the richness of the discussion. An essential element is to capture the 'whys'. The easiest way is for someone with good typing skills, who is not taking an active role in the workshop, to be taking notes directly onto a laptop. These notes need to be tidied up overnight and be printed ready for the following day. The discussion would have been so full that few people will be able to remember all the details discussed and a full note with suitable headings, for easy retrieval of the information, will be invaluable.

This process could take one or two days depending on the nature of the problem and of course the time available. When the critical co-evolving clusters have been identified, then it is time to stop and rest. The next stage will be addressing them and outlining the conditions of the Enabling Environment and participants need to be fresh when starting on that part of the exercise.

4.6   AN EXAMPLE: THE INDONESIA CASE

Details of the findings in the Indonesia case can be found in Mitleton-Kelly (2015b). This chapter focuses on the methodological process that was followed.

The process was used in Indonesia with 30 participants from a government agency responsible for forests and therefore for deforestation, which is a major problem for that particular province, for Indonesia as a whole and for the rest of the world in terms of climate change. The workshop lasted five whole days. The first day was an introduction to complexity principles to familiarise the participants with complexity theory and how to think of their organisation as a complex social system. The following two days were spent on identifying the problem space and the final two days were devoted to the Enabling Environment (EE). A weekend intervened between identifying the problem space and working on the EE and this was essential.

Identification of the problem space can be highly destabilising for an organisation. It uncovers and raises issues which are often not talked about. They can be so sensitive that if the facilitator is not highly experienced he/she can cause immense damage to the organisation. Emotions can be high and fraught and a range of distancing behaviours may emerge as the participants want to go back and delete from memories what was discussed. They feel highly vulnerable.

In this particular example after the 'high' experienced in the room, as a group, when the participants separated and went back to their rooms the phone did not stop ringing. I was supported by two local facilitators who knew the group well and had worked with them for a long time. The participants therefore approached them to complain, voicing their unease. Both facilitators came running to find me in total panic. Because the reaction was expected, I was able to reassure them that it was not out of the ordinary, but we had to calm nerves. We therefore gave up our weekend and offered all the small groups that had concerns to come and discuss them on their own. They all took up the invitation. However, by the time the meetings took place relative calm had re-established itself and we were able to listen to their concerns without too much emotion involved. But that calm has to come first from the lead facilitator.

By Monday morning the entire atmosphere had changed. We had listened to their concerns very carefully and they had a whole weekend to calm down and were ready for the creative part of the exercise. It should also be added that the entire team of 30 (plus the three facilitators), were accommodated in a hotel several hundred miles from their office. The location was beautiful and comfortable, but also rather isolated, so no one could rush back to their office for a meeting. Also no laptops were allowed in the workshop. Emails and phone calls had to be done during breaks. The team ranged from the head of the agency to very junior staff.

Culturally, it was an unusual situation as the culture is very hierarchical and junior members of staff are not normally present in the same room with senior staff. The mixture in the room was therefore unprecedented. Furthermore, the most critical comments came from the junior staff and senior members were genuinely shocked. I will give some examples to provide a fuller picture of what the workshop was like, but first of all we need to complete the exercise by looking at the Enabling Environment.

4.7 PREPARING FOR THE ENABLING ENVIRONMENT (EE)

11. The EE exercise starts with a reminder of the key issues identified in the problem space and the key clusters are revisited, so that everyone agrees that these are the clusters that need to be addressed.
12. Taking each cluster in turn, the team needs to think holistically and address all the dimensions as a whole. Focusing on one dimension will not address the issue effectively, as an intervention in one dimension, for example Finance, would impact all other related dimensions, such as the Social, Cultural, Physical, Technical, and other dimensions. Dimensions do not exist in isolation and therefore need to be seen as a whole and as part of a broader context.
13. The question is then consistently repeated: 'how will this intervention affect other parts of the organisation?' In other words the well-being of the organisation as a whole needs to be constantly, kept in mind, as solving the problems of one Department may create additional problems elsewhere.
14. The next step is to consider the enabling infrastructure. 'What needs to be put in place to enable each individual, team and the organisation as a whole to implement the EE?' That is, the support has to be provided at all scales. This support may take the form of training, new skill sets, technology, different structures or ways of working and so on, whatever may be necessary to support the new EE. If addressing the problem/ challenge is taken seriously, then changes will need to be made. Assuming that simply taking the decision to change is enough to create that change is a mistaken belief (although it is a fundamental starting point). It is essential to think through *how* that change will be implemented and *what* needs to change to make it happen.

4.8　SOME SENSITIVE CHALLENGES

Some challenges require a fundamental change in attitude, thinking, structures, and ways of working. In the Indonesia study, three of the key issues raised were corruption, nepotism and ethnic or religious groups. They impacted several dimensions: Social, Cultural, Financial, Hierarchical, Religious, Ethnic, Family and many others and they needed to be addressed at individual, team, organisational, provincial and national levels. To understand the issues some context needs to be provided. Although the main challenge was deforestation, the government agency had decided to look at their own organisation first, to enable them to address that challenge. The three issues identified above applied as much to the organisation itself, as to the broader provincial, national and international environments, which made deforestation possible.

Furthermore, we cannot assess these issues using Western values, as the cultural norms are quite different. For example the senior Indonesian male family member is responsible for the rest of the family. If he is therefore in a position of seniority or power that allows him to offer a job to his family member, he will do so and this is perceived as an honourable action, not as nepotism as we would interpret it in the West. There is, however, a distinction to be made between what is culturally expected and permissible, with the offering of that job to an acquaintance and when money changes hands. It was therefore important (a) to be aware of these differences in culture and (b) to have the courage to articulate the distinction between what was culturally acceptable and what amounted to corruption, at the workshop. Again, a researcher or facilitator that is ignorant of local values and insists on imposing his/her values will be perceived as irrelevant and will lose credibility.

Corruption is universally criticised yet it is worth understanding why it happens. Civil servants in Indonesia are not well paid, at the same time, status and indicators of status such as a large car or bigger office are important. It is therefore very difficult to resist the temptation of a bribe. In some cultures it has become so entrenched that it is expected, like giving a tip to a waiter is expected in other countries. To reduce corruption it would therefore be important to address the issue of low pay (Financial dimension), but also to see the acceptance of a bribe not as entitlement, but as contrary to the values of the organisation (Social and Cultural) and this is a significant change to be made and needs persistence and courage.

One of the senior officers told us how he had stood up against pressure not only from multinationals who wanted to cut whole forests, but also from their own ministers, who had already accepted bribes. He knew very well that his life was in danger, but he was a rare person who could see the value of the forest to his province, country and the world (deforestation affects climate change), and was prepared to take a stand. Furthermore, although the Provincial Parliament had passed a law to limit the cutting of the forest to 3 per cent, Ministers in the National Government in Jakarta had given permits to the external contractors to cut the trees.

The third issue of ethnic or religious groups was brought up by one of the most junior members of staff in the workshop and created shock waves within the room. This was not a subject open for discussion! But once it had been voiced it could not be eradicated and the facilitators recommended that it would be wise to address it, within the confines of the room. Many Indonesians are Christians, but there has been a great influx of immigrants who are Muslim. The members of the two religions support their own members and tend to exclude the others. In addition, there are many tribes and ethnic groups and again members of a tribe are loyal to their own tribe members. It would therefore be natural for those in a position to award contracts or jobs, to offer them to those of their own religion and tribe. These loyalties cannot be legislated against, but greater transparency on the allocation of contracts and jobs, can go a long way to reducing suspicion and fear and make the process less prone to favouring the 'group'.

Understanding the reasons why these things happen is not to offer an excuse, but to find a way to address them more effectively. One has to work within the cultural, moral and ethnic norms of a society and an organisation if the changes agreed have a chance of being implemented.

4.9   THE ENABLING ENVIRONMENT: PREPARING FOR ACTION

15.   The EE discussions must be recorded and all agreements documented in detail. Agreements must also be explicit and responsibility for action needs to be taken by named individuals or when everyone returns to work nothing will happen. Someone needs to be taking notes on a laptop as the discussion progresses and the agreements have to be verified by all those concerned, to carry any weight. This process needs to happen at the end of the workshop, when 'next' steps are being discussed and plenty of time has to be allowed for this exercise; it also has to happen when the entire team is present. In the Indonesian project, one of the related issues was lack of transparency and an increase in transparency was one of the actions agreed. The workshop itself had to exemplify that value.

16.   An 'action team' needs to take responsibility for implementing all the agreements or ensuring that they are implemented and that team cannot then be left unsupported. It needs continuous support and this was provided by frequent Skype calls with Mitleton-Kelly and a return visit of the team to London to discuss progress.

17.   In the year that followed the workshop, the government agency underwent some severe setbacks. They lost their Head of the Agency who was instrumental in inviting the LSE to help. The province had a new governor and a lot of new legislation from the National Parliament had to be implemented. Nevertheless the action team persisted, until the actions agreed at the workshop had been authorised by the new Head of the Agency and the new governor. They went through a great deal of opposition and resistance, but they did make changes, which were fundamental and benefited the organisation and the province. Part of their strength was derived from their direct engagement and involvement in the process and because they had defined the challenges they were facing and had also defined the actions needed to co-create an enabling environment.

Co-creation is also an essential contributor to success. Senior management will need to provide the overarching strategy and vision; fully support the changes and ensure that all facilities and training are provided, but they cannot bring about *effective* wholesale change on their own. This will come from the entire organisation becoming involved in the change process and *together co-creating the enabling environment*. This process is based on engagement of the whole work force led by local leaders, supported by the Action Team, and is predicated on the enabling of *distributed leadership*.

4.10   SOME OTHER EXAMPLES AND ESSENTIALS FOR SUCCESS

The Problem Space and EE Workshop has been used with the IT department of a pharmaceutical company in London; with an engineering company in Derby; with several groups within the NHS (National Health Service, UK) in London and Bristol; with a multi-business trust in Devon; with a European Commission agency in Helsinki; with a government agency in Indonesia and many others.

In every case, involvement and engagement of the workshop participants was essential. It is part of human nature to wish to be involved and engaged and to 'discover' the deep issues that need to be addressed as well as to decide on the actions to be taken to address the challenge. The researcher must act as a facilitator to that process and a guide and not take over the process. The workshop invariably worked best, when it had been preceded by a set of in depth interviews that included some of the workshop participants. The other essential was full support by a very senior sponsor, preferably one of the directors and the Chief Executive.

In many organisations, training of the employees to use the *EMK Complexity Methodology* was part of the project, to enable them to use it in future with other issues and to disseminate it throughout the organisation. This built capacity within the organisation and they stood a much better chance of being able to sustain the enabling environment and continue to address new problems as they arose.

4.11   CONCLUSION

This chapter has attempted to outline the *EMK Complexity Methodology* and to describe the process undertaken in the Problem Space and Enabling Environment Workshop, which has not been described in earlier papers on the methodology. What distinguishes the *EMK Complexity Methodology* and makes it so very effective with organisational, societal and global problems are the following:

a)   Identification of the multiple dimensions in the problem space;
b)   Identification of the co-evolutionary dynamics (how the different dimensions and issues interact and change each other in the process);
c)   Recognising that there is no single, simple 'solution' to a complex problem, but that the problem can be addressed effectively by setting up an Enabling Environment, which directly addresses the co-evolving clusters and evolves with the problem space.

It includes both researcher-led processes as well as a workshop, which directly involves representatives from the organisation under study. It is a robust and thorough approach, which has been proven to work effectively in many different contexts over two decades.

What has been described is however, only part of an integrated methodology, which also includes computer modelling, art (described in Chapter 8) and a tool based in psychology (Chapter 9), as well as other tools and methods. The LSE Complexity Group worked with physicists, biologists, mathematicians as well as artists and psychologists to explore the nature of complex problems and how to address them effectively. The *EMK Complexity Methodology* has emerged from those studies, has been enriched by over 40 research projects and has been tested in single organisations, with societal challenges and global issues.

In both the cases described, there were significant outcomes. In the UN Case, international policy was informed by the report on the findings and this fundamentally changed UN policy on coastal communities. The UN policy makers recognised that a single model did not represent reality and that policies had to allow for differences in context. In the second case in Indonesia, the Government Agency not only changed its practice and organisational culture fundamentally, but the understanding developed by the employees, especially by the action team, influenced policy development by the Government Agency on deforestation, in the rest of the Province. In both cases the thinking and the policies were significantly influenced.

REFERENCES

Mitleton-Kelly, E. 2003 'Complexity Research – Approaches and Methods: The LSE Complexity Group Integrated Methodology' in Keskinen, A., Aaltonen, M., Mitleton-Kelly, E. (eds), *Organisational Complexity*. Foreword by Stuart Kauffman. Scientific Papers 1/2003, TUTU Publications, Finland Futures Research Centre, Helsinki.

Mitleton-Kelly, E. 2011 'Identifying the Multi-Dimensional Problem Space and Co-creating an Enabling Environment', in Tait, A. and Richardson, K.A. (eds), *Moving Forward with Complexity: Proceedings of the 1st International Workshop on Complex Systems Thinking and Real World Applications*, London: Emergent Publications, Chapter 2, pp. 21–44.

Mitleton-Kelly, E. 2015a Report on 'Gender & Decision Making Focusing on Ocean and Coastal Management Policy' for UNEP, Jan. 2015 www.lse.ac.uk/complexity (accessed 26 July 2017).

Mitleton-Kelly, E. 2015b, 'Effective Policy Making: Addressing Apparently Intractable Problems', in Geyer, R. and Cairney, P. (eds), *Handbook on Complexity and Public Policy*, Cheltenham, UK and Northampton, MA, USA: Edward Elgar Publishing, Chapter 8.

Mitleton-Kelly, E. and Puszczynski, L.R. 2006 'An Integrated Methodology to Facilitate The Emergence of New Ways of Organising', in Bar-Yam, Y. and Minai, A.(eds), *Unifying Themes in Complex Systems*, Vol. V, Proceedings of the Fifth International Conference on Complex Systems, Springer, http://necsi.org/events/iccs/2004proceedings.html, paper #659

Tolstoy, L. [1869] 2007 *War and Peace*, translation by Anthony Briggs, London: Penguin Classics.

5. Complexity informed social research: from complexity concepts to creative applications

Dr Lesley Kuhn, Western Sydney University

5.1 INTRODUCTION

The complexity sciences provide effective ways of researching complex social phenomena where there is an extensive array of variables interacting and influencing each other in an extensive variety of ways. Extrapolating from precise scientific and mathematical approaches, the complexity sciences offer habits of thought and vocabularies for making sense that, in not being bound to linearity, more closely match the messiness of lived experience. However, choosing to pursue the challenges implicit in researching and theorizing human experience and sense making, from a perspective based on interpretations of particular findings and ideas from studies of complexity in nature, means that as a social researcher, I must translate and interpret these findings and ideas. This chapter, in describing and demonstrating a set of complexity informed qualitative social research methods, concomitantly explains the translations and interpretations made in utilizing findings and ideas from the complexity sciences (hereafter referred to as 'complexity') to effectively comprehend human social experience and meaning making.

Complexity takes a radically relational view that construes relationships as essential in the constitution of phenomena (Morin 2008). From a complexity perspective, it is through local connections or relationships that macro behaviour emerges. In describing the nature of these relationships, the ontological explanation of complexity is that 'reality' is self-organizing, dynamic and emergent (Lewin 1999; Morin 2008; Prigogine and Stengers 1984). Thus individuals, organizations, populations and environments are conceived as interrelating, self-organizing, dynamic and emergent, not reducible to component parts and mutually influential.

Likewise, epistemologically, complexity construes sense making as self-organizing, dynamic and emergent. Complexity suggests epistemological ambiguity, as both the phenomena of study and interpretation of it, are perceived as interrelating, self-organizing, dynamic and emergent, not reducible to component parts and mutually influential. Morin (2008) and others (Alhadeff-Jones 2008; Kuhn 2007) emphasize the importance of epistemological reflection and complex thinking where we remain cognisant of the biological, physical and anthropological foundations of our sense making. This view resonates with philosophical, psychological and sociological orientations (Heidegger 1969; Jung 1976; Maturana and Varela 1987; Schwartz and Ogilvy 1979) that similarly recognize the interpretive, ambiguous and uncertain character of human sense making. Complexity, in this reading, constitutes a powerful conceptual framework for sense making that is not bound to linearity or certainty.

Qualitative research spans disciplines, fields of study and subject matter. Qualitative researchers employ interpretive, naturalistic approaches to study of human behaviour and sense making so as to learn how the experience of the research participants is created and given meaning (Denzin and Lincoln 1994). Qualitative research, like complexity, takes a radically relational approach in seeking to interpret interrelationships between sense makers, fragments of knowledge, cultures, histories, futures, aspirations and so on. Complexity offers a paradigmatic orientation that can assist qualitative researchers in discerning nonlinearly-based pattern and order within the multidimensionality of existence, and specifically in terms of the meaning making of research participants. One of the advantages for qualitative researchers is that a complexity orientation assists researchers in making links between particular cases and more general situations, through understanding that it is through local connections or relationships that macro behaviour emerges.

This chapter describes and demonstrates the complexity informed qualitative social research approach, methods and techniques developed through collaboration between my colleague Robert Woog and myself. The chapter begins by presenting a methodological explanation that explicates the principles guiding the approach, methods and techniques. The relational connections between the field of investigation (that is, the practice or situation that is the focus of the research), perspective, theory, selected concepts, inquiry strategies and applied techniques are set out. *Vortical postmodern ethnography* as an inquiry approach is described along with the narrative generating method of *coherent conversations* and the narrative analysis techniques of *fractal narrative analysis* and *attractor narrative analysis*. The chapter concludes with a demonstration of these complexity informed methods as utilized in an indicative research project.

Throughout the chapter, as indicated above, I use the term 'complexity' as a shorthand means of referring to 'the complexity sciences'. Along with others (Mitleton-Kelly 2006), I take the view that there is no unified theory of complexity, but rather a diverse and contested field (Kuhn and Woog 2005). The approach taken in this chapter can be identified as drawing upon discourses on complexity that take an interpretive perspective and that highlight an integrated epistemological and ontological stance, whereby a social ecology of being and knowing is sought (Chambers 1994; Morin 2001).

5.2   INITIAL CONDITIONS

This section, in describing the initial conditions (or the starting position) out of which our complexity informed approach to qualitative inquiry emerged, indicates some of the important concepts and assumptions that shape our engagement with the complexity sciences. From a complexity perspective, initial conditions are recognized as having significant influence on shaping subsequent dynamic interactions and hence overall emergence. Outlining our initial perspectives, theoretical and method preferences and experiences provides information about the paradigmatic orientation of our complexity informed approach.

We came to the complexity sciences carrying certain epistemological, ontological and axiological preferences and assumptions and with a history of familiarity with qualitative research traditions. These constitute significant aspects of the initial conditions shaping our engagement with complexity. For many years we had engaged in transdisciplinary and interpretivist or constructivist approaches to social inquiry that concomitantly acknowledged the contingency of our own knowing, and the knowing generated through various disciplinary discourses. We supported our research practice through critical reflection on questions concerning how to understand and theorize the position of ourselves as researchers and that of the research participants, the nature of the knowledge generated through the research process and the ethics concerning our role in generating change in the situations that we had been invited to research (for example, Kuhn-White 1994).

Following Guba and Lincoln's work (1989), in identifying and naming those fundamental assumptions held by the researcher about the nature of reality and how we comprehend it that guide the researcher in all of their activities, we described ourselves as taking an interpretive and constructivist paradigmatic orientation to research (see for example Gamble, Blunden, Kuhn-White, Voyce and Loftus 1995).

We drew on a range of disciplines and knowledge traditions, while holding awareness that our knowing is shaped by our physical structure, the historic, social and cultural dimensions within which we are immersed and by our personal histories of being. This style of interpretivism, grounded in the German intellectual tradition of hermeneutics, as influenced by philosophers such as Heidegger (1969) and constructivism, as influenced by Maturana and Varela (1987), Rorty (1990) and Shotter (1993), denies the opposition of subjectivity and objectivity, and views existence itself as interpretation and construction. With this position, humans are understood to be linguistically, socially and historically constituted.

This perspective has implications for how researchers think about and engage with research methods and techniques. The interpretivist perspective on method is radically different from the conception of scientific method within logical empiricist inquiry, where method is associated with removing personal judgement and the major criteria in applying the method is correctness of application. The method itself in this instance is taken as the preeminent guide. From an interpretivist perspective, utilization of methods should be guided in the first instance by practical reasoning. This is to take what Madison (1988) describes as a normative sense of method, where, "far from supplanting personal, subjective judgement, or eliminating the need for it, [method] is meant as an aid to good judgement" (1988, p. 28).

A second influential aspect of our initial conditions was that we were engaged in systemic and participatory styles of inquiry, such as Soft Systems Methodology (SSM) (Checkland 1981), where researchers view themselves as inquiring into messy, purposeful human activity systems in order to initiate debate about change. With a SSM approach, the system is thought to learn its way forward to changes deemed by the participants as systemically desirable and culturally feasible (Checkland 1981). We viewed ourselves as inquiring into 'complex' situations, where there was little agreement as to what constituted the fundamental problem to be addressed and where it was critical that we engaged the range of values and beliefs of the participants. Through an iterative learning process, such as supported by Action Research (Reason 1991), together with the research participants we took part in a process that aimed to increase the 'complexity' by which we understood the situation.

5.3   EMERGENCE OF A COMPLEXITY INFORMED APPROACH TO QUALITATIVE RESEARCH

In the early 1990s in conversation with Vladimir Dimitrov, we began exploring the potential of the 'Chaos Sciences' to inform sociological, psychological and philosophical understandings. This evolved into critical, theoretically informed discussions about the potential of the complexity sciences to conceptually inform theorizing and practice in human inquiry and qualitative social research (for example, Dimitrov, Woog and Kuhn-White 1996; Johannessen and Kuhn 2012; Kuhn 2012 and 2009; Packman and Kuhn 2009).

Continued cycling between research practice, undertaking projects across a wide range of topics, countries and a diversity of cultures (for example, corporate organizational development (Kuhn, Woog and Hodgson 2003), health (Kuhn 2002) and community development (Gamble, Blunden, Kuhn-White, Voyce, and Loftus 1995; Khatoonabadi and Woog 1992)) and critical reflection on theory, methodology and methods (for example, Kuhn and Woog 2003), distilled into our complexity informed approach to qualitative research.

5.4   INTRODUCTION TO A COMPLEXITY PARADIGMATIC ORIENTATION

In qualitative research, identification of the research paradigm is critical as it signifies the basic principles, or set of epistemological, ontological and methodological assumptions and preferences that guide the research process and thinking about the meaning of the outcomes or findings. As indicated above, our complexity informed research approach and methods share many of the underlying assumptions that characterize an interpretivist, systemic and participative qualitative research paradigm.

However, complexity habits of thought offer distinctive ways of thinking about and describing the nature of the world and the individual's place within it. For this reason it is important to introduce the basic organizing principles or major features of the complexity ways of thinking which inform our research approach and methods.

Our approach is based on the notion of a complexity cosmography, of which the basic principles are that 'reality' is self-organizing, dynamic and emergent, comprised of individuals and various groupings of people that likewise are self-organizing, dynamic and emergent. *Self-organization* describes the capacity of complex living systems, to evolve according to internally evolving principles (Coveney and Highfield 1995; Kauffman 1995). Individuals, households, organizations, cultures, and so on are seen as involved in the ongoing metamorphosing of themselves through processes of self-organization. *Dynamism* refers to the capacity of living entities to continuously adapt, respond and influence others and the environments (social and physical) within which they exist (Waldrop 1993). The capacity to adapt highlights ability to learn from experience. *Emergence* denotes the capacity of complex living entities to exhibit unexpected and novel properties and behaviours not previously observed. With emergence comes the idea that micro-phenomena give rise to macro-phenomena, with characteristics observed in the macro-phenomena not being reducible to the micro phenomena (Coveney and Highfield 1995; Johnson 2001).

Based on these principles, we describe at the most general level, our research as inquiry into human experiential space, which we characterize as chaotic, multidimensional, non-linear and as organized around attractors.

Continuing the process of translating and interpreting findings and ideas from studies of complexity in nature, we utilize a range of specific complexity concepts as metaphors to aid in sense making. Metaphors are used in complexity to convey ideas and relationships that otherwise would take a great many words to portray. In making a metaphorical comparison we create new ways of perceiving and structuring experience. The seven complexity metaphors we have found most useful are: phase space – phrase space; communicative connectedness; sensitive dependence on initial conditions; fitness landscape; edge of chaos – chaotic edge; attractors; and fractality.

The specific characteristics of human experiential space, pertaining to the particular foci of research, may be revealed through examining the *phrase space*. Taking the notion of *phase space* (Lewin 1999), phrase space is used to describe the space wherein humans, as individuals and groups, position themselves. Whereas phase space can be mathematically represented as a plot of position and momentum variables as a function of time, phrase space brings a literary explanation. Bringing together insights from complexity, sociology, philosophy and psychology, phrase space, in focusing on narratives and discourse, is useful in describing preferred co-ordinate locations in relation to human experience and sense making. These preferred co-ordinate locations relate, in philosophical terms, to *weltanschauung* or worldview (Heidegger 1969), in sociological terms, to social constructionism (Berger and Luckman 1967; Shotter 1993) and in terms of cognitive psychology, to core values (Kelly 1955). Phase space shows that although there may be an extensive range of possibilities, an entity usually only occupies a minute proportion of its possible phase space. Likewise, phrase space shows that humans too live within a minute proportion of possibilities as individuals and as societies. The poet Rainer Maria Rilke evocatively expresses this understanding when he writes: "For if we think of this existence of the individual as a larger or smaller room, it appears that most people learn to know only a corner of their room, a place by the window, a strip of floor on which they walk up and down" (Rilke 2004, p. 51).

Qualitative research may be described as concerned with identification of certain aspects of the phrase space of a person or social group. In addition, paying attention to phrase space can help people to become aware that their views and sense of self relate to their contextual interrelatedness and thus aware that ways of knowing and doing represent limited, socially supported frameworks, within a space of infinite possibilities. In a research context this can be very important, especially where the research takes a proactive stance in seeking to bring about improvement.

Communicative connectedness (Kuhn and Woog 2006) describes the qualities of interconnections between people. As noted above, complexity construes relationships as integral to the emergence of complex phenomena and understands that it is through micro relationships that macro behaviour emerges. Human society (as well as social theorizing) can be viewed as constructed through communicative connectedness and this can be understood as generated via multiple conversations. For people, conversation forms the primary means by which communicative connectedness is created and sustained, with the types and quality of communication between people shaping phrase space. Inquiry into the nature and quality of communicative connectedness can provide critical information that assists in theorizing human experience and sense making.

Sensitive dependence on initial conditions (Gleick 1998), described colloquially as the butterfly effect, stresses the significant shaping influence of initial conditions and small perturbations in shaping the overall emergence of complex entities (such as an individual, organization, culture, society and so on). In social inquiry, beginning with the presenting circumstances, useful understandings may be garnered through tracking back into the history of the situation to reveal clues as to why things have emerged in the way that they have.

For complex entities to survive there must be a certain match, or fit, between the entity and its context (or landscape). Fitness describes the chance of survival yielded by the relationship between an entity and its context. Using the imagery of a landscape to represent fitness probabilities, depicted as peaks and valleys, *fitness landscape* describes the quality of match or fit, between a complex entity and its environment or context (Lewin 1999). The notion of fitness landscape is useful in critically appreciating histories and future possibilities in terms of contextual couplings.

The *edge of chaos* describes a dynamic zone between order and disorder within a complex entity's phase space where innovative re-arrangement and transformation take place. Complex self-organizing systems are understood to naturally gravitate to this zone (Bak and Chen 1991; Langton 1986). In researching human behaviour and sense making we find it useful to distinguish edge of chaos thinking (where being between order and disorder is thought of as an acceptable place) from *chaotic edge* thinking, where people think of themselves and their situation as full of threat rather than full of potential.

An *attractor* (Lewin 1999) functions as an organizing force that sustains behaviour (as with a point attractor) and generates change (as with a strange attractor). Holding complex entities in particular patterns, attractors in the context of theorizing human experience and sense making, may be understood as organizing principles that guide behaviour. Politics, religion, economics or environmentalism may form the attractor around which an individual or social phenomenon is organized. Identification of an attractor or attractor set assists an observer in understanding the drivers of particular behaviours.

Fractal geometry describes a family of naturally occurring irregular shapes where the degree of irregularity or fragmentation is similar (or identical) at all scales. The term fractal thus describes entities with fractal dimensions or look-alike features that are simultaneously apparent across multiple scales of focus (Mandelbrot 1977). Human society may be described as having fractal dimensions, with individuals, organizations, communities, cultures, societies and so on, exhibiting aspects of self-similarity. For example, the essayist Montaigne (in Lopate 1995) argues that each person has within him/ herself the entire human condition so that in writing about oneself, people are to some extent, writing about human experience more generally. The fractal dimensions (or self-similarity) of interest to Montaigne relate, in general terms, to the capacity of people to be self-conscious. Thinking fractally can yield new insights and recognizes that the global and local are embedded in all levels of social practice.

The key principles of fractality are that: fractals exhibit the same degree of irregularity at different scales; observing a fractal you get information proportional to the scale; and the small scale remains an equally complex microcosm of the whole.

The notion of fractality is useful for social inquiry as study of a fractal (one group or aspect of society) may lead to understanding of much larger phenomena from which the fractal is derived. In applying the concept of fractality to social research, it is necessary to recognize that the fractal relationship selected for study by the researcher is not merely a part of the whole, but that it is in a sense representative of the whole. It represents an aspect of the social system that is present at all scales of the system, and which may be examined at certain, chosen scales.

Having introduced the basic organizing principles of the complexity ways of thinking that inform our research approach and methods, the following sections introduce the complexity informed inquiry methods that we have developed.

5.5 VORTICAL POSTMODERN ETHNOGRAPHY

Vortical postmodern ethnography constitutes a complexity informed re-conceptualization of ethnography. As a form of qualitative research, ethnography seeks to develop in-depth understandings of phenomena in terms of the research participants' meaning making processes (Denzin and Lincoln 1994).

Relying largely on participant observation, classic ethnography studies the sense making and behaviour of cultural groups with the findings overwhelmingly reliant on the report of the researcher. In a classic approach, it is presumed that it is possible for people not indigenous to the cultural group or setting to gain, through direct, disciplined, appreciative and analytic involvement, valid, useful or trustworthy knowledge of other human cultural groups (Sanders 1999). Typically, ethnographic research:

1. Focuses on social behaviour within natural settings.
2. Relies on qualitative data in the form of narrative descriptions made by the researcher as observer.
3. Takes a 'holistic' perspective, where the observations and interpretations are conceived of as being made within the totality of human interactions.
4. Allows for emergence of new hypotheses and questions as the research progresses.
5. Data analysis involves contextualization of the research findings through reference to the cultural group (Atkinson and Hammersley 1994).

With postmodern ethnography, the assumed distinction between the researcher and the researched cultural group came to be viewed as less clear and more complicated. The ways by which the identities of all participants, borders between insider and outsider status and the positioning of the authoritative 'knower' (about the researched cultural group) are inevitably challenged during fieldwork came to be overtly recognized (Sanders 1999). Described in terms of the dimensions listed above (of (1) focus, (2) reliance on data, (3) overall perspective, (4) approach to hypotheses and (5) approach to data analysis), postmodern ethnography may be depicted as:

1. Focusing on discourse within settings created through the convergence of the researched cultural group and the researcher's cultural group.
2. Relying on qualitative data in the form of narrative descriptions made by the researcher and researched as discursive partners.

3. Taking a holistic perspective, where observations and interpretations are viewed as emergent from, and constitutive of, the totality of human interactions.
4. Allowing for emergence of new hypotheses and questions as the research progresses.
5. Embodying circularity in the processes of data analysis and synthesis, whereby research findings are contextualized within the discourse evoked via the research activities.

With postmodern ethnography, there is overt acknowledgement that sense of community may be expanded for both the researcher and researched (Sanders 1999). It is recognized that taking part as co-researchers may de-familiarize the commonsense reality of all involved, through momentary rupturing of everyday habits of mind and action. Taken for granted assumptions can become available for reflection, scrutiny, challenge and critique.

Undertaking ethnographic inquiry is intrinsically a challenging, highly dynamic and volatile process. Characteristic of the manifestation of vorticity is a situation where there are radically different, nonlinear dynamics occurring in close proximity to each other. The term 'vortex' describes, in a technical sense, a spinning mass of gas or liquid that pulls things into its centre. In a literary sense, vortex can be used to describe powerful feelings and situations from which one cannot escape. These descriptions of vorticity, in our view, aptly depict the experience of all actors (researchers and research participants) within the ethnographic space. For all are involved in negotiating their insider/outside status, roles, sense of identity and claims to authoritative knowledge. As Enguix (2014 p. 79), reflecting on her experience in undertaking ethnographic fieldwork, comments: "the research process becomes a 'situated' process between positions and roles that are not stable but changing . . . in research settings we encounter subjects, bodies, personalities and social positions in constant fluctuation and negotiation. All of them are embedded in dynamic categories subsumed in flexible boundaries".

As social researchers we view the activities we engage in with research participants as being not outside of either of our life experiences. We consider that our narratives become, for a time, co-emergent. Together we create a new phrase space, a new space of possibility (which may be characterized as being at the edge of chaos). Our combined discourse, although fragmentary, contributes to each other's ongoing evolution. The process of research influences key characteristics of researchers and research participants alike.

Vortical postmodern ethnography, with reference to the five aspects of ethnographic research identified above (focus, reliance on data, overall perspective, approach to hypotheses and approach to data analysis) may thus be described as:

1. Focusing on narratives, discourse and behaviour as these unfold during the research.
2. Relying on qualitative data in the form of formal and reflective narratives pertaining to the research.
3. Taking a holistic perspective, that sees the research as fractal, focused on one level, but implicating themes present at levels more macro and micro within the totality of human interactions.
4. Expecting the vorticity of the social system as generated through the research process to produce new ideas, insights, hypotheses and questions.
5. Analysing data for identification of themes that have a strong resonance for

participants and as treating data generation as a further source of vorticity that leads to new cycles of emergence.

Vortical postmodern ethnography re-conceptualizes the issue concerning the possibility of participants behaving 'normally', despite the focused gaze of the researcher. With vortical postmodern ethnography, all who are involved in the research, together with the activities engaged in, are thought to constitute the situation under investigation. 'Object', 'observer' and 'observation' are thus no longer seen as separate categorical systems, but as swirling interacting parameters of those systems. The situation is viewed as emerging from the convergence of people who may not have worked together in this way previously. Vorticity is often observed with this coming together of individuals and groups having different backgrounds, research interests and aspirations. The notion of participants behaving 'normally' may be redressed: as self-organizing beings, it is assumed that participants could not do other than behave 'normally', albeit differently as befitting novel situations.

The position of the researcher in the generation of findings is similarly redressed. The findings are not conceived of as being about an objective other, but as emergent from a particular process concerning the evolving discourse of those involved in the project.

Vortical postmodern ethnography recognizes that the vorticity or energizing of the social system, through irregular dynamics being fostered (as when different peoples come together), contributes to the emergence of new ideas and insights for both researchers and participants. The notion of vorticity applies no less to the interpretation of the research experience, which should also be understood as spontaneously co-created.

In summary, vortical postmodern ethnography, as an approach to qualitative research, leads to contextually constructed understanding of the inquiry process.

5.6   COHERENT CONVERSATIONS

Coherent conversations describe a complexity informed method of generating data (in the form of narrative accounts). Coherent conversations may take place as one-to-one interviews or group conversations. As group conversations they can be thought of as similar to focus groups, where the researcher holds a group interview with (typically) six to ten people, and where through the process of conversation, a rich account of views and beliefs may be garnered via participants reacting and responding to the views of other participants. However, whereas a focus group aims to have the conversation focus on a particular topic, coherent conversations aim for the conversation to be accepting of the breadth of topics and information brought to the conversation. With the aim of the coherent conversation being to facilitate communicative connectedness, responses from participants that may usually, in qualitative studies where the aim is to generate 'rich' descriptions of the views of participants, be deemed as non-responsive (as when the participant declines invitations to elaborate their response or asks questions of the interviewer) (Roulston 2014), are treated differently. These types of responses are valued for providing information about the concerns of the participants. It is thought that including in the conversation, as many as possible of the full range of participant' concerns, provides a richer understanding of the views of the participants.

While usually the researchers will have worked out in advance main questions and probes, each conversation will be unique. Coherent conversations aim to be critically self-reflective of the processes via which the conversation emerges. As a data gathering method, coherent conversations have the following characteristics, they:

- Are permissive not agenda bound, allowing people's priorities and own agendas to emerge.
- May reveal the way a person thinks and how they feel through what they say.
- Make the conversational dynamics and relationships as evident as what is being said.
- Are self-reflective of the conversational process.
- Are both intuitive and logical.

Coherent conversations can be used as an inquiry method that facilitates emergence. Whereas designed situation interventions are outcome driven, coherent conversations create enhanced potentiality. Coherent conversations facilitate emergence by generating conditions supportive of further developments. They do this in two important ways. First, they build coherent understanding thus strengthening communicative connectedness, and second, they reveal and help establish phrase space. The revelations carried in coherent conversations are viewed as critically significant, and sometimes as even more important than the 'answers' sought by researchers at the outset of the conversation.

In summary, a coherent conversation is a method or technique that generates as well as guides data acquisition.

Much can be learned from the narratives generated through coherent conversations by applying two complexity informed approaches to analysing narrative data: *fractal narrative analysis* and *attractor narrative analysis*.

5.7   FRACTAL NARRATIVE ANALYSIS

Fractal narrative analysis is a useful way of making sense of, interpreting or theorizing narratives (data) generated through coherent conversations, as by examining the narratives for views, or concerns that are repeatedly expressed (the fractal dimensions of the narratives) we can glimpse the macrocosm, or broader social system, despite the proportional limitation of the scale we are dealing with (as represented by the individuals or groups involved).

Viewing social and cultural groups and settings as fractally constructed, means looking for similarities that are apparent across different scales. Studying one fractal (for example, an individual or department), we can make generalizations about much larger phenomena (such as the social system or organization as a whole, or the sectors across which it operates) from which the fractal is derived.

Taking a fractal approach, we can study and make sense of small portions of social phenomena without artificially simplifying these. Smaller scale fractals remain equally complex microcosms of the whole. The capacity for self-organization and emergence of one fractal represents the dynamics and capacity for emergence of the system as a whole.

Identification of a narrative fragment as a fractal is a form of categorization and illumination of narrative data that enables and provides evidence of developing insights and exposition of meaning.

5.8   ATTRACTOR NARRATIVE ANALYSIS

Like fractal narrative analysis, *attractor narrative analysis* enables researchers to make sense of, interpret or theorize narratives (data) generated through coherent conversations without overly simplifying these. Instead, in looking for attractors, researchers aim to identify the values, issues of concern, motivators, and so on that underlie attitudes and behaviours.

Attractor narrative analysis may be recognized by social researchers as a process of formulating response category headings. Attractor narrative analysis thus constitutes a means of organizing, reducing and describing data that involves synthesizing or config-uring of the narratives into an interpretive explanation of what guides the attitudes and behaviours of the participants.

In undertaking attractor narrative analysis, it is important that the narratives have been generated through permissive, rather than tightly prescribed, conversations as to identify attractors it must be possible to glimpse something of why people hold their views and attitudes. It is in the 'why' that the attractors are revealed. Identifying the attractor or attractor set assists in creating understanding of the complex system, even beyond the group of people involved in the coherent conversation. Understanding what guides the attitudes and behaviours of those involved, we can make inferences about the self-organ-izing character of the system.

Paying attention to attractors, it is possible to gain insight about likely changes. There is some sense of predictability associated with social entities (individuals, teams, depart-ments, whole organizations, sectors and so on) undergoing critical transitions in phase space (moving from one attractor to another). Although it is hard to know when the transition will occur, symptoms of instability, such as rapidly increasing complexity and the appearance of chaotic dynamics, serve as indicators of approaching transition. As well, the emerging neophyte shape of a new attractor may be used to speculate about likely ongoing development and its ultimate character.

Use of the complexity informed research approach, methods and techniques described above, enables new insights into how the experiences of both researchers and research participants is created and given meaning. Application of these methods and techniques generate answers to questions that qualitative researchers typically ask, concerning 'what', 'why' and 'where to' regarding human experiential space. Concomitantly, engagement with the approach and methods ameliorates some commonly experienced difficulties in conceptualizing and undertaking social inquiry.

5.9   SAMPLE APPLICATION OF OUR COMPLEXITY INFORMED QUALITATIVE SOCIAL RESEARCH APPROACH, METHODS AND TECHNIQUES

This section illustrates how vortical postmodern ethnography, coherent conversations, fractal narrative analysis and attractor narrative analysis may be utilized in practice in a qualitative research project. It is intended that through describing the steps involved in undertaking a specific, but indicative research project, our approach will be rendered more accessible to others.

5.9.1   Inquiry into Community Enabling: A Review of 'The Mt Druitt Enablers Program'

This research project comprised an evaluation of a community leadership development programme entitled 'The Mt. Druitt Enablers Program'. The aim of this programme was to promote community development through assisting local people in becoming more confident as community leaders.

The evaluation was undertaken through a vortical postmodern ethnographic methodology, which focused on the views of Enablers participants and programme leaders about how the programme has enhanced the lives of participants, and enabled them to more effectively contribute to the development of the community. As well as seeking to find out what happened, we sought to inform key people about what happened. A complexity perspective suggests that the only viable long-term strategy for the development of a complex social system, such as the Mt Druitt community, is for the system to learn about itself. Informing key or relevant people is a powerful way of nudging complex social systems towards self-managed improvement.

5.9.2   Background: Specifics of the Human Experiential Space

The project was set in Mt Druitt, NSW, Australia, an area that is culturally and linguistically diverse, with a large Indigenous population and low median income (Blacktown City Council 2006).

The Enablers Program, run by Chain Reaction (a non-government organization) was begun in response to citizen requests for assistance in developing leaders and involved participants who were: engaged in the community in some way (paid or voluntary); recommended by a community organization, government agency or religious organization; between 15 and 90 years of age; committed to Mt Druitt.

The Enablers Program, based on the philosophy that self-understanding and acceptance leads to understanding and acceptance of others, involves participants meeting weekly for 25 weeks each year for two years. During the first year participants take part in a self-development programme. The second year of the programme focuses on developing awareness and understanding of community activities though involving participants in visiting different agencies to gain understanding of what the agency does and how it undertakes its activities and responsibilities.

5.9.3   Research Methodology: Vortical Postmodern Ethnography

In keeping with the features of *vortical postmodern ethnography* (as identified above), the research:

1. Focused on narratives, discourse and behaviour as these unfolded during the project (via coherent conversations with programme participants and leaders).
2. Relied on qualitative data in the form of narratives (as aurally recorded and written transcripts of the coherent conversations).
3. Took a fractal perspective, recognizing that while focused on one level, themes were implicated that are present in more micro and macro levels within human interactions (for example, between individuals and groups locally in Mt Druitt, through to national and international domains).
4. Expected the vorticity of the social system generated through the research process to stimulate new insights and questions.
5. Analysed data for identification of themes that have strong resonance for the participants.

A complexity approach complements the aim of ethnography to bring about improvement, in this case, associated with engendering a positive force to the vorticity generated within the research project. It is thought that engaging in dialogue with those involved in the situation (here the Enablers participants), about strengths, values and hopes is in itself transformational.

5.9.4   Narrative Data Generation: Coherent Conversations

A series of coherent conversations were facilitated with past and present participants along with the leaders of the Enablers Program. The guiding questions asked in the coherent conversations were:

1. In what ways do you see yourself developing through your participation in the Mt. Druitt Enablers Program?
2. Do you think that participation in the program has helped you personally?
3. Do you think participation in the program has helped your participation in the local community?
4. Would you please tell us about some of your learning in this program?
5. What have you learned about leadership from participating in this program?

The research participants comprised 14 participants from the 2004 programme, six from 2005 and two Enablers trainers. The sample consists of a representative distribution of community members, without domination of any particular ethnic grouping. Each conversation took approximately one hour, with separate conversations held for 2004 participants, 2005 participants and Enablers' trainers.

The coherent conversations were aurally recorded as narrative texts, and are understood to constitute a summary of the individual and collective experiences of those engaged with the Enablers Program. The texts were discussed, reviewed, analysed and synthesized by the researchers through many iterative cycles. It was through this review process and in conjunction with undertaking fractal narrative analysis and attractor narrative analysis that the patterns of the participants' experiences were rendered visible and we began to identify explanatory themes.

In addition to the coherent conversations described above, all research participants were invited to join a second conversation that reviewed the findings of the research, where the focus was on validating, elaborating and synthesizing the narrative.

5.9.5   Vorticity

The coherent conversations contributed to vorticity, that is, to changing the nature of the participants' phrase space. Through engaging in conversations with the researchers, the programme participants were involved in new relationally-based conversations. In critically reflecting on their experiences in the Enablers Program, the participants were engaged in conversations that were novel to them. Both these factors contributed to a dynamic situation where the common sense reality of the participants and researchers was disrupted and where taken-for-granted habits of mind and action became available for challenge and critique.

5.9.6   Data Analysis and Synthesis: Findings of Fractal Narrative Analysis

The findings were organized as fractal narratives in relation to the five questions asked in the coherent conversations. During the conversations many participants shared their views about the future of the programme and we collected these as fractal narratives under the heading 'Participants' recommendations for the Enablers Program'.

In answering the five questions and collecting recommendations for the future, a significant body of narrative was developed. Within this, certain recurrent themes or explanatory fragments of statements could be identified, which we describe as 'fractal' comments. As discussed above fractals depict similarities at different scales of reference. This means that the nature of a fractal is that the 'fragment' you are looking at is representative, not of a part, but of the whole. While the self-similar properties found at different scales of reference may be considered to express the essence or character of the fractal, the labels or detail are considered scale or level-dependent. So, in analysing the responses of participants, we judged views that were similar, albeit expressed in different ways, to be fractal in character.

In the analysis, we selected fractal comments from the narrative texts that are indicative of the character of the narratives from each respondent. We did this for each participant and for each question. We grouped together similar fractal comments and organized these in relation to a set of themes that we identified as characterizing the range of responses to each question.

In presenting our findings, we listed the themes and then the fractal comments that support identification of themes. At the end of each summary listing of fractal comments in relation to the themes identified in the responses to the question, we put together a single-voice composite, indicative or fractal, narrative that represents a broad description of the 'whole' body of narratives.

Combining the individual fractal comments into a composite fractal narrative enables the richness of detail as well as an emergent, coherent meaning to be discerned. In our view, this gives the elusive character sought by narrative analysts: that of evocative, suggestive richness along with clarity and validity of explanation.

What follows is a demonstration of this process and style of presentation, in regard to question (1). The cumulative fractal comments drawn from participant's responses are shown and following this a constructed, composite narrative response for the question is presented. The reason for constructing a composite response is to combine the fractal comments into a singular (fractal) narrative. Such a narrative represents coherence and an emergent understanding.

Question (1) In what ways do you see yourself developing through your participation in the Mt. Druitt Enablers Program?

In response to this question, we identified three themes in the narratives. Participants commented on: (1) What the program does; (2) How it made the participant feel; and (3) Where the program may lead.

Fractal comments taken verbatim from the narratives are presented below (in italics) to explain these fractal themes.

What the program does
People coming together in community and taking responsibility.
Courses often nurture intellect, this, rather, nurtured imagination.
New learning because we are so relaxed and informed.
Different ways of learning.
Sharing your life experiences.
It provides service around the area.
We all help each other.
You get better insight.
Safe environment.
It embraces diverse peoples, diverse cultures and leads to respect and diversity.
Helps and supports the marginalized.
Welcoming atmosphere . . . gave affirmation.

How it made me feel
Feel like I'm going on a holiday.
I actually feel validated.
Peaceful; good for reflection.
Accepted and credible.
Very creative.
It was like a family.
It gives confidence and self-esteem.
I felt really, really good.
Really beneficial.
A true growth experience.
Comfortable in the group.

Where the program may lead
People don't irritate me the way they used to.

Working with the marginalized and disadvantaged community.
It should be aimed at young people, particularly Year 8 and Year 9.

Drawing on these fractal comments a composite fractal narrative response to question (1) was constructed that describes the ways people see themselves developing through participation in the Program:

The program brought together a culturally diverse but like-minded group of people; people, who were interested in helping themselves and helping others and who in that way, were serving the community.
I felt safe and rewarded in working with Enablers. The program has given me a sense of purpose and I have gained self-esteem and confidence from it. I think the program has potential in supporting the individuals who participate in working with and supporting marginalized and disadvantaged members of the community. It may have potential for operating in the school system for Year 8 or Year 9 students.

This constructed, composite, fractal narrative is indicative of the responses of the individual participants and of the responses of the participants as a group. It is a fractal narrative in that it is a narrative expressed at the level of individuals and at the level of the group. In other words it depicts a similar pattern at different scales of focus.

This process and style of reporting was repeated for each of the four other questions as well as the additional topic concerning the participants' recommendations for the Enablers Program. From this we developed a composite fractal narrative that, though expressed as the voice of one person, sought to represent the views of the whole group.

Following completion of our initial analysis we met with the research participants for them to critically respond to our initial statement of findings and expression of the Enablers' composite fractal narrative. At this meeting (also conducted as a coherent conversation) the final Enablers' Fractal Narrative (comprised of all the composite fractal narratives constructed in response to each question) was endorsed by participants as representing their voices in summary form.

Enablers' fractal narrative

The program brought together a culturally diverse but like-minded group of people; people, who were interested in helping themselves and helping others, and who in that way, were serving the community.

I felt safe and rewarded in working with Enablers. The program has given me a sense of purpose and I have gained self-esteem and confidence from it. I think the program has potential in supporting the individuals who participate in working with and supporting marginalized and disadvantaged members of the community. It may have potential for operating in the school system for Year 8 or Year 9 students.

The program has been very beneficial to me in that it has made me more aware of both myself and others. It has helped me recognize my strengths and weaknesses, and from this I can recognize strengths and weaknesses in the community.

The program has improved my contact and networking with others. It has served to broaden my experiences, even my social life, and has helped me build closer relationships of trust within the family and with other people.

Enablers was definitely part of my self-development. It enabled me to really grow in confidence and I reflect on the program as a true growth experience.

Participation in the program has helped me in broadening and understanding my participation in the local community. I have built up relationships and networks with other people and with various service organizations in the Mt Druitt community. The networking has been useful in that it has increased my awareness of what services are needed and are available in the community.

My participation in the community has been enhanced because the program has given me a belief in myself and now I can do my job better. I also feel I have more control over my life and work activities. I have become more assertive in my role but, at the same time, sensitive towards the needs and circumstances of others.

I have learnt that core leadership is in the community; it is embedded within the community itself. Leadership activities develop as an evolutionary process, as one deals with individuals, groups and varying circumstances. It is necessary to understand yourself if you are to understand other people. When I understood myself and others better, I was able to be a better leader. Having a sense of leadership allowed me to feel more connected, to find a sense of place, in the Mt Druitt community.

I have also learnt that there is a responsibility to serve the community. Working and serving the community requires a sensitivity and ability to listen to people's needs. It also requires the ability to recognize what exists within the community, in terms of needs, and the resources and services that may be mobilized or accessed to provide for those needs.

I think we need to broaden and bring in more people so that Enablers becomes a broader-based and more community-centred activity. The program needs to be more inclusive and to take care in not being selective or exclusive in its participating membership. The program deserves to grow and to be extended further into the whole community.

I would like to think that the participants in Enablers can use their relationships and networks to influence government activities and government policy. I would like to see people accept roles and responsibilities in policy and government activities, such as the local council.

Again we can say that this constructed, composite, fractal narrative is indicative of the responses of the individual participants and of the responses of the participants as a group. It is a fractal narrative in that it is a narrative expressed at different levels: at that of the individual and at that of the whole group.

This particular style of recording findings is indicative of qualitative approaches to data analysis (and synthesis) and also clearly placed within a complexity epistemological framework. In qualitative research such as ethnography, the term analysis refers not only to separating out portions of the data, but also to synthesis where the data is reconfigured into a coherent whole. The constructed, composite, fractal narrative constitutes a synthesis, where units of meaning, as fractal comments, have been configured around a set of central themes. The constructed, composite, fractal narrative represents an interpretive account that draws context and expressions of sense making together in a single coherent narrative.

As is conventional in qualitative and especially participative styles of research, it is considered an important indication of validity that the participants endorse the expression of findings. The whole process, from the initial coherent conversations through to the conversations regarding participant views on the validity of the constructed, com-

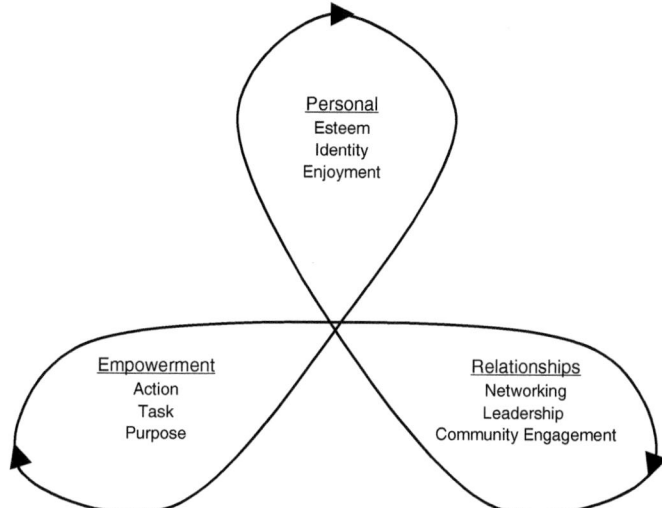

Figure 5.1   *The three interrlating attractor sets for participants in the Mt Druitt Enablers'*
            Program

posite, fractal narrative was generative of new understandings for both the participants
and researchers. In this sense the process and constructed narrative as artefact, fulfils the
accepted aim of ethnographic research of bringing about improvement for the participants. In so doing the process and constructed narrative artefact were generative of, and
borne out of, vorticity.

5.9.7   Data Analysis and Synthesis: Findings of the Attractor Narrative Analysis

Drawing on complexity depictions of the relationships between fractals and attractors,
we can state that the patterning of fractals around an attractor allows us to identify the
attractor. From the attractor, one can make inferences about form, function and processes
that have and may continue to occur. Combining the composite fractal narratives for each
question, we identified three major attractors, each made up of a number of aspects: 1.
Personal (esteem, identity and enjoyment); 2. Empowerment (action, task and purpose);
and 3. Relationships (networking, leadership and community engagement). These three
attractor sets can be depicted diagrammatically (see Figure 5.1) to show interrelation-
ship and co-dependence, which influences the organizational dynamics, value system and
purpose of the community.

The patterning of fractal narratives around particular attractors indicated that participation in the programme developed and enhanced self-esteem and a sense of personal
acceptance and that concomitantly, through this, the participants saw themselves as more
accepting of others. This enhanced acceptance of self and others meant that they were
better able to work with family and clients and to network with others active in the community. Certain tasks or priorities for action and projects were subsequently recognized
and agreed upon. The grouping of the participants' comments according to the attractor
set (personal, empowerment and relationships) shows the iterative relational nature of

these attractors. For example, personal acceptance could be viewed as leading to acceptance of others. It could also be argued that as people participated in the Enablers Program in a social way, with a diverse range of other members of the community, that their participation in a safe atmosphere supported a sense of community and acceptance of others.

We identified a sequence of activities that the Enablers Program uses to foster development of the attractors. This sequence can be described as:

- Discourse about life experience.
- Narrative events that are, in some way, stylized and formalized and may have input either from particular people or of particular content.
- Generation of co-narratives, where themes of interest and emergent, shared understanding were identified.
- Generation of narratives of complicity, that is, narratives that relate to purposive, task-related activities.
- Generation of a composite experience that becomes a narrative that guides the lives of participants.

It should be recognized that during the conduct of the programme the steps of this process, as identified above, were not strictly in sequence but followed an iterative pathway.

The participants' comments indicated that the major reason they continued their involvement was that the Enablers Program provided a safe and supportive atmosphere that engendered personal growth and awareness. Our view was that this sense of acceptance and safety provided a sense of community empowerment and self-determination within those participating in the Enablers Program.

Stated in summary form, the conclusion of our evaluation was that the Enablers Program contributed to the participant's capacity to develop autonomous agency, as well as critical autonomy, by supporting effective political participation and involvement in agencies and organizations that relate to the Mt Druitt area. We determined that, in turn, this capacity supports the participants' sense of empowerment in contributing to the self-determination of the Mt Druitt area.

5.10   CONCLUDING COMMENTS

The case study may be viewed as depicting a series of steps that are familiar to social researchers. The overall approach to social inquiry (inquiry into human experiential space) has been described as vortical postmodern ethnography with data being generated through the device of coherent conversations. Making sense of and categorization of the data was begun through fractal narrative analysis. Composite formulation of the data into a fractal narrative illuminated the data, provided evidence of developing insights and exposition of meaning and allowed inferences to be made from the particular (sense making and experiences of individuals) to the broader macrocosm or society within which they are situated. The verisimilitude of the constructed, composite, fractal narrative was developed through the process of inviting the research participants to critically review and shape the narrative so that it was held by them to have epistemological validity or to be a valid account of their experience and views. The 'reason of things', or a more in-depth analysis of the data came from the attractor narrative analysis that identified the values, issues of concern and motivators that underlie attitudes and behaviours of the research participants.

Qualitative researchers study human experience from the perspective of interacting individuals. As is recognized in qualitative research, the researcher is not an objective, authoritative observer, but rather is historically positioned, locally situated and theoretically informed. The research outcomes thus are seen as depicting emergent interpretations made through the coming together of researchers and research participants. The researcher, from a complexity perspective, is viewed as part of this self-organizing and dynamic emergence.

A complexity informed approach to qualitative research continues certain traditions in that it implicates contemporary interests, sense making preferences and interplay between and across disciplines. A complexity depiction of life as self-organizing, dynamic and emergent is well suited to inquiry into human experience as it brings a sense of openness to appreciating and studying human experience as it emerges. The outcome from complexity informed social research is that the researcher, the situation and the theory are adaptively engaged and simultaneously improved.

REFERENCES

Alhadeff-Jones, M. (2008), 'Three generations of complexity theories: nuances and ambiguities', in M. Mason (ed.), *Complexity Theory and the Philosophy of Education*, Chichester: Wiley-Blackwell, 62–78.
Atkinson, P. and Hammersley, M. (1994), 'Ethnography and participant observation' in N.K. Denzin and Y.S. Lincoln (eds), *Handbook of Qualitative Research*, London: Sage Publications, 248–261.
Bak, P. and Chen, K. (1991), 'Self-organized criticality', *Scientific American*. January 1991.
Berger, P. and Luckman, T. (1967), *The Social Construction of Reality: A Treatise in the Sociology of Knowledge*, London: Allen Lane.
Blacktown City Council (2006), *Our City* [Online] Available: http://www.blacktown.nsw.gov.au/our-city/statis tics/statistics_home.cfm (accessed 6 August 2017).
Chambers, I. (1994), *Migrancy, Culture, Identity*, New York: Routledge.
Checkland, P. (1981), *Systems Thinking, Systems Practice*, Chichester: Wiley.
Coveney, P. and Highfield, R. (1995), *Frontiers of Complexity*, London: Faber and Faber.
Denzin, N. and Lincoln, Y. (eds) (1994), *Handbook of Qualitative Research*, London: Sage Publications.
Dimitrov, V., Woog, R. and Kuhn-White, L. (1996), 'The divergence syndrome in social systems', *Complex Systems 96*, Amsterdam: IOS Press, pp. 142–146.
Enguix, B. (2014), 'Negotiating the field: rethinking ethnographic authority experience and the frontiers of research', *Qualitative Research*, **14** (1), 79–94.
Gamble, D., Blunden, S., Kuhn-White, L., Voyce, M. and Loftus, J. (1995), *Transfer of the Family Farm Business in a Changing Rural Community*, Kingston, ACT, Australia: Rural Industries Research and Development (RIRDC).
Gleick, J. (1998), *Chaos*, London: Vintage Books.
Guba, E.G. and Lincoln, Y.S. (1989), *Fourth Generation Evaluation*, Newbury Park: Sage Publications.
Heidegger, M. (1969), *Discourse on Thinking*, New York: Harper & Row.
Johannessen, S.O. and Kuhn, L. (2012), 'Ways of thinking about complexity' in S.O. Johannessen and L. Kuhn (eds), *Complexity in Organization Studies (Volumes I–IV)*, London: Sage Publications, xxv–xxxi.
Johnson, S. (2001), *Emergence*, London: Allen Lane.
Jung, C. (1976), *The Portable Jung* (ed. Joseph Campbell; trans. R.F.C. Hull), New York: Penguin Books.
Kauffman, S. (1995), *At Home in the Universe: The Search for the Laws of Self-Organization and Complexity*, Oxford: Oxford University Press.
Kelly, G. (1955), *The Psychology of Personal Constructs Vols 1 and 2*, New York: W.W. Norton and Company.
Khatoonabadi, A. and Woog, R. (1992), 'Drama as a researching tool with nomadic people', Proceedings of *The International Conference, Nomadism and Development: Survival Strategies and Development Policies*, Isfahan, Iran.
Kuhn, L (2002), 'Health promotion and complexity theory: an approach for the 21st century', in V. Dimitrov

and T. Naess (eds), *Health Ecology: Learning to Cope with Complex Realities*, NaKuHel: Asker, Norway, 135–146.

Kuhn, L. (2007), 'Why utilize complexity principles in social inquiry?' *World Futures: The Journal of General Evolution*, **63** (3–4), 156–175.

Kuhn, L. (2009), *Adventures in Complexity for Organisations Near the Edge of Chaos*, Axminster: Triarchy Press.

Kuhn, L. (2012), 'Epistemological reflections on the complexity sciences and how they may inform coaching psychology', *International Coaching Psychology Review*, **7** (1), 114–118.

Kuhn, L. and Woog, R. (2003), 'Learning our way to a better world: a complexity approach to systemic social theorizing', *Proceedings, 47th Annual Conference, The International Society for the Systems Sciences*, Iraklion, Crete, Greece, 7–11 July, CD 03-057.

Kuhn, L. and Woog, R. (2005), 'Vortical postmodern ethnography: introducing a complexity approach to systemic social theorizing', *Journal of Systems Research and Behavioural Science*, **22**, 139–150.

Kuhn, L. and Woog, R. (2006), 'Inquiry into community enabling: a review of the Mt Druitt Enablers Program', Report to Chain Reaction and the Ian Potter Foundation.

Kuhn, L., Woog, R. and Hodgson, M. (2003), 'Applying complexity principles to enhance organisational knowledge management', *Proceedings, Global Business and Technology Association Conference: Challenging the Frontiers in Global Business and Technology: Implementation of Changes in Values, Strategy and Policy*, Budapest, Hungary, 1–8 July, pp. 754–7622.

Kuhn-White, L. (1994), 'Action research still has to buy the groceries: some dilemmas in doing action research in an academic setting', in B. Neville, P. Willis and M. Edwards (eds), *Qualitative Research in Adult Education*, Adelaide: University of South Australia Centre for Research in Education and Work, Chapter 4, 38–44.

Langton, C. (1986), 'Studying artificial life with cellular automata', *Physica*, **22D**, 120–149.

Lewin, R. (1999), *Complexity*, Chicago: University of Chicago Press.

Lopate, P. (1995), *The Art of the Personal Essay: An Anthology from the Classic Era to the Present*, New York: Anchor Books, DoubleDay.

Madison, G.B. (1988), *The Hermeneutics of Postmodernity*, Bloomington: Indiana University Press.

Mandelbrot, B. (1977), *The Fractal Geometry of Nature*, New York: Freeman.

Maturana, H. and Varela, F. (1987), *The Tree of Knowledge*, Cambridge, MA: Shamhala Publications.

Mitleton-Kelly, E. (2006), 'A complexity approach to co-creating an innovative environment', *World Futures*, **62**, 223–239.

Morin, E. (2001), *Seven Complex Lessons in Education for the Future* (trans. N. Poller), Paris: United Nations Educational, Scientific and Cultural Organization.

Morin, E. (2008), *On Complexity* (trans. R. Postel and S.M. Kelly), New Jersey: Hampton Press.

Packman, A. and Kuhn, L. (2009), 'Looking at stuttering through the lens of complexity', *International Journal of Speech-Language Pathology, Special Issue, Dysfluency: Stuttering and Cluttering*, **11** (1), 77–82.

Prigogine, I. and Stengers, I. (1984), *Order Out of Chaos*, New York: Bantam Books.

Reason, P. (1991), *Human Inquiry in Action*, Newbury Park: Sage Publications.

Rilke, R.M. (2004), *Letters to a Young Poet* (trans. M.D. Norton), New York, USA: W.W. Norton and Company.

Rorty, R. (1990), *Contingency, Irony and Solidarity*, Cambridge, UK: Cambridge University Press.

Roulston, K. (2014), 'Interactional problems in research interviews', *Qualitative Research*, **14** (3), 277–293.

Sanders, C. (1999), 'Prospects for a post-modern ethnography', *Journal of Contemporary Ethnography*, **28** (6), 669–675.

Schwartz, P. and Ogilvy, J. (1979), *The Emergent Paradigm: Changing Patterns of Thought and Belief, Analytical Report 7*, Values and Lifestyles Program, Menlopark, CA: SRI International.

Shotter, J. (1993), *Cultural Politics of Everyday Life*, Buckingham, UK: Open University Press.

Waldrop, M.M. (1993) *Complexity*, St Ives: Viking.

6.   From isolation to transformation with C2

Hazel Stuteley, Institute of Health Service Research, Exeter University and Dr Jonathan Stead, C2 Connecting Communities

Figure 6.1   Giving communities back their self-belief by creating hopeful futures

During the three decades of my career in general practice, the health gap has grown wider, despite numerous costly central initiatives. This shameful state of affairs deserves a different approach rather than more of the same. (Jonathan Stead 2015)

What my long career as a nurse with a special interest in health inequalities has taught me is that it's not smoking, alcohol, obesity or substance misuse which are the greatest determinants of poor health in low income communities, but rather powerlessness, hopelessness, disconnectedness and passivity. The former are merely ways of coping with the latter, with catastrophic health consequences. (Hazel Stuteley 2014)

6.1　INTRODUCTION

It is an inescapable fact that for decades, costly interventions seeking to 'fix' Britain's most disadvantaged communities, where poverty, crime, unemployment and poor health are rife, have largely failed. The ongoing human cost is heartbreaking and the cost to the public purse is immense. The layers of challenges presented are undoubtedly complex and appear, in many cases, to be intractable. How is it possible then, to change the fortunes of these communities, with their embedded layers of learnt behaviours, aspirations and cultures, within both residents and their service providers?

This chapter describes the development and practical application of Connecting Communities: a transferable programme based on complexity principles, which consistently brings transformative change in health and social well-being, to disadvantaged communities across the UK.

6.2　C2 BACKGROUND: THE BEACON PROJECT 1995–1999

The story begins in Falmouth, Cornwall, where the Beacon Project, led by two Health Visitors, transformed an extremely challenged social housing estate (population 6,000) suffering poor health and violent crime, against a backdrop of poverty, unemployment and sub-standard housing.[1] In four years, housing stock was radically improved, crime was halved (see Table 6.1) and unemployment fell (see Table 6.2). There was no start-up

Table 6.1　Health, environment and educational outcomes between 1995–2001

Health Benefits	Environmental Outcomes	Educational Outcomes
Increased breast feeding rates by approximately 50%	£1.2 million accessed by tenants and residents + further £1m 'unlocked' as a result	On site training for tenants and residents
Postnatal depression rates down 77%	Gas central heating to 318 properties	After School Clubs
Child Protection registrations down 60%	Loft insulation in 349 houses: cavity wall in 199; external cladding to 700	Life Skills courses
Childhood accident rate down 50%		
50% reduction in incidence of asthma and schooldays lost	Fuel saving estimated at £180,306 p.a., releasing disposal income to residents	Parent and Toddler Group
78% Reduced fear of crime	£160,000 traffic calming measures	Boys & girls key stage 1 SATS up 26%
Beacon Care Centre providing on site health advice	Provision of safe play areas and Resource Centre	IT skills
Sexual health service for young people	Recycling and dog waste bins	Crèche supervisor training
All levels of crime including violent crime reduced 50%	Skateboard Park	Boys SATS key stage 1 results up 100%

Table 6.2   *Unemployment Figures: Number of adults out of work and claiming job seekers, allowance in the Penwerris ward.*

	June 95	June 96	June 97	June 98	June 99	June 2000	June 2001
Women	69	68	48	47	39	34	30
Men	356	241	208	151	172	197	103
Total	425	309	256	198	211	231	133

Source: Locally collated data from Cornwall Action Team for Jobs, Job Centre Plus, verified by the Office for National Statistics.

funding. Residents and service providers, who had collectively reached a tipping point, formed a partnership of equals, the Beacon Community Regeneration Partnership which met monthly for the duration of the project. The residents themselves, with agency support, generated monies needed to improve housing. An unexpected byproduct of this collaboration was a dramatic improvement in community physical and mental health and educational attainment. To date these outcomes have not only been sustained but, in the case of crime and anti-social behaviour, rates have actually improved year on year.

6.3   DISCOVERING COMPLEXITY THEORY TO ENABLE TRANSFERABLE SPREAD OF BEACON PROJECT: 2002–2005

The project delivered startling outcomes, which rightly resulted in national and international recognition for Beacon as a 'flagship' for community renewal. In 2001 co-author Hazel Stuteley, Health Visitor to the Beacon Estate for a decade, and co-founder of the project, was seconded to the Department of Health, which was looking for national 'spread' of the work she had led. However developing an analysis and an understanding of how this dramatic transformation came about proved difficult to articulate and even harder to identify credible, transferable methodologies. It should be remembered that the Neighbourhood Renewal Unit culture then, was very focused on process and costly, 'top down' interventions.[2] Citing trust, connectivity, listening and relationship building as being key, along with viewing residents as the 'solution, not the problem', was met with large-scale cynicism, and the process stalled.

   In the 21st century there is now compelling biological evidence that lacking control over one's immediate environment, coupled with poor social networks, causes the damaging health behaviours leading to community breakdown.[3] Our experience to date is that all this is entirely *preventable* and *treatable*.

   Then in 2002 Hazel was introduced to complexity theory by GP Dr Jonathan Stead (fellow co-author) after a chance meeting, who arranged for her to attend a complexity workshop led by Professor Eve Mitleton-Kelly, which provoked the following response:[4]

   Although the presentation was nothing to do with community development, I was mesmerized by it. What I heard being described was a process, which uncannily and exactly, mirrored the intuitive methodologies we used to lead the Beacon Project. It placed great value on widespread connectivity, the creation of new relationships and dialogue based on trust. Conversations, humility and respect, I now realised, contributed hugely to the creation of that all-important enabling environment, which released the resourcefulness of this community to become self-organizing and achieve such significant and dramatic outcomes. Sitting through that presentation was one of those rare, life-changing moments of self-enlightenment and I knew we could make all this happen again for other communities.

Transferability was now a potential reality. Viewing disadvantaged communities as complex adaptive systems was clearly key as was the *nonlinearity* of the approach used. In Beacon, transformation had occurred as a result of *resident self-organization* and *co-evolution* of trust between them, the agencies, and their environment, resulting in *'new order'*.

Further introductions to other complexity enthusiasts from Exeter University, quickly led to the foundation of the Health Complexity Group in 2002. Funding was procured in 2003 for the group to carry out retrospective research in Beacon, using complexity theory as the explanatory framework, to identify and understand transformative change factors, whilst simultaneously tracking and capturing the enablers and barriers to successful community regeneration, just beginning in nearby Redruth North. In 2004–2005 the Community Regeneration Evaluating Sustainable Transfer report[5] was published, leading to the development of the Connecting Communities programme, rapidly nicknamed 'C2' (to avoid confusion with other similarly named interventions) and the C2 seven-step delivery framework.[6]

6.4   THE C2 SEVEN-STEP FRAMEWORK: FROM ISOLATION TO TRANSFORMATION

STEP 1
C2 begins creation of enabling conditions and new relationships needed for community transformation at strategic, frontline service delivery and street levels. C2 Strategic Steering Group (SSG) established. Target neighbourhood scoped and local C2 secondee appointed. 'Key' residents identified to jointly self-assess baseline community connectivity, hope and aspiration levels.[7]

STEP 2
Establish C2 Partnership Steering Group (PSG) of frontline service providers with key residents, who share a common interest in improving the target neighbourhood. Hold connecting workshop and identify team of 6–8 members to attend two-day C2 '1st wave' Introductory Learning Programme.

STEP 3
PSG plans and hosts Listening Event to identify and prioritize neighbourhood health and well-being issues and produces report on identified issues, fed back to residents and SSG a week later. Commitment established at feedback event to form and train resident led, neighbourhood partnership to jointly tackle issues.

STEP 4
Constitute partnership which operates out of easily accessed hub within community setting, opening clear communication channels to wider community via e.g. newsletter

and estate 'walkabouts'. Host exchange visits and meetings with other local community groups and strategic organizations. Identify '2nd wave' of 6–8 new learners to C2 Experiential Learning Programme.

STEP 5
Monthly partnership meetings, providing continuous positive feedback to residents and SSG. Celebration of visible 'wins' e.g. successful funding bids which support community priorities, and promote positive media coverage, leading to increased community confidence, volunteering and momentum towards change. Partnership training undertaken to further consolidate resident skills.

STEP 6
Community strengthening evidenced by resident self-organization e.g. setting up of new groups for all ages and development of innovative social enterprise. Accelerated responses in service delivery leading to increased community trust, co-operation, co-production and local problem solving.

STEP 7
Partnership firmly established and on forward trajectory of improvement and self-renewal. Key resident/s employed and funded to co-ordinate activities. Measurable outcomes and evidence of visible transformational change, e.g. new play spaces, improved residents' gardens, and reduction in ASB, all leading to measurable health improvement and parallel gains for other public services.

6.4.1   The C2 Approach

Mirroring CREST findings in Falmouth and Redruth, the seven steps have been developed to create an enabling infrastructure to effect demonstrable behaviour change in residents and service providers, as a result of the co-evolution of new relationships within an enabling environment, with a strong emphasis on connecting and listening. New co-learning is introduced at strategic, community and frontline service delivery levels at regular intervals, providing *active feedback loops* throughout the course of delivery.

Overarching principles of C2
The C2 seven-step approach is NOT delivered as a project, but designed to bring a lasting culture shift in the way that services work with communities and vice versa. 'C2 is for life not just for Christmas!' Support and peer learning continue long after the initial step delivery, via the C2 National Network of Connected Communities (regd. charity no. 1172510), at no cost to the communities involved.

The focus at street level is on *collaboration, co-learning* and *creating new relations*, to harness the collective creative powers of residents working as equals with police, education and local authority services across the spectrum.

The starting point is a dysfunctional, fragmented, disconnected neighbourhood with low levels of community engagement and the end result is a confident, *resident-led, resilient, self-managing community* with high levels of engagement and reduced levels of disorder. The C2 'People & Provider Partnership' is the vehicle for this, providing a lasting, self-renewing mechanism of neighbourhood governance and ongoing problem solving. Usual outcomes in the short term (within a year) are a significant drop in anti-social behaviour (ASB) and greatly increased social capital and community networks. Up to four years further down the line, measurable improvements in health and well-being generally emerge.

The C2 delivery support team is drawn from a richly experienced range of practitioners from academic, NHS and community leadership roles. All have in-depth experience of the C2 seven-step approach.

The seven steps are implemented within 18–24 months, embedding the values of trust, humility, compassion and respect from 'strategic level to street level'. As its full title suggests, C2 connects communities in three different ways:

- Within themselves – creating networks and mutual co-operation.
- With local service providers -building a parallel 'community'.
- With other C2 communities across the UK, getting and giving inspiration, peer learning directly from one place to another, and exploring adjacent possibilities.

6.5   IMPLEMENTATION OF THE PRACTICAL APPLICATION OF COMPLEXITY PRINCIPLES

The foregoing has briefly described the background and development of C2. What now follows will describe and discuss each step, it's alignment with complexity theory, and some of the methodologies used. To conclude we will also explore some of the challenges faced during implementation and include a selection of impact outcomes from selected communities, successfully using this approach.

6.5.1   C2 Seven-step Framework Explained Step by Step

STEP 1: Building firm foundations and locating the energy for community change
C2 begins creation of enabling conditions and new relationships needed for community transformation at strategic, frontline service delivery and street levels. C2 Strategic Steering Group (SSG) established. Target neighbourhood is scoped and local C2 secondee appointed. 'Key' residents identified to jointly self-assess baseline connectivity, hope and aspiration levels.

This step can take up to six months and is key to success of further steps. Creating a receptive context for transformative change is essential and very much depends on readiness to change, within commissioners and their partner organizations. Getting strategic buy-in to begin to 'work differently' is the essential first step. The sponsoring organization, whether a clinical commissioning group (CCG), local authority or housing association needs to demonstrate board level support for frontline staff to begin to work differently. This extends to stakeholder partner organizations and will typically include representation from police, fire service, education and NHS if a CCG is not lead commissioner.

A strategic steering group (SSG), meeting monthly for a year, is established and they are given early learning, a clear 'road map' of the seven steps, how C2 works and what it leads to. Known within C2 as the 'dynarod' group, they act as 'unblockers' for frontline workers to work differently where resistance frequently occurs at middle and frontline management level. This group must also include one or two 'key' residents (so called because they have the potential to 'unlock' and release community capacity) to provide an expert credible 'voice' for the target community. The C2 team identifies these residents carefully as they are vital to success and are residents with energy, sense of humour and greater readiness than most to pursue improvements where they live. Described as 'gold dust' within C2 team they need to be carefully nurtured and valued by everyone.

During this step, information is gathered, mostly at street level, to identify the multiple dimensions of the *'problem space'*, that is, the background to the community's decline and the multiple factors, socially historically, culturally and economically that have led to *locked-in behaviours* of both residents and service providers in the past, which have prevented growth and kept it in this space. An example to illustrate this from Beacon:

> Behaviour becomes locked in when it seems that any potentially beneficial outcomes which might accrue from changing behaviour are outweighed by the investment that would be required to undertake the behavioural change in the first place. Typically this is because of the effort that would be required to encourage everybody else to adopt this new form of behaviour. In the case of the Beacon Estate (Falmouth) the telling evidence of locking in of behaviour was the reluctance of residents to report crimes to the police, especially cases of vandalism, because of the fear of reprisals that might follow. In a similar way, statutory agencies began to avoid the estate unless specifically called on to intervene, citing the common assumption that actively visiting the estate would only lead to more trouble and more work. The system lacked potential for any possible examples of innovative behavior to spread through it.[8]

Most importantly at this early stage, C2 must bring a spirit of hope to both that things can be different, and a sense of what could happen within the *possibility space*. This is built on further in step 2.

C2 uses a baseline 'connectivity and hope assessment scale' and scoring system, adapted from the 'Toronto indicators of community capacity'[9] which works best when used informally to stimulate discussion with mixed focus groups of providers and residents. This is repeated with the same groups 18 months on to highlight impact of increased levels of connectivity and hope.

The appointment of a local C2 community worker is made for 12–18 months, to become trained via secondment to the C2 team. This not only supports *local operational activity* via a person with local intelligence, but ensures that in-depth skills, knowledge, and *'learning by doing'* C2 in the target neighbourhood, is embedded for future sustainability of this way of working and it's future replication, using original site as exemplar for others. From this starting point C2 then sets out to address ALL of the key dimensions identified within the problem space, using the seven-step framework as vehicle to achieve this.

STEP 2: Gathering and connecting the 'journeymen'. 'What's it like to live/ work round here?'
Establish C2 Partnership Steering Group (PSG) of frontline service providers with key residents, who share a common interest in improving the target neighbourhood. Hold connecting workshop and identify team of 6–8 members to attend two-day C2 '1st wave' Introductory Learning Programme.
Neighbourhood 'walkabouts' and informal chats with residents, frontline workers and

local community groups by C2 team, accompanied by key residents, are essential in build-ing a visual picture, visceral 'feel' and a deep understanding of the 'lived experience' of what it's like to live and work there. This is often the best way to recruit membership of the Partnership Steering group (PSG) described next.

The PSG is made up of people who commit to the C2 seven steps and mirrors the stra-tegic group (SSG) except it is made up of frontline service providers and an expanding number of key residents. The group, supported by C2 team, provides steerage towards the setting up of the C2 'people and services' partnership in step 4, and will ultimately become the 'backbone' of this. A PSG typically includes representation from residents, NHS, housing, police, fire service, education and local authority as a minimum. Representation from youth, children's services and any other local organization having a prominent role within target neighbourhood, is a welcome addition. The group works best with around 20 members.

The connecting workshop

The two 'communities' of residents and service providers are often meeting for the first time. Mistrust, the 'them and us' culture, built up over many years, needs to be dissipated by creating new relationships. The first move is for the professionals to begin to behave dif-ferently and demonstrate new behaviours, by *actively listening to residents' lived experience* and visibly caring for residents, for example, by serving tea. Co-learning starts today. This early behaviour change is powerful. C2 experience bears out that when the professionals change first, residents will then reciprocate. This is the beginning of a subtle shift of power in the relationship from 'power over' to 'power with'.

This initial workshop has four purposes:

- For all members of PSG to connect, and to begin breaking down barriers by learn-ing about and seeing each other as people for the first time.
- For C2 to facilitate a shared vision and commitment to what 'people and services' partnerships can achieve together, that is, opening up 'possibility space' and to deliver a clear 'road map' of the seven-step approach and timescale.
- To identify team of 6–8 members to attend local C2 Introductory learning pro-gramme to gain new understanding around skills and mindset needed to deliver seven-step approach.
- To understand and plan the Listening Event in step 3 and collectively work together to make this happen.

The introductory co-learning programme

Connecting the SSG to the PSG. Ideally this is a two-day residential, providing opportu-nity for informal new relationships to be formed between strategic, frontline and resident stakeholders in C2. Designed for them to begin to explore the 'adjacent possibilities' offered by C2, course tutors include resident leaders from recent and long-term, trans-formed C2 sites. Site visits to local C2 communities are an integral part of this.

STEP 3: Listening together to the community

PSG plans and hosts Listening Event to identify and prioritize neighbourhood health and well-being issues and produces report on identified issues, fed back to residents and SSG a

week later. Commitment established at feedback event to form and train resident led, neighbourhood partnership to jointly tackle issues.

This is where the community begins to move from the 'problem space' to picture the 'space of possibilities'. The Listening Event provides a list of issues identified by residents, which is internalized by service providers present, identifying shared priorities for joint work, building on the new relationships and moving towards a new order. Community issues are emergent and unpredictable, but always 'do-able'. Service providers are always pleasantly surprised by the seemingly 'small' changes, which communities want. This event usually heralds a sense of growing interdependence between community and providers.

The C2 listening and feedback event

These are a fun but powerful and pivotal step towards transformative outcomes. They are specifically designed for PSG to not only create together, but to *collectively host and listen* to the community. This is based on C2 experience that communities always know what they need to 'heal' themselves. There is a great deal of detail associated with the 'lead in' to this event provided by C2 coaches, for example, how it's publicized, how to get people there and so on. A local resident paired with a service provider personally issues attendees a 'doorstep' invite. This is designed not only to embed ownership of the event across spectrum of service providers and key residents who make up PSG, but to signal to the community, who often suffer high levels of 'consultation fatigue' that something different and worthwhile is happening here. All attending are invited to attend a C2 feedback event, a week later, to receive an easy to read report, compiled by members of PSG, on what they've said and to start planning how to tackle prioritized issues via formation of a 'People and Services' partnership. This event is often a rich source of engagement of a second wave of key residents. Press releases need to be prepared for both events by PSG as this not only spreads the word but positive press coverage is helpful in deconstructing what may be a community's negative perception of itself.

STEP 4: Formalizing People and Services partnership and further exploration of 'space of possibilities'

Constitute partnership which operates out of easily accessed hub within community setting, opening clear communication channels to wider community via, for example, a newsletter and estate 'walkabouts'. Host exchange visits and meetings with other local community groups and strategic organizations. Identify '2nd wave' of 6–8 new learners to C2 Experiential Learning Programme.

Membership will be drawn from the PSG in terms of service providers but is now open to an expanded number of residents who will take on executive roles with agencies on the committee in a supportive capacity. The partnership demonstrates that power has shifted to the residents since there is a resident Chair and a majority of residents on the committee, supported by the service providers. This is a demonstration of the new order that has emerged. The partnership now jointly prepares an action plan to tackle issues identified in the Listening Event, pulling in more residents and building wider networks, and often identifying emergent leaders along the way, amongst residents and service providers. We now see spontaneous community self-organization, and evidence of taking responsibility for their neighbourhood, because they now have a leading role in becoming part of the solution to neighbourhood dysfunction and potential for improving their own environment.

Constituting the People and Services partnership
Because this is the long-term, resident-led mechanism for continued growth of the community, the partnership needs to be formally established to give it credibility and 'teeth'. It builds on the new momentum, relationships, energy and optimism developed in first three steps. C2 offers expert guidance on this and how to set up the partnership. Again a press release to publicize new partnership and public meetings leads to expanding 'ripples' of information and further community engagement.

The neighbourhood hub
It is important that these premises are visible and easily accessed, as this will not only be the HQ for the partnership meetings, but also, in time, become the 'beating heart' of the neighbourhood, offering a wide range of information and signposting services. Often a member of PSG can suggest suitable premises, which ideally can be used free of charge, at least in the short term until partnership fully established. Communicating with the wider community is essential and the 'feedback loop' can take many forms, for example, a newsletter, dedicated Facebook page, website and so on.

Exchange visits
These visits are possibly the single most powerful success factor of all the seven steps in terms of accelerating peer learning and opening up the 'possibility space' for both residents and service providers. The way it works best is to take as many members as possible from the PSG in a developing neighbourhood, to visit an established C2 site to meet residents and partners and 'see for themselves' the level of transformation and what's been achieved. The feedback is always 'If they can do it so can we'. And they do!

Then developing site, a bit further along seven steps, hosts a return visit, which is defining for them because they can then 'stock take' on their progress so far. Although included in step 4 it may be necessary to do this visit during step 1, if there is no collective sense within PSG that change can happen or indeed, what it looks like when it does.

The four-day residential Experiential Learning Programme (ELP)
Now we have reached step 4/5 it's usually the case that we now have a very committed second wave group of residents and providers, who may not have been part of step 1. Even if they were it's now timely to nominate a team of 4–6 learners to attend the Exeter ELP,[10] which could be described as C2 'immersion' and team-building course. This is a unique opportunity to meet with other teams from across the UK and learn not only the complexity theory and principles which underpin C2, but how to interpret this theory into reality during a day of visits to long established South West C2 sites, who have now been officially recognized by the University of Exeter as Guide Neighbourhoods (GNs). The GNs are a source of learning, inspiration and in complexity terms offer 'exploration of adjacent possibles' for fledgling partnerships.

STEP 5: Consolidating relationships and ongoing co-learning
Monthly partnership meetings, providing continuous positive feedback to residents and SSG. Celebration of visible 'wins', for example, successful funding bids which support community priorities, and promote positive media coverage, leading to increased community confidence, volunteering and momentum towards change. Partnership training is undertaken to further consolidate resident skills.

Positive feedback loops have the effect of reinforcing the new order, and publicizing stories in the local press changes the narrative of the community. Decades of bad publicity are being replaced by positive stories, giving a feeling of hope in neighbouring districts. Changes can begin to happen very quickly, usually in nonlinear directions.

Partnership meetings

There is absolutely no substitute for regular monthly partnership meetings. They are the 'glue' that keeps the neighbourhood on a forward trajectory by systematically tackling the issues identified at the Listening Event. Cost effective and often free solutions and early wins happen surprisingly quickly, engendered by the creativity, diversity and multiple leverage points afforded by those seated around the table. These must be publicized using range of media resources and celebrated publicly to keep that all-important positive feedback to wider community loop flowing. This is also often the point at which the service providers recognize that their workloads are easing, conversely, as a result of this extra activity and improved intelligence, for example, neighbour nuisance and ASB may be measurably reducing. By now we will also see an increase in volunteering levels.

Training opportunities

Resident collective confidence will now be increasing and this is a good stage to further consolidate and improve skills levels particularly around committee skills. There are usually training opportunities locally and links with local volunteer organizations will be able to provide contacts. There are also excellent national organizations such as Trafford Hall[11] in Cheshire that run an exciting range of community resources.

STEP 6: Residents as co-producers of services

Community strengthening evidenced by resident self-organization, for example, setting up of new groups for all ages and development of innovative social enterprise. Accelerated responses in service delivery leading to increased community trust, co-operation, co-production and local problem solving.

In all communities, new problems arise all the time. In resilient C2 communities, these problems are seen as opportunities for further self-organization and sustained transformation. These communities are no longer dependent on a few key individuals; there will be a dispersed 'army' of emergent leaders who understand the nature of and how to optimize the new relationship with the service providers.

Further community self-organization and emergence of social entrepreneurs

This is an exciting sign of community strengthening reflecting increased collective confidence and can be defined as:

- The spontaneous coming together of a group of residents to create a new activity;
- NOT directed or designed by someone outside the group;
- The group decides WHAT needs to be done, the HOW and the WHEN.[12]

Residents are now starting to take pride in and responsibility for their neighbourhood and C2 often witnesses early self-organized groups coming together during step 6 to improve green spaces, derelict land and to do neighbourhood 'tidy ups', removing rubbish and graffiti. Our consistent experience is that these activities are often the starting point for a range of social enterprise opportunities, offering employment and further education.

There will now be greater trust and more effective communications between services and people because it is visibly evident that agencies are listening and responding, so now is a good time for the Partnership to promote activities targeting poor health.

Opportunities to maximize community receptivity

Our experience suggests that most residents are completely unaware of how poor their collective health is or the differential in life expectancy between them and their more affluent neighbours and are often outraged and shocked. C2 has witnessed on many occasions the greatly increased uptake for health-promoting activities when this 'goes public'. The knock on socio-economic effect of large-scale improved health behaviours cannot be underestimated as it impacts on employability, anti-social behaviour and educational attainment.

STEP 7: Towards long-term sustainability

Partnership firmly established and on forward trajectory of improvement and self-renewal. Key resident/s employed and funded to co-ordinate activities. Measurable outcomes and evidence of visible transformational change, for example, new play spaces, improved residents' gardens, and reduction in ASB, all leading to measurable health improvement and parallel gains for other public services.

At this stage, new order is firmly established, with many stories of successful resident-led community improvement. The partnership now has much to offer in support of other disadvantaged communities to be brave enough to self-manage their neighbourhood as well. They become a source of co-learning and part of the wider national C2 learning network. This is an exciting time when the 'opportunity space' has been maximized and there is visible transformation in the way the neighbourhood looks, improving quality of life for all. Agencies are also finding their jobs easier and reinforcing interdependence, so essential to work towards long-term sustainability.

So far the Partnership will have functioned on an entirely voluntary basis but as activity and networks increase, the administration now involved will outstrip the capacity of even the most dedicated volunteers. At this point it makes sense to apply for funding to pay for a part-time key resident to co-ordinate all Partnership activities. The national C2 online webinar series offers opportunities for existing C2 partnership co-ordinators to support and share expertise with those seeking to achieve this.

6.5.2   What does a Strong Community Look Like?

All C2 Partnerships have so far stood the test of time over many years and have continued to operate this highly effective model of neighbourhood governance. Many outcomes, particularly health, will not be apparent for up to five years but our evidence shows that once transformed, neighbourhoods never slip back to the way they were, suggesting that Partnerships set up using the seven-step approach are self-renewing, with built in resilience.

6.5.3 How Will We Know When We've Achieved This?

Residents consistently define this as being where a high proportion of people:

- Are generally satisfied with their neighbourhood;
- Feel that they belong and are proud of where they live;
- Self-organize groups, events and hold budgets;
- Regularly volunteer;
- Get on well with people from different backgrounds; and
- Feel that they have influence and control in decision-making.

6.6 A SELECTION OF COMMUNITY IMPACT OUTCOMES FROM COMMUNITIES IN THE SOUTH WEST USING THE C2 APPROACH

Now that the approach has been explained, this section could be called 'so what'! What measurable difference did working in this way bring to challenged communities? A mix of quantitive and qualitative evidence of impact of this approach is therefore illustrated below, together with brief context. However we recommend reading the full stories behind these outcomes which will be available on the LSE website dedicated to this chapter.

NB: Although dates are given for initial operational project activity, most are now embedded organizations within their communities of origin, still going strong and operate as social enterprises, often with charitable status, supporting long-term sustainability. Also worth noting is that all examples had no start-up funding. The participants themselves procured whatever funding was required, often minimal, as projects unfolded.

All examples used in this chapter took place in Cornwall before national 'spread', where funded commissions began in 2010–2013. Outcomes from these are still emergent and 'hard' data is still being gathered, but all promise to be equally transformative.

6.6.1 Beacon Project: Falmouth Cornwall 1995–2000

The project that started it all! Set in severely disadvantaged social housing estate (population 6,000):

- Overall crime rate down 50%
- Unemployment down 71%
- Improvement to 1000 homes
- Educational attainment up 100%
- Child protection rates down 42%
- Post-natal depression down 70%
- Childhood asthma down 50%
- Teenage pregnancy zero.

6.6.2   REACH (Redruth Enabling Active Community Health) 2004–2006

REACH is an example of close collaboration between a community and the emergency services. It was a partnership between the resident-led Redruth North Partnership and the South West Ambulance Service. Its aim was to provide easy community access to a known and trusted practitioner (an emergency care practitioner/paramedic), while reducing the numbers of inappropriate 999 calls. The initiative won an NHS Health and Social Care Award for reducing health inequalities in July 2006. Outcomes included:

- 210 patients treated between 2004 and 2006 on site; and a
- 30% drop in incidence of under-age problem drinking and an 18% reduction in emergency call outs.

6.6.3   The Greenfingers Project: Redruth 2004–2008

'Greenfingers' was sparked by dialogue between residents and Neighbourhood Beat Manager, PC Marc Griffin. Residents and police were equally concerned by the state of many of the estate gardens and the anti-social behaviour and lack of aspiration of some young local people not in education or employment (NEET). A win–win solution was created in 'Greenfingers', literally a 'ground' breaking project.

Working in partnership with Duchy Agricultural College, it offered disaffected 16–19-year-olds the chance to access training, qualifications and earn free driving lessons, in return for completing an apprentice style course in gardening (NVQ level 1). Highly successful, it has transformed not only estate gardens but the lives of its participants, many of whom now hold a driving licence, an impossible dream prior to Greenfingers, and have since accessed further education.

Outcomes from year 1 of Greenfingers

- 10 students achieved NVQ level 1
- 2 moved on to NVQ level 2
- 16 went into full time employment
- 14 passed one day First Aid training course
- 3 passed Paediatric First Aid training course
- 15 took a course of 15 driving lessons and 4 went on to pass driving test
- 4 got LANTRA certificates in brush cutting and chainsaw (LANTRA is not an acronym but is the sector skill council for land skills)
- 3 took National Proficiency Tests Council (NPTC) in driving landscaping machinery
- 130 individual gardens were maintained for elderly and disabled
- 12 areas of open space were improved in conjunction with Kerrier District Council
- 1 new play area was created
- Support was given to a convent in landscaping their open space.

Latest 'Greenfingers' statistics as of 2014 are:

- 160 students have taken part so far;

- 154 attaining a Diploma in Horticulture or similar qualification, in the 5 years it has been running;
- Approximately 25% have gone onto employment of some nature (full/part time); and
- Further 25% moved onto a further qualification with the college.

So just over 50 per cent have gone onto employment or further education.

6.6.4   Operation Goodnight: Redruth 2008

Operation Goodnight was a groundbreaking community, police and multi-agency led, voluntary curfew scheme aimed at reducing the numbers of unsupervised children and anti-social behaviour on the streets of Redruth, Cornwall, after 9pm during the school summer holidays. Set in and around the Close Hill area of Redruth (top 2 per cent Index Multiple Deprivation Index) and as a direct response to many months of concern expressed by residents, fed up with underage drinking, swearing and vandalism, 'Operation Goodnight' focused on encouraging and supporting parental and community responsibility.

The press launch triggered a huge media response nationally and globally. Despite early concerns from residents it was highly successful, with high levels of compliance from young people and their parents. Residents describe being able to sleep properly for the first time in years, and the simple pleasure of being able to keep their windows open on summer nights!

Operation Goodnight outcomes July–September 2008

- 67% reduction in anti-social behaviour (ASB) levels;
- 64% reduction in youth-related incidents;
- 71% reduction of incidents involving 10–16-year-olds; and
- 100% reduction in youth related crime where offender is known.

6.6.5   The TR14ers Camborne: Cornwall 2005–2008

Named by the young people after their postcode, the TR14ers Community Dance Team was formed in October 2005 by the Police Neighbourhood Beat Team led by Sgt David Aynsley. It was founded in response to significant police concerns about rising levels of anti-social behaviour (ASB) and health inequalities affecting the youth of Camborne. The majority of young people that attend the TR14ers live on remote social housing estates with little social or play facilities and their families are often troubled by a raft of health and socio-economic issues.

At a C2 Listening Event in 2005, after new relationships were built between the young people and police, youngsters said that they would love to learn to dance hip-hop and street dance. The police team, young people and residents, worked together founding the TR14ers Community Dance Team, which attracted over 1000 youngsters at workshops provided free during the 'project' years, with the following measurable outcomes after three years:

- **Health and anti-social behaviour**
  - 46% reduction in anti-social behaviour
  - 60% drop in the use of tobacco, drugs and alcohol
  - 60% reduction in use of inhalers
  - 75% reduction in teenage self-harm.
- **Educational outcomes**
  - 22% increased levels of educational attainment[13]
  - 90% reduction in truancy rates
  - Weekly incidents of poor behaviour at school reduced 62%
  - 8 young people prevented from entering Criminal Justice System.

6.7　CHALLENGES ENCOUNTERED BY C2 ALONG THE SEVEN-STEP JOURNEY

6.7.1　The NHS Bio-medical Linear Approach to Health Versus C2 'Health Creation' Approach

A particular challenge for us as health practitioners seeking acceptance of our approach, has to do with NHS reluctance to 'let go' of bio-medical models of health, which have more to do with sickness than C2's model of 'health creation'. These approaches are especially problematic because they always look for direct, linear causality, which is almost impossible to find within a complex system. For commissioners, embracing complexity means being comfortable with emergence of unpredictable outcomes. A tough call for most!

We therefore discourage external evaluations of C2, as most still use a bio-medical lens through which to measure change. We prefer community and agency 'self-evaluation' as an ongoing process throughout the seven steps. As part of this, DVD clips filmed by residents or agencies provide powerful testimony to track ongoing community change. However this is sometimes viewed as 'unreliable' evidence by traditionalists. (Some C2 DVD clips have been uploaded to the LSE website dedicated to this chapter.)

6.7.2　Organizational Resistance to Change

The joys of working with this approach are many, but bringing transformation to communities often referred to as 'wicked problems', undoubtedly presents many challenges at an operational level, given their many layers and decades of embedded learned behaviours, both of the residents who live there, and the service providers who work there.

Perhaps surprisingly for the reader, by far the greatest challenge encountered during implementation comes from organizational resistance to change, from organizations threatened by the need to share power with residents, and to think and work differently. It has been our consistent experience that the 'worst' most dysfunctional and disadvantaged communities at street level invariably have a highly controlling, hierarchical, but often equally dysfunctional local authority (LA) operating at strategic level. This has no doubt evolved as a response to coping with the extremely challenging conditions encountered.

The culture of LA regeneration teams employed to 'fix' broken communities, is invariably one of 'doing to' rather than 'doing with', resulting in large-scale community passivity, which is a huge barrier to transformative change. This is why essentially we build new demonstrable and visible learning in to the seven steps, with an aim to change mindsets and bring a culture shift. However the mantra here is 'handle with care'.

Over the years the C2 team has learned to use great sensitivity and compassion when introducing the seven-step approach, to this often crowded LA 'marketplace' of service provision, all separately striving, with the best of intentions to bring improvement. The most often heard comment is 'we're already doing what you do' or 'we've already done the seven steps and it didn't work' the implication being that yet another intervention is unnecessary and unwanted. So to co-create the necessary conditions and receptive context essential for the approach to work, requires an understanding of how to change mindsets and deal with resistance.

We have found the Beckhard and Harris[14] change formula extremely useful in understanding, dealing with and assessing both the 'readiness to change' and the scale of organizational resistance. Using this scale we have also learnt to say 'no' to some commissions before they start, if the scale of resistance encountered during step 1 is deemed too great for successful community outcomes within the designated timescale. The 'readiness to change' factor is essential. Our advice to the commissioning body would be for them to work on their receptivity to new approaches and return to us at a later date.

6.7.3   'Power Crazed' Residents and Service Providers!

This happens quite a lot as a result of the delicate balancing act within the seven-step approach of redressing the loci of control within disadvantaged communities, leading to equity of influence and control between people and services. The transitional journey for a resident to make from being passive recipient to becoming a co-producer of services, is often challenging, as is the vice versa situation of sharing power for service providers and elected members (local councillors). C2 has often encountered stark personality changes from participants in both camps, who seemingly turn into mini dictators overnight! Dealing with this is always stressful and requires understanding and compassion. To hopefully prevent this, we introduce a 'C2 Code of Conduct', originally put together with residents during just such an episode, early on in steps 2–3.

Another way to minimize this is in the careful initial identification and selection of 'key' residents during step 1. So called community 'activists' often have the loudest voice but are not always helpful, as our experience demonstrates they often create a barrier to broader community engagement. Although well intentioned they frequently believe they are representing their neighbour's views, when in fact they are fixating on what is often a single issue that is not representative of what matters locally. They are often a 'turn off' for both residents and providers. As they are nearly always present, our solution is to 'dilute' them with other more representative voices, chosen as described in step 1, and they usually either respond and 'toe the line' or walk away, often to return at a later date, by which time the community voice is stronger and better able to absorb their enthusiasm.

6.7.4   Understanding the Effects of Poverty on Behaviour Change

Finally, many challenges for C2 arise as a result of the service provider's failure to understand the reality and behavioural effect of low-income living, and to recognize their own need to change their behaviours, in order for the community to change theirs. Chronic poverty causes chronic disease, educational failure and impoverished aspirations. Simply managing the state of poverty requires enormous amounts of mental energy in particular. The constant preoccupation with coping with a family on inadequate resources is enormously depleting. And yet we 'expect' residents to volunteer, become co-producers and work alongside us as equals. The fact that they do speaks volumes for human spirit! And of course, in time, their 'lived experience' changes immeasurably for the better.

In C2 we prefer to speak of 'capacity release' rather than 'capacity building', which has long been the predominant mantra for those involved in community renewal. 'Building' capacity assumes a deficit 'empty vessel' needing to be filled. Knowing capacity is already there, just needing the co-creation of enabling conditions to release it, makes for a totally different asset-based mindset from the outset. In any case it simply would not be possible to build capacity if it was not already there. In our experience it always is, even in the bleakest neighbourhoods, but it is latent, overlaid with mistrust, lack of confidence and the stress and exhaustion of coping with multiple disadvantage. To release this transformative capacity to change, requires respect, empathy, self-belief and most importantly, new relationship building at street and strategic level.

We are in no doubt that complexity science embraces all these principles, and offers a conceptual and practical framework for reversing community decline and improving health inequalities, that has eluded the UK for decades. We know it's already happening and making a difference for thousands in low-income communities across the UK.

The final challenge is for NHS and public health policy makers to embrace this approach to community wellness that we term as 'health creation'. The good news is that they are definitely listening!

We conclude with our definition, now widely quoted, that we believe describes the health phenomenon that we have been witnessing for over two decades through our practical application of complexity science.

'Health creation is the enhancement of health & well being that occurs when individuals and communities achieve a sense of purpose, hope, mastery and control over their own lives & immediate environment'.

NOTES

1. Payne, S., B. Henson, D. Gordon and R. Forrest (1996) *Poverty and Deprivation in West Cornwall in the 1990s*, http://www.bristol.ac.uk/poverty/downloads/regionalpovertystudies/Sec_1.pdf (accessed 9 June 2015).
2. New Deal for Communities, http://extra.shu.ac.uk/ndc/downloads/general/A%20final%20assessment.pdf (accessed 6 August 2017).
3. Health in Scotland (2009) Time for a change: Annual Report of the Chief Medical Officer, http://www.gov.scot/Publications/2010/11/12104010/0 (accessed 9 June 2015).
4. Durie, R., K. Wyatt and H. Stuteley 'Community regeneration and complexity' in D. Kernick (ed.), *Complexity and Healthcare Organization: A view from the Street*, London: CRC Press, pp. 279–280.
5. See https://medicine.exeter.ac.uk/research/healthserv/healthcomplexity/researchprojects/crest/ (accessed 6 August 2017).
6. See http://www.c2connectingcommunities.co.uk/ (accessed 7 September 2017).
7. Model copyright of the University of Exeter, 2003.
8. Durie, R. and K. Wyatt (2007) 'New communities, new relations: the impact of community organization on health outcomes'. *Social Science and Medicine*, p. 8.

9.   Jackson, S., S. Cleverly, B. Poland, D. Burman, R. Edwards and A. Robertson (2003) 'Working with Toronto Neighbourhoods toward developing indicators of community capacity Health Promotion International' http://heapro.oxfordjournals.org/content/18/4/339.full (accessed 9 June 2015).

10. C2 Connecting Communities Experiential Learning Programme https://medicine.exeter.ac.uk/research/healthserv/healthcomplexity/researchprojects/c2/ (accessed 6 August 2017).

11. www.traffordhall.com.

12. Mitleton-Kelly, E. (2003) *Ten Principles of Complexity and Enabling Infrastructures in Complex Systems and Evolutionary Perspectives of Organisations*, London: Elsevier.

13. 'Play it again Sir' *Times Educational Supplement* 2007.

14. Beckhard, R. and R.T. Harris (1977[1987]). *Organizational Transitions: Managing Complex Change*, Reading, MA: Addison-Wesley Publishing.

7. Using complexity principles to understand the nature of relations for creating a culture of publically engaged research within higher education institutes

Dr Robin Durie, University of Exeter, Dr Craig Lundy, Nottingham Trent University and Professor Katrina Wyatt, University of Exeter

7.1 BACKGROUND

Public engagement in higher education institutes describes the myriad of ways in which the activity and benefits of education and research can be shared with the public. Engagement is a two-way process, involving interaction and listening, with the goal of generating mutual benefit. Public engagement with and in academic research has been mandated by the Research Councils UK – the strategic partnership of the UK's seven Research Councils – which has developed a *Concordat for Engaging the Public with Research* (Research Councils UK, 2010). The *Concordat* identifies four key principles to support the embedding of public engagement with research:

1. UK research organizations have a strategic commitment to public engagement;
2. Researchers are recognized and valued for their involvement with public engagement activities;
3. Researchers are enabled to participate in public engagement activities through appropriate training, support and opportunities;
4. The signatories and supporters of this Concordat will undertake regular reviews of their and the wider research sector's progress in fostering public engagement across the UK.

Public or community engagement as such is not a new phenomenon in UK universities. A survey conducted in 2009 revealed that over 35 per cent of academics were involved in some form of community engagement (Abreu et al. 2009). Evidence suggested, however, that such engagement activity was often poorly supported and largely unacknowledged (The Royal Society, 2006; Burchell, Franklin and Holden 2009). With this in mind, the Research Councils UK, the UK Funding Councils, and the Wellcome Trust funded the creation of six Beacons for Public Engagement (BPE) each centred at a higher education institute in order to create a concerted effort to promote community–academic engagement in the UK. The BPE were given a mandate "to inspire a culture change in how universities engage with the public" (BPE, 2011). Beacons for Public Engagement were established to address five specific aims:

1. Create a culture within higher education and research institutes and centres where public engagement is formalized and embedded as a valued and recognized activity for staff at all levels and for students;
2. Build capacity for public engagement within institutions and encourage staff at all levels, postgraduate students, and undergraduates where appropriate, to become involved;
3. Ensure higher education institutes address public engagement within their strategic plans and that this is cascaded to departmental level;
4. Create networks within and across institutions, and with external partners, to share good practice, celebrate their work and ensure that those involved in public engagement feel supported and able to draw on shared expertise; and
5. Enable higher education institutes to test different methods of supporting public engagement and to share learning (Beacon Review, 2010).

Six Beacons, geographically spread across the UK, as well as the NCCPE based in Bristol, were funded. While the Beacons served as "collaborative centres, each consisting of a number of higher education institutions and partnership organizations", the role of the NCCPE was to "co-ordinate, capture, share and promote learning between the Beacons, and across UK higher education institutions, research institutes, and more widely" (BPE, 2011).

As noted by the Independent Review of Beacons for Public Engagement, each of the six BPE "evolved in different ways and have their own aims and objectives" (Beacon Review, 2010). For example, variation in the organizational and partnership structures of the six BPE was clearly evidenced, with some consisting of partnerships between several universities, whilst others were based at one; similarly, some of the BPE had structural partnerships with either public or private sector organizations, whereas for others, such collaborations arose through the work of individual projects.

As the impetus for developing community–university partnerships grows, so too has the research literature investigating the most appropriate means for understanding and modelling such partnerships. Nevertheless, research in this area remains, as yet, in its early stages of development. As Maurrasse (2001, p. 9) observed, "while practitioners have been testing the boundaries of the capacity of universities and colleges to impact communities in recent years, ironically, scholarly analyses, which address a variety of case studies representing different types of universities have not been fully developed". Maurrasse's groundbreaking study sought to apply the model of asset-based approaches to community development to community–university partnerships, focusing in particular on higher education institutions located in proximity to "poor urban neighbourhoods". Within the context of "community-building and community development movements", Maurrasse sought to determine how such higher education institutions needed "to change institutionally" in order to facilitate their involvement in community partnerships (Maurrasse 2001, p. 5). A related approach to that adopted by Maurrasse is to conceive of community–university partnerships as communities of practice. Introducing such an approach, Hart et al. (2012) argue that community–university partnerships consist in the "formation of relationships between a university and the communities within its locality, based on a principle of reciprocity". It is this reciprocity that suggests the appropriateness of the communities of practice model, where a community of practice is defined as being "a community created over time by the sustained pursuit of a shared enterprise" (Smith 2003). As Hart et al. (2012) continue, the "value in taking a community of practice approach to a community–university partnership is the focus it provides on joint enterprise, shared passion, different levels of participation and membership and the co-creation of knowledge".

Hart et al. (2012) go on to point out, however, that there remains "a lack of in-depth empirical work" in this area, and, specifically, that "evidence on the actual mechanisms of working practice" in community–university partnerships understood as communities of practice remains "emergent". Indeed, they go further, claiming that there remains "a lack of empirical data on how any communities of practice actually work in practice". We believe that this particular lack is indicative of a wider shortcoming pertaining to research in this area: whilst there is evidence in the research literature of the effectiveness of various models with which to understand the nature and structure of partnerships between communities and universities, there is little, if any, empirical research into the actual dynamic processes by which partnerships are formed between university practitioners and community partners in research projects. Similarly, there has been little work devoted to developing a theoretical understanding of the dynamic processes by which academics and community members connect with one another.

7.2   CONCEPTUAL FRAMEWORK: COMPLEXITY THEORY

Whilst there are several other theoretical approaches which have been employed to understand community engagement and community–university partnerships, such as theories of social capital (Kay 2005), asset-based community development (Maurrasse 2001), boundary-spanning (Weerts and Sandmann 2008, 2010), and communities of practice (Hart et al. 2012), in accordance with our own previous research, we adopted complexity theory as the conceptual framework for this research project. In doing so, we were seeking to build on work in which we have advanced interpretations of creating community-led partnerships for regeneration processes, involving statutory agencies and voluntary sector groups, which has been grounded in complexity (Durie and Wyatt 2007, 2013).

Complexity theory offers a series of principles for making sense of organizations such as higher education institutions, communities, and the relations between them, when these are understood as 'systems'. In particular, it offers a coherent theoretical perspective from which to make sense of the dynamic processes on the basis of which the behaviours of these systems emerge (Goodwin 1997). In simple, non-complex, systems, the behaviour of the elements of the system is the same whether the parts are taken in isolation or as parts of the whole, and the behaviour of both the system and its parts can be predicted in advance, based on a functional knowledge of the parts. Similarly, there is a linear proportionality between any change that is made to the parts and the consequent change in the behaviour of the whole system. By contrast, the distinctive behaviour of complex adaptive systems is irreducibly an effect of the nature of the relationships between the components within the system. The behaviour of the parts of complex systems is not given in advance; rather, it is determined by the relations between the parts. Similarly, the behaviour of the whole system is affected by its relations with the environment of which it forms a part, forming relations characterized as co-adaptive or co-evolving. Thus, complex adaptive systems should be understood as 'open' rather than 'closed', continually responding and adapting to changes in their environment, just as the environment itself changes and adapts to changes amongst its elements. The ongoing behaviour of the whole system, as well of its parts, remains to a greater or lesser degree unpredictable; such novel, unpredictable, behaviours – often a consequence of the 'self-organization' of the system – are thus said to be 'emergent'. Typically, such emergent behaviour tends to occur when the system is said to be 'at the edge of chaos', and when systems are at the edge of chaos, they are able to explore 'adjacent possibles', on the basis of which novel and creative systemic behaviours can emerge (Kauffmann, 2000). In order to understand the dynamic behaviour of the system, it is necessary to consider the effects of both positive and negative feedback loops within the system, and the system's sensitivity to initial conditions.

7.2.1   Rationale for a Complexity Approach to Public Engagement in Research

One of the main reasons that we have adopted complexity as the theoretical basis for our research is that it offers the potential for comparing phenomena such as networks, sustainability and resilience in biological systems with similar phenomena in social systems, and thereby opens the possibility of transferable co-learning about the causes of such phenomena. From the perspective of community engagement with academic research, the notion of co-evolution seemed appropriate for characterizing the dynamic development of the relations between communities and researchers. In particular, the notions of co-evolution and co-adaptivity seemed to offer a fruitful means for conceptualizing the dynamic processes that might underpin what Weerts and Sandmann (2010) have called the 'two-way approach' which, they argue, distinguishes community engagement with research from other university service and outreach activities. The principle of emergence offered a potentially insightful means of capturing the tendency for community engagement to yield novel and unexpected research outcomes. Most of all, the potential for phenomena such as co-evolution, self-organization and emergence stems, according to complexity theory, from the distinctively nonlinear nature of the dynamic relations between agents, and this focus on dynamic relational processes appears to offer a particularly effective means for making sense of the relations that underpin research projects involving community–university partnerships.

Our theoretical assumption at the outset of the project was that this concentration on open, fluid, nonlinearly dynamic systems, exhibiting emergent behaviours, offered a rich potential for understanding how processes of successful community engagement with research tend to occur.

In order to understand the facilitators and barriers to engaging and involving people in research to develop a process for two-way, meaningful engagement, we used a case study approach, looking at six different research projects from the Beacons.

7.3   METHODS

Our research design was explicitly informed by complexity theory. Data gathering was focused on determining the processes of engagement by which partnerships between universities and communities were formed, and the processes by which subsequent research projects were designed and delivered. We used categories and principles derived from complexity theory in order to characterize what happened within each academic–community research project, and in order to frame explanations of why this has happened. We used a case study design for the research. Case studies, in which 'the case' is examined in the context of its environment, are particularly suited as a means of investigating system inter-relationships and change (Anderson et al. 2005). Research observations that target patterns of relationships, interactions, and processes, over time, are central to gaining an understanding of the system or case (Capra 1996). By looking for patterns of behaviours across systems (cases) we can begin to see how behaviours develop and evolve, rather than merely describing the 'static behaviour' from one observation of the system (Goldstein 1999). In keeping with a complexity approach, data at both the level of the organization and that of individual projects were collected using ethnographic methods; these included initial conversations with key personnel at each site as well as interviews and focus groups. Where possible we also attended meetings and read relevant documents and reports for contextual data (Trenholm and Ferlie 2013). The goal of collecting data through a variety of means is both to enhance the theory generating capabilities of the case, and to provide additional validity to assertions made by either the researcher or the participants in the case itself (Yin 2009).Our approach also informed the inclusion of negotiated feedback sessions as part of our methodology, whereby we sought to share the findings from each study with that site in negotiated feedback sessions. Such sessions allowed for revision of any declared interpretation and validation of any emerging conclusions (Lincoln and Guba 1985).

7.3.1   Community Advisory Group

At the outset, a Community Advisory Group was established, comprising seven community partners drawn from our previous work, two practitioners with expertise in engagement work, and academics with considerable experience of involving people in research. The purpose of this Group was to inform the development and implementation of the project, and advise on the framing of research questions that would be used in the interviews and focus groups, from a community perspective. The group met three times during the course of the research and attended the final symposium. Community members were reimbursed for their time and their travel expenses to attend the meetings.

7.3.2   Identification of Case Studies

In order to identify the 'cases', a 'scoping tour' was conducted of the six Beacon sites and the National Coordinating Centre for Public Engagement. The purpose of this 'scoping tour' was to gain an understanding of the different institutional contexts developed to support community engagement in research, and to identify, in collaboration with the different Beacon teams, engagement projects to serve as case studies for further in-depth research within the project.

We used a maximum variation sampling approach when identifying possible projects for inclusion as case studies. We sought to identify a wide range of research projects on the basis both of their research question and approach, as well as nature of the community involvement. (For an indication of the diverse nature of the case study projects, see Table 7.1.)

Table 7.1   Case study projects for research

Project	Was the project jointly conceived?	What was the project's purpose?	Who participated?	Were further collaborations identified?
Project A	Jointly conceived by academics and partner organization	To work together to empower young people using theatre and to foster new relations between university and charity organization with a view to future working together	Academics and students from university, staff from Partner organization, young people and their families	Yes
Project B	Jointly conceived by academics, partner organization and community members	To enable a community to realize some of its aspirations through a series of public meetings, aided by professional facilitators	Academics and students from university, council workers, professional facilitators, community members	No
Project C	Jointly conceived by academics and community members	To facilitate a community in exploring its past through the use of rare archival footage and the production of a documentary film	Academics, professional external partners and community members	Yes
Project D	Jointly conceived by academics and community members	To creatively record and disseminate histories from an ethnic community in an urban area	Academics, community association and community members	Yes
Project E	Jointly conceived by academics and partner organization	To promote the health and well-being of young women through the shared and regular activity of growing food	Academics, mentors from partner organization, community members	Yes
Project F	Jointly conceived by academics, community liaison and community members	To inform the public about a topical science issue, to convey government funded research to a wider audience, and to increase feedback between academics and community	Academics and community members	No
Project G	Jointly conceived by academics, partner organization and community members	To explore and potentially remove barriers that inhibit the lives of adults with learning disabilities	Academics, community representatives and community members	No

In negotiation with the Beacon sites, we agreed the engagement projects which were to be the case studies for our research.

7.3.3   Case Studies

Once the case studies had been identified, a convenience sampling approach for data collection was used. For each case study we identified and interviewed the community and academic lead, and used these interviews to identify other key informants for interview or groups to conduct focus groups with. The purpose of the interviews and focus groups was to understand how the projects had been identified and developed, the nature of the community academic relationships, what effect these relationships had on the research, as well as any facilitators and barriers that had been experienced in the engagement process. As well as collecting interview and focus group data, where possible, project meetings were attended and field notes made. Where appropriate, an initial copy of the research proposal was obtained and any subsequent reports of the project's progress. Evaluation documents relating to the process and outcome of the overall Programme for each site were also scrutinized and any relevant information captured.

7.3.4   Data Collection and Analysis

A three level method of analysis was used for the data collected from the case studies.

Initial analysis: Interviews and focus groups were recorded and recordings were transcribed and then checked against the original recordings. Contemporaneous records made of conversations and during meeting observations were written up and incorporated into the analysis. Primary documents were read in close detail and memos written to capture formative ideas for more detailed analysis. The dataset for each site was then subjected to a rigorous thematic analysis (Boyatzis 1998). A thematic analysis was undertaken as it is a *process* rather than a method in and of itself and is suitable for use with many kinds of qualitative data, and is often used in conjunction with other types of data analysis, and therefore appropriate for analysing case studies. An initial within-case analysis was conducted to identify relevant themes by the researcher (CL) who was primarily involved in data collection. A convergence coding matrix was produced detailing the findings from each case which were then analysed to look for agreement, partial agreement, silence or dissonance from the different cases (O'Cathain, Murphy and Nicholl 2010). Cross-cutting themes were generated from this analysis. The validity of each site-specific themes and their possible meaning was triangulated by looking at themes from the observation of meetings, scrutiny of written documents, and by comparison with the chronologically documented sequence of relevant events at each site. A number of cross-cutting themes were identified, and these were presented to the engagement project teams during 'negotiated feedback' sessions. The resulting discussions at these sessions fed back into the analysis of the data, and led to the refinement of the cross-cutting themes.

Complexity analysis: This initial analysis was followed by a secondary analysis of the emergent cross-cutting themes from the perspective of complexity. Our provisional assumption during this level of analysis was that the cross-cutting themes should exemplify principles pertaining to complex systems. We have derived a working set of principles of complex dynamic systems from our own theoretical and primary research (Durie 2002;

Durie and Wyatt 2007); similarly, Mitleton-Kelly (2003) provides an excellent discussion of the fundamental principles of complex systems. The interpretation of the cross-cutting themes from the perspective of these principles served the function of affording theoretical insight into how and why the dynamic processes of engagement occurred in the way that they did. This analysis was undertaken by CL, RD and KW and was an iterative process, moving between the data (cross-cutting themes) and complexity principles.

Critical-reflective analysis: A third level rigorous interrogation of the complexity themes was then conducted involving the lead researcher and RD and KW. The purpose of this final iteration of analysis was to critically reflect on the appropriateness of the use of complexity theory as an interpretative framework for the primary data, and at the same time to reflect on whether this bringing together of the first two levels of analysis poses questions of complexity theory itself, and of its applicability to social systems in general, and systems of community–university research partnerships in particular. The purpose of adopting this iterative, three level, approach, was to ensure that complexity themes were not simply 'read into' the primary data, or that the application of complexity themes was not done uncritically. The third level of analysis constituted a crucial critically reflective phase, in which the theoretical underpinnings of the analyses were subjected to rigorous scrutiny, from the perspective of the data, from the perspective of the application of the complexity theory, and in respect of the relation between both of these perspectives.

A schema representing how we collected and analysed the data is presented in Figure 7.1.

7.3.5 Negotiated Feedback Sessions

All of the participants in the projects comprising the case studies were offered the opportunity of partaking in individual (case) negotiated feedback sessions. During these sessions, iterative analyses of the emerging themes from the research data were presented back to participants (and any other people they chose to invite) which allowed for revision of any declared interpretation, validation of any emerging conclusions, and consensus setting for the next part of the study. In taking part in such sessions, the researchers explicitly addressed the fact that, as researchers, they would not only be influenced by their immersion in the communities at each of these sites, but also, reciprocally, might influence some of the conversations taking place at these sites during the course of the study.

7.3.6 Symposium

The concluding activity of the project was a symposium hosted at the University of Exeter to which all participants from the project were invited, including the Community Advisory Group and other relevant stakeholders. At the symposium we presented our research findings, and the ensuing discussions tested, and ultimately confirmed, the robustness of our interpretations of these findings. Furthermore, it was from these discussions with our community and academic partners at the symposium that our new description of a multi-phased engagement cycle emerged (see below). This emergent outcome exemplifies one way in which community engagement in research can be a genuinely two-way process.

Ethics approval was secured from the University of Exeter Humanities and Social Sciences Ethics Committee in March 2011.

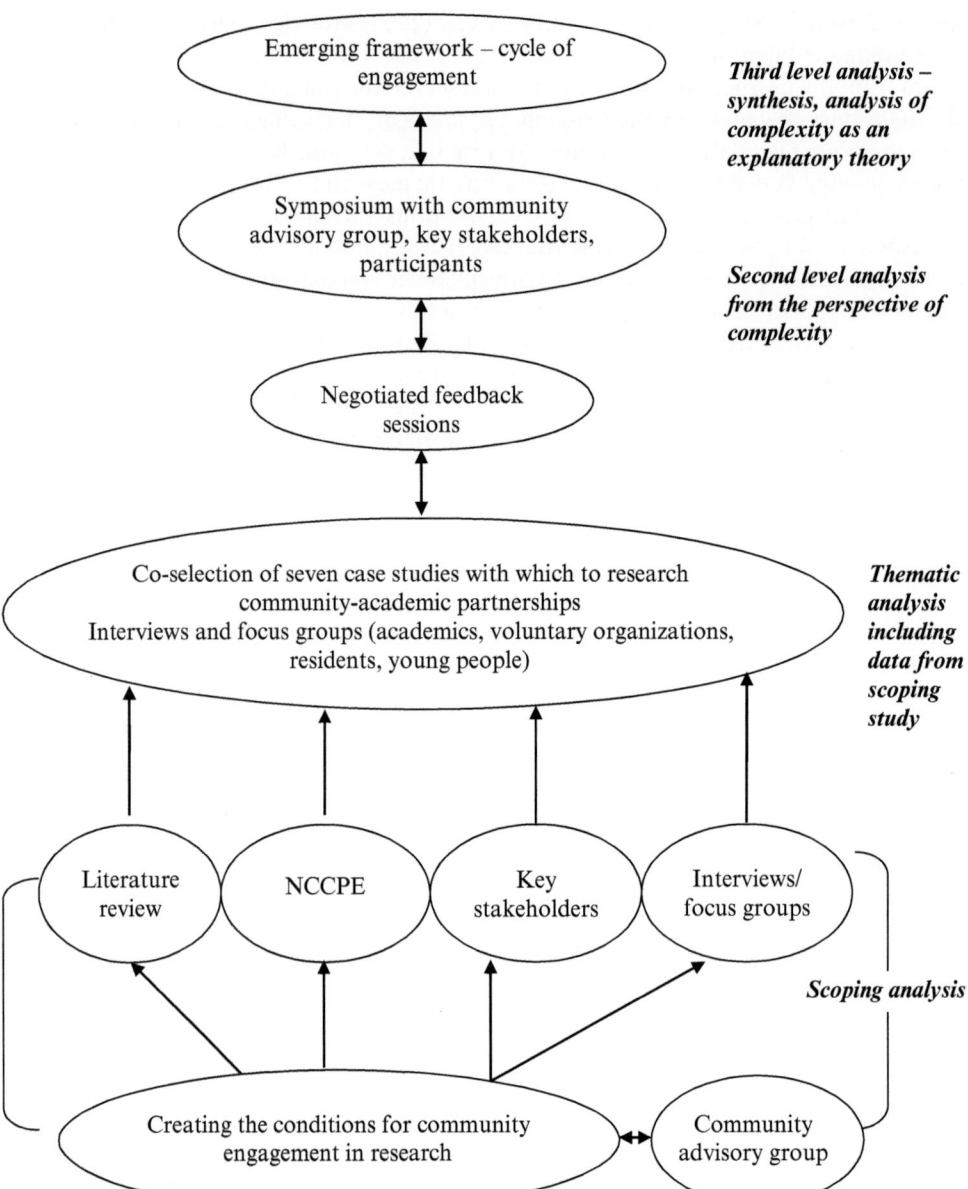

Figure 7.1   How we collected and analysed the data

7.4   FINDINGS

A total of 16 interviews and focus groups were conducted as part of the scoping study and a further 22 interviews were conducted for the case studies. Interviews and focus groups included members of the community as well as academics involved in each engagement project. Table 7.1 provides brief descriptive information about each project.

Amongst the large number of themes that emerged, six overarching themes were identified as particularly pertinent and widely cross-cutting. These overarching themes came from the data collected as part of the initial scoping tour as well as data from the case studies.

The first set of three themes relate to the conditions that support and facilitate successful engagement of communities as partners in research projects. As we discuss in section 7.4.3, these cross-cutting themes proved particularly amenable to the second level analysis from the perspective of complexity themes.

A further set of three cross-cutting themes emerged from our analysis of the data that was significant for two reasons: first, this dataset appeared to contain a series of results that offered contradictory perspectives on what seemed to be similar themes; second, these contradictions appeared to call into question a number of the fundamental assumptions about the processes of community engagement that would tend to follow from the principles of complexity theory. Specifically, there appeared to be a tension between those who considered more or less rigid, linear, systems and processes as a barrier to engagement, preferring instead the flexibility of nonlinear processes and relations; and those who emphasized the need for relatively fixed, linear, relations and processes to enable the successful functioning of engagement projects. The seemingly contradictory findings harboured within these themes therefore appeared to offer a challenge, from the perspective of our third level critical reflective analysis, to how far we could employ complexity as a theoretical lens through which to make sense of the dynamics of community engagement at the second level of our analysis.

In fact, we drew explicit attention to these issues during the project's final symposium, and then went on to present a more nuanced interpretation of the principles of complexity relative to the findings from the cross cutting themes. On the basis of this new interpretation, we also presented a new multi-phased model of the *processes* underpinning successful community engagement with academic research, which was subsequently refined and developed on the basis of discussion with our community and academic partners at the symposium.

In sections 7.4.1 and 7.4.2, we present these two sets of cross-cutting themes, and then in 7.4.3, we offer a more in-depth discussion of our findings from the perspective of complexity theory, on the basis of which we introduce this new multi-phase model of engagement processes.

7.4.1   Key Findings from the Cross-cutting Themes for Engaging Communities in Research

Time and rhythm
Many individuals who were involved in running Beacon sites, as well as those participating directly in engagement projects, discussed issues relating to time, timing and notions of rhythm. It was repeatedly suggested that co-creating projects tend to require prolonged periods of time to build robust relationships and maximize the potential of the projects. In the words of one participant from Project C, "a key difference [about this project] is just having the time to explore". Some individuals interviewed also noted, however, that

long or open-ended timescales can contribute to unproductive projects, either through the onset of inertia or the development of increased pressure to 'deliver' valued outcomes. For example, one participant from Project G recounted how the project, which was a 'flagship' project, "ran aground" and had "become this very, very, difficult process" due in part to the extensive time and space it was afforded. According to this participant, it was the implementation of time restrictions that not only facilitated the delivery of the project, but also encouraged experimentation and energized existing relations: "now we are doing some really ambitious things and it's partly because we've lost that time that we are having to be so ambitious. We're also building on a lot of relationships that . . . had existed for a long time but we're really trying to push them".

Regardless of the length of projects, several individuals involved in the creation and organization of community–academic engagement projects referred to the importance of both the 'lead-in' and 'follow-on' periods for their projects. As one academic interviewed in our initial 'scoping tour' put it, it is vital "to acknowledge both the lead-in and the follow-up [as] a legitimate part of university work". Key relationships, it was pointed out, are invariably formed *prior* to the commencement of any project, as is the exploration of possibilities for collaborative work. As for the post-project period, it was common for participants, academic researchers as well as community members, to speak of a failure to fully capitalize upon the momentum that had been built over the course of the project, following its official end date. As a participant in Project F observed, "if the project had been, you know, for a bit longer [we would have liked] to go back and then sort of really measure impact". It was therefore suggested that initial planning should also consider potential 'follow-on' activities to ensure that maximal benefit from these new relationships is more consistently realized.

Closely associated with the desired recognition of 'lead-in' and 'follow-on' periods was an insistence on the need to respect the various rhythms at play in any given engagement project. The cycles of university research, it was frequently noted, do not tend to map on to the rhythms of community life. This is particularly significant with respect to the alignment of academic funding cycles with the various non-academic patterns of activity as well as the internal rhythms of engagement projects themselves. Project work is almost inevitably discontinuous, with built-in points of finitude, in contrast with community life which is continuous and 'ongoing'.

Finally, and again following from the previous point, a precondition for the majority of projects analysed was the following question: is *now* the appropriate time for this engagement project? More specifically, as one community partner in Project D articulated, "is the community ready for this type of project?" As the data suggests, if an engagement project is to involve a genuine component of researching *with* communities, as opposed to *on*, *about* or *for* communities, this question must be answered in the affirmative.

Staying the distance

When asked to reflect on contributing factors to successful engagement projects, it was not uncommon for both academic and community partners to refer to the need for the academics involved to demonstrate a commitment to the community in question beyond their significance to achieving the academic research output. A commitment to 'stay the distance' from the outset was thus seen as highly beneficial, if not essential, for creating relationships which lead to successful engagement projects. As one community participant in Project D said: "some of the committee members were concerned regarding they [the researchers] just wanted to come in and do the project, tick their boxes and kind of leave". And in the words of another individual from that project: "people from the university are always coming in and researching what our [community] problems are and then they go away". This commitment was also recognized and considered necessary by many academic partners in order to gain the trust of communities. According to an academic partner in Project C: "if you are hit and run and you're parachuting in and out they'll never trust you". As a result, 'hit and run' research was seen by many academic partners involved in these studies to be not only potentially harmful to the communities involved, but also to the creation of meaningful community–academic relations. Both academics and communities recognized that in this type of 'hit and run' research, communities were used as a *resource* for the research rather than as *partners* in a joint endeavour. Furthermore, it was felt that academic processes, such as the funding cycles alluded to above, compounded these issues. As observed by a participate in Project E: "so much of the funding is kind of short term or is a bit like a smash and grab, like, okay you've got six months, here's a thousand pounds, run out there and, you know, find a community, whack them over the head, drive them back, stick them in a conference for half an hour, and done". Our data thus clearly resonates with, and corroborates, that of Weerts and Sandmann (2010, p. 715), which refers to such 'hit and run' academics as 'drive by' or 'wind-shield sociologists', since they "drive by the city, quickly collect their data, and eagerly leave town".

Mutual benefit
A substantial amount of the data collected reflected an unequivocal belief in the importance of 'mutual benefit' for both academics and community partners in engagement projects. Unlike some engagement projects, where the benefit to communities is largely identified by academics and justified to audiences outside of the project (such as funders), the majority of project participants we interviewed insisted upon the need for explicit and clear discussions about mutual benefit between all project participants prior to, and at every stage of, the project. As two separate individuals interviewed from Project A remarked: "I do feel very strongly that actually you do need to establish a mutual benefit"; "it's about both people trying to meet everybody's agenda and finding a workable solution, or deciding not to, but in a constructive way". While in some projects the respective benefits and outcome of the project were directly linked – that is, the outcome was one that both the academics and community partners wanted to achieve for their own purposes – participants gave examples from other projects where the beneficial outcomes for communities and academics were to a certain extent distinct, or even isolated, from one another (for example, an interesting academic finding that was of little or no benefit to the immediate concerns of the community partners; or the creation during the project of an artefact valued by the community, but which had no discernible 'academic value' in and of itself). Of course, there were occasions when the outcomes leading to such mutual benefits were not achieved; however, both academic and community participants agreed that this was of less significance than might otherwise have been the case, so long as clear discussions about desired or intended benefits and outcomes for all participating partners were maintained from the outset of the project.

7.4.2   Further Findings Requiring more Nuanced Analysis

Systems and structures
Invariably, engagement projects between academic and non-academic partners, whether they are partner organizations (governmental or non-governmental) or individuals/ groups in communities, involve a confluence of different organizational systems and structures. Because of this, it was observed by several participants that successful engagement projects require not only effective communication and understanding between the individuals involved, but also between the different systems and structures of which they are a part. As one project lead from Project A put it: "one of the biggest barriers and obstacles to come into and work out is organizations understanding other organizations and how they work and their policies and their procedures and their remit and their ethos as well". More specifically, many interviewees highlighted difficulties in negotiating and reconciling more or less rigid higher education management structures with the need for flexible and responsive organizational processes that could support engaged research. These tensions were particularly manifest in financial issues, for example the ability of HEI finance systems to pay community participants for their work in an appropriate and timely manner. In this respect, a certain degree of flexibility and openness within the implementation of University systems was seen as conducive to engaged research. One of the project case-studies (E) identified the fact that in a participating organization, "everything is quite structured really, so all of the things that they [community partners] are waiting to say . . . they can't". In an explicit effort to overcome this problem, therefore, the project partners sought to foster what they themselves characterized as 'structure-less spaces', thereby trying as far as possible to eliminate the need for relations that were predetermined by organizations.

In contrast with the experience of rigid organizational structures working as a barrier to effective engagement, for several of the projects, the establishment and implementation of predetermined and fixed systems and structures was considered an essential ingredient for their success. Pre-fabricated and non-negotiable systems and structures were thus, in some cases, the foundation upon which engagement projects were allowed to flourish, rather than an impediment. As one participant from Project A who had been involved in a number of previous engagement projects made clear, their most recent project was so successful in part because "the support network [of their organization] is already formed somewhat, all you are doing is transporting it".

Project planning and outcomes
As with the data collected on attitudes towards systems and structures, our empirical research revealed conflicting data concerning the approach to, and impact of, project planning and pre-set outcomes on engagement projects. In some cases, meticulous preplanning, with a clear articulation of project objectives and outcomes, was identified as integral to the success of the project. For one project manager interviewed from Project A, this pre-planning extended to hypothesizing possible problems and ensuring that the appropriate contingency plans had been considered: "if you don't plan and try and proactively address what issues you think are going to arise, then you are going to be reactive and firefighting and it's not going to work so well".

By contrast, open-planning and the lack of a specified set of milestones were identified by other projects as both desirable and central to effective and sustainable engagement. In the words of one participant from Project E: "the thing about [our project] is that it's kind of a wide open space . . . – you can't entirely predict it". This sentiment was especially strong in those projects that valued emergent rather than predetermined outcomes for projects. Commenting on the haphazard, yet effective, trajectory of their engagement project, one academic partner remarked: "There's not exactly a long term plan, it's evolving and emerging".

Roles (functions) and responsibilities

Data collected on the nature of roles (functions) and responsibilities within engagement projects also presented us with apparently opposed findings. Perhaps most intriguingly, however, we were presented not only with evidence of how both fixed and fluid structures and processes can contribute to successful engagement projects, but more significantly how fixed and fluid – or closed and open – processes are intrinsically related to, and rely upon, one another. On the one hand, academic and community partners from numerous projects nominated the identification of fixed and clear roles and responsibilities within the project team as a key to success. Such clarity was credited with helping to ensure the smooth running of projects, whilst also aiding in the formation of communal partnerships, comradeship and mutual respect within the project team: "I think clarity of roles is actually a great way of avoiding ambiguity about responsibility and actually makes it easier for people to have mutual respect for each other" (Project A). The evidence also suggested that fixed and clear roles and responsibilities can have a liberating effect on participants, and that, rather than constraining their behaviours, clarity of roles allowed them to experiment within the project beyond the remit of their initially determined role or function. As a result, fixed and clear roles and responsibilities were at once constraining and liberating: "we knew where responsibility lay and therefore we were actually more relaxed about missing our roles because in the end it was clear who was responsible for the role. It sounds almost contradictory but it is – it's how it worked in a strange way" (Project A).

7.4.3 A Complexity-informed Cyclical Model of Public Engagement in Research

In reflecting on these cross-cutting findings, and in particular the data which suggested explicit relations between fixity and fluidity, we were led to reconsider the initial theoretical assumptions for our research regarding the detail of whether and how complexity theory may be appropriate for explaining successful processes of community engagement with research.

In our previous research into transformative community regeneration, complexity offered a particularly powerful theory for capturing the systemic behaviours of communities as they move from cycles of decline to those of sustainable regeneration. For instance, community decline is typically manifest in the fragmentation of communities, the isolation both of individuals from one another, and from service providers and statutory agencies. By contrast, the movement of regeneration consists in the formation of multiple new relations, and the strengthening of both the community network and the networks between the community and the service providers and statutory agencies. The shift between these two phases – of decline and regeneration – can be characterized as occurring at a point of criticality, and as the community reaches what can be considered as the 'edge of chaos', self-organization starts to occur, both within the community and between the community and service providers and statutory agencies. On the basis of these new relations precipitating such self-organization, the overall behaviour of the community starts to change, and transformative regeneration starts to occur as an emergent property of the whole community (Durie and Wyatt 2007). Our theoretical assumption was that many of the features of community-led partnerships for successful change would be common to successful engagement between university researchers and community partners. In particular, we felt that success in both spheres would be a consequence of conditions that enabled the creation of dynamic, nonlinear, relations leading to unpredictable, emergent, outcomes. Whilst much of the data from the research supported this hypothesis, as we have shown, there was significant data that also challenged our theoretical assumption, and thereby made us reflect further on how complexity theory might function as a model for understanding community engagement in academic research.

These challenges prompted us to examine the data more deeply, and led us to realize that we had been positing the relations between linearity and nonlinearity, closed and open systems, and predetermined and emergent outcomes, in a way that was not sufficiently nuanced. We had fallen into the trap of thinking that these relations consisted in more or less dialectical oppositions. As such, our own thinking had been too rigid, too static – whereas complexity theory should have encouraged us to think in more fluid, dynamic, terms. We therefore sought to reconsider the apparent contradictions contained within the second set of themes from the perspective of processes, and this led us to develop a new conceptual model, informed by complexity theory, with which to characterize and make sense of the data.

Interpreting the data from a more nuanced complexity perspective, we suggest that the dynamic processes of community engagement with research could be usefully understood as phases within an engagement cycle, and that these phases have distinct dynamic characteristics. Figure 7.2 depicts the three phases and what is distinctive about each phase.

Engaging phase
The data suggest that there is an 'engaging phase' within the cycle, comprising the processes of developing relations, and building trust, with communities – the so-called 'lead-in' phase. It is within this phase that what we have called the 'initial conditions' for community engagement are created. Academics should identify community groups/special interest groups/organizations and individuals who might share a common interest in the research area. Similarly, attending meetings makes the academics visible and allows people to better understand the research which is going on at the HEI. This is very much a nonlinear phase, conversations could quickly lead to areas of mutual interest and proposed work, however it could take a considerable amount of time to build trust and develop the relationships. This may be particularly true of communities in areas of high deprivation, where their experiences with service providers and academics are often that of being 'done to', thus demanding more time and effort from all parties in the process of developing genuine relations built on trust. Similarly, meetings should be held at a time and place that is convenient for the non-academic partners which could be evenings and weekends; it should not be assumed that everyone will be able, or indeed want to, meet in

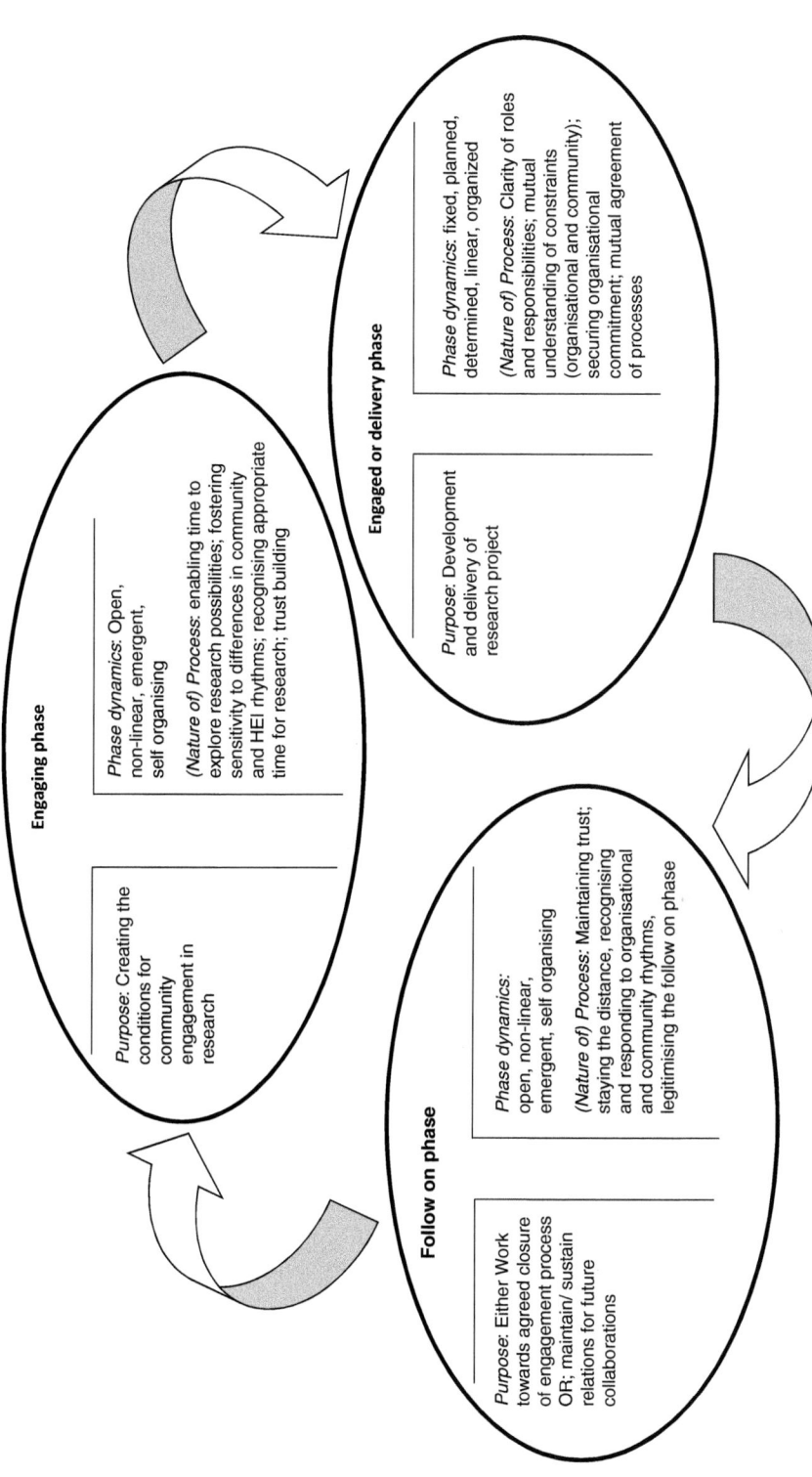

Engaging phase

Purpose: Creating the conditions for community engagement in research

Phase dynamics: Open, non-linear, emergent, self organising

(Nature of) Process: enabling time to explore research possibilities; fostering sensitivity to differences in community and HEI rhythms; recognising appropriate time for research; trust building

Engaged or delivery phase

Purpose: Development and delivery of research project

Phase dynamics: fixed, planned, determined, linear, organized

(Nature of) Process: Clarity of roles and responsibilities; mutual understanding of constraints (organisational and community); securing organisational commitment; mutual agreement of processes

Follow on phase

Purpose: Either Work towards agreed closure of engagement process OR; maintain/ sustain relations for future collaborations

Phase dynamics: open, non-linear, emergent, self organising

(Nature of) Process: Maintaining trust; staying the distance, recognising and responding to organisational and community rhythms, legitimising the follow on phase

Figure 7.2   The Engagement Cycle

a university building. Once relationships have been established, times and locations can be negotiated.

The processes characteristic of this phase of the engagement cycle tend to manifest the typical qualities of complex systems – they are open, fluid, dynamic, and lead to emergent outcomes. These emergent outcomes could be said to occur when this engaging phase reaches the 'edge of chaos', and starts to self-organize. Should any evaluation be required or conducted during this phase of the engagement cycle, we would thus suggest that it should be concerned with reflecting on the processes that facilitate or inhibit engagement, rather than focusing on the outcomes of the research, as most traditional evaluation tends to do. Moreover, such evaluation should be formative and feed back into these processes, allowing them to adapt and evolve, thereby supporting both the creation of new relations, and the identification of possible mutual benefits for communities and academic partners.

Project or delivery phase

These emergent outcomes may consist in what can be termed 'constraints' or 'parameters'. Such constraints tend to be operative within the context of projects, or pieces of work, based on community–academic partnerships that emerge from the engaging phase, and which can be interpreted as determining the form and structure of these projects. We may thus think of this as constituting a 'delivery' or 'project phase' in the cycle – the 'engaged project'. It can be postulated that such projects are the means by which the self-organized 'system' explores 'adjacent possibles'. The processes comprising this engaged phase frequently exhibit properties typically found in linear systems: for example, successful projects are often well planned, with outcomes identified in advance, on the basis of which impact evaluation can be carried out. Such properties are themselves manifestations of the emergent constraints that are determining the projects. There are several possible collaborative approaches from academics delivering the research, guided and advised by the community/non-academic personnel to community researchers delivering the research with academics providing some research knowledge as necessary. One approach is not necessarily 'better' than the other; what is important is that the approach and people's roles and responsibilities are negotiated at the outset, with opportunities to revise if necessary, during the delivery. A schedule of meetings should be agreed and joint agendas created to address governance and delivery of the research; time should be allowed to understand each other's personal and organizational constraints. Consideration of health and possible caring needs should be given to ensure participation is as equal as possible.

Follow-on phase

During the final Symposium of our research project, another phase of the engagement cycle which had been tentatively identified in interview data was confirmed and emphasized by our community and academic partners, namely a 'follow-on phase', requiring its own span of time, akin to the period following the conclusion of a run of artistic performances, during which reflection can take place and processes of planning for future work or building towards mutually acceptable closure can occur. Where possible, funding and planning for follow on meetings should occur to gain a collective sense of the process of engaging as well as agreeing next steps for the partnership. Our sense at this stage is that this 'follow-on phase' should also manifest open, fluid and dynamic relational characteristics, as well as lead to emergent outcomes, coordinate with the framework of complexity theory.

7.5 CONCLUSION

The research project sought to identify patterns of enabling behaviours for successful community engagement in research. Moreover, we sought to undertake a complexity-informed case study approach to try and identify patterns of dynamic processes by which successful engagement occurs. We have developed a multi-phased framework for understanding the distinctive dynamics involved in the processes constituting the engagement cycle. The perspective of complexity theory suggests that the focus for community engagement with academic research should be on the dynamic relations that constitute the processes of engagement, and, in particular, on how such relations might emerge, or be co-created, within engagement processes.

The development of a multi-phased engagement cycle was motivated and informed by the challenging data collected from our case study research, and through the discussion of our interpretation of this data with our community and academic partners. By developing a more nuanced application of complexity theory to the process of engaging communities with research, we were led to take into consideration more fully not only the manner in which successful projects display complex characteristics as they unfold, but even more significantly, how the phases and processes prior to and following the 'engaged project' phase also comprise critical elements in a complex understanding of engagement. This insight thereby significantly adds to the growing body of research into the appropriateness of complexity theory for understanding and promoting community engagement. By extending the application and usefulness of complexity theory in this way, we have advanced a model that similarly seeks to widen the scope for analysis of engagement projects, emphasizing above all the importance of 'lead-in' and 'follow-on' phases for maximizing the effectiveness of such projects. We believe that a framework of this kind provides a significant challenge to the dominant model of funding for academic research, given that such funding at present is almost exclusively limited to the duration of the more or less discrete project (phase two of our cyclical model). Following on from this, we believe that if genuinely effective community engagement with academic research is to be secured, there will need to be a dialogue between funders and both academics and community partners about how such 'lead-in' and 'follow-on' phases of the engagement cycle can be most appropriately supported. Finally, with respect to external evaluation of the success of community engagement work, we believe that this will have to address the means by which the potential different types of outcomes of community engagement projects can be respected and valued, where these outcomes constitute the components of the mutual benefit of the research to both academics and community partners.

REFERENCES

Abreu, M., Grinevch, V., Hughes, A. and Kitson, M. (2009) *Knowledge Exchange between Academics, and the Business, Public and Third Sectors* accessed 30 March 2015 at http://eprints.soton.ac.uk/357117/1/AcademicSurveyReport.pdf.

Anderson, R.A., Crabtree, B.F., Steele, D.J. and McDaniel, R.R. (2005) Case Study Research: The View from Complexity Science. *Qualitative Health Research* 15, 669–685.

Beacon Review (2010) Independent Review of Beacons for Public Engagement Evaluation Findings, *Final Report for RCUK, HEFCE and the Wellcome Trust*. London: People Science & Policy.

Boyatzis, R.E. (1998) *Transforming Qualitative Information: Thematic Analysis and Code Development*. Thousand Oaks, London and New Delhi: Sage Publications.

BPE (2011) *Beacons for Public Engagement Promotional Material*, accessed 30 March 2015 at www.publicen gagement.ac.uk.

Burchell, K., Franklin, S. and Holden, K. (2009) Public Culture as Professional Science: Final report of the ScoPE project – scientists on public engagement: from communication to deliberation. September, BIOS, London School of Economics and Political Science.

Capra, F. (1996) *The Web of Life*. New York: Anchor Books, Doubleday.

Durie, R. (2002) Creativity and Life. *Review of Metaphysics* 56, 357–383.

Durie, R. and Wyatt, K. (2007) New Communities, New Relations: The Impact of Community Organization on Health Outcomes. *Social Science and Medicine* 65 (9), 1928–1941.

Durie, R. and Wyatt, K. (2013) Connecting Communities and Complexity: A Case Study in Creating the Conditions for Transformational Change. *Critical Public Health*. doi:10.1080/09581596.2013.781266.

Goldstein J. (1999) Emergence as a Construct: History and Issues. *Emergence – Journal of Complexity Issues in Organizations and Management* 1 (1), 49–72.

Goodwin, B. (1997) *How the Leopard Changed Its Spots*. London: Orion Books.

Hart, A. et al. (2012) *Community–University Partnerships Through Communities of Practice*, accessed 30 March 2015 at http://www.ahrc.ac.uk/Funding-Opportunities/Research-funding/Connected-Communities/ Scoping-studies-and-reviews/Documents/Community-university%20partnerships%20through%20commun ities%20of%20practice.pdf.

Kauffman, S. (2000) *Investigations*. Oxford: Oxford University Press.

Kay, A. (2005) Social Capital, the Social Economy, and Community Development. *Community Development Journal*, 41 (2), 160–173.

Lincoln, Y. and Guba, E. (1985) *Naturalistic Inquiry*. Beverly Hills, CA: Sage Publications.

Maurrasse, D.J. (2001) *Beyond the Campus: How Colleges and Universities form Partnerships with their Communities*. London: Routledge.

Mitleton-Kelly, E. (2003) Ten Principles of Complexity and Enabling Infrastructures. In E. Mitleton-Kelly (ed.), *Complex Systems and Evolutionary Perspectives of Organisaions: The Application of Complexity Theory to Organisations*. London: Elsevier, pp. 23–50.

O'Cathain, A., Murphy, E. and Nicholl, J. (2010) Three Techniques for Integrating Data in Mixed Methods Studies. *British Medical Journal* September 17, c4587. doi: 10.1136/bmj.c4587.

Research Councils UK (2010) *Concordat for Engaging the Public with Research* accessed on 30 March 2015 at http://www.rcuk.ac.uk/pe/Concordat/.

Royal Society (2006) *Excellence in Science: Survey of Factors Affecting Science Communication by Scientists and Engineers*, accessed 30 March 2015 at https://royalsociety.org/~/media/Royal_Society_Content/policy/ publications/2006/1111111395.pdf.

Smith, M.K. (2003) Communities of Practice, in *The Encyclopedia of Informal Education*, accessed 30 March 2015 at www.infed.org/biblio/communities_of_practice.htm.

Trenholm, S. and Ferlie, E. (2013) Using Complexity Theory to Analyse the Organisational Response to Resurgent Tuberculosis Across London, *Social Science and Medicine* September 93, 229–237. doi: 10.1016/j. socscimed.2012.08.001.

Weerts, D.J. and Sandmann, L.E. (2008) Building a Two-Way Street: Challenges and Opportunities for Community Engagement at Research Universities. *Review of Higher Education* 32.

Weerts, D.J. and Sandmann, L.E. (2010) Community Engagement and Boundary-spanning Roles at Research Universities. *The Journal of Higher Education* 81 (6).

Yin, R.K. (2009) *Case Study Research: Design and Methods*. Thousand Oaks, CA: Sage Publications.

PART III

VISUAL METHODOLOGIES

Professor Alexandros Paraskevas

Visual representations in the form of illustrations, maps, models and diagrams have been widely used to describe complex systems and network structures and are well-accounted for as knowledge-elicitation methods in complexity science. The visualization of complex systems and structures offers a high-level abstraction which enables the understanding of the interdependencies and interactions of the various components of these systems from the agent (micro) to the aggregate (macro) and can provide the basis for understanding the behaviour of the entire system. This section presents three different approaches to visual representation of complexity.

Julian Burton and Sam Mockett's chapter first, describes how art can be used as a research tool for a 'Visual Dialogue' process that simplifies abstraction and jargon and enables the creation of a reflective space that can be used for the development of meaningful and focused conversations about change and the facilitation of interventions within an organization.

Central to this process is the concept that change emerges out of a constant flow of interactions, conversations and negotiations between people within the organization which is, on its own, a complex living social system. Since the quality of these conversations enables or constrains change, the 'Visual Dialogue' process is designed to positively enhance the way employees and leaders interact and relate to coordinate their activities. Barton shows how the insights from complexity thinking have influenced the development of 'Visual Dialogue' to bridge the gap between *strategic plans* and *local solutions* and to support change in large organizations.

The main components of 'Visual Dialogue' are, primarily a big picture that captures the shared story of the organization (new strategy, new structure, change programme) and, of course, the facilitation of a dialogue between groups that will enable all parties involved to make sense of this big picture. The author explains the purpose, use and benefits of each of these components (picture/dialogue) and describes step-by-step of how they are implemented in the 'Visual Dialogue' process.

The chapter ends with a real-life case study where the author illustrates main points of how this process was implemented and offers interesting insights of the people involved in the form of direct quotes.

Kate Hopkinson, in her chapter, uses the 'Landscape of the Mind' (LoM) methodology to visualize and explore the 'inner complexity' of a mind and to measure the inner skills and thinking styles that we need to live, work and play in today's complex world.

'Inner Skills' is the term that the author uses to describe a whole range of factors that affect our behaviour such as thinking, imagination, knowledge, intuition, past experience, feelings and values and so on. Hopkinson distinguishes three types of skills: *convergent inner skills* that underpin working with what we already know and understand; *evaluative inner skills* that enable us to make decisions; and *divergent inner skills* that enable out-of-the-box thinking and facilitate creativity and innovation. She uses the LoM globe to visualize the different kinds of inner skills which can be brought to bear into their three dimensions separating them in two modes, which she calls 'cool' and 'warm' (head versus heart – detached versus emotionally engaging).

The author describes in a very good level of detail the profiling process providing diagrams and illustrations and offers a number of case studies and examples to illustrate the method. This mapping of inner skills offers a situational analysis in terms of the inner skills in play at a particular time which can be used as a basis for planning and agreement for improvements in behaviour. Ultimately, the LoM visualization technique helps the participants understand their inner skills preferences and, although it is relatively unlikely that they will change their preferences over time, they can change the choices they make and the behaviours they display depending on the needs of the task or project in hand.

Whereas simple networks can be easily represented visually with mind maps and concept maps, the visualization of the structure of complex networks is far more challenging, given their large scale and their interconnected nature. In their chapter, Kurt A. Richardson and Andrew Tait argue that there is a need for tools (both mathematical and computational) that can help us make sense of these massive structures. They suggest that Dynamic Network Analysis with a focus on the simulation of dynamic networks is the preferred approach for researchers seeking to unveil the secrets networks hold. They distinguish static network topology, which does not account for the information flows around the network and how information is processed by the network nodes, from the actual active network structure that emerges when a dynamic nonlinear network follows a range of different attractors.

They use a simple Boolean 'Small World' network to demonstrate that a purely topological analysis of the original network 'graph' does not have the capability to capture all of its structural properties. This shows that studying even relatively simple dynamic networks proves to be a huge and very challenging task and that when designing network interventions, just seeing who or what is connected to who or what is not enough to gain a complete understanding of the network. Although they broadly agree with Cilliers (1998) who said that complex systems are irreducible, they propose a methodology in which they employ Power Graph Analysis (PGA) and succeed in reducing/compressing the complexity of the network without losing any information that would inhibit reliable insights and understanding of its dynamic structural properties. Clearly, the benefits of visualization by employing PGA will be more apparent in the analysis of much larger networks, since the degree of compression can be considerably higher. However, the authors also propose an alternative methodology aiming to 'slice' the network based on the Shannon Entropy (SE) of each node operating within each attractor basin. The network can therefore be layered by grouping the nodes based on their SE or activity, thus facilitating its visualisation, whilst also offering a way of obtaining an information-based network signature.

Finally, in their chapter, Göktuğ Morçöl and Sohee Kim demonstrate how they use two methodological approaches, Network Text Analysis (NTA) and Social Network Analysis

(SNA) in order to analyse complex governance networks. They define complex govern-ance networks as collaborations among multiple governmental and non-governmental actors who are self-organizing but also dynamically interrelated and interdependent and aim at solving complex social problems. Although much progress was made in the meth-odologies that explore and describe complex systems, the authors argue that none of these methods can project an accurate picture of their structural properties and the nonlinear, dynamic relationships among actors. They highlight the limitations of Agent Based Simulations (ABS) before presenting the advantages and limitations of Social Network Analysis (SNA) and its tools.

The analytical tools presented in this chapter are: Automap, a network text analysis tool based on the semantic network analysis approach which assumes that a model of networks of words and relations can be developed by extracting knowledge and language in a certain text context; and ORA which analyses the networks extracted by Automap. In essence, AutoMap develops meta-networks by linking actors and their knowledge, resources, actions, or tasks, whereas ORA analyses these meta-networks by coding the complex relationships as actors and their resources, actors and their locations, actors and their actions, and so on. This type of analysis enables researchers to visualize the struc-tural properties of the complex governance networks and the relative positions of the individual, organizational and institutional actors who operate within them.

The authors offer a range of case examples where they have used these tools highlight-ing challenges they have faced in many of them. They close by maintaining that this meth-odology can be used longitudinally to identify the evolution of the most central actors in urban governance networks and evaluate how the hypothetical removals of the central actors could impact some of the structural properties of these networks.

REFERENCE

Cilliers, P. (1998) *Complexity and Postmodernism. Understanding Complex Systems.* London: Routledge.

8. The art of complexity: using visual artefacts and dialogue to bridge the gap between strategic plans and local actions in organisations
Julian Burton, Delta7 and Sam Mockett, Visual Meaning

Figure 8.1   Command and control versus self-organisation

We will describe some of the cultural challenges of turning strategy into action that many organisations are currently facing, particularly in the context of the relationship and status gap between leaders and employees, and how difficult it can be to develop a realistic strategy without involving the people responsible for delivering it. We will also show how we have used the Visual Dialogue process as an Organisational Development intervention to address some of the key aspects of these challenges. As the process has evolved over the years we have learned that its principal impact lies in improving the quality of conversations about change between leaders and frontline employees, working to create enabling environments that foster mutual sense-making, rich connection and creativity. Finally, we will describe the component parts of Visual Dialogue and how each contributes to creating such enabling environments and supporting emergent change.

8.1   ORGANISATIONAL CHANGE

Changing the culture of a large, complicated social system is often seen as a complex, 'wicked' problem. The combination of nonlinear interdependencies, hierarchical relationships, the intransigence of human behaviour and the coexistence of many different stakeholder groups make the 'problem' impossible to resolve with traditional change management thinking and processes. Those in leadership roles focus their attention on strategic priorities, challenges and the strategic plans necessary to address them, whereas those in delivery roles attend to their local priorities, issues and solutions. Both roles, however, are interdependent and interconnected. We believe that improving the shared understanding of each other's roles, challenges and plans, and their mutual dependence, is a necessary starting point for the coordination of system-wide, local improvements.

Our objective with Visual Dialogue is to enhance the quality of mutual sense-making, listening and information-sharing between managerial and delivery-oriented groups, to enable richer interconnections and achieve closer collaboration. Helping these two groups focus on their shared purpose above personal interests can help diminish the all-too-common silo mentality and bring them together in a way that improves their effectiveness in working together to deliver strategy. For many leaders their role seems to be becoming increasingly complex. They can be under pressure to drive down costs and increase profit, and on a wider scale their organisations are asking themselves 'what is our purpose beyond profit?' or 'can we ask our people to work even harder?' and 'are we running out of road?' The need to think differently, to rethink strategy and to make decisions quickly is greater than ever. Meanwhile, employees are often bombarded with new change initiatives and strategies, without having the necessary space in which to process the new information, understand its implications for them or contribute to its

Figure 8.2   Mutually dependent yet sometimes uncoordinated roles

exploration. Employees are also expected to translate the language of change – often conveyed via jargon – into something meaningful they can act on, which can feel impossible without the opportunity to question them, voice concerns or query upward in the hierarchy. This gap between the higher managerial intent and the ground-level reality can confound efforts to engage staff, as the 'bigger picture' is not visible to them – with the result that they cannot find their place in the change landscape.

In our experience, this situation in which employees find it hard to see where they fit in can trigger an unintended reaction to change, in the form of individuals closing down and becoming disengaged and demotivated. We find that an overload of strategic communications designed to motivate people actually leads to disengagement.

Within this context, Visual Dialogue provides a powerful means of stimulating interactions around change that are free of the most typical barriers to self-organisation, which include conflicting understandings, power imbalance and lack of personal agency. It takes the naturally-occurring characteristics of social systems – co-evolution, emergence, interdependence – and works *with* them in a safe setting to loosen rigid thinking and facilitate clarity and effective action.

8.2 COMPLEXITY THEORY IN THE BUSINESS CONTEXT

Our Visual Dialogue process was inspired by the work of Professors Stacey and Mitleton-Kelly, particularly their interpretation of complexity science in its application to organisational change. It developed as a method for positively enhancing information-sharing, collaboration and the quality of team meetings – in other words, micro-interactions. Seeing an organisation as a complex, living social system of people, relationships, interactions and commitments led us to conclude that any change that happens in that system emerges through conversations. We believe that a key insight from complexity thinking is that in social systems, change, self-organisation, emergence and co-evolution are enabled or constrained by the quality of conversations that mediate relationships between people. Change emerges out of the constant flow of negotiating interests, intentions, plans and

Figure 8.3   The locus of change

expectations: the way people interact and relate to coordinate activities. We will show how the insights from complexity thinking have influenced the development of Visual Dialogue to support change in large organisations.

8.3   EMERGENCE AND THE BUTTERFLY EFFECT

One of the central ideas we use from complexity theory is that organisations can be seen as complex adaptive systems, made up of multiple interacting individuals, continuously adapting to each other and their environment, that is, co-evolving together. We find it useful to remind ourselves that we are all interdependent – in other words, we need each other and we are continuously participating in, and responsible for, creating the social world we all inhabit. Any given person is only connected to a fraction of the population, thus they have no choice but to interact locally, which can confound attempts to link cause and effect, that is, to plan and control change in organisations. This is commonly understood as a feature of the 'butterfly effect', whereby the outcome of a large number of people interacting in spontaneous, nonlinear ways cannot be planned, as tiny changes in these local interactions can escalate unpredictably, creating significantly different outcomes to those desired or intended by the leadership.

8.4   LOCUS OF CHANGE

Based on these ideas from Complexity, our work in organisations starts from the point of view that conversations are the basic unit or locus of change in organisations, and our focus is to improve the quality of sense-making in the conversations that take place. We work from the assumption that the locus of change in an organisation isn't usually within an independent or autonomous individual but rather emerges from and within the relationships between people in conversations. Interdependence, emergence and co-evolution – and thus self-organisation – are enabled or constrained depending on the quality of conversations that mediate relationships between people and guide their actions. As human beings we are constantly interacting, relating and coordinating our actions in an unpredictable flow of language, intentions, plans and expectations. To enhance the conditions in which these conversations arise, for example in team meetings, we can support leaders and employees to better achieve shared intent and outcomes.

8.5   PROTECTED TIME TO MAKE SENSE OF CHANGE AND CREATE SHARED MEANING

Therefore, to achieve change in social systems requires a unified sense of what 'change' means. Each of the two groups, the operational and strategic, needs reflective time to make sense of change in terms of the challenges faced by the other group and to create shared meaning that allows them to simultaneously address the problem. In order to create the conditions for a unified sense of change to emerge, a level of communication over and above everyday information-exchange is required. An environment conducive to mutually supportive communication requires protected time to allow people some distance from the pressures mentioned earlier, and a supportive space in which to explore challenges and possibilities. Protected time, which is an integral part of the Visual Dialogue process, offers a safe and enabling environment for teams to share ideas and barriers, their experiences and opinions, and make sense together. The safety is of particular importance as it helps diminish any resistance to the temporary suspension of hierarchy that occurs when the emphasis is on the quality of the conversation rather than the relative positions of those engaged in it.

8.6   SHARED GOALS, SYSTEMIC OBSTACLES

As argued above, in organisations people inhabit and co-create a shared cultural environment. Their roles can broadly be divided into management and delivery. They take on these roles specifically to enable the organisation to achieve its goals, but the constraining elements of the social norms, tacit assumptions and power relations can create distance between them. The two roles are both equally important and mutually dependent in terms of achieving organisational goals, as strategic decisions affect local decisions and actions, and vice versa. The content and context for each have a direct impact on the other. The potential gap between the two roles can be understood by the focus of their attention. The leadership/management group focus their attention on the strategic level, scanning the environment, making decisions on the allocation of resources, and coordinating and supporting those doing the implementation. Those with delivery roles focus their attention on the local level, that is, local challenges and innovations. We have often heard from employees that they don't grasp the strategy or new change initiative, and it leaves them confused about what exactly they are being asked to do differently.

Therefore, both groups seem to have the same high level objective, but work within a different context. Both need and desire the same 'change', but 'change' means something

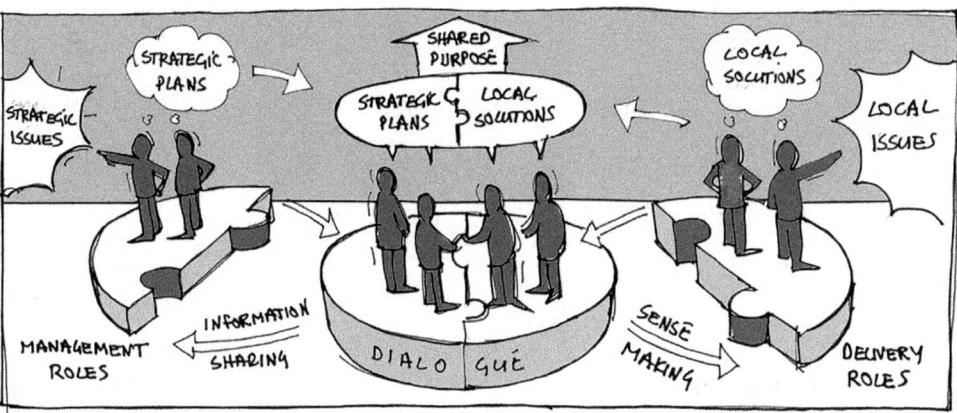

Figure 8.4   *An enabling environment to enhance the effectiveness of mutually dependent roles*

different to each group. For leaders, change is a strategic challenge at an organisational level, while for employees change means local level tasks and challenges. So it is often hard for employees to see a connection between what they need to do locally and what the strategy means to them. And conversely, the nature of a strategic plan means that it needs to have a certain level of abstraction as it describes the whole organisation, and cannot tell each individual what they need to do differently, locally.

8.7   BRIDGING THE GAP BETWEEN EMPLOYEES AND LEADERS

To bridge the gap between these two positions, a visual representation (picture) of the space in which employees and leaders operate can function as a stable reference point for a discussion through time, making it much easier to arrive at a shared meaning of what's most important for the organisation and the people in it. By using visual representation, the indirect interdependencies and interconnections within a system are arranged as relational objects. Seeing a system as a set of relational objects, rather than trying to read it as a set of described concepts, allows more space to examine the relationships between the different elements and how that translates into human experience. It also allows the strategic issues and the local issues to occupy the same space. The visual representation incorporates both sets of challenges within the same environment, which fundamentally changes the dynamic of the conversation. By showing that both sets of challenges are connected and operating within the same context and environment, it becomes easier to share information that was previously part of closed feedback loops. The visual representation

Figure 8.5   A typical communication challenge

creates the opportunity to locate oneself within the complex environment, which allows the discussion to move away from the description of the environment itself and towards the intentions, obstacles and difficulties in the *experience* of that environment. This affords leaders the opportunity to share strategy and change in a way that helps frontline employees understand the context and purpose for what is being asked of them, and gives them the opportunity to share what is needed to make local improvements, in such a way that helps leaders understand what they can do to support them.

8.8   ONE-WAY COMMUNICATION

One of the biggest challenges we see in organisations is that communication is often one-way, that is, leaders deliver new information about strategy and change to employees and can make the assumption that they will automatically understand what it means and what is expected of them. We often hear employees use the term 'command and control' to describe the sort of communication that leaves them feeling 'done-to' and disempowered.

In almost every project we work on, employees have ideas, challenges and questions about change that they do not share with leaders for various reasons, particularly as there may be a culture of silence due to the imbalance of status in the room. This situation can typically cause misunderstandings and confusion that are rarely clarified and that, in turn, can create the sort of organisational distress that inhibits service delivery and improvement. I know from my own experience that whenever work is very busy, I can sometimes forget the wider purpose of my endeavours, the reason why I am doing what I am doing and how my colleagues and I depend on each other. In this state my anxieties and worries will likely inhibit my performance and ability to collaborate effectively. In one organisation we worked in, at an employee session where we were testing the strategy big picture, one employee said "at team briefings our manager is so busy telling us what we need to do, there's no space to talk about anything, it's just one long download from him". Feeling ignored, overwhelmed, stressed or tired can often reduce communication to the level of a transaction, when in fact a mutually supportive level of communication is more beneficial. Because of the lack of adequate space to make sense of the chaos at work, problems often go unvoiced and are compounded by jargon and abstractions.

8.9   DIFFERENT LANGUAGES

This communication gap is often also caused by the use of different language. Any effort at communication is futile when talking different languages, but the pressure that drives that effort (trying to achieve the shared organisational goals in a complex environment) inevitably results in rising frustration on both sides. In most organisations we work with, there can be intense pressures on employees: to cut costs, continually audit or measure, and always improve performance. There are corresponding pressures on leaders: to reduce budgets, improve quality and efficiency, and enable the continued progress and increased profitability of the organisation. This can have a significant negative impact on the quality of learning and sense-making, constraining creativity and innovation. Factors such as culture, status and history shape the behaviours of each group within itself and

Figure 8.6   Both sides often contribute to building silos

towards others, which introduces constraints on both sides. These constraints inhibit the ability to communicate effectively to share information, which has a significant impact on the ability to achieve shared goals. In one recent example from an organisation we have worked with, a senior team communicated to employees that they were all to "live the values". We heard from employees that they were not clear exactly what they needed to do differently in practice, and wanted leaders to show them by example. However, this response went unvoiced as they believed that to speak up would be a "CV Moment"!

Therefore, informal, honest and grounded conversations undertaken in an appropriate space for reflection are the key to making sense of what is going on and can lead to the clarity of shared meaning necessary for making creative responses to change.

We have discovered over the years that there are three vital ingredients to achieving a rich and enriching conversation: informality, honesty and groundedness. By informality we mean that when there is little or no status being played out in the conversation, people feel safe and trust each other, there is a relaxed, friendly, or unofficial style to the conversation. In this atmosphere, people tend to be more honest and forthcoming about how they are feeling, what is going on for them and what is getting in the way. In this kind of setting, plain, simple language can be a helpful bridge between differing ideas, interpretations and realities. By groundedness, we mean that people talk of their own experience using the first person (I, me or my). This is more meaningful for people when they want to go beyond the conceptual, to arrive at a mutual understanding that is grounded in personal, lived examples, feelings and reflections. It is therefore the best antidote to corporate jargon.

8.10   WHAT IS DELTA7'S VISUAL DIALOGUE?

Visual Dialogue is an organisational development process that uses a Big Picture to represent a new strategy or change programme, and facilitated dialogue to support employees in small groups to make sense of it. A Big Picture is a two-metre colour illustration that represents the organisational strategy (and scenarios deriving from it) in story form. Participants in the Visual Dialogue session are then able to have conversations about the complex or difficult issues represented in the picture, and about how they can take ownership of their role in resolving them. The key elements of Visual Dialogue are a big picture that captures the shared story of the organisation and dialogue facilitation. We will explain the purpose, use and benefits of each, beginning with pictures, and then show the process of how we use them.

Visual Dialogue sessions are designed to achieve the following three things: share important information about where the business is going, why and what needs to change; help colleagues make sense of the change and what it means to them personally through discussion with their peers and how to affect the necessary changes; and create useful insights, learning and ideas to enable the change process. In a Visual Dialogue session, a leader or manager, trained in dialogue, facilitates using the Big Picture to share the story the change journey. He/she shares a story about where the business has come from, where it's going and how it's going to get there. Once they have shared the story, the facilitator then invites the group to discuss the issues raised, using the different parts of the picture as prompts.

Because it represents the cares and concerns of most employees, everyone can see their experience represented in the Big Picture, which helps to create a sense of shared meaning. Issues that might be too difficult to talk about become easier to discuss because they are externalised in picture form. The Big Picture typically engages even those people who start off resistant because there is always something in it they care about and have a useful opinion about. The Big Picture is a powerful catalyst for conversation and we find that most people want to talk about what they see in it. In a story sharing session, people learn more from each other than from the picture itself or the facilitator running the session. The leader's role is to set up and manage the discussion to encourage that learning. The Visual Dialogue process and the feedback it generates can act as a bridge between the workforce and the senior leadership team, helping each to better understand concerns or tensions. Part of the leader's role as a facilitator is to help build that bridge.

Poor communication, and hence coordination, in organisations is currently a major barrier to successful change. We often hear from our clients that their employee surveys are indicating worryingly low levels of engagement and particularly that employees don't understand a new strategy or change initiative, do not know what is being asked of them, and are not sure how their role fits into the larger scheme of things. What leaders want is for employees to have a clear 'line of sight' between their local role and tasks, and the overall purpose and direction of the organisation.

Figure 8.7   A typical Delta7 strategic narrative Big Picture

Figure 8.8   Abstract words point to concrete things

8.11   WHY PICTURES?

Pictures and visual metaphors can help leaders create meaning; they can offer senior teams and employees a different way of exploring complex problems, of understanding the bigger picture and generating a richer kind of meaning, knowing and understanding. Using visual metaphors to represent complex strategic issues can bring more areas of the brain into play and help to create a shift in how the problem is seen, talked about and made sense of. We will look at how pictures differ from words, then examine two ways that pictures can act as a leadership tool: first, as a sense-making tool for the leaders themselves, and second as a way of engaging the rest of the organisation with change.

8.11.1   Pictures as a Tool for Sense-making and Meaning-making

The dominant representational system in organisations today is verbal. Leadership teams generally make sense together and communicate strategic change using spoken words. Words are a powerful communication tool, yet they do have their limitations. First, they are processed by the left, more rational, side of the brain. Second, they are abstract: it is only by arbitrary convention that words, spoken or written, represent the objects in the world that they point to.

Finally, they are linear in that they can only describe one thing at a time. As the world becomes more complex they can become inadequate for representing emotionally charged, complex, or multidimensional problems. Imagine trying to explain how something as complicated as the London underground system looks, with all its stations and their interrelationships, using words alone!

But the problem is not just in language itself, but in the *type* of language that is typically found in organisations, as argued previously. In many client change projects we have worked on, we have found a consistent problem to be that the words used to communicate

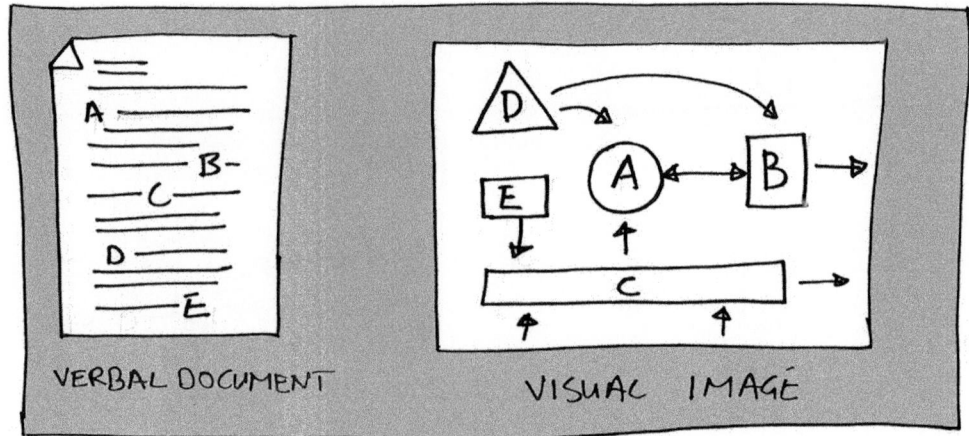

Figure 8.9   Visual images can show complex nonlinear and causal relationships unavailable to verbal language

change can be highly abstract and distant relative to the experience of employees. Real and complex challenges can get smothered in abstractions and generalisations.

The result is that a new change strategy or a critical directive that may be coherent and obvious to a leadership team may feel very disconnected from the employees' lived experience if communicated in traditional ways. And when the strategy is communicated one-way, with little space for local sense-making, there is likely to be a lack of clarity, coupled with confusion and subsequently little or no action. As Stacey would say "the meaning is in the response". Difficult but important subjects are avoided, or made safe and unthreatening, so the truth gets hidden away and lost. This way of communicating is so ingrained that people are barely aware they do it. So if communication through verbal language has its limitations, what are the alternatives?

There is a growing body of research that advocates the use of visualisation as a meaning-making tool that can help senior teams make sense of, and articulate, strategy, in a way that employees and other stakeholders can understand. Pictures have been used for thousands of years to tell stories, create meaning and represent what is most important for people in communities. The visual medium is able to capture even the most complex interrelationships in an instant, which means that the observer can have an overview of the whole. Pictures are memorable and tangible, and allow us to re-present complex and tacit information in visceral, meaningful ways. Pictures can help individuals locate themselves in relation to their world, help them to see connections, can be non-threatening and can appeal to the imagination.

The Western corporate world's reliance on linguistic and rational intelligences is perhaps nowhere more evident than in the boardroom, where it is clear that conventional thinking about change is no longer enough. Professor Keith Grint at Warwick Business School proposed the idea of 'wicked problems', which have no linear causes or linear solutions, and it is these kinds of challenges that business leaders are increasingly facing. Rational processes are simply not robust enough to address wicked problems. A different level of thinking is needed – a different way of knowing. It is through this 'different way of knowing' that visual metaphors can help leadership teams discover different ways of looking at the world or new ideas to solve the problems represented, bringing more creativity to the boardroom in the process. Visual art, used properly, is a particularly powerful way of representing important themes symbolically, non-verbally and aesthetically. Visual images can present knowledge in a more concrete form than words alone, stimulating more spontaneous and grounded conversations, the sort in which new meaning or knowing can arise naturally within the group.

In our work with leaders to help them visualise a new strategy, we create reflective spaces out of which new ideas emerge. We ask them open questions to expand their current thinking, and make draft representations of the concepts that allow a more thorough engagement of their intelligence as they see how their different perspectives and ideas fit together. With the help of our artists we can then create a synthesised view of the big picture as it emerges. Leaders can also use the picture-making process for their own internal development, to create a shift in their own thinking. These pictures may be particularly intimate, expressing vulnerabilities and personal weaknesses as well as unspoken dreams and aspirations. They become powerful anchors that remind individual leaders of the shifts they need to make personally before they can successfully create shifts in the organisation.

8.12   USING PICTURES TO ENGAGE STAFF IN TURNING STRATEGY INTO ACTION

The traditional practice of communicating strategy seems to be based on the assumption that by delivering information one-way into people's heads they will naturally act on it. Too often corporate communication has the opposite effect – creating confusion and lack of meaning for people that can lead to de-motivation. Ultimately no one can dictate, demand or force others to do things they do not understand or that have no obvious meaning for them. These days the reality for leaders is that they can only set the stage for people to want to act on their own. This was highlighted recently when a senior leader mentioned to me that in the past when he asked people to "jump" they asked "how high?" whereas nowadays they ask "why should I?". For leaders to help employees to understand a change strategy, they need to ground the meaning of that strategy in the personal experience of the employees. Turning words into pictures is a very effective way of doing this. For words to be acted-upon – for language to do its job properly – the benefits of change have to be personally meaningful. The more vague and abstract words are, the more distant from personal experience they become, and the less likely they are to influence personal behaviour.

When leaders fail to engage frontline staff in a new change initiative, it may be because it does not have any real meaning for them as they have been given precious little space to make sense of it for themselves. This can be highly de-motivating as people feel that change is being *done to* them, not understanding its significance or rationale. We hear from employees that when they are communicated-*at*, there is little space to respond and they feel that their leaders rarely listen to what they have to say.

Eliciting group thinking in a visual format can be a powerful approach to closing the gap that separates words from experience. Grounding the meaning of their words is the

Figure 8.10   How some employees feel about corporate values

first step to turning them into action. It can help create the shared meaning needed to align and motivate people towards a common goal. Complex or difficult situations can be described in more concrete, memorable and highly discussable forms than words alone. Pictures bring ideas to life. It is essential for leaders to articulate their strategy in a way that is meaningful to other people, and to create space for them to discuss what is important to them as individuals. Using visual art as a tool to assist leaders helps because it conveys different meanings to different people, stimulating a dialectical movement in their understanding of each other and the world. Pictures can stimulate meaningful conversations where continuous and local interpretation of experiences can keep meanings moving and flowing.

8.12.1   The Benefits of Pictures

- **Shared story** – they bring together the concerns of the leadership and the workforce into a single, engaging colourful artefact that can provide a framework for rich discussion. This can allow comprehension of complex states of affairs outside of personal experience, yet will affect the conditions of the viewer.
- **Overcome limitations of verbal language** – Pictures are time independent in that they are available for continuous scrutiny, relieving the burden on short-term memory. They are instantly referable and available.
- **Line of sight** – they help people locate themselves and their day-to-day experience in relationship to the wider organisation and its journey, giving them a clearer line of sight between their role and the organisation's purpose.

- **Concrete ideas** – they make important concepts 'concrete' and real – people can point to them and talk about them.
- **Universal language** – everyone can understand a picture, and they are very effective at communicating across boundaries, be they organisational, cultural or linguistic. This reduces the risk of costly misunderstandings.
- **Unspokens** – they can help people talk about the 'elephant in the room'. Making a picture of what people find difficult to talk about encourages honest inquiry into the real issues that are blocking change. Pictures can be a safer way for people to talk about their experiences and feelings about change in a more honest, constructive and motivating way.
- **Create new meaning** – they bring the meaning of words to life. We translate jargon, often meaningless to employees, into visual metaphors that can ground the meaning of words in people's experience, increasing understanding, giving employees a tangible and memorable representation that anchors new understanding of how the business actually works.
- **Show complex interrelationships** – when lots of factors need to be considered, a picture can give people an overview that clearly shows the complex interrelationships that are so difficult to explain in words, because verbal language can only show linear relationships.
- **Collective thinking** – they pull together a team perspective. Pictures can represent a group perspective that transcends personal differences. When developing a new strategy, drawing out what you're thinking makes it easier to discuss as a team and helps get you all clear and aligned around the same thing. It can help teams get a grip on complex situations, and bring together their collective thinking.

To summarise, using visuals to help create meaning offers many benefits that can complement the more traditional or verbal approaches to the sense-making and communication that are crucial to change. Complexity theory has helped us realise that change can be enabled by creating better, more productive conversations, and visuals are a powerful way to represent information in a way that can focus and stimulate those conversations.

8.13   WHY DIALOGUE?

One of the biggest inspirations for Delta7 when it was being formed was the dialogical approach to leadership and change developed by William Isaacs (1999) and David Kantor (1994). A dialogical approach means that conversations are facilitated in a way that is more meaningful and productive for all participants, not only for those who usually dominate them. Dialogue is a conversation that surfaces new ideas, perceptions, meanings and understanding, which happen when people explore uncertainties and questions that no one person has the answer to. For this to happen it needs to be a safe environment of trust and respect where people can think together for a shared purpose and mutual benefit.

With these ideas in mind, we had a language that helped us observe strategic meetings in organisations, and we found that often these meetings were conversations in which people tended to hold on to and defend positions and attempt to control the content for their own interests, possibly unconsciously, or from positions of status. Of course there needs to be

Figure 8.11   Kantor's four dialogical positions

structure in meetings, yet too much control of the flow of a conversation significantly constrains the level of connectivity, listening, creativity and mutual sense-making. With a more facilitative approach to meetings using dialogue as their model, leaders can create space for conversations that can be more free-flowing, enabling participants to reach a common understanding, experiencing everyone's point of view fully, equally and non-judgementally. This can lead to new and deeper understanding.

Another inspiration for us was David Bohm (1996), for whom dialogue has no prede-fined purpose and no agenda, other than that of inquiring into the movement of thought, and exploring the process of 'thinking together' collectively. This can allow group partici-pants to surface their assumptions and blind spots, as well as to explore the more general movement of thought. By simply exploring and learning together without an agenda or fixed objective there can be space for something new to emerge: new meaning, new ideas, new possibilities and new relationships.

Movers are typically leaders who initiate ideas that they want to put into action. Followers are those who support what movers say and can put it into action. Opposers challenge what is being said and bring a different perspective into the room. Bystanders bring a 'helicopter view' to what is going on. Conversations become dialogue when all four positions are being voiced, and if some of the positions are silent for too long there can be imbalance.

Someone showing dialogical leadership can support the conversation to happen in a more balanced, dynamic way. They can create a safe space for people's genuine voices to be heard. They can listen deeply to what is being said. They can also hold space for – and respect as legitimate – other people's views which oppose their own. They can also keep people focused on what they have all contracted to do, for example, work together and collaborate better towards a shared goal or purpose. When some of these qualities are missing the leader can bring that role themselves or encourage others to do so.

This dialogical approach to meetings, briefings and workshops can be very counter-cultural to many organisations steeped in a command and control culture. Once we train leaders to run Visual Dialogue, they experience the value of learning to really listen and connect to their teams in a completely different and more constructive way.

8.13.1   The Benefits of Dialogue

In our experience of implementing Visual Dialogue in many large organisations, our clients have experienced these benefits from the dialogue component:

Increases business literacy: Dialogue brings together many different perspectives leading to a more sophisticated and mature understanding of the business context and greater understanding of the strategic decisions being made.

Makes time for sense-making: Dialogue creates space for everyone to work out what's going on and what needs to happen.

Gives employees a voice: By offering a safer space in which to speak, Dialogue opens up a channel for feedback, ideas and insights that can be actioned locally or at leadership level.

Helps to break down silos: Dialogue plays a vital role in breaking down silos by increasing levels of understanding between different and hierarchically separate parts of the business, reducing the 'us and them' dynamic.

Evolves a common language: Through dialogue, people create a shared language for the changes the business is going through.

Fosters a meta-perspective: One of the consequences of an iterative dialogue process in which all participants share a common (visual) focus is that employees can understand the strategy and environment. As the scope and content of the Big Picture evolve, with input from all, the dialogue begins to gravitate toward a shared understanding of the larger patterns and forces at play within and beyond the organisation. This is invaluable in terms of organisational learning, problem-solving and capacity-building. The blend of dialogue and visual representation facilitates the implementation of change by supporting people to self-organise around the new ideas and new connections, process feedback together and adapt their behaviours and actions accordingly.

8.14   THE VISUAL DIALOGUE PROCESS

The first step in the Visual Dialogue process is to the senior team who own the strategic plan. Interviews with team members are then done individually and are taken as a chance for individual reflection on the challenges that may be faced in achieving the organisation's outcomes. Following those, a first draft picture is created that captures the story elements. Next this draft picture is tested with a cross-section of the organisation and their feelings, experiences, ideas and feedback are woven into the story and picture. Multiple perspectives are essential to ensuring the credibility of the story and for representing all perspectives. The mixing-up of roles and levels works best because it connects people who may otherwise not get the chance to meet, and to check sense-making across the organisation. This space is created to get people thinking about what the vision for the future would look like or what it would take to get them there, thus ensuring that the gap between employees and leaders is bridged. This feedback is not only about picture changes but also about testing the 'feeling in the room', how employees honestly feel about the strategy and change and whether the picture creates an enabling environment for them to give their input. The final draft of the picture is then agreed and painted in watercolour.

The next stage involves training groups of leaders and managers to facilitate dialogue sessions and share the story with small groups of about 6–8 employees. The trainees learn how to share the story in an engaging way and create a safe space for honest and open discussion. We pay particular attention to listening out for, and visually capturing, the cares and concerns of employees as this makes the picture a shared story and helps people locate themselves in it. Once the initial Big Picture is complete, Visual Dialogue is then typically facilitated during a 90-minute session for 6–9 employees. These sessions are often run throughout an entire organisation so that all employees have an opportunity to talk about the changes in question in a relatively safe environment.

CASE STUDY

This anonymised case study comes from an interview with a HR manager we worked with on a recent change project.

We are a multinational engineering business of over X000 people and have recently been created from a merger of three very different businesses, very different backgrounds and different perspectives. So when we came together it was quite disjointed and it felt like that right through the organisation. We had formed the new business and we didn't know what kind of business we were. People didn't know how they fitted in and they didn't know how the components fitted in. So nobody really understood how it all fitted together. Even the management team didn't understand how it fitted together and they were quite a disconnected team, all with great intentions but moving in different ways. Our first challenge was how can we to get the senior team onto the same page so they're all saying, "this is the direction we want to take" and also how do we cascade that down to employees?

The management team got together and did a fabulous strategy. It was a PowerPoint presentation. It had so many boxes, the whole thing was complex. Unless you were the one who wrote it, you wouldn't understand it. It wasn't hearts and minds stuff. At the first employee brief that I was at, two people fell asleep. We realised we needed something different, then we thought that perhaps big pictures could be a good way to start a dialogue around the new strategy, and ask each other "actually what is this business about?" We saw big pictures as a dialogue tool, primarily. At first people thought it was just a picture. But as we started to develop the picture with leaders and staff in conversation they really liked the whole process. People could personalise what they saw in the picture, personalise their role in the story, they saw in it things that they could relate to and connect with. Also it created space for managers to have a real dialogue with their teams. One employee said that the Visual dialogue session was the first time that a manager had actually had dialogue with him. He'd been with the company a year, yet no one had asked him his thoughts until then. Because the manager had asked his opinion and listened, he was fully involved in that session, he felt like he owned the picture after that. I think the Visual Dialogue process was powerful because we got so many people involved in it. It was powerful because we could get everyone's views into it. It was really valuable because of time and effort that went into it, and there was a lot of thought and real testing of the story. I think even today it still endures because it is a conversation tool. It's about sparking that debate. It's not giving you the answers, it's a space to ask questions and listen to each other.

When we rolled out the Visual Dialogue sessions we shared the story which sparked debate with all our employees. What the process showed me was that people were really interested and actually they did want to know where we're going and it was a tool that gave people space to do that. It captured people's hearts and minds, and actually a lot of those people inputted into those sessions as well, so they saw themselves in the picture. Using the process the management actually heard the employees for the first time. It was the first time we had asked employees, "What's it like to work here?" And people weren't afraid to say what they thought, which was perfect. Also they weren't afraid to challenge each other. So I ran some sessions and one of the sessions that stood out for me was one group in there was saying how bad it was and how they couldn't get anything done in this place, and one group was saying, "Oh, actually, I don't find that an issue". We had a great debate across these two groups about, "So why is your experience so different from this experience?" And I think the guys that could actually make a change sparked ideas in the others. "Is it because my mindset says we can't make change?" I think it tested some of those myths.

As a result many people realised they've got to change. A lot more people now know where we're going, it's something that generates a conversation. It's been very useful as a common tool that anyone can pick up and have that conversation with. It's also been valuable for personal development. For example someone said, "I would have never have usually done the Visual Dialogue training, but because I'm talking about something I know and love and I'm passionate about it".

I'd say it's not an easy journey [the Visual Dialogue process]. It tests you. It really makes you think about what you want. I think it tests all levels of the organisation. It's not just a picture. It gets people talking, no matter where you are. It tests your logic. It tests your reasoning. I think it tests peoples assumptions, because I think in a group you assume that you know, but actually, when you see it in black and white, you realise you don't! I think that was proved when we went through so many iterations, what they thought they were saying, when it's in black and white in front of you as a picture, "Oh, that's not what we're saying". So I think it really helps people and brings clarity about "What I'm really thinking." It brings that all to-life, really, and we use the word 'bringing strategy to life' and I think that is really what it's done for us. You look at the money that we spent on it and the individuals it touched, what programmes do you get that value for money from? We spent £x and we touched thousands of people. How many programmes do that? Everybody had an opportunity to experience and contribute in the story sharing sessions and have a chance to actually share their voice.

8.15   CONCLUSION

We believe there is a pressing need to create more safe spaces in organisations for people to reflect on the overwhelming chaos and confusion they face on a daily basis and make sense of it. Having a space for informal, open and honest conversations is fundamental for sense-making and relationship building which are the foundation for successful change. Used properly, visual representations and dialogue training can be a valuable tool to help leaders learn to create this kind of space. Visual representations of a shared story are a powerful frame for the kind of grounded, specific and owned conversations that are at the heart of any organisational transformation. Visual Dialogue can therefore make a valuable contribution to organisational development by bringing leaders and employees together to turn strategy into action and equipping them with the means to co-create a shared language, a shared context for change, and build relationships. Using the picture as a shared clarification tool, participants from all hierarchical levels within the organisation are able to transcend those barriers that typically hinder relationship building and the improvement of collaboration and coordination need to get from strategy to action.

Visual Dialogue respects the organisation as a complex adaptive system and works to create a space in which the inherent characteristics of emergence, co-evolution and interdependence can be enabled. Using complexity theory has given us a useful model to create an enabling environment that can generate the common ground needed to build the good relationships which are critical for thriving and successful organisations.

REFERENCES

Bohm, D. (1996), *On Dialogue*. London: Routledge.
Isaacs, W. (1999), *Dialogue: The Art of Thinking Together*. New York: Doubleday (Random House).
Kantor, D. and Lonstein, N.H. (1994), 'Reframing Team Relationships: How the Principles of "Structural Dynamics" Can Help Teams Come to Terms with Their Dark Side.' In Peter Senge et al. (eds), *The Fifth Discipline Fieldbook*. New York: Doubleday, pp. 407–416.

9. Inner complexity: using *Landscape of the Mind* to catalyse change in organisations
Kate Hopkinson, Inner Skills Services Ltd

9.1 INTRODUCTION

Most social science research which applies complexity principles is an attempt to get to grips with important aspects of the outside world. There has been much less emphasis on the intrinsic complexity human beings bring to any situation, by reason of being human.

Yet multifariousness and unpredictability of response is a key factor in human social systems in general, and organisations in particular (look how often 'human error' is identified as a cause, when complex systems fail).

Landscape of the Mind (LoM) is a model and a methodology designed to take account of our inner complexity (Shaw and Frost 2015, pp.638–641) and to illuminate how it interacts with the complexity of the world around us. This in turn has significant implications for management at all levels.

9.2 THE CONCEPT OF INNER SKILLS

We begin from the concept of *inner skills*. Every individual brings an extraordinary range of inner gifts and qualities to every situation. These include experience, logic, imagination, intuition, feelings and values. Somehow, we orchestrate all these intangible capabilities (Hailey 2015) into coherent, if sometimes unexpected, action in the world (hence, inner *skills* (see Figure 9.1)).

Generally speaking, everyone is equipped with all inner skills, though we use them differently. On the basis of our research over 30 years, several characteristics of inner skills have become clear:

- There are significant individual, team and corporate differences in preferences.
- Preferences are reflected in performance.
- Some inner skills strategies seem to be much more enabling of some outcomes than others. For instance, navigating successfully in uncertainty, complexity and turbulence seems to depend on using more of some types of inner skills, rather than others.

By preference being reflected in performance, I mean that if you know someone's pattern of inner skills preferences, you can usually make some quite good predictions about where their time, energy and attention will go, and vice versa. This is probably one of our most

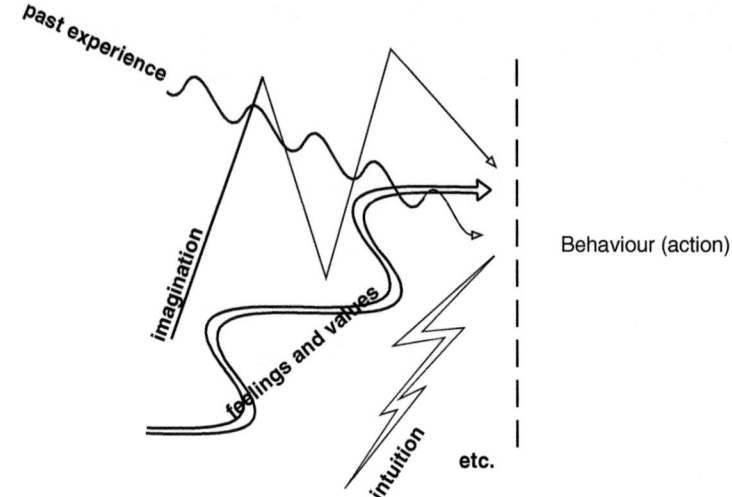

Behaviour (action)

Note: 'Inner skills': how we orchestrate the different aspects of our inner lives - thinking, imagination, knowledge, intuition, feelings, values, perceptions, past experience, etc. - into a process which culminates in effective ACTION.

Source: © 2014 Kate Hopkinson. All Rights Reserved.

Figure 9.1   What are inner skills?

important findings, as it highlights the immense practical consequences which follow from apparently nebulous, invisible inner activity.

What is unusual about LoM, is that, as well as using it to understand people, you can also apply the concepts to work. Different tasks, projects, and jobs require a different mix of inner skills to perform successfully. This doesn't mean there is only one 'right' way to do something – on the contrary – but there are sequences of inner skills which will not get you to where you want to go. LoM is something you do *with people*, not *to* them (Argyris and Schon 1989, pp.612–623), so LoM projects are collaborative explorations, out of which emerge chosen courses of action which have been generated by, and are owned by, the participants. This is facilitated by LoM providing a language and framework for exploring both what they are trying to do, and what, in inner skills terms, they are bringing to the task.

9.3   DIFFERENT KINDS OF INNER SKILLS

Most everyday managerial work in all fields draws mainly on convergent and evaluative inner skills. It involves collecting and organising facts and figures, and drawing logical conclusions based on the evidence. It also involves talking and listening to others, and having a drive to overcome obstacles and achieve outcomes. By exercising these inner skills (and other related ones) appropriately, individuals and organisations can achieve and sustain considerable success. But there is one crucial criterion which must be met: for this kind of strategy to work, the organisation's operating environment needs to be relatively stable. The more turbulent and unpredictable the environment, the less will a combination of convergent and evaluative inner skills, support continuing success. Many highly successful companies have gone out of business doing what they had always done best, because the fitness landscape (Kauffman 1995, pp.166–168, 184, 193, 204) around them had changed. There is widespread recognition of this – hence repeated calls for organisations to become more flexible, agile and resilient.

However, it is one thing to identify the need, and another to change behaviour to meet it. LoM is very good at catalysing this shift in awareness and behaviour. Later in this chapter, I will be presenting a number of case studies illustrating its use, but first we need to look more closely at the model, and then the methodology.

9.4   THE *LANDSCAPE OF THE MIND* (LoM) MODEL

If we are interested in how to help organisations navigate in ambiguity and uncertainty, and shape a future in a continually changing context, we need to understand and mobilise another major type of inner skill, which we have, but largely do not value. Convergent inner skills underpin working with what we already know and understand. Evaluative inner skills enable us to make choices, judgments and decisions. But we also come equipped with a whole range of inner skills for moving away from what we currently know and understand, out into the unknown. We call these *divergent inner skills*[1] (Hudson 1966) – and we now know that they are crucial to navigating successfully in turbulence.

Although we all arrive with these divergent inner skills (it's hard to see how we could accomplish growing up without them), they are not adequately supported by our education system, or the culture at large. So by the time most people are adult, they have learned not to use divergent inner skills, and are often rather out of touch with these inner gifts.

LoM can:

1. Identify individuals and groups who are still in touch with their capacity to diverge from the known.
2. Help individuals and organisations to develop both competence and confidence in doing this, by using divergent inner skills, and
3. Link them to the other dimensions of inner skills which are essential for implementing ideas in a dynamic, constantly co-evolving process.

To capture the universe of possibilities of all the different kinds of inner skills which can be brought to bear, we map them onto a globe. Just as with a geographical globe, by drawing imaginary lines on the LoM globe, we can work out both where we are (in the sense of which inner skills are in play at a particular time), and agree where we want to get to.

Our basic globe is shown in Figure 9.3. As you can see, as well as the distinctions between divergent, convergent and evaluative inner skills, we also make another

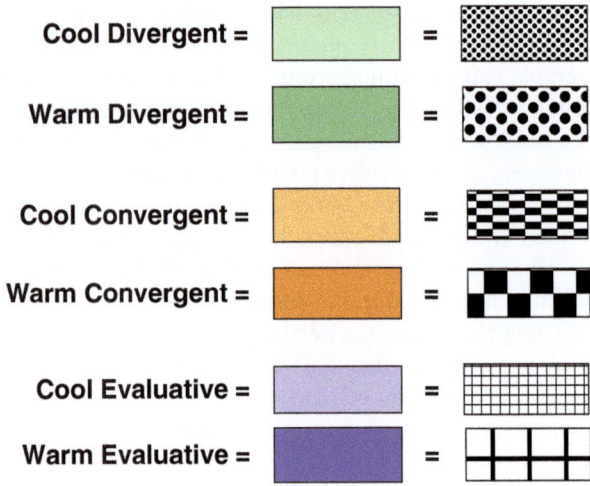

Figure 9.2   Key to the figures

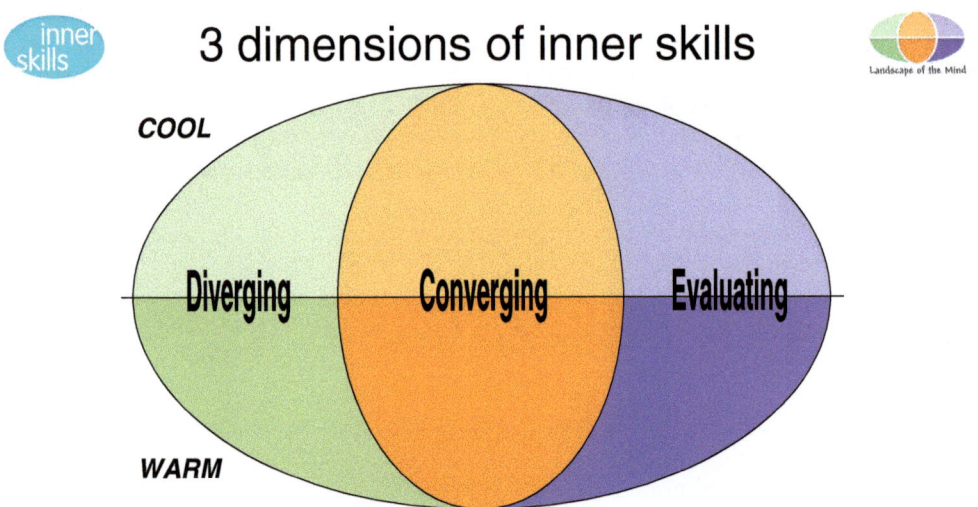

Figure 9.3   Basic First Level globe

distinction, orthogonal to those, between cool and warm inner skills. Cool inner skills (those above the 'equator') are detached, and seem to stand on the outside looking in. Warm inner skills, on the other hand (those below the 'equator') are emotionally engaged and on the inside looking out, as it were (Cox, Hopkinson and Rutter 1981, pp.283–291; Hopkinson, Cox and Rutter 1981, pp.406–415).

To close this section, Figure 9.4 provides another version of the LoM globe, populated with some examples of behaviour which are primarily underpinned by each kind of inner skill. Complex tasks always involve more than one 'colour'/type of inner skill, so this globe is a simplification for familiarisation purposes only.

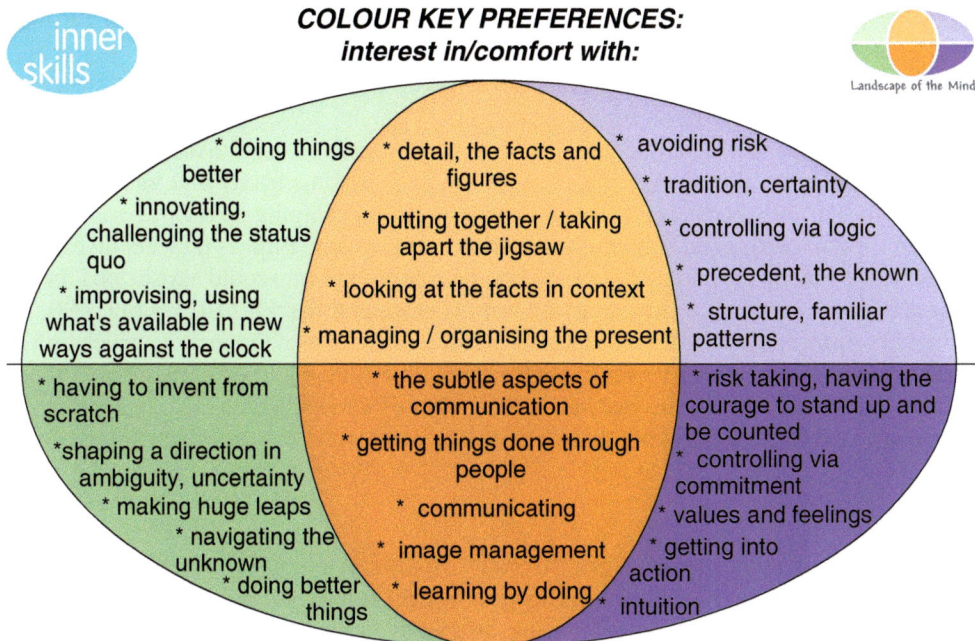

COLOUR KEY PREFERENCES:
interest in/comfort with:

inner skills

Landscape of the Mind

* doing things better
* innovating, challenging the status quo
* improvising, using what's available in new ways against the clock

* detail, the facts and figures
* putting together / taking apart the jigsaw
* looking at the facts in context
* managing / organising the present

* avoiding risk
* tradition, certainty
* controlling via logic
* precedent, the known
* structure, familiar patterns

* having to invent from scratch
*shaping a direction in ambiguity, uncertainty
* making huge leaps
* navigating the unknown
* doing better things

* the subtle aspects of communication
* getting things done through people
* communicating
* image management
* learning by doing

* risk taking, having the courage to stand up and be counted
* controlling via commitment
* values and feelings
* getting into action
* intuition

Figure 9.4   Colour key globe

9.5   THE PROFILING PROCESS

As a consultant working with organisations, it helps to have a simple method for collecting a lot of useful data quickly, without putting heavy demands on the client. The LoM electronic profiling process enables this. Triangulating the results with other sources of information (Keskinen, Aaltonen and Mitleton-Kelly 2003, pp.65–67) also allows us to populate the LoM globe with actual examples from the client's own story and experience. This 'grounds' the 'theory' so it becomes real and relevant. It is less the absolute scores and more the relative scores, the patterns of use in practice, and the choreography with others' contributions which are interesting and helpful to people. This is why feedback is a face-to-face process where the meaning is an emergent aspect of the conversation between participants and consultant. The LoM globe and the profiling results usually make sense intuitively to participants, sometimes after a little time to reflect, particularly when participants themselves begin to give telling examples from their experiences. The LoM framework is then fairly readily adopted and applied, sometimes more skilfully than others, of course.

Having introduced the LoM model, we are moving to black and white figures. Figure 9.5 shows the three First Level LoM profiles. The two profiles on the left-hand side of Figure 9.5 are very common, in the sense that there are many examples of patterns of strong preference for these inner skills on our database. The one on the right, on the other hand, is very unusual: only a small proportion of profiles on the database display the characteristic that divergent inner skills dominate. Crucially, these are the people who navigate very easily and fluently in turbulence and uncertainty, and who come up with sometimes radically novel ideas and solutions to problems (I describe one striking example, Carole, later in the case studies section).

However, they don't fit easily or comfortably in large organisations, and often either jump or are pushed, especially when companies are trying to become leaner and meaner. This results in the curious spectacle of these same organisations ramping up the rhetoric about the need for flexibility, creativity, and innovation – at the same time as they are divesting themselves of exactly the potential trail blazers and role models who could have helped them achieve the changes they say they need and want (Lettice and Parekh 2010, pp.139–158).

First Level profiles provide an introductory, broad brush picture of the territory which LoM covers. There is another much more detailed analysis which puts a particular pattern of inner skills preferences under a microscope, as it were, and brings into focus each separate inner skill within each of the families of inner skills. Figure 9.6 provides an example of a Depth profile.

9.5.1   Depth Profiling

Depth profiling is a very powerful tool, especially with senior executives. From this vantage point you can see the individual inner skills within each 'family'. Interestingly, there can be substantial differences in preference for using inner skills within the same family. Since every Depth profile on the planet is likely to be different, this goes some way to throwing light on the unique flavour which characterises each of us. It also means that the only category which *Landscape of the Mind* ultimately wedges an individual into, is the one with their name on it.

Depth analysis includes exploring which inner skills an individual uses to 'dance' through their profile to achieve their aims (there is a temporal flow from left to right, though it is rarely linear, in reality). Depth profiling also brings into focus gaps, biases and blind spots, thus providing a rich picture of how the use of their inner resources serves them well, and where the inner skills strategies they choose may be tripping them up.

Describing these 'mental muscles' and their interactions in detail would require a chapter in itself, so for our purposes here, we will concentrate on First Level Analysis. This brief overview of Depth analysis is merely included to alert the reader to the level at which *Landscape of the Mind* can engage with inner complexity. Lastly, here are answers to some frequently asked questions about LoM.

Some points of clarification:

- Everyone uses all the colours, over time and in different circumstances.
- The concept of inner skills applies to tasks as well as people.
- The model is fractal-like, in that patterns appear at individual, team, organisational and social eco-system levels.
- Stronger preferences are not necessarily 'better': it depends what you are trying to do, and who you are trying to do it with.

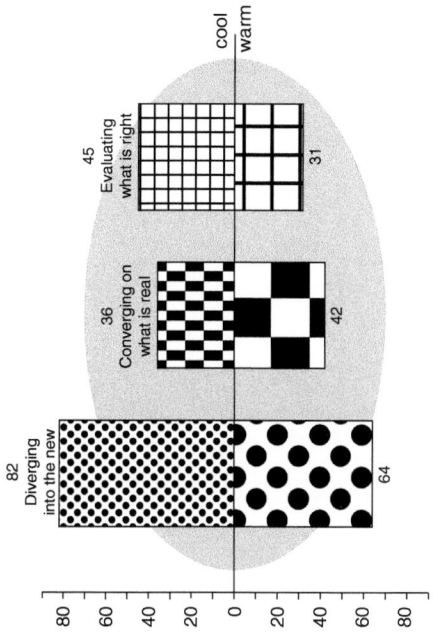

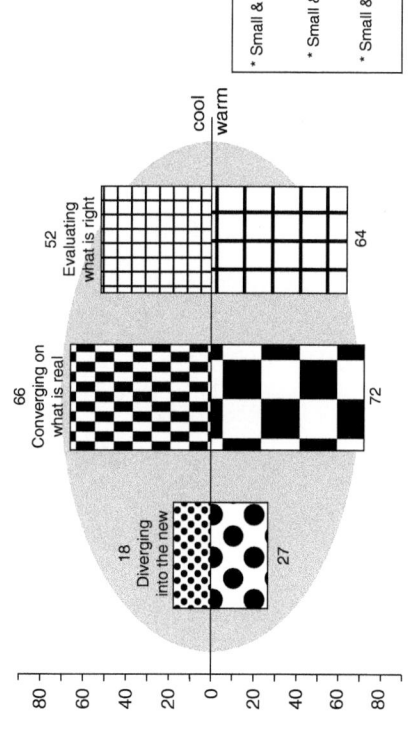

* Small & large CIRCULAR hatching ▓ = DIVERGENT inner skills = light (cool) & dark (warm) GREEN colour in the model

* Small & large CHECKERBOARD hatching ▓ = CONVERGENT inner skills = light (cool) & dark (warm) GOLD colour in the model

* Small & large SQUARE hatching ▦ = EVALUATIVE inner skills = light (cool) & dark (warm) BLUE colour in the model

Source: © 2014 Kate Hopkinson. All Rights Reserved.

Figure 9.5 Three First Level profiles

LANDSCAPE OF THE MIND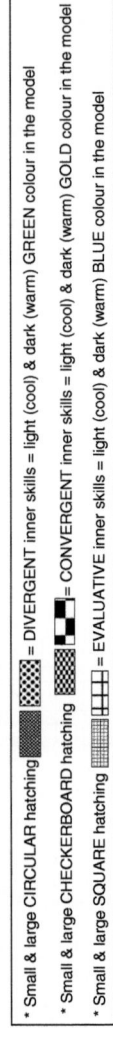

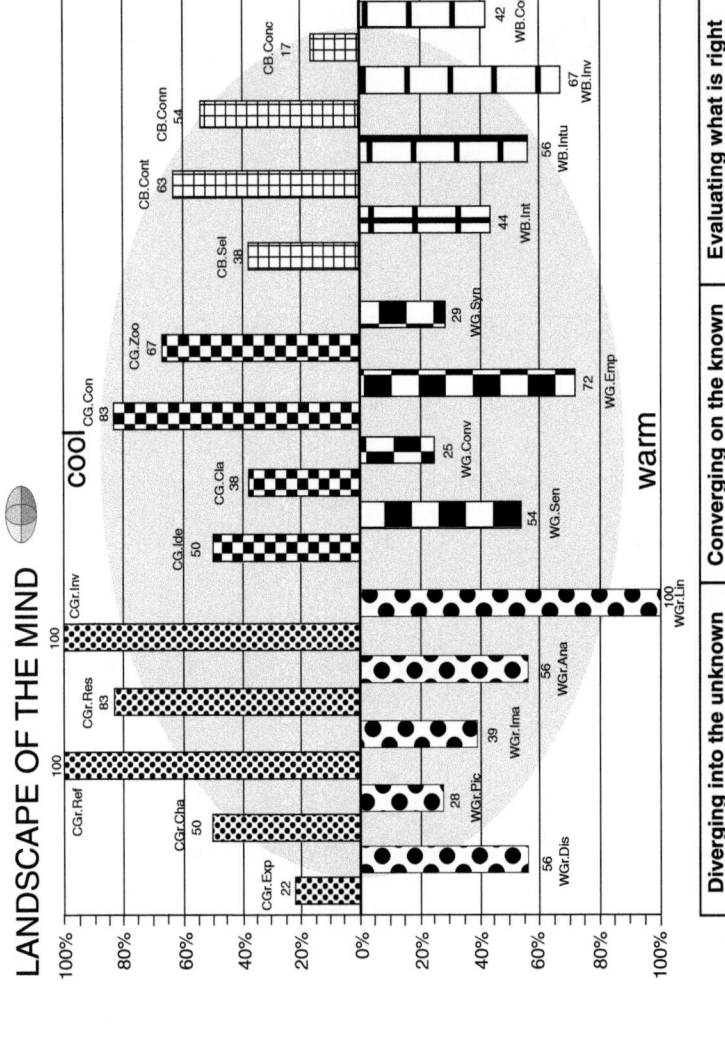

Diverging into the unknown	Converging on the known	Evaluating what is right

Source: © 2014 Kate Hopkinson. All Rights Reserved.

Figure 9.6   Depth profile

164

- Preference is not necessarily the same as competence: many people through practice, have become competent at using inner skills which are not strongly preferred.
- Divergence is not the same as creativity: the use of divergent inner skills is a necessary but not sufficient condition for creativity to occur. Creativity is a complex process – in both senses – which involves all the colours/kinds of inner skills.
- Scores are not comparable across the 3 main LoM dimensions. The database averages are much higher for convergent and evaluative inner skills, than for divergent ones.
- Not all colours/inner skills get on equally well – some have a natural affinity (e.g. cool gold and cool blue), while others have difficulty valuing each other (e.g. cool blue and warm blue).
- People are context sensitive, so will vary the inner skills they use, depending on the context. But this influence is reciprocal, as what individuals do (especially if they are in positions of power) will in turn, change what inner skills others contribute. One way of thinking about organisational culture, is that the 'flavour' of a department or unit, or a whole organisation, is a function of the dominant inner skills preferences in it.
- Colours/kinds of inner skills become relevant in different ways at different stages of a process, so a dynamic 'matching' is needed as the task or project unfolds.

Another recurrent question concerns what changes and what doesn't:

- Our experience suggests patterns of preference for using different kinds of inner skills are fairly stable over time.
- But *behavioural choices* change as a result of the fresh perspective which LoM provides.

For instance, individuals and teams

(i)   Recognise when a task needs their less preferred inner skills.
(ii) Develop competence in using non-preferred inner skills, and in timing contributions.
(iii) Develop tolerance to allow others to contribute inner skills which are not preferred by the observer(s).

Finally, because of the way complex systems work, very small changes in inner skills use can result in substantial changes and practical benefits to the system as a whole.

9.6   *LANDSCAPE OF THE MIND* IN ACTION – THEORY INTO PRACTICE

It has been said that there is nothing so practical as a good theory, but how do LoM's theoretical predictions compare with what we find on the ground? Simply on the basis of the LoM model, it's possible to predict that units and organisations which are primarily focused on delivery (whether of products or services), are likely to have cool gold, cool blue and warm gold as their dominant preferences. In other words, the majority of their time, energy and attention will be devoted to producing whatever they produce, to standard, on time and within budget. This gives us what we call the *Delivery* pattern. 'Improver' groups, on the other hand (that is, units or groups tasked with bringing about innovation and significant change), would be expected to show stronger preferences for the inner skills which underpin those kinds of activities. This gives the *Discovery* pattern. The discovery pattern includes all green inner skills, because these are the ones which enable us to diverge from the status quo; plus warm blue because these inner skills are implicated in having the courage to take personal risks, to stand up for what you believe in, and trial new options.

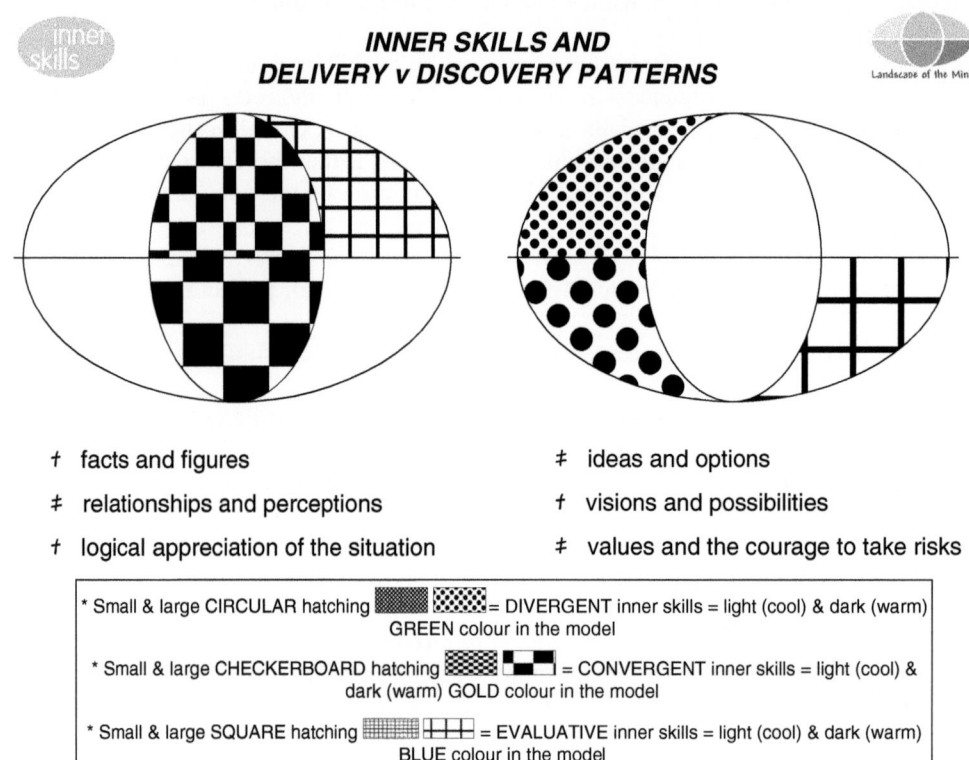

INNER SKILLS AND DELIVERY v DISCOVERY PATTERNS

† facts and figures

‡ relationships and perceptions

† logical appreciation of the situation

‡ ideas and options

† visions and possibilities

‡ values and the courage to take risks

* Small & large CIRCULAR hatching = DIVERGENT inner skills = light (cool) & dark (warm) GREEN colour in the model

* Small & large CHECKERBOARD hatching = CONVERGENT inner skills = light (cool) & dark (warm) GOLD colour in the model

* Small & large SQUARE hatching = EVALUATIVE inner skills = light (cool) & dark (warm) BLUE colour in the model

Source: © 2014 Kate Hopkinson. All Rights Reserved.

Figure 9.7   Delivery versus Discovery patterns

Is this difference reflected in real organisations? Well, it certainly seems to be. Here is an example, taken from the NHS as it is very easy to classify different groups into primarily *deliverers* or *discoverers*.

Figure 9.8 shows the rank ordered preferences for 130 senior managers in Acute Hospital Trusts, with for contrast, the rank ordered preferences for the top 60 managers in the NHS Modernisation Agency (since closed down). The Modernisation Agency was set up to do exactly that – support innovation and modernise the NHS.

There are many other ways of interpreting LoM data, especially at the Depth level,

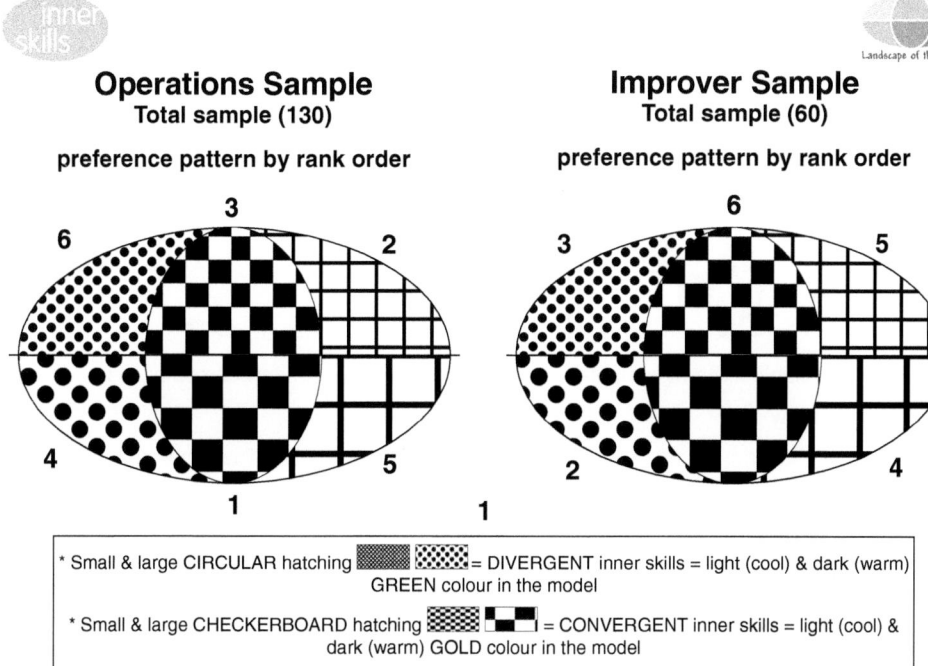

Operations Sample
Total sample (130)

preference pattern by rank order

Improver Sample
Total sample (60)

preference pattern by rank order

* Small & large CIRCULAR hatching = DIVERGENT inner skills = light (cool) & dark (warm) GREEN colour in the model

* Small & large CHECKERBOARD hatching = CONVERGENT inner skills = light (cool) & dark (warm) GOLD colour in the model

* Small & large SQUARE hatching = EVALUATIVE inner skills = light (cool) & dark (warm) BLUE colour in the model

Source: © 2014 Kate Hopkinson. All Rights Reserved.

Figure 9.8 Operational sample (Deliverers) versus Improvers sample (Discoverers)

but rank ordered preferences are useful because they are very accessible (6 = strongest preference 1 = least preferred option). In each case, the improvers sample's preference for the inner skills involved in the discovery pattern, is higher than for their colleagues in the delivery sample. This is only the beginning of illuminating the not always comfortable dynamic between these two groups/tasks, because there are potentially strong tensions inherent in the juxtaposition of the delivery and discovery patterns. Figure 9.9 provides a few indicators of where the gremlins lurk.

The catch is that organisations need to be doing both at once, with the balance progressively shifting further towards allocating more resources to discovery relative to delivery, the more turbulent their operating environment becomes. But once the issues of when and how to actually do this are opened up to constructive practical debate, instead of an endless exchange of conflicting opinions (which themselves reflect different patterns of inner skills preferences), progress can usually begin to be made.

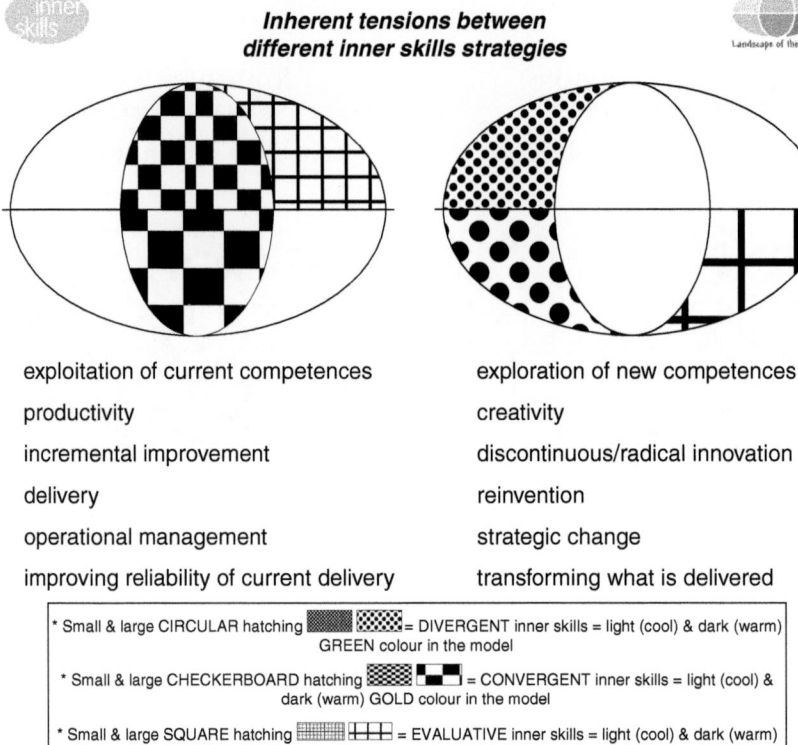

Inherent tensions between different inner skills strategies

exploitation of current competences	exploration of new competences
productivity	creativity
incremental improvement	discontinuous/radical innovation
delivery	reinvention
operational management	strategic change
improving reliability of current delivery	transforming what is delivered

* Small & large CIRCULAR hatching = DIVERGENT inner skills = light (cool) & dark (warm) GREEN colour in the model

* Small & large CHECKERBOARD hatching = CONVERGENT inner skills = light (cool) & dark (warm) GOLD colour in the model

* Small & large SQUARE hatching = EVALUATIVE inner skills = light (cool) & dark (warm) BLUE colour in the model

Figure 9.9   Inherent tensions between different inner skills strategies

CASE STUDIES AND EXAMPLES TO ILLUSTRATE THE APPLICATION OF *LANDSCAPE OF THE MIND*[2]

An easy way to engage participants with LoM findings, is to use rank ordered preferences, mapped onto a globe. Dominant preferences usually 'run the show', tending to take up most of an individual or group's time, energy and attention. Figure 9.10 shows two members of a senior team, who have 'opposite' dominant preferences. We can predict that they may find it hard to work together, even with good will on both sides.

There will be a strong temptation for AN, whose top three preferences are all cool, to regard AW as a time-wasting chatterbox, caught up in emotions and gossip; while AW will likely see AN as cold, aloof, possibly arrogant and certainly not a team player. Being able to look at their differences as a function of their inner skills preferences, and not malice or insensitivity, is a good starting point to explore and discuss which other inner skills sequences they could use, which might improve their working relationship.

Figure 9.11 provides an example where the dominant preferences for a whole top team, are highly discrepant with those of their new CEO.

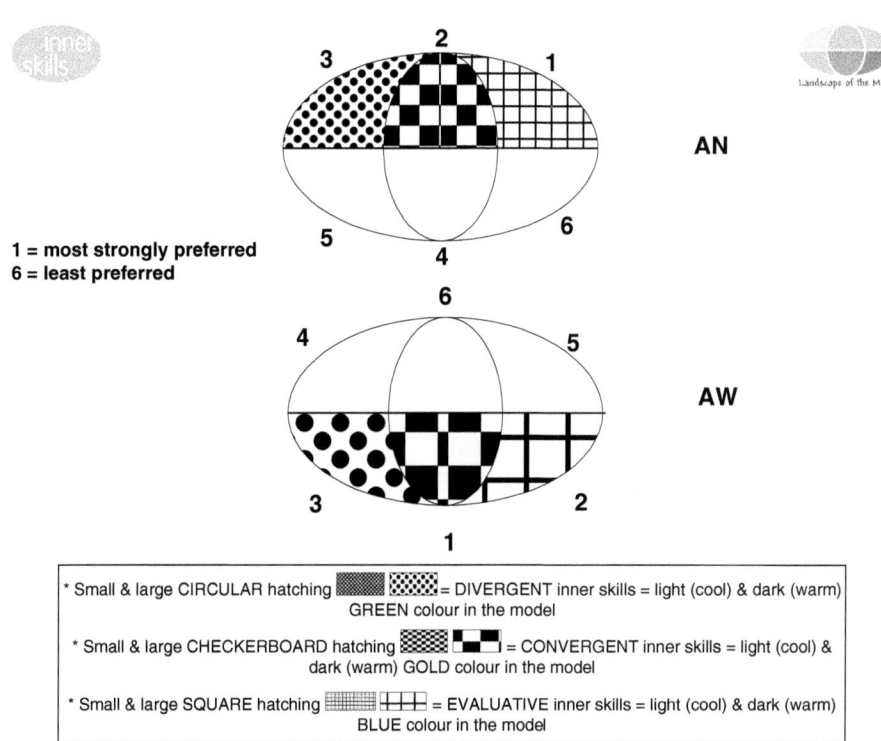

1 = most strongly preferred
6 = least preferred

* Small & large CIRCULAR hatching ▨ ▩ = DIVERGENT inner skills = light (cool) & dark (warm)
GREEN colour in the model

* Small & large CHECKERBOARD hatching ▦ ▰ = CONVERGENT inner skills = light (cool) &
dark (warm) GOLD colour in the model

* Small & large SQUARE hatching ▤ ╫ = EVALUATIVE inner skills = light (cool) & dark (warm)
BLUE colour in the model

Source: © 2014 Kate Hopkinson. All Rights Reserved.

Figure 9.10   Comparison of rank ordered preferences (AN and AW)

This kind of situation is common, when a new CEO has been brought in to break up the old ways of doing things, and bring about radical change. It can have predictable consequences too, when the situation blows up completely and the new CEO leaves. An expensive and time consuming maladaptive walk for all concerned. But this outcome is not inevitable. Once everyone in the situation can see the gulf in LoM terms, they can (and do) set about figuring out how to 'bridge the gap' and work successfully together (even if, as in this case, the team referred to their CEO as 'The Tornado' behind her back).

What about the reverse case? Figure 9.12 shows a senior team and team leader where the dynamics are quite different.

This too can be tricky to manage, but is much more likely to be successful if there can be open and honest discussion of what it is like for participants. The regional centres were a new venture for the organisation, so they needed directors who could innovate, improvise and start something up from scratch. This is reflected in their higher than average scores on divergence. But their boss's two most dominant preferences are both blue (evaluative). Strong 'blue' preferences tend to shut down green divergence, frustrating the divergers, and depriving the organisation of the innovation at the periphery which it needs. This configuration risks the regional centre directors being the direction-finders, with their boss always saying 'no' and blocking needed innovation/variation.

Incidentally, the LoM model is significantly 'no blame'. It is not about the right way or the wrong way; or the people who don't see it as you do, being either mad or bad or both. The relevant questions are about what we're trying to do together, and what kinds of inner skills patterns this will need to support it, to be assessed against what kinds of sequences we're currently using.

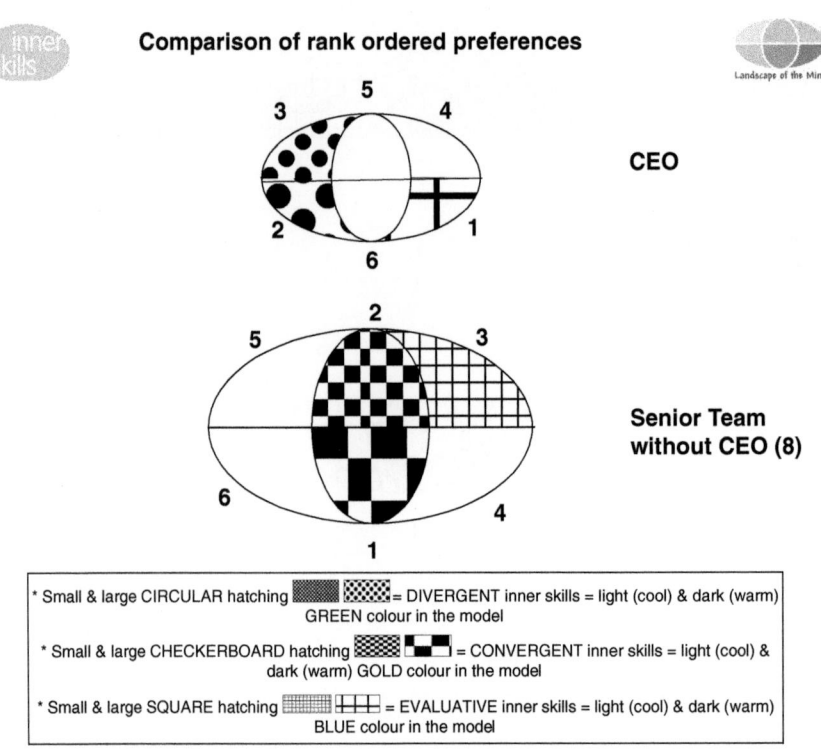

Figure 9.11   The new broom

Inner Skills Patterns Across Management Levels

Figure 9.13 shows a slightly more complicated example, looking at the top three levels of a world class commercial manufacturing company. This organisation, although it was at pole position internationally in its field, recognised that simply carrying on doing what it did best was not going to save it, in a rapidly changing fitness landscape. So there was good theoretical recognition of the need for increased agility, flexibility, and innovation. Practically everyone was talking the talk. But when we looked at the dominant patterns of preference for each of the top three levels of management, Figure 9.13 shows what we found.

At each level, the pattern was different; but for all three levels, diverging from what they were currently doing was not getting a lot of time, energy or attention, in spite of all the aspirational rhetoric swirling around. This raised serious questions about whether this group were well placed to lead the kind of fundamental changes which they recognised intellectually they needed to design and implement. This in turn led to the incorporation of LoM workshops and profiling in their development activities, both for senior executives and also for young high flyers.

Working with the Less-than-Thrilled

As already mentioned, LoM is an approach which you do *with* people, not *to* people. So how does it work if the prospective participants are not keen? Here is a case study about a global, world class, household-name company and their use of LoM to address a strategic issue they faced. The head of their IT

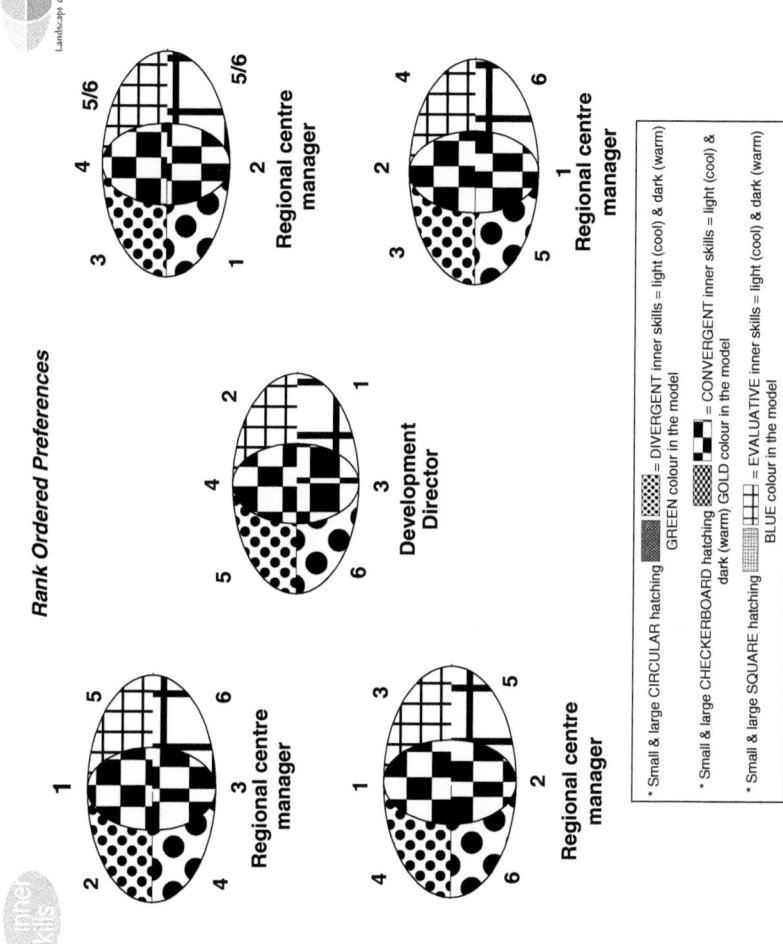

Rank Ordered Preferences

Regional centre manager

Development Director

Regional centre manager

Regional centre manager

* Small & large CIRCULAR hatching = DIVERGENT inner skills = light (cool) & dark (warm) GREEN colour in the model

* Small & large CHECKERBOARD hatching = CONVERGENT inner skills = light (cool) & dark (warm) GOLD colour in the model

* Small & large SQUARE hatching = EVALUATIVE inner skills = light (cool) & dark (warm) BLUE colour in the model

Source: © 2014 Kate Hopkinson. All Rights Reserved.

Figure 9.12   Development Director and regional centre managers – national not-for-profit organisation

Rank ordered preference aggregated by team

Most preferred by level

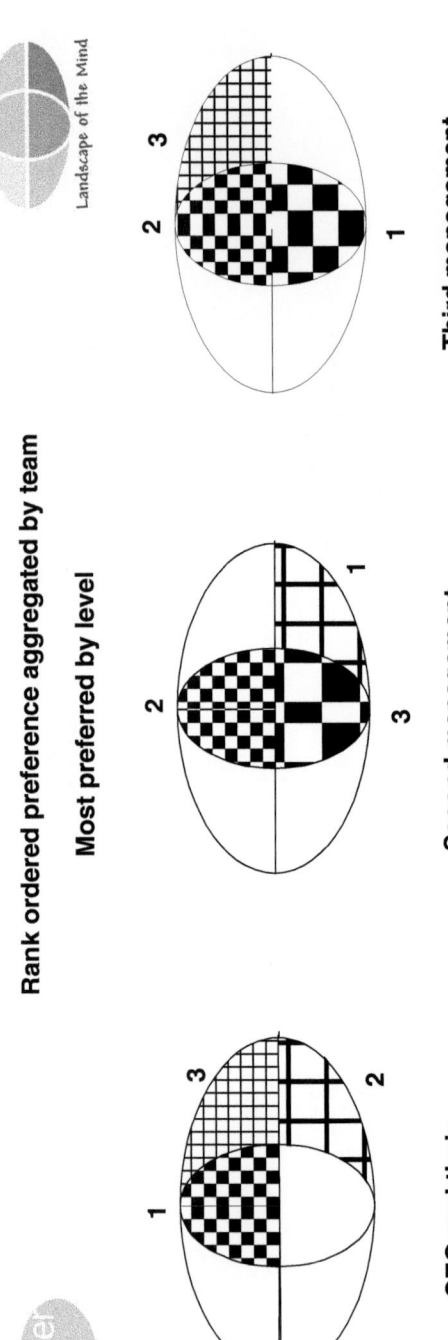

CEO and their direct reports

Second management level

Third management level

Landscape of the Mind

inner skills

* Small & large CIRCULAR hatching ▨ = DIVERGENT inner skills = light (cool) & dark (warm) GREEN colour in the model

* Small & large CHECKERBOARD hatching ▨ = CONVERGENT inner skills = light (cool) & dark (warm) GOLD colour in the model

* Small & large SQUARE hatching ⊞ = EVALUATIVE inner skills = light (cool) & dark (warm) BLUE colour in the model

Figure 9.13   Three levels of leadership

function, Paul, was concerned that this side of the business risked being out sourced. He saw this as having serious unintended consequences for the company, as well as for the individuals concerned.

He then took an unusually fresh, thoughtful look at how his IT experts added value to the company (his ability to do this is reflected in his own LoM profile). They themselves assumed their value lay in their technical skills and leading edge software know-how. But actually, their irreplaceable contribution lay in the quality of the relationships they built with their internal clients, so that they became trusted advisors to line managers, not just 'fix-it' merchants and fire fighters when company systems went down. And this was what would be lost if the IT function was out sourced, so the challenge was to strengthen their capability to initiate and sustain these high quality relationships, where they really understood the needs of the business and were seen to, by their internal customers.

Many 'techies' have difficulty seeing the value of what they regard as this touchy feely stuff, and definitely don't recognise it as fundamental to their work. In fact, we were told that if we even used the word 'relationships' in the title of the project, we'd be dead in the water. It also became clear that, in some cases because of less than good experiences in the past, they would be very resistant to individual profiling.

So at the front end, the project was described as being about 'network development', and built around jointly exploring the aggregate LoM data, and mapping that onto their work and how it was evolving. Meanwhile, the scaffolding for the whole project was four goals which had already been decided and which they had to meet anyway – so it wasn't 'extra' on top of what was already going on, but an enabler for current commitments. This noticeably increased participants' motivation to take part. In the first workshop, we presented the aggregate findings regarding LoM preferences for the whole department (Figure 9.14).

Comparison of rank ordered preferences
Dept. without MT (48)

Landscape of the Mind

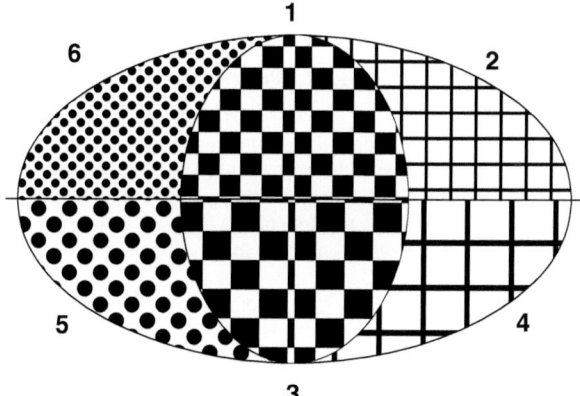

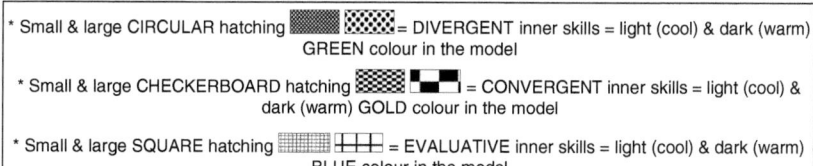

* Small & large CIRCULAR hatching ▨ ▦ = DIVERGENT inner skills = light (cool) & dark (warm) GREEN colour in the model

* Small & large CHECKERBOARD hatching ▦ ◼ = CONVERGENT inner skills = light (cool) & dark (warm) GOLD colour in the model

* Small & large SQUARE hatching ▦ ╬ = EVALUATIVE inner skills = light (cool) & dark (warm) BLUE colour in the model

Figure 9.14   Whole group (48 people) without Management Team and Head of Department

Comparison of rank ordered preferences
Management Team without H of D (5)

Landscape of the Mind

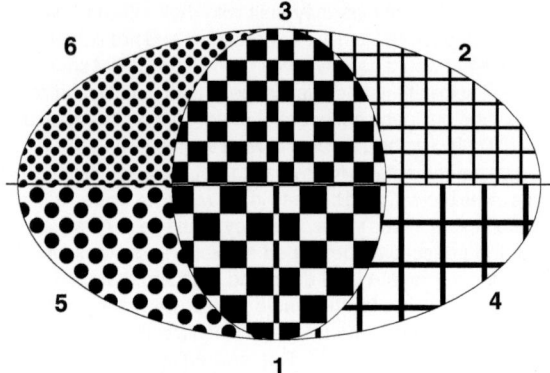

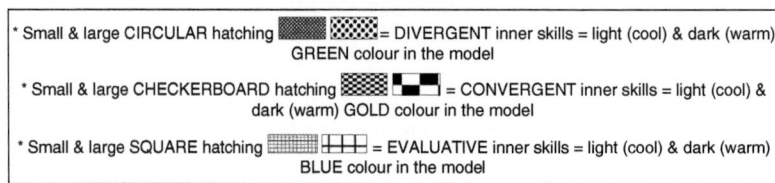

* Small & large CIRCULAR hatching [◼◼][▨▨] = DIVERGENT inner skills = light (cool) & dark (warm) GREEN colour in the model

* Small & large CHECKERBOARD hatching [▨▨][▣▣] = CONVERGENT inner skills = light (cool) & dark (warm) GOLD colour in the model

* Small & large SQUARE hatching [▦▦][╫╫] = EVALUATIVE inner skills = light (cool) & dark (warm) BLUE colour in the model

Figure 9.15   IT Management Team (5 people) without Head of Department

This exactly mirrored predictions, based on the demands of their jobs. But we also showed aggregate figures for the management group in the department – which were a bit different (Figure 9.15).

The Head of Department had specifically chosen this group for their higher loading on warm gold (the inner skills particularly implicated in relationships). This was simply presented as information, but since they all knew this group, it wasn't hard for them to see the relevance, and the difference which this altered weighting made.

The final plank in the initial sub-project to gain credibility for the LoM model, was, with his permission, to present the individual profile results for the Head of Department. Again with his prior consent, this was introduced with the observation "so if you always thought Paul was from another planet, you were absolutely right!" (Figure 9.16).

Since they had indeed thought this, as we unpacked what these differences meant on the ground (Paul was present in the workshop), while there was a good deal of laughter, there was also a dawning recognition that the methodology could accurately surface important but intangible issues – as well as providing a language and a framework for discussing them. And it could tell them something useful as well as intriguing, such as how to manage their boss better.

It was at this point that the participants insisted they wanted their own individual profiles – which then had to be built in to the project. Finally, the *before* and *after* measures by an independent consultancy, did indeed show significant improvement on exactly the relationship-building skills which the project had been set up to support. (Hopkinson 2015, part 2).

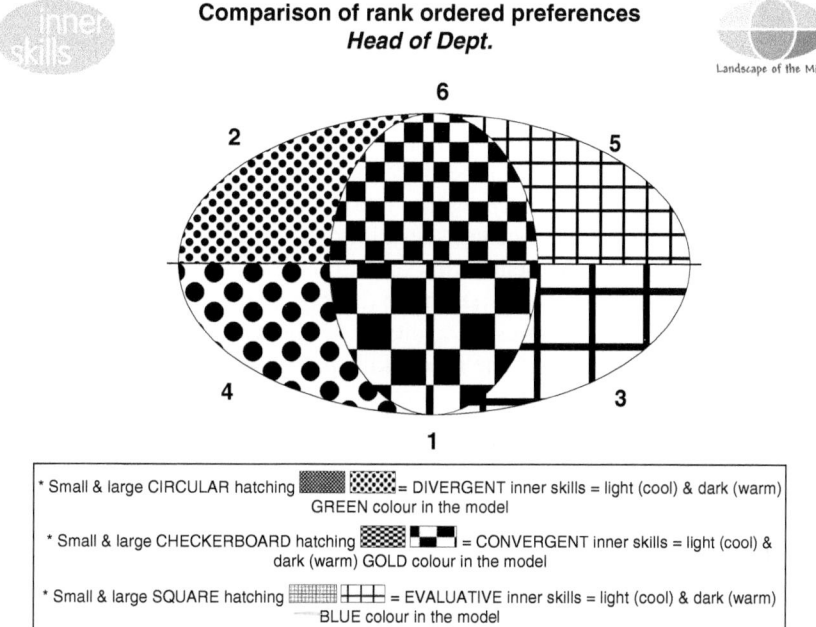

Comparison of rank ordered preferences
Head of Dept.

Landscape of the Mind

* Small & large CIRCULAR hatching ▓▓ ▒▒ = DIVERGENT inner skills = light (cool) & dark (warm) GREEN colour in the model

* Small & large CHECKERBOARD hatching ▓ ◼ = CONVERGENT inner skills = light (cool) & dark (warm) GOLD colour in the model

* Small & large SQUARE hatching ▦ ╫ = EVALUATIVE inner skills = light (cool) & dark (warm) BLUE colour in the model

Source: © 2014 Kate Hopkinson. All Rights Reserved.

Figure 9.16   Head of Department's rank ordered preferences

Serendipitous Research Findings

Another characteristic of the LoM approach is that it can turn up completely unexpected phenomena. Not just interesting oddities, but findings which may turn out to have direct consequences for a company's bottom line. Here is one such case study. In the course of profiling a group of about 60 people whose dominant preferences were strongly cool gold and cool blue (convergent and evaluative inner skills) we noticed a few outliers, who had more green (divergence) in their profiles than the majority – and one in particular, Carole, who had a significant loading on divergence.

Although not covered by the project, I arranged to give Carole feedback on her profile. This was personally life-changing, as it not uncommonly is. Since the management team in her department was familiar with the LoM model, I asked Carole if she was willing for me to show them her profile. She agreed. The reaction of the management team was interesting: they immediately recognised that here was someone with highly discrepant preferences to most of their staff, and someone with something distinctive and important to contribute – but they had no idea what to do with her.

This again, is common, even when the air is thick with exhortations about the need for flexibility and innovation. I was able to suggest that they take her off the very run-of-the-mill stuff she was currently doing; give her a project *no one* knew how to do; and resist the temptation to micro-manage her while she tackled it. So no fortnightly review meetings, where she would have to account for how she'd spent her time, and which sub-targets she'd hit. When she thought she'd found something interesting, the idea was that she should put herself on the agenda for a management team meeting, and come and tell them about it. That's what they did. Within a matter of weeks, she had saved the company £1/4million.

The reaping of a substantial *divergence dividend* is not peculiar to this project. On the contrary, it is a potential source of advantage which is present in most circumstances – just unrecognised.

9.7   CONCLUSIONS

What we see repeatedly in our work is people, individually and collectively, largely trapped inside their inner skills preferences, whether or not these reflect the real needs of the situation. Using LoM is unlikely to change their preferences, but it often changes the behavioural choices they make: they start to become 'inner skills globetrotters', moving fluently and appropriately around the LoM globe, depending on task needs.

In the case of improving flexibility, agility, innovation and resilience, how might this help? Well, it moves us beyond the generic descriptions, to look more carefully at what is happening in a specific case, and hence what directions of movement might be beneficial *for this organisation*, at this point in time, and in this operating environment. The enquiry is thus grounded practically in present realities. At the same time fresh and relevant adaptive walks are signposted, and the inner skills patterns which would support them, identified.

Thus without relying on simplistic 'answers', LoM provides both a conceptual and practical enabling environment for continuing co-evolution into a turbulent and unpredictable future (Mitleton-Kelly 2011, pp.21–44) – a future perhaps a little less threatening when viewed through the lens of LoM. There is much further research to be done on the LoM methodology and its potential. I hope this chapter will stimulate interest in doing it.

For those readers interested in learning more about the methodology, there is a film about *Landscape of the Mind* which can be streamed free from www.innerskills.co.uk and also from the LSE website (Hopkinson 2015). The film is in three parts. The first part introduces the concepts and the model (including a small pilot study we carried out using fMRI brain imaging, with the Institute of Cognitive Neuroscience). The second part offers a wide variety of case studies of real projects. And the third part consists of interviews with users, about their experiences of LoM.

Of course, it is not just organisations which are struggling with the turbulence of our times. So are governments and the international community. We have used LoM as a lens to make sense of social ecosystems, and because it is visual and accessible, perhaps it could in future make a small contribution to navigating successfully through the uncertainties which face us all.

ACKNOWLEDGEMENTS

I would like to acknowledge the help of the following people in the preparation of this chapter: Suzanne Bramham, Simon Carruth, Billie Edwards, Debbie Fresco, Duncan Frowde and Alexandra Hopkinson.

NOTES

1. I do not use the concept of divergence in the same way as Hudson, but I acknowledge his influence on my thinking.
2. Some case study details have been changed to protect participants but all are based on real projects.

REFERENCES

Argyris, C. and D.A. Schon (1989), 'Participatory Action Research and Action Science Compared. A Commentary: The Dilemma of Rigor or Relevance', *American Behavioral Scientist.* **32** (5), 612–623.

Cox, A., K. Hopkinson and M. Rutter (1981), 'Psychiatric Interviewing Techniques II. Naturalistic Study: Eliciting Factual Information', *The British Journal of Psychiatry*, **138** (4), 283–291.

Hailey, J. (2015), 'Rethinking NGO Leadership: Strategic Capabilities for the New Future', paper presented at an IMA International Event, New Economics Foundation, 24 June.

Hopkinson, K. (2015), 'Working with Inner Complexity', film accessed 31 July 2015 at www.innerskills.co.uk/film-introduction/.

Hopkinson, K., A. Cox and M. Rutter (1981), 'Psychiatric Interviewing Techniques. III. Naturalistic Study: Eliciting Feelings', *The British Journal of Psychiatry*, **138** (5) 406–415.

Hudson, L. (1966), *Contrary Imaginations: A Psychological Study of the English Schoolboy*. London: Methuen.

Kauffman, S. (1995), *At Home in the Universe: The Search for Laws of Self-Organization and Complexity*. New York: Oxford University Press.

Keskinen, A., M. Aaltonen and E. Mitleton-Kelly (2003), *Organisational Complexity*. Helsinki, Finland: Finland Futures Research Centre.

Lettice, F. and M. Parekh (2010), 'The Social Innovation Process: Themes, Challenges and Implications for Practice', *International Journal of Technology Management*, **51** (1), 139–158.

Mitleton-Kelly, E. (2011), 'Identifying the Multi-Dimensional Problem-space and Co-creating an Enabling Environment', in Andrew Tait and Kurt A. Richardson (eds), *Moving Forward with Complexity. Proceedings of the 1st International Workshop on Complex Systems Thinking and Real World Applications*. AZ, USA: Emergent Publications, pp. 21–44.

Shaw, R. and N. Frost (2015), 'Breaking out of the Silo Mentality', *The Psychologist*, **28** (8), 638–641.

10. On the visualization of dynamic structure: understanding the distinction between static and dynamic network topology

Dr Kurt A. Richardson, Emergent Publications and Andrew Tait, Decision Mechanics

10.1 INTRODUCTION

Understanding the structure of complex networks and uncovering the properties of their constituents has been, for many decades, at the center of study of several fundamental sciences, such as discrete mathematics and graph theory. In the past decade there has been an explosion of interest in complex network data, especially in the fields of biological and social networks. Given the large scale and interconnected nature of these types of networks there is a need for tools (both mathematical and computational) that enable us to make sense of these massive structures.

The field of network theory/analysis has emerged in the last decade or so, offering tools that allow analysts to begin to comprehend, and therefore intervene in, complex networks.

Social Network Analysis (SNA) (Borgatti, Everett, and Johnson 2013) has grabbed much of the headlines (for example, *Economist* 2015) – in no small part due to the growth of social media. However, SNA focuses primarily on the extraction and analysis of *static* networks. This limits the understanding of how networks evolve or how the structure of the network shapes behavior. Dynamic network analysis (DNA) is essential if we wish to gain comprehensive access to the secrets networks hold. A National Academy of Sciences workshop on SNA for building community disaster resilience concluded that "Building resiliency into social networks requires an understanding of the dynamic nature of networks and of how positive changes that prevent network failure during a disaster may be promoted" (Magsino 2009, p. 50).

> Dynamic network analysis (DNA) is an emergent scientific field that brings together traditional social network analysis (SNA), link analysis (LA), social simulation and multi-agent systems (MAS) within network science and network theory. There are two aspects of this field. The first is the statistical analysis of DNA data. The second is the utilization of simulation to address issues of network dynamics. (https://en.wikipedia.org/wiki/Dynamic_network_analysis, accessed August 7, 2017)

Statistical analysis of static networks is by far the more mature of these two aspects of DNA. In this chapter we focus more on the simulation of dynamic networks. In so doing, we illustrate the distinction between static network topology and the actual active network structure that emerges when a dynamic nonlinear network follows a range of different attractors. In this way we illustrate, for a fairly simple Small World network, that when it comes to designing network interventions, just seeing who or what is connected to who or what is insufficient to obtain a complete understanding of the network.

This chapter will explore network dynamics using a simple Boolean network model. Using this framework, the concepts of information barriers and emergent effective dynamic structure will be explicated. We will also consider the concept of 'network compressibility' as an approach to simplifying networks and examine the extent to which it can be applied to Boolean networks. We also present some initial speculations as to how this work might inform the analysis of real-world dynamic networks.

10.2   BOOLEAN NETWORK ANALYSIS

As mentioned above, a major limitation of traditional network analysis tools (such as SNA) is that they focus primarily on the static topology (structure) of the network of interest. There is little to no accounting of how information flows around the network *and* how information is processed by the network nodes. Boolean networks are dynamic, and thus can be used to provide insights into how the flow of information and its transformation impacts the effective (dynamic) network topology.

As a trivial example, imagine a fully interconnected network where each node is connected, in both directions, to every other node. Additionally, each node transforms incoming information by ignoring it, that is, the node contains a set of rules whose net result is that none of the incoming information affects the node's state; the node's state remains the same regardless of incoming signals. From a conventional network/graph-based analysis, the network would appear to have many hundreds/thousands of feedback loops – such a high level of connectivity itself is of interest. However, from dynamical considerations we would see that this network is really very dull. In the language of nonlinear dynamics theory, this network has many (equal to the number of possible network states, which for a Boolean network is 2^N, where N is the number of nodes) point attractors which completely define its qualitative dynamics; basically the network's state does not change from the initial conditions used to 'seed' it – not very interesting at all. Although this example is trivial, it does highlight in an extreme way how excluding dynamical considerations can lead to very misleading conclusions about the significance of the connectedness of some networks.

It seems self-evident that static networks are going to provide a limited perspective on most social systems. Companies, for instance, are swirling masses of blossoming and decaying relationships (edges).

The aim of this section is to consider how information flows determine effective network structure. We won't go into much detail here about how a formal Boolean Network is defined. The interested reader is directed to Wikipedia (https://en.wikipedia.org/wiki/Boolean_network). In the opinion of the authors Boolean network analysis provides a relatively simple tool for exploring and developing the fundamental 'physics' of complexity theory.

A Boolean network is simply a set of connected nodes. Each node can adopt one of two states: 0 or 1 (hence 'Boolean'), and has associated with it a rule that relates the next state of the node from the current states of the nodes that are connected to that node. Connections in Boolean networks have direction. If A is connected to B this means that

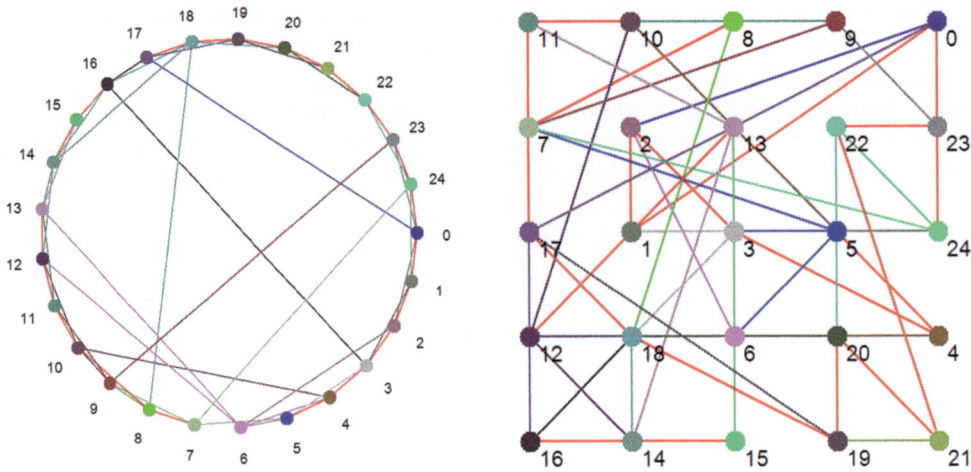

*Note: The red connections indicate that the nodes are connected in both directions, otherwise the color of an
edge is determined by the color of the node it comes out of.*

Figure 10.1   Boolean 'Small World' network used for this study

information can flow directly from A to B. It does not follow that information can flow
directly from B to A, unless B is also connected to A. This does not exclude the possibility
that B is part of a feedback loop that eventually feeds-back to A. Figures 10.1a and 10.1b
show pictorially the Boolean network that is used throughout this study. The two figures
represent the same network, just drawn differently.

For Figure 10.1b, the nodes were randomly allotted to a point on a simple grid.
The total distance between all the nodes was then calculated; literally the length of
all the connections was totaled. The nodes were then reallocated and the calculation rerun. This process was repeated 100,000 times, and the layout with the shortest
total distance was deemed the most efficient one. This is the layout shown in Figure
10.1b.

Figure 10.1 just illustrates the network's static topology. Table 10.1 shows the same
network in tabular form, but also includes the all-important Boolean rule that is associated with each node, and which is used to process the information (the collection of states)
from incoming connections. We can see here that the number of incoming connections to
any one node ranges from 1 (nodes 6, 7, and 14) to 5 (node 12).

To understand how the Rule determines a node's next state, let's consider node 10,
which has two inputs (or neighbors – nodes 8 and 11), whose states are combined via the
Rule 0xB (hex for 1011 in binary). Table 10.2 shows how different combinations of the
node 10's inputs are processed.

The rule value represents the outputs of a truth table where the inputs are all the possible combinations of the input nodes. The input columns, in this case, are ordered by
node number and the rows are ordered according to the binary number that the inputs
represent – that is, 00, 01, 10, 11.

Table 10.1 Lists the Boolean rule and set of incoming connections from other nodes for each node in the network

Node	Rule	Number of Incoming Connections	Incoming Connections				
0	6A96	4	24	1	2	23	
1	1E	3	24	0	2		
2	5	2	1	3			
3	DE36	4	1	2	4	16	
4	5A35	4	3	5	6	10	
5	B493	4	3	4	6	7	
6	2	1	2				
7	2	1	8				
8	5B	3	7	10	18		
9	A6	3	7	8	10		
10	B	2	8	11			
11	F0	3	9	10	12		
12	54FD75C8	5	11	13	14	10	6
13	E6CA	4	11	12	14	6	
14	2	1	15				
15	D3	3	13	14	16		
16	38	3	14	15	18		
17	99	3	16	18	0		
18	81	3	17	19	14		
19	A6	3	17	18	20		
20	2D	3	18	19	21		
21	4D	3	19	20	22		
22	23	3	20	21	23		
23	87	3	22	24	9		
24	6B8E	4	22	23	0	7	

Table 10.2 Rule table for node 10

Node 8 state	Node 11 state	Resulting Node 10 state
0	0	**1**
0	1	**0**
1	0	**1**
1	1	**1**

Note: The rows of the third column combined are 1011, or 0xB. This is the Boolean rule for node 10.

10.2.1   Network Structure

Before looking at the dynamics of this network in detail there are number of measurements we can calculate that characterize the network's static topology. Table 10.3 list a number of important metrics that characterize this network.

Table 10.3   Topological metrics for the network of interest

Number of nodes	25
Number of edges	75
Average outdegree	3
Average indegree	3
Number of structural feedback loops	2,345
Average feedback loop length	14.38
Longest structural feedback loop	22
Number of qualitatively different structural feedback loops	21
Average clustering	0.27
Diameter	8
Average path length	3.55
Reciprocity	0.125

Most of these metrics will be very familiar to those studying networks. At the simplest level we have a network containing 25 nodes and 75 connections between those nodes. The network has an average outdegree (outgoing connections from any given node) of 3 and an average indegree (incoming connections to any given node) of 3. We can also see that this network can be best described as a 'Small World' as the average path length (the average of the shortest distance between any two nodes) is only 3.55 – compared with a diameter (shortest distance between two most distant nodes) of 8. The number we'd like to highlight is the number of structural feedback loops: 2,345 – a surprisingly high number perhaps to those not familiar with the challenges of understanding network dynamics. Even relatively simple networks, such as this, can exhibit very complex dynamic behavior.

Most readers will be familiar with the three-body problem in classical mechanics (https://en.wikipedia.org/wiki/Three-body_problem). Essentially, the nonlinear mathematics for a gravitational system containing only two bodies is relatively trivial; although nonlinear, it is *analytically* solvable. Adding just one more body, therefore creating a three-body system, changes all this and all of a sudden the mathematics is far from trivial and the system's dynamics becomes chaotic (in the formal mathematical sense) and qualitatively rich. As you'd expect, the problem is exacerbated as more bodies are added to the system. In short, the three-body problem is an example of the challenges of predicting the 'paths' of systems containing more than two non-linearly connected bodies/nodes. The challenge of prediction for chaotic systems is well-known.

The reason we bring this up is simply to illustrate the challenge of understanding the dynamics of even simple dynamic networks. The three-body problem is a three-node network containing only four feedback loops; the network of interest here has 2345 – what chance do we have of understanding this? For completeness the full structural feedback profile of the network above can be written as:

20P2, 16P3, 7P4, 4P5, 10P6, 29P7, 40P8, 51P9, 61P10, 75P11, 137P12, 254P13, 366P14, 401P15, 340P16, 243P17, 153P18, 82P19, 38P20, 15P21, 3P22.

'20P2' says that there are 20 structural feedback loops with length 2.

As the discussion above suggests, studying the dynamics of even relatively simple dynamic networks is an enormous challenge. What we need are tools and methods for somehow reducing/compressing the complexity and yielding reliable insights about the whole by studying subsets of that whole. Seems reasonable, until we add into the mix the observation that complex systems are themselves (by definition) irreducible (Cilliers 1998).

Taken literally, Cilliers's observation would suggest that there is little we can do to overcome the fundamental limitations of simplifying complexity – bearing in mind that Science itself is really nothing more than constructing and testing simplifications of observable reality, we seem to have reached a point where Science itself cannot proceed (in the sense of discovering 'absolute' truths). However, it has been shown (for example, Richardson 2005) that reliable and robust, and even insightful, simplifications of complex systems can indeed be extracted, as long as we remember that any conclusions drawn from those simplifications must be applied with care; all conclusions/insights must be regarded as context dependent and provisional.

Returning to our Boolean network, our next step is to explore the *dynamics* of our network of interest. The Boolean network analyses were conducted using a tool developed by one of the authors over a number of years that runs on Microsoft's .NET framework.[1]

Table 10.1 shows the rules used by each node to process incoming information (other node states). By randomly seeding each node with 1 or 0 at *time* = 0, the network is simulated forward in time (setting *time* = *time* + 1 at each step) by applying the rule for each node as illustrated in Table 10.2. If the macroscopic network state is taken to simply be the result of concatenating the states for each of the nodes, we can easily visualize the time-dependent (space-time) behavior of the network. Figure 10.2 is an example of the time-dependent behavior of this network given a random initial condition.

The initial state, the first row of pixels, is shown in red. This is the state of the network at *time* = 0. Each of the subsequent rows shows the state at the next epoch. So, for example, node 0 switches from 0 to 1 at *time* = 2.

Given that the state space of Boolean networks is discrete and finite (2^{25} = 33,554,432 states in the case of a 25 node Boolean network) the network will eventually get back to a state it has visited before, and the network's behavior is said to have reached an attractor on which its behavior will repeat forevermore (unless perturbed externally).

The dynamic (phase) space, as opposed to the topological (structural) space, of any given (Boolean) network is characterized by a number of dynamic cycles, or attractors, that is, all possible network states can be mapped onto an attractor in phase space. For this particular network we find that there are 2p3, 10p4, 3p6, 3p12, 1p18, 1p42 attractors. This expression is read much the same way as the structural feedback loop expressions given above, for example, 2p3 represents 2 phase space attractors with period-3.

Figure 10.3 shows six qualitatively different phase space attractors for the network of interest, and their respective state-time diagrams. At the center of each attractor is the attractor basin, and connected to each basin state are transient branches that show how the network progresses from any state to eventually 'fall' into one of the basins.

The number and type of attractors is a way of visualizing and characterizing the 'function' of the network. For example, in biological regulatory networks the genotype is the structural network itself (and the associated node rules), and the phenotype is the various

Note: Node 0 is the right-most column, with node 24 being the left-most.

Figure 10.2   The state-time (or, space-time) diagram for the network of interest

attractors that 'emerge' from the genotype (https://en.wikipedia.org/wiki/Genotype-phenotype_distinction, Kauffman 1993). Alternatively, we have a way here to associate form (structure) with function (dynamics).

Table 10.4 provides a summary of the phase space characteristics of this network. In short, there are 20 quantitatively different attractors, and 6 qualitatively different attractors (that is, attractors with different periods) with periods ranging from 3 to 42. There is a single 18 period attractor that accounts for 58.4 percent of all phase space states, and the longest transient, that is, the largest number of steps between a state not on an attractor basin (that is, at the far end of an attractor transient branch), and a state on an attractor basin, is 117.

An interesting metric in Table 10.4, which we won't discuss much here, is Dynamic Robustness (DR). A DR of 0.594 says that, for nearly 60 percent of all network states, an external perturbation that flips the state of just one of the nodes will not result in a qualitative change in network behavior. Conversely, for just under 40 percent of all network states, a small external force that flips the state of one node will push the network into a different attractor basin, that is, the external force will result in a *qualitative* change in network behavior. DR is discussed in detail in Richardson (2009).

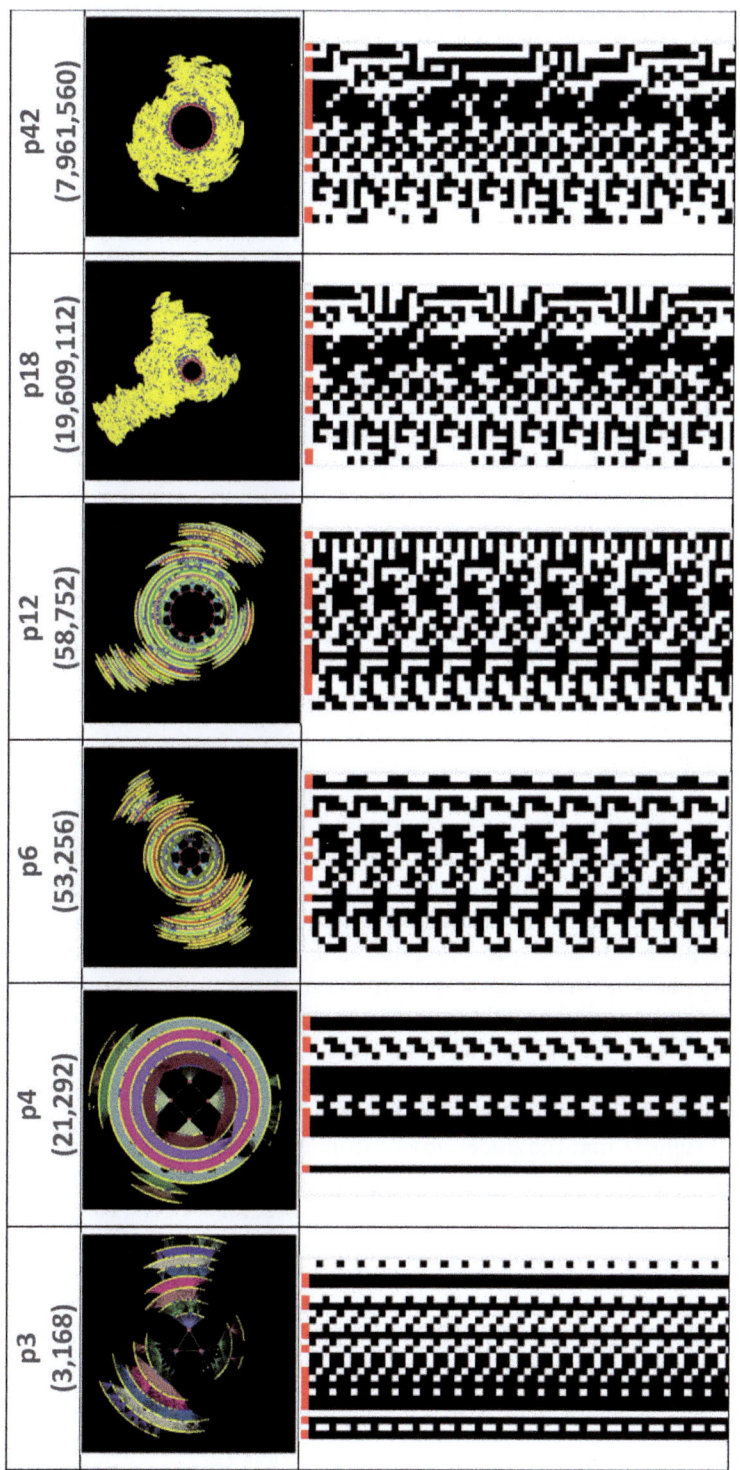

Note: Node 0 is the right-most column, with node 24 being the left-most.

Figure 10.3   *The phase space and state-time diagrams for 6 of the 20 attractors for the network of interest*

Table 10.4   Summary of phase space characteristics of given network

Number of phase space attractors	20
Average period of phase space attractors	8
Longest period	42
Number of qualitatively different phase space attractors (different periods)	20
Number of qualitatively different phase space attractors (different structures)	6
Longest transient length	117
Largest number of pre-images	208
Average number of pre-images	9.7027
Garden-of-Eden percentage	89.69
Dynamical robustness	0.594
Size of largest attractor (relative to total size of phase space)	0.5844
Period of largest attractor	18

10.2.2   Dynamic Structure

You may have already noticed, from Figure 10.3, that some of the nodes seem to be fixed in the state-time diagram, that is, their state does not change at all whilst in certain attractor basins. A fixed state suggests that no information is flowing through that node, or to put it another way, the multiple flows of information around the network cooperate to 'freeze' some of the nodes. These nodes effectively become 'walls of constancy', or information barriers, through which no information flows.

An interesting characteristic of Boolean networks, and perhaps this extends to all complex networks, is that, as far as network function is concerned, any nodes that are frozen/inactive *in all attractors* can safely be removed without affecting the *function* – the *qualitative* structure of phase space – of the network (Richardson 2005). For some networks (but not the one under study herein) the emergence of frozen nodes can result in the modularization of a network, in which the network is effectively carved-up into dynamically independent sub-networks. The point is that for any given attractor not all the nodes contribute, and so there is a sub-structure that is primarily responsible for the observed dynamics that is a subset of the overall network structure – static structure and effective dynamic structure can be very different indeed.

To illustrate this distinction further Figure 10.4 shows the effective dynamic structure for each of the attractors (functions) shown in Figure 10.3.

It is tempting to suggest that the nodes absent from each view contribute little to the network's dynamical behavior. However, not only is it important to realize that a node that plays a small role in one function, may actually play a very special role in another (and all the possibilities in between), it is also essential to remember the importance of a 'container' for emergence to occur within.

In Glenda Eoyang's *Conditions for Self-Organizing in Human Systems* (2001) thesis, she presents her CDE Model (Container/Difference/Exchange), in which the "Container bounds the system of focus and constrains the probability of contact among agents." The inactive nodes in the network under study represent such a container within which the other nodes can interact and exchange information that contribute to the particular function that emerges. Notably, even though the whole network itself can be seen as a container (the very

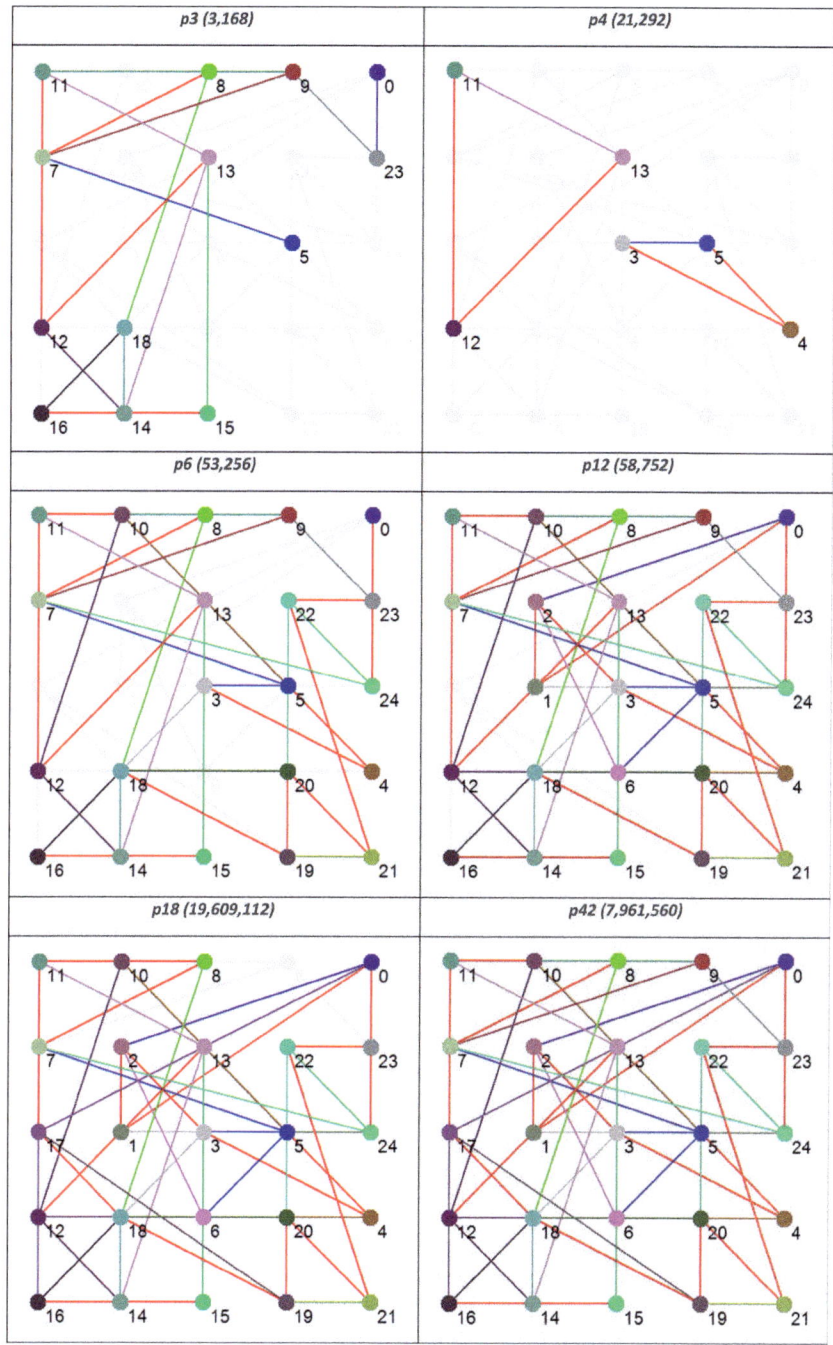

Note: The number in brackets after the attractor type is the number of states in phase space that map to that particular attractor. Red connections are bi-directional.

Figure 10.4   Effective dynamic structures for selected attractors

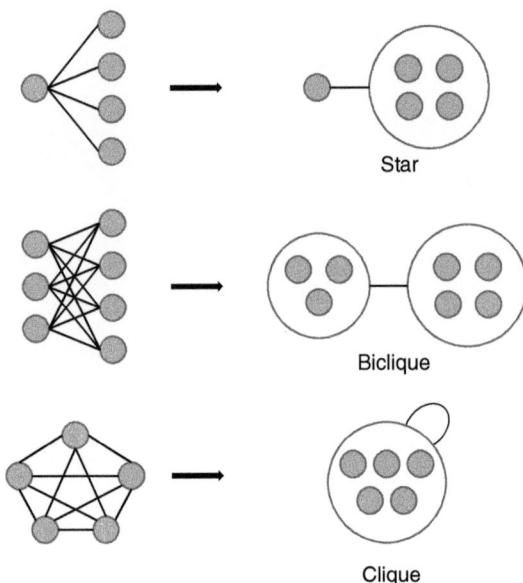

Figure 10.5   Power graph analysis primitives

fact it is a network 'contains' the system considerably), the dynamic shape of that container is co-dependent on the attractor (function) being exhibited – the container itself is emergent; a node is 'frozen' as a result of the net effect of the 2,345 interacting feedback loops.

Ensuring that removal of the frozen nodes left the compressed network functional equivalent to the original network would require checking that the node was frozen in every attractor. Even then, removal of the frozen nodes would still reduce the DR of the compressed network. So, compression needs to be considered in context.

10.3   POWER GRAPH ANALYSIS (PGA)

Power graphs are novel representations of graphs, developed within the field of computational biology, that rely on two abstractions: power nodes and power edges. Power nodes are sets of nodes brought together and power edges connect two power nodes thus signifying that all nodes contained in the first power node are connected to all the nodes contained in the second power node. These language primitives allow for the succinct representation of stars, bicliques and cliques (see Figure 10.5).

As Figure 10.5 shows, a star is expressed as a node connected via a power edge to a power node, a biclique is expressed as two power nodes connected by a power edge, and a clique is a power node connected to itself by a power edge. PGA does not concern itself with the direction of a connection – only that a connection exists.

Figure 10.6 shows the result of applying power graph analysis (PGA) to a simple graph. The power graph representation reduces the number of edges needed to represent the network, groups together highly connected nodes as well as nodes having similar neighbors, and this without any loss of information. In the following, we will often use

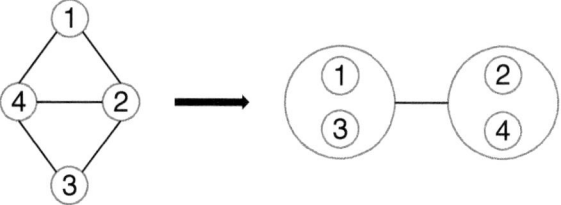

Figure 10.6   Simple power graph analysis example

the notion of edge reduction, that is, the proportion of edges that are abstracted from the original network in the power graph representation.

PGA is the computation and analysis of power graphs. Based on the work of Royer et al. (2008), an algorithm that computes power graphs is presented. Node clustering, module detection, network motif composition, network visualization, and network models can be recast in terms of PGA. PGA can be thought of as a lossless compression algorithm for graphs.

Compression levels of up to 95 percent have been obtained for complex biological networks. Microsoft and a team at Monash University have also looked at PGA (along with other edge compression techniques) as a way of simplifying complex software dependency graphs (Dwyer et al. 2013a, Dwyer et al. 2013b). We will now examine the use of PGA to compress Boolean networks.

10.3.1   PGA of Boolean Networks

Figure 10.7 re-visualizes the diagrams in Figure 10.4 using PGA. The diagrams were generated using the CyOog plugin for Cytoscape (CyOog, 2012). Small grey numbered circles represent single nodes, black circles are power nodes and green power nodes are cliques. Although PGA is most effective on vast networks containing thousands of nodes we can see some simplifications when it is applied to these smaller graphs.

The benefits of revisualization using PGA are more obvious with much larger networks, primarily because the degree of compression is considerably higher. With the relatively small network used for this study, the level of compression is minimal, but these alterative views do offer some structural insights that are not readily apparent from the traditional graph view. For example, although it would be difficult to say that the PGA view of the p3 (3,168) basin is any simpler than the equivalent traditional view offered in Figure 10.4, the fact that node groups 7, 8, 11, 23 and 12, 13, 15, 16, 18 represent two connected power nodes (in the language of PGA) may be of consequence.

We say 'may' because within the abstract framework we're working within here, understanding the *significance* of these power nodes is problematic. That being said, it would be interesting to see if targeting members of these power nodes leads to an increased probability of qualitative behavioral change within the network. This is a potentially interesting area for investigation in follow-up work.

We can also see, from a brief examination of p3 (3,168) and p6 (53,256) that nodes 9 and 14 are highly connected, and at the top of a hierarchy of power nodes. While the number of connections could have been obtained from a degree analysis, the

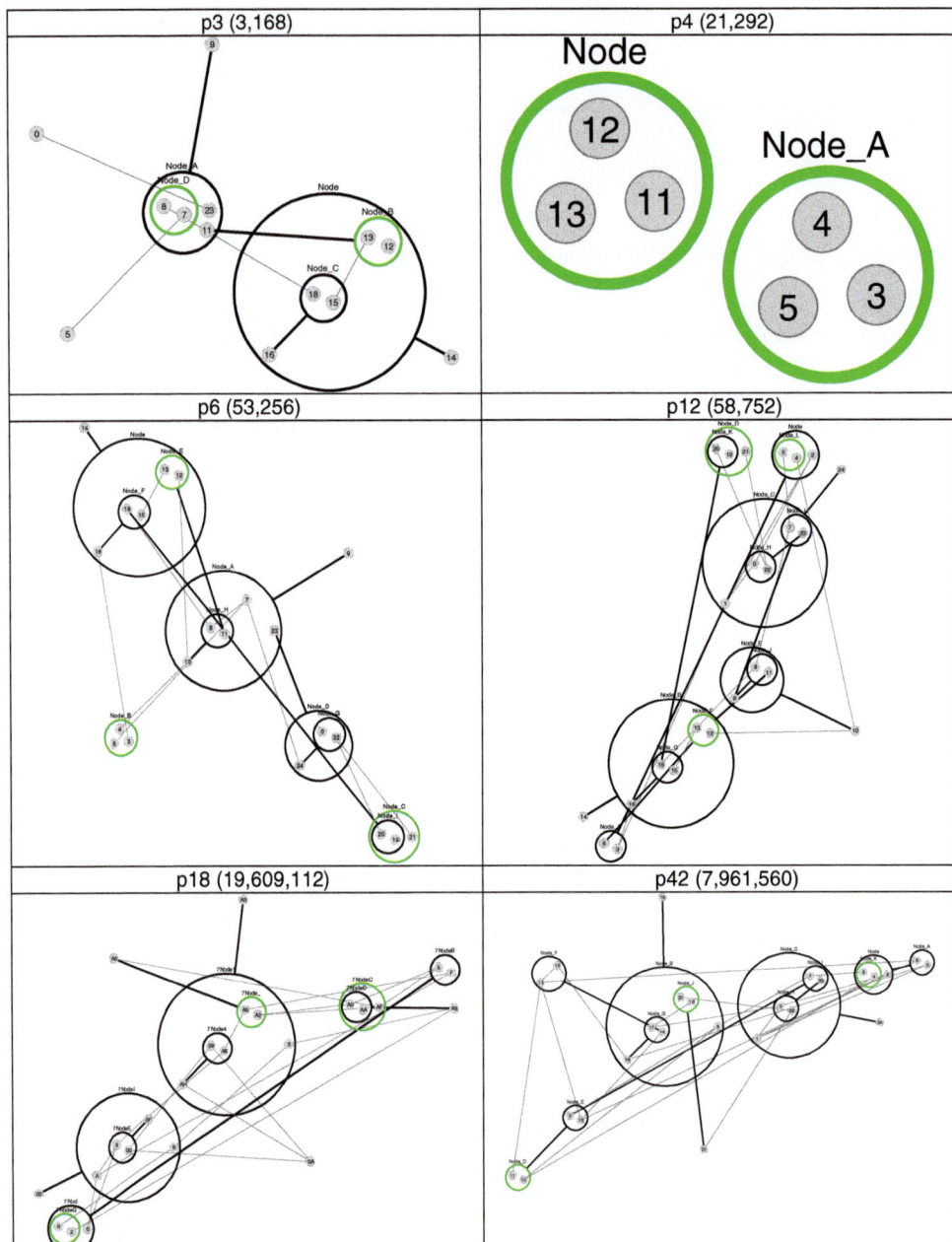

Figure 10.7   Power graph analyses of effective dynamic structures for selected attractors (see Figure 10.4)

relationships between the subgroups would have been lost. Nodes 8 and 9 both have degree 4, but it is clear that they are playing different structural roles. Also, the PGA makes it clear that node 14 has links to a group of nodes with internal structure – the biclique of 16, 15 and 18.

It may be that nodes such as 9 and 14 act as levers to push the system into other attractors, but this would require further investigation. In addition, it is also notable that the set of nodes at the top of the power hierarchy varies between the different structural attractors (for p3 they are nodes 9 and 14; for p12 they are nodes 10, 14, and 24, and so on), suggesting that the nodes which are the most significant when considering interventions that might have the largest impact on dynamics, depends on which attractor basin the network is currently in – the best levers are determined by dynamic context. Looking at the dynamic robustness with particular focus on these nodes would also be of considerable interest.

In short, PGA has the potential to highlight areas of interest within a dynamic network that would benefit from deeper consideration.

10.4   SHANNON ENTROPY AS A MEASURE OF ACTIVITY

Another way to partition (and so simplify) the network is to 'layer' it by considering the activity of each node. As a metric for activity we require a measure that is zero if the node state was 'frozen', and close to unity when the node is very active. The Shannon Entropy (https://en.wikipedia.org/wiki/Entropy_%28information_theory%29) of the state sequence for each node provides the appropriate activity profile – it provides a measure of the information contained in the node activity. The node state sequence is simply the state of a given node as it changes in time. So for a frozen node this might be '0,0,0,0,0,0,0,. . .'; this contains no information, and above is referred to as an information barrier.

The Shannon Entropy, H, of a particular node state sequence is calculated as:

$$H(x) = \int P(x) I(x)\, dx = -\int P(x) \log_b P(x)\, dx$$

Where $P(x)$ is the proportion of 1s (or 0s) in the state sequence. Table 10.5 shows the Shannon Entropy for each node when the network is in each one of the attractor basins. The cells have been colored to indicate nodes with high Shannon entropy and those with low; the lowest being the red 'frozen' nodes. The table rows have been ordered by each node's total entropy across all attractors, and the columns have been ordered by the number of red 'frozen' nodes in each attractor.

Using the data presented in Table 10.5 allows us to 'layer' the network now in a different way, by grouping the nodes based on their Shannon Entropy, or activity. The group number is given in the last, rightmost, column. Similar to what is shown in Figure 10.4, Figure 10.8 provides six different views that, starting with group 6 (nodes 11, 12 and 13 – the most active group), sequentially adds each group to the network structure until we arrive at the whole original network.

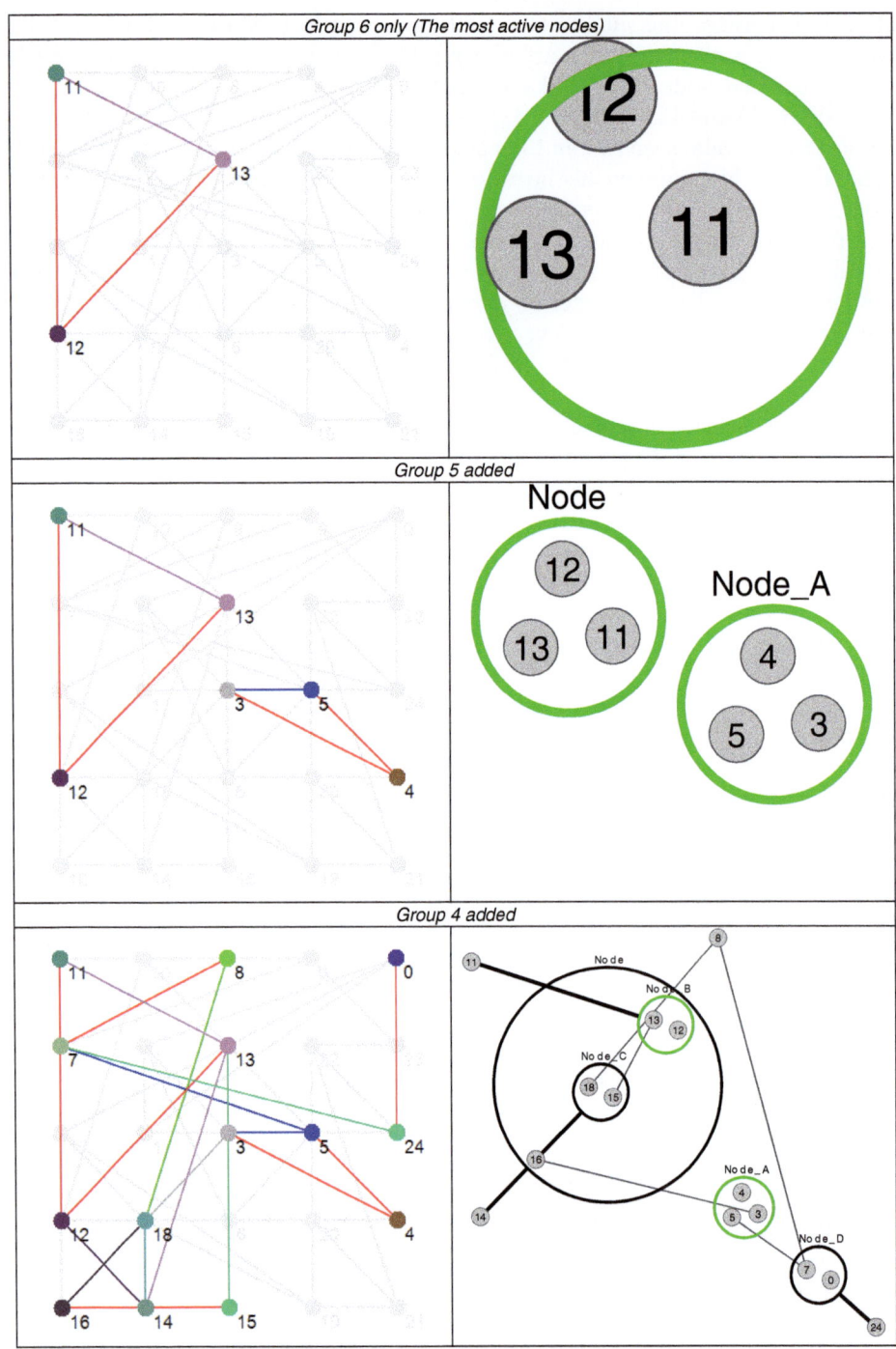

Figure 10.8   Effective dynamic layers based on node Shannon Entropy group

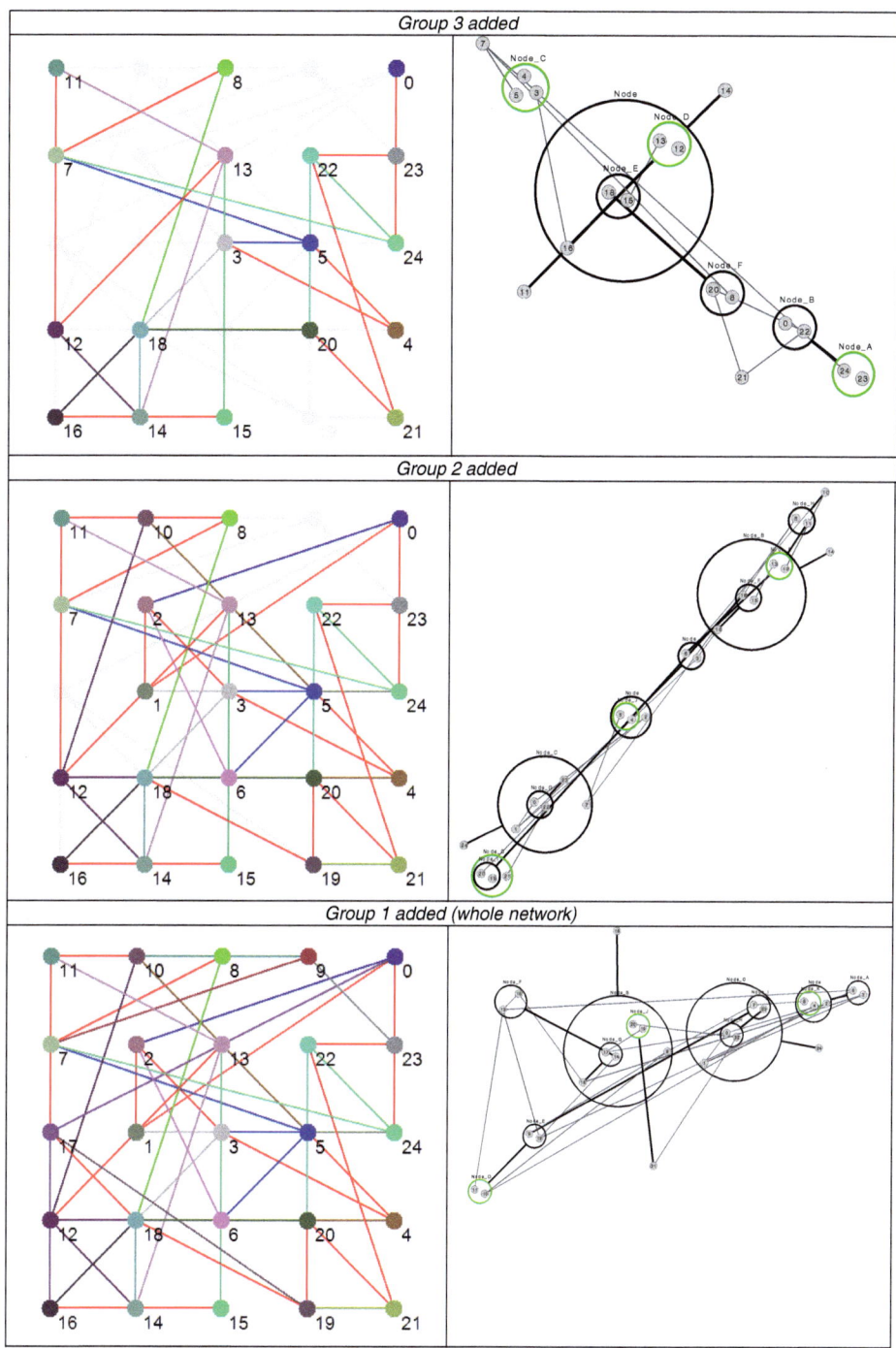

Figure 10.8 (continued)

Table 10.5   Shannon Entropy for each node operating within each attractor basin

Node	p12 (58,752)	p42 (79,615,060)	p12 (578,032)	p18 (19,609,112)	p6 (93,504)	p12 (675,816)	p6 (53,256)	p6 (2,946,648)	p3 (1,720)	p3 (3,168)	p4 (23,768)
12	0.30	0.30	0.30	0.30	0.30	0.30	0.30	0.28	0.28	0.28	0.30
11	0.30	0.30	0.30	0.30	0.30	0.30	0.30	0.28	0.28	0.28	0.30
13	0.28	0.30	0.28	0.30	0.30	0.29	0.30	0.30	0.28	0.28	0.30
5	0.30	0.30	0.29	0.26	0.20	0.24	0.28	0.28	0.28	0.28	0.24
3	0.29	0.27	0.29	0.29	0.30	0.20	0.28	0.20	0.00	0.00	0.30
4	0.28	0.30	0.29	0.28	0.20	0.28	0.28	0.00	0.00	0.00	0.30
14	0.30	0.28	0.28	0.28	0.28	0.28	0.30	0.28	0.28	0.28	0.00
15	0.30	0.28	0.28	0.28	0.28	0.28	0.30	0.28	0.28	0.28	0.00
0	0.20	0.28	0.28	0.29	0.30	0.29	0.30	0.28	0.28	0.28	0.00
16	0.28	0.28	0.28	0.28	0.28	0.28	0.28	0.28	0.28	0.28	0.00
7	0.28	0.23	0.20	0.20	0.28	0.24	0.28	0.30	0.28	0.28	0.00
8	0.28	0.23	0.20	0.20	0.28	0.24	0.28	0.30	0.28	0.28	0.00
24	0.28	0.29	0.20	0.29	0.30	0.30	0.30	0.28	0.28	0.00	0.00
18	0.28	0.20	0.20	0.20	0.28	0.20	0.28	0.28	0.28	0.28	0.00
22	0.30	0.29	0.28	0.30	0.30	0.28	0.28	0.00	0.28	0.00	0.00
20	0.28	0.30	0.30	0.30	0.28	0.28	0.28	0.00	0.28	0.00	0.00
21	0.28	0.30	0.30	0.30	0.28	0.28	0.28	0.00	0.28	0.00	0.00
23	0.28	0.21	0.28	0.20	0.20	0.24	0.00	0.30	0.00	0.28	0.00
1	0.30	0.28	0.30	0.26	0.28	0.20	0.00	0.20	0.00	0.00	0.00
2	0.30	0.28	0.30	0.26	0.28	0.20	0.00	0.20	0.00	0.00	0.00
6	0.30	0.28	0.30	0.26	0.28	0.20	0.00	0.20	0.00	0.00	0.00
10	0.28	0.20	0.20	0.20	0.28	0.20	0.28	0.20	0.00	0.00	0.00
19	0.20	0.28	0.28	0.28	0.28	0.28	0.20	0.00	0.00	0.00	0.00
9	0.20	0.08	0.00	0.00	0.00	0.12	0.20	0.30	0.28	0.28	0.00
17	0.00	0.30	0.30	0.30	0.00	0.30	0.00	0.00	0.00	0.00	0.00
Total Entropy	6.63	6.61	6.48	6.35	6.29	6.28	5.82	4.97	4.42	3.59	1.75

10.4.1   Shannon Entropy Signatures (SESs)

Another way to visualize the dynamic behavior of the network of interest is to construct their Shannon Entropy (SE) profile or signature. For each attractor the SES is simply a column chart showing the SE for each node. Figure 10.9 shows the SESs for four of the network's attractors.

p4 (31,392)	p4 (148,956)	p4 (34,412)	p4 (175,400)	p4 (173,700)	p4 (33,728)	p4 (178,968)	p4 (106,304)	p4 (676,336)	Total Entropy	Group
0.30	0.30	0.30	0.30	0.30	0.30	0.30	0.30	0.30	5.04	6
0.30	0.30	0.30	0.30	0.30	0.30	0.30	0.30	0.30	5.04	6
0.30	0.30	0.30	0.30	0.30	0.30	0.30	0.24	0.24	4.95	6
0.24	0.24	0.24	0.24	0.24	0.24	0.24	0.00	0.00	3.76	5
0.30	0.30	0.30	0.30	0.30	0.30	0.30	0.00	0.00	3.67	5
0.30	0.30	0.30	0.30	0.30	0.30	0.30	0.00	0.00	3.43	5
0.00	0.00	0.00	0.00	0.00	0.00	0.00	0.00	0.00	1.96	4
0.00	0.00	0.00	0.00	0.00	0.00	0.00	0.00	0.00	1.96	4
0.00	0.00	0.00	0.00	0.00	0.00	0.00	0.00	0.00	2.02	4
0.00	0.00	0.00	0.00	0.00	0.00	0.00	0.00	0.00	1.94	4
0.00	0.00	0.00	0.00	0.00	0.00	0.00	0.00	0.00	1.85	4
0.00	0.00	0.00	0.00	0.00	0.00	0.00	0.00	0.00	1.85	4
0.00	0.00	0.00	0.00	0.00	0.00	0.00	0.00	0.00	1.75	4
0.00	0.00	0.00	0.00	0.00	0.00	0.00	0.00	0.00	1.77	4
0.00	0.00	0.00	0.00	0.00	0.00	0.00	0.00	0.00	1.43	3
0.00	0.00	0.00	0.00	0.00	0.00	0.00	0.00	0.00	1.40	3
0.00	0.00	0.00	0.00	0.00	0.00	0.00	0.00	0.00	1.40	3
0.00	0.00	0.00	0.00	0.00	0.00	0.00	0.00	0.00	1.49	3
0.00	0.00	0.00	0.00	0.00	0.00	0.00	0.00	0.00	0.92	2
0.00	0.00	0.00	0.00	0.00	0.00	0.00	0.00	0.00	0.92	2
0.00	0.00	0.00	0.00	0.00	0.00	0.00	0.00	0.00	0.92	2
0.00	0.00	0.00	0.00	0.00	0.00	0.00	0.00	0.00	1.14	2
0.00	0.00	0.00	0.00	0.00	0.00	0.00	0.00	0.00	1.02	2
0.00	0.00	0.00	0.00	0.00	0.00	0.00	0.00	0.00	1.17	1
0.00	0.00	0.00	0.00	0.00	0.00	0.00	0.00	0.00	0.60	1
1.75	1.75	1.75	1.75	1.75	1.75	1.75	0.85	0.85		

Again, the dominant nodes in the p4 attractors are readily apparent.

If, via a small external perturbation, the network was pushed from a p4 attractor to a p42 attractor for example, we could say that the network has moved from a *modular, low information* (0.85) mode to a *distributed, high information* (6.61) mode. In this way, it may be possible to determine different operational modes in real networks, once a proxy for SE is determined.

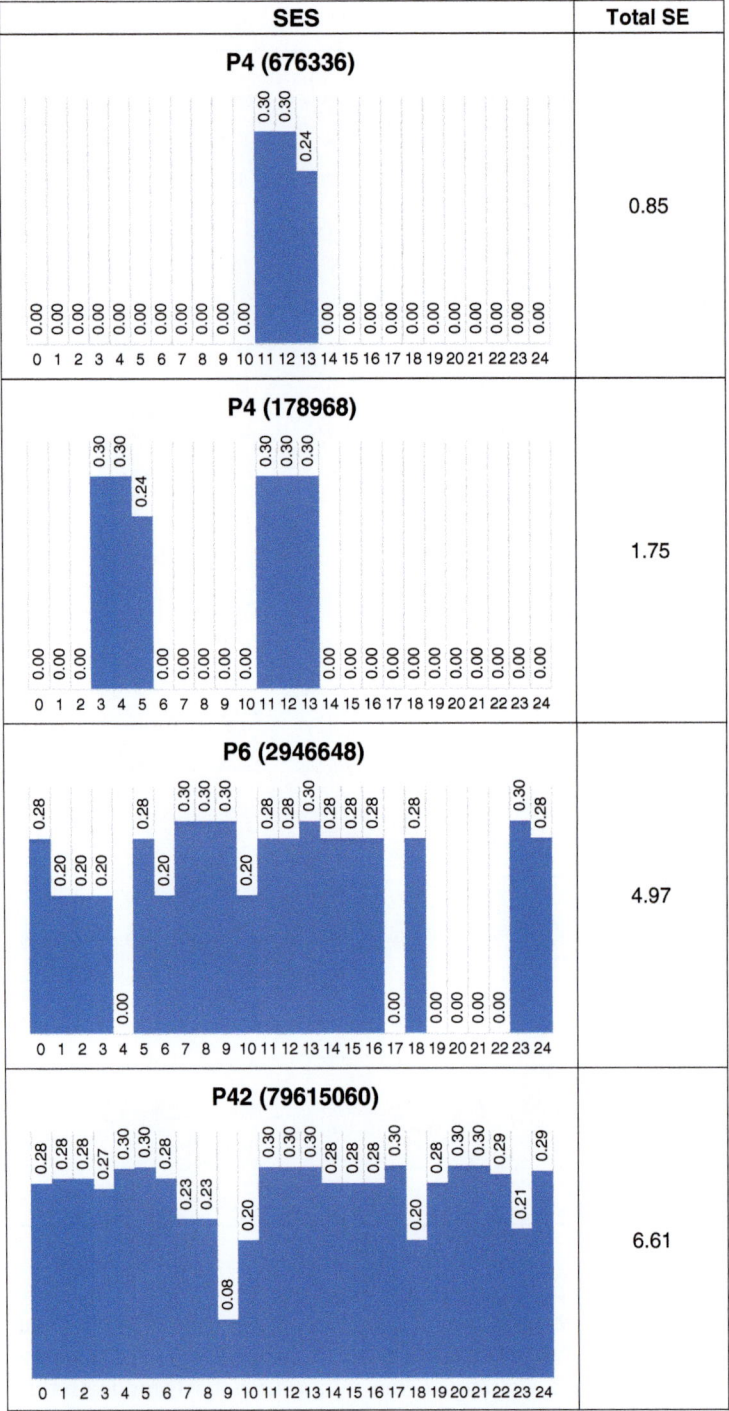

Figure 10.9   Shannon Entropy Signatures for four attractors

Furthermore, rather than developing a set of archetypes for the different 'connectivity modes' and 'information levels' that the current network behavior might be mapped to, the focus would be on how the SES changes over time. In this way, we can monitor for qualitative changes in the SES, wave a flag to announce such a change, and then consider the significance within the network of interest of such an event.

10.5   DISCUSSION AND CONCLUSIONS

The aim of this chapter in some ways is to make the obvious point that a pre-occupation with network structure in the absence of dynamic considerations (as is typical of traditional SNA) can lead to very misleading insights into how one might interact with a real-world complex network. In the presentation above it is clear that a purely topological analysis of the original network 'graph' would not capture some of the structural properties that only become apparent when dynamical, functional, and phase space analysis is performed.

Boolean networks were compressed by identifying their frozen nodes. We then looked at the use of power graph analysis (PGA) as another way to compress these networks. While PGA seems to achieve some simplification of network structure, the Boolean networks studied here were too small to clearly illustrate the full capabilities of PGA for network simplification. That being said, although the PGA views appear as complicated as the more traditional graph views, they do highlight different structural elements, such as stars and cliques more readily, as well as hierarchically significant nodes.

Finally, a Shannon Entropy measure was applied to the networks, allowing them to be layered (from dynamical considerations) to aid visualization, while also providing a way of obtaining an information-based network signature. Further work will be required to see how these approaches can be applied to real-world dynamic networks.

10.5.1   Practical Relevance

Analyzing static network structures is akin to staring at a photograph, looking for insights, while ignoring the full-length documentary that's available on the subject. When one pauses to think about real-world networks it's almost always the dynamics that are of interest. What can we say about the suitability of an empty road network? We can analyze the route efficiency, for example, comparing point-to-point and hub-systems. However, it's obvious that it's only when you put traffic on the roads that today's problems come to light. A road network that is efficiently traversable at 2am on Monday morning can be gridlocked six hours later.

Financial meltdowns are important because of the change. You can't have a market crash without change. Similar, when you consider social network analysis – somewhat of a poster-child for network analysis – the companies who run these networks are focused on change: increasing and decreasing numbers; getting the right type of users; adding new services, and so on.

Networks without dynamics are merely the skeletons of past scenarios. However, dynamics are difficult to analyze. The sheer volumes of data produced by small RBNs quickly overwhelm the trendiest of 'big data' platforms. Hence, in this chapter, we've started to explore ways of simplifying dynamic networks – making them easier to understand and reason about. Given the importance of dynamic networks we urgently need tools that can uncover their secrets.

10.5.2   Future Work

The authors plan to develop this research in a number of areas. First, they intend to port elements of the Boolean dynamic network analysis toolkit to a cloud-based web service. This will allow the study of larger networks, but also make the toolkit available to all academic researchers.

Once we have results for larger networks, PGA will be revisited to see if it is more effective in simplifying larger networks, and also to understand the identification of power nodes and the role they may play in network dynamics and robustness. We are also planning to further study the Shannon Entropy signature of various networks in an attempt to identify both valid proxies for the node state used herein, significant changes in network behavior. The authors believe that analysis tools based on this concept could be invaluable in areas such as security analysis and news reporting.

Finally, we hope to be able to start testing some of these approaches on real-world networks. As mentioned in the introduction, some of the ideas herein are somewhat speculative, but we believe that by using insights from Boolean network analysis, robust general-use dynamic network tools can be developed.

NOTE

1.   Contact Kurt Richardson (kurt@exploratorysolutions.com) for details.

REFERENCES

Borgatti, S.P., Everett, M.G., and Johnson, J.C. (2013). *Analyzing Social Networks*, London: Sage Publications.
Cilliers, P. (1998). *Complexity and Postmodernism: Understanding Complex Systems*, London: Routledge.
CyOog (2012). http://www.biotec.tu-dresden.de/research/schroeder/powergraphs/download-cytoscape-plugin.html (accessed August 7, 2017).
Dwyer, T., Riche, N.H., Marriot, K., and Mears, C. (2013a). *Edge Compression Techniques for Visualization of Dense Directed Graphs*, Microsoft Research, http://research.microsoft.com/en-us/um/people/nath/docs/edge-compression_infovis2013.pdf (accessed August 7, 2017).
Dwyer, T., Mears, C., Morgan, K., Niven, T., Marriot, K., and Wallace, M. (2013b). Improved Optimal and Approximate Power Graph Compression for Clearer Visualisation of Dense Graphs, Paper presented in: Visualization Symposium (PacificVis), 2013 IEEE Pacific Conference Proceedings, March, pp. 105–112, https://arxiv.org/pdf/1311.6996.pdf (accessed July 2017).
Economist (2015). The Nuke Detectives, *Economist*, September 5, 2015.
Eoyang, G. (2001). *Conditions for Self-Organizing in Human Systems*, http://www.hsdinstitute.org/about-hsd/dr-glenda/glendaeoyang-dissertation.pdf (accessed August 7, 2017).
Kauffman, S. (1993). *The Origins of Order: Self Organization and Selection in Evolution*, Oxford: Oxford University Press.
Magsino, S.L. (2009). *Applications of Social Network Analysis for Building Community Disaster Resilience: Workshop Summary*, Washington, DC: The National Science Academies Press.
Richardson, K.A. (2005). Simplifying Boolean Networks, *Advances in Complex Systems*, 8(4): 365–381.
Richardson, K.A. (2009). Robustness in Complex Information Systems: The Role of Information 'Barriers' in Boolean Networks, *Complexity*, 15(3): 26–42.
Royer, L., Reimann, M., Andreopoulos, B., and Schroeder, M. (2008). Unraveling Protein Networks with Power Graph Analysis, *PLoS Comput Biol*, 4(7): e1000108.

11. Network text analysis and social network analysis in investigating complex governance networks: applications of AutoMap and ORA

Professor Göktuğ Morçöl, Penn State Harrisburg and Dr Sohee Kim, Penn State Harrisburg

11.1 INTRODUCTION

It is an increasingly common understanding among social theorists that the problems of contemporary societies are complex and they cannot be solved by single actors using simple tools. This understanding requires new and more fitting conceptualizations. In this chapter we first briefly describe such a conceptualization: the complex governance networks conceptualization. Then we discuss why it is necessary to use new tools to investigate these networks. After this background information, we turn to the primary focus of this chapter: applications of network text analysis and social network analysis in investigating complex governance networks. We describe these tools, illustrate their applications, and conclude the chapter with a discussion of the strengths and limitations of such applications.

There is no overarching theory of complex systems, nor is there a single method that would help us to gain a complete knowledge of these systems. Complex systems should be conceptualized for circumscribed realms of reality. In this chapter we focus on the realms that are traditionally known as public policy and public administration processes (public policymaking and public service delivery). We adopt particularly the complex governance networks conceptualization (Morçöl 2014).

The methods we describe in this chapter – the network text analysis (NTA) and social network analysis (SNA) combination – is suitable for investigating some aspects of complex governance networks. Despite the major methodological advances in recent decades, neither these nor any other methods are capable of depicting complete pictures of nonlinear and dynamic relationships among actors/agents and the emergent structural properties of complex governance networks. NTA and SNA can still be useful tools when they are applied with an understanding of their strengths and limitations.

Unlike traditional social science research methods in which individual attributes or behaviors are used as units of analysis, in SNA relations among nodes (actors) are used as units. This emphasis on relations among actors, rather than attributes of atomized individuals, makes SNA philosophically compatible with complex system conceptualizations and suitable for investigating complex social systems. An in-depth discussion of this compatibility and suitability is beyond the scope of this chapter, but in the next section we will highlight the importance of using relations as unit of analyses in investigating complex governance networks.

A brief summary of how NTA and SNA can be useful in investigating complex governance networks. NTA and SNA can help us identify the central actors in governance networks. The centrality of an actor in a network may be interpreted as his/her popularity (Kadushin 2012, p.31) or prestige (Wasserman and Faust 1994, pp.174–175), which in turn may be an indicator of his/her social capital (Borgatti, Everett, and Johnson 2013, p.164) or influence (power) in the network. Using SNA methods a researcher can investigate whether and to what extent the centrality (popularity, prestige, or power) of an actor can affect the macro properties of a network, as we illustrate in this chapter.

Complex systems are dynamic. There are methods that can be used to understand system dynamics, such as agent-based simulations, whose strengths and limitations we describe briefly later in this chapter. Social network analyses are not dynamic – they provide only snapshots of the network relations in given time periods – but they can be used longitudinally to gain insights into system dynamics, to a certain extent. In this chapter we will discuss and illustrate how NTA and SNA can be used in such longitudinal studies.

Established theoretical frameworks of policy processes, such as the Advocacy Coalition Framework, suggest that policy actors coalesce around particular idea clusters (worldviews and more specific policy positions that are associated with them). As we illustrate with a case study in this chapter, the NTA and SNA combination can be used to reconstruct the structural properties of the complex governance networks by analyzing and mapping the relative positions of individual and organizational actors and their links to ideas and resources in these networks.

11.2   COMPLEX GOVERNANCE NETWORKS

There is a growing recognition among social theorists and researchers that contemporary human societies are complex and this creates governance challenges. This complexity and such challenges result from the increasingly multi-centered nature of our societies (Castells 2000; Jessop 1991). No governmental actor, or a private actor, has the capacity to control these multiple centers or to solve the complex and dynamic social problems of societies singlehandedly (Kettl 2002; Koliba, Meek, and Zia 2011; Kooiman 1993; Rhodes 1997; Torfing et al. 2012). The diffusion of political and economic power in contemporary societies poses new challenges in solving 'collective action problems' (Feiock 2013), but the increasingly 'polycentric' new order of societies may also provide opportunities for self-organization and self-governance in human communities (McGinnis 1999).

A detailed discussion of the merits and demerits of the literature on the complexity of governance processes is beyond the scope of this chapter. But a clarification and a discussion of the key concepts and conceptualizations the theorists use is necessary to make the case for the investigative tools we will present in the following sections. We want to stress first the importance of the key observation that the primary reason for the increased complexity of social and policy processes is that societies are multi-centered. This observation leads to the conclusion that the methods that were developed to investigate simple causal relations cannot be as useful.

Most of the traditional methods of investigation, like experimental designs and linear regression models, were built on the assumption that causal relationships between independent and dependent variables could be isolated. This assumption is suitable for the traditional conceptualization of public policy, which is expressed succinctly in Dye's (2013) classic textbook definition: Public policy "is concerned with what governments do, why they do it, and what difference it makes" (p.3). In this definition governmental action is the independent variable and a social problem is the dependent variable. There is a series of implicit or explicit assumptions under this definition. It is assumed that 'government' is a unified actor that represents the public interest (or the collective will of the society) and it can solve social problems. It is also assumed that a government is a rational actor that can make cohesive decisions and take actions on behalf of the society. It is also assumed that the government's decisions can be implemented as intended, because there is a linear and knowable causal link between governmental actions and their 'outcomes.' Finally, it is assumed that the government's analysts can determine objectively whether its actions reached their intended goals (that is, impact assessment).

These assumptions have been criticized from various theoretical angles. For example, rational choice theorists (Mueller 1997) argue that governments are not unified actors that represent public interests. Pressman and Wildavsky (1984) empirically demonstrated that actually policies are not implemented as intended or designed. Radin (2013) notes that in the evolution of policy analysis since the 1960s these assumptions have been either abandoned or at least modified. During this evolution policy theorists and analysts have been compelled to recognize the "pluralistic and fragmented policy world that operates in a society characterized by diversity" (p.x).

In the governance networks conceptualization, social problems can be solved not solely and independently by governmental actors, but through collaborations among multiple governmental and non-governmental actors (Koliba, Meek, and Zia 2011; Provan and Kenis 2007). These actors are self-organizing, but they are also interdependent. They share information and resources. These interdependent relationships are also dynamic. These systems of relationships are called 'governance networks.'

Complexity theory suggests that social problems cannot be solved by single (for example, governmental) actors because the actions of multiple self-organizing actors constitute processes from which complex systems emerge. "Because complex systems are self-organizational and composed of self-organizing elements (actors, agents), they cannot be controlled or directed centrally, nor can their 'problems be solved' centrally" (Morçöl 2014, p.10). A key implication of complexity theory is that the relations between the policy actions of governments and 'their outcomes' are not easily predictable or controllable because 'policy outcomes' actually emerge from the nonlinear interactions of self-conscious and self-organizing actors (Salzano 2008). Policy interventions (that is, governmental actions) can at best be used to 'nudge' systems toward socially desirable states (Zia et al. 2014), or to enable the capacity of a social system to self-organize in solving a social problem (Stewart and Ayres 2001, p.87).

Then how do macro social structures (for example, 'policy outcomes') emerge from the interactions of self-organizing individual and organizational actors? Also, how can one study the processes and structures of complex governance networks and the roles of individual and organizational actors in them? There are some developments in answering these questions (for example, Koliba, Meek, and Zia 2011; Morçöl 2012), but more conceptual refinement is needed. Detailed discussions of these conceptual developments and the needs are beyond the scope of this chapter. We will instead focus on a group of methods that can be used to investigate the evolutions of the relations between individual and organizational actors and macro-level in complex governance networks.

11.2.1   Methods of the Complex Governance Networks Conceptualization

It is important to note that despite the major advances in the methods that have been developed to investigate complex systems, none of these methods is capable of depicting complete pictures of nonlinear and dynamic relationships among actors/agents and the emergent structural properties of these systems. Each method has its strengths in capturing some aspects of these systems, however. In this chapter we will discuss the strengths and weaknesses of the methodological combination of network text analysis and social network analysis that we propose.

Agent-based simulations (ABS) are a powerful set of methods of investigating the dynamics of micro–macro relationships and transformations in complex systems. They have been used in the studies on governance networks (Johnston et al. 2008; Zia et al. 2006). ABS can help researchers capture the dynamism of the interactions among actors and understand the emergence of macro structures. These methods have a limitation: A researcher has to make some abstractions from what is known about the behaviors of actors and then enter them into their simulations. Because of these abstractions, applications of ABS are somewhat 'artificial' (ABS methods are alternatively called 'artificial life simulations'; see Epstein and Axtell 1996).

Social network analysis (SNA) methods have strengths and limitations, as well. Because SNA researchers use data about real actors, not abstracted behavioral profiles, they are not artificial and they reflect better the characteristics of the networks they are applied to study. SNA is not a dynamic methodology, however: An SNA researcher can only take 'snapshots' of reality in given time periods. Complex systems are dynamic; therefore ABS is a better method to study this dynamism. SNA can still be useful in investigating complex systems dynamics. The snapshots of networks that are taken in successive time periods can be studied longitudinally to investigate system dynamics. We illustrate how this can be done in the following sections.

11.3   SOCIAL NETWORK ANALYSIS AND NETWORK TEXT ANALYSIS

The general principles and specific methods of social network analysis are described and discussed in several other sources (for example, Freeman, White, and Romney 1992; Knoke and Yang 2008; Wasserman and Faust 1994). We provide a brief summary of its relevant aspects here.

The history of social network analysis can be traced back to the sociometric models of Moreno and others in the 1930s (Wasserman and Faust 1994, pp.9–17). These researchers mapped various forms of relations and interactions among individuals (for example, friendship, information exchanges, and economic exchanges) and described the social structures of these individuals. Several social science schools adopted and refined sociometric conceptualizations (for example, Gestalt psychology, the Manchester anthropologists, and the Harvard structuralists; see Berry et al. 2004). Major methodological advances in graph theory, statistical and probability theory, and algebraic models helped investigate these conceptualizations empirically.

At its most basic level, a social network is composed of individual actors (nodes) and relations among those actors (links, ties) (Wasserman and Faust 1994). In a study of social networks, a node may be an individual person or a group of individuals, such as a formal organization. Unlike traditional social science research methods that use quantified individual attributes or behaviors as units of analysis, SNA uses relations as units. More specifically, SNA utilizes relational data – data that describe the relationships between actors embedded in the network.

Relational data can be collected using surveys or archival texts. The most commonly used method of collecting social network data is surveys (Milward et al. 2010; Sandström and Carlsson 2008; Wasserman and Faust 1994). Surveys are useful when a researcher studies people as network actors and their relationships in the network. However, the survey method may be problematic in longitudinal investigations of networks. Informant bias is always a problem in survey research, and this can be an even bigger problem in longitudinal studies. Informants are not likely to recollect past events accurately (Freeman, Romney, and Freeman 1987; Knoke and Yang 2008, p.35).

A researcher can avoid these biases by using secondary data, such as archived written texts (Borgatti, Everett, and Johnson 2013, pp.29–30). Newspaper reports, government publications, blogposts, social media posts, and tweets can be used as archival data in SNA (for example, Human and Provan 2000; Kapucu 2006, 2009). A principal advantage of using archival information is that it provides a researcher with the opportunity to study the past. Because the texts and data were recorded in the past, closer to the events, there is no problem of inaccurately recollecting past information. Additionally, compared with other sources of data such as surveys, experiments, or field studies, archival data sources are relatively easier to access, which makes using archival data a relatively cost-efficient way (Singleton and Straits 2010). The increased availability of archival sources in electronic formats and the new techniques and tools to retrieve data from them made these sources more popular in recent years (Knoke and Yang 2008; Wasserman and Faust 2009).

Using archival data sources for SNA has its potential problems as well. One of the problems is that the information recorded in journalistic accounts may be inaccurate because of the biases of the journalists who collected them. However, this data collection method still have more validity compared to asking informants in surveys to recollect events that happened in a distant past; to the extent that the journalistic accounts of events were screened through serious editorial processes, they can be valid sources of information. So when archival data are used carefully and with some caution, they can be useful sources for social network analyses.

There is a long tradition of using archival data, or text data in general, in content analyses (Krippendorff 2004). Among the methods of content analysis, particularly semantic network analysis is important to highlight here, because it can be used as a form of social network analysis (Atteveldt 2008; Atteveldt, Kleinnijenhuis, and Ruigrok 2008; Carley, Bigrigg, and Reminga 2009; Lim, Berry, and Lee 2015). A researcher who conducts a semantic network analysis attempts to extract relations among concepts in a text and maps them using specialized software. Researchers who use semantic network analysis assume that they can find components of social structures, such as relations among people, organizations, resources, events, and locations in the texts they study.

Network text analysis (NTA) is a form of semantic network analysis (Carley 1997; Diesner and Carley 2004; Popping 2003). The AutoMap software, which we discuss in the following sections, was developed for NTA. Using the NTA tools, a researcher can map a variety of links among the concepts extracted from a text, classify them into meaningful groups, and analyze the mathematical properties of the linkages and groups (Carley, Columbus, and Landwehr 2013).

11.4　AUTOMAP AND ORA AS NETWORK TEXT ANALYSIS AND DYNAMIC (SOCIAL) NETWORK ANALYSIS SOFTWARE

In this section we describe the software system that was developed by Kathleen Carley and her colleagues at Carnegie Mellon University (Carley, Bigrigg, and Reminga 2009; Carley, Columbus, and Landwehr 2013; Diesner and Carley 2005) and discuss its strengths and limitations. This software system has two components: AutoMap, which extracts networks from texts (http://www.casos.cs.cmu.edu/projects/automap), and ORA, which analyzes networks extracted by AutoMap (http://www.casos.cs.cmu.edu/projects/ora). The applications of Carley's software are illustrated with two cases in the following sections. A protocol of using the software is described in the Appendix of this chapter.

AutoMap can extract networks from texts and analyze not only the relations among network actors (individual and organizational actors) but also the 'resources' and 'knowledge' that are associated with them. AutoMap extracts networks based on the co-occurrence of, or proximities between, words or word combinations in a pre-determined 'window size' in the analyzed text (for example, sentence, paragraph, or whole text) (Atteveldt 2008, p.38). Using AutoMap, a researcher can identify the concepts used in a text, code them into concept categories (for example, actors, resources, and knowledge), and extracts a model of links and relations among the concepts, which is called a 'meta-network' (Carley, Columbus, and Landwehr 2013).

The meta-networks that are extracted by AutoMap can be read into ORA, dynamic meta-network analysis software, which can be used to conduct traditional social network analyses, such as the ones that can be conducted using UCINET – a more popular software (for applications in UCINET, see Borgatti, Everett, and Johnson 2013) – but ORA has additional capabilities. ORA is useful particularly when a researcher aims to analyze multiple sets of network relations. For example, ORA is capable of analyzing and mapping not only the networks of actors, but also the links between actors and their actions, tasks, knowledge, resources, or locations (Carley et al. 2013; Diesner and Carley 2004). Similar analyses can be conducted using UCINET, but in a more limited fashion.

One of the major advantages of using the AutoMap–ORA system is that the meta-networks that are extracted by AutoMap are multi-modal (they readily include the links between the actors and the other characteristics, such as actions, tasks, knowledge, resources, and locations) can be read into ORA easily. Using ORA a research can analyze these meta-networks to map the relative positions of individual and organizational actors and their links to ideas and resources. This capability is particularly useful in testing the propositions of established theoretical frameworks of policy processes, such as the Advocacy Coalition Framework (ACF) and the Resource Dependency Theory (RDT). ACF posits that policy actors coalesce around particular idea clusters (worldviews and more specific policy positions that are associated with them). RDT posits that they coalesce around the economic and natural resources available to members of societies. Using the AutoMap–ORA system, a researcher can extract the links between idea clusters and/or resources and groups of actors from archival texts. We will illustrate an application of AutoMap–ORA on textual data in testing the predictions of the ACF in this chapter.

The method of extracting such links from texts has limitations as well. In any kind of content analysis of written texts, there is a common problem: how to ensure the validity of the relationships (links) extracted from texts. In the semantic network analysis literature, this problem is cited as the problem of validity in the identifications in texts of the 'objects' (that is, concepts in texts). More specifically, the literature suggests, the researcher faces problems in identifying the 'ontological categories' of the concepts extracted from texts (that is, whether they should be categorized as actors, resources, tasks, and so on), and the hierarchies among those categories (that is, which concept categories are inclusive of others) (for details, see Atteveldt 2008, chapter 2). In the end, the ontological and hierarchical categorizations are dependent on the researchers' interpretations and/or the agreements among the members of a research community. Therefore some 'subjectivity' in the categorizations is inevitable and researchers are primarily responsible for their interpretations.

This is true for the applications of the AutoMap–ORA system: Although AutoMap has the capability of automatically detecting concepts, the researcher has a major responsibility in identifying their ontological categories, selecting the most appropriate methods of analyses for these categories in ORA, and interpreting the results. For example, the researcher is responsible for defining the ranges of texts where the concepts could be found (for example, sentence, paragraph, or whole text) and for coding the concepts into categories using AutoMap's tools. In the Appendix we illustrate the kinds of decisions a researcher would have to make in the process of using the AutoMap–ORA system.

A major problem in any network text analysis is that some words in texts may not be directly relevant to the problem under study or they may have different meanings in different contexts. There are validity problems particularly in identifying individual and organizational actors. There is a possibility of double-counting of actors when the name of an actor who represents an organization and the organization are cited in the same text, for example. The researcher has to identify possible cases of double-counting and remedy them using the tools available in AutoMap. The researcher also has to make decisions on how to classify the concepts AutoMap identifies as 'actors,' 'tasks,' 'resources,' and so on.

Another potential problem is how to interpret the network analysis results that are generated using ORA. To ensure the validity in identifying the structural relations in the ORA networks, a researcher needs to supplement the information in them by directly interpreting the original texts and understanding the contexts.

11.5   ILLUSTRATIONS OF AUTOMAP AND ORA APPLICATIONS

11.5.1   Identifying Central Actors in an Urban Governance Network: The Longitudinal Case Study of the Center City District

There are multiple individual and organizational actors in the complex governance networks conceptualization we described earlier in this chapter. The social networks literature has demonstrated that some of these actors are more central (more visible, influential, or powerful) in these networks and these central actors are not necessarily governmental actors, which are granted legal authorities and expected to be more influential and powerful. Then it is the task of the researcher to identify the actors that are actually more central in governance networks and how their positions and relations with other actors affect the network structures and processes.

The centrality measures in SNA are used to measure the degrees popularity, prestige, and/or social capital, as we noted earlier. These measures are used by some of the researchers of urban governance processes in their investigations of the types of relationships among the actors in urban governance networks (for example, De Socio 2010).

Morçöl, Vasavada, and Kim (2011, 2014) applied the AutoMap–ORA software in their longitudinal case study of some of the characteristics of the urban governance network in Center City, Philadelphia. They investigated the centralities of the governmental and non-governmental actors in the network and what specific policy areas these actors played roles in. In this chapter we summarize the AutoMap and ORA methods these researchers used and illustrate their findings in answering the question on the centrality of the actors in the network.

Morçöl and his colleagues (2011, 2014) gathered the data for their network text analyses from 354 newspaper articles published between 1990 and 2009. The authors first grouped the articles according to the years in which they were published and subsequently conducted longitudinal analyses on these grouped texts. Then, using AutoMap, they identified the individual and organizational actors mentioned in the articles and organized them in two thesauri: a 'generalization thesaurus' and a 'meta-network thesaurus.' Using the information in these thesauri, AutoMap generates a 'meta-network' for each year between 1990 and 2009, which included the matrices of actors, resources they used, actions they took, their locations, and so on. The authors analyzed only the actor–actor matrices to generate the results we describe in this chapter.

The validity and reliability of the identifications of the nodes and their ontological categories are important problem areas in network text analyses, as we mentioned earlier. In the Center City case study, the authors faced particularly the problem of how to properly identify the names of potentially central actors (avoiding miscounting and double-counting), such as mayors who served in successive terms and those who served in multiple capacities (city council member, mayor, and governor) in the period they studied (1990–2009), and how to categorize them properly for the subsequent ORA analyses. The details of how they addressed these problems are discussed in Morçöl et al. (2014).

Morçöl and his colleagues entered the meta-networks that AutoMap generated for each year into ORA. They conducted total degree centrality analyses to determine which actors were most central in each year's network (that is, which actors had the highest number of

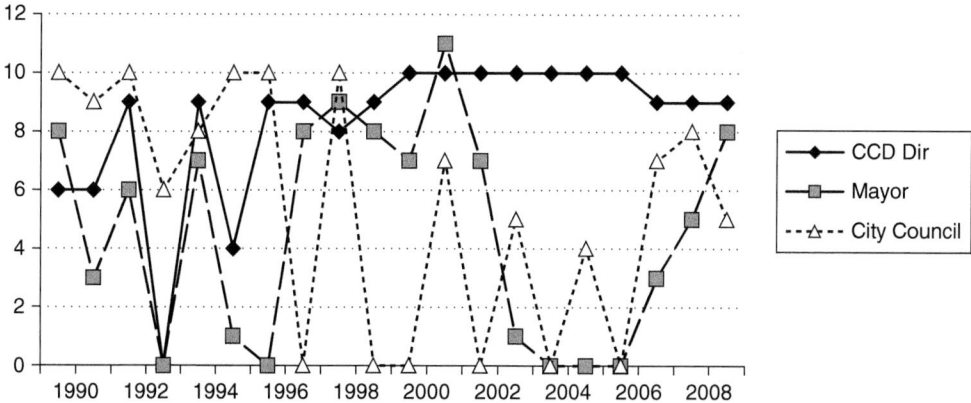

Figure 11.1 Rankings of top three actors over time in the CCD study

connections to the other actors in the texts). For details of the characteristics of degree centrality as an SNA measure, see Wasserman and Faust (1994, pp.178–180), and Knoke and Yang (2008, pp.63–65).

The ORA analyses identified the 10 most central actors in each of the 20 years between 1990 and 2009. Morçöl and his colleagues (2014) found that the director of the Center City District (CCD), the business improvement district of the urban area they studied, was the most central actor in general over the 20 year period. The two governmental actors one would expect to be central in the network, the mayor and the city council, were actually not as central as one would expect from the legal authorities they were granted. The evolutions of the rankings of the centrality scores of these three actors are illustrated in Figure 11.1. (The rankings are converted to ratings to make their plots intuitively meaningful.)

Figure 11.1 indicates that the ranking of the CCD director increased over time, with some fluctuations between 1990 and 2000 and after that he stayed either the first or second ranked (central) actor in the network. The centrality rankings of the city council and the mayor fluctuated over the years, but in general they stayed below those of the director. The consistent centrality of the CCD director over the years was not accidental, nor was it a methodological artifact, as the further interpretations of the texts by Morçöl and his colleagues revealed. They found that the director's involvement in the governance of Center City became deeper and wider, as he took on policy and advocacy roles that were more diverse and more frequently cited over the years (for the details, see Morçöl et al. 2014).

Morçöl and his colleagues (2011) conducted other analyses with the same Center City data. For these additional analyses, they used the two procedures that were available in ORA: immediate impact analysis and sphere of influence analysis. The immediate impact analysis allows the researcher to examine the hypothetical impact of removing an actor from the network on the some of the structural properties of the network (clustering coefficient, density, and so on). ORA calculates the percentage changes in several measures after a node is removed. The sphere of influence analysis can be used to determine the size of the ego network of each individual actor.

For the immediate impact analyses, the researchers calculated the arithmetic means

Table 11.1   Arithmetic means of percentage changes in key indicators after immediate impact analyses in the center city case

	Percent Change After CCD Director Removed	Percent Change After the City Council Removed
Clustering Coefficient	66.98%	29.60%
Density	65.44%	44.69%
Number of Isolated Agents	36.83%	53.93%

of the percentage changes in three key whole network characteristics in the 20 years they studied: clustering coefficient (degree to which nodes in a graph tend to cluster together), social density (number of links/number of potential links), and the number of isolated agents. They removed the CCD director and the city council, separately from the networks to investigate the impacts of each removal on the structural properties in each year. They did not conduct immediate impact analyses for 'the mayor' because multiple mayors served between 1990 and 2009, and some of them served in multiple capacities. This problem could be solved in the analyses Morçöl and his colleagues (2014) conducted, but it would create additional validity problems for the immediate impact analyses. They calculated the arithmetic averages of the changes caused by removing the CCD director and the city council in each of the 20 years. The results are presented in Table 11.1.

The percentage changes presented in the table show that the director's removal had considerable effects on these indicators. In average, the agents in the networks are almost 70 percent more clustered, the density of the network increases by 65 percent, and the number of isolated agents increases by almost 37 percent after the removals. These three measures indicate that the director was indeed a key actor in the network. Without him the network would be more splintered into groups (more clustered), the relations among the agents would be less dense (more distant), and over a third of the agents would not be connected to the network at all. The removal of the city council had similar effects, but the magnitudes of the changes are smaller on the clustering coefficient and density, but larger on the number of isolated agents. This means that there were more agents in the network who were connected solely to the city council.

Using the sphere of influence analysis option in ORA, a researcher can determine the size of an ego network. ORA allows researchers to choose different numbers for the radius of an ego network. Morçöl and his colleagues (2011) selected the radius of 2 (that is, all the actors who are separated from the ego by one and two degrees) for their analyses for the director and the city council. Then they conducted sphere of influence analyses for them for every year between 1990 and 2009. The results of our analyses are shown in Figure 11.2.

The trend lines in Figure 11.2 indicate that the sphere of influence of the CCD director was above that of the city council in general. These results were corroborated by the Pearson's r correlations between the total number of agents on the one hand and the numbers for the director and the city council on the other; these correlations were 0.95 and 0.68 respectively. These results indicate that the director had a wider sphere of influence over the years, compared to that of the city council.

These brief examples and discussions of the analyses in the Center City case study

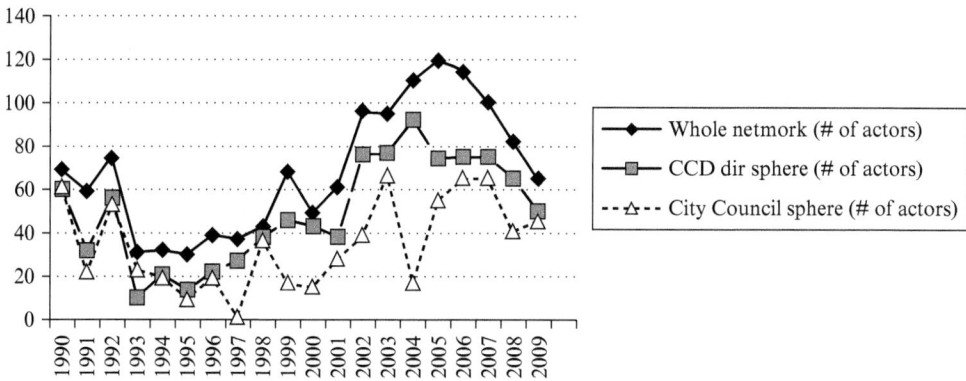

Figure 11.2   Sphere of influence analyses in the center city case

illustrate how AutoMap and ORA can be used to investigate the centralities of actors in governance networks (degree centrality scores and sphere of influence analyses) over time. The methods like hypothetical removals of central actors and sphere of influence analyses can be used to confirm the central position of these actors. The case illustrations in this section show how influential central actors can be on the macro properties of a network and how SNA methods can be used to study the micro–macro relations in networks.

11.5.2   Identifying the Actor Coalitions in Policy Issue Networks: the Illinois SCHIP Case

Complex governance networks do not include only one micro and one macro levels. There are groupings of individuals and organizations that relate to each other in multiple configurations and at multiple levels. A group of scholars of policy processes developed refined conceptualizations of governance networks in response to the increased realization of the complexities in the policy implementation processes and the intergovernmental relations since the late 1970s and early 1980s (for a detailed history, see Ferlie, Lynn, and Pollitt 2005; Morçöl 2014; Weber and Khademian 2008). An important example of these conceptualizations is the Advocacy Coalition Network (ACF) framework (Sabatier and Jenkins-Smith 1993). ACF suggests that policy networks are composed of advocacy coalitions, which are defined by shared beliefs and some level of coordinated behavior among policy actors.

Then it is important to develop methods of identifying advocacy coalitions, which are subgroups in policy networks, and the shared beliefs among the members of these coalitions. To identify the policy coalitions and the shared policy beliefs that are associated with them, Kim and Morçöl (2015) used network text analysis and social network analysis. In their case study of the State Children's Health Insurance Program (SCHIP) policy processes in Illinois, the researchers collected archival data from newspaper articles and used AutoMap and ORA to trace the evolution of the policy networks in this process. They applied the procedure described in the Appendix of this chapter.

Kim and Morçöl (2015) selected the state of Illinois for their study because it is relatively representative of the policy processes in more populous states (that is, California, New York, Pennsylvania, and so on), where there are complex policy environments, including urban and rural interests and policy coalitions and a variety of policy perspectives and approaches. The Illinois SCHIP case was suitable also because the history of the policy process was long enough to track its evolution and there was a wide coverage of SCHIP in the media, which provided the researchers with sufficient information resources to conduct this study.

Kim and Morçöl analyzed the contents of 261 newspaper articles that included references to SCHIP between 1997 and 2007. They grouped the newspaper articles by year during this period and applied AutoMap to extract one meta-network for each year, using the methods described in the previous section, and the protocol described in the Appendix. They then entered these meta-networks into ORA to conduct a series of cluster analyses.

Cluster analysis is used to group individuals or entities into clusters so that entities in the same cluster are more similar to one another than they are to the ones in other clusters (Hair et al. 2009). Kim and Morçöl used the cluster analysis options in ORA to develop some measures of similarity among the actors to identify the coalitions in the SCHIP policy process. More specifically they used the 'Locate Subgroups Report' option in ORA to detect the clustering of actors and the beliefs that those actors held. ORA has a few clustering algorithms available; the researchers selected the Newman algorithm, which is one of the hierarchical clustering methods. The Newman algorithm is useful to find clusters in networks, specifically large networks like the ones in the Illinois SCHIP case (Carley et al. 2013).

ORA generates traditional cluster analysis outputs, such as dendograms (hierarchical clustering diagrams), as well as network maps of clusters, such as the ones presented in Figure 11.3a, 11.3b, and 11.3c. Kim and Morçöl used both dendograms and maps to decide which clustering step is to select as the most appropriate one for each of the years between 1997 and 2007. Figure 11.3a, 11.3b, and 11.3c display the clustering maps for three selected years (1997, 2002, 2007) for illustration purposes. Due to the limited space, we present only the results from selective years that are considered as milestone years in the SCHIP processes, 1997 (SCHIP Authorization), 2002 (IL SCHIP *KidCare* Expansion) and 2007 (SCHIP Reauthorization).

The cluster analyses the researchers conducted confirmed that the policy actors in each cluster were similar to each other, compared to those in other groups in terms of their beliefs, as predicted by the Advocacy Coalition Framework. The cluster analyses in ORA identified three or four groups of individuals during the 11 years they analyzed. Overall, the policy networks of the SCHIP policy process involved actors from all levels of governments as well as non-governmental actors. The results of the ORA analyses show that the central actors in the IL SCHIP policy processes were governors, state agencies, the groups of the members of IL state congress (for example, IL GOP and IL Democrats), and non-profit organizations. The federal government was often involved as an actor (either as a supporter or a challenger) in the SCHIP authorization and reauthorization processes.

As shown in Figure 11.3a, 11.3b, and 11.3c, the actors are clustered around beliefs in the time periods displayed. The cluster analyses identified policy issue coalitions in the SCHIP policy subsystem based on shared policy belief system: pro-SCHIP coalition and against-SCHIP coalition during the time period of 11 years. It is noteworthy that one hegemonic pro-SCHIP coalition existed in the initial year, 1997. That is because there was a consensus among the policy actors in Illinois on SCHIP authorization. As controversial

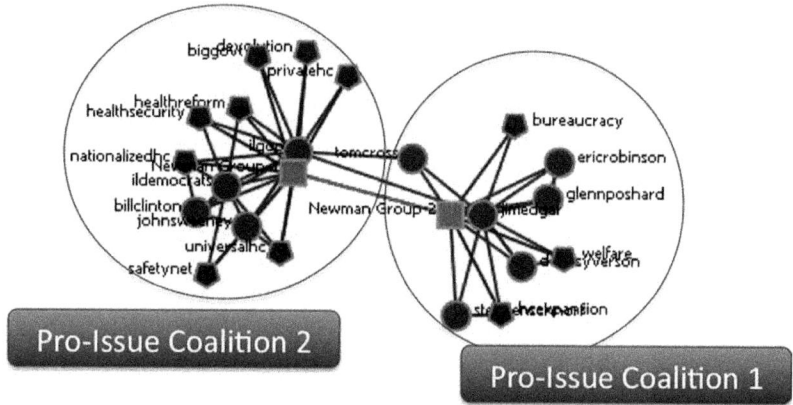

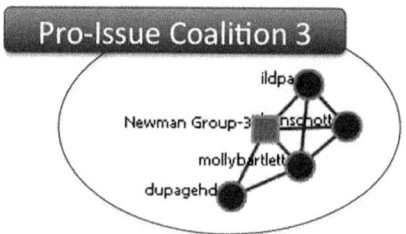

Figure 11.3a   Clusters of actors and beliefs in the 1997 IL SCHIP case

issues (for example, SCHIP expansion) appeared in the years 2002 through 2007, competing pro- and against-SCHIP coalitions were formed over the issues.

A neutral coalition can be observed in 2002 – as well as in other years that are not shown in Figures 11.3a, 11.3b, and 11.3c. These neutral coalitions may be methodological artifacts: residual groups that cluster analyses could not group otherwise. Alternatively they may be interpreted as groups of actors who were in the state government bureaucracy and therefore they were not associated with particular policy positions.

The figures presented in this chapter and the more detailed analyses in Kim and Morçöl (2015) indicate that the configurations of SCHIP coalitions evolved over the years, as a series of controversial policy issues arose. As the figures illustrate, the pro-SCHIP coalitions were split into two to four sub-coalitions based on their different beliefs. Policy actors in each coalition consisted of diverse policy actors and these actors frequently switched their allies or membership over the years.

The Illinois SCHIP case study illustrates how NTA and SNA can be used to identify groupings in networks (for example, advocacy coalitions) and the beliefs associated with them. The application of these methods in a longitudinal study like this one also can help researchers trace the evolution of a governance network. These applications will not yield complete understanding of complex systems, but only partial understanding of a circumscribed system, such as a governance network.

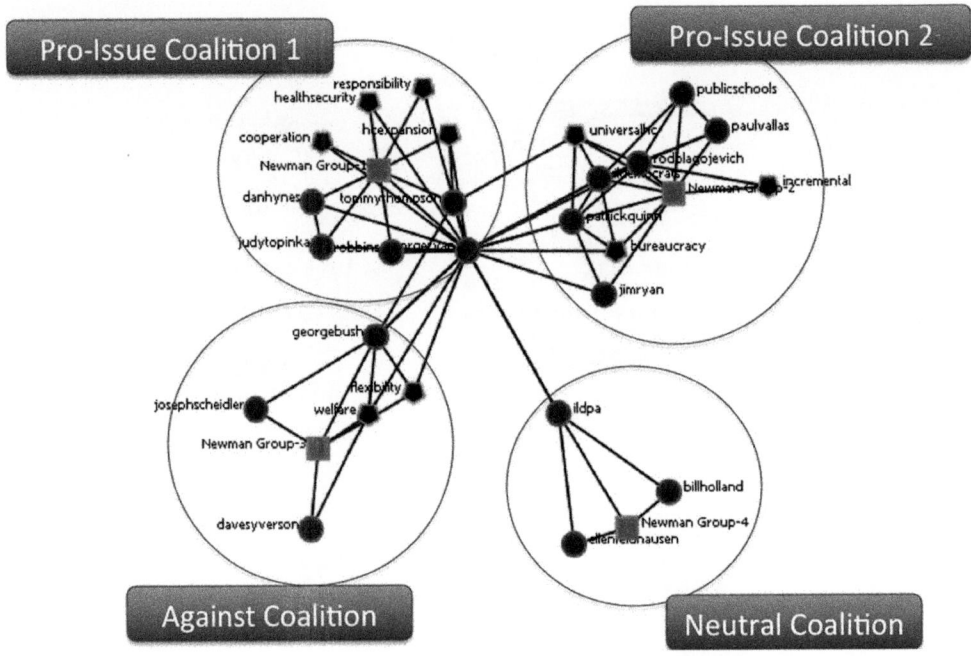

Figure 11.3b   Clusters of actors and beliefs in the 2002 IL SCHIP case

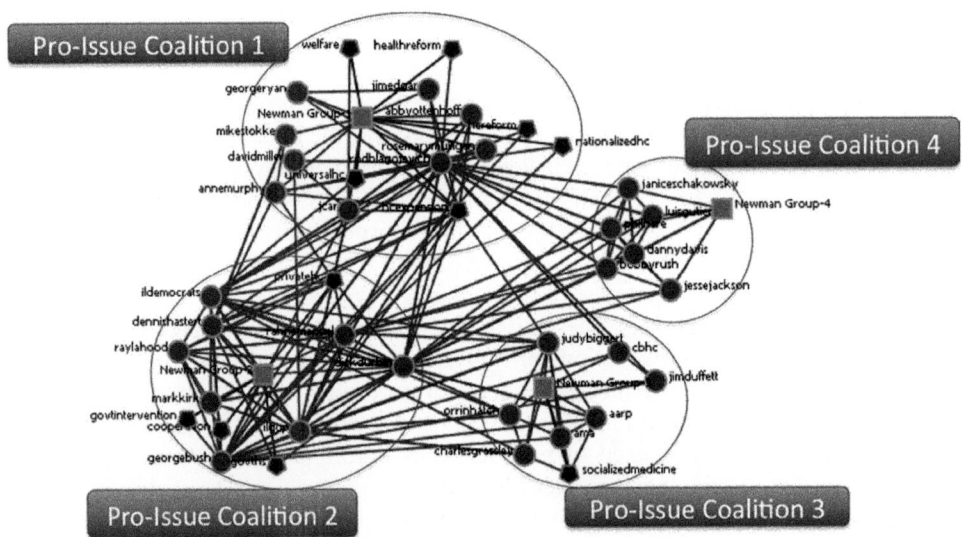

Figure 11.3c   Clusters of actors and beliefs in the 2007 IL SCHIP case

11.6   CONCLUSIONS

The illustrations and discussions in the previous sections show how AutoMap and ORA can be used to investigate parts of the processes in complex governance networks. Although social network analyses, including the ones conducted using ORA, are basically static (they provide snapshots of reality at a given time), the longitudinal analytical approaches in the cases presented in this chapter illustrate that this analytical approach can be used in a dynamic way as well.

Morçöl, Vasavada, and Kim's (2011, 2014) studies illustrate how AutoMap and ORA can be used to identify the evolution of the most central actors in urban governance networks and how the hypothetical removals of the central actors could impact some of the structural properties of the networks (micro–macro relations). Obviously these illustrations are not complete depictions of the micro–macro relations in networks or the evolutions of networks, but they provide information that can be used to make inferences. Kim and Morçöl's (2015) study demonstrates that using AutoMap and ORA a researcher can identify the coalitions of actors in governance networks and trace the evolution of the coalitions (their membership and the values, beliefs associated with the coalitions).

There are some general and specific methodological problems in the applications of AutoMap and ORA, as we discussed in this chapter. The case-specific problems, like how to properly identify the mayors over the years in the Center City case, may be dealt with depending on the context of each study. More general potential validity problems are worth reiterating here. There is the problem of how to interpret textual information in a valid manner in all kinds of content analyses, including network text analysis. AutoMap automates some functions in enumerating and collating concepts in texts, which helps speed up analyses, but researchers should still be cognizant of potential problems of validity and make informed decisions when selecting the options provided by the software.

The protocol we present in the Appendix of this chapter may help future users of AutoMap and ORA to avoid potential pitfalls. However, the protocol should be used only as a list of suggestions and reminders, not as definitive guidelines. As the AutoMap and ORA system evolves, parts of the protocol may become irrelevant and the system is likely to become easier to use.

REFERENCES

Atteveldt, van W. (2008). *Semantic network analysis: Techniques for extracting, representing, and querying media content* (Doctoral dissertation, Vrije Universiteit, Amsterdam, 2008).

Atteveldt, van W., Kleinnijenhuis, J., and Ruigrok, N. (2008). Parsing, semantic networks, and political authority using syntactic analysis to extract semantic relations from Dutch newspaper articles. *Political Analysis, 16*(4), 428–446.

Berry, F.S., Brower, R.S., Choi, S., Goa, W.X., Jang, H., Kwon, M., and Word, J. (2004). Three traditions of network research: What the public management research agenda can learn from other research communities. *Public Management Review, 64*(5), 539–552.

Borgatti, S.P., Everett, M.G., and Johnson, J.C. (2013). *Analyzing social networks*. Los Angeles: Sage Publications.

Carley, K.M. (1997). Network text analysis: The network position of concepts. In C.W. Roberts (ed.), *Text analysis for the social sciences: Methods for drawing statistical inferences from texts and transcripts* (pp. 79–100). Mahwah, NJ: Lawrence Erlbaum Associates.

Carley, K.M., Bigrigg, M., and Reminga, J. (2009). *Basic lessons in ORA and AutoMap 2009*. Retrieved on August 8, 2017 from http://www.dtic.mil/cgi-bin/GetTRDoc?Location=U2&doc= GetTRDoc. pdf&AD=ADA501738.

Carley, K.M., Columbus, D., and Landwehr, P. (2013). *AutoMap user's guide 2013*. Retrieved on August 8, 2017 from http://www.casos.cs.cmu.edu/publications/papers/CMU-ISR-13-105.pdf.

Carley, K.M., Pfeffer, J., Reminga, J., Storrick, J., and Columbus, D. (2013). *ORA user's guide 2013*. Retrieved on August 8, 2017 from http://www.casos.cs.cmu.edu/publications/papers/CMU-ISR-13-108.pdf.

Castells, M. (2000). *The rise of the network society*. Oxford, UK: Blackwell.

De Socio, M. (2010). Marginalization of sunset firms in regime coalitions: A social network analysis. *Regional Studies*, *44*(2), 167–182.

Diesner, J., and Carley, K.M. (2004). Revealing social structure from texts: Meta-matrix text analysis as a novel method for network text analysis. In V.K. Narayanan and D.J. Armstrong (eds), *Causal mapping for research in information technology* (pp. 81–108). Harrisburg, PA: IGI Global.

Diesner, J., and Carley, K.M. (2005). Revealing social structure from texts: Meta-matrix text analysis as a novel method for network text analysis. In V.K. Narayanan and D.J. Armstrong (eds), *Causal mapping for research in information technology* (pp. 81–108). Harrisburg, PA: IGI Global.

Dye, T.R. (2013). *Understanding public policy* (14th edition). Boston: Pearson.

Epstein, J.M., and Axtell, R. (1996). *Growing artificial societies: Social science from the bottom up*. Washington, DC: Brookings Institution Press.

Feiock, R.C. (2013). The institutional collective action framework. *Policy Studies Journal*, *41*(3), 397–425.

Ferlie, E., Lynn, L.E., and Pollitt, C. (2005). Networks and interorganizational management: Challenging, steering, evaluation, and the role of public actors in public management. In E. Klijn (ed.), *The Oxford handbook of public management* (pp. 257–281). New York: Oxford University Press.

Freeman, L.C., Romney, A.K., and Freeman, S.C. (1987). Cognitive structure and informant accuracy. *American Anthropologist*, 89, 310–325. doi:10.1525/aa.1987.89.2.02a00020.

Freeman, L.C., White, D.R., and Romney, A.K. (eds) (1992). *Research methods in social network analysis*. Fairfax, VA: George Mason University Press.

Hair, J.F. Jr., Black, W.C., Babin, B.J., and Anderson, E. (2009). *Multivariate data analysis*. Upper Saddle River, NJ: Pearson Prentice Hall.

Human, S.E., and Provan, K.G. (2000). Legitimacy building in the evolution of small-firm networks: A comparative study of success and demise. *Administrative Science Quarterly*, *45*,327–365.

Jessop, B. (1991). *State theory: Putting the capitalist state in its place*. State College, PA: The Pennsylvania State University Press.

Johnston, E.W., Nan, N., Zhong, W. and Hicks, D. (2008). Between implementation and outcomes, growth matters: Validating an agent-based modeling approach for understanding collaboration process management. *The Innovation Journal: The Public Sector Innovation Journal*, *13*(3), 1–31.

Kadushin, C. (2012). *Understanding social networks: Theories, concepts, and findings*. New York: Oxford University Press.

Kapucu, N. (2006). Interagency communication networks during emergencies: Boundary spanners in multiagency coordination. *American Review of Public Administration*, *26*(2), 207–225.

Kapucu, N. (2009). Interorganizational coordination in complex environments of disasters: The evolution of intergovernmental disaster response systems. *Journal of Homeland Security and Emergency Management*, *6*(1). Retrieved on August 8, 2017 from http://www.bepress.com/jhsem/vol6.

Kettl, D.F. (2002). *The transformation of governance: Public administration for twenty-first century America*. Baltimore, MD: The Johns Hopkins University Press.

Kim, S., and Morçöl, G. (2015). *Policy issue networks in the State Children's Health Insurance Program (SCHIP) in Illinois: A longitudinal case study*. Paper presented the 76th National Conference of the American Society for Public Administration, Chicago, IL.

Knoke, D., and Yang, S. (2008). *Social network analysis*. Thousand Oaks, CA: Sage Publications.

Koliba, C., Meek, J.W., and Zia, A. (2011). *Governance networks in public administration and policy*. Boca Raton, FL: CRC Press.

Kooiman, J. (1993). Social-political governance: Introduction. In J. Kooiman (ed.), *Modern governance: New government–society interactions* (pp. 1–6). London: Sage Publications.

Krippendorff, K. (2004). *Content analysis: An introduction to its methodology* (2nd edition). Thousand Oaks, CA: Sage Publications.

Lim, S., Berry, F.S., and Lee, K-H. (2015). Stakeholders in the same bed with different dreams: Semantic network analysis of issue interpretation in risk policy related to Mad Cow Disease. *Journal of Public Administration Research and Theory*. doi: 10.1093/jopart/muu052. First published online: January 20, 2015.

McGinnis, M.D. (ed.) (1999). Introduction. In M.D. McGinnis (ed.), *Polycentricity and local public economies: Readings from the workshop in political theory and policy analysis* (pp. 1–27). Ann Arbor, MI: The University of Michigan Press.

Milward, H.B., Provan, K.G., Fish, A., Isett, K.R., and Huang, K. (2010). Governance and collaboration: An evolutionary study of two mental health networks. *Journal of Public Administration Research and Theory*, *20*(suppl 1), i125–i141.

Morçöl, G. (2010). *AutoMap and ORA instructions for the center city district network analysis*. Unpublished manuscript.

Morçöl, G. (2012). *A complexity theory for public policy*. London: Routledge.

Morçöl, G. (2014). Complexity of governance networks: An assessment of the state of the literature and proposals for future research. *Complexity, Governance & Networks, 1*(1), 5–16.

Morçöl, G., Vasavada, T., and Kim, S. (2011, September). *Evolution of an urban governance network: The case of Center City, Philadelphia*. Annual Meeting of the American Political Science Association, Seattle, WA.

Morçöl, G., Vasavada, T., and Kim, S. (2014). Business improvement districts in urban governance: A longitudinal case study. *Administration and Society, 46*(7). 796–824.

Mueller, D.C. (1997). Public choice in perspective. In D.C. Mueller (ed.), *Perspectives on public choice: A handbook* (pp. 1–18). Cambridge, UK: Cambridge University Press.

Popping, R. (2003). Knowledge graphs and network text analysis. *Social Science Information, 42*(1), 91–106.

Pressman, J.L., and Wildavsky, A.B. (1984). *Implementation: How great expectations in Washington are dashed in Oakland; or, why it's amazing that federal programs work at all* (3rd edition). Berkeley, CA: University of California Press.

Provan, K.G., and Kenis, P. (2007). Modes of network governance: Structure, management, and effectiveness. *Journal of Public Administration Research and Theory, 18*(2), 229–252.

Radin, B.A. (2013). *Beyond Machiavelli: Policy analysis reaches midlife* (2nd edition). Washington, DC: Georgetown University Press.

Rhodes, R.A.W. (1997). *Understanding governance: Policy networks, governance, reflexivity and accountability*. Maidenhead, UK: Open University Press.

Sabatier, P.A., and Jenkins-Smith, H.C. (1993). *Policy change and learning: An advocacy coalition approach*. Boulder, CO: Westview Press.

Salzano, M. (2008). Economic policy hints from heterogeneous agent-based simulations. In K. Richardson, L. Dennard, and G. Morçöl (eds), *Complexity and policy analysis: Tools and methods for designing robust policies in a complex world* (pp. 167–196). Goodyear, AZ: ISCE Publishing.

Sandström, A., and Carlsson, L. (2008). The performance of policy networks: The relation between network structure and network performance. *Policy Studies Journal, 36*(4), 497–524.

Singleton, R.A., Jr., and Straits, B.C. (2010). *Approaches to social research*. Oxford: Oxford University Press.

Stewart, J. and Ayres, R. (2001). Systems theory and policy practice: An exploration. *Policy Sciences, 34*(1), 79–94.

Torfing, J., Peters, G.B., Pierre, J., and Sørensen, E. (2012). *Interactive governance: Advancing the paradigm*. Oxford, UK: Oxford University Press.

Wasserman, S., and Faust, K. (1994). *Social network analysis: Methods and applications*. Cambridge: Cambridge University Press.

Wasserman, S., and Faust, K. (2009). *Social network analysis: Methods and applications* (9th edition). New York: Cambridge University Press.

Weber, E.P., and Khademian, A.M. (2008). Wicked problems, knowledge challenges, and collaborative capacity builders in network settings. *Public Administration Review, 68*(2), 334–349.

Zia, A., Norton, B.G., Noonan, D.S., Rodgers, M.O., and DeHart-David, L. (2006). A quasi-experimental evaluation of high-emitter non-compliance and its impact on vehicular tailpipe emissions in Atlanta, 1997–2001. *Transportation Research Part D, 11*, 77–96.

Zia, A., Kauffman, S., Koliba, C., Beckage, B., Vattay, G., and Bomblies, A. (2014). From the habit of control to institutional enablement: Re-envisioning the governance of social-ecological systems from the perspective of complexity sciences. *Complexity, Governance & Networks, 1*(1), 79–88.

APPENDIX: A PROTOCOL FOR AUTOMAP AND ORA ANALYSES

The following is a summary of the protocol that was developed by Morçöl (2010). It was used in Morçöl, Vasavada, and Kim (2011, 2014) and Kim and Morçöl (2015). More detailed information about the steps in this protocol is available from the authors of this chapter upon request.

STEPS IN AUTOMAP ANALYSIS

I.   Extracting Meta-Networks using AutoMap

1.   Develop/revise/update a generalization thesaurus and a meta-network thesaurus
Note: Generalization thesaurus matches words and word groups with their 'synonyms.' AutoMap detects these words and word groups in the texts loaded into it and converts them to their synonyms, as instructed in the generalization thesaurus (for example, in Morçöl, Vasavada, and Kim's (2014) analyses, 'Center City District' was converted to "CCD"). Using a meta-network thesaurus, AutoMap generates 'meta-networks,' which include matrices of a variety of nodes: actors, resources they use, actions they take, their locations, and so on.

The validity and reliability of the identifications of the nodes and their ontological categories are important in generalization and meta-network thesauri. A potential problem is that both the proper name of an actor who represents an organization and the organization may be cited back-to-back in the same text, which may cause double-counting. Morçöl, Vasavada, and Kim (2014) noticed, after their initial analyses, that the proper name of the executive director of the BID and the name of the BID were frequently cited in close proximity or back-to-back. To prevent double-counting, they included only the name of the director of the BID in their analyses and excluded the name of the BID.

2.   Preparation of the generalization thesaurus file

2.1   Read each news articles and add category lists.
The categories will include individual agent, collective agent, organization, resource, knowledge, and action.
2.2   Convert this file into a CSV file (for example, 'SCHIP Generalization Thesaurus. csv').
2.3   Edit the generalization thesaurus file in AutoMap:
Tools
Thesauri Editor
Under 'File':

- Open file.
- Check the file visually. (Is there any item on the third column? If yes, make the correction or remove it.)

Under 'Procedures':

- Check thesaurus for missing entries.
- Check thesaurus for duplicate entries.
- You may want to use 'Sort thesaurus file,' but this is tricky; it sorts some of the entries in a way that you may not want.
- Once these are done, save the file.

2.4 Open the generalization thesaurus (csv) file in Excel and double-check the corrections and revisions you made using the thesaurus editor in AutoMap.

2.5 Once the corrections are made in the generalization thesaurus, add the categories in the original Excel file ('SCHIP Generalization Thesaurus.xls') and save it under the name of this Excel file.

3. Prepare the meta-network thesaurus file

3.1 Remove the first column of entries in this Excel file. (The categories in the last column should remain.)

3.2 Remove the duplicate entries in this file. (This is labor intensive. This step may not be necessary. ORA seems to be tolerant of duplicate entries. If it is, then there is no need to do this.)

3.3 If you see errors in the generalization thesaurus and the Excel file, correct them.

3.4 Save this file as a meta-network thesaurus file.

3.5 If for some reason you need to develop your own thesaurus from scratch using AutoMap, here is how:

   Generate

   MetaNetwork

   Suggested MetaNetwork Thesauri

3.6. Once AutoMap creates a "suggested thesaurus" for you, you will need to revise it.

4. Prepare the file(s) for AutoMap analysis

4.1 Make sure that each newspaper article is directly relevant to the key word you study (for example, SCHIP).

4.2 Save each newspaper article separately as a text file (.txt, Windows Default). Name each article using the following coding scheme:

   CT19970207.txt

   Abbreviation for the newspaper's name (for example, Chicago Tribune is named as 'CT' and date of publication in the following format: year (four digits), month (two digits), day (two digits)

4.3 Load the text file(s) into AutoMap.

   File

   Import Text Files

- Load all the articles published in the same year together.
- Select all of them from the directory and load all at once.

4.4   Develop a meta-network:
a.   Preprocess data to clean up:
Note that whether you remove symbols, punctuation, and numbers will have consequences. Also whether you choose the 'replace with space' option in the following operations will have consequences. These choices should be coordinated with the wording used in the generalization thesaurus. The most straightforward method may be not to remove symbols, punctuation, or numbers and apply generalization thesaurus directly.

- Remove symbols
- Remove punctuation
- Remove numbers.

b.   Apply the generalization thesaurus ('SCHIP Generalization Thesaurus.csv' or the most recent version of the thesaurus file).
- *Select 'Use thesaurus content only: Yes.'*
- [This is a potentially problematic issue. Carley (personal communication) states that this creates sparsely populated networks. That is a problem, but this method still yields the most valid results.]
- *Select 'type of delete processing: Rhetorical.'*
- *'Would you like to include an exception list: No.'*

c.   Generate a *meta-network*:
Generate
MetaNetwork
MetaNetwork DyNetML (Per Text) or (Union Only)

- Select (or create) a directory for the meta network to be created.
- Then select the meta-network thesaurus to be used in the meta-network generation (that is, select 'SCHIP Meta Network Thesaurus.csv').
- Select *directionality*: unidirectional.
- Select *window size*: a number between 20 and 40.
- Select *stop unit*: 'All.'
- Select *stop unit value*: none [This is not applicable when 'all' is selected. If 'sentence' is selected, this value represents the number of sentences to be included in each step of the analyses.]
- 'Confirm.'

II. Prepare Meta-network Files for ORA

- AutoMap saves meta-network files for each year.
- Rename the meta-network files by adding the year and window size to easily recognize them when you load the files to ORA.

Steps in ORA Analyses

[Note the following does not include all the options in ORA. It is meant to provide a basic outline of some of the methods that can be used in ORA.]

1. **Read the meta-network into ORA**
   In ORA:
   File:
   >   Open meta-network.

2. **Analyze the meta-network in ORA**

   a. Visualize the network first.
      This creates both the meta network map and the semantic network map.
   b. Generate reports:
      To generate reports:
      Analysis
      Generate Reports

   - Select 'Standard Network Analysis.'
   - Select the number of ranked nodes to display [we choose the default value of 10 or insert a larger number here].
   - Select an attribute and value [We choose 'frequency' and '1'].
   - Select 'agent*agent' network.

The Standard Network Analysis report generates group level measures (for example, number of articles, number of agents, density, degree centralization, network levels, clustering coefficient, network fragmentation, Krackhardt connectedness, Krackhardt hierarchy) and individual level measures (for example, top ranked agent charts, total degree centrality, eigenvalue centrality, betweenness centrality).

- Select 'Locate Subgroups' report for cluster analysis.
- Select the input network: agent * agent, agent * knowledge [For the purpose of your study, you can modify the analytical options].
- Select a 'Grouping Algorithm': Newman.
- Select 'Automatically Compute Number of Groups.'

PART IV

MODELLING AND STATISTICAL ANALYSIS OF EMPIRICAL DATA

Professor Bill McKelvey

Phase 1 of complexity science began back in the 1950s when Ilia Prigogine started writing about phase transitions just beyond the '*edge of order*' (the *1st critical value*) in physical systems resulting from the reduction of imposed tensions (Prigogine 1955, 1962, 1997; Prigogine and Stengers 1984; Nicolis and Prigogine 1989). Starting in the late 1980s, scientists at the Santa Fe Institute of Complexity Sciences shifted complexity science to the study of emergent behaviour in living biological and social systems (Anderson 1972; Gell-Mann 1988, 2002; Holland 1988, 2002; Kauffman 1993; Arthur 1994).

Phase 2's focus was on heterogeneous agents interacting just before the '*edge of chaos*' (the *2nd critical value*) of imposed tensions. In between the '*edges*' of *order* and *chaos* is the *Region of Emergent Complexity*, what Kauffman terms the '*melting*' zone (1993, p. 109). Bak (1996) argued that to survive, organisms have to have a capability of staying within the melting zone, maintaining themselves in a state of '*self-organized criticality*,' that is, adaptive efficacy. Holland (2002) defined emergent phenomena as *multi-level hierarchies*, *intra- and inter-level causal processes*, and *nonlinearities*.

Phase 3 focuses on nonlinearity, scalability, power laws (PLs), the 'butterfly effect',[1] scalability and fractals. Though beginning many decades ago with Pareto (1897), Auerbach (1913) and Zipf (1949), Phase 3 re-focused attention to PL phenomena (Newman 2005; Andriani and McKelvey 2007, 2009), and eventually includes econophysics (West and Deering 1995, Mantegna and Stanley 2000). Econophysics began with Benoit Mandelbrot's focus on stock market crashes (1963). While crashes are negative extreme events, their showing of the PL signature indicates that the markets were free to go up or down without restraint. PLs often appear as indicators of self-organization, emergence-in-action, self-organizing economies (Krugman 1996), and the growth of firms, cities and economies (Stanley et al. 1996; Axtell 2001).

Various descriptions of how complexity science has been applied to organizations appear in Maguire et al. (2006) Hazy et al. (2007), Allen et al. (2011), Strathern and McGlade (2014). These existing views and theories about how complexity science thinking and concepts apply to organizations and management are more explicitly tested and elaborated in the chapters comprising this Section.

Crawford and McKelvey (Chapter 12) focus on statistical tests of whether power laws (PLs) actually exist in industry distributions. Especially important is their use of the x_{min} concept to identify the point where normally-distributed phenomena shift into PL-distributed phenomena as some firms start growing out toward the 'stochastic frontier' (Aigner et al. 1977; Koop et al. 1999; Kumbhakar and Knox Lovell 2000; Lensink and Meesters 2014; Ravishankar and Stack 2014). Brown et al. (Chapter 13) focus on multifractal empirical analyses of 'web-traffic' connections to gender terms used in Wikipedia. Dister et al. (Chapter 14) conduct an empirical analysis of multiple cases (power-grid segments in the US state of Ohio) that uses both qualitative, quantitative, and statistical methods available in the SACS Toolkit created by Castellani and Rajaram (2012). Wolf-Branigin et al. (Chapter 16) use two NetLogo (Wilensky and Rand 2015) agent-based computational models (ABMs) to study the relative impacts of policy alternatives such as 'respite care, tax incentives, work-place policies and adult day-care services' on caregiver stress. Hazy and Wolenski (Chapter 15) use a canonical mathematical model – Goldstein et al.'s *Cusp of Change* model (2010) – to study the impact of degrees of freedom and internal and external complexity conditions on human interaction dynamics. Pourbohloul et al. (Chapter 17) focus on the use of complexity-relevant mathematical network modelling methods that would significantly improve the management of emerging disease outbreaks here and there around the world.

Crawford and McKelvey (Chapter 12) begin by explaining the basic causes of skew distributions. "These processes are generated by scale-free mechanisms – the same cause at multiple levels of analysis – that result in self-similar fractal structures, that is, power-law distributions (PLDs) within firms and other social entities". They then briefly describe how Bak's (1996) 'self-organized criticality' can be "used to explain how and why social and organizational entities maintain viable new-order creation as they coevolve" with competing firms. This discussion provides context for their data analysis and subsequent interpretation of the results. The authors use a bootstrapped maximum likelihood estimation (MLE) technique developed by Clauset et al. (2009) using three longitudinal datasets of entrepreneurial firms. They use "these variables and samples because any hypothesis about any mechanism driving the emergence of PLDs in a research domain requires a significant empirical finding (that is, the distribution is actually a PLD)". They "then draw conceptual links from the individual parameters of the MLE to the generative mechanisms". Most specifically, they identify the "critical point in a distribution where systems change from linear to nonlinear". They explain how each of the parameters generated by their statistical techniques can be interpreted via complexity science concepts.

Brown et al. (Chapter 13) begin by using gender studies – such as black versus white or male versus female – to illustrate how the traditional focus on binary examples supports the traditional deeply ingrained desire for logical formalisms and conceptually dynamic models of systems. The traditional way of avoiding math-driven simplifications was to use case studies so as to be able to describe the richness of human and social realities. In contrast, the authors focus on what is now termed 'intersectionality theory', which recognizes that many people have complex multiple memberships in various biological, behavioural, and social categories. They introduce multifractal analysis as a means for studying 'cascade dynamics', for example, defined as how DNA becomes more differentiated and complex as a family grows from two original parents to many distantly-related offspring. The authors suggest that "cascade dynamics and multifractal analysis can provide a logical formalism and statistical framework to make intersectionality a quantitatively tractable" representation of gender development. They review recent cognitive-science advances in which multifractal analysis identified key features of the cascades affecting cognitive performance. Using "web-traffic data for gender terms on Wikipedia", they demonstrate how similar cascade structures in gender dynamics develop by using multifractal analysis. They conclude by proposing that "cascade formalisms and multifractal analysis may provide new avenues for gender studies that balance both logical formalisms and dynamic concepts".

Dister et al. (Chapter 14) observe that "if one is to improve reliability and resilience in infrastructures . . . it is necessary to adopt a 'complex, smart territory' modelling strategy, particularly one that gives particular attention to the importance of social complexity. To test the veracity" of their argument, they "conduct a case study on a segment of the United States power grid". Their goal is simple: they "seek to create a first proof-of-concept sufficient to show, in the simplest of cases, how thinking about infrastructures in 'complex systems' terms, primarily in terms of their social aspects, can prove beneficial". For their case study, they use "the SACS Toolkit: a new method for quantitatively modeling complex social systems, based on a case-based, computational approach to data analysis" (quote taken from Castellani and Rajaram, 2012: p. 153), "which is part of the new approach to modelling complex systems, called *case-based complexity*. As a technique, the SACS Toolkit is a computationally grounded, case-comparative, mixed-methods platform for modelling complex systems as sets of cases." It integrates "case-based reasoning with complex critical realism and the latest developments in complexity theory". They provide a "basic overview of their research process, ending with a summary of novel insights".

Hazy and Wolenski (Chapter 15) advocate a mathematical research agenda to categorize a theoretic representational framework to 'connect all scientific approaches to human interaction dynamics (HID), which seek to "build a cumulative base of knowledge to inform individual choice and behavior". They illustrate their approach by using the *Cusp of Change* canonical mathematical model (Goldstein et al., 2010) that describes the potential for first order-phase transitions. They suggest that the "degrees of freedom that are active in the action orientation of a stable organizing state may be a useful order parameter and further that indexes reflecting the internal and external complexity conditions that are confronting the population can be used as control parameters". Their approach takes the position that action orientation gathers its potency via "gradients of differences in predictability and uncertainty regarding relevant events. As agents seek to reduce the unpredictability and thus their cognitive load by moving along this gradient, informational-influence forces set up a potential field under which organizing for action occurs among independent semi-autonomous agents. By using the information in socio-technical structure that can be recognized and decoded, agents can ease their cognitive load while maintaining a perceived level of predictability about events in the environment."

Wolf-Branigin et al. (Chapter 16) note that social planners "create innovative interventions to address the diverse and expanding needs of vulnerable populations. Over time, these social innovations require an increasing high level of flexibility and adaptability to remain effective in addressing the continually changing, disparate, and incompatible preferences of clients, funding sources, and other stakeholders". The authors apply complexity science as a social programme evaluation methodology and then test programme and policy options using agent-based models. They then discuss how complex adaptive systems (CAS), and its related components are beneficial to social programme evaluators and researchers. Their use of complex adaptive systems (CAS) facilitates bridging complexity science to social programme evaluation; several characteristics of complex adaptive systems have direct implications to programme evaluation, which "include non-linearity, emergence, being adaptive, having uncertainty in that estimating values cannot be exact, and being dynamical and co-evolutionary". The authors use ABMs to evaluate social programmes framed as complex adaptive systems. Foremost is the ability to forecast outcomes (represented as emergent behaviours) over time because the running of an ABM assumes the presence of an iterative process. They note that problems remain, however, in sufficiently matching modelling schemes and social realities. The ABM approach appears to function well when the individual agents are empowered and have the ability to make choices from the programmes from which they receive information and services.

Pourbohloul et al. (Chapter 17) use mathematical network models to describe various kinds of contacts among people that can cause quickly spreading disease outbreaks here and there around the world that are a special concern of health care providers and public health officials. They also note that further research is needed to better understand how infectious-disease spreads occur in hospitals, so as to better identify which individual(s) or group(s) of healthcare workers in hospitals who are most likely to foster the spreading of infectious diseases, for example, trainees, senior clinicians and so on. Strategies for more effectively requiring self-protection for these highly network-connected individuals (for example, wearing surgical masks during their working hours, wearing disposable gloves, and washing hands more often) may effectively reduce disease spread in hospitals. By incorporating the heterogeneous behaviour of individuals as well as the heterogeneity in disease transmissibility, complexity science tools allow them to accurately pinpoint optimal and economical intervention strategies, thus contributing to evidence-based public health research and decision-making, and also to the health of people worldwide. They emphasize one reassuring fact: the sooner control measures are implemented appropriately, the lower the rate of infection across many populations.

NOTE

1.   The so-called '*butterfly effect*' stems from Lorenz's 1972 paper: 'Does the flap of a butterfly's wings in Brazil set off a tornado in Texas?' These are Holland's (2002) 'tiny initiating events' that scale up to extreme outcomes.

REFERENCES

Aigner, D.J., Lovell, C.A.K. and Schmidt, P. 1977. Formulation and estimation of stochastic frontier production function models. *Journal of Econometrics* **6**(1): 21–37.

Allen, P., Maguire, S. and McKelvey, B. 2011. *The Sage Handbook of Complexity and Management*. London: Sage Publications.

Anderson, P.W. 1972. More is different. *Science* **177** 393–396.

Andriani, P. and McKelvey, B. 2007. Beyond Gaussian averages: Redirecting organization science toward extreme events and power laws. *Journal of International Business Studies* **38** 1212–1230.

Andriani, P. and McKelvey, B. 2009. From Gaussian to Paretian thinking: Causes and implications of power laws in organizations. *Organization Science*, **20**(6): 1053–1071.

Arthur, W.B. 1994. *Increasing Returns and Path Dependence in the Economy*. Ann Arbor, MI: University of Michigan Press.

Auerbach, F. 1913. Das Gesetz d er Bevolkerungskoncentration. *Petermanns Geographische Mitteilungen* **59** 74–76.

Axtell, R.L. 2001. Zipf distribution of US firm sizes. *Science* **293** 1818–1820.

Bak, P. 1996. *How Nature Works: The Science of Self-Organized Criticality*. New York: Copernicus.

Castellani, B. and Rajaram, R. 2012. Case-based modeling and the SACS Toolkit: A mathematical outline. *Computational and Mathematical Organization Theory*, **18**(2): 153–174.

Clauset, A., Shalizi, C.R. and Newman, M.E.J. 2009. Power-law distributions in empirical data. *SIAM Review* **51**(4): 661–703.

Gell-Mann, M. 1988. The concept of the Institute. In D. Pines (ed.), *Emerging Synthesis in Science*. Boston, MA: Addison-Wesley, pp. 1–15.

Gell-Mann, M. 2002. What is complexity? In A.Q. Curzio and M. Fortis (eds), *Complexity and Industrial Clusters*. Heidelberg, Germany: Physica-Verlag, pp. 13–24.

Goldstein, J., Hazy, J.K. and Lichtenstein, B.B. (2010), *Complexity and the Nexus of Leadership: Leveraging Nonlinear Science to Create Ecologies of Innovation*. Englewood Cliffs, NJ: Palgrave Macmillan.

Hazy, J.K., Goldstein, J.A. and Lichtenstein, B.B. (eds) 2007. *Complex Systems Leadership Theory*. Mansfield, MA: ISCE Publishing.

Holland, J.H. 1988. The global economy as an adaptive system. In P.W. Anderson, K.J. Arrow and D. Pines (eds), *The Economy as an Evolving Complex System*, Vol. 5. Reading, MA: Addison-Wesley, pp. 117–124.

Holland, J.H. 2002. Complex adaptive systems and spontaneous emergence. In A.Q. Curzio and M. Fortis (eds), *Complexity and Industrial Clusters*. Heidelberg, Germany: Physica-Verlag, pp. 24–34.

Kauffman, S.A. 1993. *The Origins of Order*. Oxford, UK: Oxford University Press.

Koop, G, Osiewalski, J, and Steel, M. 1999. The components of output growth: A stochastic frontier analysis. *Oxford Bulletin of Economics and Statistics* **6**(1): 455–487.

Krugman, P. 1996. *The Self-Organizing Economy*. Malden, MA: Blackwell.

Kumbhakar, S.C. and Knox Lovell, C.A. 2000. *Stochastic Frontier Analysis*. New York: Cambridge University Press.

Lensink, R. and Meesters, A. 2014. Institutions and bank performance: A stochastic frontier analysis. *Oxford Bulletin of Economics and Statistics* **76**(1): 67–92.

Lorenz, E.N. 1972. Predictability: Does the flap of a butterfly's wings in Brazil set off a tornado in Texas? Paper presented at the 1972 meeting of the American Association for the Advancement of Science. Washington, DC.

Maguire. S., McKelvey, B., Mirabeau, L. and Öztas, N. 2006. Complexity science and organization studies. In S. Clegg, C. Hardy and T. Lawrence (eds), *Handbook of Organizational Studies* (2nd edition). London: Sage Publications, pp. 165–214.

Mandelbrot, B.B. 1963. The variation of certain speculative prices. *Journal of Business* **36**: 394–419.

Mantegna, R.N. and Stanley, H.E. 2000. *An Introduction to Econophysics*. Cambridge, UK: Cambridge University Press.

Newman, M.E.J. 2005. Power laws, Pareto distributions and Zipf's law. *Contemporary Physics* **46** 323–351.

Nicolis, G. and Prigogine, I. 1989. *Exploring Complexity: An Introduction*. New York: Freeman.

Pareto, V. 1897. *Cours d'Economie Politique*. Paris: Rouge & Cie.

Prigogine, I. 1955. *An Introduction to Thermodynamics of Irreversible Processes*. Springfield, IL: Thomas.

Prigogine, I. 1962. *Non-Equilibrium Statistical Mechanics*. New York: Wiley-Interscience.

Prigogine, I. and Stengers, I. 1984. *Order Out of Chaos: Man's New Dialogue with Nature*. London: Fontana Paperbacks.

Prigogine, I. and Stengers, I. 1997. *The End of Certainty*. New York: Free Press.

Ravishankar, G. and Stack, M.M. 2014. The gravity model and trade efficiency: A stochastic frontier analysis of Eastern European countries' potential trade. *World Economy* **37**(5): 690–704.

Stanley, M.H.R., Amaral, L.A.N., Buldyrev, S.V., Havlin, S., Leschhorn, H., Maass, P., Salinger, M.A. and Stanley, H.E. 1996. Scaling behaviour in the growth of companies. *Nature* **379** 804–806.
Strathern, M. and McGlade, J. (eds) 2014. *The Social Face of Complexity Science*. Litchfield Park, AZ: Emergent Publications.
West, G.B. and Deering, B. 1995. *The Lure of Modern Science: Fractal Thinking*. Singapore: World Scientific.
Wilensky, U. and Rand, W. 2015. *An Introduction to Agent-based Modeling: Modeling Natural, Social, and Engineered Complex Systems with NetLogo*. Cambridge, MA: MIT Press.
Zipf, G.K. 1949. *Human Behavior and the Principle of Least Effort*. New York: Hafner.

12. Using maximum likelihood estimation methods and complexity science concepts to research power law-distributed phenomena

Assistant Professor G. Christopher Crawford, Ohio University and Professor Bill McKelvey, UCLA

12.1 INTRODUCTION

Life is not normally distributed – we live in a world of extreme events that highly skew what we consider 'average.' In many cases, scholars use a complexity science perspective to explain the emergence of such events. However, complexity science scholars have long been criticized for evoking too many metaphors and not enough quantitative rigor in building and testing theory on the emergence and creation of new order. This is an especially important problem in social sciences because explaining and predicting emergence is vital to pedagogy, policy, and practice. Nonetheless, with its assumptions of nonlinearity, indeterminism, postmodern perspectives, and the potential for interdependent agents to cause large events, studies of the emergence of complex adaptive systems require methodological techniques that are often outside of what traditional science would consider 'normal'. Indeed, to remain theoretically consistent, methods to build and test complexity-based hypotheses require statistical techniques with the same underlying assumptions.

Complexity science makes a fundamental transition from the math-required *i.i.d.* (independent, identically distributed) assumptions that physicists and economists make about the entities they study – atoms, molecules, planets or people, firms, and economies – to the reality of entities (agents) that typically have idiosyncratic attributes and various patterns of interacting with other agents. Instead of the traditionally assumed Gaussian distributions of homogeneous agents (where means dominate), complexity assumes that agents exhibit heterogeneous attributes and interaction patterns, which result in inputs and outcomes that consistently display highly right-skewed, heavy-tailed distributions (where outliers dominate).

In most emerging social systems, these heavy-tailed distributions are power laws. Over the last several decades, there have been significant conceptual and empirical efforts by complexity-science scholars in their search for a universal mechanism to explain the emergence of these distributions. Andriani and McKelvey (2009) summarize this body of research, identifying 15 mechanisms (that is, causes) of emergent new-order creation in social, organizational, and financial systems. Within power law distributions (PLDs), the outliers in the tail exert a disproportionate influence on the statistical and behavioral characteristics of the other observations in the population. As evidence, Crawford et al.'s (2015) longitudinal study of four samples of more than 12,000 entrepreneurial ventures, at different states of organizational emergence, found PLDs in all measures of *inputs* – including human, social, and financial capital at the individual and team level, as well as expectations for future growth and environmental resources – and all measures of *outcomes* – including revenue, number of employees, and growth. Our chapter suggests that systems (for example, individuals, teams, ventures, and so on) with inputs and outcomes in the tail of the distribution have the highest potential to emerge successfully and have the greatest potential to exert co-evolutionary influence on higher levels of analysis (for example, community, industry, economy). Outliers in these distributions exhibit nearly infinite variance, novelty, and co-evolutionary influence – indeed, outliers are worthy of study on their own. However, without the identification that an outlier came from a PLD, traditional research methods suggest that outliers are simply random events that cannot be explained or predicted. Complexity science, however, posits that there is an underlying pattern of emergence that generates PLDs and the outliers therein. Thus, our chapter focuses on the questions: *What causes the emergence of PLDs? How do we know if it's really a PLD? At what point in the distribution do observations begin to shift from a normal distribution to a PLD? How influential are PLD outliers on the rest of the distribution? How can we test what mechanism(s) might drive PLDs?* We address each question both theoretically and empirically.

Our chapter begins with a brief explanation of the basic causes of skewed distributions. We follow with a section on horizontal scalability processes, including the search for and acquisition of resources that sometimes turn small Ma&Pa stores into giant firms like Walmart. These processes are generated by scale-free mechanisms – the same cause at multiple levels of analysis – that result in self-similar fractal structures (that is, PLDs) within firms and other social entities. We also briefly describe Bak's (1996) explanation of how 'self-organized criticality' works in nature, which can also be used to explain how and why social and organizational entities maintain viable new-order creation as they coevolve with the evolution of competing firms. This discussion provides context for our data analysis and subsequent interpretation of the results.

Our method incorporates a bootstrapped maximum likelihood estimation (MLE) technique developed by Clauset, Shalizi, and Newman (2009) on *input* and *outcome* variables from three longitudinal datasets of emerging organizations from Crawford, McKelvey, and Lichtenstein (2014) and Crawford et al. (2015) to answer the research questions. We utilize these variables and these samples because any hypothesis about any mechanism driving the emergence of PLDs in a research domain requires a significant empirical finding (that is, the distribution is actually a PLD). We then draw conceptual links from the individual parameters of the MLE to the generative mechanisms identified in section 12.5. Most specifically, we focus on identifying the critical point in the distribution where systems transition from linear to nonlinear, from normal to novel. We explicate the parameters generated by our statistical techniques and explain, in relatively simple terms and with clear examples, how each can be interpreted with complexity science concepts.

In this chapter, while we describe and demonstrate the MLE method, we also stress the importance of good theory development, where theoretical and methodological assumptions are aligned. Hence, throughout, this chapter provides scholars with the conceptual-empirical link for moving beyond loose qualitative metaphors to rigorous quantitative analysis so as to enhance the generalizability and utility of complexity science theories.

12.2 BASIC CAUSES OF SKEWED DISTRIBUTIONS

The fundamental concepts and dynamics of complexity science have been described many times previously (see for example: Allen, Maguire and McKelvey 2011; Hooker 2011; McKelvey 2013a, 2013b, 2013c, 2013d, 2013e; plus chapters in this volume). In this section, we define key indicators of measurable variables that can represent data of high face validity for testing complexity dynamics relevant to organizations and/or management. After a brief introduction of relevant theory, each section ends with a table presenting a list of key complexity concepts that are essential for valid research about complexity dynamics. Example definitions of each variable and implicit concept are also included.

12.2.1 Adaptive Tension: Phase 1

Phase 1 in the development of complexity science emphasizes imposed tension (heat), the 1st critical value, and dissipative structures. It is based on the works of Prigogine (1955, 1962, 1997), Haken (1983), Prigogine and Stengers (1984), Nicolis and Prigogine (1989), and Mainzer (1994/2007), among many others. It begins with the Bénard (1901) process in which a critical value indicates where a phase transition occurs, for example, water turns from boiling water into steam at 100°C. An organizational example occurs when a firm is impacted by the competitive tension imposed by a fast growing competitor or one that creates a dramatically different new product. A dramatic recent example is the impact of Apple's iPhone, iPad, smartphones in general, and Microsoft's Cloud computing that has resulted in reduced production of laptop computers by Dell and Hewlett Packard. In the iPhone example, the phase transition is from computing via laptops to computing via smartphones and the Cloud. Prigogine termed the new phenomena '*dissipative structures*' because the phase transition causes the phenomenon to appear in a form of new order (for

Table 12.1   Phase 1: Variables related to tension effects

---

1   ***Tension*** variables of forces causing adaptation: e.g., indicators of tensions imposed by Jack Welch's phrase at GE: 'Be #1 or 2 in your industry, or else. . ..' Or: We want 25% new products every 5 years. Or: Top management demands such as: increase efficiency; cut costs; produce higher profits; speed up rate of new technology development, etc. Or: Could be the indicators of: new product concepts, more efficient production methods, off-shore production that has lower costs; a new competing firm; new government regulations, etc. This type of information could be derived from specific verbiage in a company's mission or vision statements within annual reports, or via a content analysis for some of the terms listed above.

2   ***1st critical value*** (edge of order) variables: e.g., an indicator of the level of tension at which new order appears. Indicators showing when employees are more creative, come up with new ideas and products; work together to create new product ideas, create new groups or departments, etc.

3   ***Dissipative structure*** (phase transition) variables: e.g., indicators could be firms resulting from M&As; new groups, departments, units, divisions, etc., within an organization; could be new ventures resulting from entrepreneurs trying to reduce the tensions between lack of supply to meet demand.

Table 12.2   Phase 2: Variables related to the emergence of new forms of order

1   *2nd critical value* (edge of chaos) variables: e.g., indicators of the edge of chaos; two different kinds of force: (1) when there are too many different kinds of tension imposed on an agent or system at the same time that it can't respond is effectively to any of them or (2) if there so much of one kind of tension imposed on an agent or system that it becomes dysfunctional; so many different kinds of imposed demands that employees don't know how to respond in a timely and effective manner.

2   *Region of emergence* (melting zone) variables: e.g., indicators of the 1st & 2nd critical values (edge of order and edge of chaos): lies between the 1st & 2nd critical values; how wide is the Region? Do people respond to tension quickly, easily, and effectively? People make effective responses to multiple tensions without getting freaked out or not knowing what to do next.

3   *Heterogeneous Agents* (many kinds) as variables: e.g., entities such as employees, groups, departments, subsidiaries, networks, ideas in a firm and different kinds of activities; competing firms, governments, economies, resources. The indicators of each are whatever descriptive elements are identifiable and measurable.

4   *Self-organization* variables: indicators occur only when agents themselves become motivated to change – there is no 'global controller' as Holland (1988) puts it – agents don't need to be told to start changing; they just do it because of some individual or collective motivation. The minimum ingredients for self-organized new order to emerge are tension, connectivities among agents, and agents' motivations to adapt to the imposed tension (or negation of imposed tensions); indicators are new ideas, behaviors, groups, and networks.

5   *Tiny initiating event* variables: There are many (tens, hundreds, thousands) of seemingly meaningless incidents or changes in any given firm over the course of a year. Most are just random events that quickly disappear. All indicators appear as random, novel events to begin with. BUT, some repeat and start growing/repeating, thereby becoming the beginnings of networked behavior, agreements, groups, and eventually cause significant changes.

6   *Connectivity* variables: Indicators: e.g., various kinds of connectivities such as: influences one agent has on another; influences affecting many agents in the same way and/or at the same time; influences that change because of other influences or changing agent attributes.

7   *Motives to connect (link)* variables: Indicators are various needs and actions to influence others.

8   *Motives to survive and grow* variables: Employees have all kinds of motivations, but they can enter a firm or be trained or incentivized to become passive-dependent, loners, and maintain the status quo OR they can learn, change, interact, motivate others, innovate, and adapt to and survive changing competitive environments. Some people are strongly self-motivated but managers and/or fellow employees may lead or stimulate them in either direction (i.e., toward passive dependence or innovation and change).

9   *Bottom-up emergence* variables: Newly emergent ideas and intellectual capital (IC) may be intangible and based on tacit knowledge, but even so, emergent developments in IC are there to be found. Emergent networks, groups, and hierarchies are more tangible and hence more easily discovered kinds of emergence. Implicit in the foregoing is to what extent a firm tolerates, punishes, or rewards people generating emergent behaviors and structures. They may be treated as deviations from approved behavior or treated as developments at least worthy of further study and potentially worthy of value and further stimulation.

10   *Upward and downward streams of influence* variables: Indicators: various information and influence flows. All firms show top-down influence from the CEO down through the management hierarchy. Firms show much more variance in whether or not vibrant bottom-up influence exists. Some firms show one or more layers of middle management that act as blocks to either kinds of influence; e.g., such a layer exists at GM (personal information source).

Table 12.2   (continued)

---

11 ***Haken's enslaving principle*** variables: Many employees become more or less enslaved by other more personal or more immediate personal or business tensions: pleasing the local boss, trying to leverage some other new project to get a promotion, searching for a new job, finding a better school for their kids, buying a new car, getting ready for the annual ski trip as winter approaches, on and on. Hence, many employees slowly become enslaved by various other tensions. Consequently, as the phase transition develops, there are usually only a few remaining highly networked individuals who actually determine the nature of new order. This could be good or bad. Indicators: a lot of phase transitions (emergent dissipative structures) in a firm, but they always appear to be created by a small subset of employees acting at the last minute. Most employees that should be involved seem too distracted by other issues to pay close attention to the issue at hand.

12 ***Co-evolution*** variables: Indicators: Changes in one entity force responsive adaptive changes in a 2nd entity. Changes the 2nd entity then force the 1st entity to make further changes; and then the 2nd entity makes even more changes; the 1st entity responds to these. Positive feedback results. This is one way we see tiny initiating events scaling up into noticeable emergent new order.

---

example, boiling water turns into steam; or smartphones and the Cloud replace laptops) that reduces the tension at the 1st critical value (people increasingly wanted to search the Internet via a hand-held smartphone and then do further computing if necessary).

12.2.2   Bottom-up Emergence: Phase 2

Bottom-up emergence emphasizes agents' self-organization absent outside influence. Its advocates consist largely of scholars associated with the Santa Fe Institute (Holland 1988, 1995; Anderson, Arrow and Pines 1988; Cowan, Pines and Meltzer 1994; Bak 1996; Arthur, Durlauf and Lane 1997). While Phase 1 focuses mostly on dramatic phase transitions at R_{c1} – the '*edge of order*', Phase 2 complexity scientists focus mostly on R_{c2} – the '*edge of chaos*' (Lewin 1992; Kauffman 1993) – chaos occurs where there are so many tensions occurring at the same time that agents can't agree about which tension to respond to. The *Region of Emergence* exists between the two Edges. Focusing on living systems (Gell-Mann 2002), Phase 2 emphasizes the spontaneous co-evolution of entities (usually termed '*agents*') in a complex adaptive system. Agents restructure themselves continuously, leading to new forms of emergent order consisting of patterns of evolved agent attributes and hierarchical structures displaying both upward and downward causal influences. The signature elements within the melting zone are self-organization, emergence and nonlinearity. Kauffman's (1993) '*spontaneous order creation*' begins when three elements are present: (1) heterogeneous agents; (2) connections among them; and (3) motives to connect – such as mating, improved fitness, performance, learning, and so on. Remove any one element and nothing happens. According to Holland (2002) we recognize emergent phenomena as *multiple level hierarchies*, *bottom-up and top-down causal effects*, and *nonlinearities*. Nonlinearity often stems from scalability reflected as power-laws (see below). Although Holland (1988) said there is no 'global controller', Westerman et al. (2014) note that employees in the middle of an organization 'have strong rolls to play' but 'leaders at the top must actively drive the effort.'

12.2.3   Self-organized Criticality and Adaptive Variability

In his now classic book, *How Nature Works*, Bak (1996) explained power-law (PL) distributions by looking at how sandpiles build up: falling grains of sand are allowed to slowly accumulate in a pile. Eventually the sandpile becomes high enough and its slope steep enough to trigger sand avalanches of varying sizes. These restore stability to the slope. The steepness of the slope depends on two elements: (1) *gravity* and (2) the sharp *irregular* shape of the individual sand grains. Take away gravity and there is no force causing the grains to slide down past each other – call the influence of this force the *tension* effect. Take away the irregular shape of the individual grains, on the other hand, and they become frictionless, unable to resist the downward force exerted by gravity – somewhat like smooth M&M candies, they will then scatter, unable to cohere enough to build up a pile. Call the influence of the friction the *connectivity* effect. Bak observed that sand-grain movements varied from the frequent but barely perceptible movement of a few isolated grains to the rare but gigantic avalanches in which thousands of sand grains move in unison. The size and frequency of sand grain avalanches are PLD (Bak, Tang and Wiesenfeld 1987).

The nonlinear tensions and connectivities that lead to extreme outcomes (the largest avalanches) are key elements of complexity science. Bak labeled the results of the nonlinear interplay of tension and connectivity, '*self-organized criticality*' – when the force of gravity encounters the friction-induced resistance of irregularly shaped grains of sand, these will move so as to maintain the sandpile's slope in a precarious state of equilibrium. The rate and volume of sand moving at any given instant is (1) nonlinear, (2) unpredictable, and (3) occasionally results in extreme events. In addition to the normally-distributed phenomena characterizing much of physical, social, and organization science – and described by Gaussian statistics (data points assumed to be *i.i.d.*).

Many researchers, however, have discovered an ever-increasing number of phenomena – from physical to biological to social to organizational and financial – that are best described by skew distributions and more specifically PL distributions. These skew distributions begin from random '*tiny initiating events*' [what Holland terms 'small inexpensive' inputs or 'lever-point phenomena' (2002)], a few of which grow into rank/ frequency distributions of outcomes that are PL distributed (Andriani and McKelvey 2007) and often can be explained by scale-free theories (Newman 2005; Andriani and McKelvey 2009). These tiny initiating events, as we interpret them, exhibit novelty (that is, non-normal) characteristics, similar to outliers. When the accumulation of many small initiating events are interconnected, thereby creating a network, they can reach a critical threshold, and often trigger unpredictable and extreme outcomes – such as giant companies like Apple, ExxonMobil, Carrefour, Walmart, Tata, and so on. PLs are indicators of emergent co-evolution and networks among companies comprising an industry that has become non-normally distributed and exhibits one or more extreme outcomes. Since PLs mostly appear to be the result of self-organization, they often – if not always – signify active processes that indicate some kind of self-organized criticality – meaning that a company keeps showing its ability to stay ahead of its competitors and keep coevolving and growing in its competitive ecosystem. This calls for a Method that accomplishes the following:

Table 12.3   Self-organized criticality

---

1   ***Tension*** variables equivalent to the 'gravity' effect: Indicators: any of the tension indicators mentioned in Tables 12.1 and 12.2.

2   ***Connectivity*** variables equivalent to the 'irregularities among sand grains' that cause their connectivities. Indicators: any of the various kinds of connectivities mentioned in Tables 12.1 and 12.2.

3   ***Slope adaptivity*** outcome equivalents: The slope of a sandpile changes because of the effects of gravity and sand-grain connectivities; in organizations the equivalent to 'slope' is a firm's adaptivity to changing internal and/or external conditions (co-evolution effects) Indicators: Any of the adaptivity and/or coevolving changes mentioned in Tables 12.1 and 12.2.

M1: A Method that validly identifies when a PL distribution exists.

We note, especially, that the key forces defined by the concepts, variables, and indicators stemming from complexity science – briefly described in Tables 12.1, 12.2, and 12.3 – all combine in various ways that result in fractals and PL distributions. Consequently, we next briefly describe fractals, PLs, and rank/frequency distributions. Reality is that PL distributions are far more ubiquitous in the real world than statisticians and econometricians (for example, Greene 2011) are willing to admit. While statisticians may keep themselves happy by assuming reality is normally distributed, managers have to perform in a world of PL distributions.

12.2.4   Fractals and Power Laws

Consider Microsoft's ecosystem – described by Iansiti and Levien (2004); it includes 28 categories of companies: some of which are: 7,752 'systems integrators' (the smallest firms); 2,252 'small specialty firms'; 1,253 'Internet service providers'; 653 'consumer electronics companies'; 238 'media stores'; 46 'e-tailers'; 13 'office superstores'; 5 'applications integrators'; and 3 giant firms out at the extreme end of the distribution (at the right-hand end of the X-axis in Figure 12.1). Despite increasingly larger size, each company buys/makes and sells software products – all in the same ecosystem and more or less competing with each other (especially in each niche). The technology- and competition-based causes of repetitive formation are the same at each level and hence are explained by a '*scale-free*' theory. This feature defines the ecosystem as '*fractal*'. The numbers of the firms in the 28 categories show a traditionally-defined PL distribution (that is, a straight line of the data when plotted on log-log scales) of the total number of Microsoft's 38,233 ecosystem companies.

Fractals are most often shown to result from mathematical formulas – as in Mandelbrot's '*Fractal Geometry*' (1983, 1997/2009). However, fractal structures also originate from adaptive processes – like the cauliflower – in biological, social, organizational, and financial-economic contexts. In these fractal structures, the same co-evolutionary adaptation dynamics appear at multiple levels. For example, McKelvey, Lichtenstein and Andriani (2010) cite 19 studies showing biological adaptation-based predator/prey fractal dynamics. Also, Zanini (2008) argues that the same effects hold for merger and acquisition activities in business niches, which Park et al. (2009) empirically confirm with a century's worth of data.

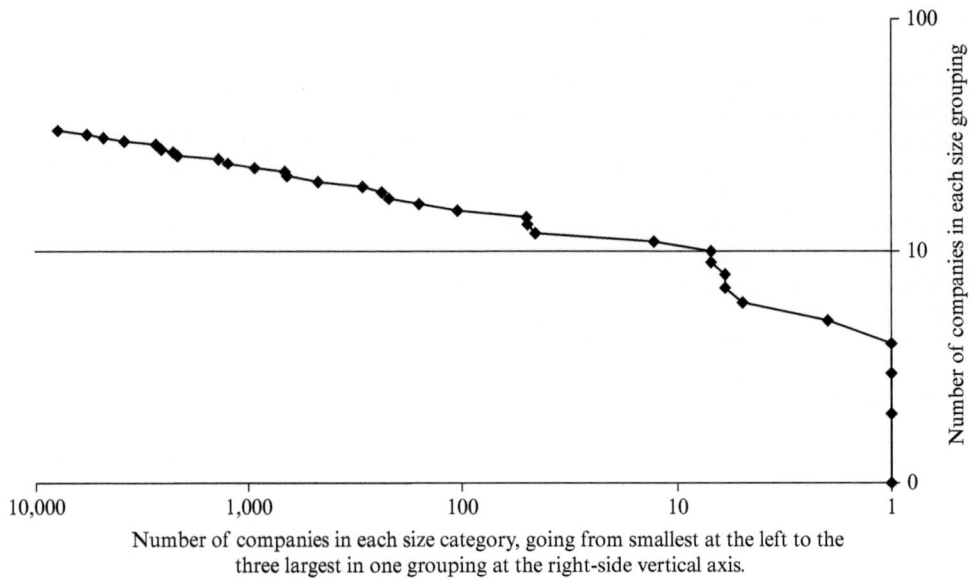

Number of companies in each size category, going from smallest at the left to the
three largest in one grouping at the right-side vertical axis.

Note: The data underlying this figure are from Iansiti and Levien (2004), p. 71.

Figure 12.1   PL Distribution of all the companies in Microsoft's ecosystem

PLs act as good indicators of fractal structures. A well-formed Pareto R/F distribution
plotted using double-log scales appears as a PL distribution – an inverse sloping straight
line, as stylized in Figure 12.2b. PLs often take the form of rank/size expressions such as
$F \sim N^{-?}$, where F is frequency, N is rank (the variable) and $?$, the exponent, is constant.
In a typical 'exponential' function, for example, $p(y) \sim e^{(ax)}$, the exponent is the variable
and e is constant. The now famous PL 'signature' dates back to the early 20th century
(Auerbach, 1913; Zipf, 1929, 1949). Andriani and McKelvey (2009) list ~140 kinds of
PLs in physical, biological, social, and organizational phenomena, all of which are good
indicators of fractal geometry. McKelvey and Salmador Sanchez (2011) list another
60+ specifically in financial economics. See also a variety of additional firm-related PL
studies (Newman 2005; Stanley, Amaral and Plerou 2000; Glaser 2013; Crawford et al.
2015).

   In these distributions, the data are highly skewed to the right. Figure 12.2a depicts a
typical PLD when data are plotted on normal scales. Here, the majority of observations
(values high on the Y-axis) are at the lowest values of the X-axis, while very few observations (values low on the Y-axis) are out at the end of the X-axis – this makes the distribution look like a playground slide. Figure 12.2b is the same data, plotted on log-log scales.
Extant research demonstrates that few phenomena adhere to a PL over all values; instead,
the PL most often applies for values greater than some minimum or threshold, represented
by the dashed vertical line separating the Gaussian and Paretian worlds in Figure 12.2b.

   PL distributions also characterize nearly all network data. Barabási (2002) connects
scalability, fractal structure, and PL findings to social networks. He shows how networks
in the physical, biological and social worlds, are fractally structured such that there is a

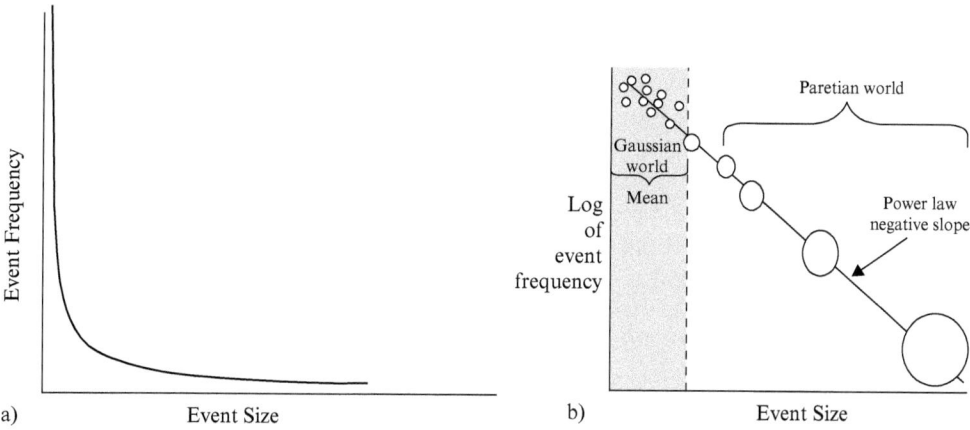

Source: Figure 2 in Boisot and McKelvey (2010), p. 417.

Figure 12.2 a) *Rank/frequency distribution on linear scales; b) Distribution plotted on log-log scales*

rank/frequency effect – an underlying Pareto distribution showing many sparsely connected nodes at one end and one very well connected node at the other. Many post 2002 studies show PL-distributed networks (McKelvey 2015).

12.2.5 Adaptive Rank/Frequency Distributions

So far, our focus has been on how order creation actually unfolds once the forces of *emergent order creation* by *self-organizing agents* – such as biomolecules, organisms, people, or social systems – are set in motion. *New order often appears as rank/frequency distributions*; the latter date back to Pareto (1897). The *outcomes* of self-organization and emergent new order often appear as rank/frequencies. According to Holland (2002) we recognize emergent phenomena in multi-level hierarchies, in intra- and inter-level causal processes, and in nonlinearities. Nonlinearity incorporates two key ideas: *butterfly events* and *scalability*. Tiny butterfly-events and scalability produce nonlinearities that may extend across multiple levels within organisms or organizations or across multiple species or firms within an ecosystem. Extreme outcomes, long-tailed Pareto distributions and PLs (Zipf 1929, 1949; Newman 2005), scalability (Brock 2000), and scale-free causes (Zipf 1949; West and Deering 1995; Andriani and McKelvey 2009) often result.

In his opening remarks at the founding of the Santa Fe Institute, Gell-Mann (1988) emphasizes the search for scale-free theories – simple ideas that explain complex, multi-level phenomena. Brock (2000) goes so far as to say that '*scalability*' is the core of the Santa Fe vision – no matter what the scale of measurement, the phenomena appear the same and result from the same causal dynamics. Gell-Mann (2002) concludes his chapter, 'What is Complexity?' with a focus on scalability in living systems. Key parts of this third phase are *fractal structures*, *PLs*, and *scale-free theory*.

Mandelbrot applied fractal geometry and later PLs to economic rank/frequency distributions (1963, 1997/2009; Mandelbrot and Hudson 2004), Econophysicists start with early foci on rank/frequency distributions of: *returns in financial markets, income and wealth, economic shocks and growth-rate variations, firm sizes and growth rates,* and *scientific discoveries* (Rosser 2008). Econophysicists often begin with a focus on Lévy skew distributions and applications of statistical-physics methods to economic phenomena (Mantegna and Stanley 2000; Chakrabarti, Chakraborti and Chatterjee 2006; Chatterjee and Chakrabarti 2006; Sinha et al. 2011). As Rosser (2008; online) notes: "A common theme among those who identify themselves as econophysicists is that standard economic theory has been inadequate or insufficient to explain the non-Gaussian distributions empirically observed for various of these phenomena, such as 'excessive' skewness and leptokurtotic 'fat tails' (McCauley 2004)."

12.3   IDENTIFYING WHERE A POWER LAW DISTRIBUTION SEPARATES FROM A NORMAL DISTRIBUTION

As traditionally portrayed, the Gaussian region of a distribution represents normal outcomes; that is, there is independence, stability and order. The Paretian region of the distribution represents nonlinear outcomes, that is, extreme events. Theoretically, complexity-science scholars use the concepts of *phase transition, threshold, bifurcation point,* and *critical value* to identify the point where a system changes from an *i.i.d.* state into a skew and/or PL state; this state is demonstrably different from the *i.i.d.* characterization of normal distributions (Prigogine 1980; Dooley and Van de Ven 1999; McKelvey 2004; Lichtenstein 2011).

It is also worth noting that the tip of the downward-pointing bracket in Figure 12.2b. While the point implies that this is the mean of the distribution, it is probably closer to the median. In skewed distributions like the ones we study here, extreme values on the right often pull the mean beyond the lower bound of the PL tail – we will revisit the practical aspects of this when we discuss the results. Moreover, as the value of α decreases, the tail becomes heavier, where the total value of all observations in the tail increases as a percentage of the entire distribution. In other words, as α moves closer to one, it indicates that the outliers in the distribution become more influential on the statistical and behavioural properties of other observations.

Firms beyond (to the right of) this critical point are operating in an interdependent, highly scalable, nonlinear environment and have the potential to influence outcomes that cascade at multiple levels and change the entire landscape; firms below this point have little ability to cause nonlinear outcomes within their existing environment. The dashed line separating the Gaussian and Paretian Worlds in Figure 12.2b indicates this point. Thus, there is a point beyond which the nonlinearity of events can be assumed, that is, the critical threshold. We show the location of this critical point in the distribution, labeled the 'Critical Threshold' (x_{min}) point, in Figure 12.3. In the following section, this critical point is calculated as the parameter x_{min}. Later in the chapter, in Figures 12.4 and 12.5, we reveal the actual data plots and empirical parameters from our analysis. These figures are conceptually important and provide a foundation for discussing additional descriptive statistics and parameter estimates. The above calls for the following:

Complexity science proposes that outcomes are an emergent result of lower-level aggregation of inputs. The self-similar nature of PL distributions suggests that sub-distributions will exhibit visually similar patterns.

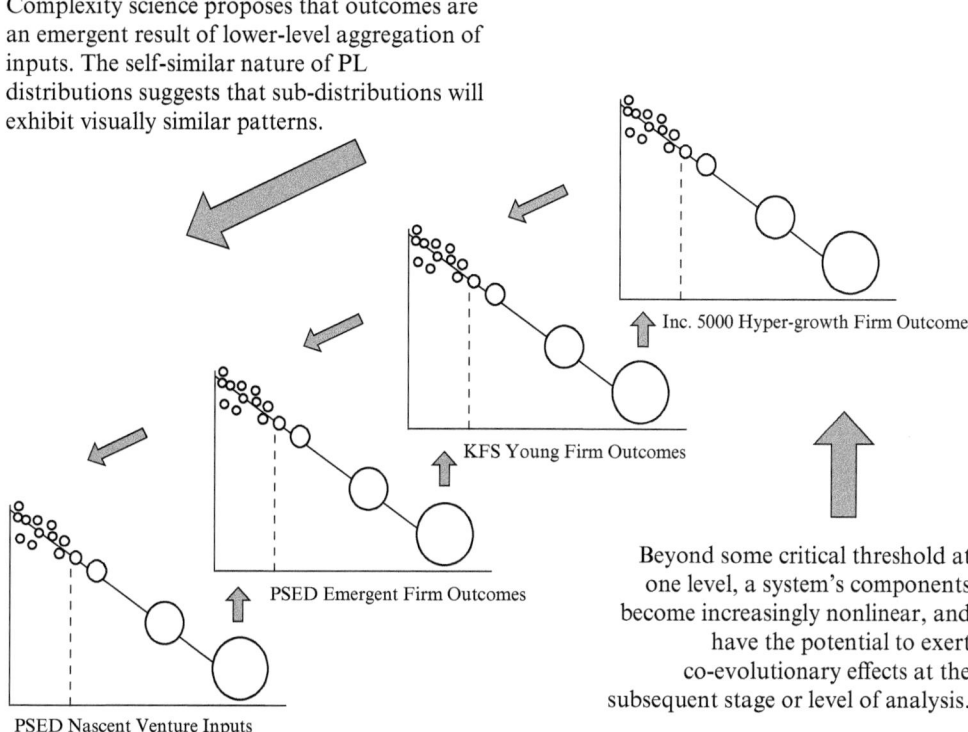

Inc. 5000 Hyper-growth Firm Outcomes

KFS Young Firm Outcomes

PSED Emergent Firm Outcomes

PSED Nascent Venture Inputs

Beyond some critical threshold at one level, a system's components become increasingly nonlinear, and have the potential to exert co-evolutionary effects at the subsequent stage or level of analysis.

Figure 12.3   *Conceptual co-evolutionary causation inherent in PL distributions and self-organized criticality*

M2: A Method that identifies the critical threshold in the distribution, where data tip from linear to nonlinear, and effects become co-evolutionary.

12.4   METHOD

This section describes a robust maximum likelihood estimation (MLE) method for finding whether or not a given distribution is actually a PLD and, if it is, identifies where the data tip from a normal to a PL distribution (that is, the x_{min} point). Using this method, we are able to calculate the relative influence that outliers have on the rest of the distribution. More explicitly, we utilize the parameter estimates generated by the analysis to identify basic propositions about self-organized criticality and other foundational complexity concepts. Specifically, we propose that when the value of a variable approaches the tail of the distribution, the variable becomes critical, and has a disproportionate potential to influence nonlinear growth at multiple levels of analysis. Below, we briefly outline the samples and data analysed, then describe the MLE techniques that allow researchers to quantitatively test basic complexity science propositions.

12.4.1   Linking Complexity-science Concepts to Empirical Samples

We use data from both Crawford, McKelvey, and Lichtenstein (2014) as well as Crawford et al. (2015) – conducted in the domain of entrepreneurship – to demonstrate the potential emergence, growth, and extreme outcomes of organizations at different stages of development. The basic premise of these two studies is that a new-venture founder's inputs have the potential to cause nonlinear interactions with would-be stakeholders; above some critical threshold, *inputs* get positive feedback, become recursive and, subsequently, exhibit self-similar patterns of *outcomes* at multiple stages of venture development.

We draw from three longitudinal samples: The Panel Study of Entrepreneurial Dynamics II (PSED), the Kauffman Firm Survey (KFS), and the Inc. 5000 Fastest Growing Companies in the United States (INC). The PSED is a representative sample of 1214 entrepreneurial ventures that was collected from random-digit dialling of US households. The PSED tracks nascent ventures' (that is, prior to their official beginning or 'start-up' because they have not, for example, filed a Schedule C, applied for an Employer Identification number, or earned revenue) inputs and outcomes over five years; the KFS is a panel study of 4928 newly operating firms; the INC is taken from Inc. Magazine's annual special issue of America's Fastest Growing Private Companies in 2010 based on three years of data. From a theory-based perspective, our samples are consistent with complexity-science assumptions: non-determinism (ventures, especially at their early stages, may fail); potential nonlinearity (ventures emerge from non-existence to existence, with some becoming giants, for example, Walmart, Carrefour, Apple, Google); potential co-evolutionary effects (once a firm gets large enough, it can influence the next higher level of analysis); and potential for large outcomes (disproportionately influential firms).

From the PSED, we analyse variables pertaining to the founder's *inputs*, including her/his *net worth, strong network ties, number of employees supervised,* and *total hours worked on the venture* from the first wave of data collection. Also, from all three samples, we analyse identical measures of ventures' *outcomes* – annual revenue and number of employees – using the PSED's fifth year of data collection, the KFS's fourth year of data collection, and the 2010 INC data. Since we are most interested in the positive-feedback and/or co-evolutionary processes that cause PLDs, as demonstrated in Figure 12.3, we focus on finding the x_{min} point and then analyse firms in the long tail on the X-axis, that is, the firms that show more extreme (non-Normally distributed) growth resulting from self-organized criticality, a phase transition, a contagion burst, or an episode of spontaneous order creation. The two key elements of our Method are:

M3: Describe, authenticate, and demonstrate a Method that validly identifies when PL distributions exist.

M4: Describe, authenticate, and demonstrate a Method that validly identifies the x_{min} tipping point at which a Normal distribution changes to a PL distribution extending out on the X-axis.

In the past, distributions were considered to follow a PL if a histogram of the data exhibited a straight line when plotted on log-log scales – this traditional method can be seen in the data from Iansiti and Levien's (2004); shown in Figure 12.1. Without a quantitative measure of goodness-of-fit, however, it is difficult to assess how well data approximates a PL. Moreover, a quantitative goodness-of-fit enables the identification of possible interesting phenomena that could be causing the distribution to deviate from a

PL. Since simple visual-based techniques for fitting a PL distribution, grounded on linear fitting of log-log transformations and data binning, tend to provide biased estimates for the exponent (α, the scaling parameter), we utilize the MLE method, which produces more accurate and robust estimates (Mitzenmacher 2005). We enhance the readability and relatability of our chapter by explaining the mathematical calculations therein and provide descriptions of the data analysis process in the Appendix.

12.5   ILLUSTRATING HOW THE METHODS ARE USED

Before displaying the results, we have a few cautionary words about the process of analysing and interpreting the data. First, be aware that estimates become biased when there are fewer than 50 data points analysed (Clauset, Shalizi, and Newman 2009), so interpretations of the results of small samples should include an explanation of how this might limit the robustness of the analysis. Second, this method is very computationally taxing on a computer. Quite often, runs on large sample sizes (~2000+) or on variables with many extremely large values (for example, millions and billions) can crash the computer or take more than 30 hours to complete, even on a relatively updated system (in our case, an Intel i7 quad-core 3.1 GHz 64-bit CPU with 12GB of memory). We recommend that, unless you have significantly more processing power on your primary computer, you should use a separate computer for word processing, spreadsheets, music, and the like to avoid potential catastrophic outcomes. Finally, while the MLE method can analyse continuous or discrete (integer) data, it is not appropriate for other types, like Likert scales or data where there is some known constraint on the maximum outcome. This constraint could be something similar to an organizational policy, a legislative law, an operational process that reduces variability, or an employee performance rating that is categorical (not continuous). We suggest, therefore, researchers obtain appropriate knowledge about how the data are collected and how each variable is actually measured.

12.5.1   Results and Interpretations

Table 12.4 shows results from the fitting of a PL form of distribution to the input and outcome variables, using the procedures and format described by Clauset, Shalizi, and Newman (2009) in the Appendix. A variety of descriptive statistics are included, including the number of observations (n), median (med), mean, skewness (skew), standard deviation (sd), and the maximum value in the data (max). These values can be calculated in Excel; we use MATLAB (R2013b) to calculate the remaining parameters, which are italicized to indicate that they are estimates. Table 12.4 also includes x_{min} (the critical threshold) and α (the negative slope of the PL's tail); n_{tail} indicates the number of observations in the tail of the distribution (and identifies the percentage of the total observations that are in the tail). The model's goodness of fit, calculated by the Kolmogorov–Smirnov test (K-S), measures how well the observed data fit to a hypothesized power law distribution. As Clauset, Shalizi, and Newman (2009, p.11) explain "if this value is suitably small we can say that the power law is a plausible fit to the data." Consistent with other studies, we propose a suitably small K-S value to be *below* 0.10, while values *above* 0.10 rule out a PL explanation for the data. Finally, as described at length in Clauset, Shalizi, and Newman

Table 12.4 Power law distribution parameter estimates and model significance – nascent inputs and emergent outcomes

	n	med	mean	skew	sd	max	x_{min}	β	n_{tail}	K-S	p-value
PSED Input Variables											
Individual Net Worth ($000)	892	152	667	23	5,807	153,000	835	2.44	159 (18%)	**0.04**	**0.64**
Strong Network Ties	1,214	0	1	10	2	50	2	2.61	231 (19%)	**0.02**	**0.66**
Employees Supervised	1,179	4	20	12	83	1,500	28	2.12	329 (28%)	0.07	0.01
Total Hours Worked	1,211	400	1931	7	5,681	73,000	4,000	2.16	177 (15%)	0.07	0.05
Outcome Variables											
Number of Employees											
PSED	68	0	30	12	122	1,500	2	1.73	49 (72%)	**0.06**	**0.19**
KFS	1,462	1	4	13	14	320	22	2.56	40 (3%)	0.07	**0.14**
INC*	5,000	51	364	40	3,769	194,000	244	1.95	906 (18%)	**0.02**	**0.80**
Annual Revenue ($000)											
PSED	145	42	354	7	1,400	12,000	600	1.78	64 (44%)	**0.06**	**0.62**
KFS	1,901	125	849	25	4,500	160,000	1,900	2.19	295 (16%)	**0.05**	**0.13**
INC**	5,000	10,700	73,135	36	631,131	30,700,000	15,800	1.79	449 (9%)	0.03	0.01

Notes:

n=number of observations; **med**=median; **skew**=skewness; **sd**=standard deviation; **max**=score with the largest value (maximum); x_{min}=estimated minimum value of power law tail, where data change from linear (additive) to nonlinear (multiplicative); α=estimated scaling exponent (i.e., slope) of the power law curve [the lower the value, the more of the total distribution resides in the tail]; n_{tail}=estimated number of observations in the tail of the distribution, i.e., the number of outliers; the percentage of the sample's total observations in the tail are in parentheses. **K-S**=estimated Kolmogorov-Smirnov goodness-of-fit statistic, which compares observed data to hypothesized power law distribution [K-S values that are suitably small suggest that a power law is a plausible fit to the data – we use values below 0.10 as a cutoff]; **p-value**=estimated fit between maximum likelihood estimation (MLE) of perfect PL tail and empirical observations [a 'perfect' empirical tail would have a p-value of 1.00]. Significant p-values *greater than* or equal to 0.10 are bolded; values less than 0.10 suggest that an alternative heavy-tailed distribution (e.g., exponential or log-normal) might characterize the data more accurately. Plots of INC Employees* and INC Revenue** are shown in Figures 12.4 and 12.5, respectively.

(2009), the *p*-value is the fit between all empirical data points in the tail of the distribution and a perfect maximum likelihood estimation line based on synthetically generated data – we explain this in more detail later. Below, we describe some ways to interpret the results of Table 12.4.

The presence of PLDs can be initially detected by comparing the mean and median of each distribution. As shown in all of Table 12.4's variables, the mean is several orders of magnitude greater than the median. A similar indicator of potential PL data occurs when the standard deviation is significantly higher than the mean, as is the case with all of our variables. We also point out the significant skewness – the measure of asymmetry about a distribution's mean – for all the variables. Greene (2011) suggests that data must be statistically manipulated to maintain adherence to Gaussian assumptions if a distribution's skew is above 3. Our calculations show the smallest skew of 7, the largest skew of 40, and the average skew of 22 among our distributions. This, along with maximum values of each distribution that exhibit nearly infinite variance from the 'average' observations, validates the foundational premise that Gaussian methods – and the assumptions that underlie them – are not likely to yield accurate descriptions of the complex phenomena they study.

As confirmation, Table 12.4 shows that all ten of our *K-S* statistics are significant, that is, below 0.10, suggesting that both *inputs* and *outcomes* of the venture emergence process could be plausibly described as power law distributed; as well, indirectly, these findings suggest that the data are most likely not able to be described as normal, Gaussian distributions. This result is consistent with the parameter estimates for mean, median, standard deviation, skewness, and maximum value described above: Pareto-based methods like our MLE technique are more likely to provide more appropriate theory-testing techniques than those based on Gaussian assumptions.

Interpreting *p*-values, seven of the ten variables have estimates above the 0.10 threshold to support a PL explanation for the data (Clauset, Shalizi, and Newman 2009). Based on the calculated *K-S* statistic, the *p*-value quantifies the fit of the empirical data to a 'perfect' PL, one that is synthetically generated from 10,000 Monte Carlo bootstrapped distributions with the same α and x_{min} (the process for which is described in the Appendix). As shown in the parameter values for INC 5000 Employees, there is a significant *K-S* statistic of 0.02 and a significant *p*-value of 0.80; this is evident in Figure 12.4, where almost all of the observations (in blue circles) fall directly on the MLE line. In contrast, the *p*-value for INC 5000 Annual Revenue 0.01 is not significant. As reflected in Figure 12.5, there is a large percentage of the observations that fall considerably far from the MLE line. When the *K-S* is significantly small, but the *p*-value is not, it suggests that an alternative heavy-tailed, non-power law distribution (like a log-normal or exponential) would more accurately describe the empirical data. In many cases, when the *p*-value is not significant, there is likely to be some sort of constraint on the system. In this case, with significant estimates for Employees, but not Revenue, an example of a constraint for the fastest growing firms in the US could be a federal corporate tax policy that provides loopholes or incentives for adding employees without similar policies for growing revenue.

The parameter x_{min} is the tipping point between the Normally-distributed data and the PL-distributed data, that is, the point of criticality, the critical threshold. As noted by Miller and Page (2007), a system self-organizes when an aggregation of individual action produces an organized pattern at the macro-level, and a system is critical when small events can trigger large cascades. Our x_{min} estimates make intuitive sense as they relate

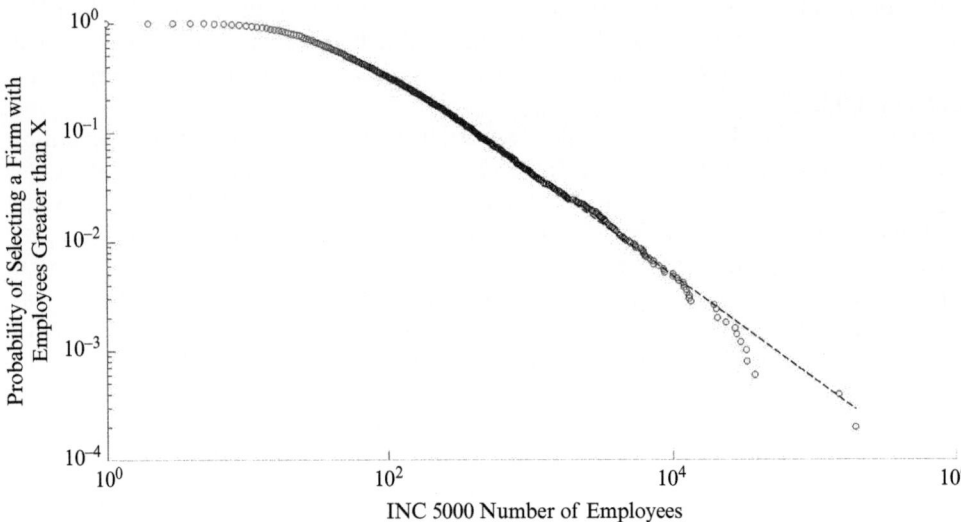

Note: In this sample of 5000, the slope (α) of the dotted MLE line is 1.95. The critical threshold (x_{min}) is 244, as indicated by the left edge of the MLE dotted beginning slightly beyond 10^2 on the X-axis. The *p*-value is significant at 0.70, as demonstrated by the substantial number of observations (blue circles) directly on the MLE line.

Figure 12.4   *PLD plot of INC 5000 number of employees (on logarithmic scales)*

to the co-evolutionary process we illustrate in Figure 12.3. The *input* variables are highly consistent with the emergence of a new venture and its potential to influence stakeholders at higher levels of analysis, either *when the venture starts with outlier endowments* (like a founder with net worth above $835,000), or *once it emerges to some critical threshold of interactions* (like a founder putting in 4,000 hours of work). Moreover, the *outcome* variables demonstrate that it takes more employees and more revenue at each subsequent level of emergence (that is, PSEDàKFSàINC) to create co-evolutionary effects. We discuss this emergence more extensively in the following section.

The parameter α indicates the negative slope of the MLE line. The lower the α, the longer the tail of the distribution, the more of the total value of the distribution resides in the tail, and the more influential each outlier becomes on the statistical and behavioral aspects of other observations in the distribution (Crawford et al. 2015). Whereas scaling parameters typically exist between the range 2 < α < 3, exponents lower than this are usually indicators of early emergent dynamics in newer systems (Kohli and Sah 2006; White, Enquist and Green 2008; Zanini 2008). Given our attention to different stages of firm development, we would expect nascent entrepreneurial ventures to exhibit exponents below 2; for a sample of operational firms, we would expect most of the distributions to exhibit exponents between 2 and 3 where, as Newman (2005) suggests, black-swan (Taleb 2007) type (for example, rare, extreme, unpredictable) outcomes are more likely to occur due to each firm's relatively larger size and higher degree of interconnectedness with other firms.

Andriani and McKelvey (2009) suggest that *alphas* below 2 are critical to understanding

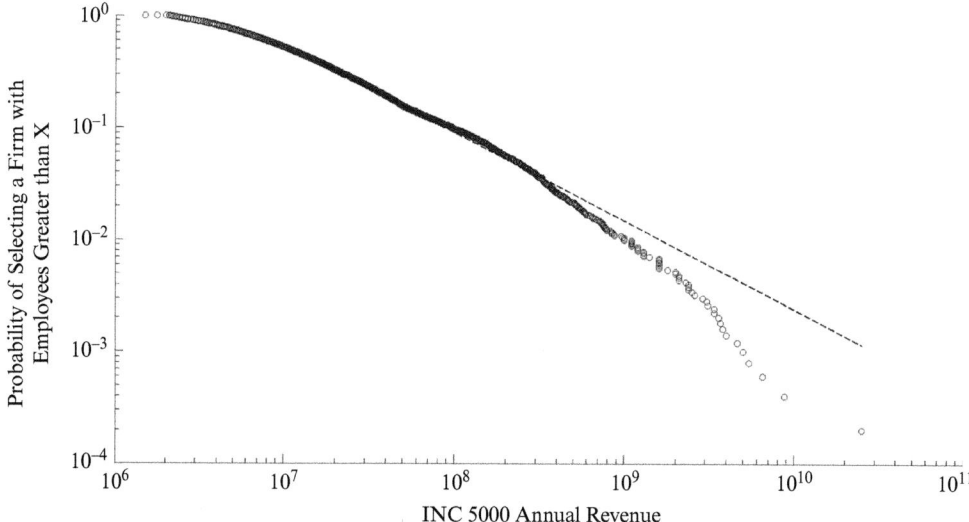

INC 5000 Annual Revenue

Note: In this sample of 5000, the slope (α) of the dotted MLE line is 1.79. The critical threshold (xmin) is $15,800,000, as indicated by the left edge of the MLE dotted beginning slightly beyond 107 on the X-axis. With a very small K-S value of 0.03, it is plausible that the distribution could be adequately described as a power law. However, since the p-value of 0.01 is insignificant (as demonstrated by the substantial number of observations (blue circles) that fall below the MLE line), the distribution is more likely to be characterized as some other heavy-tailed distribution.

Figure 12.5   PLD plot of INC 5000 annual revenue (on logarithmic scales)

the potential scalability of ventures. Additionally, Thietart (2015), agreeing with Morel and Ramanujam (1999, p.282), proposes that an *alpha* less than two is "the most conspicuous signature of a self-organized criticality phenomenon." As shown in the *alphas* of outcomes of the PSED (newly emerging ventures) and INC (fastest growing companies in the US) – those most likely to have a co-evolutionary effect – are all below 2. We posit that the alphas above 2 for both employees and revenue for firms in the KFS sample are indicators that they are more stable than those in the PSED and less likely to experience a catastrophic event (that is, go bankrupt and fail).

The n_{tail} parameter indicates the number of observations in the tail of the distribution. In normal distributions, all of these observations would be considered outliers. Keep in mind that many studies have shown that just *one* outlier in the distribution can cause significant errors in analysing and interpreting the data, including the misrepresentation of direction and relative effect of causal relationships among constructs (c.f., O'Boyle and Aguinis 2012). As a reminder, Gaussian statistics suggest that outliers are random and the probability of an outlier event is 0.03 percent. As Table 12.4 shows, the distribution of each variable has exponentially many more outliers (on average ~24 percent of all observations) than traditional statistics would lead us to believe (0.03 percent of all observations). Thus, coupled with α, the n_{tail} parameter can lend insight as to how influential outliers are on statistical and behavioural properties of the rest of the observations in the distribution.

Next, we link some of the parameter estimates from our empirical analysis to seminal concepts from complexity science. We begin by continuing our discussion on self-organized criticality, emergence, and nonlinear, co-evolutionary effects, as demonstrated by the process in Figure 12.3.

12.5.2   Linking Some Empirical Results to Complexity Concepts

As Sornette (2006, p.396) notes: "A sub-critical process has exponentially decaying activity, always dying out." In other words, if a system is below a critical threshold (that is, below x_{min}), it is less likely to survive than systems above the threshold. Looking at outcome variables from Table 12.4, in the PSED, the critical threshold for number of employees is 2. How could this have nonlinear effects, and how might this facilitate a venture's survival? Even with one employee, a founder's efforts can become more focused and more scalable. If it is an administrative or labour employee, the founder has the potential to put more attention on activities – like sales or networking – that generate immediate revenue or procure future sales contracts. Present and future cash flows can pay bills, fund expansions, or upgrade facilities, among a plethora of other things that could increase the legitimacy and probability of survival for a new venture.

Additionally, in small work environments, a founder's relationship with the employees becomes interdependent (Wiklund, Davidsson and Delmar 2003), where any decision to expand or cease operations has the potential to influence everyone in the company. When a business has *no* employees, however, there is little opportunity for nonlinearity and little possibility for adaptive efficacy. Without a single employee, to either pick up the slack (for example, continue administrative work if the founder is sick for a few days) or to provide some additional slack (for example, answer a company phone call while the founder greets a new client at the door), all of a founder's attention must be focused on proximate tasks and her ability to efficaciously adapt is reduced. While having two employees may not appear to be an 'extreme' event, it is important to note that the median and the mode of the distribution are *zero*.

Consistent with seminal and recent complexity science research (Bak and Chen 1991; Benbya and McKelvey 2011), emergent outcomes beyond the x_{min} tipping point also present the possibility of an unexpected *negative* extreme event. For number of employees, criticality for a new venture in the PSED begins at 2. Though these employees can help scale the business, venture resources must dissipate to sustain them (Davis, Eisenhardt, and Bingham 2009). For example, if a venture has a critical (that is, nonlinear) number of employees and only linear revenue (the mean revenue in the fifth year of the PSED is only $30,000), it may be financially impossible to support these employees. This is similar to a "decreasing share of the accessible energy differentials" for the founder (Levie and Lichtenstein 2010, p.334), which could result in lay-offs or firm closure, unless there is a sufficient base of alternate venture resource endowments (for example, external funding, a cache of cash). A more appropriate benchmark of criticality for revenue might be the PSED's threshold of $600,000, an amount where both the entrepreneur and the employees could conceivably have the opportunity for adequate compensation.

In the KFS sample, as a venture emerges to a conceptually more established, dynamic state (Levie and Lichtenstein 2010), there is an increased potential for co-evolutionary effects and nonlinear cascades of environmental influence, because the critical threshold for number of employees is 22. Here, there appears to be a different underlying dynamic (as evidenced by a significantly different *alpha* than the outcomes from PSED and INC outcome variables), and the influence of new business operations goes beyond the family relationships described within the PSED. This would seem to confirm the results of the Wiklund, Davidsson, and Delmar (2003) study of Swedish companies with more than 20 employees, which finds that the primary component of a manager's willingness to grow is his/her concern about the potential reduction of a 'family-like' relationship with employees. With fewer than 22 employees, those relationships may still come into play. However, if a new business creates more than 22 jobs, though the strength of the tie between a founder and each employee may be reduced, the influence of the firm's outcome cascades to a different environment. New job creation of this magnitude could affect a local economy and potentially revitalize a small town; as well, a firm of this size may extract premium offers from corporations looking to acquire it – all of these could be considered positive extreme events. As mentioned, once a system reaches a critical state (that is, above x_{min}), there is also a greater probability of a negative extreme event. Therefore, with more than 22 employees, as tie strength and interdependence are decreased between a founder and each employee, there is less relational embeddedness and an increased incentive for the founder to actually sell the firm, thereby indirectly affecting the employees. Thus, given complexity science's postmodernist epistemology (Cilliers 1998), selling the firm could be viewed as a negative extreme event from an employee's perspective.

For the INC sample, criticality begins at 244 employees. Again, private firms larger than the threshold have the potential to exert a disproportionate, co-evolutionary influence on the environment. Stating the obvious: there are firms in the tail – like HCA Healthcare with 194,000 employees – that have underlying interaction dynamics that are qualitatively and quantitatively different than smaller companies. Firms in the tail have qualities similar to gravity: they pull in additional resources. Similarly, empirical studies of elite performers support the view that outlier entrepreneurs are like stars: once they emerge beyond a critical threshold in the distribution, their gravity starts to attract resources and produce co-evolutionary effects on the environment (Aguinis and O'Boyle 2014). Thus, beyond the threshold of 244 employees, a firm could effectively influence a city's tax infrastructure and attract employees from outside the proximate geographical region.[1]

12.5.3   Testing for Mechanisms that Drive PLDs

Given our explanations above about using x_{min}, we could make the case that self-organized criticality is the mechanism that drives PLDs in entrepreneurship. However, any conclusions about whether a specific mechanism drives PLDs in a specific domain must examine and consider the full realm of statistics drawn from the MLE method we describe here. While we use x_{min} to inductively interpret emergence and criticality and the co-evolutionary effects of outliers, the parameter α can also lend insight about the generative mechanisms that drive the emergence of PLDs. As we mention in the previous section, α is a measure of the underlying dynamics of the observations in the tail. When there are consistent alphas across datasets that measure the same variable at different levels of analysis, this could be an indication of a consistent mechanism driving the emergence of PLDs. To support any proposition of *one* generative mechanism, the distribution of variables under study must have significant *p*-values. As Clauset, Shalizi, and Newman (2009, p.22) proffer, "If . . . our goal is to infer plausible mechanisms that might underlie [power law distributed phenomena], then it may matter greatly whether the observed quantity follows a true power law or some other form."

We point out that there is significant undercurrent in PL research and theorizing as to whether a particular distribution is a 'pure' or 'true' PL, as opposed to an alternative distribution, such as a log-normal or exponential distribution. Some research suggests that, even when PLDs are shown to be plausible explanations for the data (with significant K-S statistics and *p*-values), if an alternative distribution has a higher *p*-value, then there is less support for a PL hypothesis and, thus, no support for a proposed mechanism. One example of this is work by Brzezinski (2014) that studies whether wealth distributions are PL distributed. Though he finds *p*-values above 0.10 for the entire world and three countries across almost three decades, he rejects a PL explanation because, in some of the data, alternative heavy-tailed distributions have slightly higher *p*-values. As Leek and Peng (2015, p.612) note, "Arguing about the *p*-value is like focusing on a single misspelling, rather than on the faulty logic of a sentence."

Consistent with Leek and Peng's (2015) concept of a *summary statistic*, we propose that finding suitably small *K-S* values and significantly large *p*-values provide strong support for a PL explanation, even if an alternative distribution's *p*-value may be higher. The authors note that in between the summary statistic and the calculation of the *p*-value is 'inference'. Without reporting values for *alphas* or x_{min}, or discussing the possible mechanisms that might be driving the PL distributions, the analysis becomes strictly focused on numerical analysis. If, for example, *alphas* were consistent across behavioural and outcome variables at multiple *levels of analysis* (as the Bettencourt et al. (2010) findings of α ~ 2.15), proposing a mechanism like self-organized criticality might be plausible, provided each distribution has a *p*-value above 0.10. Or, for example, if all input variables have significant *p*-values across multiple *units of analysis* then it could be justified to propose interacting fractals as a generative mechanism. Alternatively, if all variables that measured a system's *initial conditions* have significant *p*-values and similar alphas *over time*, a preferential attachment model (that is, cumulative advantage) might be plausible. In all cases, for theory and research purposes, careful logical explanations – with sufficient counterfactual arguments against competing mechanisms – are essential.

12.6   DISCUSSION AND CONCLUSION

At the beginning of the chapter, we asked the questions that guided our content and method. We now briefly answer these questions before we move on to our conclusions.

What causes the emergence of power law distributions? Mechanisms where interdependent tensions drive repeated, self-similar interactions among connected agents.

How do we know if it's really a PLD? It is really a PL distribution when the maximum-likelihood bootstrap analysis estimates the *K-S* value to be suitably small, *below* 0.10, and the *p*-value to be *above* 0.10.

At what point in the distribution do observations begin to shift from a normal distribution to a PLD? The data tip from normal to a PLD at the critical threshold in the distribution; in our method, the parameter for this critical point is x_{min}.

How influential are PLD outliers on the rest of the distribution? The influence of outliers can be measured with the parameter α (when the closer it gets to 1, the more of the total distribution is from those in the tail) and with the parameter n_{tail} (which identifies how many outlier observations are in the distribution).

How can we test which mechanism(s) might cause PLDs? With all of the proposed mechanisms, the *p*-value of each variable's distribution must be significant (above 0.10) and specific parameter estimates should be nearly identical. When proposing a mechanism that spans across *similar levels* of analysis and *multiple units* of analysis, scholars need only to have significant *p*-values. When proposing a mechanism that spans across *multiple levels* of analysis with *similar units* of analysis, scholars should look for consistent values of *alpha* among all variables.

While symmetrical, bell-shaped Gaussian distributions help characterize data when there are reasons for limited divergence from a norm, like a person's height or analysing team-based interactions, most phenomena in social systems have no such limits. Preexisting limits don't exist for things like a company's revenue or an entrepreneur's wealth and, subsequently, interdependent phenomena like these can result in extreme outcomes. When PLs are present, we know that the data in the Paretian tail have a disproportionately large effect on the entire ecosystem the firms are in and, contrarily, that most of the data in the body of a Gaussian distribution of firms have a negligible effect (Sornette 2006). By formally identifying the x_{min} point we extend complexity theory by including *both* linear and nonlinear data into a single framework. Rather than focusing on one to the exclusion of the other, our chapter shows how both can be incorporated into a single framework, highlighting the broad value of complexity science for explaining all aspects of social phenomena.

While we feel our MLE method and interpretation of results are robust, they could be enhanced with nonlinear correlation analysis (for example, Spearman's Rho) and agent-based modelling. As Sornette (2006, p.223) explains, "correlations are the signatures that inform us about the underlying mechanisms"; and, as Axtell (1999) notes: "You have to grow it to know it", suggesting that agent-based computational modelling (Epstein 2006; McKelvey 2013d), with consistent experiments and validation with empirical data, is the only way to truly build theory about generative mechanisms and the emergence of new order.

In conclusion, PLs in particular and complexity science in general provide tools for uncovering the linear and nonlinear dynamics of social systems. In combination, they can explain the entire spectrum of human interactions and outcomes, offering increased validity and relevance to our research and productivity to practice. From a Darwinian selectionist perspective, the theory with the most utility – that which most accurately describes and reliably predicts – wins. With its robust and growing cadre of methods (in this chapter and the rest of this book), complexity science is making a case for the theoretical perspective with the highest utility for describing generative mechanisms, extreme events, and the creation of new order – the changes that drive the evolution of social systems.

The thoughts above, however, must be taken with a humble pill. In the design of future research projects, keep in mind this important observation: Without proper alignment of theoretical and methodological assumptions; without internal consistency among concepts, variables, and data analysis; without communicating esoteric concepts in a manner that potential stakeholders can comprehend and utilize, complexity science often appears as a very large hammer looking for nails. We hope that this chapter provides a solid foundation for moving complexity scholars from loose qualitative metaphors to robust qualitative analysis.

ACKNOWLEDGEMENTS

We thank Aaron Clauset and Didier Sornette for comments and recommendations on Crawford (2012), the paper upon which this chapter is founded. We appreciate discussions with Jaehu Shim and Per Davidsson, especially their suggestions to build on, differentiate from, and clarify potential inconsistencies with Crawford et al. (2015). We are also sincerely grateful to the Ewing Marion Kauffman Foundation for funding portions of this research.

NOTE

1. It is interesting to note that Walmart – a company that has survived numerous lawsuits about gender discrimination and fair wages, has withstood attempts at unionization with nearly 2,200,000 employees and 8,970 locations, and has been an exemplar case study for both the revitalization and the destruction of many small towns – has an average of 234 employees per store (en.wikipedia.org/Walmart accessed April 14, 2015).

REFERENCES

Aguinis, H. and E.H. O'Boyle (2014), 'Star performers in twenty-first-century organizations', *Personnel Psychology*, **67** (2), 313–350.
Allen, P., S. Maguire and B. McKelvey (eds) (2011), *Handbook of Complexity and Management*, London: Sage Publications.
Anderson, P.W., K.J. Arrow and D. Pines (eds) (1988), *The Economy as an Evolving Complex System*. Proceedings of the Santa Fe Institute, Volume V. Reading: Addison-Wesley.Andriani, P. and B. McKelvey (2007), 'Beyond Gaussian averages: Extending organization science to extreme events and power laws', *Journal of International Business Studies*, **38** (7), 1212–1230.
Andriani, P. and B. McKelvey (2009), 'From Gaussian to Paretian thinking: Causes and implications of power laws in organizations', *Organization Science*, **20** (6), 1053–1071.
Arthur, W.B., S.N. Durlauf and D.A. Lane (eds) (1997), *The Economy as an Evolving Complex System II*. Proceedings of the Santa Fe Institute, Volume XXVII. Reading: Addison-Wesley.
Auerbach, F. (1913), 'Das Gesetz der Bevolkerungskoncentration.' *Petermanns Geographische Mitteilungen*, **59**, 74–76.
Axtell, R.J. (1999). 'The emergence of firms in a population of agents: Local increasing returns, unstable nash equilibria, and power law size distribution', Brookings Institution Discussion Paper: *Center on Social and Economic Dynamics*. Washington, DC.
Bak, P. (1996), *How Nature Works: The Science of Self-organized Criticality*, New York: Copernicus.
Bak, P. and K. Chen (1991), 'Self-organized criticality', *Scientific American*, **264** (1), 46–54.
Bak, P., C. Tang and K. Wiesenfeld (1987), 'Self-organized criticality: An explanation of $1/f$ noise', *Physical Review Letters*, **38** (4), 381–384.
Barabási, A.-L. (2002), *Linked: The New Science of Networks*, Cambridge, MA: Perseus.
Bénard, H. (1901), 'Les tourbillons cellulaires dans une nappe liquide transportant de la chaleur par convection en régime permanent', *Annales de. Chimie et de Physique*, **23**, 62–144.
Benbya, H. and B. McKelvey (2011), 'Using power-law science to enhance knowledge for practical relevance', *Best Paper Proceedings*, Academy of Management Conference, San Antonio, TX, August.
Bettencourt, L.M, J. Lobo, D. Strumsky and G. West (2010), 'Urban scaling and its deviations: Revealing the structure of wealth, innovation, and crime across cities', *PLoS ONE*, **5** (11), 1–9.
Boisot, M. and McKelvey, B. (2010), 'Integrating modernist and postmodernist perspectives on organizations: A complexity science bridge', *Academy of Management Review*, **35**, 415–433.

Brock, W.A. (2000), 'Some Santa Fe scenery', in D. Colander (ed.), *The Complexity Vision and the Teaching of Economics*, Cheltenham, UK and Northampton, MA, USA: Edward Elgar Publishing, pp. 29–49.

Brzezinski, M. (2014), 'Do wealth distributions follow power laws? Evidence from "rich lists"', *Physica A*, **406** (C), 155–162.

Chakrabarti, B.K., A. Chakraborti and A. Chatterjee (eds) (2006), *Econophysics and Sociophysics: Trends and Perspectives*, Weinheim, Germany: WILEY-VCH.

Chatterjee, A. and B.K. Chakrabarti (2006), *Econophysics of Stock and other Markets*, Berlin: Springer-Verlag Italia.

Cilliers, P. (1998), *Complexity and Postmodernism: Understanding Complex Systems*, London: Routledge.

Clauset, A., C.R. Shalizi and M.E.J. Newman (2007), 'Power-law distributions in empirical data', *arXiv pre-print arXiv*:0706.1062, 64, 1–26.

Clauset, A., C.R. Shalizi and M.E.J. Newman (2009), 'Power-law distributions in empirical data', *SIAM Review*, **51** (4), 661–703.

Cowan, G.A., D. Pines and D. Meltzer (eds) (1994), *Complexity: Metaphors, Models, and Reality*, Proceedings of the Santa Fe Institute, Vol. XIX. Reading, MA: Addison-Wesley.

Crawford, G.C. (2012), 'Emerging scalability and extreme outcomes in new ventures: Power law analyses of three studies', in Leslie A. Toombs (ed.), *Proceedings of the Seventy-Second Annual Meeting of the Academy of Management*, ISSN 1543-8643.

Crawford, G.C., B. McKelvey and B. Lichtenstein (2014), 'The empirical reality of entrepreneurship: How power law distributed outcomes call for new theory and method', *Journal of Business Venturing Insights*, **1** (1–2), 3–7.

Crawford, G.C., H. Aguinis, B. Lichtenstein, P. Davidsson and B. McKelvey (2015), 'Power law distributions in entrepreneurship: Implications for theory and research', *Journal of Business Venturing*, **30** (5), 696–713.

Davis, J., K. Eisenhardt and C. Bingham (2009), 'Optimal structure, market dynamism, and the strategy of simple rules', *Administrative Science Quarterly*, **54** (3), 413–452.

Dooley, K.J. and A.H. Van de Ven (1999). 'Explaining complex organizational dynamics', *Organization Science*, **10** (3), 358–372.

Epstein, J.M. (2006), *Generative Social Science: Studies in Agent-Based Comnputational Modeling*, Princeton, NJ: Princeton University Press.

Gell-Mann, M. (1988), 'The concept of the institute', in D. Pines (ed.), *Emerging Synthesis in Science*, Boston, MA: Addison-Wesley, pp. 1–15.

Gell-Mann, M. (2002), 'What is complexity?' In A.Q. Curzio and M. Fortis (eds), *Complexity and Industrial Clusters*, Heidelberg, Germany: Physica-Verlag, pp. 13–24.

Glaser, P. (2013), 'Inequality in an increasing returns world: US stock market capitalizations from 1930 to 2008', in B. McKelvey (ed.), *Complexity: Crucial Concepts*, Vol. 5: *Power-law Distributions in Society and Business*, London: Routledge, pp. 410–423.

Greene, W.H. (2011). *Econometric Analysis* (7th edition), Englewood Cliffs, NJ: Prentice Hall.

Haken, H. (1983), *Synergetics, An Introduction* (3rd edition), Berlin: Springer-Verlag.

Holland, J.H. (1988), 'The global economy as an adaptive system', in P.W. Anderson, K.J. Arrow and D. Pines (eds), *The Economy as an Evolving Complex System*, Vol. 5. Reading, MA: Addison-Wesley, pp. 117–124.

Holland, J.H. (1995), *Hidden Order: How Adaptation Builds Complexity*, Reading, MA: Addison-Wesley.

Holland, J.H. (2002), 'Complex adaptive systems and spontaneous emergence', in A.Q. Curzio and M. Fortis (eds), *Complexity and Industrial Clusters*, Heidelberg, Germany: Physica-Verlag, pp. 25–34.

Hooker, C. (ed) (2011), *Philosophy of Complex Systems*, Amsterdam, the Netherlands: Elsevier.

Iansiti, M. and R. Levien (2004), 'Strategy as ecology', *Harvard Business Review* (March), 68–78.

Kauffman, S.A. (1993), *The Origins of Order*, New York: Oxford University Press.

Kohli, R. and R. Sah (2006), 'Some empirical regularities in market shares', *Management Science*, **52** (11), 1792–1798.

Leek, J. and R. Peng (2015), 'P values are just the tip of the iceberg,' *Nature*, **520** (April 30), 612.

Levie, J. and B.B. Lichtenstein (2010), 'A terminal assessment of stages theory: Introducing a dynamic states approach to entrepreneurship. *Entrepreneurship Theory and Practice*, **34** (2), 317–350.

Lewin, R. (1992), *Complexity: Life at the Edge of Chaos*, Chicago, IL: University of Chicago Press.

Lichtenstein, B. (2011), 'Complexity science contributions to the field of entrepreneurship', in S. Maguire, P. Allen and B. McKelvey (eds), *Sage Handbook of Complexity and Management*, London: Sage Publications, pp. 471–493.

Mainzer, K. (1994), *Thinking in Complexity*, New York: Springer-Verlag [5th edition published in 2007].

Mandelbrot, B.B. (1963), 'The variation of certain speculative prices', *Journal of Business*, **36**, 394–419.

Mandelbrot, B.B. (1983), *The Fractal Geometry of Nature* (2nd edition), New York: Freeman.

Mandelbrot, B.B. (1997), *Fractals* and *Scaling* in *Finance: Discontinuity, Concentration, Risk*, New York: Springer [2nd edition published in 2009].

Mandelbrot, B.B. and R.L. Hudson (2004), *The (mis) Behavior of Markets: A Fractal View of Risk, Ruin, and Reward*, New York: Basic Books.

Mantegna, R.N. and H.E. Stanley (2000), *An Introduction to Econophysics: Correlations and Complexity in Finance*, Cambridge, UK: Cambridge University Press.

McCauley, J.L. (2004), *Dynamics of Markets: Econophysics and Finance*, Cambridge, UK: Cambridge University Press.

McKelvey, B. (2004). 'Toward a complexity science of entrepreneurship', *Journal of Business Venturing*, **19** (3), 313–341.

McKelvey, B. (ed.) (2013a), *Origins of Order-Creation Science: Complexity Science From Basic Disciplines*, Routledge Major Works Series: Complexity: Critical Concepts, London: Routledge.

McKelvey, B. (ed.) (2013b), *Self-organization, Emergence & Self-organized Criticality*, London: Routledge.

McKelvey, B. (ed.) (2013c), *Organization and Management Complexity* Dynamics, London: Routledge.

McKelvey, B. (ed.) (2013d), *Socio-economic Agent-based Models*, London: Routledge.

McKelvey, B. (ed.) (2013e), *Power-Law Distributions in Society & Business*, London: Routledge.

McKelvey, B. (2015), 'Why *i.i.d.* economics is increasingly obsolete'. Working Paper, UCLA Anderson School of Management, Los Angeles, CA.

McKelvey, B. and M.P. Salmador Sanchez (2011), 'Explaining the 2007 bank liquidity crisis: Lessons from complexity science and econophysics', Working Paper, Universidad Autónoma de Madrid, Spain.

McKelvey, B., B.B. Lichtenstein and P. Andriani (2010), 'When systems and ecosystems collide: Is there a law of requisite fractality imposing on firms?' In M.J. Lopez Moreno (ed.), *Chaos and Complexity in Organizations and Society*, Madrid, Spain: UNESA, pp. 153–191.

Miller, J.H. and S.E. Page (2007), *Complex Adaptive Systems: An Introduction to Computational Models of Social Life*, Princeton, NJ: Princeton University Press.

Mitzenmacher, M. (2005), 'A brief history of generative models for power law and lognormal distributions', *Internet Mathematics*, **1** (2), 226–251.

Morel, B. and R. Ramanujam (1999). 'Through the looking glass of complexity: The dynamics of organizations as adaptive and evolving systems', *Organization Science*, **10** (3), 278–293.

Newman, M.E.J. (2005), 'Power laws, Pareto distributions and Zipf's law', *Contemporary Physics*, **46** (5), 323–351.

Nicolis, G. and I. Prigogine (1989), *Exploring Complexity: An Introduction*, New York: Freeman.

O'Boyle, E.H. and H. Aguinis (2012), 'The best and the rest: Revisiting the norm of normality of individual performance', *Personnel Psychology*, **65** (1), 79–119.

Pareto, V. (1897), *Cours d'Economie Politique*, Paris: Rouge & Cie.

Park, J.W., B. Morel and R. Madhavan (2009), 'Riding the wave: Self-organized criticality in merger and acquisition waves', Working Paper, Katz Grad. School of Business, University of Pittsburgh, Pittsburgh, PA.

Prigogine, I. (1955), *An Introduction to Thermodynamics of Irreversible Processes*, Springfield, IL: Thomas.

Prigogine, I. (1962), *Non-Equilibrium Statistical Mechanics*, New York: Wiley Interscience.

Prigogine, I. (1980), *From Being to Becoming: Time and Complexity in the Physical Sciences*, San Francisco, CA: Freeman.

Prigogine, I. (with I. Stengers) (1997), *The End of Certainty*, New York: Free Press.

Prigogine, I. and I. Stengers (1984), *Order Out of Chaos: Man's New Dialogue with Nature*, New York: Bantam.

Rosser, J.B. Jr. (2008), 'Econophysics' (article), *The New Palgrave Dictionary of Economics* (2nd edition) http://www.dictionaryofeconomics.com/article?id=pde2008_E000253&edition=current&q=econophysics&topicid=&result_number=1 (accessed April 14, 2015).

Sinha, S., A. Chatterjee, A. Chakraborti and B.K. Chakrabarti (2011), *Econophysics: An Introduction*, Weinheim, Germany: WILEY-VCH.

Sornette, D. (2006), *Critical Phenomena in Natural Sciences: Chaos, Fractals, Self-Organization, and Disorder: Concepts and Tools*, Berlin: Springer-Verlag.

Stanley, H.E., L.A.N. Amaral and V. Plerou (2000), 'Scale invariance and universality of economic fluctuations', *Physica A*, **283** (1–2), 31–41.

Taleb, N.N. (2007), *The Black Swan*, New York: Random House.

Thietart, R.A. (2015), 'Strategy dynamics: Agency, path dependency, and self-organized emergence', *Strategic Management Journal*, **36** (early online edition).

West, B.J. and B. Deering (1995), *The Lure of Modern Science: Fractal Thinking*, Singapore: World Scientific.

Westerman, G., D. Bonnet and A. McAfee (2014), *Leading Digital: Turning Technoloy into Business Transformation*, Boston, MA: Harvard Business Review Press.

White, E.P., B.J. Enquist and J.L. Green (2008), 'On estimating the exponent of power-law frequency distributions', *Ecology*, **89** (4), 905–912.

Wiklund, J., P. Davidsson and F. Delmar (2003), 'What do they think and feel about growth? An expectancy-value approach to small business managers' attitudes toward growth', *Entrepreneurship Theory and Practice*, **27** (3), 247–270.
Zanini, M. (2008), 'Using power curves to assess industry dynamics', *McKinsey Quarterly* (November), 1–6.
Zipf, G.K. (1929), 'Relative frequency as a determinant of phonetic change', *Harvard Studies in Classical Philology*, **40**, 1–95.
Zipf, G.K. (1949), *Human Behavior and the Principle of Least Effort*, New York: Hafner.

APPENDIX

This appendix briefly describes the procedures to assess the presence of a PL in the data. We use MATLAB software (R2013b) and follow the protocol and techniques for calculating PL fit as described by Clauset, Shalizi, and Newman (2009). Given the research community's interest in PL phenomena, we note that this article is one of the most influential in the last decade, with more than 3,600 citations as of October 2015. We encourage the reader to explore this article, the descriptions of the computations at www.santafe.edu/~aaronc/powerlaws/, as well as our online companion. The mathematical formulae that underlie each technique, as well as the computational scripts for MATLAB and R can be found that the previous link. Below, we describe the mathematics and scripts using MATLAB software.

The method described by Clauset, Shalizi, and Newman (2009) includes the following five-steps:

1. Determine the best fit of the power law (PL) to the data, estimating both the scaling parameter *alpha* and the cutoff parameter x_{min}.
2. Calculate the Kolmogorov-Smirnov statistic (K-S) for the goodness of fit of the best-fit power law to the data and calculate the standard errors of the estimates in Step 1 using a semi-parametric bootstrap analysis.
3. Generate a large number of synthetic datasets using the procedure above and calculate the K-S statistic for each fit.
4. Calculate the *p*-value as the fraction of the K-S statistic for the synthetic datasets whose value exceeds the K-S statistic for the real data.
5. A PL is a plausible fit to the data if the K-S statistic is "suitably small" (we suggest *below* 0.10); however, if the *p*-value is small (*below* 0.10), the PL distribution can be ruled out.

To set up MATLAB most efficiently, we use the command matlabpool to use all available cores for processing.

For Step 1, we use the script *plfit.m* (accessed April 14, 2015) to reliably estimate α. The plfit script estimates *xmin* and *alpha* according to the goodness-of-fit based method described in Clauset, Shalizi, and Newman (2007). *x* is a vector of observations of some quantity to which we wish to fit the PL distribution p(x) ~ x^-alpha for x >= x_{min}. plfit automatically detects whether x is composed of real or integer values, and applies the appropriate method. For discrete data, if min(x) > 1000, plfit uses the continuous approximation, which is reliable in this regime.

As a rejoinder, a quantity mathematically obeys a PL if it is drawn from a probability distribution

$$p(x) \sim x^{-\alpha} \tag{1}$$

The first step utilizes equation (2) to estimate α

$$\hat{\alpha} = 1 + n \left[\sum_{i=1}^{n} \ln \frac{x_i}{x_{min}} \right]^{-1} \tag{2}$$

And equations (3) and (4) to estimate x_{min}

$$\ln = Pr(x|x_{min}) \simeq \mathcal{L} - \frac{1}{2}(x_{min} + 1)\ln n \qquad (3)$$

where $\mathcal{L}$ is the value of the conventional log-likelihood at its maximum. This type of approximation is known as a Bayesian information criterion (BIC), where the maximum of the BIC with respect to x_{min} then gives the estimated value, and

$$D = \max_{x \geq x_{min}} |S(x) - P(x)|. \qquad (4)$$

which is the K-S statistic that minimizes the distance between the cumulative distribution functions of the data and the fitted model. The estimate of x_{min} is the value that minimizes D.

For steps 2, 3, and 4, we use the parplva2.m script takes a and x_{min} from step 1, and computes the corresponding K-S statistic and p-value. The fitting procedure generates the synthetic datasets in MATLAB with the command [alpha, x_{min}, n_{tail}] = plvar(x) using equation

x = (1−rand(10,000,1)).^(−1/(2.5−1)) $\qquad (5)$

which generates 10,000 synthetic datasets using Monte Carlo simulations on randomly generated power law distributions with the same alpha (in this case 2.5); this command also estimates the number of observations in the tail of the distribution, n_{tail}.

Then, the command [p, gof] = parplva2 (x, x_{min}, alpha) provides the results of the p-value and K-S statistic that minimizes the distance between the empirical data and the bootstrapped value of all the Monte Carlo results of all the synthesized datasets. Plots, like the ones we display in Figures 12.4 and 12.5, can be generated using the command h=plplot (x, x_{min}, alpha).

13. Multifractal signatures of intersectionality: nonlinear dynamics permits quantitative modeling of hierarchical patterns in gender dynamics at the cultural level

Hannah L. Brown, Grinnell College, Chase R. Booth, Grinnell College, Elizabeth G. Eason, Grinnell College and Assistant Professor Damian G. Kelty-Stephen, Grinnell College

Gender catches traditional scientific wisdom by surprise. Identities and ideologies surrounding gender emerge at complex intersections of social-cognitive constructs and biological inheritance, tied to both but belonging to neither (Fausto-Sterling 2012a, 2012b). Long-standing cultural forces left white, heterosexual men plenty of space to build a science unable to anticipate the currently evolving understanding of gendered experience. Science has, until very recently, been at odds with any formulation of gender not adhering to a Mendelian binary of male/female or not evoking evolution-themed apologies for male dominance (Fausto-Sterling, Gowaty, and Zuk 1997; Ah-King and Nylin 2010). Modern gender theory challenges the very notions of 'natural' binaries or dominance (Pinker and Spelke 2005; Fausto-Sterling 1992; Liesen 2007), and historical analysis show that notions of 'natural' binaries written into genetics is cultural rather than objectively chemical (Richardson 2013).

Whether it is our reasoning about sexual differences (that is, genitalia dimorphism) or about how these differences intertwine with social roles to constitute gender, psychological study of individual experience sits uncomfortably within strictures of traditional wisdom. We inherit ancient Greek notions of mind and thought as abstract, mathematical phenomena, separate and private from material phenomena (Gare 2008). The Aristotelian idea of 'five senses' remains so popular that psychology textbooks still include disclaimers to the contrary (for example, Schacter, Gilbert, and Wegner 2011, p.127). Modern computational approaches to mind have recourse to, strangely enough, similar Mendelian and Darwinian mechanisms vindicating traditional gender binaries and norms (Carruthers, Laurence, and Stich 2006).

Whether we consider two genders or five senses, taxonomies show themselves to be ill-suited to the intersections of real-world phenomena. Intersectionality theory highlights the failure of simple decompositions to fully encompass the diversity of human experience, reflecting a hierarchical interweaving of causal influences that make traditional linear modeling inapplicable and irrelevant (Bowleg 2008). The current recommendation for understanding how we think about and undergo our gendered experience is to apply qualitative analysis rather than quantitative analysis to gendered behavior (Cole 2009; Davis 2008). This chapter reviews recent multifractal advances in cognitive science that may provide new expression to intersectionality theory when studying gender. In this chapter, we review multifractal analysis and use it to demonstrate 'cascade dynamics' in web traffic for Wikipedia articles describing gender concepts and issues. In sum, we propose that cascade dynamics offers a mathematical foundation that may give renewed impact to intersectional discourse.

13.1 AGAINST THE SOCIOCULTURAL DISAPPEARING ACT: INTERSECTIONAL DISCOURSE PROTECTS INDIVIDUAL EXPERIENCE FROM FALLING BETWEEN THE CRACKS OF LINEAR MODELING

Some magicians make rabbits appear from hats and doves disappear underneath handkerchiefs. But what about the disappearance of an individual's experience? To make an entire group of people's experience disappear from history into thin air is a far greater feat, one which our legal system performed with ease several decades ago. Imagine a company that hired black men and white women but not black women. Does this recruiting method strike current moral sensibilities as fair hiring? In 1976, the American justice system told five women that a mass layoff of mostly black women was within the limits of the law. Their experience of discriminatory firing disappeared as mysteriously as a dove under the magician's handkerchief.

The secret to this magic trick was additive thinking. The courts required the women to file their suit taxonomically, that is, according to predetermined social categories, first by race and then, only after race, by gender. A wave of the legal wand kept racial discrimination undetected so long as black men remained on the payroll. Similarly, gender discrimination remained invisible while the positions of white women remained secure. When we expect social structures to decompose along straightforward fault-lines, the end product is usually a linear model insensitive to the intertwining of identities and obfuscating the overlapping experience of oppression.

If linear models are the problem, then interactive frameworks are the solution. This legal case, involving General Motors and five women, caught the attention of legal and gender theorist Kimberlé Williams Crenshaw (Crenshaw 1997). Crenshaw gave name to a framework that gender theorists and black feminists had recognized for years: intersectionality. The novel term quickly spread throughout gender theory, social sciences, and even humanities. However, the discourse behind this label goes back much farther, long before organized feminism to when Sojourner Truth highlighted the discrepancies between her experience as a woman and the experiences of white feminists in her classic "Ain't I a Woman" speech. Intersectionality originally referred to a matrix of intersecting oppressions constricting gendered and racial experience and the theory offers a non-additive framework in which social constraints (oppressive or otherwise) meet and interact. Although the appreciation of such troublesome intersections existed long before the General Motors legal case, coining the term 'intersectionality' helped focus disparate debates in public policy and social sciences by addressing the conundrum of a single person's multiple memberships within different behavioral, social, and even biological categories. Intersectionality entails that the individual's experience unfolds through a hierarchy of overlapping constraints and boundaries built at different scales of social and biological variation (Bowleg 2008; Cole 2009). This concept operates outside classic additive models and provides social sciences a dynamical framework of the human experience of oppression.

13.2   INTERSECTIONALITY IN THE SCIENCES: GENDER THEORY AND THE DIVERSITY OF PSYCHOLOGICAL EXPERIENCES

The term 'intersectionality' has grown beyond its original definition, and we follow others in applying this term to a broader class of interacting socio-behavioral constraints. Intersectionality has inspired new interest in the complexity of our gendered behavior as social and cognitive beings navigating dynamic, ever-changing contexts. This interest faces a twofold challenge: the first obstacle is traditional dominance of linear modeling in scientific research, and consequently, the other hurdle is the explicit unsuitability of treating behavior as the sum of component factors (Bowleg 2008). Intersectionality theory promises to loosen sexual and gender-based stereotypes and to ground the diversity of an individual's social, sexual experience in the complexity science boiling upwards from physical sciences. We think that intersectional discourse highlights precisely what has been so appealing about complexity-based approaches to psychology. The hierarchy of constraints shaping an individual's experience of gender may share similar architecture with the hierarchy of constraints shaping psychological experience.

13.3   MULTIFRACTAL INSIGHTS INTO THE HIERARCHIES DRIVING PSYCHOLOGICAL EXPERIENCE

Thoughtful behavior reflects interactions across scales (Van Orden, Holden, and Turvey 2003). For instance, decisions, memories and thoughts wax and wane over the long-term time scales and refer to the big-picture concerns of our day to day life ("Is it time for a career change?" or "Do I have time to stop for coffee?"), but these mental events rest upon the brief, rapid firing of extremely small neurons. Interactions across scales produce events whose probability relates to their magnitude (Figure 13.1; top-left panel). When we take the logarithm of probabilities and the logarithm of magnitudes, the power-law function takes on a linear shape, growing or decaying invariantly over all scales (Figure 13.1; top-right panel). This growth or decay depends on a potentially fractional power-law exponent. The fractional aspect of this exponent has been one motivation for calling power-law-distributed fluctuations 'fractal.'

How we understand and interpret these power-law forms depends on how much we need to parcel the measurement into separate classes of event magnitude. Small events might relate only to other small events, and big events might only depend on other small events. This view (Figure 13.1; bottom-left panel) respects a simple decomposition. However, scale-invariant behavior raises the possibility that processes unfolding at many scales are shaping one another (Figure 13.1; bottom-right panel). That is, the less-frequent, higher-magnitude events impact the more-frequent, lower-magnitude events. Much less intuitively, the more-frequent, lower-magnitude events that seem so miniscule

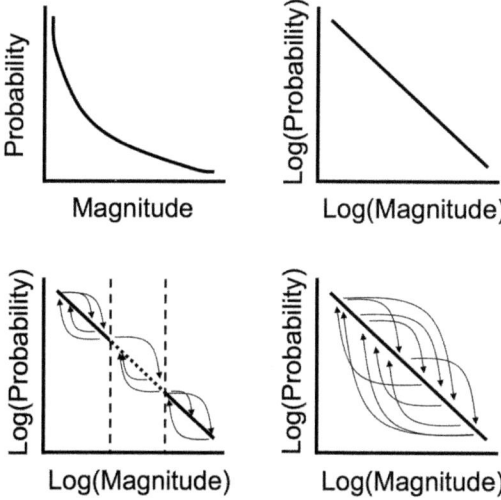

Figure 13.1   *Power law distributions as a skewed distribution on standard axes (top-left panel) and on double-logarithmic axes (top-right panel). Bottom panels elaborate the top-right panel with arrows suggesting the extent to which events of different magnitude might influence each other*

compared to larger, less-frequent, catastrophic events actually eventually foster new catastrophic change. This latter case appears in everyday discourse sometimes as 'the straw breaking the camel's back' or, less often but more poetically, the suggestion that the beat of a butterfly's wing might stir up a storm halfway across the world. Scale-invariant behavior across all available scales suggests that the distributions have no lower or upper bound – that there are no discernible limits to how fast, slow, or large or subtle cognition can be. Perceptual learning, cognitive change and integration of information from different sensory streams may all be predictable from the changes in fractal indicators of how these many different scales interact (Dixon et al. 2012). We expect that the same interactions across scales give rise to gendered experience.

We propose that intersectionality may be conceptually analogous to the formalism of 'cascade dynamics.' Much in the same way that the term 'intersectionality' has grown beyond its original meaning, 'cascades' have grown beyond simply denoting the natural phenomenon of waterfalls. In mathematical terms, cascades comprise a way of generating series of measurements as though each individual measurement was a distant grandchild split apart from many generations of parent measurements. As in the waterfall, the mathematical example begins at the top with one aggregate pooling together of all measured activity. Coherent streams moving down a precipice splinter and fracture away from each other to give way to more disordered, more variable streams and uneven splits at one generation feed down through the generations to all of the child cells, as the pooled measurements are spread apart into smaller bins. This heredity across generations of splitting produces nonlinear contingencies across scales that bring the cascade beyond the scope of linear description. There's precious little guarantee that any linear decomposition aligns with rippling seams spreading across scales, rooting one process in many others of different sizes.

13.3.1   Multifractal Paths Forward

Power law structure abounds in cognitive and perceptual dynamics down to its physiological roots (Kello 2011; Van Orden 2010). Cognition and perception are not simply drawing pesky power-law forms to light, but power-law forms are predicting how well cognition and perception go about their business (Stephen, Arzamarski, and Michaels 2010; Stephen and Hajnal 2011; Kelty-Stephen and Dixon 2014). This recent evidence regarding multiple fractional exponents governing their power-laws leads us to the possibility that cognition and perception depend on what are known as multifractal fluctuations (Dixon et al. 2012).

We suspect that a similar nesting of effects has plagued attempts to mathematize gender studies: chromosomal differences may shape sexual dimorphism, but pre-symbolic parent–child interactions in early life, family environments in later life, and social and cultural factors beyond the home all conspire to embellish and shape the gender constructs that emerge in the social adult. Intersectionality's complexity has often made quantitative studies challenging, but it may not so much prohibit mathematical quantification as it may reflect cascade dynamics and so demand the application of multifractal methods for proper quantification.

13.4   MULTIFRACTAL DIAGNOSIS OF INTERACTIONS ACROSS SCALES

We have outlined the various steps of multifractal analysis and its use in diagnosing interactions across scales in a series of figures. Keeping along the stream of water metaphors, let us suppose that civil engineers are interested in the rainfall across a span of paved road (for example, Figure 13.2; top-left). What follows involves 'binning' the measurement series: here, 'binning' involves liquid in a container. Figure 13.2 schematizes the binning of the measurement series into many non-overlapping windows, using progressively smaller window sizes. For smaller window size L, the average proportion P of the total rainfall will decrease. Smaller windows cover necessarily less of the measurement series. However, the variability of rainfall in each window will increase as we apply smaller windows. Shannon entropy quantifies this variability (Figure 13.2, bottom right). Both P and entropy usually distribute as power laws as a function of window size, and so, we can estimate power-law exponents for each α and f respectively.

In the figure, we intentionally drew the curve to ensure heterogeneous rainfall across the same stretch of pavement. Multifractal analysis assesses how much these power-law relationships in Figure 13.2 (right) change as we emphasize relatively dense and relatively sparse regions of our measurement series. Figure 13.3 shows how we can selectively emphasize different parts of the same measurement. An exponent q applied to individual windows' proportions of total rainfall produces a 'mass' term represented by the symbol μ. For each value of q, we attempt to fit a new power-law function for both α and f. Each pair of values α and f for which the functions are in fact power laws adds a single point to a convex, single-humped curve on the axes for $\alpha(q)$ and $f(q)$. This curve is the multifractal spectrum. For diagnosing interactions across scales, we are most concerned with its width, that is, the range between the greatest α and the least α.

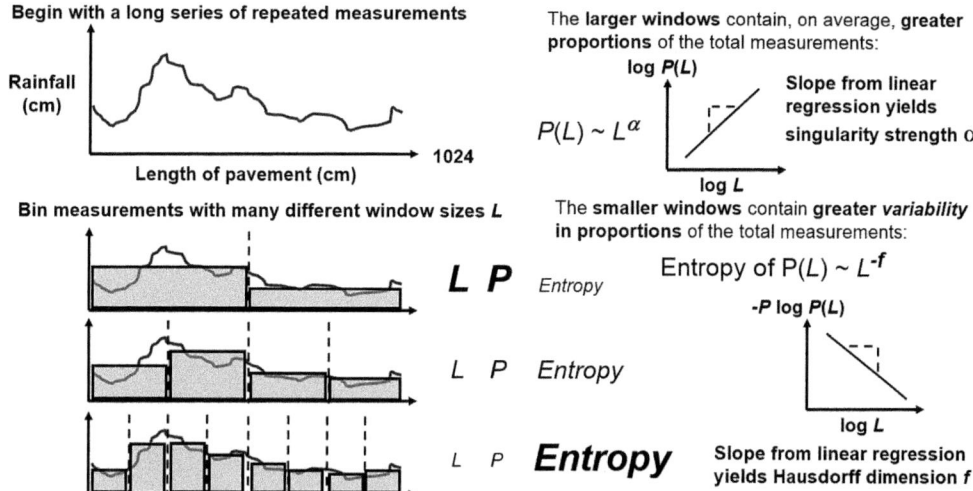

Figure 13.2   Schematic of initial premise of multifractal analysis. Multifractal analysis begins with a measurement series (top-left) and bins the measurement at many scales (that is, 'window sizes'; bottom-left). Proportion per window and entropy of window proportions vary directly and inversely, respectively, with window size (top-right and bottom-right, respectively)

Once we have a multifractal spectrum for our measurement series, we have two options for interpreting the multifractal spectrum widths: either the multifractal-spectrum width is trivially attributable to linear properties of the measurement series (Option A in Figure 13.4), or multifractal-spectrum width reflects the nonlinear contingencies across scales as we find in cascades (Option B in Figure 13.4). Figure 13.4 schematizes the hypotheses for comparison to what are called 'surrogate data' series. We submit surrogate data mimicking these linear properties to multifractal analysis and test the null hypothesis that the original series has a multifractal-spectrum width within the 95 percent confidence interval of surrogates' spectrum widths. If the original series has a multifractal-spectrum width greater than zero but outside the 95 percent confidence interval of the surrogates' spectrum widths, we reject the null hypothesis and conclude that the multifractal-spectrum width reflects nonlinear cascade structure and interactions across scales (for example, Ihlen and Vereijken 2010; Kelty-Stephen et al. 2013).

13.5   MULTIFRACTAL ORGANIZATION OF PRESYMBOLIC DEVELOPMENT: FROM COGNITIVE SCIENCE TO GENDER STUDIES

Gender studies has recently made strides towards elaborating its explanation of gendered experience to encompass the multi-scaled diversity in evidence. Specifically, gender studies has recently begun to take on the challenging position of embracing science but rejecting old-fashioned computational logic. For too long, a gender binary seemed the requisite

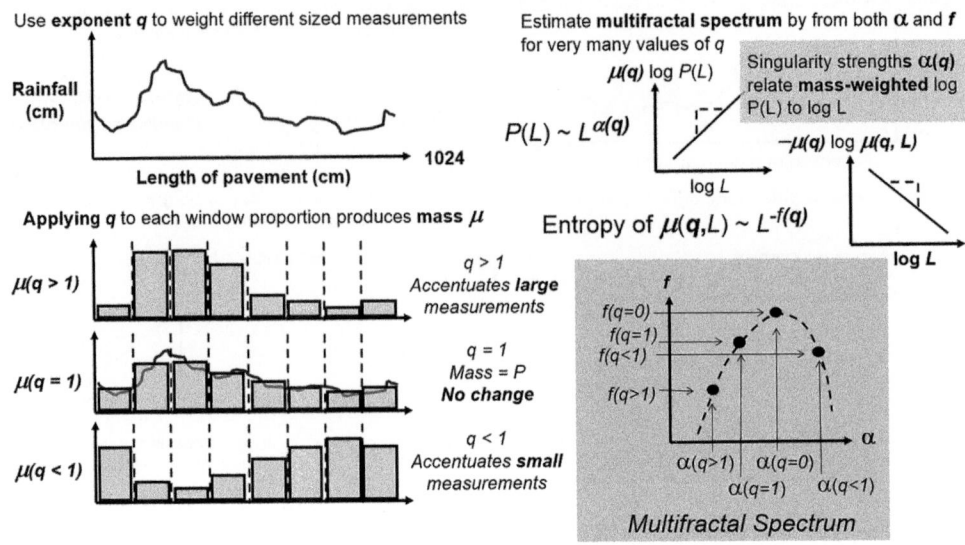

Figure 13.3   *Schematic of 'mass' term produced by applying q exponents to window proportion (bottom-left). New power-law relationships for each value of q are potentially produced by mass-weighted proportion and entropy (top-right). Each pair of power-law exponents for a given q provides a two-dimensional point. All such pairs compose the multifractal spectrum (bottom-right)*

consequence of genetic coding. Current work looks past old, limiting intuitions about genetics and now anchor gender diversity in the richly complex sensorimotor experience before infants have discovered or invented symbols that define gendered adult life (Fausto-Sterling 2012a, 2012b). Dynamical-systems approaches to infant development have unearthed a fertile ground in this presymbolic stage of the developmental trajectory from which the growing infant can fashion its own concepts through its own experiences (Martin and Ruble 2013; Thelen and Smith 1994). One day, the infants' habits and distinctions stabilize and invite the shorthand description as codes or symbols for social-cognitive computation. Until then, during the presymbolic stage, development follows a relatively plastic, nonlinear trajectory in which genetic, cellular, tissue, and behavior cooperate across time to generate new structure across all scales (Gottlieb 2000). This interaction-driven gendered experience even manifests itself through fractal patterning of variability well past infancy and through the lifespan (DiDonato et al. 2012; DiDonato, Martin, and England 2014).

This so-called presymbolic stage need not be a chaotic free-for-all. Rather, work in cognitive science has found that the discovery of novel cognitive structures may follow from multifractal processes. Symbolic-seeming structures such as rules and mathematical strategies occur to a perceiving-acting adult human in a flash of insight – like an 'aha!' moment. In many cases, hand movements provide participants a means for interacting with their task environment (for example, tracing static displays of gear systems to simulate force exchange between interlocking gears) and provide experimenters a window

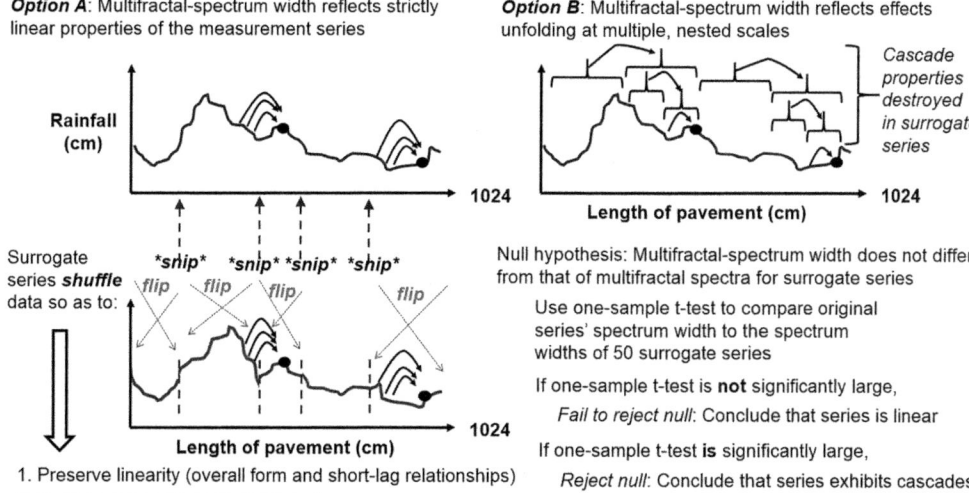

Option A: Multifractal-spectrum width reflects strictly linear properties of the measurement series

Rainfall (cm)

→ 1024

Surrogate series **shuffle** data so as to:

snip *snip* *snip* *ship*

flip   flip   flip   flip

→ 1024

Length of pavement (cm)

1. Preserve linearity (overall form and short-lag relationships)
2. Destroy any nonlinearity in original sequence

Option B: Multifractal-spectrum width reflects effects unfolding at multiple, nested scales

Cascade properties destroyed in surrogate series

→ 1024

Length of pavement (cm)

Null hypothesis: Multifractal-spectrum width does not differ from that of multifractal spectra for surrogate series

Use one-sample t-test to compare original series' spectrum width to the spectrum widths of 50 surrogate series

If one-sample t-test is **not** significantly large,

Fail to reject null: Conclude that series is linear

If one-sample t-test **is** significantly large,

Reject null: Conclude that series exhibits cascades

Figure 13.4   *Schematic of comparison to linear surrogates preserving a series' linear structure (that is, mean, variance, and autocorrelation; top-left). Cascade dynamics reflects contingency across multiple scales of measurement (top-right). Surrogate series mimic linear structure of original series but destroy original sequence, destroying nonlinear contingence of potential cascade structure (bottom-right). The null hypothesis is that original series' multifractal-spectrum width is not different from surrogates' multifractal-spectrum widths. Rejecting this null hypothesis entails concluding that original series exhibits cascade process*

on dynamic thought processes (for example, sorting cards into piles according to an uninstructed, yet-to-be-induced dimensional rule; Stephen, Anastas, and Dixon 2012). Analysis of the fractal structure of fluctuations in hand movements and eye movements during spatial reasoning or categorization tasks have shown that these 'aha!' moments often follow a gradual rise and fall in power-law exponents. Participants who did not experience the same 'aha!' moment showed no such variety of power-law structure (Stephen, Dixon, and Isenhower 2009; Stephen and Dixon 2009; Stephen et al. 2009). So, multifractal organization of presymbolic stages of performance may provide insight into the interactions across scales supporting development towards new cognitive symbols when that cognition addressed gender.

13.6   CONCLUDING DEMONSTRATION: INTERACTIONS ACROSS SCALES IN EXPLORATION OF GENDER TOPICS

For the present demonstration, we take a coarse measure of gendered experience. We downloaded daily visit counts on web traffic to Wikipedia sites on gender-related topics from January 1, 2008 to May 31, 2014 (time series length $N = 2345$ points), including

Table 13.1   Multifractal spectrum widths for original series WOrig, average width for their corresponding surrogates WSurr (and standard error SE), and t statistics representing their differences

Page title	W_{Orig}	W_{Surr} (SE)	T	p
Transgender	.0922	.0980 (.0013)	−4.45	< .0001
Transsexualism	.0973	.0845 (.0016)	7.86	< .0001
Transvestitism	.0983	.0606 (.0011)	33.18	< .0001
Intersex	.1074	.0854 (.0012)	18.71	< .0001
Womyn	.1058	.0670 (.0019)	20.24	< .0001
Ze	.1084	.0922 (.0029)	5.59	< .0001
Zer	.1774	.2356 (.0076)	−7.69	< .0001
Gender	.0867	.0958 (.0016)	−5.82	< .0001
Genderqueer	.1207	.1307 (.0038)	−2.61	< .05
Sex	.1947	.1850 (.0029)	3.34	< .01
Crossdressing	.0979	.1027 (.0021)	−2.34	< .05
Dragking	.0993	.0945 (.0015)	3.29	< .01
Dragqueen	.0880	.1027 (.0009)	−17.18	< .0001
Masculinity	.1157	.1101 (.0011)	4.94	< .0001
Femininity	.1041	.1220 (.0025)	−7.06	< .0001

traffic counts for various gender topics from arguably traditional to much less traditional topics (Table 13.1; Wiki Trends, 2014; http://wikipediatrends.com). These series reflect a measure of popular interest in specific gender-related topics well beyond presymbolic stages: computer users surfing the Internet have already taken on sufficient understanding of the gender-related symbols to produce a relevant letter string or to direct a mouse-cursor click on a linked letter string that would bring them into contact with information about the symbol. We were curious whether users' ability to manipulate these symbols might show multifractal results revealing the interaction across scales.

We submitted these daily series of web traffic to multifractal analysis, using windows sized 4 to 586 (that is, $N/4$) points (that is, days) long and incrementing q by .1 from −50 to 50. We produced 50 surrogate series from each original series using an iterative algorithm that reordered the daily traffic counts to another position in the 2345-point-long series while preserving the sum of all available sinusoids (that is, regular rise-and-fall patterns) at all possible frequencies (that is, speed of rise-and-fall patterning) that, together, add up to the observed series (Schreiber and Schmitz 1996). Preserving these sinusoids is mathematically equivalent to preserving the autocorrelation. We submitted surrogate series to multifractal analysis to generate 95 percent confidence interval for multifractal-spectrum widths based purely on linear structure. We tested for differences between multifractal-spectrum width for original series and for corresponding surrogate series. All series bore significantly different multifractal width than linear surrogates (Table 13.1). Figure 13.5 shows Wikipedia traffic for 'Transsexualism.'

All Wikipedia traffic series exhibited multifractal evidence of interactions across scales. Hence, even in the symbol-heavy dynamics of gender, we can find that Wikipedia's freely available information regarding the public's gender-related web-browsing exhibits

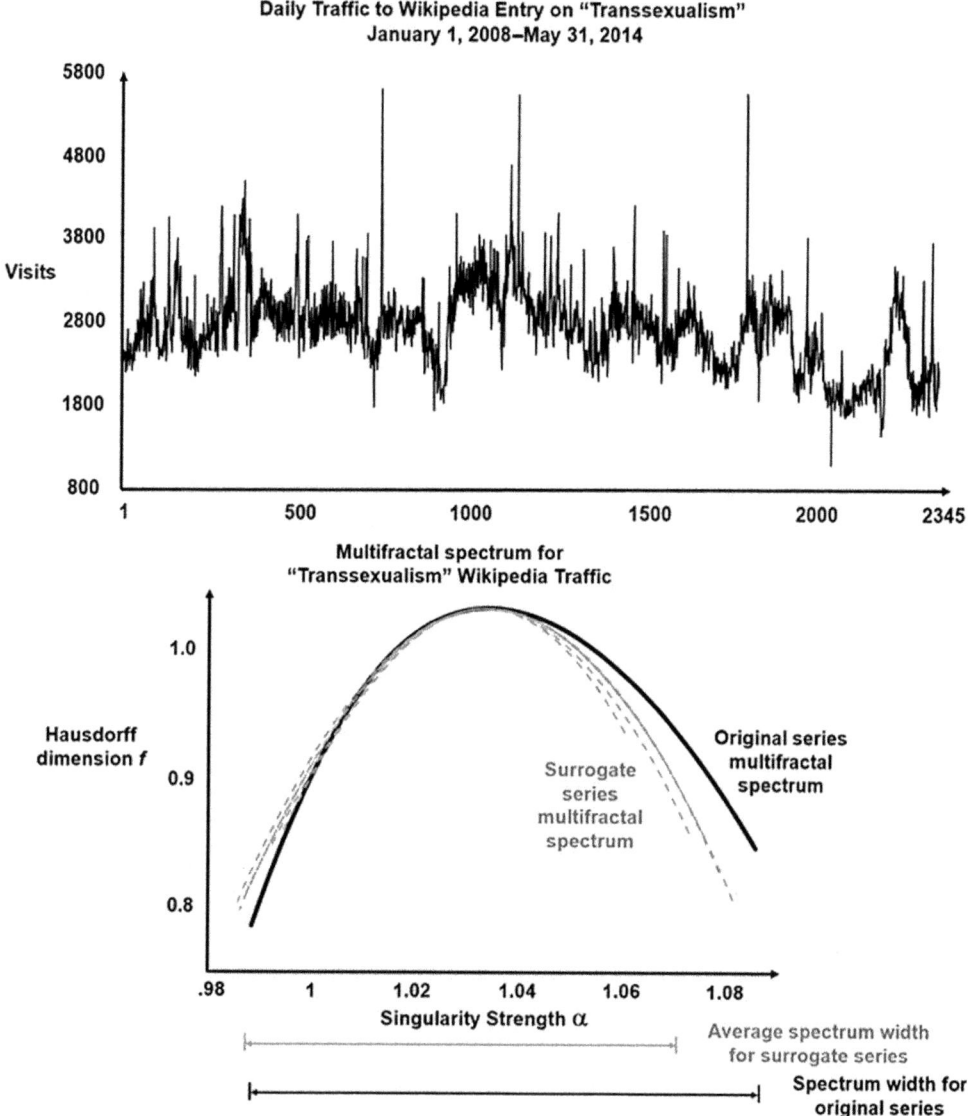

Figure 13.5   Example series of daily Wikipedia traffic, as well as corresponding multifractal spectrum (solid black) for the original series and a sample of multifractal spectra (dashed grey) for five surrogate series

interactions across scales consistent with cascade patterns. Likely factors in these cross-scale interactions include day-to-day social transactions, week-to-week changes in media trends and cultural goings-on making specific gender outcomes more or less publicly salient, month-to-month changes in social policy, and year-to-year changes in political administrations.

13.7   IMPLICATIONS AND FUTURE DIRECTIONS

The eventual promise for multifractal approaches to gender is offering quantitative metrics for sensitivity to or flexibility of gender symbols that follow from intersectional logic. It is the clumsiness of gender binaries that taught gender studies to limit itself to qualitative research, and cascade dynamics offers a breath of fresh air. A skeptical newcomer might find this approach at first cluttered by an embarrassment of multifractal riches – indeed, many things appear to be fractal or multifractal, leaving many to wonder at the usefulness of descriptions applying to everything (Likens et al. 2015). However, in those domains where multifractal results fail to differ from that of corresponding surrogates (that is, a null result), we may interpret diffuse and imprecise protestations of intersectionality as newly suspect.

Beyond adjudicating the validity of cascade-laden rhetoric, this approach may progress towards using the width of the multifractal spectrum (or the marginal difference of its width compared to that for surrogates) as an index of how flexibly or inflexibly a person or group of people view gender symbols. For instance, present work has begun to examine the multifractal structure of word-reading times in a story that begins with an ambiguously named protagonist exhibiting stereotypically 'male' behaviors but reveals, for half of the participants, that the protagonist is female. Early findings with a sample of 34 readers may indicate that the marginal difference between original and surrogate spectra predicts the startle effect on word-reading times just following the gender-related surprise, suggesting that multifractal structure may index participant attention to gender information in discourse (Booth et al., 2016). Such relationships between multifractal analysis and the wide-ranging attention needed to cultivate something so intersectionally loaded as gender expectations might suggest new ways of elaborating existing relationships between fractal scaling and gender flexibility (DiDonato et al. 2012). Similarly to Dixon and colleagues' use of multifractal statistics to quantify cross-scale interactions supporting the cognitive development of individual human participants (Dixon et al. 2012) and the aggregation of multicellular slime molds (Dixon and Kelty-Stephen 2012), gender studies might exploit multifractal statistics to elucidate cross-scale interactions during critical periods in gender development, whether for single individuals' experience or entire Wikipedia-using cultures' growing receptivity to novel gender outcomes.

ACKNOWLEDGMENTS

The authors thank Anne Fausto-Sterling for her helpful comments on an earlier draft.

REFERENCES

Ah-King, M., and Nylin, S. (2010). Sex in an evolutionary perspective: Just another reaction norm. *Evolutionary Biology*, *37*, 234–236.
Booth, C.R., Brown, H.L., Eason, E.G., Wallot, S., and Kelty-Stephen, D.G. (2016). Interactions across the hierarchical scales of discourse processing: Multifractal evidence from self-paced word-reading times predicts both short- and long-range effects of reader surprise at finding gender stereotypes unfulfilled. *Discourse Processes*. doi: 10.1080/0163853X.2016.1197811.

Bowleg, L. (2008). When black + lesbian + woman ≠ black lesbian woman: The methodological challenges of qualitative and quantitative intersectionality research. *Sex Roles, 59,* 312–315.

Carruthers, P. Laurence, S., and Stich, S. (eds) (2006). *The Innate Mind: Culture and Cognition.* New York: Oxford University Press.

Cole, E.R. (2009). Intersectionality and research in psychology. *American Psychologist, 64,* 170–180.

Crenshaw, K.W. (1997). Intersectionality and identity politics: Learning from violence against women of color. In M.L. Shanley and U. Narayan (eds), *Reconstructing Political Theory: Feminist Perspectives* (pp. 178–193). University Park, PA: Pennsylvania State University Press.

Davis, K. (2008). Intersectionality as buzzword: A sociology of science perspective on what makes a feminist theory successful. *Feminist Theory, 9,* 67–85.

DiDonato, M.D., Martin, C.L., and England, D. (2014). Gendered interactions and their consequences: A dynamical perspective. In P.J. Leman and H.R. Tenenbaum (eds), *Gender and Development* (pp. 20–42). New York: Psychology Press.

DiDonato, M.D., Martin, C.L., Hessler, E.E., Amazeen, P.G., Hanish, L.D., and Fabes, R.A. (2012). Gender consistency and flexibility: Using dynamics to understand the relation between gender and adjustment. *Nonlinear Dynamics, Psychology, and Life Sciences, 16,* 159–184.

Dixon, J.A., and Kelty-Stephen, D.G. (2012). Multiscale interactions in *Dictyostelium discoideum* aggregation. *Physica A, 391,* 6470–6483.

Dixon, J.A., Holden, J.G., Mirman, D., and Stephen, D.G. (2012). Multifractal dynamics in the emergence of cognitive structure. *Topics in Cognitive Science, 4,* 51–62.

Fausto-Sterling, A. (1992). Building two-way streets: The case of feminism and science. *National Women's Studies Association Journal, 4,* 336–349.

Fausto-Sterling, A. (2012a). Not your Grandma's genetics: Some theoretical notes. *Psychology of Women Quarterly, 36,* 411–438.

Fausto-Sterling, A. (2012b). The dynamic development of gender variability. *Journal of Homosexuality, 59,* 398–421.

Fausto-Sterling, A., Gowaty, P.A., and Zuk, M. (1997). Evolutionary psychology and Darwinian feminism. *Feminist Studies, 23,* 402–417.

Gare, A. (2008). Approaches to the question 'What is life?': Reconciling theoretical biology with philosophical biology. *Cosmos and History: The Journal of Natural and Social Philosophy, 4,* 53–77.

Gottlieb, G. (2000). Environmental and behavioral influences on gene activity. *Current Directions in Psychological Science, 9,* 93–97.

Ihlen, E.A.F., and Vereijken, B. (2010). Interaction-dominant dynamics in human cognition: Beyond $1/f^\alpha$ fluctuation. *Journal of Experimental Psychology: General, 139,* 436–463.

Kello, C. (2011). Intrinsic fluctuations yield pervasive $1/f$ scaling: Comment on Moscoso del Prado Martin (2011). *Cognitive Science, 35,* 838–841.

Kelty-Stephen, D.G., and Dixon, J.A. (2014). Interwoven fluctuations during intermodal perception: Fractality in head sway supports the use of visual feedback in haptic perceptual judgments by manual wielding. *Journal of Experimental Psychology: Human Perception & Performance, 40,* 2289–2309.

Kelty-Stephen, D.G., Palatinus, K., Saltzman, E., and Dixon, J.A. (2013). A tutorial on multifractality, cascades, and interactivity for empirical time series in ecological science. *Ecological Psychology, 25,* 1–62.

Liesen, L.T. (2007). Women, behavior, and evolution: Understanding the debate between feminist evolutionists and evolutionary psychologists. *Politics & the Life Sciences, 26,* 51–70.

Likens, A., Fine, J., Amazeen, E., and Amazeen, P. (2015). Experimental control of scaling behavior: What is not fractal? *Experimental Brain Research.* [Epub ahead of print] doi: 10.1007/s00221-015-4351-4.

Martin, C.L., and Ruble, D.N. (2013). Patterns of gender development. *Annual Review of Psychology, 61,* 353–381.

Pinker, S., and Spelke, E. (2005). A conversation with Steven Pinker and Elizabeth Spelke: The science of gender and science. Presented at the Harvard University. Harvard University Mind/Brain/Behavior Initiative. http://edge.org/3rd_culture/debate05/debate05_index.html.

Richardson, S. (2013). *Sex Itself: The Search for Male and Female in the Human Genome.* Chicago: Chicago University Press.

Schacter, D.S., Gilbert, D.T., and Wegner, D.M. (2011). *Psychology* (2nd edition). New York: Worth.

Schreiber, T., and Schmitz, A. (1996). Improved surrogate data for nonlinearity tests. *Physical Review Letters, 77,* 635–638.

Stephen, D.G., and Dixon, J.A. (2009). The self-organization of insight: Entropy and power laws in problem solving. *Journal of Problem Solving, 2,* 72–101.

Stephen, D.G., and Hajnal, A. (2011). Transfer of calibration between hand and foot: Functional equivalence and fractal fluctuations. *Attention, Perception & Psychophysics, 73,* 1302–1328.

Stephen, D.G., Anastas, J., and Dixon, J.A. (2012). Scaling in executive control reflects multiplicative cascade dynamics. *Frontiers in Physiology, 3,* 102.

Stephen, D.G., Arzamarski, R., and Michaels, C.F. (2010). The role of fractality in perceptual learning: Exploration in dynamic touch. *Journal of Experimental Psychology: Human Perception & Performance*, *36*, 1161–1173.

Stephen, D.G., Dixon, J.A., and Isenhower, R.W. (2009). Dynamics of representational change: Action, entropy, and cognition. *Journal of Experimental Psychology: Human Perception & Performance*, *35*, 1811–1822.

Stephen, D.G., Boncoddo, R.A., Magnuson, J.S., and Dixon, J.A. (2009). The dynamics of insight: Mathematical discovery as a phase transition. *Memory & Cognition*, *37*, 1132–1149.

Thelen, E., and Smith, L.B. (1994). *A Dynamic Systems Approach to Cognition and Action*. Cambridge, MA: MIT Press.

Van Orden, G. (2010). Voluntary performance. *Medicina (Kaunas)*, *46*, 581–594.

Van Orden, G., Holden, J.G., and Turvey, M.T. (2003). Self-organization of cognitive performance. *Journal of Experimental Psychology: General*, *132*, 331–350.

14. Modeling social complexity in infrastructures: a case-based approach to improving reliability and resiliency

Carl J. Dister, ReliabilityFirst, Professor Brian Castellani, Kent State University and Dr Rajeev Rajaram, Kent State University

14.1 YES, INFRASTRUCTURES ARE SOCIALLY COMPLEX!

We wrote this chapter to address a major limitation in the current literature: the continued and significant failure to address the profound but oft-hidden role that complexity and, more specifically 'social complexity' play in the reliability and resiliency of various infrastructures. In doing so, we follow a 'small but growing trend' in several interconnected literatures, ranging from systems engineering and engineering infrastructures to globalization studies and urban design to green architecture and social policy to ecology and sustainability, which seek to understand infrastructures from a *complex social systems perspective* (for example, Braha et al. 2006; Byrne 2013; Byrne and Callaghan 2013; Capra and Luisi 2014; Gerrits 2012; Gerrits and Marks 2015; Haynes 2015; Pagani and Aiello 2013, 2015; Teisman, van Buuren and Gerrits 2009).

By way of definition, infrastructures are, traditionally speaking, the core service systems in a country, region, city or community, without which it would be near impossible for such places to function, grow or sustain themselves. Such systems range from transportation and emergency services to the power grid and water supply to telecommunications and cyber-infrastructure. More recently, in the wake of rapid globalization, these services have grown exponentially in the speed and range of their spatial importance and reach, as well as their degree of complexity and interconnectedness, including global air-traffic networks, the world wide web, nationally and internationally distributed supply and power networks, and global social media (for example, Castells 2011; Giddens 2002; Urry 2007).

When viewed on their own or (more importantly) in combination, these systems are, by definition, complex; that is, "[they] are composed of many heterogeneous subsystems and are characterized by observable complex behaviors that emerge as a result of nonlinear spatio-temporal interactions among the subsystems at several levels of organization and abstraction" (Braha et al. 2006). Equally important, many of these subsystems are social, insomuch as: (a) they are not self-steering (that is, while strictly technical in composition, they require humans to manage them) or (b) they are explicit forms of social organization (Pagani and Aiello 2013, 2015). As an illustration, the power grid in many countries is a complex web of evolving, dynamic, spatially distributed and interdependent social and technical networks – power stations and cyber-infrastructures, power companies and distributors, users and regions. In turn, this web is situated within and impacted by the larger residential, political, economic and environmental systems of the regions or countries of which they are a part (Pagani and Aiello 2013, 2015).

For Giovannella (2013, 2014) and others (for example, Pagani and Aiello 2013, 2015), the combined sociotechnical complexity of infrastructure suggests that such systems are, perhaps, better viewed as 'smart' complex systems. Let us explain.

14.1.1   Steering Infrastructures as Smart Territories

For complexity scientists, when viewed together as a bounded system, complex infrastructures, given their social complexity, constitute a 'smart territory' or 'smart grid' (Pagani and Aiello 2013, 2015). A *smart territory/grid* is a data-driven system (think Big Data!) whose 'smart' benchmarking and management requires the integration and cooperation of a variety of often competing or countervailing social actors, including public and private organizations (Giovannella 2013, 2014). Such data-driven 'smart' management includes the sharing and exchange of targets and growth strategies, innovation and development, and monitoring and regulation; and it is particularly important in terms of overseeing reliability and resiliency (Pagani and Aiello 2013, 2015). In other words, how infrastructures, as smart territories, are steered, has a significant impact on their corresponding reliability and resiliency. The main challenge, however, is that, currently, such a 'smart territory' approach to steering infrastructures lags behind significantly, due to the failure to embrace a complex systems approach to modeling.

14.1.2   Reliability and Resiliency

In the new cyber-security-concerned, big-data, digitally-saturated, hot-flat-and-crowded globalized world in which we live today, maintaining the reliability and resilience of societal infrastructures is front and center (Castells 2011; Friedman 2008; Giddens 2002; Urry 2007). In fact, along with the environment and inequality (health, political, economic), it is one of the most pressing issues we currently face. How, for example, do we ensure daily power, water and food to our ever-growing super cities? And what about social mobilities, how do we sustain increasing global travel and its demands? In turn, how do we make sure our infrastructures can withstand attack, be it cyber-based or environmental? And what about the additional myriad of social factors impacting our services, including the sheer interdependent complexity of it all?

By way of definition, reliability of an infrastructure is a measurement of how well it serves its intended function over time. Resiliency, in turn, is defined as an infrastructure's ability to recover from adverse events. A solid risk management process – such as the International Organization for Standardization's ISO31000 set of standards, for example – helps to deliver the appropriate level of these two important properties of the system. However, traditional risk management programs continue to employ classical, linear methods from the natural sciences to measure reliability and resiliency in order to inform decision-making. Unfortunately, these methods are not robust enough to account for the complexity that exists in many infrastructures – in large measure because (1) humans are an integral part of these systems; and (2) because external forces (for example, weather, economics) cannot be easily linearized or ignored.

As illustration, see Figure 14.1 for a segment of the United States power grid. The top

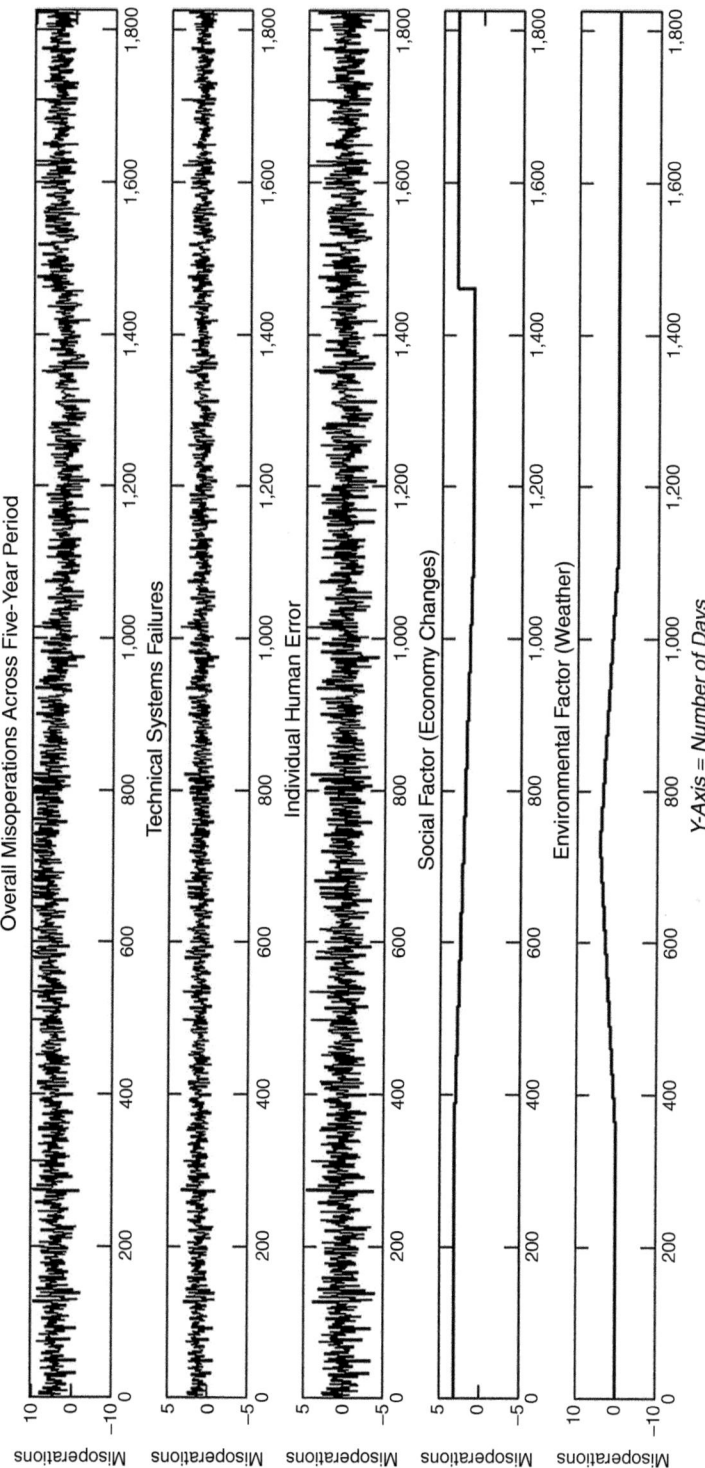

Brief Descrioption: The top signal represents total misoperations as a function of time. All of the lower signals are components that makeup this total misoperations signal. As a waveform goes up, that is an indication that the number of misoperations is increasing due to that factor. For the random signals (Equipment Failure and Human Error) these go up or down rather randomly each day. For the social and environmental factors, they move a little more slowly, following longer term trends.

Figure 14.1   Decomposing misoperations in the power grid: an example of a complex, interdisciplinary, multi-level approch

269

signal demonstrates a set of misoperations (for example, transmission line drops, natural gas line breakages, water piping issues) experienced over a five-year period in a generic infrastructure. The second signal shows that part of the composite signal due to technical equipment failure. However, the lower three signals show contributions from human factors, economic factors and weather. Key to a complex systems perspective, is that, embedded in these dynamic signals are not just the hidden patterns of reliability (a low number of misoperations in this case) but also the hidden patterns of recovery (resiliency). The question is how to get at them and model them correctly?

This question brings us to the heart of the problem we seek to address in this study: at present, the majority of methods used to model infrastructures are highly linear and mechanistic in focus. And, even when grounded in a complex systems (networks) perspective, they tend to focus on an infrastructure's network of technical subsystems, at the expense of the social (for example, Braha et al. 2006; Gerrits 2012; Gerrits and Marks 2015; Haynes 2015; Pagani and Aiello 2013, 2015; Teisman, van Buuren and Gerrits 2009). The challenge, then, as Linkov et al. (2013) and others have pointed out (Mingers and Gill 1997; Pachauri et al. 2014; Rosa et al. 2013), is to build new risk management models that allow experts to quantitatively or qualitatively identify how social and technical factors, in combination, influence key outcomes. Or, as Braha et al. state, "Understanding, designing, building and controlling such complex systems is going to be a central challenge for engineers in the coming decades" (Braha et al. 2006).

14.1.3   Purpose of Current Study

Our thesis is that, if one is to improve reliability and resilience in infrastructures, particularly in the new globalized society in which these infrastructures must effectively sustain us, it is necessary to adopt a 'complex, smart territory' modeling strategy, particularly one which gives full truck to the importance of social complexity.

To test our hypothesis, we will conduct a case study on a segment of the United States power grid. Our goal, however, is simple: we seek to create a first proof-of-concept (not a full-blown study) sufficient to show, in the simplest of cases, how thinking about infrastructures in 'complex systems' terms, primarily in terms of their social aspects, can prove beneficial. As demonstration, our goal is to show how our approach leads to novel insights that regulators do not know about or, presently, do not effectively measure or model or predict.

For our case study, we will employ the SACS Toolkit, which is part of the new approach to modeling complex systems, called case-based complexity; which comes from the work of David Byrne and Charles Ragin (2009) and, more specifically, Byrne and Callaghan's (2013) integration of *case-based reasoning* with *complex critical realism* and the latest developments in complexity theory. Extending the work of Byrne and colleagues, the SACS Toolkit is a computationally grounded, case-comparative, mixed methods platform for modeling complex systems as sets of cases (Castellani and Rajaram 2012; Rajaram and Castellani 2012, 2014). For those interested in a complete overview of the SACS Toolkit, including its mathematical underpinnings, see Castellani and Rajaram (2012), Rajaram and Castellani (2012, 2014) and Castellani et al. (2015a, 2015b).

For those new to the SACS Toolkit, several quick comments are necessary. While most statistical techniques focuses on the average, aggregate behavior of samples, the SACS

Toolkit (similar to such statistical techniques as growth mixture modeling and cluster analysis) focuses on cases, specifically how they aggregate and cluster into similar groups (Castellani and Rajaram 2012; Castellani, Schimpf and Hafferty 2013). The strength of the SACS Toolkit is its ability to: (1) identify sub-group differences (*typologies*) amongst highly complex data and (2) explore the *different* causal pathways of these typologies, based on differences in their profiles of key variables.

To date, we have used the SACS Toolkit to model a variety of complex systems, including (a) allostatic load (Galen Buckwalter et al. 2016), (b) depression trajectories (Castellani et al. 2015b), (c) community health (Castellani et al. 2015a), (d) international, global health trajectories (Rajaram and Castellani 2014) and (e) medical professionalism (Castellani and Hafferty 2006; Hafferty and Castellani 2010). And, as we will illustrate with the current case study, we are beginning to use the SACS Toolkit to model the complexity of infrastructures, including the power-grid and cyber-security for infrastructure. However, while the majority of our work, to date, has been computational in focus, the SACS Toolkit (given its mixed-methods approach) is equally facile with existing qualitative and statistical techniques. To demonstrate, for the current study we will develop the qualitative and statistical side of the SACS Toolkit. With this general framework in mind, we turn now to our case study: a segment of the United States Power Grid.

CASE STUDY: COMPLEXITY OF THE ELECTRIC GRID

In the United States, the complexity of the power grid, as a smart territory, begs for a more complex regulatory strategy [Ashby – Law of Prerequisite Variety – 1950s]. However, historically, the dominant approach to regulation has been and continues to be insufficiently static and simplistic in its conceptualization of this problem. A bit of background:

As the electric grid expanded after the Second World War, it started to exhibit large-scale blackouts (for example, New York City 1965), prompting groups of experienced technical specialists to codify their *Lessons Learned*. In response, the US government sought to implement the *U.S. Electric Power Reliability Act of 1967* to improve grid reliability, but it did not pass. Instead, the legislation motivated the Electric Power Industry to create NERC (National Electric Reliability Council), forming regions within the United States that adopted the *Lessons Learned* from the experts and began planning power exchange more reliably. However, major blackouts in the States continued to occur, every 10–20 years, and more pressure was put on the government to take legal action. *The Energy Policy Act of 2005* was successfully passed, and made the previous *Lessons Learned* federal law. The NERC and its designated regions were given power to monitor compliance to the new reliability standards. However, once again, blackouts, like the 2008 Florida blackout and the 2011 Southwest blackout continued to occur. Finally, in 2014, the United States government launched its *Reliability Assurance Initiative*, which attempted to look at a wider set of issues. Regulators began digging deeper into the *Internal Controls and Management Practices* (ICMP) that make up the power companies operating the grid. Still, in terms of addressing the issues of complexity and, more specifically, social complexity, the ICMP's approach remains significantly challenged. Here is a quick list of some of the most important issues – which the reader can think of as an insider's view, based on the first-hand experience in the field by the first author:

The Challenges of Complexity to Grid Reliability

A. Time and change
To begin, the velocity of change in the power industry is rapid. Although compliance audits in the current regulatory scheme take place on a multi-year cycle, and new requirements take several

years to adopt, the actual regulated business are experiencing mergers and acquisitions, climate variations, equipment failures, and economic fluctuations at a time constant of a year or less, instead of multiple years. In other words, slow regulation is aiming at a moving target. Speed of response in regulatory influence needs to be increased.

Second, the time perspective is also slightly misaligned. Regulatory standards based upon lessons learned have a historical viewpoint. Management practice evaluations examine the present state of organizations, but neither viewpoint looks ahead, in a predictive fashion. Prediction is sorely needed.

Third, regulators are not spending enough time at the power companies. Audits or appraisals, for example, take approximately one week. Outside of these activities, immersion in the power company is hardly existent, although it is hoped to have some understanding of the organization's culture in such a short time. Immersion in the culture is required more on an ongoing basis to better steer influential regulatory action.

Fourth, while current regulations cover requirements from short-term control room switching (minutes) to long-term maintenance intervals (years), the regulatory cycle for this wide variation in system performance is relatively slow. The pace of response to risks discovered through the current regulatory process in the form of penalties, through mitigating process improvement efforts are on the order of a year or more. A faster regulatory scheme in the delivery of influential steering is required.

B. Organizational modeling

The next issue has to do with organizational modeling. Current models of power industry behavior are mostly based on the human advice of experts, sometimes supported by their profession's classic computer tools. These experts are typically from the fields of Power System Engineering, Cyber and Physical Security, or Legal Professions. Yet a large portion of the problems facing the grid involves human behavior issues, more commonly addressed in the social sciences. More involvement from those working in the complexity sciences, from the fields of sociology, for example, are needed in the regulatory scheme.

Second, the level of mathematics used in understanding the power industry is typically binary (black and white, pass/fail interpretations of standards), or simple algebra (classic risk management weighting with accumulative multiplication and addition of risk likelihoods). Tools that look at the dynamics of the large complexity, fusing in statistics and probability are seldom considered. As such, experts in the regulatory field require a higher proficiency in computational mathematics and the tools of complexity theory.

Third, the models used by industry experts have a prescriptive and normative architecture. Even those models used in addressing organizational practices by systems experts are often prescriptive and normative. The grounded, empirical models common in the social sciences are seldom used. The use of social scientific modeling also needs to be increased.

Finally, although recent model variables have evolved from counts of violations and events (all historical) to a better focus on metadata surrounding Root Causes and Risk Rankings, the measurements rarely include human and social factors. Human and social factors need to be included.

C. Smaller and larger social factors

The next issue concerns the need to address a wider set of microscopic and macroscopic social factors. For example, the power and influence components of social behavior in organizations are only talked about behind closed doors, and not made visible. The view of the system is closer to maps of components in concrete walls (cyber and physical security perimeters) than the interactions between people inside those walls. When discussions are performed at higher levels between regulators and power company officials outside of audits and penalty discussions, they often take place by business executives with little experience in the data gathering discipline known by sociologists and anthropologists. At best, the executives may bring in an attorney, but attorneys are not trained in sociology.

D. Regulatory influences

The final issue is regulatory influences. Although improvements have been made to focus more on Reliability and Resiliency than Compliance, Regulators are still working in a system that uses a proxy for the real concern: for the power companies to optimize uncertainty in all aspects of their business, from reliability, to profitability, to environmental sustainability. To turn a back to all rationalities used to govern a power industry organization is to miss crucial data that can help steer influence. The regulator must be a part of the overall decision-making of the companies served. Also, communications between regulators and the regulated entities, although evolving to be more reward-based than penalty-based, is still mistaken with the issue of control. The realization that there is only hope of communicating, and not really a hope of control, is missing in the influential actions of the regulator. Regulators will need to view themselves as elements inside a larger, open system, including those they regulate.

Given the above concerns, for the current proof-of-concept we sought to see if it could address one or two of these issues. Here is how we proceeded.

Analytic Procedure

As a first proof of concept, our goal was to address some of the above issues concerning the complexity and, more specifically, social complexity of the United States power grid. To conduct our study, we employed the following steps, based on the SACS Toolkit:

1. Identify the boundary of the system of interest
2. Select cases
3. Conduct qualitative interviews: grounded theory and snowballing
4. Determine larger sample via cluster analysis
5. Identify measurable variables
6. Conduct scatterplot analysis
7. Run vector quantization
8. Examine time domain waveforms – dynamic observations
9. Draw-up conclusions and communicating results
10. Begin next round of study.

1. Boundary definition. Or, why did we study Ohio?

Despite their increasing independence and interconnectivity, smart territories (as the name implies) have boundaries. The question, however, is how to define and identify them? To demonstrate the applicability of the SACS Toolkit to grid reliability and resiliency, we decided to narrow our study to a small geographic portion of the Power Grid. The State of Ohio was selected, by convenience, for the physical boundary, and the time period from 2009 to 2014 was selected for the temporal boundary. Also, although there are various data feeds that correlate to grid reliability and resiliency (for example, Violation History, Events, Power Studies. . .) we decided to limit the scope of variables studied to misoperations, as misoperations are a good earlier indicator of larger scale reliability events, as mentioned earlier, and a good start for a proof-of-concept study.

2. Selection of cases

After selecting our system boundary (that is, the power grid in Ohio from 2009 to 2014), the next step was to select the cases to be studied. A Case has a clear meaning in case-based modeling, and therefore in the SACS toolkit. A Case is the unique group being studied, including the internal and external variables that describe the group. To select the Cases for this study, several options were considered: Cases as individual Power Companies; Cases as groupings of Power Grid Assets; Cases as Counties in the State of Ohio; Cases as different categories of Employees in the Power Industry, Trade Union geographic regions, and so on. Because the SACS toolkit is looking at social factors, it did not make sense to select Cases that were grounded in equipment. (There are several techniques to model the complex interactions between various grid assets, including Power Flow

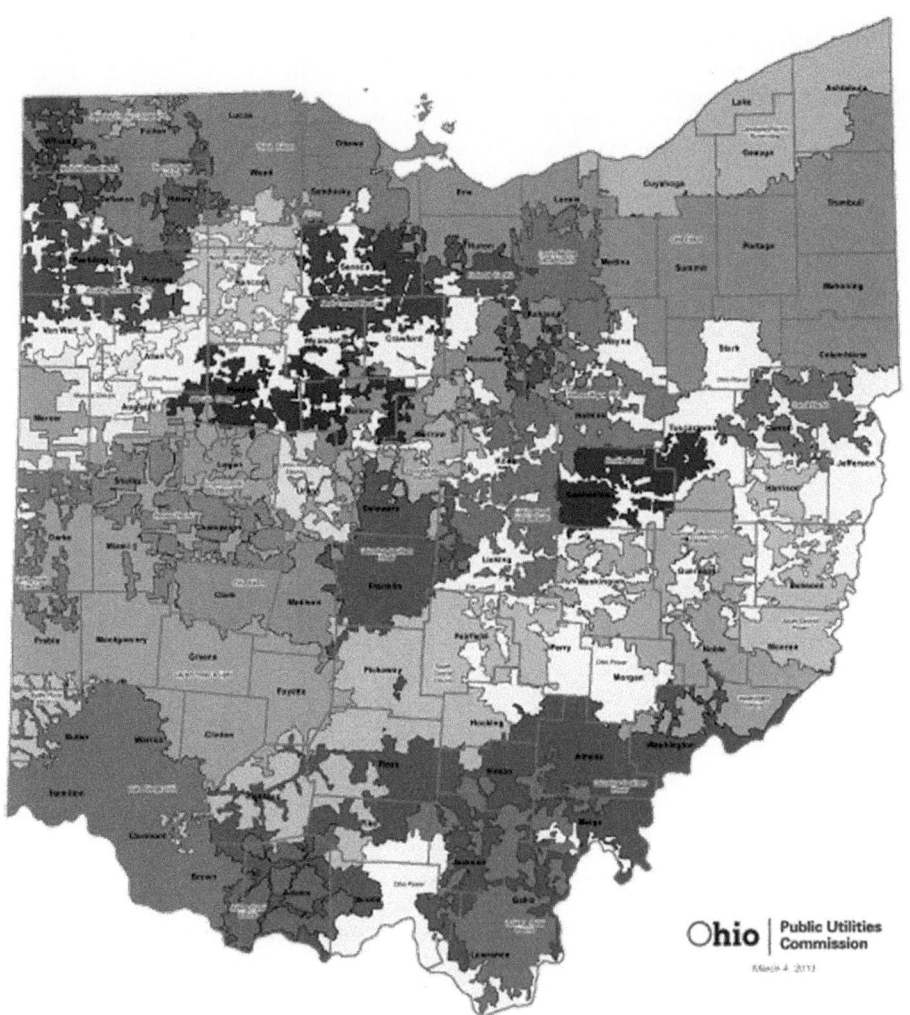

Figure 14.2   Electric service areas of Ohio

analysis.) As such, we decided to define our cases geographically, based on different areas of service. We therefore used the geographical divisions of Service Areas for Ohio. The Service Areas have the unique qualities of being geographically grounded, yet distributed within the context of electrical power. Figure 14.2 shows the Service Areas in Ohio.

3.   Qualitative interviewing: grounded theory and snowballing
After selecting the Boundary (Ohio) and Cases (Service Areas), the next step was to determine which social factors had the strongest influence on the study goal (grid reliability and resiliency, specifically, misoperations). The list of possible social factors to choose from was endless. Therefore, a technique was required to narrow down the list of factors to only those most impactful. For this study, grounded theory was chosen as the method (Glaser and Strauss 1967), using the Snowballing (or Chain Sampling) technique during interviewing (Coyne 1997).

Grounded Theory method is a popular sociological research method. It makes the assumption that the people interviewed are the experts on the topic and therefore can be interviewed to determine which issues are the most important, primarily by looking for themes across the interviews, and using each interview to drive the questions and issues to be explored in the next interview, and so forth, until a consensus about one's data is achieved. There are several variations to the Grounded Theory method. For example, a decision needs to be made regarding how much information is allowed to pass from one interviewer to the next. In purely quantitative statistical surveys, independence may be desirable (since in quantitative surveys, there may be a preconception of the model or behavior underlying the social factors). In the current study, while the investigators had a working model of what the key social factors might be, those were held to the side in order to allow the people interviewed to provide their own working models. Also, there is not endless time or budget to perform hundreds of interviews on independent subsamples. Therefore, the snowballing technique was selected. In snowballing, any factors uncovered in the first interview are fed into the next interview. Therefore the list of factors, although perhaps prioritized differently from interview to interview, continues to grow. In fact, when the growth of the number of factors begins to stabilize, this is an indication of what grounded theorists call 'saturation', the point when further interviewing does not yield significantly more illumination on the study.

4. Determining interviewing sample via cluster analysis

Determining the number of people to be interviewed for the larger study sample is not trivial, primarily because, prior to the interview, the number of factors and ranking of those factors is unknown. How the factors will be distributed across a Service Area, for example, is impossible to discover until interviewing takes place. However, there are some known issues that can be factored into the analysis. For the current study, prior to the interviewing, the number of misoperations in Ohio and their approximate geographic location was known. This information was used to determine the initial interview sampling.

First, misoperations needed to be connected to the Service Areas. To aid in this study, each misoperation in the database was approximately located on a Map of Ohio using textual information provided in the misoperation description (for example, the name of the substation where the misoperation took place). Next, these locations were located by approximate longitude and latitude (GPS coordinates were not available) in the Matlab® Mapping Toolbox. The Public Utility Commission of Ohio (PUCO) provided shape-files for the 32 Ohio Service Areas, and the Mapping Toolbox functions were utilized to determine how many misoperations occurred in each Service Area during each of the five years of the study. Also, the number of grid assets contained in the Regional Map (Figure 14.2) was encircled using the Mapping Toolbox to estimate an approximate number per Service Area.

After completing the mapping, the next step was to determine how to structure the interviews in some effective manner. Initially this 'structuring' was performed by clustering cases according to misoperations, with the idea that this would provide meaningful differences between the various service areas. The misoperations-per-year in each Service Area, along with the number of grid assets per Service Area were used to create common clusters in Ohio.

Cluster analysis is a vector quantization technique that attempts to group data by common patterns. This is a tremendous asset in a study like the current one, as no mathematical models are available to predict patterns. The cluster analysis is semi-supervised. The user needs to input the number of clusters; then the software shows how well the clustering algorithm finds a fit. The clustering used was the K-Means Clustering algorithm. This algorithm shows membership of each Service Area to the proposed clusters, and provides statistical metrics (for example F-Statistics) to show how strongly the cluster contains the Service Areas in its boundary. Visual inspection determined the selection of five clusters as the best fit. Therefore, interviewing was performed in those five clusters of Service Areas, shown in the Ohio map, Figure 14.3.

After selecting the clusters, a series of fifteen interviews (roughly 20 minutes in length) were conducted throughout the clustered Service Areas in Ohio using the Grounded Theory methodology, the Snowballing Technique to determine the strongest sample. It is helpful to have a multi-level

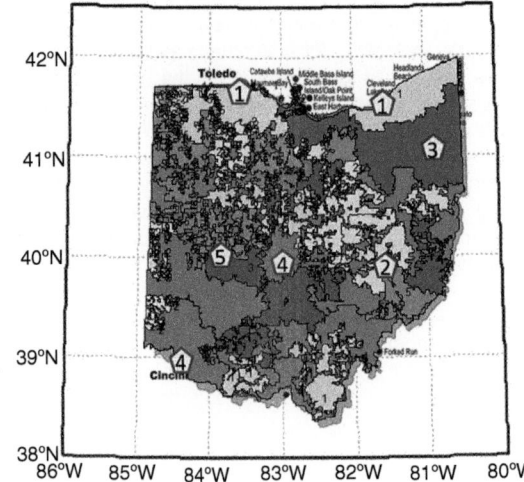

Social Factor ⏷	COUNTI ↲
Weather	11
Economy	7
Public Policy	5
Reorganizations	5
Contractors	4
Unions	3
Demographics	3
Leadership	2

Figure 14.3   Clusters of service areas targeted for interviewing and the corresponding highest ranking social factors from interviews

approach in the diversity of experts in the interviewing. The interviewees ranged from utility line workers to high-ranking executives, each bringing a different expertise and therefore perspective to the interview. Figure 14.3 also shows the result of the coding analysis from the interviews conducted.

Circled in Figure 14.3 are the two top factors, Weather and Economy. Public Policy and Reorganizations were next on the list, however, reviewing the interview transcripts, varying beliefs about the impact of these were provided. For example, some felt that only State of Ohio public policy influenced reliability. Others felt Federal policy influenced reliability. Therefore, although Public Policy ranked relatively high in the list, there was not consensus on which policies influenced behavior. The majority selecting public policy as a factor thought State of Ohio rate case and siting decisions by the Public Utility Commission of Ohio were the stronger public policy impact upon grid reliability. Also, reorganizations were high on the list; but there was not consensus on which type of reorganizations actually drove reliability. In Ohio, during the time period covered in the study, Dayton Power and Light and First Energy experienced reorganizations, but neither were considered the type that would be impactful in the State of Ohio. Therefore, given that we were conducting a case study to generate a first proof-of-concepts, we chose the top two factors: weather and economy.

5.   Identifying measurable variables

Even after the social factors of interest are determined in a sociological study, quantitative data are not always easy to obtain or readily available. The researcher is then faced with finding data that serve as a close proxy for the measurable variables of interest. In this study, Weather information was readily available by county using the publically available National Oceanic and Atmospheric Association (NOAA) database. The categories of storms with high winds and icy conditions were selected as most correlated with electric grid reliability, which was determined by comparing Department of Energy data on storm types for major outages to the NOAA database categories of storms.

Economic data most mentioned during the interview process were unemployment rates. Once again, these data were available from the Federal Bureau of Labor Statistics by county. Shape-files for Ohio Counties were used to partition the Weather and Economic information into Service Areas instead of Counties. Weather data and Economic data came from the Sources by County in Ohio. Therefore, a County-to-Service-Area conversion was required. Using the Matlab Mapping Toolbox intersect and union commands, an estimate was made for the percentage of each county in contact with a service area. Finally, quantitative data was available for Electricity Rates by service areas for the years 2009–2012 (by year). Since this is an incomplete dataset for the five-year period and only a proof-of-concept study, it will not be analyzed quantitatively, but used qualitatively in final cluster descriptions.

6. Scatterplot analysis

Before performing further clustering and analysis of the variables for the study, a scatterplot of these data was examined to see if any large-scale patterns existed. Figure 14.4 shows an interesting correlation, in general between misoperations and weather events.

These results are interesting because weather related events are not counted as a Misoperation. However, weather somehow seems to be influencing the number of misoperations. This will be explored in more depth during further analysis later in our case study.

Figure 14.5 shows another interesting correlation between misoperations and unemployment. The number of misoperations seemed to be rather high when unemployment is in the middle of extremes. These findings correspond to our qualitative interviews.

As one respondent stated: "When the economy is great, errors go down because people are motivated, when the economy is poor, errors go down because people are alert in fear of job loss – it's the middle ground that is of concern for human error."

Although the scope of this study was limited to the correlations between social factors and misoperations, resiliency was examined. However, not enough data were available to trace the resiliency metrics by service area or county. With variable data available for misoperations, unemployment and weather, further analysis was performed in hopes of illuminating correlations, described in subsequent sections.

7. Vector quantization

With our preliminary clusters identified (based strictly on misoperations) and our expert interviews conducted, we had arrived at a working model of two of the key factors our study sought to explore: weather and unemployment. The next step, then, was to re-run the cluster analysis with a more developed profile. With these new discoveries, the methodology leads to clustering again, this time to be performed on the full set of variable data.

The full set of variable data is known as longitudinal data. For this part of the study, the statistical software package, SPSS was used for the clustering analysis on the longitudinal data, due to the more stable results and ability to iterate/converge compared to Matlab's K-Means clustering. Longitudinal data contains full time sampled data: each time stamp (each month in the five years of data) for each of the three variables (misoperations, weather, and unemployment) was considered as a separate trajectory. This resulted in a large dataset (over 5,000 elements) for each of the 32 service areas.

Clustering is a key concept in this style of case-based methodology. Using cases, we help mitigate the chances of missing important details due to regression of the means. In fact, it is the differences between clusters that often lead to insight, and provide directions for further study and enlightenment.

The optimum number of clusters was experimented with, looking at the F-Statistics and distance between means for each step from 1 to 32. A four-cluster solution was determined to be optimum – see Figure 14.6. These algorithmically determined clusters were then compared against the qualitative interview data and overall statistical summary information to determine the attributes that can be stated regarding these clusters. The following descriptions of the four clusters are summarized below:

Cluster 1, No Big Players: This grouping has a very low number of weather events and misoperations. Also, since none of the 'big 4' power companies are in this cluster, it is comprised of an assortment of rural electrics. Interestingly, none of the service areas in this cluster are in 'Appalachian Ohio' (the counties that are in the Eastern part of Ohio, where the terrain is hilly, and are designated by the federal government to be Appalachian). None of the Rural Electrics in this cluster are registered

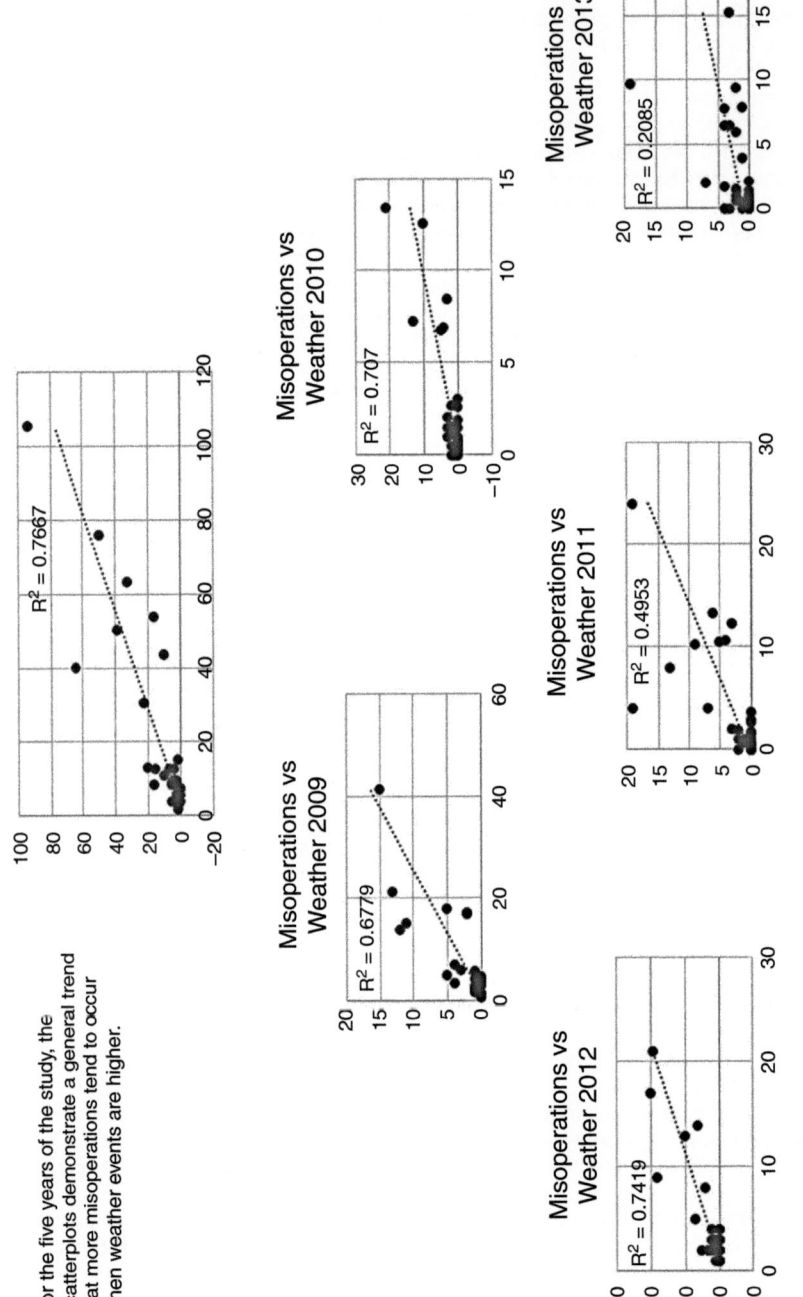

For the five years of the study, the scatterplots demonstrate a general trend that more misoperations tend to occur when weather events are higher.

Figure 14.4   Correlation of misoperations and weather

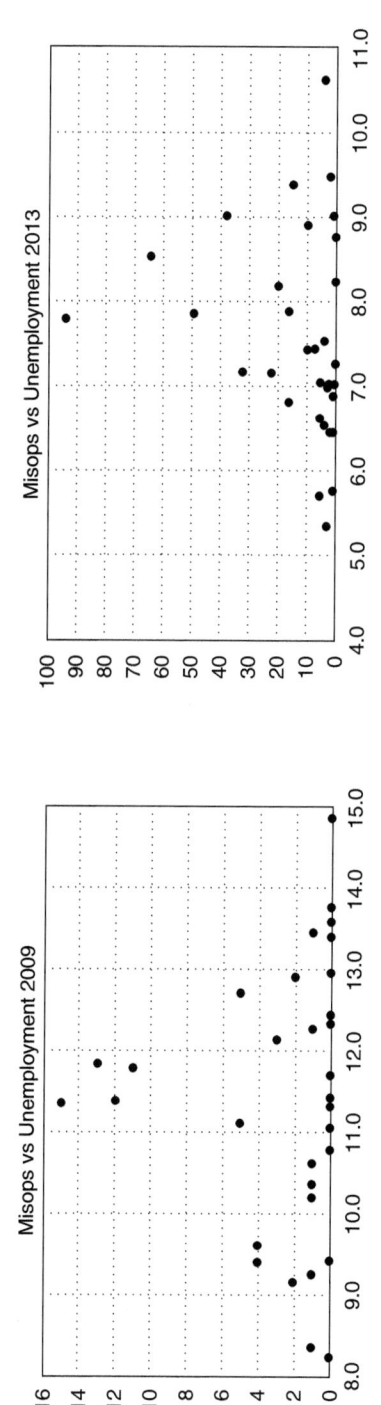

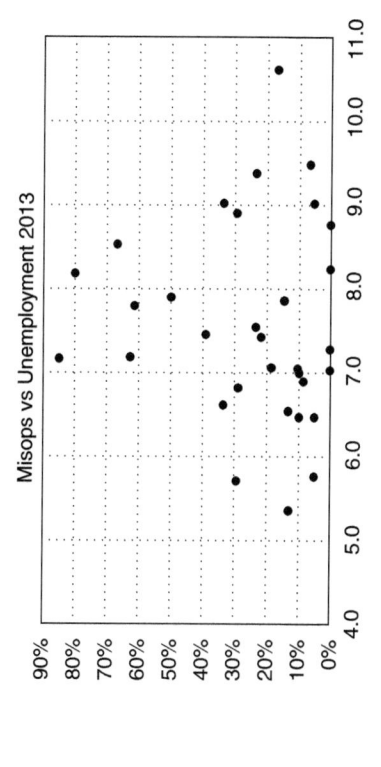

Figure 14.5 Correlation of misoperations and unemployment

as a governmental entity. Finally, this contains, on average, the lowest Electricity Rates of the four clusters. Interview Quote:

> Keep in mind that Ohio is two states – West of 71 and along the lake, and East of 71 and south (Appalachia). The crews in the west are used to flat land working out of buckets. The crews around here in the east have to hike and climb poles. The roads west are nice and north/south/east/west. The roads out east follow old Indian trails and are up/down/winding. Big difference when servicing equipment and working day to day. The storms are different as well in the two regions.

Cluster 2, Lake Erie Shoreline: This cluster is mostly along the shore of Lake Erie along with one other western Ohio service area. Once again, none of the service areas in this cluster are in 'Appalachian Ohio.' Also, due to proximity to Lake Erie, there are a relatively higher number of storms compared to other clusters. None of the Rural Electrics in this cluster are registered a NERC entity. Finally, this cluster, like cluster four, contains the mid to high range of the Electricity Rates.

Cluster 3, Frequent Weather Events: This cluster contains some high weather event areas and high misoperations. This grouping has a very large number of assets. Several of the Rural Electrics in this cluster are registered as NERC entities. Finally, this cluster consistently contains, on average, the midrange Electricity Rates of the other clusters. Interview Quote: "And, even though misop-cause codes [the interviewee means misoperation codes] don't include weather, during a storm there are more switch operations, therefore, more likelihood of misops."

Cluster 4, Low Misoperations per Asset: This grouping has a large number of assets, and a fairly high number of weather events, but has a much lower number of misoperations than Cluster 3. Some of the Rural Electrics in this cluster are registered as NERC entities. There is a concern in this Cluster of the impact in the future that the retirement of Coal plant may come to bear on misoperations in the future. Finally, this cluster, like cluster two, contains the mid to high range of the Electricity Rates. Interview Quote: "Weather is the biggest changing factor over the last 5 years (increasing) and EPA Ruling [United States Environmental Protection Agency] will be impactful around the Southwest part of Ohio near the river."

The four clusters for the State of Ohio map are shown in Figure 14.6. The validation of the algorithmic determination of the clusters, as those shown in Figure 14.7, is not simple. After all, the computer optimized the solution based upon thousands of pieces of information, rather a difficult task for the human brain to perform. However, the SACS Toolkit recommends testing the results of the clustering analysis using a Kohonen Self Organizing Map. The same data fed into the cluster analysis were entered into the Matlab SOM Toolkit, and after the 30-node neural network training was complete, the SOM produced the following analysis of the four clusters (Figure 14.7 contains the SOM U-Matrix showing in a graphic form how the neurons of the neural net clustered).

Notice in Figure 14.7 that the majority of the service areas in the clusters determined by SPSS were in similar groupings to the SOM analysis, validating the computer selected clusters by two separate methods, and two different software packages.

8.   Time domain waveforms – dynamic observations

Time domain waveforms for misoperations, weather events, and unemployment rate for each service area in each of the four clusters was available for detailed study. It is important not to assign 'causation' from weather and economy to misoperations because of the other factors contributing to the complex overall misoperation signals.

However, there are interesting correlations discovered through detailed inspection of the dynamic waveforms. First, Service Areas experiencing higher number of weather events also have higher misoperations. These misoperations are not aligned in time with the misoperation. For the five-year timescale of the study, visual inspection suggests an approximately six months or greater delay from the weather event to the next misoperation. Second, service areas experiencing few or no weather events often show correlation between changes in the unemployment rate and the occurrence of misoperations. Finally, while we cannot make any strong causal statement regarding the link between weather and misoperations, given the rich data we had, we decided to split our database

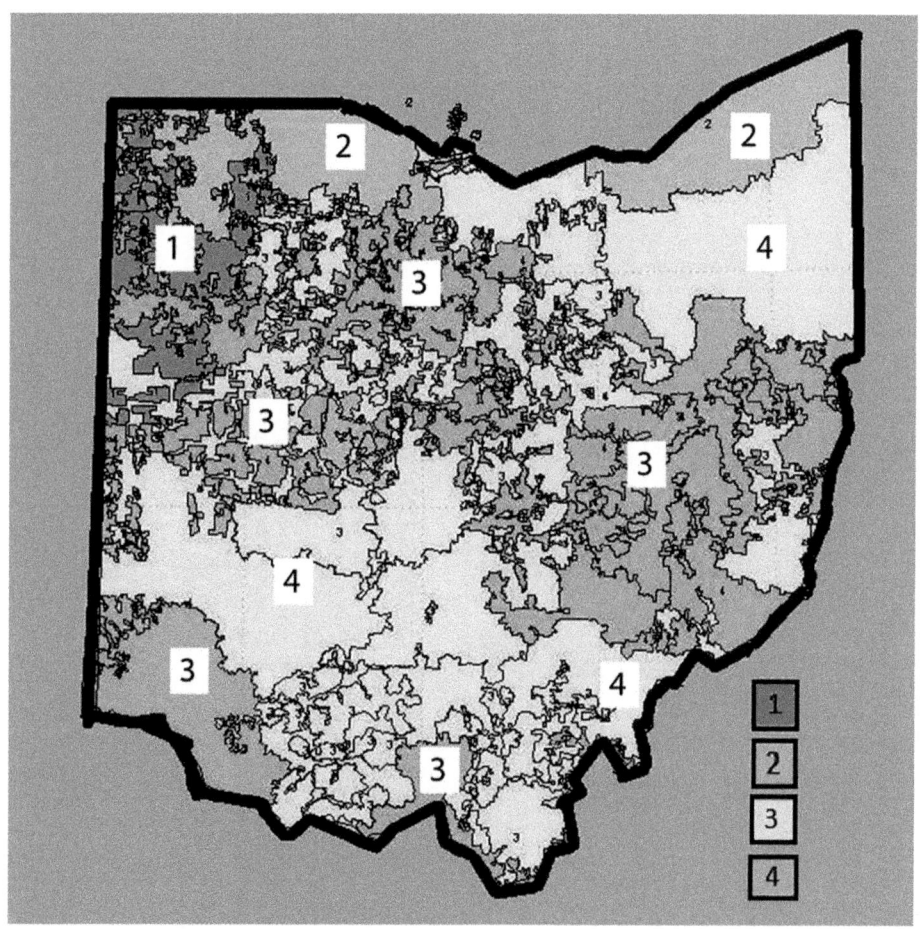

Figure 14.6 Four main clusters of Ohio

and perform a test-retest hypothesis. We used the first half of the database to determine the lag between weather events and misoperations; then, we used the second half of the database to test our hypothesis about this data-derived lag. This can be visualized by comparing the left and right half of each waveform.

9. Conclusions and communicated results

With the analysis complete, the last step is to write up one's report, as we did here, and communicate results. Here is a list of key suggestions we made for the current study: (a) for all clusters, continuously collect weather, economic, electricity rate, and misoperations data by service area and alert if storms increase, unemployment rates change or approach the mean, and if electricity rates change. Compare these against the baseline misoperations for each service area; (b) *Cluster 1 – 'No Big Players'* do not take any measures unique for this cluster unless misoperations increase well beyond normal; (c) *Cluster 2 – 'Lake Erie Shoreline'* monitor storm activity carefully. Consider advising entities owning assets in this cluster to increase to higher operational contingencies (N–2 or above) for a period of six months after storms occur. Consider influencing the Public Utility

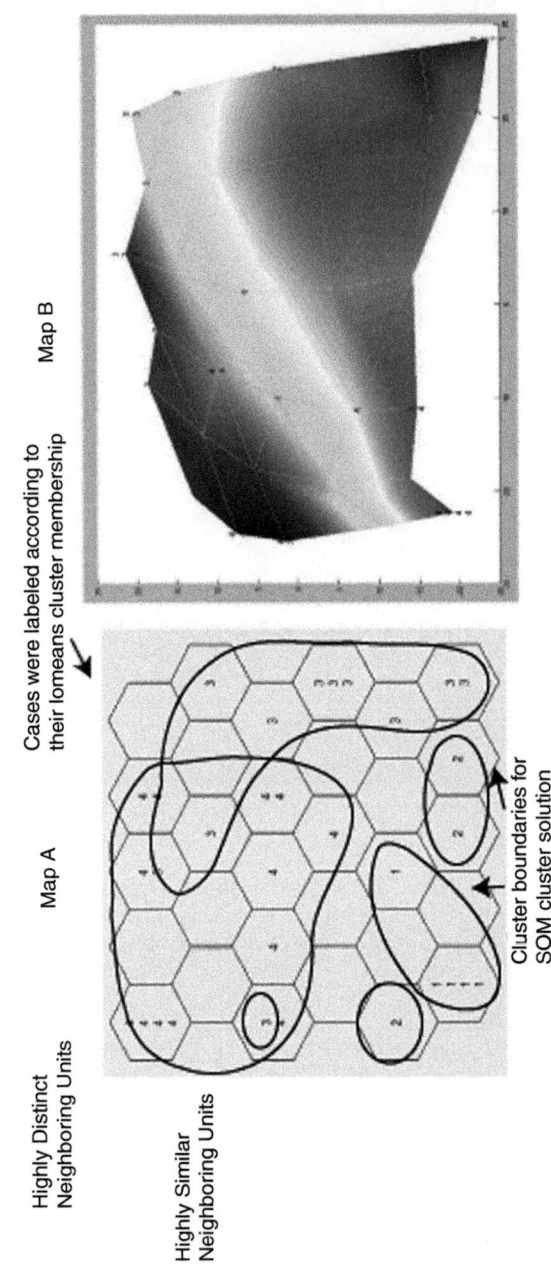

Highly Distinct
Neighboring Units

Map A

Cases were labeled according to
their Iomeans cluster membership

Map B

Highly Similar
Neighboring Units

Cluster boundaries for
SOM cluster solution

Map A and *Map B* are graphic representations of the cluster solution arrrived at by self-Organizing Map (SOM) Neural Net, referred to as the U-Matrix. In terms of the information they provide, *Map B* is a three-dimensional (topgraphical) u-matrix for it, the SOM adds hexagons to the original map to allow for visual inspection of the degree of similiarity amongst neighboring map units; the dark areas indicate neighbor-hoods of acases that are highly similar, in turn, bright areas, as in the middle of the map, indicate highly defined cluster boundaries. *Map B* is a two-dimensional version of *Map A* that allows for visual inspection of how the SOM clustered the individual cases. Cases on this version of the u-matrix (as well as *Map A*) were labelled according to their k-means cluster membership (The 4 cluster solution show in Figure 14.6) to see if the SOM would arrive at a similar solution.

Figure 14.7   U-matrix for four cluster solution

282

Commission of Ohio (PUCO) to consider higher rates in this cluster unless misoperation rates change or economic conditions change rapidly; (d) *Cluster 3 – 'Frequent Weather Events'* and *Cluster 4 – 'Low Misoperations per Asset'* monitor economic activity carefully; and consider advising entities owning assets in this cluster to increase to higher operational contingencies (N–2 or above) for a period of six months after large changes in economic rates occur. Consider influencing the PUCO to maintain rates in this cluster unless misoperation rates change or economic conditions change rapidly.

10. Return to study

With the analyses complete and recommendations made, the last step is to begin again with the first step – seeking to learn new things about the power grid – based on new concerns that regulators have.

14.2 CONCLUSION

As this study has hopefully demonstrated, the optimal decision-making process for addressing infrastructure risk and resiliency requires a dynamic, evolving 'complex systems' model, which fuses current infrastructure models (that is, engineering, control theory, and so on) with those found in social sciences, all grounded in evolving temporal big data. Only then can our models inform the most probable course of action for organizational decision makers, in the face of high uncertainty, allowing them to apply influential communications to steer the sociotechnical systems they manage to provide higher reliability and resiliency, year after year.

Furthermore, as we also hopefully demonstrated, the SACS Toolkit is a highly useful modeling platform to support these efforts, as demonstrated by the case study of the State of Ohio Misoperations illustrated in this chapter. The SACS Toolkit offers an effective multi-method approach that can address several of the key challenges associated with such an empirical study, as well as overcome many of the limitations of conventional modeling, which do not do a very good job modeling complexity or complex systems.

REFERENCES

Braha, D., Suh, N., Eppinger, S., Caramanis, M., and Frey, D. (2006). *Complex Engineered Systems* (pp. 227–274). Berlin Heidelberg: Springer.

Byrne, D. (2013). Evaluating complex social interventions in a complex world. *Evaluation* 19(3): 217–228.

Byrne, D., and Ragin, C.C. (2009). *The Sage Handbook of Case-based Methods*. London: Sage Publications.

Byrne, D., and Callaghan, G. (2013). *Complexity Theory and the Social Sciences: The State of the Art*. London: Routledge.

Capra, F., and Luisi, P.L. (2014). *The Systems View of Life: A Unifying Vision*. Cambridge: Cambridge University Press.

Castellani, B., and Hafferty, F.W. (2006). The complexities of medical professionalism. In J. Spandorfer, Charles A. Pohl, Susan L. Rattner, and Thomas J. Nasca (eds), *Professionalism in Medicine* (pp. 3–23). New York: Springer.

Castellani, B., and Rajaram, R. (2012). Case-based modeling and the SACS Toolkit: A mathematical outline. *Computational and Mathematical Organization Theory*, 18(2): 153–174.

Castellani, B., Schimpf, C., and Hafferty, F. (2013). Medical sociology and case-based complexity science: A user's guide. In C. Martin and J. Sturmburg (eds), *Handbook of Systems and Complexity in Health* (pp. 521–535). Berlin: Springer.

Castellani, B., Rajaram, R., Gunn, J., and Griffiths, F. (2015b). Cases, clusters, densities: Modeling the nonlinear dynamics of complex health trajectories. *Complexity, Early View* http://onlinelibrary.wiley.com/doi/10.1002/cplx.21728/full.

Castellani, B., Rajaram, R., Buckwalter, G., Ball M., and Hafferty, F. (2015a). *Place and Health as Complex Systems: A Case-Based Study and Empirical Test*. In *SpringerBriefs in Public Health Series*. New York: Springer.

Castells, M. (2011). *The Rise of the Network Society: The Information Age: Economy, Society, and Culture* (Vol. 1). London: John Wiley & Sons.

Coyne, I.T. (1997). Sampling in qualitative research. Purposeful and theoretical sampling: merging or clear boundaries? *Journal of Advanced Nursing*, 26(3): 623–630.

Friedman, T.L. (2008). *Hot, Flat, and Crowded: Why we need a Green Revolution – and how it can Renew America*. London: Macmillan.

Galen Buckwalter, J., Castellani, B., McEwen, B., Karlamangla, A.S., Rizzo, A.A., John, B., . . . and Seeman, T. (2016). Allostatic load as a complex clinical construct: A case-based computational modeling approach. *Complexity*, 21(S1), 291–306.

Gerrits, L.M. (2012). *Punching Clouds*. Litchfield Park, Arizona: Emergent Publications.

Gerrits, L., and Marks, P. (2015). How the complexity sciences can inform public administrastion: An assessment. *Public Administration*, 93: 539–546.

Giddens, A. (2002). *Runaway World: How Globalisation is Reshaping our Lives*. London: Profile books.

Giovannella, C. (2013). 'Territorial smartness' and emergent behaviors. In *Systems and Computer Science (ICSCS), 2013 2nd International Conference on* (pp. 170–176). IEEE.

Giovannella, C. (2014). Smart territory analytics: Toward a shared vision. In *47th SIS Scientific Meeting of the Italian Statistical Society* (pp. 10–14).

Glaser, B.G., and Strauss, A.L. (1967). *The Discovery of Grounded Theory: Strategies for Qualitative Research*. Observations. Chicago: Aldine Publishing.

Hafferty, F.W., and Castellani, B. (2010). The increasing complexities of professionalism. *Academic Medicine*, 85(2): 288–301.

Haynes, P. (2015). *Managing Complexity in the Public Services*. London: Routledge.

Linkov, I., Eisenberg, D.A., Bates, M.E., Chang, D., Convertino, M., Allen, J.H., . . . and Seager, T.P. (2013). Measurable resilience for actionable policy. *Environmental Science and Technology*, 47(18): 10108–10110.

Mingers, J., and Gill, A. (1997), *Multimethodology: The Theory And Practice of Integrating Management Science Methodologies*. Chichester; New York: Wiley.

Pachauri, R.K., Allen, M.R., Barros, V.R., Broome, J., Cramer, W., Christ, R., . . . and van Vuuren, D. (2014). Climate change 2014: Synthesis report. Contribution of Working Groups I, II and III to the Fifth Assessment Report of the Intergovernmental Panel on Climate Change.

Pagani, G.A., and Aiello, M. (2013). The power grid as a complex network: A survey. *Physica A: Statistical Mechanics and its Applications*, 392(11): 2688–2700.

Pagani, G.A., and Aiello, M. (2015). A complex network approach for identifying vulnerabilities of the medium and low voltage grid. *International Journal of Critical Infrastructures*, 11(1): 36–61.

Rajaram, R., and Castellani, B. (2012). Modeling complex systems macroscopically: Case/agent-based modeling, synergetics and the continuity equation. *Complexity*, 18(2): 8–17.

Rajaram, R., and Castellani, B. (2014). The utility of non-equilibrium statistical mechanics, specifically transport theory, for modeling cohort data. *Complexity*, Early View.

Rosa, E., McCright, A., and Renn, O. (2013). *The Risk Society Revisited: Social Theory and Risk Governance*. Philadelphia, PA: Temple University Press.

Teisman, G., van Buuren, A., and Gerrits, L.M. (eds) (2009). *Managing Complex Governance Systems*. London: Routledge.

Urry, J. (2007). *Mobilities*. London: Polity.

15. Phase transitions and social contagion as enabling mechanisms for coordinated action in populations: a mathematical framework

Professor James K. Hazy, Adelphi University and
Professor Peter R. Wolenski, Louisiana State University

15.1 INTRODUCTION AND OVERVIEW

This chapter advocates a research agenda that builds on prior work that uses mathematics to understand social behavior (Von Neumann and Morgenstern 1944; Axelrod 1997; Lasry and Lions 2007). It targets canonical modeling of how human organizing changes over time within populations and how these properties emerge. It argues that the genesis of organizing occurs within the individual, the agent, and spreads to other agents through contagion. The framework relates to how each agent engages the complexity that it observes; how it models or represents this complexity; and how it uses these representations to navigate its local environment to survive, thrive and ensure its genetic and cultural fitness. Taken together, we refer to these fitness enhancing outcomes as the agent's 'welfare.'

To further its welfare each agent acts in response to various observable informational influences available to it at its particular place and time. These informational influences are associated with complex external demands in the ecosystem as well as complex local interaction dynamics internal to the population that it must navigate along with others. Taken together, these two interacting sources of information involve a stream of events that although relevant to the agent, can reflect a very high level of computational and statistical complexity (Crutchfield, Ellison, and Mahoney 2009). This complexity must be modeled or represented by the agent if that agent is to be able to predict and thus anticipate events to further its welfare.

Through social learning, these models, or aspects of them, spread through the population by the mechanism of social contagion. As order spreads, information about the environment is embedded or stored in socio-technical structure (effectively as the agent's external memory). As a result, agents who have been primed to access and decode this information can use it to guide their actions. This reduces the load on the agent's scarce internal memory and cognitive processing capacity. The cost to the agent of this individual benefit is that the agent loses control of the memory that has been stored. It becomes a shared resource. This chapter suggests that, by assuming each agent is a 'black box,' these processes can be described using nonlinear mathematics of phase transitions and contagion models that are in consilience (Wilson 1998) with models used in the natural sciences.

15.1.1   Overview of the Cusp of Change Framework

To illustrate the above general argument, this chapter looks in detail at *phase transitions* and *contagion* in human interaction dynamics (HID) to explore how conditions external to the agent (that is, 'context') impact the potential for discontinuous change in organizing structure. To do this, it describes a canonical fourth degree *potential function* (Thom 1989) that is used in the natural sciences to model *potential fields* (Haken 2006). This chapter argues that it can also be applied to social systems assuming the requisite parameters can be identified and measured. The approach effectively sets aside individual difference occurring inside an agent's 'black box' and treats each agent as a particle of an 'ideal population' (like the 'ideal gas' in thermodynamics) subject to informational influence 'forces' and contagion effects during local interactions.

Called the cusp catastrophe (Thom 1989; Zeeman 1977), Haken (2006) shows how this static model describes the potential for discontinuous change in the value of a single order parameter. As two *control parameters* (coefficients in the equation) are allowed to vary, discontinuous change is predicted to occur in the vicinity of *critical points*. These are the points on a smooth function where the derivative is zero (Arnold et al. 1993). These points can be identified for all values of the control parameters when a *potential surface* is plotted in the relevant *parameter space*.

In the context of human organizing, this potential surface embedded in parameter space has been called as the Cusp of Change model (Goldstein, Hazy, and Lichtenstein 2010). Metaphorically, this model has been averred to describe the canonical potentials for dynamical stability of an ordered regime within HID where the potential for stability can be indexed by a single variable to reflect the level of order within the population. This chapter defines an application of this model that we argue extends beyond mere metaphor (Hazy 2014).

To make this case under a general framework that assumes fields of informational influences acting on individual agents, the control parameters in the Cusp of Change model are taken to index the complexity conditions that each agent must confront. (One may think of this as observations by the agent that include both information and surprise.) These complexity conditions suggest informational influence pressures that are acting on individual agents who are seeking to enhance their welfare, perhaps by organizing with others.

The first control parameter indexes complexity conditions that are external to an agent's local social environment. High levels of surprise in the broader ecosystem – for example with regards to the availability of resources beyond the local population and inter-group competition – implies conditions where, by organizing with others, individuals might more effectively discover, gather, use, and accumulate resources that are necessary to further their welfare. The second control parameter indexes interaction complexity within the local population. High levels of surprise associated with the human interactions that are unfolding locally in real-time – for example the computational complexity required for an agent to predict relevant peer-to-peer interactions and their relative payoffs – implies conditions where individuals face local challenges when accessing resources available within the local population. Internal complexity would seem to be greater in more densely populated localities than in sparsely populated ones, assuming that neither is organized.

For both of these control parameters, as the value of the parameter increases, the agent is forced to navigate and model increasing complexity in an effort to predict events that might impact its welfare. Absent opportunities to organize which might reduce perceived complexity, an ever increasing probability for surprise would seem to consume ever greater cognitive load and internal memory capacity within each agent. The challenge presented by increased complexity in an agent's local informational environment is whether and if so how organizing with others might reduce the agent's cognitive load and do so without increasing risk to the agent's individual welfare.

15.1.2   The Cognitive Benefits to Organizing

The number of degrees of freedom that an agent must navigate is a reflection of the computational or statistical complexity of the informational environment which that agent must model to effectively predict events. This implies that more memory and processing power (more cognitive load) is required of the agent as complexity increases. When informational conditions become too complex, we posit that an agent can reduce the number of degrees of freedom that it must navigate – and thus its requisite cognitive load – by organizing with other agents as a mechanism to 'share the load.'

As the level of predictability or 'order' that is observed in social interactions increases, one would expect that the number of degrees of freedom for choices that the individual agents who participate or 'enroll' in the requisite order regime would need to model would decrease. Thus, the presence of social order constrains (or limits) the number of degrees of freedom that must be modeled for autonomous action by the enrolled agents. This reduces cognitive load. To gain these benefits, agents must 'follow the rules.' 'Order' enacted in this way has the effect of encoding, or 'storing' information in these 'rules' that are external to the agent's memory, reducing the participating agent's cognitive load. When a pattern of social interaction is observed, a predetermined response can be enacted with minimal cognitive load. But this is only possible if the agent is privy to the decoding algorithm that unlocks the payoff structure of the game. Only agents equipped to decode these micro-enactments, as they have been called (Hazy and Silberstang 2009; Silberstang and Hazy 2008), can access information encoded in the social environment to anticipate events.

15.1.3   Order and Degrees of Freedom

Just as *observing* an ordered environment enacted by others helps each enrolled agent reduce cognitive load, each enrolled agent must also reciprocate for others who are likewise observing social patterns by *enacting* an ordered environment for their benefit. Conversely, within a population, if the average agent operates with fewer degrees of freedom by conforming to norms that promote order, then any enrolled agent needs to model fewer degrees of freedom than are actually present in the ambient context. Thus, enacting social order simplifies the modeling of experience for an enrolled agent who individually no longer needs to model the full complexity of macro and micro conditions. However, because the agent is no longer independently modeling the full complexity of the environment, its welfare is dependent on the efficacy of social structure that has been adopted.

To preserve cognitive load, enrolled agents lean on the collective for their welfare. As such, organizing is essentially a social heuristic, a simplified view of reality that may or may not provide accurate prediction of events in a specific instance. Although there appears to be more order, and thus less surprise during local interactions, the potential for surprise in the events within the broader ecosystem has not changed. Stated differently, although there is greater 'order' locally because more information is embedded in social structure and available to enrolled agents, this local order may, but does not necessarily, provide welfare benefits to the agent. If it does, it offers what evolutionary biologists who favor group selection call 'nesting safety' (Nowak, Tarnita, and Wilson 2010). We therefore propose that an *order parameter* in HID can be indexed as the mean number of degrees of freedom that encode the game for an average agent in an organizing system.

15.1.4   Phase Transitions in HID

By analogy with phase transitions in natural systems, herein we describe an order parameter in a particular representation of events within a closed HID system (Hazy and Backström 2013a). Note that the representation is not intended to completely reflect all aspects of ordering that might be observed in HID, and there may be many more active degrees of freedom than are observed in physical systems. Any particular closed representation that uses the Cusp of Change is limited to representations having a single order parameter. Examples might include: phase transitions among stable levels of utility maximization, cost minimization, or employee engagement metrics. However, only one order parameter can be predicted with the Cusp of Change model. Further, not any representation will do. The representation must be constructed according to mathematical requirements of Category Theory (Mac Lane 1978; Spivak 2014).

We aver that many problems in social science relate to cases where the number of active (that is, nonzero) degrees of freedom that are being decoded by individual agents within a stable instance of an organizing system (as indexed by an order parameter) undergoes a sudden, discontinuous change. Such analysis can indicate that there are distinct stable values in this order parameter that are analogous to the 'phases' indicated by the term 'phase transition' in the natural sciences.

For any specific parametric state, these phases might suggest a fixed point equilibrium, or perhaps in some cases, bi-stability with two stable equilibria separated by an unstable equilibrium between them. The stable levels for the order parameter suddenly change as the system moves along the potential surface by varying one or both of the two control parameters.

15.1.5   Summary of the Argument

In the first main section that follows we suggest that the notions of potential surfaces and phase spaces that are widely used in the natural science can also be used as an underlying representational framework to describe the dynamics of human organizing. The presentation herein draws from algebraic geometry rather than analysis. Therefore, the results it describes relate to general abstract relationships among variables rather than specific solutions to particular problems or equations. Although for any particular application – a cost or utility function, for example – a specific instance must be carefully defined, the category theoretic schema presented here is internally consistent, and we suggest canonical, and therefore could be a useful starting point from which to begin deductive arguments.

The second section describes the Cusp of Change framework as a canonical fourth degree equation that is often used in the natural sciences to model *phase transitions* (Haken 2006). In the third and final sections before the conclusions, we explore the interaction mechanism of contagion among agents interacting within a potential field. We posit that social organizing contagion operates locally within complex adaptive systems (Holland 1995) of human interaction dynamics (Hazy and Backström 2013a, 2013b) to enable the spread of ordered structure that stabilizes collective action flows in a given parametric situation. We conclude with suggestions for future research directions.

15.2 INFORMATIONAL INFLUENCE AS THE CONTEXT FOR ORGANIZING

The methodological approach we suggest incorporates a static representation of the potential for dynamic stability in HID and for phase transitions between dynamically stable states. Further, it posits that there are putative informational influence 'forces' that push the actions of individuals to maximize the potential benefits that accrue from stable organizing as reflected in the value of the order parameter. Organized stability implies order that agents can recognize and use to reduce their cognitive load.

Thus, this section proposes that the potential to organize in HID can be represented by a mathematical category in parameter space which forms a potential surface within a conservative field that is enacted through informational influences. By identifying stable potential states as well as rates of change at other points, the representation describes imputed 'informational influence forces' that are applied locally on each agent to alter an agent's *action vector* along the relevant degrees of freedom. In many cases, informational influence forces result in the potential for action along a particular degree of freedom to go to zero. For example a hard surface or wall in physical space constrains action along that dimension to zero. Likewise, a legal construct such as a trade embargo would truncate action in the direction of trade with the embargoed entity. These informational forces constrain action and can reduce the active degrees of freedom available to each agent. This simplifies the representation and reduces cognitive load for agents.

15.2.1 Action Vectors in Phase Space

The simplest object within the representation is an *action vector*, $x = (x_1, x_2, \ldots, x_n)$ which is constructed to represent action along N selected dimensions. A dimension might be cost cutting activity, for example. Each component of this vector suggests two degrees of freedom, the magnitude of action, x_i, along that dimension, and the rate of change, dx_i/d_t in the magnitude of that dimension over time. Thus, when treated as a black box, each object, each agent, can be represented by a point in *2N* dimensional phase space, S_N.

More specifically, let $D \leq N$ be the number of nonzero or active dimensions of action vector x. Then $2D$ identifies the number of degrees of freedom and thus the number of dimensions in the relevant instance of phase space such that, $S_D \subseteq S_N$. Each phase space generated in this way – formed from a set of possible active dimensions of an action vector (of an agent) – is defined to be an object. These objects along with the preorder defined by the inclusion morphisms among them form the category of possible phase spaces for an agent or a set of agents in a population. This category, denoted C_p, represents the potential dimensions for organized agent action.

15.2.2   Categorical Representations of Action Vectors

When phase spaces are treated as the objects in the category C_p, the identity and inclusion morphisms between objects, as well as the composition of maps must also be defined. In this context, one can further construct a map via functors F_i from C_p to the category of sets, Set_p composed of sets of agents generated by the population, P. This new category, called C_p-Set_p, reflects an *instance* of the phase space represented in the population. One would say that for a given F, the image set of agents in Set_p operate within the pre-image phase space in C_p.

One can define the category of *action vectors* for the agent a_i as the functor F_j which maps objects in C_p (phase spaces) to objects (sets) in Set_p in which a_i is an element, and within which each active dimension in objects of C_p and its rate of change reflect two degrees of freedom for the set of agents of which it is a member – the agent a_i is in the image object in Set_p. For example, if the object in C_p implies nonzero attention to 'cost cutting action' as a dimension for action, then the agents in the image set would have 'cost cutting action' as an active dimension upon which to focus their attention. The magnitude and the rate of change in magnitude of activity toward 'cost cutting action' would be the relevant degrees of freedom for this dimension of the instance.

Any object in C_p with a single active dimension has two degrees of freedom and thus forms a two-dimensional phase space. This is the smallest non-trivial object within C_p because the only other object in C_p that is included within it is the null set. Objects in C_p with multiple active dimensions (where inclusion relationships are the morphisms in the category) include other objects that have a single active dimension. These relationships are relevant in the context of the preordering among all possible phase spaces.

15.2.3   Categorical Representations of Organizing Systems

More generally, define C_p-Set_p to be the category of functors F_i: $C_p \rightarrow Set_p$ (the objects are these functors) mapping C_p to Set_p where Set_p is the category generated by the power set of the population P and the inclusion morphism is a preorder that forms a category Ord_p of preordered sets in Set_p. An *instance* is an object in C_p-Set_p, that is, a functor F mapping C_p (the *phase space*) to an object (the set) in Set_p. Because the objects in Set_p are sets of agents in the population, an instance may include a mapping of the *phase space* to a set containing one or many agents.

When multiple agents are in the instance, they are represented by the same phase space and therefore have the same action dimensions. Taken together, the action vectors of the individual agents in the instance form an *action flow* of a set of agents. If the action flow in an instance tends to a limit set, the instance is called *stable*. This implies that the agents who are elements of the image of F in the codomain Set_p as defined by the pre-image of F in the *phase space* domain C_p have action vectors that tend to a limit set. These agents can better predict the future state of events that tend to the limit, and this reduces their cognitive load. Any non-empty closed subcategory of C_p-Set_p with at least one nonempty stable instance is called an *organizing system*. Further, agents who participate in an organizing system perceive a more predictable local environment in at least one instance, and as a result, they benefit from a reduced cognitive load.

15.2.4   The Order Parameter as an Equivalence Class among Stable Instances

To formally measure the level of organizing observable in a subpopulation, P^S, we define the *order parameter* to be the mean number of degrees of freedom for which the individual agents in P^S are able to decode the correct choice within a stable instance in a subcategory of preordered sets in $C_p\text{-}Set_p$. Each stable instance of the subcategory has a corresponding value for its order parameter. However, since in general $C_p\text{-}Set_p$ is a preorder and not a total order, there is not a single path of inclusion from an instance involving a particular agent to the instance involving the maximal set in Set_p. Likewise, multiple distinct stable instances that involve the same agent can have the same order parameter.

With respect to levels of organizing, because the Integer values of the order parameter form a total order, the order parameter can be used to define equivalence relationships. This means that from an order parameter perspective, for three agents x_a, x_b and x_c, the functor F_b mapping a phase space with two degrees of freedom to the set $\{x_a, x_b\}$ is equivalent to the functor F_c mapping the same phase space to the set $\{x_a, x_c\}$ in the sense that in both cases the value for the order parameter is equal to two. Informally, this means that from the order parameter perspective of two degrees of freedom, x_a being aligned with x_b, is the same as or equivalent to x_a being aligned with x_c and so on. This correspondence remains true even if the active dimensions are quite different. In the former case, the degrees of freedom might involve attention to cost cutting activities and change in the levels of those activities. In the latter, the two degrees of freedom might involve spending level to stimulate sales growth and its rate of change. In either case, the value of the order parameter for each instance is equal to two.

If one assumes that there are two phase spaces in C_p and that each maps through a functor to the same object, a set of agents, in Set_p, then there are two instances in $C_p\text{-}Set_p$ with the same set of agents in the codomain such that each agent has two action vectors in common with others member agents, each with a different dimension and each with two degrees of freedom. Both of these would be included in a higher order instance of $C_p\text{-}Set_p$ with the four degrees of freedom. These agents could use this information in the ecosystem to decode cooperative choices along four degrees of freedom and benefit from a further reduced cognitive load. A simple example of this is the case where one active dimension reflects a nonzero magnitude of cost cutting activity to minimize costs while a second active dimension might be a nonzero magnitude of spending activity (perhaps through bonuses and commissions) to increase sales. Together these imply four degrees of freedom. We will refer to this example as the Cost Containment versus Revenue Growth model in the remainder of this chapter.

Inclusion mapping can continue in this way into higher numbers of dimensions and thus degrees of freedom, and the active dimensions can vary across the population. As these examples suggest, active dimensions within an instance can impact a single agent, multiple agents or the whole population, and they can complement or conflict with one another.

15.2.5   Stability in Organizing Systems

We posit that all dimensions and degrees of freedom in any instance of $C_p\text{-}Set_p$ are subject to informational influences. The resulting potential field impacts the agents (that is, their action vectors) as each is independently influenced by conditions that are both external to and internal within the population. These informational 'forces' can change both the magnitude and the direction of an agent's action vector. Taken together, agents' action vectors reflects collective *action flow* within a chosen instance of a subcategory of $C_p\text{-}Set_p$.

Some instances are dynamically stable which means that all agent trajectories in the instance tend to a limit set. Others diverge for all agents in all dimensions. Still others might converge to quasi-periodic or to strange attractors for some or all agents. Because the inclusion morphism in $C_p\text{-}Set_p$ is a preorder rather than a total order, not all stable instances can be reached from another stable instance by a direct morphism. Sometimes it may be necessary to compose morphisms by passing through unstable instances, moving from one stable instance to another first increasing degrees of freedom and accepting instability before again decreasing degrees of freedom along a different chain of inclusion and projection morphisms to regain stability.

15.2.6   Informational Influence and the Order Parameter

Potential fields that have been posited to arise from informational influence forces that enable dynamically stable organizing systems include cost minimization to maintain profits, shareholder value maximization to maintain access to capital, economic utility maximization to further self-interest, and climbing fitness landscapes for survival. The first of these is assumed to exert an informational influence force such that costs are minimized, whereas the latter three represent informational influence pressure to maximize the measured value. In each case, activity that occurs in these fields suggests that there is an imputed informational influence force that changes the values of an agent's action vector along a relevant dimension. Thus, action flows that include many agents within an instance of the organizing system are likewise subject to this 'force.' The active dimensionality and thus the implied degrees of freedom within a given instance determine the value of the *order parameter* for that instance.

For example, information forces might be observed to influence agents in an instance of the organizing system to develop an active positive dimension of cost cutting where none existed before. This change would be represented as a category map from the then current instance in the organizing system, through the inclusion morphism of the preorder, to a higher dimensional instance of the organizing system. This higher order instance would completely include the earlier instance and also preserve its structure. However, the domain of the new instance would be the phase space with two additional degrees of freedom, a magnitude of cost cutting activity and a rate of change in that magnitude.

Likewise, informational influence forces could be observed to influence agents in an instance of the organizing system to erase a previously active positive dimension toward spending for sales growth. These forces would take the magnitude and rate of change of spending for growth to zero. For this to be captured in the representation, this change would be represented as a category map from the then current instance of the organizing system, through a projection to a lower dimensional instance of the organizing system.

This lower order instance would preserve the relevant structure in the lower level instance. However, the domain of the new instance would be the phase space with two less degrees of freedom, the magnitude of spending for sales growth and its rate of change in that magnitude would have gone to zero. Taken together, the composite of this mapping within an organizing system with the one in the prior paragraph would effectively replace the focus on spending on sales growth with a focus on cost cutting while keeping all other active action dimensions unchanged.

15.2.7   An Integrated Analytical Framework: The Value Potential Function

This chapter proposes to generalize various ad hoc constructions that are common in management and social science into a category theoretic representation of informational influence forces as reflected in *value potential fields*. This general representation would be well-specified in each instance according to the category theoretic requirements of the general theory. This value potential function, V, generalizes the role currently played by utility functions, fitness or performance landscapes, cost functions, and so forth, and would be assumed to exert putative informational influence forces that act locally on individual agents as well as at other levels of order and scale.

   This overall blueprint is intended to be consistent with mathematical concepts contained in category theory. As a result, the nature of these potential functions and how organizing states are recognized in phase space could be described rigorously through category theoretic relationships (Mac Lane 1978; Spivak 2014) and would be consistent with analogous models of potential fields in the natural sciences. The theory we propose draws upon axiomatic work pioneered by Whitney (1955), Thom (1989), and Arnold (1991), and now is mature enough to suit our needs. Of particular interest is the manner in which potential functions and fields include critical points that anticipate and describe discontinuous phase transitions. The resulting possibilities of these phase transitions in organizing patterns suggest relevant hypotheses that can be tested empirically to shed light on organizational change dynamics in complex ecosystems.

15.3   PHASE TRANSITIONS BETWEEN ORGANIZING STATES

In this section we explore a mathematical framework to describe phase transitions between instances of stable action flows within organizing systems. In an earlier section, an organizing system was defined to be a subcategory of instances that could be observed in a population. As this section describes, organizing states are defined as the stable instances of an organizing system that are relatively larger than related instances in that they have more dimensions in the phase space of their domain and more agents in their codomain set that other related instances included within them.

15.3.1   Organizing States and Relationship among Them in Organizing Systems

For a given set of parametric conditions, some instances (more precisely, some of the action flows within instances) are dynamically stable. An action flow that is in an instance of an organizing system is stable when it tends to a limit set and there exists a threshold value below which small perturbations from outside the system or fluctuations from within it do not permanently change the qualitative dynamic behavior of the action flow. Other instances of the organizing system may not be stable.

Phase transitions in the organizing system are morphisms from one stable instance to another. For example, consider the collection of instances in an organizing system with a given phase space as the domain. Inclusion maps in a preorder among sets of agents in the codomain would each incorporate additional agents into an image set in the codomain. Similarly, instances in the organizing system are also related by preordering with inclusion that maps each instance into instances with more agents for a given phase space.

A new order regime with stable action flows (among agents) within a given phase space can be represented by constructing a new category that is a partially ordered subcategory of the organizing system generated by a phase space. The category has stable instances of the organizing system as objects and inclusion maps as its morphisms. By composing morphisms (in this case the inclusion morphisms) it makes sense to speak of 'an order regime as spreading through the population' as morphisms reach codomains that are larger instances in the preorder. When the terminal object of these maps approximates the organizing system, this category represents what is called a second-order phase transition. Likewise, phase spaces can map into one another as a preorder with the same set of agents as the codomain. Further, both of these inclusion maps might be represented as objects in the same larger category.

Discontinuous phase transitions, sometimes called first-order phase transitions, are compositions of morphisms passing through unstable instances of the organizing system before arriving at a new stable instance. This can mean compositions of morphism along paths that pass through other stable instances or in some cases unstable ones. A social contagion mechanism that is represented by maps between instances is described in the next main section. Before this, however, in this section we explore the dynamics of phase transitions more closely.

The terminal instance as organizing state

To explore phase transitions in HID, we make more precise the way that an organization can be represented as an organizing system. In particular, we explore how to identify a specific relevant instance of the organizing system as opposed another, that is, we define an organizing state. To do this, the one must define the organizing system such that an order parameter can be identified and measured. To complicate matters, as Karl Weick (1979) has noted, a basic premise of human organizing (which we assume) is that organizations are dynamic and are always evolving. It is more accurate and evocative of the situation to describe how organizing states change rather than an 'organization' as existing in a 'state,' per se. Thus, we represent an organizing system as cycling through (via morphisms) instances that map relevant phase spaces to sets of agents and their action flows in the context of a parameterized potential field.

A dynamically stable organizing state is defined to be an abstract category theoretic representation of an instance within an organizing system that is a terminal object in the sense that it is the terminal codomain of a composition of inclusion morphisms between instances of an organizing system such that its action flows tend to a limit set. Further, more practically, an abstract representation of a dynamically stable organizing state can be taken to have semantic meaning in the sense that it can be taken to represent the behavior of agents organizing within a complex adaptive system and tending toward dynamically stable action flows as reflected in the relevant phase space.

Note that this definition is entirely generic and might be applied to all organizing. Further, there may be many such organizing states in an organizing system if the category that represents the organizing system is not closed. Our notion of an organizing system is thus a *trajectory* (or composition of morphisms) among instances in phase space that progresses from one dynamically stable organizing state to another. The chapter posits that conditions or contexts for organizing that enable stable organizing states can be parameterized.

Order parameters and organizing systems

The potential surface for an organizing system represents the stable organizing states in phase space as control parameters vary. One selects the *order parameter* for the organizing system to reflect the number of degrees of freedom in the instances of phase space being observed. Some of these are stable (the action flows in the instance converge to a limit). Others are quasi stable, and still others diverge. An organizing system includes all possible instances (objects) and paths (morphisms) between them. For example, inclusion and projection morphisms allow instances of increasing and decreasing dimensions of phase space as well as instances with a larger and smaller set of agents within each phase space projection. A stable value for the order parameter is the number of degrees of freedom of an organizing state (which by definition is stable) with the organizing system. Note that a stable value for the order parameter of the organizing system may or may not include all of the possible degrees of freedom included in objects of the organizing system. Likewise, it may or may not include all of the agents who are included in objects in the organizing system. It simply reflects the value of the order parameter for a stable subcategory of the organizing system.

Hazy and Boyatzis (2015) described an illustrative model for the case of emotional orientation while individuals are organizing. If one assumes that sets of individuals, as the objects in an organizing state category, are in one of either a positive emotional state that is open and collaborative, or a negative emotional state that is defensive and self-interested, then emotional contagion dynamics can be observed as a recursive mapping process along the preorder to sets that includes an increasing number of agent with either a positive or negative state along the dimension of interest. The potential to reach stability in either the majority positive or negative state would occur as these maps among instances terminate in a stable organizing state.

15.3.2   Changing Organizing States on a Potential Surface

A potential surface in parameter space represents the values of the order parameter for which stable organizing states are possible within an organizing system. Thus, it identifies how stable values of the order parameter ('phases') of an organizing system change as the control parameters vary. As conditions change in the ecosystem, for example, stability may occur at higher or lower level of the order parameter. This approach seeks to understand and predict changes to the values for these phases for a given set of parametric conditions. It does this by assuming that each agent is represented by an action vector and that action vectors interact in a field of informational influence forces that nudge them toward a point or points of stability on the potential surface. This would mean that action flows in the organizing state tend to a limit set.

A more rigorous representation of a potential surface for an organizing system of generalized action vectors could assist management as they seek to anticipate how ecosystem forces that are influencing actors and their organizing either toward or away from a desirable organizing state. Of course, for any given application (for example cost containment or revenue growth) the specific action vectors and the interaction affects between them would need to be specified. We will take this up in the next major section on social contagion mechanisms. First, however, we propose a general model to describe an approach to parameterizing change.

Assuming a choice of sign such that informational influence forces draw behavior to lower potentials, a potential field could be specified to include one or more relative minima, that is, points on the potential surface. Each relative minimum represents a stable state, and the potential field would reflect influences on action vectors by quantifying ambient social pressure that pushes the organizing system toward one of various organizing states on the surface.

In the Cost Containment versus Revenue Growth example, the potential field might include macro-information about conditions that would influence the firm to prioritize cost controls over spending for revenue growth for example. One might imagine ecosystem conditions where there is information that suggests an impending economic recession. In this parametric situation, where unpredictability and the possibility of surprise in the ecosystem have increased, it would be destabilizing for the firm to continue to pursue both cost cutting and aggressive spending on revenue growth. Thus the organizing system would tend toward a point of stability – an organizing state – that is primarily focused on cost containment. One can imagine this as a continuous move along the potential surface to a new point of stability that matches the new parametric conditions.

In other situations, however, the parametric conditions might suggest strong underlying economic growth along with restricted capital. In this case, the parameterization of the potential function should demonstrate that it might indeed be the best overall strategy to pursue both cost control and growth because stability would be maintained while incremental benefits are also being accrued. This would be seen in the model where the two competing basins of attraction merge into a single stable state as shown on the left side of Figure 15.1.

15.3.3   Phase Transitions between Organizing States – An Illustration

The approach proposed in this chapter highlights the need to define a measureable order parameter for organizing systems in human interaction dynamics. The measure would reflect the number of degrees of freedom of the action flows that are observed in the instance of the system that is being observed. This would measure the number of degrees of freedom needed to describe the action flows in the organizing state of the organizing system, as indexed by the average number of active degrees of freedom per agent in the organizing state. As such, a particular instance of the organizing system, its 'organizing state' is represented as the largest (in the sense of the preorder) stable instance with stable action flows in phase space.

A generic example of an order parameter for an organizing system is the number of

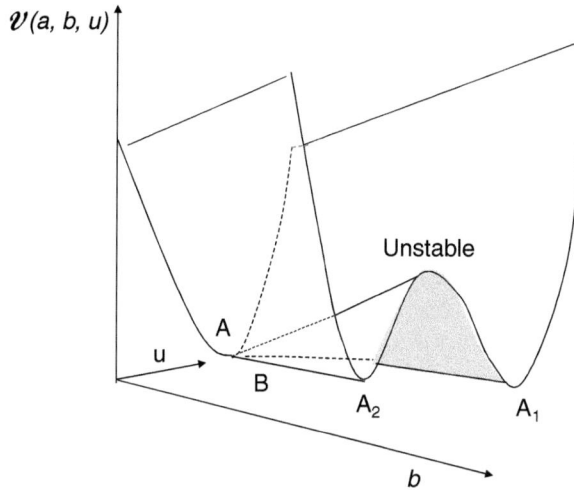

Note: With the parameter a = c_{int} held constant at the value where bi-stability with approximately equal attraction is reflective of ecosystem conditions, the parameter b= c_{ext} varies along the lower axis varies to reflect an increasing level of Opportunity/Threat Tension. Two deepening attractor basins imply an increasing tendency to align oneself with one or the other of two survival strategies – in this case, the choice between attachment to product innovation or to building and maintaining strong customer relationships by offering valuable services – with little hope for stability in between.

Figure 15.1   As External Complexity b increases, minimal potential A bifurcates into A$_1$ and A$_2$

degrees of freedom in the organizing state within which a critical mass of the population aligns. More specifically, this would be the number of degrees of freedom in the organizing state that includes a threshold set of agents that itself includes (in the preorder) a well-specified subset of agents that one might call a governing coalition. Note that this is a general approach that assumes nothing about structure or lack of structure among the agents, nor about hierarchical power or other relationships among agents. The threshold set might include a majority of agents acting autonomously or it might be a complicated set of sets of particular agents or groups regardless of plurality or majority.

For instance, on one hand, assume there is evidence of positive action orientation for the cost cutting priority in an organizing-state observed in every sample drawn from a threshold subset of the population. In this case, the probability of observing this dimension in the organizing state approaches 1. There are two degrees of freedom, this value and its rate of change. For simplicity, assume that in the stable state, there is a fixed point limit in phase space such that the rate of change is zero. This leaves only one degree of freedom and the order parameter is assigned the normalized value 1. If on the other hand, however, the cost cutting priority is never observed and its probability approaches 0, then spending continues, and there are two degrees of freedom in the spending direction. If we assume that for the stable state the rate of change is zero, then for simplicity we assign the order parameter the normalized value −1.

Conditions between the extremes, where the order parameter is between −1 and 1, would imply different mixtures of the population detecting the relative presence of the focal organizing state in a sample. Thus, one could assume that a mean value of 0 for the order parameter would imply an even mix of 'cost cutting' and 'continued spending' action states among agents within the population, that is, the probability of the organizing system forming a 'cost cutting' orientation as an organizing state would be 0.5. Thus, there is no action vector (for the instance that includes all of the agents) pointing definitively in either direction, and thus there is no 'order' in the action flows and only noise along that cost management dimension.

A simple illustration of a construction with two mutually exclusive states is a large stadium filled with people where a value of 1 is assigned to a standing position and −1 is assigned to a seated position. When there is no activity on the field that acts as a priority which would draw everyone's attention, people might stand or sit, almost at random, a value that approximates 0. As the game begins, however, one might expect a potential function to emerge such that everyone feels social informational influence in the direction of sitting to maximize visibility and comfort for all individuals who seek to observe the action on the field. These parametric conditions would imply an informational influence field which would push toward a fixed point stable attractor state, 'sitting' in the dynamical system. A majority of individuals could be observed to be seated at any point in time (absent perturbation or fluctuation that the reader can easily imagine). Sometimes, however, people could be observed to stand in unison due to local contagion dynamics when a certain play injects nervous energy into the population. This is an example of an external perturbation that arises from uncertainty and complexity in the ecosystem. At times, however, through internal forces alone, the potential field might include a wave function that changes over time leading to the ubiquitous 'stadium wave' phenomenon that is often observed at sporting events (Lasry and Lions 2007). This is an internal fluctuation that arises from uncertainty and complexity inside the organizing system itself.

There may be many parameters in general, and these may be defined in many ways. However, mathematics of Whitney (1955), Thom (1989) Arnold (1991) and others have shown that for cases with five or less control parameters and for two or less order parameters, there are canonical representations of geometric relationships. It is these that we call potential functions. The challenge, of course, is identifying what these parameters represent empirically in the case of human interaction dynamics. Although the relationships among these parameters are well studied, the mathematics itself says nothing about how to identify the parameters for empirical study (Arnold 1991).

15.3.4 The Potential Function for the Cusp of Change

In the Cusp of Change approach proposed herein, we examine the simplest case of interacting control parameters and the critical points that result from this interaction. This is the case where the potential value of a single order parameter is shaped by the interacting values of two control parameters such that the relative maxima or minima indicate points of relative stability to form a potential surface. This relationship is described canonically by a fourth degree equation (see equation 1) and by its third degree derivative equation (see equation 2) that together are called the cusp catastrophe. Specifically, the potential V is a function of the internal order parameter of the system u and two control parameters c_{int} and c_{ext} that can be written as:

$$V\left(c_{int}, c_{ext}, u\right) = \tfrac{1}{4}\, u^4 + \left(c_{int}/2\right) u^2 + c_{ext}\, u \tag{1}$$

This potential function is posited to be a general model of the potential environment within which organizing systems might form for the case with two control parameters, called c_{int} and c_{ext} and a single order parameter, u. The potential that individual action within a population will form into an organizing system, as measured in reference to achieving a stable value of the chosen order parameter u, is reflected by the presence or absence of maximum or minimum of the potential function described by equation 1.

Several cases of this function for different values of c_{int} are shown in Figure 15.2 which assumes that potentials tend to be minimized. As shown, there can be two distinct stable states under certain parametric conditions. By letting the parameters c_{int} and c_{ext} vary, all of the possible stable instances of this system can be found by differentiating equation 1 to find the points where $dV/du = 0$. This equilibrium surface is shown in Figure 15.2 and described by:

$$0 = dV / du = u^3 + c_{int}\, u + c_{ext} \tag{2}$$

The specific equilibrium points shown as minima in Figure 15.2 are identified in the equilibrium surface shown in Figure 15.3 to indicate a series of values for the parameter $a = c_{int}$ along a single path when the parameter $b = c_{ext}$ is held constant.

To summarize, the nuances of the examples we use aside, in this chapter we limit the model discussion to cases with a single organizing system represented by one order parameter that measures the mean probability that an agent in a population will decode and implement the requisite choice along a single action dimension. This might, for example, be a measure of whether cost containment action is consistently observed during agent interaction within the population. As shown in Figure 15.3, a stable value of the order parameter is assumed to vary according to the values of two constraining factors (control parameters). These are:

1. The external complexity of the ecosystem in which the population that is forming an organizing system is situated, and
2. The internal complexity of the social networks (Burt 1982), social interactions (Dodds and Watts 2004), emotional displays (Fowler and Christakis 2008; Hatfield, Cacioppo, and Rapson, 1993; Hill et al. 2010), communicative acts (West 2012) and competitive dynamics (Axelrod 1997; Lasry and Lions 2007) inside the population.

Future study may further parse this framework to specify an organizing system more precisely and by doing so identify other parameters. However, singularity theory suggests that these more highly specified representations would further unravel aspects of the same essential phenomena, that is, the nature of discontinuity at a mathematical singularity on the potential surface (Arnold 1991). For our purposes, the case with two control parameters, also called the cusp catastrophe model (Zeeman 1977; Thom 1989; Arnold 1991; Guastello 2002), is both nonlinear and complex, but without being too intractable. As such we describe this case in detail. Note that models with a single control parameter and a single order parameter are special cases of the two control parameter analysis where the other control parameter is held constant. Next, we describe these two parameters in more detail.

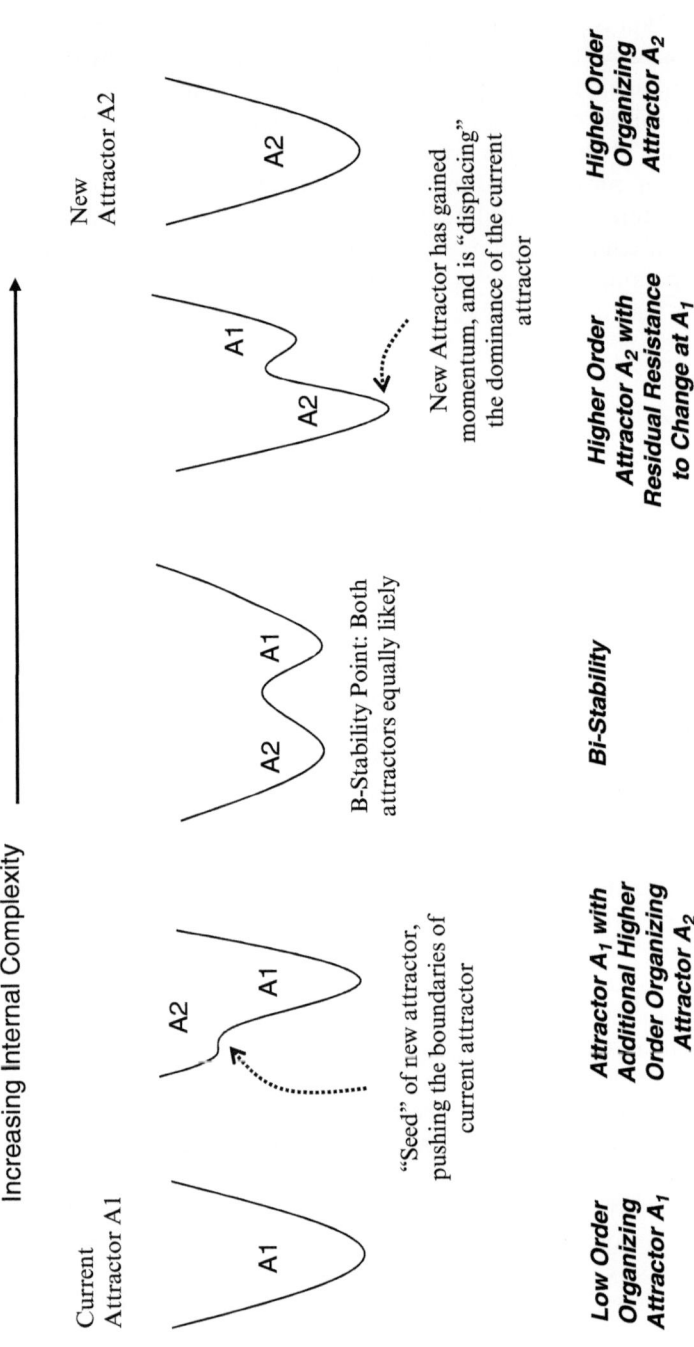

Increasing Internal Complexity

Current Attractor A1

New Attractor A2

A1

A2

A2 A1

A2 A1

A1

A2

A2

"Seed" of new attractor, pushing the boundaries of current attractor

B-Stability Point: Both attractors equally likely

New Attractor has gained momentum, and is "displacing" the dominance of the current attractor

Low Order Organizing Attractor A₁

Attractor A₁ with Additional Higher Order Organizing Attractor A₂

Bi-Stability

Higher Order Attractor A₂ with Residual Resistance to Change at A₁

Higher Order Organizing Attractor A₂

Note: Changing 'attractor basins' around stability points with one parameter constant and the second parameter a five distinct values imply distinct shapes for the value potential function.

Figure 15.2 As internal complexity increases, minimum potentials vary at A₁ to A₂

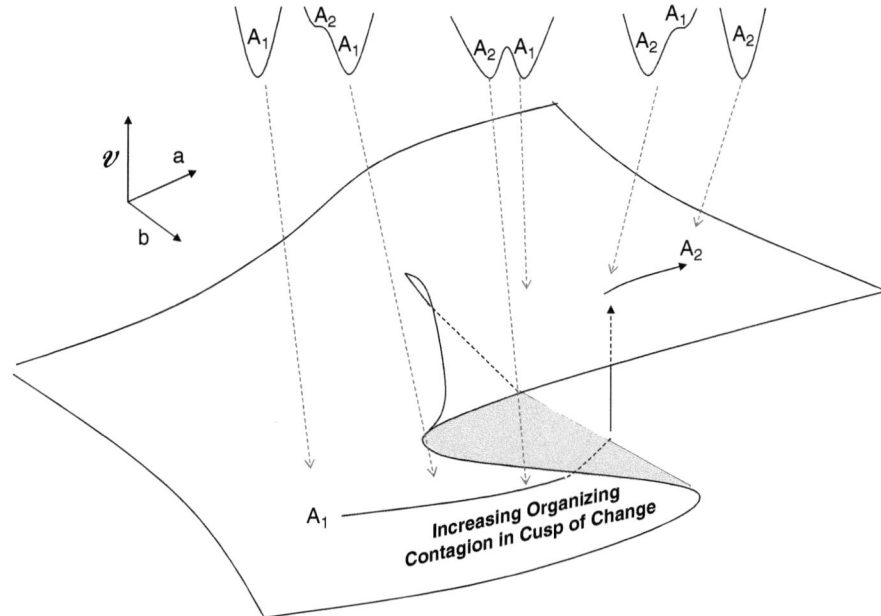

Note: The equilibrium surface of stable points for the order parameter u (when dV/dt = 0) as the values for the control parameters a = c_{int} and b = c_{ext} that describe the ecosystem conditions vary.

Figure 15.3   Stability of the order parameter varies with external and internal complexity

15.3.5   The First Control Parameter – External Complexity

To be more specific, we define the first control parameter, c_{ext}, to reflect the *external complexity* of the ecosystem. This parameter is intended to reflect the impacts on the dynamics of an organizing system (the action flows within and maps between various organizing states) that emanate exogenously from beyond the organization's boundary. This parameter has been called opportunity/risk tension between the organization and its environment. This is because it reflects the complexity of environmental conditions that the population must face as it organizes to exploit opportunities or respond to risks. In business contexts, this might involve the complexity of predicting the presence and characteristics of resources and markets that define conditions within which requisite organizing is needed to acquire these resources – such as a continuing supply of raw materials – or serve these markets – such as accessing customers – as well as other risks or surprises such as unexpected competitive pressures, regulatory changes or resource scarcities.

One can imagine why varying the values of this control parameter might imply the existence of a critical point in the order parameter and thus the possibility of bi-stability on the potential surface that shapes organizing dynamics. For example, consider a firm that uses inputs from the ecosystem to produce a product that it then sells into the ecosystem to accumulate the resources that it needs to sustain its various stakeholders. This process must drive enough cash flow through the firm such that its organizing system remains stable day-in and day-out for months, years or even decades. One would expect that there would be a cost point for inputs beyond which the organizing system would be unsustainable. This would reflect a critical point on the potential surface.

As long as access to ecosystem inputs remains predictable, the uncertainties inherent in the internal complexity of the system might allow local inefficiencies and fluctuations to creep into the internal process. For example, a flu epidemic might temporarily cause a significant portion of the workforce to be out sick. But even so the relationships among the various organizing states that might be taken on by the system are such that the organizing system as a whole remains stable (that is, it 'finds' a stable organizing state). This continues to be true even as uncertainties of external complexity are manifested as demand and supply prices ebb and flow. When such minor perturbations do arise, the organizing system accommodates them by cycling through a constellation of stable organizing states. As an example, the system might activate dynamic capabilities such as budget tightening or the use of resources held in reserve to provide the organizing system with resilience that eventually regains stability.

Now assume that a firm, an organizing system, is operating and is stable with a positive mean value in the cost containment action orientation dimension forming an attractor of a potential field, perhaps during a recession. Assume further that the cost of a supply (from the ecosystem) of an important input to the firm, say oil, increases beyond a critical point and that the cost remains beyond the point where existing inventories and other resilience mechanisms are able to sustain stability through its constellation of known stable organizing states. This might mean that many of the previously accessible organizing states – for example, temporarily operating at a loss, relying on existing inventories, cutting costs elsewhere in the system – are no longer accessible. The maps among alternative organizing states become more constrained and stability is therefore more difficult to regain.

Access to alternative organizational states that support resilience to perturbation in the face of external complexity is effectively equivalent to the organizing system having to access additional active degrees-of-freedom or having access to states with fewer active degrees of freedom to conserve resources. When an organizing system has adequate degrees-of-freedom to recover from a perturbation, the organizing system is dynamically stable. However, when its access to additional degrees of freedom is constrained or it is unable to deactivate some of its degrees of freedom, the organizing system can become unstable. This is true with respect to fluctuation in internal and external complexity conditions. This condition has been called 'disequilibrium' (cf. Goldstein, Hazy, and Lichtenstein 2010).

When new or different degrees of freedom become available as new organizing states for an organizing system, the 'organizing system' itself has changed. It now has (it maps to) new and different instances with different active degrees of freedom. The map from the old organizing system to the new one (as objects in a larger category) might achieve stability in a different attractor basin as implied by a different position on the potential surface, perhaps with the same parametric conditions. If this occurs, there are conditions of bi-stability.

When parametric conditions are such that bi-stability becomes a possibility, informational conditions are more complex. This can be seen in Figure 15.1 as a single minimum on the left of the graph becomes dual minima on the right. From the point of view of the organizing system, it takes a larger algorithm, a more involved categorical construction that embodies more queries or questions and more logical steps to predict unfolding events (Crutchfield, Ellison, and Mahoney 2009). In part, the value of the second parameter, which measures internal complexity, determines the direction that informational influence forces will push.

15.3.6   The Second Control Parameter – Internal Complexity

The second parameter, c_{int}, is chosen to reflect the *internal complexity* within an organizing system. These are the complexities of prediction and surprise that are associated with social networks (Burt 1982), differences in rank and influence (Hazy 2012), the quality of social encounters, exchanges and interactions (Dodds and Watts 2004), emotional displays (Fowler and Christakis 2008; Hatfield, Cacioppo, and Rapson 1993; Hill et al. 2010), and communicative acts (West 2012) inside the population when it is in a particular organizing state, or when it is transitioning from one state to the other.

These complexities impact the potential for the organizing population to create consistent interaction structures that agents can decode to predict relevant events within an 'organizing system' in the context of its environment. This information is needed to effectively address the complexities in the local interaction environment for the benefit of the organizing system and the individuals in the population that are the objects within it.

The level of internal complexity within an organizing system influences the potential for the emergence of properties within the population. These complexities have been called informational differences. This is because this parameter reflects the capacity of the CAS and its internal structure to sense and locally absorb information from many points inside the organizing system and then to process and store it in 'organizing structures' (Hazy 2012). These structures enable the accessing, processing and use of stored information (in a manner analogous to encoded information stored in computer programing code) that is retrieved and decoded to execute the properties of an organized system. Drawing on the HID model (Hazy and Backström 2013a) we informally call this the *autonomy/integration tension* parameter.

15.3.7   Requisite Complexity among Control Parameters

Together, the interactions between internal and external complexity are represented as a potential surface. This surface in parameter space reflects stable values of the order parameter under various combinations of control parameter values. The internal complexity inside the organizing system helps the population to organize in the context of the external complexities that are apparent in the ecosystem in order to exploit opportunities or respond effectively to risks. Together these parameters interact to determine the potential level of 'order' in an *organizing system* given parametric conditions. This level of order is a surrogate for the quantity of information about the environment that is stored in socio-technical structure to be decoded and used by agents to simplify their choices and reduce cognitive load.

Following Boisot and McKelvey (2011), the modeling approach being proposed can be seen to be a conceptual restatement of Ashby's (1956) 'law of requisite variety.' If a population is to sustain itself by controlling the opportunities and risks in its ecosystem, it must have the internal capacity (that is, the information stored in its structure) to express at least the number of degrees of freedom in its stable organizing states that are present in the environment.

If the internal complexity of human interaction dynamics is not sufficient to facilitate the requisite organizing which would effectively integrate individual actions into the informational structures of organizing states that would predict success, individuals in the population would be left to react independently to events in environment as individuals or in small, local groups. This sub-system-level agency within the organizing system may eventually move the organization toward an organizing system of greater effectiveness. However, it could also fracture the organizing system into smaller entities by undermining the effectiveness of the structural properties that would have otherwise emerged at the population level to form an integrated complex adaptive system.

15.4   A MODEL OF CONTAGION IN ORGANIZING SYSTEMS

In the context of phase transitions, it is necessary to also understand how change occurs through interactions among members of the population. There are multiple studies of contagion that explore the complexities associated with social encounters, exchanges and interactions (cf. Dodds and Watts 2004). For simplicity, we explore the conditions whereby one or the other of two opposing action orientations (positive or negative) along a particular dimension can become 'contagious.'

15.4.1   Modeling the Dynamics of Action Orientation Contagion

In this subsection and the next main section, we describe how aligning action along a dimension of agents' action vectors so that the component vector is either predominantly positive or predominantly negative and thus nonzero creates an active component of action in the organizing state. This mechanism for collective organizing, which we call action orientation contagion (AOC), can be observed when individuals who are in either a positive or negative orientation influence others to synchronize with them through interactions.

This transfer of state through interaction is a type of information transit (about their respective action orientation) since what had been a random distribution becomes ordered. An information signal is encoded into the orientation of a population and this can be observed by agents equipped to decode it. For these participating agents, a signal about the potential for specific collective action replaces noise and confusion. The end result may be a population that is all synchronized or 'ordered,' a signal in the informational environment that is distinct from the background. Alternatively, a fluctuation might occur and be recognizable within the population and then dissipate into a disordered or noisy background. The order can be retained if the benefit it brings outweighs the costs of sustaining it in the face of dissipating forces (Hazy 2013).

The transit of information in an organizing system through the activation of AOC within a population can occur unconsciously and quickly. It is effectuated by a fluctuation in the current organizing system. If a fluctuation or perturbation is great enough, the system may escape its prior basin of attraction and fall into a new one. This indicates a transition from its prior stable state (as described by the potential function) to a different stable state (on a potential surface). A sudden change in the stability dynamics of the aggregate state is called a *phase transition* (West 2012).

The flow of information, including a signal about the dominant action orientation across the organizing system, can vary depending upon the internal complexity of the system, its social networks and its management processes (Dodds and Watts 2004). One factor might be the probability that during a given interaction an action orientation might transfer from one party to the other. Does the first individual even bother to try to convince the other? Does the second party even receive a 'dose' with the new action orientation 'infection?' Another factor might be the strength of the argument. In infectious disease terms, what is the dosage level? A third factor might be how broadly the information is disseminated. Does everyone hear about the disturbance or is the distribution of that information limited? Finally, are individuals able to recognize the message, decode it and understand how to react, thus changing their state to synchronize with it? All of these questions must be explored when trying to understand contagion dynamics.

15.4.2   The Dodds and Watts General Model

To explore these questions, we draw on prior analysis (Hazy and Boyatzis 2015) to examine the generalized model of contagion developed by Dodds and Watts (2004) as an extension to the standard epidemiology models. For our analysis, we say that an 'infection' describes a positive action orientation along some action dimension (positive magnitude along the component vector). The relevant dimension might be a cost control imperative, for example. (Note that although we choose to focus on the positive orientation, the negative orientation could just as easily become infectious.)

As Hazy and Boyatzis (2015) describe, Dodds and Watts have developed a model for generalized contagion that is defined as follows. A population of N individuals is divided into three subpopulations of individuals with three different states, S (susceptible), I (infected) or R (removed). Note that $I + S + R = N$, [and in our case, I is the proportion of the population with a positive action orientation state]. At each time step, t, each individual $i \in N$ comes into contact with another individual $j \in N$ which is randomly selected from the population. Infection occurs as follows. If i is susceptible and j is infected [with a positive action orientation], then i receives a dose $d_i(t)$ (of positive action orientation along a particular dimension) with probability p drawn from a distribution $p \in f(d)$. If i is not susceptible or is removed, $d_i(t) = 0$. Each individual maintains a memory (that is, neurological or psychological memory) of doses received over the last T time steps, remembering a cumulative dose of $D_i(t) = \sum_{t'= t-T+1}^{t} d_i(t')$. Individuals become infected when $D_i(t) > d_i^*$, i's dose threshold. Each d_i^* is drawn randomly at time t_0 from the distribution $g(d)$. This means that each individual can take only so many 'doses' from others during an interval of time (which reflects memory of prior events) before he or she is 'infected' by the positive action orientation state. Further, this threshold might vary among individuals.

Dodds and Watts (2004) continue their model to include situations where individuals recover from the 'infection' (and are 'Removed' from the infected population) if their cumulative dosage falls below their threshold. They also might become at risk for reinfection (perhaps with a new threshold) if they interact with an infected individual again. If the dosing (of positive action orientation state) that an individual is receiving from others drops below a threshold, then the person will return to his or her normal or neutral state.

Dodds and Watts (2004) point out that these dynamics can be quite complex. However, for the simplified case, where the rate of removal, $r = 1$, and the rate of becoming re-susceptible $\rho = 1$, the system can be solved analytically. This means that falling below the infection threshold returns the individual to the susceptible population so that $R = 0$ and the population $N = I + S$. [For our purposes, N = I + S is the number of individuals, I, who are a positive state, plus S, the number in a neutral state.]

Dodds and Watts (2004) report a series of interesting results from their model even for this simplified case. In particular, they show that the contagion dynamics within a population can vary considerably depending upon several exogenously determined conditions such as: the probability of receiving a positive dose, the distribution of dosage size, and the distribution of individual dose-size-thresholds. Varying these factors individually and in combination can be used to replicate many models of social contagion that have been used in other studies. Because we are interested in social contagion that results in aligned action orientation in response to conditions in the ecosystem, it is important to note that the flexibility of this model derives from the introduction of individual memory and thus an accumulated dosage calculation that relates to an individual's threshold of infection. Memory to store information about prior interactions enables many possible contagion relationships.

By comparing the probability P_1 that an individual is infected on the first contact with the probability P_2 that an individual will be infected on the second contact, and where T is the memory parameter that describes the time interval wherein dosage is accumulated for effect, Dodds and Watts (2004) divide the dynamic course of contagion into three classes.[1] Although the different classes identified by Dodds and Watts (2004) point to the richness of the possibilities for future research, herein we limit our discussion to just two cases.

The analysis suggests circumstances where two distinct dynamic situations exist depending upon parametric conditions. On the one hand, if the probability that an individual will assume the action orientation state of another with the first interaction, is greater than half the probability that it will assume it on the second interaction, we say that there is no relevance to further social reinforcement and assume that these individuals offer little resistance to infection to action orientation exchange and are likely to be infected upon first encounter.

On the other hand, if the probability of adopting the action orientation state of the other on the first interaction *is less* than half the probability of adopting it after the second interaction, then there is an apparent need for additional reinforcement. This suggests that there is resistance to independent influence. This latter case exhibits the conditions where external complexity in the ecosystem and internal complexity inside the organizing system interact to enable a discontinuous phase transition in action orientation state. Under these conditions multiple interaction with others are needed before an individual synchronizes its state with others.

15.5   A SIMPLE MODEL OF PHASE TRANSITIONS IN ORGANIZING SYSTEMS

In this section we describe a simple model of a phase transition in action orientation states from a neutral action orientation to a positive action orientation in human interaction dynamics (HID). To explore this dynamic, we build upon the Cusp of Change Model described by Goldstein, Hazy and Lichtenstein (2010) which uses the canonical mathematical model for one order parameter, ξ, and two control parameters, c_{int} and c_{ext}. In our case, we define the parameters c_{ext} and c_{int} to be the probability of surprise in the ecosystem, what Goldstein, Hazy, and Lichtenstein (2010) called *opportunity/risk tension* and the probability of surprise in internal interactions, what they called *autonomy/integration tension* respectively. Together, these parameters describe complex organizing system dynamics. This is shown in Figure 15.3. Further, for $c_{ext} > 0$ and for values of c_{int} between the curves of the cusp, there are conditions (that is, inside the cusp-shape) where the order parameter experiences bi-stability.

We assume that for a given organizing system, action orientation contagion occurs when the order parameter for the organizing system changes significantly and becomes stable at a new level, an event that signals to observers that a phase transition is occurring. The signal is observable because one can recognize that the locally synchronized action orientation states within the subpopulation have become differentiated from the background and stabilize at that differentiated level. As this occurs, a synchronized state of individual action orientation spreads through the organizing system and establishes the organizing system as persistently different than the background at least in this one aspect.

15.5.1   Second-order Phase Transition: Smooth Action Orientation Contagion

For simplicity, we initially assume from the Dodds and Watts (2004) model that the probability of infection from the first interaction P_1 is high and (specifically that $P_1 \geq P_2/2$ where P_2 is a second interaction). This is the case when the opportunity/risk tension parameter $c_{ext} < 0$ which is outside the cusp of change in Figure 15.3. Dodds and Watts show that these conditions imply a deterministic threshold (DT) model of social contagion. In this case, as disturbances occur in the environment, synchronized action orientation states spread by contagion and are globally correlated across the entire organizing system. In this case, no clumping would be observed and no organizing would occur in subpopulations because no subgroups would deviate significantly from the changing substrate.

15.5.2   First Order Phase Transition: Bi-stability in Action Orientation Contagion

In cases identified by Dodds and Watts (2004) where P_1 is lower (specifically that $P_2 > P_1/2 \geq 1/T$ where P_2 is a second interaction and T is the length of memory) two stable states might be possible within various regions or subpopulations of the overall population. (From Figure 15.3 note that this is where the *opportunity/risk tension parameter* $b = c_{ext} > 0$ is in the cusp of change model and where the *autonomy/integration tension* parameter c_{int} is between two threshold or 'tipping point' values.) In these cases, path-dependence affects tend to make action states 'sticky' in these subpopulations as a means to sustain organizing state potentials.

In these cases, phase transitions in the broader population can occur if subgroups or other objects first synchronize locally into clusters and only then do clusters synchronize with one another across the entire population. During these transitions, only local correlations are observed. Self-similar contagion would likely be observable at higher levels of scale indicating scale-free dynamics are at work. These interaction dynamics are assumed to relate to social or action orientation contagion processes internal to the population, and these may not be homogeneous. For these cases, Dodds and Watts (2004) work shows that the specific dynamics of contagion in the broader population can be quite complex. Thus, the exact nature of the course of the 'infection' through the population will ultimately be explored via empirical studies and improved modeling techniques that operate across scale.

However, to advance this thinking initially, we propose the Cusp of Change Model (Goldstein, Hazy, and Lichtenstein 2010) with well-specified assumptions as a starting point – a complement to the general model of Dodds and Watts (2004). This model describes how action contagion can serve as a mechanism that enables contagion among stable instances of the organizing system thus enabling clusters of stable organizing states to emerge along certain action orientation dimensions under specific parametric conditions. In addition, this model might also shed light on when action orientation alignment remains stable as subpopulations change their constituency over time (as occurs in firms and other organizations). Such a model would describe the formation of dynamically stable organizing systems.

15.5.3   The Mechanism of Action Orientation Contagion

To recapitulate, action orientation contagion is the process where positive or negative action orientation states along specific action dimensions spread through and synchronize within a population via social interaction. It can occur gradually in response to changing opportunity/risk conditions – a Second-Order Phase Transition – or it can be sudden, a First-Order Phase Transition. The latter case is enabled when two factors interact to create the requisite enabling conditions as shown in Figure 15.3. These conditions exist when:

1. The parameter c_{ext} or *opportunity/risk tension* in the ecosystem is such that a threshold is crossed whereby the complexity of the situation is recognized by agents to indicate that as isolated individuals it would be difficult to respond absent interaction with others, and
2. The parameter c_{int} or *autonomy/integration tension* in the organizing system within the population is in a range whereby the population is ordered enough such that individuals identify with the action responses of others and are able to recognize and decode these to use the information embedded in them to simplify their own choice of action orientation and reduce cognitive load. However, the complexity is also low enough that they do so with skepticism and do not immediately adopt the action orientation of another after only a single interaction.

Under these parametric conditions, the process unfolds as follows: a subset of individuals who find themselves in a position to recognize the opportunity or risk directly in the environment assume varying action orientation states in response, consciously or unconsciously. Through interactions with others, these action orientation states can be imitated and thus spread to others who did not directly observe the event. The likelihood of 'infection' with an action orientation depends upon each individual's susceptibility to socio-emotional influence from a particular interaction including relative reputation and status (Hazy 2012). Local stability is enabled when local organizing conditions support local synchronization of action flows as well as resistance to outside influences.

Citing Dodds and Watts (2004), Hazy and Boyatzis (2015) argue that when individuals interact the likelihood that the two will synchronize is positively related to: (1) the strength of the argument or 'dosage' exchanged in the interaction, (2) the total impact of the arguments or dosage that has accumulated from recent interactions, (3) the number of recent interactions that are accumulated by the alter agent in 'memory', and (4) whether a level that exceeds that agent's threshold for infection is achieved. Further they suggest that there is a range of bi-stability called the 'cusp of change' between two tipping points where an organizing system can suddenly switch between two stable organizing states during a strong perturbation on or fluctuation within the organizing system.

15.6 CONCLUDING THOUGHTS

This chapter makes a conceptual argument for a mathematical research agenda that extends and subsumes other mathematical and computational work, for example, models such as the Dodds and Watts (2004) general model of social contagion, to provide a consistent framework that connects all complexity informed social science. Until such a framework exists, knowledge is difficult to accumulate, compare across disciplines or even within disciplines, and test.

The methodological approach we propose would require the construction of a category theoretic representational framework similar to what is suggested herein to connect all scientific approaches to human interaction dynamics (HID) which seek to build a cumulative base of knowledge to inform individual choice and behavior. In the middle sections of the chapter, the approach is illustrated using a model that describes the potential for discontinuous or first order phase transitions in the context of a single order parameter and two control parameters. This approach is called the Cusp of Change model (Goldstein, Hazy, and Lichtenstein 2010).

The mathematical relationships in the Cusp of Change model are well understood. Here we take the additional step and argue for a canonical parameterization of the Cusp of Change. In this regard, we suggest that the mean value of activity along action orientation dimensions within a stable organizing state may be a useful order parameter. We further suggest that two indexes, reflecting respectively the internal and external complexity conditions confronting the population, can be used as control parameters.

This approach implicitly takes the position that action orientation, that is, agency, gathers its potency through gradients of differences in predictability and uncertainty with respect to relevant events. As agents seek to reduce the unpredictability and thus their cognitive load by moving along this gradient, informational influence forces set up a potential field under which organizing for action occurs among independent semi-autonomous agents. By using the information in socio-technical structure that can be recognized and decoded, agents can ease their cognitive load while maintaining a perceived level of predictability about events in their environment. It is our hope that by using an approach that formalizes what has been described herein, eventually even subtle and complex forms of organizing behavior can be modeled from within a common framework.

NOTE

1. More specifically, in Class I, where $P_1 \geq P_2/2$, the probability of infection on the first interaction is high enough such that the model implies behavior similar to *epidemic threshold models* also called *deterministic threshold models*. This means that as the probability that an interaction results in a positive dose increases, the steady-state percentage of infection at which the population stabilizes – called the fixed point attractor for a fixed value of the parameter – increases as the value of the parameter setting increases. Further, this relationship has a positive slope – as the parameter setting increases, the stabilizing point also increases. In Class II, where $P_2 > P_1/2 \geq 1/T$, the probability of infection after a second interaction is high enough relative to that from the first interaction and for a given length of memory that there are cases wherein a critical mass might be necessary to achieve a stable fixed point attractor level of infection within the population. This behavior is similar to *vanishing critical mass models*. In Class III, where $1/T > P_1$, the probability of infection on the first interaction is so small that a preexisting critical mass is always necessary if the infected population is to stabilize at a level greater than zero – whatever the probability that an interaction results in a positive dose of the infection – to imply a stable fixed point attractor level of infection. This means that the model implies behavior similar to *pure critical mass models*.

REFERENCES

Arnold, V.I. (1991), *Catastrophe Theory* (3rd edition), Berlin: Springer-Verlag.

Arnold, V.I., Goryunov, V.V., Lyashko, O.V., and Vasil'ev, V.A. (1993), *Singularity Theory I*, Berlin: Springer-Verlag.

Axelrod, R. (1997), *The Complexity of Cooperation: Agent-Based Models of Competition and Collaboration*, Princeton, NJ: Princeton University Press.

Boisot, M., and McKelvey, B. (2011), 'Complexity and Organization – Environment Relations: Revisiting Ashby's Law of Requisite Variety', in P. Allen, S. McGuire, and B. McKelvey (eds) *The Sage Handbook of Complexity and Management*, Los Angeles: Sage, pp. 279–298.

Burt, Ronald S. (1982), *Toward a Structural Theory of Action*, New York: Academic Press.

Crutchfield, J.P., Ellison, C.J., and Mahoney, J.R. (2009), 'Time's barbed arrow: Irreversibility, crypticity, and stored information', *Phys. Rev. Lett* **103**, 094101.

Dodds, P.S., and Watts, D.J. (2004), 'Universal behavior in a generalized model of contagion', *Physical Review Letters*, **92**(21), 218701.

Fowler, J.H., and Christakis, N.A. (2008), 'Dynamic spread of happiness in a large social network: Longitudinal analysis over 20 years in the Framingham Heart Study', *British Medical Journal*, **337**(dec04 2), a2338.

Goldstein, J., Hazy, J.K., and Lichtenstein, B. (2010), *Complexity and the Nexus of Leadership: Leveraging Nonlinear Science to Create Ecologies of Innovation*, Englewood Cliffs, NJ: Palgrave Macmillan.

Guastello, S.J. (2002), *Managing Emergent Phenomena*, Mahwah, NJ: Lawrence Erlbaum.

Haken, H. (2006), *Information and Self-organization: A Macroscopic Approach to Complex Systems* (3rd edition), Berlin: Springer.

Hatfield, E., Cacioppo, J.T., and Rapson, R.L. (1993), *Emotional Contagion: Studies in Emotion and Social Interaction*, New York: Cambridge University Press.

Hazy, J.K. (2012), 'Leading large organizations: adaptation by changing scale-free influence process structures in social networks', *International Journal of Complexity in Leadership and Management*, **2**(1/2), 52–73.

Hazy, J.K. (2013), 'Complexity mechanisms in human interaction dynamics and the organizing functions of leadership', in Leslie A. Tombs (ed.), *Proceeding of the Seventieth Annual Meeting of the Academy of Management* (CD), ISSN 1543-8643.

Hazy, J.K. (2014), 'Is this "thermodynamic inquiry" metaphor or more?', *Journal of Consulting Psychology*, **66**(4), 200–305.

Hazy, J.K., and Silberstang, J.K. (2009), 'Leadership within emergent events in complex systems: Micro-enactments and the mechanisms of organisational learning and change', *International Journal of Learning and Change*, **3**(3), 230–247.

Hazy, J.K., and Backström, T. (2013a), 'Human interaction dynamics (HID): Foundations, definitions, and directions', *Emergence: Complexity and Organization*, **15**(4), 91–111.

Hazy, J.K., and Backström, T. (2013b), 'Editorial: Human interaction dynamics (HID): An emerging paradigm for management research', *Emergence: Complexity and Organization*, **15**(4), i–ix.

Hazy, J.K., and Boyatzis, R.E. (2015), 'Emotional contagion and proto-organizing in human interaction dynamics', *Frontiers in Psychology*, 6:806. doi: 10.3389/fpsyg.2015.00806.

Hill, A.L., Rand, D.G., Nowak, M.A., and Christakis, A. (2010), 'Emotions as infectious diseases in large social networks', *Proceedings of the Royal Society Biology* (published online July 7, 2010).

Holland, J.H. (1995), *Hidden Order: How Adaptation Builds Complexity*, Reading, MA: Perseus Books.

Lasry, J.M., and Lions, P.L. (2007), 'Mean field games', *Japanese Journal of Mathematics*, **2**, 229. doi:10.1007/s11537-007-0657-8.

Mac Lane, S. (1978), *Category Theory for the Working Mathematician* (2nd edition), New York: Springer.

Nowak, M., Tarnita C.E., and Wilson, E.O. (2010), 'The evolution of eusociality', *Nature*, 466, August 26, 2010, p. 1057. doi: 10.1038/nature0920.

Silberstang, J., and Hazy, J.K. (2008), 'Toward a micro-enactment theory of leadership and the emergence of innovation', *The Innovation Journal: The Public Sector Innovation Journal* **13** (3), Article 5.

Spivak, D.I. (2014), *Category Theory for the Sciences*, Cambridge, MA: MIT Press.

Thom, R. (1989), *Structural Stability and Morphogenesis: An Outline of a General Theory of Models*, Reading, MA: Addison-Wesley.

Von Neumann, J., and Morgenstern, O. (1944), *Theory of Games and Economic Behavior*, Princeton, NJ: Princeton University Press.

Weick, K. (1979), *The Social Psychology or Organizing* (2nd edition), New York: McGraw-Hill.

West, B.J. (2012), 'Consensus control and minority opinion', Paper presented at the Society for Chaos Theory in Psychology and Life science, 22nd Annual Conference, held at the Johns Hopkins University in Baltimore, Maryland.

Whitney, H. (1955), 'Of singularities of mappings of Euclidean Spaces. I. Mappings of the plan into the plane', *Annals of Mathematics, II.* Ser. 62, 374–410.

Wilson, E.O. (1998), *Consilience: The Unity of Knowledge*, New York: Alfred A. Knopf.

Zeeman, E.C. (1977), *Catastrophe Theory:* Selected Papers 1972–1977, Reading, MA: Addison-Wesley.

16. Applying complex adaptive systems to agent-based models for social programme evaluation

Professor Michael E. Wolf-Branigin, George Mason University, Dr William G. Kennedy, George Mason University, Dr Emily S. Ihara, George Mason University and Dr Catherine J. Tompkins, George Mason University

Within the field of human services, social planners and evaluators with varying levels of success create innovative interventions to address the diverse and expanding needs of vulnerable populations. Over time, these social innovations require an increasing high level of flexibility and adaptability to remain effective in addressing the continually changing, disparate, and incompatible preferences of clients, funding sources, and other stakeholders. The programmes and schemes we devise to improve peoples' quality of life are typically well intentioned; however, forecasting the outcomes of these interventions can be difficult to justify to key stakeholders. Understanding the logic and assessing the impact behind the intervention can be difficult, time consuming, and inconclusive. This occurs because commonly used tools such as logic models are based on a set amount of input, throughput, and output in which it is believed that interventions will operate in a linear fashion; whereby a set effort results in a set result (Knowlton and Phillips 2009). The monitoring and evaluation of these interventions use traditional research methods focusing on group comparisons that contrast pretest to posttest scores or an experimental group versus a control group prevail when determining the effectiveness of the intervention, rather than understanding the environment in which the intervention operates. This chapter seeks to facilitate the use of agent-based modelling by providing the requisite background for evaluators and researchers to frame their evaluation efforts as complex adaptive systems. As will be discussed, this is requisite in moving to the next step of developing ABMs.

To address the schism between attempting to measure linear (a specific unit of input results in a specific unit of output) and more environmentally oriented nonlinear outcomes, we apply complexity science as a social programme evaluation methodology by discussing the components of complex adaptive systems, and then building on this knowledge by applying it to test programmes and policy options using agent-based models. Within this method, we view complexity as a mathematical field where the relations between inputs and outputs cannot be described in a closed linear form, but rather are better understood through a simulation. We address both qualitative and quantitative aspects of complexity through two applications of agent-based modelling that considers policy issues. The first application is in relation to caregiving stress for those who are caregivers for persons with Alzheimer's disease. The second application takes a geographic approach that inquires into the patterning of where persons with intellectual and developmental disabilities live relative to several variables.

Research methods focusing on the third wave of science – complexity – provides improved insights to understanding effectiveness of programme efforts (Patton 2011). Complexity science, and its undergirding theory as operationalised through complex adaptive systems (CAS), provides a promising organising framework for social programme evaluators by providing a structure of complex systems to social service practice. We argue that complexity theory provides a promising approach in social work as researchers and evaluators begin examining and quantifying nonlinearity and emergence.

Viewed as the third wave of systems thinking, complexity science has antecedents in both systems theory and cybernetics (Williams and Imam 2006). As a discipline, social work has relied heavily on systems theories. The first wave – general systems theory – included the refinement of a developmental ecosystems approach (Bronfenbrenner 1994). Continuing in this first wave, the refinements made through ecosystems theory builds upon social work's person-in-environment theoretical approach in that people have mutually beneficial interdependencies with other individuals and their environment (Germain and Gitterman 1980). Ecosystems theory uses a developmental approach relying on the importance of interactions with micro-systems or settings in which the person lives (Bronfenbrenner 1994), whilst stressing the importance of interpersonal learning environments (Germain 1978). The second wave of systems thought – cybernetics – uses feedback to inform key stakeholders about their future actions and decisions. These two waves lead to the third wave, complexity. Social work as a discipline was an early adopter of systems theories, and has applied cybernetics in practice through the maintenance and improvement of social service organisations; however, compared to other disciplines it has been late to adopting complexity as a primary organising framework.

Warren, Franklin, and Streeter (1998) introduced chaos and complexity theory to social work researchers and evaluators. Since then social work researchers have begun applying these theories to see complexity arising from the interactions of agents from the bottom up rather than from the top down. Complex systems are in a continual state of dynamic equilibrium as these systems navigate being in rigid order and chaos. These systems are operating by balancing between two the forces of order and chaos. Such systems operate according to a set of simple rules, yet patterns emerge from these simple interactions without a predetermined template (Mankiewicz 2001). Applications have also addressed social policy issues (Harvey 2001) to forecast how changes in policy can affect the decisions and outcomes of differing levels of intervention.

Whilst still limited in application to social workers, it leads to a discussion of how complex adaptive systems (CAS), and its related components are beneficial to social programme evaluators and researchers. Limited use of and specific examples of complexity, and the related computational methodology of agent-based modelling (ABM), exist in the social work literature (Ihara, Horio and Tompkins 2012; Woehle 2007; Woehle et al. 2009; Wolf-Branigin 2009). Applications in health and conceptual social science applications appear more common including works describing public health (Gorman et al. 2006), racial segregation (Schelling 1978), behavioural and ecological interactions (Epstein and Axtell 1996), and drug epidemics (Agar 1999).

Applying complexity science to social programme evaluation can be confusing. To simplify, we propose using complex adaptive systems (CAS) to represent complexity as it facilitates one approach of bridging complexity science to social programme evaluation. To begin, evaluators need to identify and describe how seven components of CAS – are agent-based, have self-organisation, have boundaries, provides heterogeneous options from which the agents choose, uses feedback, adapts based on the feedback provided, and produces an emergent behaviour – will apply to the programme they intend to evaluate. In order to assist the reader, we apply these components of CAS to two recently developed agent-based models. The first simulates caregivers' stress when supporting persons with Alzheimer's disease. This application has particular relevance because it demonstrates policy analysis and decision-making within the evaluation research method. The second model inquired into how persons with intellectual and developmental disabilities selected their housing options.

Coming from a human services background, our professional social work orientation focuses on the larger environment and how people interact with other individuals, groups, and communities whilst realising the value of observing people in the larger environment and within systems of all scales. Person-in-environment is a basic tenet to social workers (Green and McDermott 2010; Wolf-Branigin 2006). Intuitively we understand the roles of interconnectedness and interdependencies aid clients in developing more robust systems to support themselves and others. Complexity appears at several levels within social work and provides a promising approach for conducting programme evaluations (Westley, Zimmerman and Patton 2006). For example, people with diverse backgrounds seek support from others with a similar need; the concerns of a residents may lead neighbourhood citizens who band together in order to resolve community problems. What these and other social service examples have in common is the presence of self-organisation and emergence. As we will demonstrate, applying a complexity model for social programme evaluation facilitates the analysis of the collective behaviour of agents that reflect interconnectedness, rather than simply the aggregate outcomes of individuals. It provides an undergirding organising framework to evaluators given the increased availability of evaluation tools, computational power, and speed to analyse qualitative and quantitative data (Israel and Wolf-Branigin 2011).

Besides the seven components, several characteristics of complexity science, and more specifically complexity adaptive systems, have direct implications to programme evaluation; these include nonlinearity, emergence, being adaptive, having uncertainty in that estimating values cannot be exact, and being dynamical and co-evolutionary (Patton 2011). Nonlinearity implies sensitivity to initial conditions which assumes that the initial state of a CAS has a large influence on how the system will evolve, or being cautious of early unexpected influences or programme effects as these may carry forward and confound expected outcomes. Emergence involves the patterns in behaviours or outcomes that result from individual agents self-organising. Being adaptive includes the continual use of feedback to change in relation to their interactions with others and the larger environment. Uncertainty implies that processes and outcomes within complex systems likely will be unpredictable; addressing uncertainty aids in keeping programme efforts on track. Being dynamical involves understanding that interactions within systems, with all of its changing parts, can often become unstable; therefore tracking changes not only when they occur, but also how and why. Finally, co-evolutionary involves how as agents within a system make their connections and develop their networks, the larger system also changes.

To put all of this in context, the three objectives are designed to assist evaluation researchers – interested in utilising complexity science – frame their efforts. This will be achieved by:

- Introducing the components of complex adaptive systems (CAS) by using brief prompts for identifying their application to the components.
- Identifying and applying CAS components to two established agent-based models for the purpose of aiding readers understand the link between theory and application.
- Discussing the benefits and disadvantages of using a complexity evaluation scheme for decision-making and policy analysis.

16.1 COMPONENTS OF COMPLEX ADAPTIVE SYSTEMS

Understanding complexity science is confusing given the broad field of contributing disciplines. Creating a generalisable scheme through which different programmes can apply complexity likewise is difficult given that individuals who have used complexity theory as an undergirding conceptual framework do not necessarily agree on what *complexity* means. To address this dilemma, we begin by taking a short-cut that frames complexity through the lens of complex adaptive systems (CAS). In social work this becomes useful because the current person-in-environment framework can become static time even though its interventions are evolutionary and process oriented. Understanding the components underlying an intervention as viewed from the lens of a complex adaptive system can therefore aid in visualising the process oriented intended effects.

To begin, we frame a CAS by deconstructing the social service intervention/programme into several components. The components of a CAS represent a process of interacting pieces (Mitchell 2009) and are discussed below. These include: (1) agents having a choice of options, (2) defining boundaries, (3) occurring within an environment/organisation, (4) allowing for self-organisation, (5) having different options from which to choose, (6) using feedback to adapt as interactions evolve along with the larger environment, and (7) producing an emergent behaviour (Johnson 2002; Wolf-Branigin 2013). Patton (2011) in discussing developmental evaluation that applies the concepts of complexity science, recommends a continual questioning process in understanding a programme's operation. Using this logic and to aid the reader, prompts to ask when deconstructing the intervention into the components of a CAS are provided.

Agent-based – Traditional research methods tend to be top down in which the central tendencies of the groups are measured. Within a complexity oriented methodology, a vital difference from traditional methods is that the model is agent-based. This assures that the framework evolves from the grassroots in which the decisions, actions, and behaviours of individuals are taken into account. The agents are the unit of analysis. Within social service organisations, the individuals, families, small groups, or communities are the *agents* within a CAS framework. They are the interacting parts at the most basic level and make decisions from their available and perceived options. Agents interact with one another, they have the capacity to decide whether it is in their best interest to act alone or to work cooperatively in order to take full advantage of benefits. This builds upon social work's person-in-environment perspective by encouraging interdependencies with individuals and their environment (Germain and Gitterman 1980). When seeking to identify the agents within a social programme evaluation, possible questions to ask when identifying the agents include "*Who are the primary actors in the system and who are the intended recipients of the intervention*" and "*Is the unit of analysis individual or groups?*"

Attraction and Self-organisation – This component represents the array of options in first attracting and then maintaining the agent's continued participation in an intervention. Attraction reflects upon consumer empowerment, and concerns itself with the complete set of variables affecting agent decisions and behaviours (Halmi 2003). Attraction and self-organisation are what draw and encourage agents to initiate and continue participation. Attraction represents the clients' initial and continued programme participation. It facilitates an agent's initial interest. This encourages interdependencies, dignity and the worth of human relationships striving to promote future cooperation by creating a network of agents involved in: (1) enlarging future impacts, (2) changing the payoffs, (3) teaching people to care about each other, and (4) encouraging reciprocity (Axelrod 1984). Initial evaluation prompts to consider when identifying the component of attraction/self-organisation include "*Why are clients seeking services?*" and "*What keeps clients motivated to continue or complete services?*"

Boundaries – Complex adaptive systems have boundaries that may be imposed by restrictions in the system representing limited governmental funding or local statutes. Boundaries are determined often from legislation, court decisions or administrative rules supporting policies, funding availability or incentives. Therefore, boundaries play an essential role in understanding and developing a CAS approach to social programme evaluation. Within a CAS scheme these boundaries include the rules agents follow that eventually lead to an emergent behaviour, with the possible questions of "*What rules and laws that affect the agents' behaviour?*" and "*Under what funding influences the agents' behaviour?*"

Heterogeneity – The fourth component of a CAS refers to the options from which the agents choose. Without agents having the ability to select the option that they believes or perceives to best meet their needs or preferences, a system cannot be complex. In the realm of social work this includes the different service options that are available. A prompt to ask when identifying heterogeneity is "*What options are available from which the agents can choose?*"

Feedback – Using feedback remains an essential characteristic of organisations seeking to continually improve their operations. The use of organisational feedback focuses on the iterative and continual use of information (Proehl 2001). Most social service organisations have at least one form of continuous feedback; these may include data systems required for external accreditation purposes or for governmental or non-governmental contract conformance. Failing to use feedback results in an organisation maintaining the status quo whilst not responding to the emerging challenges presented to them by the community and the agents receiving services from them. These organisations fail to use external data to inform their decisions. This can discourage experimentation, continual improvement, and potential innovations. An evaluation prompt may include "*Does the organisational array of services reflect their clients' changing preferences over time?*"

Adaptation/Use of Feedback – As agents in models use information in an iterative way garnered from feedback, adaption to the situation occurs. This process refers to the systems' ability to evolve based on the preferences of the individual agents. A prompt to ask when identifying adaption is "*Does the available options to clients change over time?*"

Emergent Behaviour – This is less a component than a result or dependent variable of the working of the other components of a CAS. It represents the outcomes resulting from interventions provided and identifies how agents with similar interests, strengths or needs create an interconnected result (Hudson 2000; Hudson 2004). In traditional research methods the outcomes of an intervention often identifies linear patterns. Because a CAS takes a bottom-up approach, we are better able to identify nonlinear patterns of outcomes. A potential prompt to ask for identifying the emergent behaviour is "*Do client outcomes reflect any pattern?*"

The above provides a brief overview of the components underlying complex adaptive systems. In order to build upon this scheme and to continually improve, understand, and define a CAS, as will be evident in the following two models, an important aspect is to continually ask questions.

16.2   APPLICATIONS OF AGENT-BASED MODELS

We consider the application of complexity science to be a mathematical field where the relations between inputs and outputs cannot be described in a linear of closed form. Currently the best way known to understand the relationship between inputs and outputs rests through a simulation; particularly some form of an agent-based model (ABM). Developing an appropriate ABM that sufficiently represents the phenomena of the environment we are seeking to represent is a difficult process as it involves programming skills, solid knowledge of one of the modelling software packages, and most importantly the salient variables within the environment we are attempting to replicate.

16.2.1   Model One: Caregivers for Persons with Alzheimer's Disease

For our first model, we use a recent example that describes the components of a CAS for an ABM developed to experiment on differing levels of policy support for caregivers of persons with Alzheimer's disease. This was done in order to forecast whether the caregivers stress levels could be reduced. We used *NetLogo* software (Wilensky 2015) a freely distributed programme and agent environment package.

To describe the decision-making involved, we use a mixed approach to create the agent-based model (Gilbert 2008) in which the interactions of individual agents were built on system dynamics models of their health and stressors. The model simulated the impacts of policy alternatives which included increased respite care, tax incentives, work place policies, and adult day services in order to reduce caregiver stress (Ihara et al. 2015). Experiments demonstrated that policy options providing support to caregivers could reduce their stress.

The model had 100 agents representing older adults and 60 agents representing their caregivers. This model assumed that 40 of the older adults provided their own care. The caregivers were assumed to be family members (spouse, adult daughter, or other kin), professional caregivers, or institutions. Each iteration (step) of the model represents one year. At each iteration, the physical and behavioural health of the older adults could decline. If conditions change, the provider of the care may change from self to family, from family to a professional, or from a professional to an institution. Changes in conditions are based on the health of the older adult or the perceived stress of the caregiver. The older adult or the caregiver may also pass away. New older adults are added in each step to keep the population of older adults at 100 agents. The mix of care providers is driven by the health of the associated older adult. As described below, this application represents the flexibility of the complexity framework as applied to social programme evaluation, with each of the components of a CAS are identified.

Agent-based – In our first application, there were two different sets of agents. The first set were those representing the *older adults* in the system. These had variables to represent their age, physical health, mental health, and who was their caregiver. These agents were initialised randomly; they did have assigned behavioural characteristics in order to replicate the population statistics mean and standard deviation. These descriptive statistics were appropriate for the simulated age of the older adult agents. The second set represented as agents in the system, were the *caregivers*. Caregiver agents had characteristics describing their capabilities and motivation to provide care to the older adults and level of stress and limits of their levels of stress.

Attraction and Self-organisation – The model did not involve complicated attraction or self-organisation as a variable. The results were simply described as population means and standard deviations. Although attraction and self-organisation was not specifically a variable, the shared interests of the caregivers on a common interest with others and benefitting from the variety of policy options conceptually represented the attraction and self-organisation in the model.

Boundaries – The current policies on financially supporting programmes and services provided the primary boundaries for the model that include increasing options that will support family caregivers. Policy options such as increased respite care availability, tax incentives, work place policies, and adult day health services may support aging-in-place (Chen 2014). In the United States, several of these options are available through the Family Medical Leave Act (P.L. 103–3), provisions under Title III, Part E of the Older Americans Act related to the National Family Caregiver Support Program (P.L. 109–365), and the Lifespan Respite Care Act (P.L. 109–442) (Ihara et al. 2012).

Until recently, in-home and community-based services are often not accessible to low income older adults and for those individuals not qualifying for publicly funded services. Provisions for long-term care under the 2010 Patient Protection and Affordable Care Act included several expansions of home- and community-based services (HCBS) under state Medicaid programmes including the Community First Choice state plan option, Balancing Incentives Program, and the home health state plan option (O'Shaughnessy 2013). These and other programmes such as the Community Innovations for Aging in Place Program help promote aging in place (Greenfield 2012); however, the growing need for services may not match the availability or ability of state and local communities to meet all of the demand.

One example of a policy acting as a boundary comes from a National Alliance for Caregiving (2009) study indicating that over half of the responding caregivers, when rating six potential policies or programmes, would choose a $3,000 tax credit as their first or second preferred option. Ihara et al. (2012) tested this policy option by developing an ABM to forecast and explore the likelihood that grandchildren would decide to become the primary caregiver for their grandparent needing significant support. Results indicated that by using a targeted-policy scenario in which high-income families who did not receive a tax credit, middle-income families who receive a $3,000 tax credit, and low-income families who receive a higher tax credit had better results for motivating grandchildren to become caregivers than the universal policy of a flat tax credit across the income distribution.

Heterogeneity – In this model, the options from which the agents can choose (the amount of support the caregivers choose) represents heterogeneity. The above policy options underlie and represent the agent level decision-making process of an older adult and his/her family regarding the choice of living situation. The range of options from which to choose ranged from having the older adult living independently to nursing home placement with several options in-between.

Feedback – Similar to the delivery of other social services, the model of adult caregiving is an iterative process. Feedback comes back to the agents over time from the variables at each stage. These variables within the ABM included the increasing ages of both the caregivers and the older adult, and the death of an older adult in the model. This aggregate feedback from each of the caregivers modified their behaviours and decisions as represented in the forecasted stress levels. At each stage, individual agents react to the policy options presented to them and make their decisions accordingly.

Adaptation/Use of Feedback – The experiment conducted with the model demonstrated that providing services and supports to caregivers can reduce their stress by reducing their direct contact time, this reduction potentially helps the caregiver to continue to care for their loved ones within their home for a longer period of time. To model the effects of this, we presumed that the relief would reduce the stress proportional to the amount of the time relief relative to the total time, in which taking care of the older adult for M hours a day every day causes the stress.

Emergent behaviour – The experiments revealed that benefits beyond anticipated cost saving. Furthermore, other studies demonstrate that using adult day centres generate beneficial effects for individuals with dementia and their caregivers on the days the individual attended the adult day centre. Benefits included fewer behaviour problems, better sleep, and decreased caregiver stress, improved cortisol levels, and reduced depression (Zarit et al. 2011; Zarit et al. 2014; Zarit et al. 2003). In order to better understand the output generated from the model, please refer to Ihara et al. (2015).

16.2.2 Model Two: Housing Patterns of Persons with Disabilities

In our second example we look at housing patterns for persons with intellectual and developmental disabilities. Wolf-Branigin, LeRoy and Miller's (2001) evaluation on community inclusion of people with intellectual and developmental disabilities provided a framework and data or constructing an agent-based model by using a quantitative geography approach, or spatial analysis, to evaluate the efficacy of community inclusion. In the study a random sample was drawn from 2,300 people who had transitioned from institutional residences to community residences such as small group homes, foster homes, and supported/shared living units. The successful inclusion of people with disabilities into the community was in part the result of support by individuals' families and advocates over several years.

In evaluating the inclusion of the people with disabilities (the agents), we used five explanatory variables: (1) degree of involvement by families and friends in the planning process; (2) number of unrelated persons the individual resides with; (3) proportion of earned income to total income; (4) level of disability; and (5) level of mobility. Below the application of the components of a CAS are defined. The hypothesis for the experiments conducted on the AMB is that individual support and encouragement of persons with disabilities is essential to community inclusion, and that becomes evident through the emergence of agents preferences. It was assumed that a more random pattern of housing locations indicated better inclusion because clusters of individuals with disabilities would be limited. Similar to Model One, this agent-based model was created with *NetLogo* software.

Agent-based – The evaluation from which the model was developed involved a sample (n = 150) that included data on the five variables. The different sets of agents represented in the model include the persons with a disability, their supporters who were either family members or friends who participated in their annual planning meetings (allies), and staff employed by the different organisations. The gender ratio of the samples was consistent with the gender ratio of the entire population. In creating the model, 150 agents are created. The numbers of agents with disabilities was set at 24 to provide a population percentage similar to the number of people with disabilities in the general population. These agents were linked to varying numbers of agents that represented their allies. Numbers of allies (those who attended the annual planning meeting who were not employees of the agency) ranged from 0 to 3 per person with a disability agent. In linking the agents, they were able to influence each other's movements throughout the model in the same manner that the decisions allies and people with disabilities influence each other's actions.

Attraction/Self-Organising – The agent-based model assumed that a random pattern of residential locations of the persons with an intellectual or developmental disability represented a higher level of community inclusion (Wolf-Branigin 2006). This assumption was based on the likelihood that persons with a disability would have a more robust and established support system compared to others and therefore led to their ability to become randomly dispersed in the community. This could also indicate a greater level of inclusiveness as these individuals would benefit from the natural supports provided by family members and other allies rather than paid staff.

In order to assess attraction and self-organising behaviour spatial autocorrelation (Moran's I, a measure of spatial randomness) and spatial regression statistics were used. The design included collecting data on the individual consumer level, mapping the data, assessing spatial randomness using Moran's *I* for each explanatory variable, creating local indicator of special associations for regression analysis, performing regression of the five explanatory variables, and assessing influence of control variables in regression procedures. Secondary data were collected from agency records and annual planning meetings that included allies of the person with a disability, agency staff, and the person with the disability. The residential location of individuals was represented in a two dimensional space with longitude and latitude coordinates. In the model, the agents are designated by different colours.

Boundaries – The primary boundaries in the model was whether the person with a disability was eligible for eservices, and the availability of housing options. To illustrate, in the model the role of boundaries on a community, all agents formed groups or neighbourhoods with other agents. An additional boundary specific to this location was a local ordinance that required a minimum of 500 feet between the locations of each group homes.

Heterogeneity – The different housing options from which the person with a disability in conjunction with the feedback from his/her allies could choose represented heterogeneity. These options included larger group homes where any were where from four to 16 persons with a disability also lived, semi-independent living, and independent living.

Feedback – the information provided at the annual planning meetings attended by the person with the disability, family members, other allies, and paid organisational staff served as the main primary feedback mechanism. In partnership with consumers and their allies, the organisation continually located and developed housing and support services options based on the available information. The staff members were assumed to have high levels of adherence to the values of inclusion in the same degree as the persons with a disability and their allies did. This facilitated the ability to create options (Balcazar et al. 1998).

Adaptation/Use of Feedback – For these agents, having a disability limited their movement within the model. These agents could increase their mobility by organising in groups containing five or more agents who represent people without disabilities. The selection of the housing option represented the adaptation and use of feedback. A key aspect in the running of the model was that as time progressed, those with a lesser level of disability were more likely to live near public transport lines so they could more independently travel to their training programmes or employment. Those with a higher level of disability were more likely to live in more isolated environments, such as group homes, and were likely transported by their agency van.

Emergent Behaviour – In all runs of the model similar results occurred. As the agents sought out groups to join, both person with a disability agents and ally agents were near the largest numbers of agents at approximately 15 iterations. However, the agents' numbers remained at the same level with some fluctuations. Based on these results it appears that agents with allies in the community experience greater community inclusion than agents without allies.

Placement of homes likewise may have resulted from political and zoning issues that led to larger residences being located in less populated areas. This dispersion may be interpreted as isolation as opposed to the random dispersion of smaller homes. People living alone or with fewer unrelated persons appear to benefit from residing in more populated areas of the region. The results of several runs of the model indicated that the greater degree of community inclusion among cyan agents and red agents resulted from the community links between cyan agents and green agents as well as the reciprocal energy relationship between the cyan and green agents. For a more detailed explanation of the model and the output generated, please see Wolf-Branigin (2013, pp.157–163).

16.3 BENEFITS AND LIMITATIONS OF AGENT-BASED MODELLING OF COMPLEX ADAPTIVE SYSTEMS

As the above examples of models demonstrate, using agent-based models to evaluate social programmes framed as complex adaptive systems derives benefits. Foremost is the ability to forecast outcomes (represented as emergent behaviours) over time because the running of an ABM assumes the presence of an iterative process. In Model One, the ABM provided a vehicle for modelling and in our circumstance understanding how using differing levels of intervention, within a bounded system in which we can modify policy options, would yield nonlinear outcomes as represented by an emergent behaviour. The ABM was a simulated environment that, to the best of our abilities, represented an actual environment to the level of resolution necessary. Using an ABM approach had the advantage that given a well-defined environment, of being able to perform experiments within this simulated environment that reasonably gauge the impact of the various policy options.

Programme evaluation in contemporary social service practice increasingly exhibits several attributes found in complex systems theory. These include: (1) decisions come from the client level or grass roots level and demonstrated in empowerment evaluation (Fetterman 2001), (2) attractors are instrumental in clients maintaining interest and completing interventions, (3) limits, laws and rules set boundaries under which systems operate, (4) client and organisational feedback are vital to improving outcomes, and (5) self-organisation leads to an emergent behaviour. Although the application of complexity science is far from being consistently adopted across the evaluation field, it does provide a valuable alternative from which to conduct evaluation studies. Using the methods we suggested above, we can envision how deconstructing social interventions into the components of CAS facilitates the use of complexity science. As noted above, we were able to manipulate and test policy options in order to forecast the effects of the policies through the simulated environment.

The advantages and disadvantages of applying agent-based modelling to a complexity framework to policy analysis are numerous. Among its advantages, complexity framework builds upon Patton's utilisation-focused evaluation (2008). It best performs when empowered agents have the ability to select services and length of participation from programme options based on information they receive (Patton 2011). Because of advances in recent knowledge garnered on complex networks (Barabasi 2002), there are suggested methods for identifying and improving the interconnectedness amongst participants in 'small world' networks by adding additional linkages to others with similar interests in their communities. These additional linkages make paths to accessing services shorter (Watts and Strogatz 1998).

ABMs further aid in the detailed policy analysis in the era of big data (Couldry and Powell 2014; Pentland 2014) because the collective effects of individual level data that includes the choices made over time can be incorporated into the model. Problems remain, however, in sufficiently matching modelling schemes and social realities (Miller 2014) and in this sense, it can be overly reductionist. The ABM approach appears to function well when the individual agents are empowered and have the ability to make choices from the programmes from which they receive information and receive services (Pentland 2014). Such innovation suggests that organisational planners and staff implement ideas beyond the status quo that encourage unique solutions to their clients' demonstrated needs and capabilities. Agent-based modelling matters to social programme evaluators because it provides a bottoms-up approach, rather than top-down testing of models.

Whether at the macro level (Tenkasi and Chesmore 2003) or micro level (Hudson 2004), as demonstrated on Model One, creating robust social support systems have the capacity to facilitate less stress, and assist in individuals becoming more resilient to ecosystem threats (Newman 2010; Watts and Strogatz 1998). Across different evaluation situations, ABMs provide a lens for visualising interconnectedness and interdependencies. Given a person-in-environment foundation, the potential use of agent-based modelling respects the viewpoints and decisions made at the client level, whilst simultaneously understanding that these activities occur within an organisational body. Potential applications within the social services cover a broad range. These applications may range from conducting historical analyses, visualising emerging patterns in social phenomena, testing social policy options, understanding indigenous populations, to applying trajectory growth curves in clinical trials (Wolf-Branigin 2013).

Methodologically, evaluators may explore the use of agent-based modelling (ABM) for creating predictive simulations of social service phenomena. ABMs can provide a sensitive method for investigating individuals' responses to social interventions. Investigating the properties of robustness (resiliency), scalability in generalist social work practice (for example, use of goals and objectives at both the individual client and programme levels), and the diversity of agents (Miller and Page 2007) are promising lines of inquiry to develop. Because the modelling of complexity uses an iterative process and a priori information contained in management information systems, it generates multidimensional perspectives of social service delivery and resulting outcomes. Future efforts to develop complexity applications in social programme evaluation include further defining a concise set of prompts for defining the components, framing of the social programme components, and creating viable procedures and checklists for programme planners and evaluators to use.

Challenges associated with integrating complexity theory into social service programme planning and evaluation remains (Fish and Hardy 2015). Utilising ABMs as a proxy for complex adaptive systems in order to measure the collective behaviour – rather than simply the aggregated outcomes of intended programme participants – remains difficult. Having the requisite skill to sufficiently model a reasonably simulated environment that adequately represents the real world is difficult. Problems of addressing boundaries, modelling the various properties and variables, keeping the model agent-based, and determining when to use a CAS approach will be difficult (Epstein 2007). Challenges exist when applying CAS to measure the collective behaviour of agents rather than the simple aggregated outcomes of programme targets. These problems must address defining the boundaries of models, verifying the reliability and validity of models, keeping the model agent-based, and deciding when a CAS approach best fits (Epstein 2007).

16.4 CONCLUSION

The rational planning, implementation, delivery, and evaluation of social innovations involving policy options usually remains a complicated endeavour. Before an evaluator can adequately develop an agent-based model, however, we suggest that they first should frame the phenomena to be evaluated as a complex adaptive system. In addressing these difficult issues, we sought to introduce an alternative approach in which the components that comprise complex adaptive systems are considered, defined, and applied to two situations that inquired into policy options of caregivers to persons with Alzheimer's disease and housing patterns of persons with disabilities. Benefits and disadvantages of using this approach were discussed. Key aspects to remember when framing as a CAS, is that the policy options typically are the boundaries and that most social innovations have some form of continuous feedback that informs the evolution of the system. As such, we view complexity as a mathematical field where the relations between inputs and outputs could not be described in a closed form and the best currently way known is through a simulation, particularly some form of ABM.

We did not attempt to provide the reader with a detailed approach to agent-based modelling, but rather sought to provide a succinct introduction to applying the components of complex adaptive systems to the preparation of an ABM. For more detailed procedures for creating ABMs the reader should explore the *NetLogo* website (ccl.northwestern.edu/netlogo/) where software, existing models, information on user groups, sample coding, and other procedures can be found.

The study of complexity science includes a wide variety of thought, definitions, methods, and theoretical orientations. In time, this will become further complicated as the desire to analyze big data becomes more pronounced. Agent-based modelling provides one powerful tool to those interested in complexity because it allows for simulated environments to be created that replicate a social situation. Within these simulated environments, experiments can be conducted to test various options, including policy changes. However, to sufficiently develop such experiments in the simulated environments the process of framing as a complex adaptive system aids programme planners and evaluators in assuring that the necessary components are taken into account. Developing and programming the simulated environments can be difficult; therefore having a group in which there is at least one person with these skills will be beneficial.

REFERENCES

Agar, M. (1999). Complexity theory: An exploration and overview. *Field Methods*, 11, 99–120.

Axelrod, R. (1984). *The Evolution of Cooperation*, pp. 109–168. Jackson, TN: Basic Books.

Balcazar, F.E., Mackay, M., Keys, C.B., Henry, D. and Bryant, F.B. (1998). Assessing perceived agency adherence to the values of community inclusion: Implications for staff satisfaction. *American Journal of Mental Retardation*, 102, 451–463.

Barabasi, A.-L. (2002). *Linked: The New Science of Networks*. Cambridge, MA: Perseus.

Bronfenbrenner, U. (1994). Ecological models of human development. In *International Encyclopedia of Education*, Vol. 3 (2nd edition). Oxford, England: Elsevier Science, pp. 1643–1647.

Chen, M.-L. (2014). The growing costs and burden of family caregiving of older adults: A review of paid sick leave and family leave policies. *The Gerontologist*, gnu093. doi:10.1093/geront/gnu093.

Couldry, N. and Powell, A. (2014). Big data from the bottom up. *Big Data & Society*, 1(5), 1–5.

Epstein, J.M. (2007). *Generative Social Science: Studies in Agent-based Computational Modeling*. Princeton, NJ: Princeton University Press.

Epstein, J.M. and Axtell, R.L. (1996). *Growing Artificial Societies: Social Science from the Bottom Up*. Cambridge, MA: MIT Press.

Fetterman, D.M. (2001). *Foundations of Empowerment Evaluation*. Thousand Oaks, CA: Sage Publications.

Fish, S. and Hardy, M. (2015). Complex issues, complex solutions: Applying complexity theory in social work practice. *Nordic Social Work Research*, 5(supl. 1), 98–114.

Germain, C. (1978). General-systems theory and ego psychology: An ecological perspective. *Social Service Review*, 52(4), 534–550.

Germain, C. and Gitterman, A. (1980). *The Life Model of Social Work Practice*. New York: Columbia University Press.

Gilbert, N. (2008). *Agent-based Models*. Thousand Oaks, CA: Sage Publications.

Gorman, D., Mezic, J., Mezic, I. and Gruenewald, P. (2006). Agent-based modeling of drinking behavior: A preliminary model and potential applications to theory and practice. *American Journal of Public Health*, 96(11), 2055–2060.

Green, D. and McDermott, F. (2010). Social work from inside and between complex systems: Perspectives on person-in-environment for today's social work. *British Journal of Social Work*, 40, 2414–2430.

Greenfield, E.A. (2012). Using ecological frameworks to advance a field of research, practice, and policy on aging-in-place initiatives. *The Gerontologist*, 52(1), 1–12.

Halmi, A. (2003). Chaos and non-linear dynamics: New methodological approaches in the social sciences and social work practice. *International Social Work*, 46(1), 83–101.

Harvey, D. (2001). Chaos and complexity: Their bearing on social policy research. *Social Issues,* 1(2). Complexity Science and Social Policy. http://www.whb.co.uk/socialissues/indexvol1two.htm (accessed 10 August 2017).

Hudson, C.G. (2000). At the edge of chaos: A new paradigm for social work? *Journal of Social Work Education*, 36(2), 215–230.

Hudson, C.G. (2004). The dynamics of self-organisation: Neglected dimensions. *Journal of Human Behavior in the Social Environment*, 10(4), 17–37.

Ihara, E.S., Horio, B.M. and Tompkins, C.J. (2012). Grandchildren caring for grandparents: Modeling the complexity of family caregiving. *Journal of Social Service Research*, 38(5), 619–636.

Ihara, E.S., Kennedy, W.G., Tompkins, C.J. and Wolf-Branigin, M.E. (2015). Long-term dementia care: Modeling the decision process. *Proceedings of the 24th Annual Behavior Representation in Modeling and Simulation Conference (BRiMS)*, 59–62.

Israel, N. and Wolf-Branigin, M. (2011). Nonlinearity in human service evaluation: A primer on agent based modeling. *Social Work Research*, 35(1), 20–24.

Johnson, S. (2002). *Emergence: The Connected Lives of Ants, Brains, Cities, and Software*. New York: Simon & Schuster.

Knowlton, L.W. and Phillips, C.C. (2009). *The Logic Model Guidebook: Better Strategies for Great Results*. Los Angeles: Sage Publications.

Mankiewicz, R. (2001). *The Story of Mathematics*. Princeton, NJ: Princeton University Press.

Miller, J.H. and Page, S.E. (2007). *Complex Adaptive Systems: An Introduction to Computational Models of Social Life*. Princeton, NJ: Princeton University Press.

Miller, K.D. (2014). Agent-based modeling and organisational studies: A critical realist perspective. *Organisational Studies*, 36(2), 175–196.

Mitchell, M. (2009). *Complexity: A Guided Tour*. New York: Oxford.

National Alliance for Caregiving (2009). Caregiving in the US available online: http://www.caregiving.org/pdf/research/CaregivingUSAllAgesExecSum.pdf (accessed 22 August 2017).

Newman, M.E (2010). *Networks: An Introduction*. New York: Oxford University Press.

O'Shaughnessy, C.V. (2013). *Medicaid home- and community-based services programs enacted by the ACA: Expanding opportunities one step at a time* (Background Paper No. 86). National Health Policy Forum. http://www.nhpf.org/library/background-papers/BP86_ACAMedicaidHCBS_11-19-13.pdf (accessed 10 August 2017).

Patton, M.Q. (2008). *Utilization-focused Evaluation* (4th edition). Los Angeles: Sage Publications.

Patton, M.Q. (2011). *Developmental Evaluation: Applying Complexity Concepts to Enhance Innovation and Use*. New York: Guilford Press.

Pentland, A. (2014). *Social Physics: How Good Ideas Spread- the Lessons from a New Science*. New York: Penguin.

Proehl, R.A. (2001). *Organisational Change in the Human Services*. Thousand Oaks, CA: Sage Publications.

Schelling, T. (1978). *Micromotives and Macrobehavior*. New York: Norton.

Tenkasi, R.V. and Chesmore, M.C. (2003). Social networks and planned organisational change: The impact of strong network ties on effective change implementation and use. *The Journal of Applied Behavioral Science*, 39(3), 281–300.

Warren, K., Franklin, C. and Streeter, C.L. (1998). New directions in systems theory: Chaos and complexity. *Social Work*, 43(4), 357–372.

Watts, P.J. and Strogatz, S.H. (1998). Collective dynamics of 'small world' networks. *Nature*, 393, 440–442.

Westley, F., Zimmerman, B. and Patton, M. (2006). *Getting to Maybe: How the World is Changed*. Toronto: Random House Canada.

Wilensky, U. (2015). *NetLogo*. Center for Connected Learning and Computer-Based Modeling, Northwestern University: Evanston, IL. http://ccl.northwestern.edu/netlogo/ (accessed 10 August 2017).

Williams, B. and Imam, I. (2006). *Systems Concepts in Evaluation: An Expert Anthology*, 3–16. Point Reyes, CA: EdgePress.

Woehle, R. (2007). Complexity theory, nonlinear dynamics, and change augmenting systems theory. *Advances in Social Work*, 8(1), 141–151.

Woehle, R., Jones, G., Baker, T. and Piper, M. (2009). Theory and modeling of emergent behaviors: The effects of intervention on social and cultural capital. *Social Development Issues*, 31(2), 43–56.

Wolf-Branigin, M. (2006). Self-organisation in housing choices of persons with disabilities. *Journal of Human Behavior in the Social Environment*, 13(4), 25–35.

Wolf-Branigin, M. (2009). Applying complexity and emergence in social work education. *Social Work Education: The International Journal*, 28(2), 115–127.

Wolf-Branigin, M. (2013). *Using Complexity Theory in Research and Program Evaluation*. New York: Oxford University Press.

Wolf-Branigin, M., LeRoy, B. and Miller, J. (2001). Physical inclusion of people with developmental disabilities:

An evaluation of the Macomb-Oakland Regional Center. *American Journal on Mental Retardation*, 106(4), 368–375.

Zarit, S.H., Kim, K., Femia, E.E., Almeida, D.M. and Klein, L.C. (2014). The effects of adult day services on family caregivers' daily stress, affect, and health: Outcomes from the Daily Stress and Health (DaSH) Study. *The Gerontologist*, 54(4), 570–579.

Zarit, S.H., Stephens, M.A.P., Townsend, A., Greene, R. and Femia, E.E. (2003). Give day care a chance to be effective: A commentary. *The Journals of Gerontology Series B: Psychological Sciences and Social Sciences*, 58(3), P195–P196.

Zarit, S.H., Kim, K., Femia, E.E., Almeida, D.M., Savla, J. and Molenaar, P.C.M. (2011). Effects of adult day care on daily stress of caregivers: A within-person approach. *The Journals of Gerontology Series B: Psychological Sciences and Social Sciences*, 66B(5), 538–546.

17. Complexity, the bridging science of emerging respiratory outbreak response

Dr Babak Pourbohloul, University of British Columbia,
Dr Krista M. English, University of British Columbia and
Dr Nathaniel Hupert, Cornell University

Due to the increasing incidence of emerging infectious diseases (EIDs), significant time and resources have been allocated to the development and implementation of pandemic preparedness plans worldwide. From the evolution of novel influenza strains to the emergence of new lethal pathogens, such as Severe Acute Respiratory Syndrome (SARS) and the recent Middle East Respiratory Syndrome (MERS), infectious diseases remain a key public health concern of the 21st century. The World Health Organization has characterized the influenza virus in particular as 'sloppy, capricious and promiscuous' (World Health Organization 2005), characteristics that contribute to its constant pandemic threat. However, it is the social context derived primarily from the behaviors of its human hosts that enable the potential lethality of this virus and others. Incorporating this social perspective is important to understand and control the potential impact of viral disease not only within individuals, but also within and between populations.

Public health plans designed to prevent or mitigate the impact of an EID pandemic were put to the test with the emergence of SARS in 2003 and a novel influenza strain in 2009. These experiences demonstrated that despite the best efforts to prepare for such events, real-time management of emerging disease outbreaks is often marked by confusion and uncertainty. During these outbreaks, decision makers were challenged to make impactful decisions with little time and incomplete information. The recent 2014–2015 Ebola outbreaks in West African countries and the spread of Zika virus in the Western Hemisphere serve as urgent examples of the need to evaluate health systems' capabilities to respond to emerging diseases.

To date, the scientific approach of health authorities to such threats has primarily focused on evaluation of the safety and effectiveness of interventions at the level of individuals, that is, regarding the safety and efficacy of particular vaccines and antivirals. Such science, however, does not make clear *how* these countermeasures should then be used to optimally benefit society as a whole (that is, at the level of population health as opposed to individual medical care). Once a pandemic is declared, policy makers must respond quickly; to the extent possible, their decisions should be informed by both pre-event and real-time analyses in order to minimize associated morbidity, mortality and disruption at the societal level. Decisions around how to best use a potentially limited supply of pharmaceutical or disruptive social distancing interventions in these situations require a unique combination of scientific fields, and integrating these fields in real time requires a bridging science. Mathematical modeling of these complex socio-epidemiological systems – or 'complexity science' for short – represents that bridging science.

The spread of communicable diseases is a dynamical process and, as such, the understanding and control of infectious disease outbreaks and epidemics is pertinent to the temporal evolution of disease propagation. Historically, this aspect of disease transmission has been studied using *compartmental models*, which are meant to be mathematical representations of the population in question. In these models, the population is divided into a small number of coarse epidemiological states (also referred to as 'compartments' or 'classes'). This approach – and its more detailed variants – have been instrumental to understanding several features of infectious diseases over the past three decades. Yet, the entire approach is a significant over-simplification. The main assumption is that the population is 'well mixed', that is, every infectious individual has an equal opportunity to infect others. However, this is in contrast with reality where individuals heterogeneously contact each other based on their demographic profile and movement history. Mathematical epidemiology has been advanced by the recent development of detailed community contact network models. This advancement enables the more complex study of random virus behavior within the important context of non-random human behavior.

In several areas of science, technology, industry and logistics, quantitative frameworks based on advanced mathematical and computational techniques are employed to aid decision makers tailor evidence-informed decisions at times of crises. While modeling has a strong foothold in many other disciplines, its incorporation as an *integrated* decision-support tool within health systems decision-making and policy design has been slow and unevenly adopted. This may be in part because, historically, modeling methodologies emerged from disciplines outside the health sciences – mainly within physics, computer science, mathematics, engineering, and related fields. Despite this, there has been progress over the past decade within both public health and mathematical modeling communities by members who have remained committed to an ongoing dialogue. Furthermore, there have been sporadic attempts to encourage collaboration between senior-level policy makers and modelers in workshops intended to bring key players together. These efforts have highlighted a recurring set of questions raised by enthusiastic senior policy makers who would like to capitalize on this approach: 'How can complex systems modeling inform public health decision-making at times of crisis?'; 'What is the best example of successful integration that I can replicate in my jurisdiction?'; and 'What processes are required to establish an integrated decision-support framework?'

In this chapter, we discuss the conceptual design and structure of complex network models, within the context of transmission dynamics of respiratory-borne pathogens in human populations. Application of network models to address the spread of other communicable diseases, such as vector-borne, sexually-transmitted, or zoonotic infections deserve separate treatment. We believe that assembling outbreak response teams with wide expertise – not just infectious disease epidemiologists but also experts in contact networks and mathematical modeling, virology, immunology and emergency management, are necessary to ensure valuable quantitative decision-support tools to assist policy makers at the time of crisis.

The application of complex systems tools in public health policy design may support decision-making in five major ways, by including:

1. Understanding and quantifying the network on or through which the pathogen is transmitted.

2. Quantifying the transmission dynamics potential within the community at large, in specific subpopulations such as children, or within healthcare facilities and other smaller settings.
3. Estimating the initial real-time rate of disease spread in the community when the biological characteristics of the emerging pathogen are largely unknown.
4. Assessing the impact of different logistically attainable intervention strategies on the propagation of the pandemic within the community at large and/or in select locations.
5. Evaluating the potential for nonlinear and/or emergent or unintended consequences from intervention strategies.

The number of peer-reviewed scientific papers that self-reported the application of complexity science to infectious disease modeling remains small: fewer than 10 such papers were published annually as recently as 2012. However, more than five times that number were published that year incorporating contact network methods into disease epidemiology (without explicitly labeling the resulting methodologies as 'complexity science'), and that number is now rapidly increasing.

17.1 MODELING INFECTIOUS DISEASES

Mathematical models have progressed dramatically over the past four decades leading to several mathematical approaches that explicitly consider sociological factors. In the past, lack of reliable large-scale data about social contacts (be it proximity contacts for respiratory-borne infections, or sexual contacts for sexually-transmitted infections) has been an obstacle to having a realistic picture of disease-specific contact networks. In the absence of such data, the best assumption was that people could be grouped based on their infection status (for example, susceptible, infected, recovered) and/or their basic demographic characteristics, such as age. Historically, this aspect of disease transmission has been studied using a coarse representation of population dynamics, known as compartmental models.

Figure 17.1 is a schematic representation of a simple **SEIR** compartmental model that may be compatible with natural history of some real-life communicable diseases: at any given time, individuals (circles) can be part of the population within any of the following compartments: <u>S</u>usceptible (not infected but could potentially become so upon interaction with infectious individuals), <u>E</u>xposed (carries the disease, but cannot transmit it), <u>I</u>nfectious (can transmit the disease) or <u>R</u>emoved (can neither transmit the disease nor become newly infected). Individuals within each of these compartments are indistinguishable (that is, fully interchangeable) and transition rates between compartments are fixed, meaning that flow among compartments depends solely on the total population size of the compartments. More complex structures can be built by adding more compartments to account for more specific groups of individuals within a population, which allows more detailed dynamics. In these models, the temporal variation in population size in each epidemiological/demographic class is described by a set of differential equations.

In the absence of large-scale contact data, this approach, and its more complex variants, has been instrumental in understanding numerous features of infectious disease transmission. Yet, it has a major simplifying assumption, namely that every infectious individual

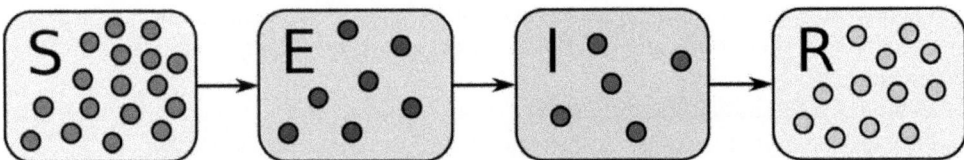

Figure 17.1   A simple SEIR compartmental model. At any moment, individuals (circles) can be part of the population within any of the following compartments: Susceptible (not infected but could potentially become so upon interaction with infectious), Exposed (carries the disease but cannot transmit it), Infectious (can transmit the disease) or Removed (cannot transmit the disease nor become newly infected). Individuals within a compartment are indistinguishable and thus flow among compartments depends solely on the total population of compartments. More complex variations exist where, for example, these compartments are divided into sub-compartments that define more specific groups of individuals, which allows more detailed dynamics

has an equal opportunity to infect others. Postulating such homogeneous, or 'well-mixed,' populations is valid in the broader context of population biology, but human populations tend to contact each other in a heterogeneous manner based on age, profession, socio-economic status, geographic location, behavior, and other measurable factors. The 'well-mixed' approximation therefore may not accurately portray disease spread among humans (Song et al. 2010; González et al. 2008).

Apart from these classical compartmental models, other mathematical frameworks have been employed to address infectious disease transmission dynamics, including: stochastic compartmental models (Bailey 1957); branching process models (Becker 1976; Farrington, Kanaan and Gay 2003); dyad models (Groendyke et al. 2012; Keeling et al. 1997; Ferguson and Garnett 2000); Reed–Frost chain-binomial models (Lefevre and Picard 1989) allowing more exact predictions of the size and probability of epidemics; and 'individual-based modeling' (Germann et al. 2006), a primarily computational approach based on following the contact and infection histories of simulated individuals, yielding detailed statistical predictions about disease outcomes. Several individual-based models assume predefined simple contact patterns such as regular lattices (Durrett 1999; Kleczkowski and Grenfell 1999; Britton and O'Neill 2002; Sander et al. 2002), although there have been efforts to consider more realistic contact patterns (Van der Ploeg et al. 1998; Chowell, Fenimore et al. 2003; Chowell, Hyman et al. 2003; Mossong et al. 2008; Eubank et al. 2010).

Recent advances in network and percolation theories have paved the way for physicists to offer a new perspective for understanding disease spread. Over the past decade, seminal work by Watts and Strogatz on small-world networks (Watts and Strogatz 1998; Watts 1999), Albert and Barabási on scale-free networks (Albert and Barabási 2002), and Dorogovtsev and Mendes (Dorogovtsev and Mendes 2003) and Pastor-Satorras and Vespignani (Pastor-Satorras and Vespignani 2007), among others, on the dynamics of networks, have shed light on a number of intriguing aspects of epidemiological processes. In particular, groundbreaking work by Newman et al. (Newman et al. 2001; Newman 2002, 2003) has provided a strong foundation for the formulation of epidemiological problems on a broad range of network structures using tools developed by physicists.

Thanks to the increasing availability of mesoscale (that is, larger than individual person-to-person but smaller than national-level) contact data and new methods for their storage, analysis, and visualization, it is now possible to develop realistic network-based models to assist public health decision-making.

17.2 CONTACT NETWORK MODELING

Contact network epidemiology is an analytical framework that explicitly captures the realistically-diverse interactions that underlie the spread of diseases (Longini 1988; Sattenspiel and Simon 1988; Mollison 1995; Ball et al. 1997; Diekmann et al. 1998; Lloyd and May 2001; Newman 2002; Keeling et al. 2003; Meyers et al. 2005; Pourbohloul et al. 2005). The first step in contact network epidemiology is to construct networks based on information about real-life contacts between individuals. One can then analyze these networks to determine their crucial topological features and apply analytical methods to make epidemiological predictions and intervention recommendations. The two primary advantages of this approach are that (1) it makes no a priori assumptions about the overall network structure; and (ii) the mathematical analysis allows one to bypass computationally-intensive simulations.

Contact network models attempt to characterize every interpersonal contact that can potentially lead to disease transmission in a community. These contacts may take place within households, schools, workplaces, hospitals, and so on (Figure 17.2). Each person in a community is represented by a node (vertex) in the network, and each contact between two people is represented by an edge (link) connecting their nodes. The disease can only be transmitted to neighbors of an infectious node, with a probability between 0 and 1, depending on disease contagiousness; 0 meaning no transmission, and 1 meaning transmission with certainty. Neighbors do not need to be geographical neighbors. For instance, an individual may not have adequate contact with his/her household neighbors to pass on the disease, but does have such contact with someone else in a shopping mall. Therefore, one may infect or be infected by *topological* neighbors. The number of edges emanating from a node, or the number of potentially disease-causing contacts a person has, is called the *degree* of the node. The distribution of the numbers of contacts is a fundamental quantity in network theory, known as the *degree distribution*.

Different transmission routes may lead to different network structures; thus different degree distributions. Some, that are analytically tractable (that is, they have solutions that can be mathematically derived), have been extensively studied in the literature, such as those networks following a *Poisson degree distribution* (Erdos and Rényi 1961), or *scale-free distribution* (Pastor-Satorras and Vespignani 2001; Boguñá et al. 2003). While some real-life networks may be very well-approximated by these approaches, others may not.

Data may be affected by incompleteness and inaccuracies due to the difficulty of detecting infections, especially if symptoms are mild (for example, as with the 2015–2016 Zika virus outbreak), if ill individuals do not seek medical care, or if the infection causes non-specific symptoms that could also be due to other causes (as is common with influenza). Figure 17.3 illustrates this concept. Suppose the network in panel (a) represents a totally susceptible population (blue) prior to the invasion of the pathogen; panel (b) denotes the individuals' various epidemiological states (different colors) after the epidemic is over.

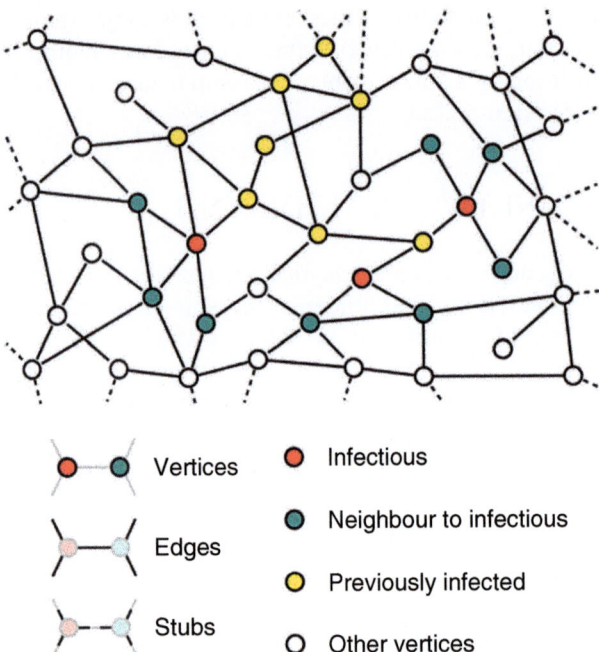

	Vertices	● Infectious
		● Neighbour to infectious
	Edges	
		○ Previously infected
	Stubs	
		○ Other vertices

Figure 17.2   *A network model. Each individual is represented by a node (vertex) in a network. Neighboring vertices are vertices linked by an edge. A stub is the half of an edge that is connected to a given vertex. The disease can only be transmitted to neighbors of an infectious vertex (but maybe not to all of them). Neighbors do not need to be geographical neighbors. For instance, if an individual does not come into contact with his/her household neighbors, but does come into contact with someone in a shopping mall (not necessarily in the household neighborhood) he/she may infect these 'topological' neighbors*

Excluding un-infected nodes (still susceptible) from panel (b) leaves all infected nodes in panel (c). In reality, a fraction of these nodes (light green) may have been asymptomatic – they may not represent any signs or symptoms associated with the disease despite being infected. Among the remainder of the symptomatic nodes (panel (d)) some may report to health authorities to seek care, while others (dark green) may not. Finally, the causative agent of infection for a fraction of the symptomatically-reported cases may be confirmed by laboratory tests (red), while disease diagnosed for others (purple) might be attributed to the same agent (different strain) due to similarity of clinical symptoms but without laboratory confirmation.

The latter group may or may not have been infected with the emerging pathogen. It is worth noting that not all laboratory-confirmed cases are necessarily linked to one another, making it near impossible to reconstruct the transmission chain for all of these cases, let alone the underlying contact network for the entire population. The fraction of symptomatic, asymptomatic and laboratory-confirmed cases depends on the severity of disease and availability of diagnostic tests for the novel pathogen.

Contrasting examples from the last decade include SARS (for which almost all true cases were symptomatic) and the 2009 pandemic H1N1 influenza (which typically presented with mild symptoms in adults). All these issues are particularly relevant at the very onset of the outbreak, when the detected number of cases remains small and biases in detection rates can lead to a serious underestimation of key epidemiological characteristics of disease.

17.3   TRANSMISSION DYNAMICS

Contact network epidemiology allows us to assess the vulnerability of a population to an infectious disease on the basis of the structure of the network (for example, its degree distribution) and on the average transmissibility (T) of the disease (Newman 2002; Meyers et al. 2005). T is the average probability that transmission will occur from an infected person (node) to an uninfected person during the entire duration of infectiousness, and the average is taken over all transmissibility values, T_{ij}, between each pair of individuals i and j who are connected by an edge. This parameter summarizes multiple aspects of transmissibility, including the contact intensity between persons, duration of infectiousness, and the host's susceptibility to the infectious pathogen.

A critical transmissibility value, T_c, exists in any given network and indicates whether a large-scale epidemic is probable (Figure 17.4). Any disease with average transmissibility < Tc cannot cause sustained transmission within a population and will thus be limited to small outbreaks. Such disease outbreaks are likely to die out because of the probabilistic nature of transmission before the disease has a chance to spread to the population at large. On the other hand, diseases with average transmissibility > T_c may spark large-scale epidemics (but not with certainty). As disease transmissibility exceeds beyond T_c, large networks may coalesce to harbor one 'giant component' linking together a very large number of individuals. And while many individuals within the giant component may interact with only a few other people, they can have a significant risk of infection due to the global property of the network (Newman et al. 2001). The value of T_c depends on the structure of the network that is created by the contact patterns within a community. Roughly speaking, when abundant opportunities exist for transmission, disease will spread easily, and the epidemic threshold will be low.

In this Figure, **A** and **B** represent two different network structures and therefore have two different epidemic thresholds, even for the same hypothetical disease (T_c^A for **A**, T_c^B for **B**). Differences in network structure could correspond to two different communities, or to a single community before and after a contact-reducing intervention. A contact-reducing intervention, such as self-isolation or school closure may result in removing a fraction of nodes and/or edges in the network, generally leading to an increase in the epidemic threshold.

This difference between the two communities in Figure 17.4 is entirely the result of difference in their network structure. Thus, although a disease with transmissibility X can ignite an epidemic in community A ($X > T_c^A$), it will only cause a small outbreak in community B ($X < T_c^B$). Also, although intervention I_1 can reduce the size of a small outbreak in community B, its implementation avoids an epidemic in community A by bringing the transmissibility below threshold.

By analogy, transmission-reducing interventions such as I_2 could bring down the

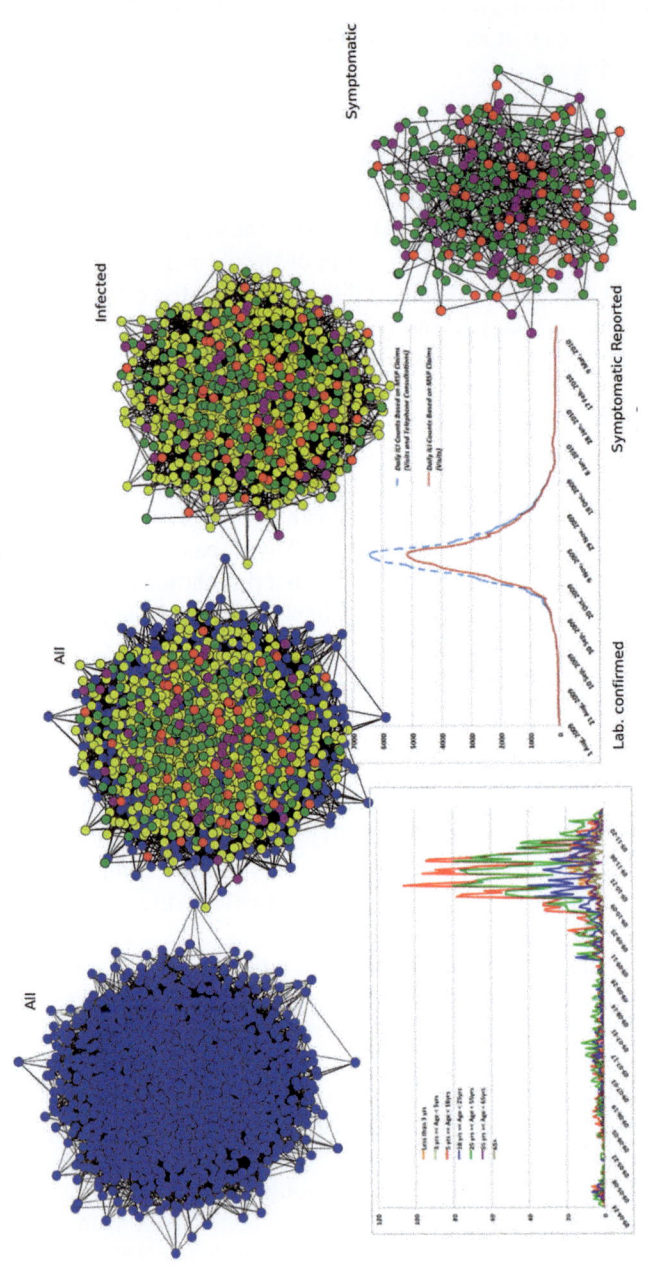

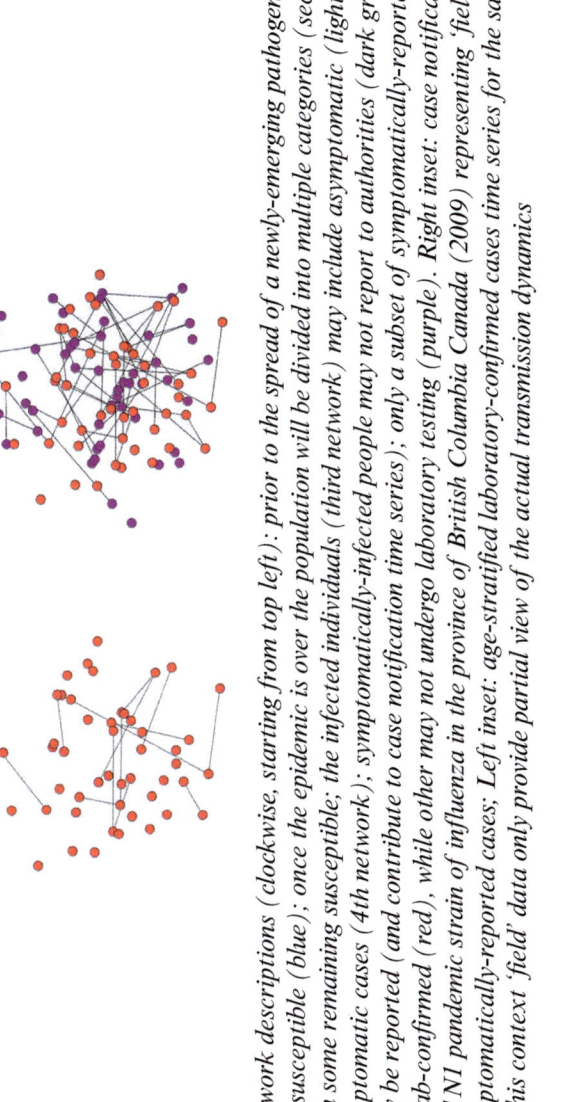

Figure 17.3 Network descriptions (clockwise, starting from top left): prior to the spread of a newly-emerging pathogen all individuals are susceptible (blue); once the epidemic is over the population will be divided into multiple categories (second network), with some remaining susceptible; the infected individuals (third network) may include asymptomatic (light green) and symptomatic cases (4th network); symptomatically-infected people may not report to authorities (dark green), or they may be reported (and contribute to case notification time series); only a subset of symptomatically-reported cases may be lab-confirmed (red), while other may not undergo laboratory testing (purple). Right inset: case notification data for pH1N1 pandemic strain of influenza in the province of British Columbia Canada (2009) representing 'field' data for symptomatically-reported cases; Left inset: age-stratified laboratory-confirmed cases time series for the same epidemic. In this context 'field' data only provide partial view of the actual transmission dynamics

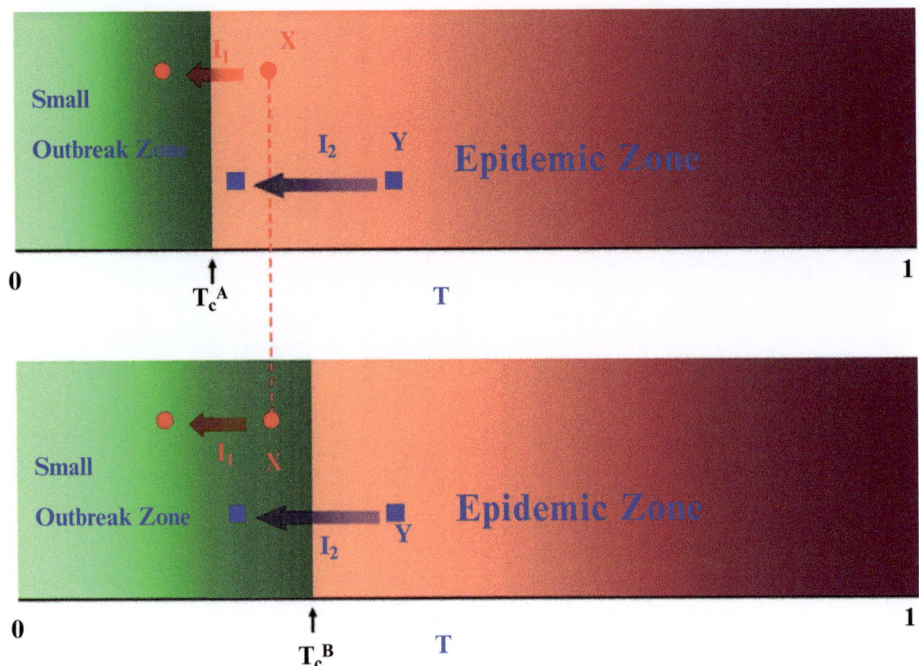

Figure 17.4   *[Figure reproduced with author's permission from (Pourbohloul and Brunham 2004)] Epidemic threshold: The network structure of each community sets a unique threshold value that determines its vulnerability to infectious disease. The horizontal axis represents the transmissibility of disease varying between 0 and 1. Every disease can be uniquely identified on this axis for a specific network. A and B represent two different network structures and therefore have two different epidemic thresholds, even for the same hypothetical disease (T_c^A for A and T_c^B for B). Differences in network structure may correspond to two different communities, or may correspond to a single community before and after a contact-reducing intervention. If the disease transmissibility is smaller than the epidemic threshold (X in the bottom panel), it dies out after a small-scale outbreak (green areas); but if it is above threshold (X and Y in the top panel and only Y in the bottom panel), then it could potentially ignite a large-scale epidemic (red areas). The size of outbreaks is shown by different green shades, and it increases (darker shades) as disease transmissibility approaches the epidemic threshold from below. Also, the probability of the occurrence of an epidemic is shown by different red shades, and it increases (darker shades) as disease transmissibility approaches 1*

transmissibility of a relatively rapid spreading disease like **Y** in community **B** to a level below its threshold value, but it cannot be as successful in community **A** and will only reduce the probability of an epidemic without being able to avoid it (because it still remains in the epidemic zone).

17.4   MODES OF TRANSMISSION

An agent's mode of transmission affects its transmissibility and spread across the network (Tang et al. 2009; Hui et al. 2015; Tang et al. 2013). For most respiratory diseases, it is widely accepted that droplets, which have been coughed, sneezed, or simply exhaled into the air, are the most important means of transmission. Large droplets carry the most infectious particles. They are affected by air motion, however the dominant physical force acting on them is gravity; thus, they settle out of the air quickly. Large droplets are effectively trapped in the upper airways and are not deposited in the lower respiratory tract. In contrast, very small particles, which carry a much smaller dose of infectious agents, quickly transform into 'droplet nuclei' through evaporation and remain suspended in air for prolonged periods of time and can penetrate deeper into the lower respiratory tract.

The relative contribution of these different modes of transmission to the spread of the influenza virus remains unresolved. Almost all countries have incorporated means to prevent direct and indirect transmission of influenza via droplet and interpersonal contact in their pandemic plans (for example hand washing, wearing gloves where appropriate); however, the debate continues about the need for airborne precautions (Tellier 2006; Brankston et al. 2007; Tellier 2007; Tang and Li 2007; Lee 2007). Based on epidemiological patterns of disease transmission, droplet transmission has been considered by many to be the primary route of influenza transmission (Bridges et al. 2003). However, several studies in the scientific literature support the importance of the contribution of airborne influenza transmission by aerosols (Tellier 2006; Riley 1974; Salgado et al. 2002). An accurate assessment of the mode of transmission has obvious implications for pandemic preparedness and is essential for making rational decisions about infection control measures, such as intervention strategies, screening, patient placement, staff cohorting, use of personal protective equipment, antivirals, and others (Pan-Canadian Public Health Network 2015; US Department of Health and Human Services 2012; Department of Health, Commonwealth of Australia 2014; Department of Health, England and Health Departments of the Devolved Administrations of Scotland, Wales and Northern Ireland 2012; Pandémie grippale 2011). Additionally, the duration of contact and proximity to an infected person will strongly predict the likelihood of further spread. Obtaining a thorough understanding of patterns of contact (direct or indirect) along with the different modes of viral transmission (droplet and airborne [short-range or long range aerosolized particles]) is advantageous when designing infection control strategies (Tang et al. 2006; Atkinson and Wein 2008; Li et al. 2007; Ip et al. 2007). Contact network modeling facilitates the study of these factors and their effect on eventual outbreak sizes and probabilities of initiating an epidemic (Meyers et al. 2005).

Experiments with the influenza A virus show that both large droplets (that is, intranasal drops) and aerosols cause infections in human volunteers. Therefore, proximity and duration of contact may be used as surrogate indicators for the probability of infection and/or transmissibility, although being at a certain distance from an infectious person for a specific time does not necessarily guarantee that transmission of infection will occur. This suggests that the notion of probabilistic population-based transmissibility, T, encapsulates multiple probabilistic components of infection across multiple scales of biosocial systems (for example, cellular defenses, interpersonal contact). The analytical framework

A)   B)

C)   D)

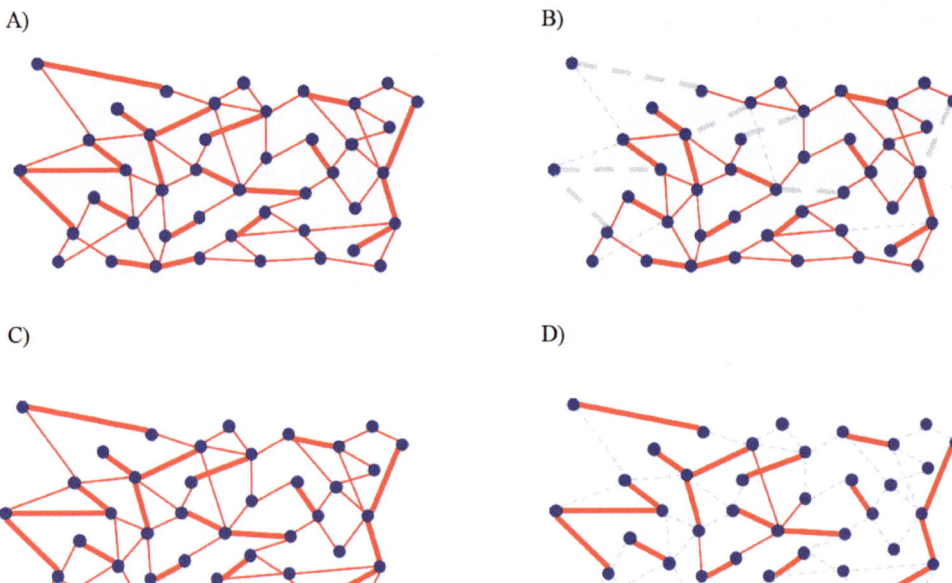

Figure 17.5   *A hypothetical contact network assembled using real-time data. Depending on the range and duration of contact required for the effective transmission of a pathogen, different 'transmission contact networks' A–D (from top to bottom) may arise from the same "physical contact network". In these figures the length of the edge corresponds to the actual distance between individuals and the edge thickness correspond to the minimum duration required for effective transmission. Red lines correspond to existing edges in each case and gray lines correspond to non-transmitting edges, as they do not satisfy the range/minimum duration required for effective infection transmission*

provided by contact network epidemiology can support these multi-scale risk calculations that result from the probabilistic nature of transmission.

Making assumptions about the maximum distance and minimum duration required for transmission permits creation of different risk assessment profiles for the impact of various intervention strategies. Figure 17.5 illustrates this concept. Panel (a) shows a hypothetical outcome of our contact network obtained from real-time data. The edges are drawn proportional to the 'actual distance' and the thickness of each edge is proportional to the 'duration of contact'. Panel (b) shows the situation where the cough/sneeze/exhale profile only permits short-range transmission of a disease that typically requires a short duration of contact (for instance, less than a minute). Although the 'physical contact network' remains the same as in panel (a) (both solid red and dashed gray edges), the 'transmission contact network' looks different (only solid red edges), because only edges satisfying the proximity/duration criteria have the potential to transmit the infection. The latter network is comprised of only those edges that are within the range required for effective transmission (since we assumed that transmission can occur in a short period of time, both thin and thick edges (short- and long-duration contacts) are accepted as long as individuals are spatially close to one another). Panel (c) has a disease with a longer range and similar duration of infection, and panel (d) has an agent with an even longer range that requires a longer time for potential transmission. In this case, only thick edges (long duration of contact) can effectively transmit the infection; thin edges (short duration of contact) are not sufficient for transmission even if individuals are spatially close to one another. Therefore, by making various assumptions about the proximity and duration of contact, which correspond to various modes of respiratory infection transmission, network models will provide invaluable risk assessment profiles for different modes of transmission.

17.5 ESTIMATING THE BASIC REPRODUCTION NUMBER R_0

The epidemic potential of disease is commonly estimated by using the basic reproduction number, R_0, the number of secondary infections arising from a single infection in a relatively naïve population. This quantity is linearly related to the transmissibility of the disease, that is, $R_0 = \gamma T$, where γ depends on the structure of the network. When T is at the epidemic threshold ($T = T_c$), then $R_0 = 1$. Public health interventions aim to reduce the number of new infected cases, ideally decreasing the *effective reproduction number* of the disease below the epidemic threshold, $R_{eff} < 1$.

The difference between the average transmissibility T and the basic reproduction ratio R_0 is important. While both have threshold values that distinguish epidemic from non-epidemic scenarios ($R_0 = 1$ and $T = T_c$), T is determined by the transmission characteristics of the pathogen and the nature of human interactions, but not affected by the numbers of contacts in a community, whereas R_0 is influenced by all of these factors, particularly on the numbers of interactions within the community. For example, consider a single airborne pathogen spreading through a hospital, where there are an abundance of close contacts, and through a rural community, where close contacts are rare. The transmission (T_{ij}) probability per contact may be similar in these settings because it is determined by the particular host-pathogen interactions specific to the disease agent in question, while the numbers of contacts are different. Therefore, the average transmissibility T will be similar in the two locations, while R_0 will be substantially higher in the hospital than in the rural setting. As such, one should be very cautious in generalizing the reported estimates of R_0 in the literature to populations and settings with different network structures.

17.6 ESTIMATING R_0 IN REAL-TIME

In order to determine the appropriate intervention, the primary objective at the onset of an emerging epidemic is to estimate its basic reproduction number, R_0. Generally, the key epidemiological characteristics of a disease are inferred from the analysis of case-notification data that are collected during the early phase of the outbreak, when surveillance systems can implement active case-finding by contact tracing. The real-time estimation of the reproduction number is then obtained by combining the information gathered in the early phase with the epidemic curve reported by surveillance systems, generally based on cases seeking care.

In recent years, several statistical methods have been developed in order to obtain a reliable estimate of the reproduction number at a given time period, $R(t)$, from a temporal sequence of incident case data (Wallinga and Teunis 2004; Forsberg White and Pagano 2008; Valleron et al. 2010). Some of these methods have been proposed as real-time estimators of the reproduction number, and have usually been tested on synthetic epidemic curves or historical outbreak data (Forsberg White and Pagano 2008; Fraser 2007; Cauchemez 2006a; Garske et al. 2007; Hall et al. 2006; White and Pagano 2008; Nishiura, Chowell et al. 2009; Davoudi et al. 2012). Among these methods are (1) the likelihood-based procedure originally proposed by Wallinga and Teunis (Wallinga and Teunis 2004) and modified as a real-time estimator by Cauchemez et al. (Cauchemez 2006b); (2) the branching process likelihood-based method proposed by White and Pagano (Forsberg White and Pagano 2008); and (3) the method based on a contact network model as proposed by Davoudi et al. (Davoudi et al. 2012). These methods generally provide estimates for the value of the *effective reproduction number* at any given time t, $R(t)$. This is different from the basic reproduction number R_0, since the population can only be fully susceptible at the beginning of the epidemic, if no pre-existing immunity is present.

All these methods need at least two data inputs for the real-time estimation of the reproduction number. The first input must be the epidemic curve, which represents the number of reported cases by date of symptom onset. The second input can be either the *generation interval distribution*, which is the distribution of time intervals between two successive infections (Fine 2003; Wallinga and Lipsitch 2007), or the *distribution of time to removal*, which is the probability that each infected individual is removed from the infectious population after a certain time, t (Davoudi et al. 2012). In this context, we refer to individuals as removed if they are removed from the population as a result of isolation, quarantine, vaccination, or death. Both quantities represent fundamental information needed to characterize the infectivity profile of infected cases.

The first estimation method, originally proposed by Wallinga and Teunis (Wallinga and Teunis 2004), is based on a statistical likelihood approach that infers the transmission tree from the observed dates of symptom onset, as provided by the epidemic curve. In order to estimate the reproduction number at time t, this method assumes knowledge of the full epidemic curve, to the end of the epidemic, which of course is not available in real time. For this reason, a number of modifications have been proposed to make it suitable for real-time estimations of the reproduction number, such as Bayesian modification proposed by Cauchemez et al. (Cauchemez 2006b) – the Bayesian transmission tree.

The second method, proposed by White and Pagano (Forsberg White and Pagano 2008), is based on a statistical Bayesian approach to the classic branching process estimator (Becker 1977) and defines a maximum-likelihood estimator for the reproduction number at time t.

Finally, a method was proposed more recently by Davoudi et al. (Davoudi et al. 2012) to produce an early estimation of the reproduction number of the pandemic 2009 H1N1 in North America (Pourbohloul et al. 2009). This method is significantly different from the above-mentioned approaches as it includes the structure of the contact network underlying the transmission process into the reproduction number estimation algorithm – called the contact network estimator. This method focuses not on the generation interval distribution, but rather on the time-to-removal distribution, whose expression can be more

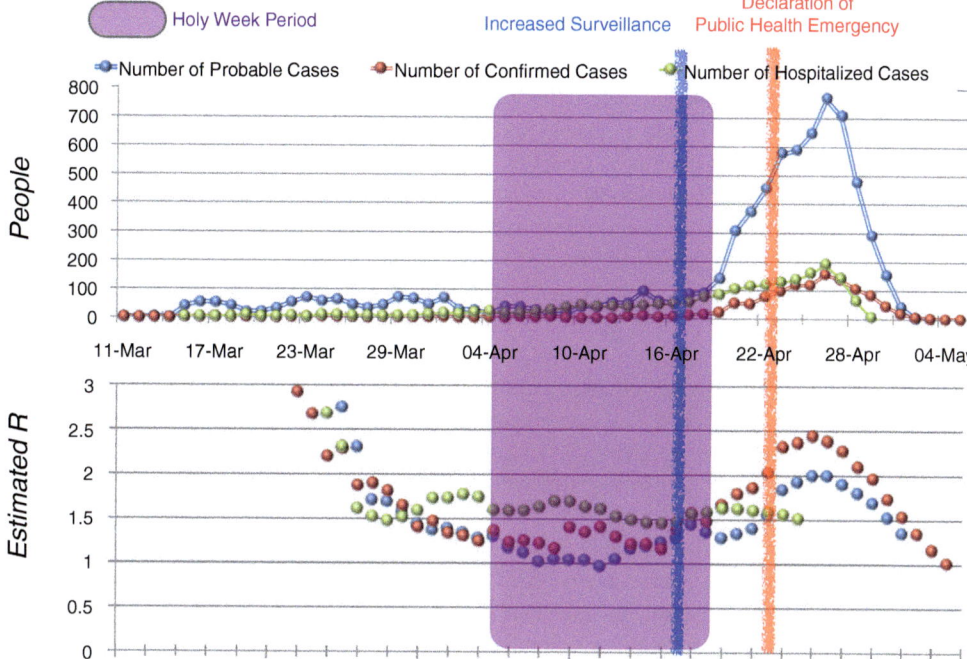

Figure 17.6   *pH1N1 in Mexico City, Mexico, 2009: Top panel: The three time series*
shown in this figure correspond to the daily incidence based on self-reported
onset of symptoms dates, also known as probable cases, provided by Direccion
General de Epidemiologia, Secretaria de Salud, Mexico (DGE) (purple
curve), the number of hospitalized cases due to acute pneumonia (green
curve) and the number of confirmed cases (red curve) between March 11 and
May 4, 2009, in Mexico City. While the first two time series were available
in real-time, the time series for confirmed cases became available only after
the first epidemic wave was over, due to delays in developing a diagnostic test
for a new pandemic strain of influenza, and in collecting laboratory results.
Bottom panel: estimates for the initial reproduction number R corresponding
to the three time series plotted above (depicted with the same color). While
the initial reproduction number corresponding to the number of hospitalized
cases remain more or less stable before and after the public health decree, the
R corresponding to the two other time series was influenced by social factors
depicted in the figure by vertical blue and red lines and the return of Mexico
City residents after 'Holy Week' holidays, who generally take this period as
vacation to travel elsewhere

easily related to the expected transmissibility. This approach enabled policy makers to
estimate the rate of influenza spread in Mexico, where the pandemic strain was originated,
remarkably early on – when relatively few cases overall were reported to the authorities
(Figure 17.6).

17.7   ASSESSING THE IMPACT OF INTERVENTION STRATEGIES

Mathematical models of the spread of the 2009–2010 pandemic influenza (pH1N1) virus across the globe played a prominent role in the assessment of the pH1N1 pandemic risk and in the evaluation and design of intervention and control strategies. During the early stages of the pH1N1 pandemic, mathematical analyses of the initial data from Mexico and other countries allowed researchers to estimate the transmissibility of the pH1N1 virus, as measured by its basic reproduction number (Fraser et al. 2009; Nishiura, Castillo-Chavez et al. 2009; Pourbohloul et al. 2009; Yang et al. 2009). As the pandemic progressed, many modeling studies investigated the impact of different kinds of containment strategies such as social distancing (Gojovic et al. 2009), vaccination (Yang et al. 2009; Chowell et al. 2009), and the use of antivirals (Dimitrov et al. 2011; Moghadas et al. 2009).

Vaccination is an important influenza control measure and was a key component of many countries' pandemic preparedness plans. Over the summer of 2009, much remained unknown about the effect of timing and prioritization of vaccination against pandemic (pH1N1) 2009 virus on health outcomes. Production of the pH1N1 vaccine began soon after the pandemic potential of pH1N1 was recognized during its Spring wave. However, the early arrival of the second wave of pH1N1 in many regions of the northern hemisphere, combined with production delays, resulted in the implementation of vaccination programs in populations already experiencing moderate to high incidence of pH1N1, a sequence of events expected to reduce the ultimate population impact of immunization. Quantifying this reduction and determining how it might have been mitigated through alternative dispensing schemes motivated a number of modeling efforts.

Seasonal influenza vaccination campaigns have historically targeted those at greatest risk of the severe outcomes of influenza– notably the very young, the elderly, any individual with underlying medical conditions – or their close contacts, as well as healthcare workers (Public Health Agency of Canada 2016). It has been hypothesized that vaccination of schoolchildren might be a more effective strategy (Halloran and Longini 2006; Longini 2012). Younger age groups are disproportionately responsible for influenza transmission, and therefore targeting them would indirectly protect at-risk groups (Bansal et al. 2006; Galvani et al. 2007; Kwong et al. 2008). Some regions – notably the province of Ontario in Canada – have adopted a universal influenza immunization program whereby seasonal influenza vaccine is provided free to all citizens over the age of six months (Kwong et al. 2008).

In the province of British Columbia, Canada, health officials adopted a provincial-level contact network model to study the potential impact of different pH1N1 vaccination campaign logistics on influenza morbidity and mortality. In the Fall of 2009 it became clear that misalignment between vaccine availability and the peak of the second wave of infection required prioritization of vaccine recipients. Greater infection risk in younger individuals supported targeted vaccination of younger age groups. Conversely, while older individuals were at decreased risk of infection with pH1N1, they were, according to British Columbia outcome surveillance data, experiencing higher rates of severe outcomes, including mortality. Additionally, per-laboratory confirmed-case hospitalization and fatality rates were greatest in older adults, with substantial increase beginning at age 50 (Louie et al. 2009; Vaillant et al. 2009; Tuite et al. 2010). The pH1N1 vaccine

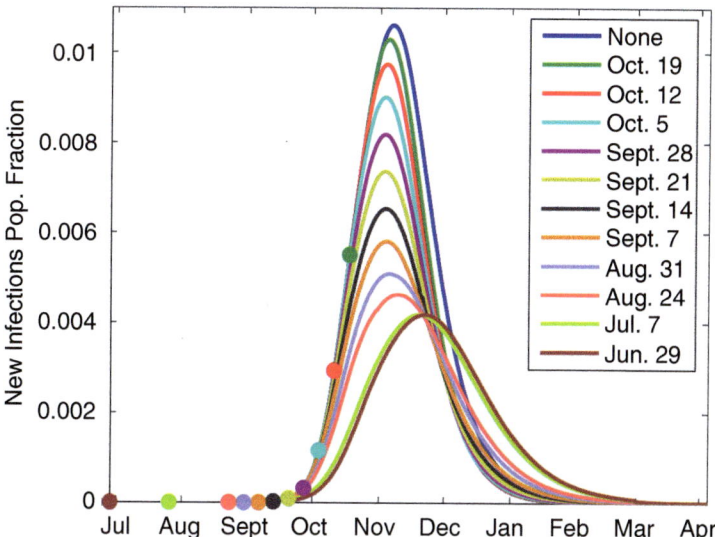

Figure 17.7   *Evaluating potential impact of vaccination strategies against the spread of*
a pandemic strain of influenza. Epidemiological characteristics similar to
the pH1N1 (2009) influenza pandemic were used to produce this outcome.
Epidemic spread was evaluated using a contact network for the Vancouver
population. The actual epidemic in British Columbia, Canada started during
the first week of September 2009. The circles show different start dates
for a vaccination campaign, each of which may change the profile of the
no-intervention epidemic curve (dark blue) to various degrees resulting in
new epidemic curves with the same color.

prioritization schemes adopted by many countries required a balance of these competing
considerations and ultimately differed from seasonal influenza recommendations as a
result of vaccine delay and unique pandemic patterns of age-related risk.

Figure 17.7 demonstrates how delay in vaccination program implementation during
pandemics may impact the epidemic curve. The epidemic in Vancouver, British Columbia
began during the first week of September 2009. The modeling results (Figure 17.7), which
is based on contact networks for the Vancouver population (comprising of ~2.4 million
nodes) suggest that if implemented prior to or at the onset of a pandemic, vaccination
programs could have significant impact on the progression of the epidemic. However,
this beneficial impact is rapidly and severely diminished when implementation is delayed.

Delays in vaccine production due to technological or logistical barriers may reduce
potential benefits of vaccination for pandemic influenza, and these temporal effects can
outweigh any additional theoretical benefits from population targeting. Careful modeling
may provide decision makers with estimates of these effects before the epidemic peak to
guide production goals and inform policy. Integration of real-time surveillance data with
mathematical models holds the promise of enabling public health planners to optimize the
community benefits from proposed interventions before the pandemic peak.

Results of modeling around pH1N1 provided important information to policy makers in Canada and many other countries, demonstrating the value of mathematical modeling as a risk assessment tool for potential interventions in responding to this global pandemic.

17.8   ANTIVIRAL OPTIMIZATION ON SIMPLE NETWORKS

As stated earlier, delays in production of a viable vaccine to offset the spread of an emerging pathogen are highly likely, if not inevitable. Therefore, alternative non-vaccine-based interventions, such as stockpiling and dispensing antiviral medications for influenza, represent an important second line of population defense. Recently-developed scenario optimization algorithms can rapidly and systematically analyze numerous policy options (Coquelin and Munos 2007; Azadivar 1999; Even-Dar et al. 2002). Using these algorithms with carefully developed models, policy makers can map disease intervention strategies onto policy trees that are coupled to specific stochastic epidemic models. For example, during the pH1N1 pandemic we used an optimization algorithm (Dimitrov et al. 2011), previously used with favorable convergence properties on large policy trees, to evaluate influenza response strategies (Coquelin and Munos 2007; Gelly and Wang 2006). The algorithm preferentially samples sub-trees of the policy tree that have performed well in the past. The algorithm performs best when all of the policies within a single sub-tree of the policy tree perform similarly; it can then effectively determine the 'goodness' of any sub-tree by sampling it only a few times. To achieve algorithmic efficiency, one should therefore use expert knowledge and intuition to structure the policy tree in this way. If there is a single optimal solution in a sub-tree surrounded by many poorly performing solutions, then the algorithm may require many simulations to find it (although it is guaranteed to eventually do so).

We performed this type of analysis in the summer of 2009, as the pH1N1 pandemic was unfolding, in response to questions posed to us by public health agencies regarding the effective use of antivirals prior to the availability of pH1N1 vaccines. The analysis suggested that pH1N1 might have been slowed with targeted, aggressive dispensing of antivirals. Yet, the impact of such a policy would have been highly uncertain in countries such as the United States or Canada where pandemic antiviral distribution policies are primarily decided at the federal level, but the stockpiles are distributed to local public health agencies without the ability to ensure distribution to targeted populations or to redirect supplies to meet shifting demand. In the United States in 2009, for example, 31 million doses were made available at the state level in the US as soon as the pandemic was declared (roughly a quarter of the total national antiviral stockpile), but little is known about the fate of those medications after states "signed off" on receipt of their pro rata shipment.

Our modeling found that simple strategies involving regular fixed releases perform as well as more complex optimized strategies. For example, for a pandemic strain with $R_0 \sim 2$, a monthly distribution of 10 million regimens divided proportionally among the US population consistently matches or outperforms other policy options, regardless of the uptake levels, potential misallocation or loss of medication – which are all likely in a complex emergency response setting. Slight variations on this policy, such as regular distributions of 5 or 25 million doses are predicted to perform significantly worse across a large range of uptake values. For more contagious pandemic strains (with higher reproduction numbers), use of an optimized distribution schedule with the possibility of redistribution would significantly improve the intervention outcome.

In the Canadian context, the model included a realistic network of the 141 largest Canadian cities (totaling 25.6 million people) and was based on detailed domestic air and ground travel data. Disease spread occurred between cities via the travel of infected individuals and within cities according to mass action models. Using transmission rates estimated for 2009 pH1N1 flu, the model predicted that, without any intervention, the outbreak would infect approximately 16.5 million people in Canada. We can use this model to evaluate the efficacy of various intervention strategies for ongoing and future epidemic and pandemic outbreaks of flu. For example, we used the model to evaluate various distribution policies for the national antiviral stockpile. It predicted that, under an optimal distribution schedule, aggressive use of antivirals might avert up to 75 percent of cases. However, under more realistic treatment rates, even the best policies will only moderately mitigate transmission. Figure 17.8 shows snapshots taken from one round of computer simulations to evaluate the impact of antiviral distribution with a 25 percent uptake on influenza spread across Canada.

From a public health perspective, the best policies are those that perform robustly in a variety of scenarios. However, a single policy may not be robust for all potential scenarios. These policies should be periodically re-evaluated to consider distribution wastage (which we conservatively estimated at up to 10 percent of distributed amounts) and antiviral resistance, which may develop in currently circulating strains of seasonal influenza (and are in fact commonly seen against the most commonly used anti-influenza antiviral, oseltamivir) (Hurt et al. 2009).

The effectiveness of any antiviral policy will depend critically on the extent to which antivirals reduce the severity and transmission of flu. Most studies in the literature (Lee and Chen 2007; Lee et al. 2012; McCaw and McVernon 2007; Lipsitch et al. 2007) assume maximum likelihood-based estimates of antiviral efficacies calculated by Longini et al. (Longini et al. 2004) using data from a clinical study by Welliver et al. (Welliver et al. 2001). More recent clinical trials indicate that the odds of a secondary infection in individual contacts decreases by approximately 50 percent when antivirals are used on the day of onset (OR: 0:5, 95 percent CI: 0:17, 1:46) (Ng et al. 2010; Goldstein et al. 2010). The results using either assumption will produce very different outcomes.

17.9   THE PROMISE OF MULTI-TYPE NETWORKS

Network modeling allows for rational design of infection control interventions and can play a crucial role in identifying optimal intervention strategies in healthcare settings. These strategies should target transmission-reducing interventions (to bring the transmissibility of infection, $T_{disease}$, below the threshold value, T_c) or contact-reducing interventions (changing the network structure to increase the threshold level, T_c), either of which serve to prevent more contagious infections from spreading (Meyers et al. 2006). Unlike most simpler network models referenced in the published literature, employing analytical and computational descriptions of 'multi-type network' models to accurately study transmission patterns within healthcare settings is a promising avenue (Allard et al. 2009; Allard et al. 2015). In this type of model, several major types of nodes can be

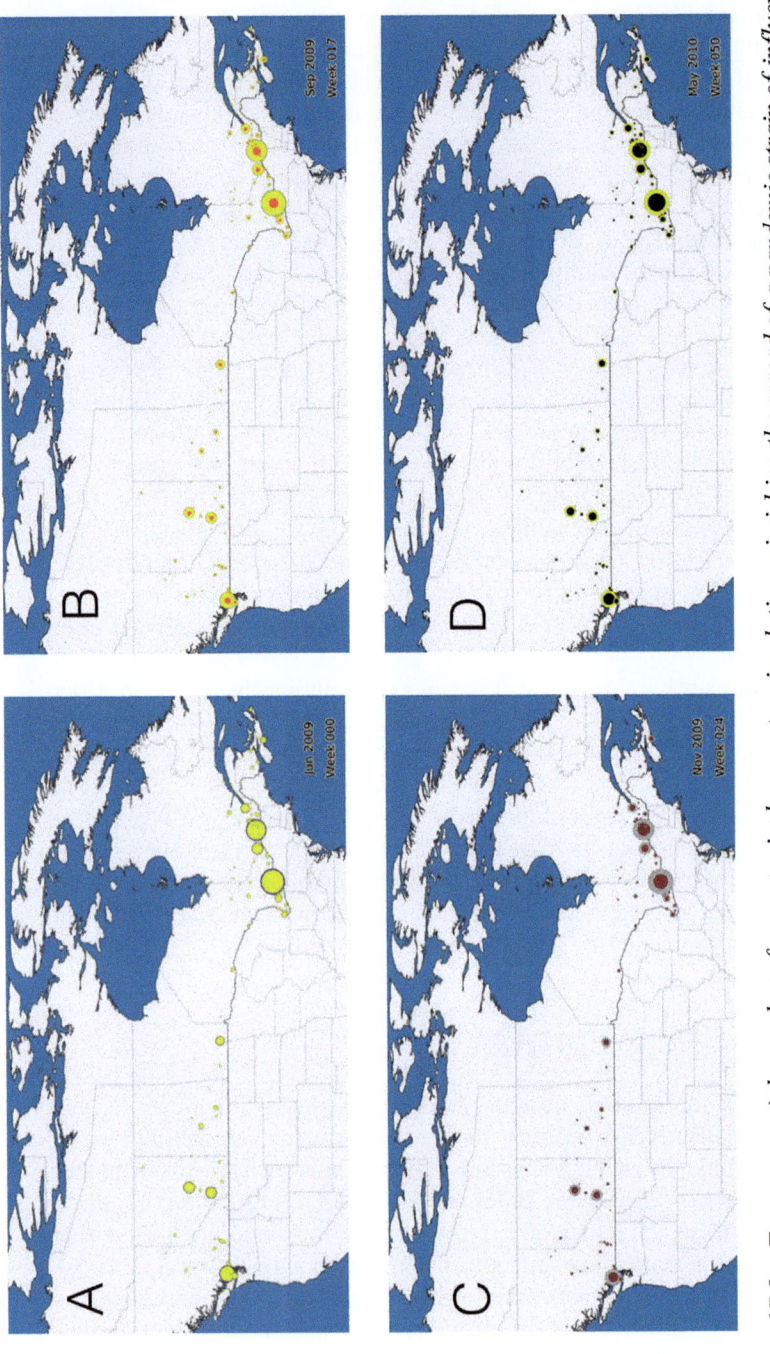

Figure 17.8 Four sequential snapshots from a typical computer simulation mimicking the spread of a pandemic strain of influenza across Canada. The areas of the circles representing cities are proportional to their population sizes. Inside each circle, a smaller circle represents the cumulative number of infections within each city. Around each circle, a blue ring represents the number of antiviral courses available in each city. Both the inner infection circle and the outer antiviral ring have areas proportional to the respective number of individuals (or courses). The color of the inner circle changes from red to black as the effective R in each city drops below 1. At this point, disease spread in that city has reached its peak and the disease can be expected to die out shortly. The color of each city changes from yellow to gray as the city's local cache of antivirals is depleted, i.e., the fraction of antiviral courses over city population drops below 0.1 percent. This particular simulation run models an intervention policy at a plausible uptake value of 25 percent.

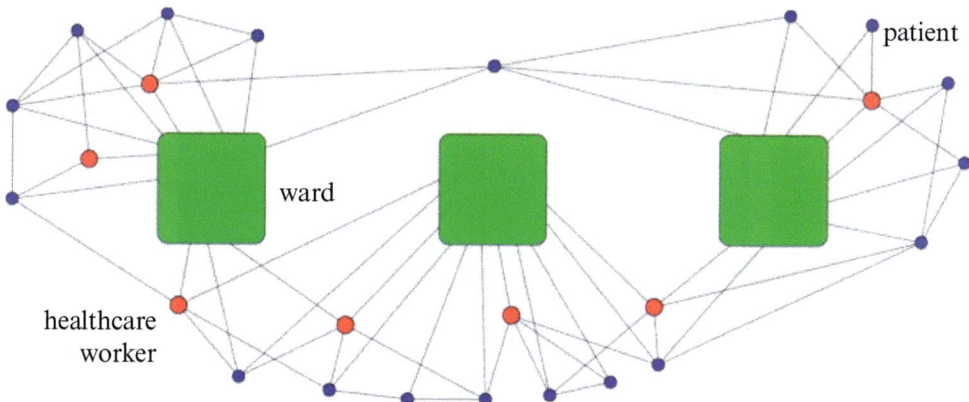

Figure 17.9   *Multi-type networks. In this 3-type network, three distinctive sets of nodes interact within and between each other. Green squares, red circles and small blue circles represent hospital wards/units, caregivers and patients, respectively. Since the nodes (and edges connecting these nodes) can be studied separately, intervention strategies targeting each category can be tested and compared with one another in more detail*

distinguished within the network; for instance, the nodes that represent patients, health-care workers (HCWs) and space units (which may represent wards or smaller units within a ward) are all distinct subgroups (see Figure 17.9). In addition to modifying key structural aspects of the network (for example, disconnecting large components), complex intervention strategies, such as the use of personal protective equipment (PPE) among healthcare workers combined with antiviral prophylaxis in work and home settings can also be modeled.

The advantage of this approach is that the role and impact of contacts within and between patients and caregivers on the spread of disease can be analyzed, separately or in combination. This, in turn, permits evaluation of various intervention strategies designed for specific subgroup(s), separately or in combination, within the same quantitative framework. It may also be extended to capture indirect transmission for evaluating situations where the pathogen can survive on surfaces (beds, door knobs, and so on) or may have been aerosolized in poorly ventilated areas.

In order to make logistically-realistic advance preparations to respond to emergent diseases, public health agencies need a fundamental understanding of how new pathogens may move along the existing contact networks in their jurisdictions. Mapping these contact networks in the pre-outbreak period is akin to understanding coastal evacuation routes before a storm: they are a necessary but not sufficient part of the analytical toolbox that response agencies will use to understand why events are playing out the way they are, and what can be done to alter them. Depending on the effort invested in investigating these person-to-person interactions in the community and especially in healthcare settings, models will yield varying levels of understanding of how disease spreads. Therefore, one important preparatory step in outbreak response is to assess the quality of available contact network mapping prior to the next outbreak.

To illustrate this, recall that mathematical models of infectious disease depend critically on estimates of the risk of transmission associated with a sick contact. Previous models, primarily designed for large populations, have used rather crude definitions of a 'contact' (for example, two people have been presumed to be in contact if they work together in the same environment), which requires only generic empirical data from workplace studies. However, models that investigate the impact of specific behavioral changes require either controlled studies (which would be impractical for an emergent epidemic) or a pre-existing detailed understanding of the fundamental physical processes by which infectious agents travel from an infected individual to a susceptible individual. The challenge is to identify the types of contacts that pose a risk and determine how to quantify that risk.

17.10   CASE STUDY: PANDEMIC INFLUENZA AS A HOSPITAL– ACQUIRED INFECTION

The need to diminish infection transmission within a hospital at the onset of an influenza pandemic is dramatically illustrated by Blumenfeld et al. They described the onset of the 1957 pandemic in a New York hospital ward with a total of 65 patients and staff (Blumenfeld et al. 1959; Louria et al. 1959). One patient with influenza who was admitted to that ward initiated an outbreak that over the next 12 days caused 30 symptomatic cases and 8 asymptomatic infections (documented by seroconversion); of the remaining 26 uninfected persons, 12 had been vaccinated previously and may not have been susceptible.

The expected burden of pandemic influenza will be added to the routine burden of nosocomial, or hospital-acquired infections (HAI). In Canada, more than 200,000 patients acquire a nosocomial infection annually, and more than 8,000 die as a result (Zoutman et al. 2003). Estimates in the United Kingdom indicate at least 100,000 cases of HAIs in England, causing approximately 5,000 deaths and costing England's National Health Service (NHS) as much as £1 billion a year (Anon 2000). Approximately 4.1 million patients are estimated to acquire a HAI in the European Union each year. The number of deaths occurring as a direct consequence of these infections is estimated to be at least 37,000 and these infections are thought to contribute to an additional 110,000 deaths each year (European Centre for Disease Prevention and Control (ECDC 2008). In the United States, HAIs are estimated to affect two million people every year and are associated with 100,000 deaths, adding $30.5 billion to the nation's healthcare tab (Scott 2009; Centers for Disease Control and Prevention 2016; Magill et al. 2014).

Public health agencies, acute care hospitals, and other healthcare settings have drafted pandemic plans to prepare for and manage influenza or other respiratory pandemic outbreaks. In institutional healthcare settings, these plans may include modifying emergency room patient flow, establishing dedicated wards for respiratory illnesses, and/or using alternative treatment sites to discourage patients from seeking care in emergency rooms. These new care patterns are likely to come with new and unpredictable webs of interpersonal contacts, which may in turn impact nosocomial transmission risk. To ensure that pandemic plans do not inadvertently worsen outbreaks within healthcare settings, planners should define methods to quickly measure and analyze these new interpersonal contact networks in real time. Such emergent contact information can then be incorporated into previously mapped network design to further understand and limit infectious disease transmission.

17.11   MAPPING HOSPITAL CONTACT NETWORKS IN PRACTICE

Almost a decade ago, we attempted to create such site-specific contact network maps at three urban university-affiliated tertiary care Canadian hospitals, using surveys, architectural maps, and floor plans. Anyone working (or volunteering) in these hospitals was eligible to participate. We used anonymous paper-based surveys attached to employee paystubs to inquire about interpersonal contacts during the workday.

The surveys collected demographics, spatial movement, and patient interaction (contact) data. We defined direct contact as two or more individuals coming within one meter (3 feet) of each other for two minutes or more. This proximity and duration are deemed necessary but not sufficient for respiratory infection transmission (Xie et al. 2007). An indirect contact was defined as two or more individuals co-locating in the same room but not closer than 1 meter.

For each location (patient rooms, administrative area, coffee break room, and so on) within the facilities surveyed, we plotted the number of visits per week and mean hours spent per location, which differed significantly by location type. This survey provided one of the most comprehensive portrayal of hospital-wide contact networks yet reported, providing an evidence base to develop novel strategies for the prevention and control of nosocomially-transmitted pathogens. Figure 17.10 summarizes a typical contact network resulted from this study, where different occupations are depicted by different colors. The figure captures the heterogeneity of contacts between individuals within a healthcare setting. It is worth noting that data from participating hospitals yielded different structures, supporting the view that hospital infection control strategies in different settings might lead to different outcomes, depending on the structure of their contact network.

Despite this achievement, the technology to overcome one of this study's main limitations – that is, the ability to track a target individual's actual location in relation to their contact – has only recently been applied within a healthcare setting. With the emergence of Radio Frequency Identification (RFID) technology and its variants, including Ultra Wide Band (UWB) and infrared (IR) technology, it is now feasible to know the location, within a few decimeters, of every person at all times while inside a given facility (Salathé et al. 2010). The advantage of electronic tracking technology is that we are able to identify exactly when the participant is in a particular location and how many contacts they have. For example, the paper survey would have indicated a participant spent five hours in the cafeteria during a typical work week; however, there is no way to determine which five hours were spent in the cafeteria or who else this person came in contact with while there. Electronic tags may demonstrate that a volunteer participant, for example, arrived in the cafeteria at noon on Monday, spent 45 minutes within two meters (acceptable range for influenza transmission) of two other participants and stopped at the hospital gift shop for three minutes before all three individuals return to their respective departments and had further contacts with x patients, y physicians and nurses and z other healthcare workers.

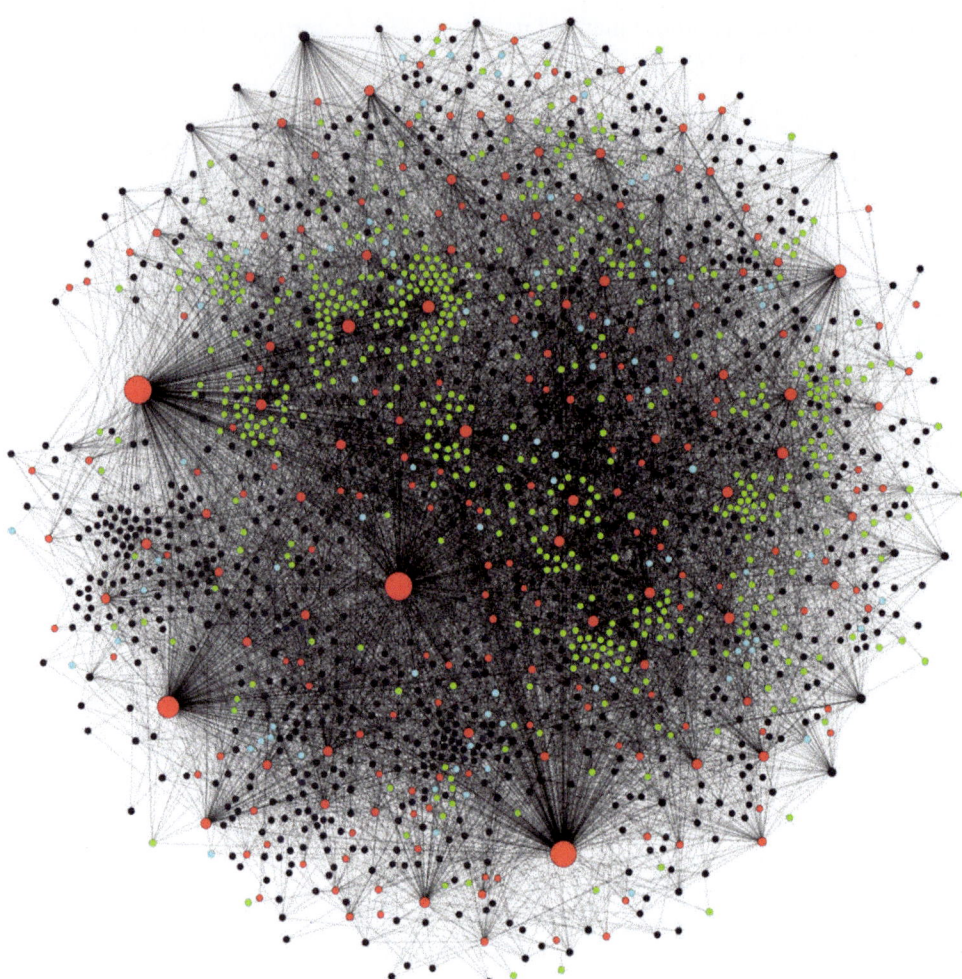

Figure 17.10   *A typical contact network generated from the questionnaire-based survey
data described in the text. Node colors: turquoise: staff physician; green:
nurse, black: other HCWs; red: location. Several networks, whose degree
distributions are identical, may be generated using the same dataset. As
such, all analyses should be performed on an ensemble of such networks*

Gathering data about the proximity of contacts and how healthcare workers interact with
one another, as well as patients, allows us to develop effective containment strategies at the
sub-ward level. More importantly, these datasets will allow us to incorporate representations of different modes of transmission – droplets, aerosolized particles, and surfaces –
or a combination of these mechanisms into future models and to estimate the potential
outcomes of spread via routine contact or so-called 'super-spreaders.'

17.12   CONCLUDING REMARKS: COMPLEXITY, MODELING, AND POLICY

Policy inference from realistically detailed contact models is an active ongoing area of research and implementation science that requires close collaboration between mathematical modelers, healthcare providers, public health officials, and related occupations. Future work in the study of nosocomial infections may help to elucidate which individual(s) or group(s) of healthcare workers, for example, trainees, senior clinicians, and so on, provide the greatest opportunity for infection transmission within the hospital. Strategies for prioritizing self-protection for these 'nexus individuals' (for example, wearing surgical masks during their working day) may prove efficient means of reducing disease spread in these environments.

As described in the introduction, the network theory approach to analyzing disease spread is advantageous in allowing one to model and evaluate a broad range of intervention strategies without sacrificing detailed information about interpersonal contacts. By incorporating the heterogeneous behavior of individuals as well as the heterogeneity in disease transmissibility, complexity science tools will allow us to accurately pinpoint optimal and economical intervention strategies, thus contributing to evidence-based public health decision-making and to the well-being of individuals at the societal level.

Despite all of the challenges that face researchers and policy makers dedicated to the difficult task of designing pandemic and other emergency preparedness plans, there is one reassuring fact – the sooner control measures are implemented appropriately, the lower the rate of infection in the population (Markel et al. 2007; World Health Organization Writing Group 2006; World Health Organization Writing Group 2012; Ferguson et al. 2006). This implies that the development of advanced, rapidly implementable quantitative methods to understand the pattern of disease spread – and systems to quickly analyze and present that information to policy makers – is a high priority for overall public health preparedness. Developing the means to identify detailed interpersonal contact information in various settings, both non-emergent and during emergency events, should now be considered a core public health function, along with such classic tasks as disease surveillance, community partnerships, and environmental health services.

ACKNOWLEDGEMENTS

BP and KE wish to acknowledge the Canadian Institutes of Health Research (CIHR) and Michael Smith Foundation for Health Research (MSFHR), both of whom have provided generous support through grants which facilitated the research and collaborations culminating in this book chapter. NH is supported by the Cornell Institute for Disease and Disaster Preparedness, which is jointly funded by New York Presbyterian Hospital and Weill Cornell Medicine.

REFERENCES

Albert, R. and Barabási, A.-L. 2002. Statistical mechanics of complex networks. *Reviews of Modern Physics*, 74(1), 47–97.

Allard, A. et al. 2009. Heterogeneous bond percolation on multitype networks with an application to epidemic dynamics. *Physical Review E*, 79(3), 036113–9.

Allard, A. et al. 2015. General and exact approach to percolation on random graphs. *Physical Review E*, 92(6), 062807–15.

Anon, 2000. *UK National Audit Office Report: The Management and Control of Hospital Acquired Infection in Acute NHS Trusts in England (HC 230 1999/00) – Full report [1,016 Kb]*.

Atkinson, M.P. and Wein, L.M. 2008. Quantifying the routes of transmission for pandemic influenza. *Bulletin of Mathematical Biology*, 70(3), 820–867.

Azadivar, F. 1999. *Simulation Optimization Methodologies*, New York: ACM.

Bailey, N.T.J. 1957. *The Mathematical Theory of Epidemics*, London: Hafner.

Ball, F., Mollison, D. and Scalia-Tomba, G. 1997. Epidemics with two levels of mixing. *The Annals of Applied Probability*, 7(1), 46–89.

Bansal, S., Pourbohloul, B. and Meyers, L.A. 2006. A comparative analysis of influenza vaccination programs. *PLoS Medicine*, 3(10), e387.

Becker, N. 1976. Estimation for an epidemic model. *Biometrics*, 32(4), 769.

Becker, N. 1977. Estimation for discrete time branching processes with application to epidemics. *Biometrics*, 33(3), 515–522.

Blumenfeld, H.L. et al. 1959. Studies on influenza in the pandemic of 1957–1958. I. An epidemiologic, clinical and serologic investigation of an intrahospital epidemic, with a note on vaccination Efficacy. *Journal of Clinical Investigation*, 38(1 Pt 1–2), 199–212.

Boguñá, M., Pastor-Satorras, R. and Vespignani, A. 2003. Absence of epidemic threshold in scale-free networks with degree correlations. *Physical Review Letters*, 90(2), 028701–4.

Brankston, G. et al. 2007. Transmission of influenza A in human beings. *The Lancet Infectious Diseases*, 7(4), 257–265.

Bridges, C.B., Kuehnert, M.J. and Hall, C.B. 2003. Transmission of influenza: Implications for control in health care settings. *Clinical Infectious Diseases*, 37(8), 1094–1101.

Britton, T. and O'Neill, P.D. 2002. Bayesian inference for stochastic epidemics in populations with random social structure. *Scandinavian Journal of Statistics*, 29(3), 375–390.

Cauchemez, S. 2006a. Estimating in real time the efficacy of measures to control emerging communicable diseases. *American Journal of Epidemiology*, 164(6), 591–597.

Cauchemez, S. 2006b. Estimating in real time the efficacy of measures to control emerging communicable diseases. *American Journal of Epidemiology*, 164(6), 591–597.

Centers for Disease Control and Prevention, 2016. National and state healthcare associated infections. Progress Report 2016, pp. 1–147.

Chowell, G. et al. 2009. Adaptive vaccination strategies to mitigate pandemic influenza: Mexico as a case study, H.E. Gendelman (ed.), *PLOS ONE*, 4(12), e8164–e8169.

Chowell, G., Fenimore, P.W. et al. 2003. SARS outbreaks in Ontario, Hong Kong and Singapore: The role of diagnosis and isolation as a control mechanism. *Journal of Theoretical Biology*, 224(1), 1–8.

Chowell, G., Hyman, J.M. et al. 2003. Scaling laws for the movement of people between locations in a large city. *Physical Review E*, 68, 066102.

Coquelin, P.-A. and Munos, R.E. 2007. Bandit algorithms for tree search. *Uncertainty in Artificial Intelligence*, pp.1–9.

Davoudi, B. et al. 2012. Early real-time estimation of the basic reproduction number of emerging infectious diseases. *Physical Review X*, 2(3), 031005–15.

Department of Health, Commonwealth of Australia, 2014. Australian health management plan for pandemic influenza, pp.1–232.

Department of Health, England and Health Departments of the Devolved Administrations of Scotland, Wales and Northern Ireland, 2012. UK pandemic influenza communications strategy 2012, pp.1–26.

Diekmann, O., de Jong, M.C.M. and Metz, J.A.J. 1998. A deterministic epidemic model taking account of repeated contacts between the same individuals. *Journal of Applied Probability*, 35(2), 448–462.

Dimitrov, N.B. et al. 2011. Optimizing tactics for use of the US antiviral strategic national stockpile for pandemic influenza, B.J. Cowling (ed.), *PLOS ONE*, 6(1), e16094–10.

Dorogovtsev, S.N. and Mendes, J.F.F. 2003. *Evolution of Networks: From Biological Nets to the Internet and WWW (Physics)*, New York: Oxford University Press.

Durrett, R. 1999. Stochastic spatial models. *SIAM Review*, 41(4), 677–718.

Erdos, P. and Rényi, A. 1961. On the evolution of random graphs. *Bull. Inst. Internat. Statist*, 38(4), 343–347.

Eubank, S. et al. 2010. Detail in network models of epidemiology: are we there yet? *Journal of Biological Dynamics*, 4(5), 446–455.

European Centre for Disease Prevention and Control (ECDC), 2008. *Annual epidemiological report on communicable diseases in Europe 2008*.

Even-Dar, E., Mannor, S. and Mansour, Y. 2002. PAC bounds for multi-armed bandit and markov decision processes. In *Computational Learning Theory*. Lecture Notes in Computer Science. Berlin, Heidelberg: Springer Berlin Heidelberg, pp. 255–270.

Farrington, C.P., Kanaan, M.N. and Gay, N.J. 2003. Branching process models for surveillance of infectious diseases controlled by mass vaccination. *Biostatistics*, 4(2), 279–295.

Ferguson, N.M. and Garnett, G.P. 2000. More realistic models of sexually transmitted disease transmission dynamics: Sexual partnership networks, pair models, and moment closure. *Sexually Transmitted Diseases*, 27(10), 600–609.

Ferguson, N.M. et al. 2006. Strategies for mitigating an influenza pandemic. *Nature*, 442(7101), 448–452.

Fine, P.E.M. 2003. The interval between successive cases of an infectious disease. *American Journal of Epidemiology*, 158(11), 1039–1047.

Forsberg White, L. and Pagano, M. 2008. A likelihood-based method for real-time estimation of the serial interval and reproductive number of an epidemic. *Statistics in Medicine*, 27(16), 2999–3016.

Fraser, C. 2007. Estimating individual and household reproduction numbers in an emerging epidemic. *PLOS ONE*, 2(8), e758.

Fraser, C. et al. 2009. Pandemic potential of a strain of influenza A (H1N1): Early findings. *Science*, 324(5934), 1557–1561.

Galvani, A.P., Reluga, T.C. and Chapman, G.B. 2007. Long-standing influenza vaccination policy is in accord with individual self-interest but not with the utilitarian optimum. *Proceedings of the National Academy of Sciences*, 104(13), 5692–5697.

Garske, T., Clarke, P. and Ghani, A.C. 2007. The transmissibility of highly pathogenic avian influenza in commercial poultry in industrialised countries, E. Scalas (ed.), *PLOS ONE*, 2(4), e349–7.

Gelly, S. and Wang, Y. 2006. Exploration exploitation in Go: UCT for Monte-Carlo. In the Annual Conference on Neural Information Processing Systems. Last accessed on September 10, 2017 at https://hal.inria.fr/hal-00115330/document.

Germann, T.C. et al. 2006. Mitigation strategies for pandemic influenza in the United States. *Proceedings of the National Academy of Sciences*, 103(15), 5935–5940.

Gojovic, M.Z. et al. 2009. Modelling mitigation strategies for pandemic (H1N1) 2009. *Canadian Medical Association Journal*, 181(10), 673–680.

Goldstein, E. et al. 2010. Oseltamivir for treatment and prevention of pandemic influenza A/H1N1 virus infection in households, Milwaukee, 2009. *BMC Infectious Diseases*, 10(1), 211–217.

González, M.C., Hidalgo, C.A. and Barabási, A.-L. 2008. Understanding individual human mobility patterns. *Nature*, 453(7196), 779–782.

Groendyke, C., Welch, D. and Hunter, D.R. 2012. A network-based analysis of the 1861 Hagelloch Measles data. *Biometrics*, 68(3), 755–765.

Hall, I.M. et al. 2006. Real-time epidemic forecasting for pandemic influenza. *Epidemiology and Infection*, 135(3), 372–385.

Halloran, M.E. and Longini, I.M. 2006. Public health: Community studies for vaccinating schoolchildren against influenza. *Science*, 311(5761), 615–616.

Hui, D.S. et al. 2015. Airflows around oxygen masks. *CHEST*, 130(3), 822–826.

Hurt, A.C. et al. 2009. Emergence and spread of oseltamivir-resistant A(H1N1) influenza viruses in Oceania, South East Asia and South Africa. *Antiviral Research*, 83(1), 90–93.

Ip, M. et al. 2007. Airflow and droplet spreading around oxygen masks: A simulation model for infection control research. *American Journal of Infection Control*, 35(10), 684–689.

Keeling, M.J. et al. 2003. Modelling vaccination strategies against foot-and-mouth disease. *Nature*, 421(6919), 136–142.

Keeling, M.J., Rand, D.A. and Morris, A.J. 1997. Correlation models for childhood epidemics. *Proceedings of the Royal Society B: Biological Sciences*, 264(1385), 1149–1156.

Kleczkowski, A. and Grenfell, B.T. 1999. Mean-field-type equations for spread of epidemics: The 'small world' model. *Physica A: Statistical Mechanics and its Applications*, 274(1–2), 355–360.

Kwong, J.C. et al. 2008. The effect of universal influenza immunization on mortality and health care use. *PLoS Medicine*, 5(10), e211.

Lee, R.V. 2007. Comments on: Transmission of influenza A in human beings. *The Lancet Infectious Diseases*, 7(12), 760–761– author reply 761–763.

Lee, V.J. and Chen, M.I. 2007. Effectiveness of neuraminidase inhibitors for preventing staff absenteeism during pandemic influenza. *Emerging Infectious Diseases*, 13(3), 449–457.

Lee, V.J. et al. 2012. Economics of neuraminidase inhibitor stockpiling for pandemic influenza, Singapore. *Emerging Infectious Diseases*, 12(1), 95–102.

Lefevre, C. and Picard, P. 1989. On the formulation of discrete-time epidemic models. *Mathematical Biosciences*, 95(1), 27–35.

Li, Y. et al. 2007. Role of ventilation in airborne transmission of infectious agents in the built environment: A multidisciplinary systematic review. *Indoor Air*, 17(1), 2–18.

Lipsitch, M. et al. 2007. Antiviral resistance and the control of pandemic influenza. *PLoS Medicine*, 4(1), e15.

Lloyd, A.L. and May, R.M. 2001. Epidemiology: How viruses spread among computers and people. *Science*, 292(5520), 1316–1317.

Longini, I.M. 1988. A mathematical model for predicting the geographic spread of new infectious agents. *Mathematical Biosciences*, 90(1–2), 367–383.

Longini, Jnr. Ira M. et al. 2004. Containing pandemic influenza with antiviral agents. *American Journal of Epidemiology*, 159(7), 623–633.

Longini, I.M. 2012. A theoretic framework to consider the effect of immunizing schoolchildren against influenza: Implications for research. *Pediatrics*, 129(Supplement), S63–S67.

Louie, J.K. et al. 2009. Factors associated with death or hospitalization due to pandemic 2009 influenza A(H1N1) infection in California. *JAMA*, 302(17), 1896–1902.

Louria, D.B. et al. 1959. Studies on influenza in the pandemic of 1957–1958. II. Pulmonary complications of influenza. *Journal of Clinical Investigation*, 38(1 Pt 1–2), 213–265.

Magill, S.S. et al. 2014. Multistate point-prevalence survey of health care – associated infections. *New England Journal of Medicine*, 370(13), 1198–1208.

Markel, H. et al. 2007. Nonpharmaceutical interventions implemented during the 1918–1919 influenza pandemic. *JAMA*, 298(19), 2260.

McCaw, J.M. and McVernon, J. 2007. Prophylaxis or treatment? Optimal use of an antiviral stockpile during an influenza pandemic. *Mathematical Biosciences*, 209(2), 336–360.

Meyers, L.A. et al. 2005. Network theory and SARS: Predicting outbreak diversity. *Journal of Theoretical Biology*, 232(1), 71–81.

Meyers, L.A., Newman, M.E.J. and Pourbohloul, B. 2006. Predicting epidemics on directed contact networks. *Journal of Theoretical Biology*, 240(3), 400–418.

Moghadas, S.M. et al. 2009. Post-exposure prophylaxis during pandemic outbreaks. *BMC Medicine*, 7(1), 73–10.

Mollison, D. 1995. *Epidemic Models: Their Structure and Relation to Data*, Cambridge: Cambridge University Press.

Mossong, J. et al. 2008. Social contacts and mixing patterns relevant to the spread of infectious diseases, S. Riley (ed.), *PLoS Medicine*, 5(3), e74.

Newman, M.E.J. 2002. Spread of epidemic disease on networks. *Physical Review E*, 66(1), 016128–11.

Newman, M.E.J. 2003. The structure and function of complex networks. *SIAM Review*, 45(2), 167–256.

Newman, M.E.J., Strogatz, S.H. and Watts, D.J. 2001. Random graphs with arbitrary degree distributions and their applications. *Physical Review E*, 64(2), 026118–17.

Ng, S. et al. 2010. Effects of Oseltamivir treatment on duration of clinical illness and viral shedding and household transmission of influenza virus. *Clinical Infectious Diseases*, 50(5), 707–714.

Nishiura, H., Castillo-Chavez, C. et al. 2009. Transmission potential of the new influenza A (H1N1) virus and its age-specificity in Japan. *Euro Surveillance: bulletin Europeen sur les maladies transmissibles = European Communicable Disease Bulletin*, 14(22).

Nishiura, H., Chowell, G. et al. 2009. The ideal reporting interval for an epidemic to objectively interpret the epidemiological time course. *Journal of The Royal Society Interface*, 7(43), 297–307.

Pan-Canadian Public Health Network 2015. Canadian pandemic influenza preparedness: Planning guidance for the health sector, pp. 1–60.

Pastor-Satorras, R. and Vespignani, A. 2001. Epidemic spreading in scale-free networks. *Physical Review Letters*, 86, 3200–3203.

Pastor-Satorras, R. and Vespignani, A. 2007. *Evolution and Structure of the Internet: A Statistical Physics Approach*, Cambridge: Cambridge University Press.

Plan national de prévention et de lutte (Pandémie grippale) document d'aide à la préparation et à la décision 2011, pp.1–78. Last accessed on January 25, 2017 at http://social-sante.gouv.fr/IMG/pdf/Plan_Pandemie_Grippale_2011.pdf.

Pourbohloul, B. and Brunham, R.C. 2004. Network models and transmission of sexually transmitted diseases. *Sexually Transmitted Diseases*, 31(6), 388–390.

Pourbohloul, B. et al. 2005. Modeling control strategies of respiratory pathogens. *Emerging Infectious Diseases*, 11(8), 1249–1256.

Pourbohloul, B. et al. 2009. Initial human transmission dynamics of the pandemic (H1N1) 2009 virus in North America. *Influenza and Other Respiratory Viruses*, 3(5), 215–222.

Public Health Agency of Canada 2016. An Advisory Committee Statement (ACS), National Advisory Committee on Immunization (NACI), pp. 1–61.

Riley, R.L. 1974. Airborne infection. *The American Journal of Medicine*, 57(3), 466–475.

Salathé, M. et al. 2010. A high-resolution human contact network for infectious disease transmission. *Proceedings of the National Academy of Sciences*, 107(51), 22020–22025.

Salgado, C.D. et al. 2002. Influenza in the acute hospital setting. *The Lancet Infectious Diseases*, 2(3), 145–155.

Sander, L.M. et al. 2002. Percolation on heterogeneous networks as a model for epidemics. *Mathematical Biosciences*, 180(1–2), 293–305.

Sattenspiel, L. and Simon, C.P. 1988. The spread and persistence of infectious diseases in structured populations. *Mathematical Biosciences*, 90(1–2), 341–366.

Scott, D. 2009. *The Direct Medical Costs of Healthcare-Associated Infections in US Hospitals and the Benefits of Prevention*, Division of Healthcare Quality Promotion, National Center for Preparedness, Detection, and Control of Infectious Diseases, Coordinating Center for Infectious Diseases, Centers for Disease Control and Prevention.

Song, C. et al. 2010. Limits of predictability in human mobility. *Science*, 327(5968), 1018–1021.

Tang, J.W. and Li, Y. 2007. Comments on: Transmission of influenza A in human beings. *The Lancet Infectious Diseases*, 7(12), 758, author reply 761–763.

Tang, J.W. et al. 2006. Factors involved in the aerosol transmission of infection and control of ventilation in healthcare premises. *The Journal of Hospital Infection*, 64(2), 100–114.

Tang, J.W. et al. 2009. A Schlieren optical study of the human cough with and without wearing masks for aerosol infection control. *Journal of The Royal Society Interface*, 6(Suppl. 6), S727–S736.

Tang, J.W. et al. 2013. Airflow dynamics of human jets: Sneezing and breathing – potential sources of infectious aerosols, E. Subbiah, ed. *PLOS ONE*, 8(4), e59970–7.

Tellier, R. 2006. Review of aerosol transmission of influenza A virus. *Emerging Infectious Diseases*, 12(11), 1657–1662.

Tellier, R. 2007. Comments on: Transmission of influenza A in human beings. *The Lancet Infectious Diseases*, 7(12), 759–760, author reply 761–763.

Tuite, A.R. et al. 2010. Estimated epidemiologic parameters and morbidity associated with pandemic H1N1 influenza. *Canadian Medical Association Journal*, 182(2), 131–136.

US Department of Health and Human Services 2012. 2009 H1N1 influenza improvement plan, pp.1–51.

Vaillant, L. et al. 2009. Epidemiology of fatal cases associated with pandemic H1N1 influenza 2009. *Euro Surveillance: bulletin Europeen sur les maladies transmissibles = European Communicable Disease Bulletin*, 14(33).

Valleron, A.-J. et al. 2010. Transmissibility and geographic spread of the 1889 influenza pandemic. *Proceedings of the National Academy of Sciences*, 107(19), 8778–8781.

Van der Ploeg, C.P.B. et al. 1998. STDSIM: A microsimulation model for decision support in STD control. *Interfaces*, 28(3), 84–100.

Wallinga, J. and Teunis, P. 2004. Different epidemic curves for severe acute respiratory syndrome reveal similar impacts of control measures. *American Journal of Epidemiology*, 160(6), 509–516.

Wallinga, J. and Lipsitch, M. 2007. How generation intervals shape the relationship between growth rates and reproductive numbers. *Proceedings of the Royal Society B: Biological Sciences*, 274(1609), 599–604.

Watts, D.J. 1999. *Small Worlds: The Dynamics of Networks Between Order and Randomness*, Princeton, NJ: Princeton University Press.

Watts, D.J. and Strogatz, S.H. 1998. Collective dynamics of 'small-world' networks. *Nature*, 393(6684), 440–442.

Welliver, R. et al. 2001. Effectiveness of Oseltamivir in preventing influenza in household contacts: A randomized controlled trial. *JAMA*, 285(6), 748–754.

White, L.F. and Pagano, M. 2008. Transmissibility of the influenza virus in the 1918 pandemic. *PLOS ONE*, 3(1), e1498.

World Health Organization, 2005. *Avian influenza: assessing the pandemic threat*. Last accessed on September 10, 2017 at http://www.who.int/influenza/resources/documents/h5n1_assessing_pandemic_threat/en/. Also, pdf available at: http://apps.who.int/iris/bitstream/10665/68985/1/WHO_CDS_2005.29.pdf.

World Health Organization Writing Group, 2006. Nonpharmaceutical interventions for pandemic influenza, international measures. *Emerging Infectious Diseases*, 12(1), 81–87.

World Health Organization Writing Group, 2012. Nonpharmaceutical interventions for pandemic influenza, national and community measures. *Emerging Infectious Diseases*, 12(1), 88–94.

Xie, X. et al. 2007. How far droplets can move in indoor environments – revisiting the Wells evaporation–falling curve. *Indoor Air*, 17(3), 211–225.

Yang, Y. et al. 2009. The transmissibility and control of pandemic influenza A (H1N1) Virus. *Science*, 326(5953), 729–733.

Zoutman, D.E. et al. 2003. The state of infection surveillance and control in Canadian acute care hospitals. *American Journal of Infection Control*, 31(5), 266–273.

PART V

MULTI-LEVEL NETWORKS

Professor Patrick Beautement

As Johnson, Fortune and Bromley state in their chapter "Almost all social systems have many levels of organization, from micro to macro levels, and multi-level structure is fundamental to their dynamics. Part-whole structures play a major role in multi-level systems, where intermediate wholes may themselves be parts in higher level structures."

But what is really meant here by 'multi-level'? Patrick Beautement defines it in broad terms in Chapter 20 (Employment of Tools and Models Appropriate to Complex, Real-world Situations). He says that these levels are certainly *not* hierarchies of homogenous entities in a reductionist breakdown (as in parts of a machine where subsystems and sub-subsystems have fixed and predictable properties and relationships), nor even fractal self-similarities.

Instead, he draws on Cohen and Stewart's view[1] that these multi-levels of networks are so dissimilar that phenomena, information and activities at one level may have little meaning or equivalence at another. The reason for this is that the emergent properties of one level can be the influences on another level and as such, those emergent properties are hidden from the level that generates them.

Indeed, even the term 'level' is misleading as it implies hierarchy. For example, take the nervous, endocrine, immune and autonomic systems of mammals. They are distinct and operate concurrently within the bodies of creatures yet no one of them is at the top or the bottom, nor are they bounded and disjoint, there is overlap and inter-twingling of functionality throughout.

To address these, Beautement's chapter first provides a model of practice, and a framework for judging appropriateness of tools based on that model (with three examples of the framework in use). The chapter then offers a critique of two sets of example tools: examining the applicability of autonomous agents and multi-agent systems to a range of situations; and explaining how to employ multi-modal, multi-level influence networks to bring about ongoing change. Finally, the chapter presents a list of principles of practice, drawn from experience in the field, to be used to inform real-world decision- and policymaking. Beautement asserts that, to employ the multi-level networks approach in analysis, modelling or as a practical tool, one must take account of the fact that they:

- Are *made up of a number of heterogeneous network structures* each of which have distinct characteristics (not defined by humans) and that function differently;

- *Exist at many different scales and rates* – from micro to macro (such as from the virus to human societies), and over timescales from the immediate to the geological;
- Are *inter-dependent, yet loosely coupled, and interact via a variety of modes* (for example, using chemical or electrical signalling; by passing information in a number of forms; or though exchanging resources or by modifying the environment, as in stigmergy);
- Have *modes of interaction that are not homogenous* between all networks. Instead, there is a 'soup' of signalling in which they exist from which networks use the modes which are appropriate to their scale and function (and all other modes are ignored or are transparent);
- Have *effects and influences which extend and may cascade beyond the immediate bounds* of the networks' structures and may occur 'up', 'down' or 'sideways'. For example, the firing of synapses in the brain releases neurotransmitters (hormones) which may affect other organs, the whole body or even social behaviours beyond the person's body – such as the reaction to a shock making the person feel ill and anxious and so causing concern for relatives;
- Have *transitions between the networks which can be emergent* in that activity within one network may generate emergent phenomena which may be inputs to other networks (for example, the processes within the mitochondria in cells are driven in a manner which seems largely decoupled from the activities of the whole organism – yet there *are* cross-network influences, even though the mitochondria will not be 'aware' of their source[2]);
- *Exchange information, much of which will be unknowable or unobservable* by human beings and/or our sensing devices, partly owing to the vast range of types of 'information';
- They *exist regardless of whether or not humans can detect them*. It is not relevant to their functioning whether or not we can identify and label the network structures, their nature, bounds or influences – indeed, science tries to find those bounds and categorize them.

This implies that, for real-world applications, research methods and tools should at least consider these characteristics of multi-level networks – though, in practice, this list is very exacting and demanding. In Johnson, Fortune and Bromley's chapter (Multilevel Systems and Policy) they take a pragmatic approach and present a simple tool for bounding what might otherwise be an unmanageable research task.

Johnson et al. indicate that anything that impacts on what is of particular interest will be included in the representation of the system and its environment. Thus:

- Anything the system can control directly is inside the system;
- Anything else that affects the system or its elements but cannot be controlled by the system is in the system environment;
- Everything else is outside the representation of the system.

Johnson et al. then develop the defined 'system' through taxonomic aggregation techniques. These are of the form of reductionist decompositions – an approach which would be at odds with Érdi[3] who would claim that there still remains the issue of bounding the depth of decomposition that is appropriate – which Johnson et al. then address. For example, in a health system, can one stop at the patient? What about the role of the epidemic that is bringing the patient to the doctor's door (and of bacteria that caused it)? Cohen and Stewart would also point out that each subsystem is itself distinct and bounded without any ambiguous blurring. The tool to apply here is, of course, expert judgement which can be the final arbiter of whether an item is included or not.

As any system is dynamic (even a dead one is decaying), Johnson et al. then examine the traffic through the system using the taxonomy as a backdrop. This technique enables the researcher to begin to identify patterns of, for example, conflict in the use of resources, or overloads, blockers or enablers. Dynamic phenomena, such as beneficial or damaging oscillations may also be evident.

However, even within these bounds, if results are to be valid and insights relevant to practice there is a but, and it is a big but . . . how to obtain suitable data which spans these heterogeneous multi-levels of networks for use in research work.[4] This is a significant issue. Is the data needed observable, detectable and meaningful – is its significance something that we can even begin to appreciate? Johnson et al. attack this issue, showing the value of using micro-data from synthetic micropopulations which, within the context of the system under investigation, can be derived from detailed models.

Finally, a rich model of multi-level systems is developed based on a Formal System Model (the FSM)[5] and Johnson et al. use this to analyse one aspect of the current crisis in the United Kingdom's National Health Service – that of a resource clash between emergency and social care. The analysis shows clearly that the mismatch between the responsibilities and scope of authority of both systems generates mal-adapted behaviour – one symptom of which is patients unable to be admitted to hospital wards. Finally, they recommend that agent-based modelling be employed to explore alternative arrangements which could inform policy.

In Palit, Banerjee and Mukherjee's chapter (Complex Scenarios in Socio-Economic Data: A Comprehensive Analytical study), they start by stating that "Complex systems are those which are composed of many particles, or objects, or elements of same or different kinds. The elements may interact with each other in a more or less complicated manner by various nonlinear couplings. The global human society, especially the economy, with its numerous participants – managers, employers and consumers, its financial systems, its capital goods, its natural resources, and its traffic probably form the most complex system in our world."

They go on to show that, when comparing countries, their GDP (Gross Domestic Product) and population dynamics are key indicators. To analyse these, the team employed three nonlinear tools: recurrence rate, mean conditional recurrence (MCR), and complex networks (CN). These analyze country level GDP and population data, and have successfully validated the derived results with the standard conclusions based on general theories of economics as follows:

- Recurrence rate is used to show how two non-identical systems get synchronized through their phase spaces;
- MCR detects the driver and response system in synchronized states; and
- CN reflects the overall scenarios of the complex systems by its various statistical measures.

To address the challenge of obtaining relevant data (mentioned above), the whole data are collected from a data centre in NASA's Earth Observing System Data and Information System hosted at Columbia University and are downscaled projected based on Special Report on Emissions Scenarios (SRES).

An important part of Palit et al.'s chapter is clarifying the nature of the types of synchronization that can occur between non-identical systems at different levels. This clarity is necessary as it determines the complex influences that the systems may have on each other. They cite three important ones: Complete Synchronization (CS), Generalized Synchronization (GS) and Phase Synchronization (PS). Obviously though, if there is no direct synchronization there will be little effect – though influence may still manifest itself through intermediate, indirect or underlying routes. In the chapter, the synchronization tools have been applied in a number of ways, such as to analyze the worldwide population data so as to verify whether the population data of a particular year is synchronized with that of the previous years.

As part of Palit et al.'s analysis the nature of the phase space and how it is reconstructed is examined in detail. A number of nonlinear tools are described and a food web example is employed to show their utility. In addition, measures such as bifurcation and sensitivity analysis are discussed. Palit et al. then raise an important issue: "constructing a proper model for a complex phenomenon and observing their nonlinear behaviour is one of the efficient ways to predict the long term dynamics of the system. However, if the parameters and variables of the system are not properly taken, [the] mathematical model does not reflect the real scenario". One might add that, at the limit, Gödel's Incompleteness Theorem[6] may make the model misleading unless expert judgement is applied.

To address this they cite the 'Takens Model' which is used to construct attractor spaces which are topologically equivalent to given attractors. They then go on to describe how to construct a recurrence plot and to show how these are used to develop phase synchronization indicators. Between countries, these form networks of influences and socio-economic GDP networks are then analysed in the chapter.

Following this, standard network analysis tools (such as degree distribution) are employed to derive an important insight – that the GDP network is scale free and so there are huge differences between the GDP characteristics of countries across the globe. These GDP networks are then further analysed for their: small-world properties (which shows that they are a small world); transitivity or clustering (which indicates a recent reduction in the tendency to form cliques); assortativity (where it is shown that there are mixed types of association); and centrality (which shows that high GDP countries tend to be involved with low-GDP countries).

In conclusion, Palit et al. summarize the value of using the analysis tools associated with complex multi-level networks to investigate real-world issues – in this case of economic and population data. The results are striking and insightful and are an excellent case study of what can be achieved by using the tools of complexity science in an appropriate way.

The last chapter in this Multi-Level Networks Section of the Handbook is from Michael Gabbay. His chapter (Leadership Network Structure and Influence Dynamics) describes a quantitative methodology for the analysis and modelling of leadership networks which leverages research in complex systems, in particular nonlinear dynamical systems theory and network science. A prototype software package, PORTEND, is introduced which implements the methodology using data from expert analysts in order to help assess policy and factional outcomes with respect to the internal dynamics of a system of political actors.

Gabbay starts by highlighting the importance of being able to understand, if one is going to negotiate effectively with them, the dynamics of leadership groups and of their networks of allegiances and inter-dependencies.

He then introduces a software package called PORTEND (Political Outcomes Research Tool for Elite Network Dynamics). This integrates quantitative techniques from nonlinear systems theory and network science to aid the analysis of policy and factional outcomes with respect to the internal dynamics of a system of political actors. The political actors may be individual leaders or organizations within a government or movement. The outcomes of concern may be policy decisions, winning and losing factions, the positions of individuals, or the potential for issues to cause dissension or factional realignment.[7]

His case study involves Iran and he obtains relevant data[8] from a survey of two experts in Iranian matters. They not only gave information about the characteristics of fifteen leaders but also provided data under the following headings:

- Liberalism (LIB): The proper role for Western culture, Islam, media sources, and democratic institutions.
- Economic Reform (ECON): Whether economic policies should benefit the current elites or a wider set of interests.
- Arab States (ARAB): Whether Iran's peers in the Arab world are potential allies or enemies.
- Syrian Regime (SYR): Whether the Assad regime in Syria should be supported.
- US/Israel (USISR): The extent to which Iran should confront the US and Israel.
- Nuclear Issues (NUKE): The extent to which Iran should develop nuclear technology.
- IRGC Influence (IRGC): The appropriate role for the Islamic Revolutionary Guard Corps (IRGC).

As part of this analysis the experts generated an influence network indicating the nature and strength of connections between the various actors selected. From this and the other data collected it was possible to generate a matrix of the range of attitudes to a topic for each actor. Plots of these opinions of graphs showed various degrees of clustering of opinions from which one could infer that certain actors were aligned in their opinions (which might indicate which actors would be compatible within coalitions on that matter). Further analysis was applied using Principal Components Analysis (PCA) which seeks to represent a data matrix by a series of coordinate vectors, known as principal components (PCs), each of which corresponds to a pattern of covariation in the data. PORTEND then applies a network analysis tool and displays the results in the visualizations shown in Gabbay's chapter.

He then describes the Nonlinear Social Influence Simulation built into the tool. In the model, an actor's position changes under the influence of two separate forces: the 'self-bias force' and the 'group influence force'. For example, for self-bias force, each actor is assumed to come to the debate with an initial issue position given by his natural preference (also called the natural bias) which reflects the actor's underlying beliefs, attitudes, and worldview pertinent to the issue. The group influence force is the total force acting to change an actor's position due to the persuasive efforts of the other actors in the group. It is assumed to operate in a pairwise manner so that an actor – the message receiver – experiences a persuasive 'coupling force' from another actor – the message sender – to whom he is connected (and vice versa). The results of the simulation in respect to one actor are described in detail.

Gabbay completes his chapter by discussing possible further research. One area could involve the investigation of whether automated content analysis of actor rhetoric could be a viable input source for either the structural analysis or the simulation. Another area could be extending the social influence model to a multidimensional issue space in order to allow issues to trade off against each other. Additionally, complexity research on adaptive networks could be used to develop an issue-network coevolution model in which both issue positions and network ties would interact and change dynamically, thereby explicitly modelling alliance formation processes, a capability not present in the current model.

In summary, the chapters in this Multi-Level Networks Section of the Handbook provide pragmatic examples of the use of such networks as an analytical approach. The Section:

- Defines the nature and characteristics of multi-level networks.
- Illustrates the range of tools and techniques that are available to represent different areas of real-world phenomena, to analyse them and support decision-making.
- Examines some of the issues which could inhibit the use of the approach (such as the need for relevant data) and offers solutions and insights.
- Provides a number of 'industrial weight' case studies to illustrate that this approach has real relevance to policy makers and those seeking to tackle real-world global challenges.

NOTES

1. Cohen, J. and Stewart, I. (2000) *The Collapse of Chaos. Discovering Simplicity in a Complex World*, London: Penguin Books.
2. See Lane, N. (2005) *Power, Sex, Suicide: Mitochondria and the Meaning of Life*, Oxford: Oxford University Press, and especially the work by Dr Peter Mitchell on 'proton pumping' cited in this book.
3. Erdi, P. (2008) *Complexity Explained*, pp.328–351, Berlin: Springer-Verlag.
4. As Dr Robert Myers said in the panel discussion (on the Complexity of Global Change) at ECCS Complex'09 "... there is no shortage of irrelevant data, but a serious lack of relevant information" (see: http://www2. warwick.ac.uk/fac/cross_fac/complexity/newsandevents/events/archive/2009/eccs09/publicsession, accessed 13 August 2017).
5. Readers may wish to compare the FSM with Stafford Beer's Viable Systems Model (VSM), for example, at: https://en.wikipedia.org/wiki/Viable_system_model, which places emphasis on viability within the wider environment and therefore on the adaptive and self-regulatory capabilities of the parts of the system.
6. For a detailed discussion of the Incompleteness Theorem see Hofstadter, D.R. (1983) *Gödel, Escher, Bach: an Eternal Golden Braid*, London: Penguin Books.
7. A related technique is by Howard, N., Bennett, P. and Bryant, J. (2001) 'Drama Theory and Confrontation Analysis'. In Rosenhead J.V. and Mingers J. (eds), *Rational Analysis for a Problematic World Revisited: Problem Structuring Methods for Complexity, Uncertainty and Conflict*, London: Wiley. Also see so-called 'Ripeness Theory', Zartman, I. William (2002) *Power and Negotiation* (edited by I. William Zartman and Jeffrey Z. Rubin), Ann Arbor: University of Michigan Press.
8. Readers may wish to consider how the validity of such data (and hence of the outputs of a model) can be established given that the publicly stated and privately held views of all of us are usually different.

18. Multilevel systems and policy

Professor Jeffrey Johnson, Open University,
Professor Joyce Fortune, Open University and
Dr Jane Bromley, Open University

18.1   INTRODUCTION

Making multilevel systems well defined is essential for the implementation of computer models to investigate the multilevel consequences of policy. This chapter shows that systems thinking can provide practical guidance to those building models of complex multilevel social systems in order to inform policymaking. Part–whole aggregation and taxonomic aggregation are described as methods of representing multilevel structure, and it is shown how they are interleaved in the construction of vocabulary to describe multilevel systems. This enables complex nested structures to be represented as a kind of backcloth that supports patterns of aggregate and disaggregate numbers that describe the day-to-day traffic of people, resources and responsibility that are essential for systems to function.

Decision-making in policy involves examining complex systems and evaluating the possible outcomes of interventions. This can be done in an open way so that different stakeholders can agree on what might happen, even if they disagree on what should happen.

Almost all social and economic systems have many levels. At the highest levels, every ministry in every country is responsible for a multilevel system, for example, health, agriculture, transport, justice, defence, education, and so on. Similarly, all but the smallest companies and enterprises in the private sector have many levels of organisation. To illustrate this, consider the National Health Service (NHS) in the UK. For England this has organisational structure: 1. Secretary of State > 2. Department of Health > 3. NHS England > 4. Clinical Commissioning Groups > 5. Planned hospital care, rehabilitative care, urgent and emergency care, community health services, mental health services (NHS 2016). This five-level organisational scheme would have even more levels if it went down to the level where individual patients receive treatment. As an example from the private sector, the mining company RioTinto divides its operations into twenty countries with varying numbers of companies (RioTinto 2016). Each of the companies has its own internal organisational structure. In both these examples, the vocabulary and the organisational structure it is expressing are essential to represent and manage the organisation at the different levels of aggregation.

The science of complex systems has developed new computational methods to investigate possible system behaviours and policy outcomes. These include building computational models that give replicable and sometimes unanticipated insights into system behaviours. The great advantage of computer modelling is that the representation of the system and its dynamics are explicit and open to question.

This chapter analyses the properties of multilevel systems and presents a method of building the multilevel vocabulary necessary to represent their different levels of organisation. It introduces a way of representing multilevel systems at *all* levels in a well-defined way. The presentation will be qualitative through examples and diagrams. A more comprehensive technical account can be found in Johnson (2006, 2014).

18.2   SYSTEMS THINKING, MODELLING AND POLICY

In the natural sciences it is assumed that there is an external reality that can be studied objectively to discover immutable scientific truths. The phenomena observed seem to be indifferent to being studied and who studies them. The study of social systems is not so straightforward. It is a commonplace that people see the same situation differently. When people act together to achieve a shared goal their views need to be coordinated. One way to do this is construct formal *models* of systems as the basis for consensus. Model construction is a social process with the model evolving through social interaction. *Systems thinking* provides methodologies for building models of social systems. The starting point here will be the four part definition of a system given in Bignell and Fortune (1984) which states that a system is:

1.   An assembly of components, connected together in an organised way;
2.   The components are affected by being in the system and the behaviour of the system is changed if they leave it;
3.   The organised assembly of components does something; and
4.   The assembly has been identified as being of particular interest.

The first problem that any investigator has to answer is *where is the boundary of the system*? In the physical sciences it has been very easy to put boundaries around the system studied. The physicist can shut the door of the laboratory and ignore everything that happens outside. Even inside the laboratory much can be ignored. When looking at social systems everything can seem to affect everything else and there are no clear boundaries. However, it is impracticable to include everything in a description of a social system so some things must be ignored or left out but it is really important to know what can be omitted and what it is important to include. When things are left out that do impact on the behaviour of a system it may act in unpredictable ways, and policies may have unintended consequences. The fourth part of the definition of system provides guidance on this when building policy-oriented models of multilevel systems. The system does something that is of particular interest (to the analyst) and will be analysed in this light. Anything that impacts on what is of particular interest will be included in the representation of the system and its environment. Thus:

1.   Anything the system can control directly is *inside* the system.
2.   Anything else that affects the system or its elements but cannot be controlled by the system is in the *system environment*.
3.   Everything else is *outside* the representation of the system.

The four part definition is explicitly tied to a particular stakeholder or stakeholder group and allows that different stakeholders may have different views of the same system.

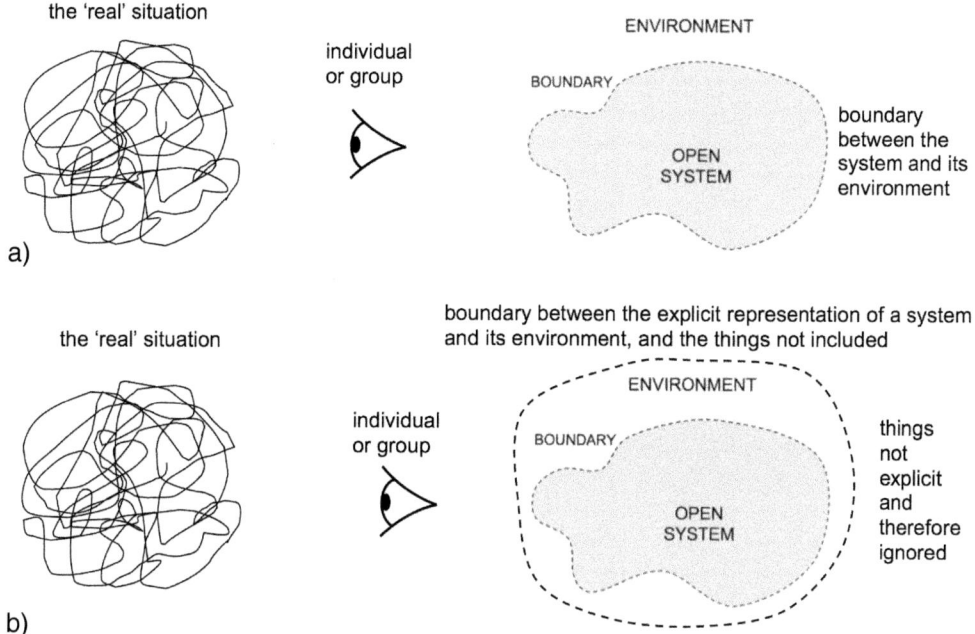

Figure 18.1   Situations and individual systems representations of them

For example, Figure 18.1a shows an individual or group's view of a system that may differ from another individual or group's view of the 'real' situation. Figure 18.1b shows the boundary to the formal representation and those things not explicitly included in the system or its environment.

Models are used to explore the behaviour of systems. All policy is based on models, whether or not they are formalised and based on evidence. Although all models are imperfect representations of the 'real' situation, an advantage of using formal models is making explicit what is under consideration and what is not, and what is known and what is not. A *model* of a system is an explicit list of its components and the components of its environment and an explicit description of how they are connected in an organised way. The model will state explicitly what the system does and the transformations it makes to itself and its environment through time, and how it responds to the environment and changes in the environment through time.

18.3   REPRESENTING MULTILEVEL PART–WHOLE SOCIAL STRUCTURES

Any system has at least two levels – the whole system at the highest level and the component parts at the lowest level. For example, Figure 18.2 illustrates a UK General Practitioner (GP) system. A patient with a health problem wanting to see a doctor contacts the receptionist to request an appointment, which either results in an appointment being made, or

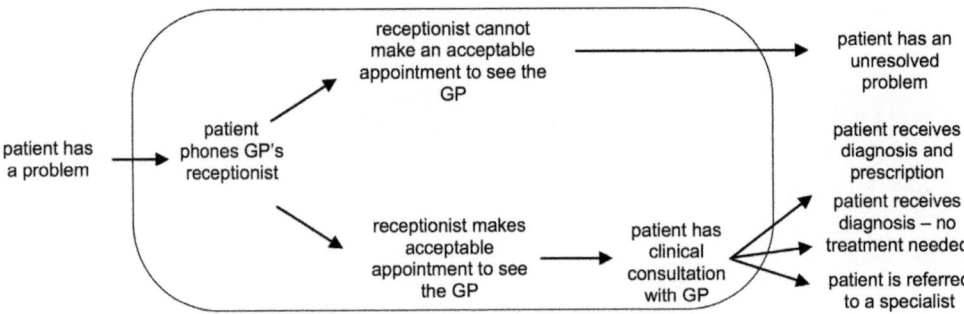

Figure 18.2   An action-flow description of the General Practitioner system in the UK

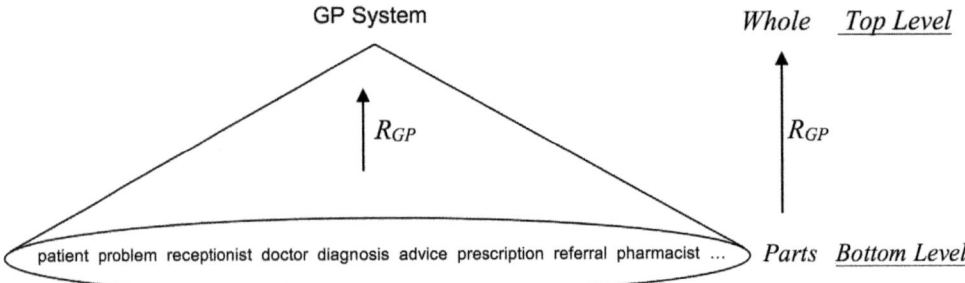

Figure 18.3   The GP system as a part–whole cone

the patient being left with their problem unresolved. In the former case the patient has a clinical consultation with the doctor resulting in a diagnosis with or without a prescription for medicinal treatment, or it may result in a referral to a specialist physician.

Figure 18.3 shows the GP system drawn as a *part–whole cone*. At the lowest level are all the parts, and at the highest level the apex represents the whole system. These are called the *Top Level*, and the *Bottom Level*. The ellipse at the bottom of the cone is called its *base*. We write *base*(GP System) = {patient, problem, receptionist, doctor, diagnosis, advice, prescription, referral, pharmacist, . . .} to show the set of parts. The symbol R_{GP} signifies the relational structure between the parts. Thus R_{GP} maps the base of the cone at one level to its apex at a higher level, as shown by the arrow between the base and apex of the cone.

However, the GP system is just one subsystem of the whole NHS. There are other subsystems such as the Hospital and Ambulance Systems. These all have to fit together to make the whole, and there will be legal and administrative documents defining this. Let the symbol R_{NHS} represent all the relational information that says how these subsystems work together.

For example, in Figure 18.4 the Ambulance subsystem is connected to the Emergency Phone Line System (not shown) used by members of the public for emergency access the Police, Fire and Ambulance systems. Given the necessary information, an ambulance is dispatched in the Ambulance system and either treats the patient where they are or transports them to a hospital, where the Ambulance subsystem hands

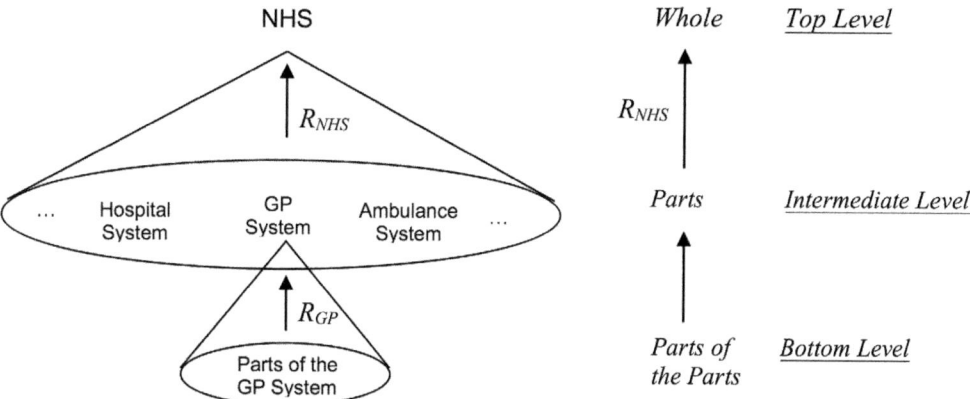

Figure 18.4   Intermediate levels between the top and bottom levels

over responsibility for the patient to the Hospital System. This is part of the R_{NHS} relationship.

Now, as shown in Figure 18.4, the representation has three levels: the whole system at the *top level*, the subsystems of the system at an *intermediate level* and the parts of the subsystems at the *bottom level*.

The part–whole relationship by assembly is particularly important in multilevel systems because constructed wholes are never parts of their parts, for example, a car is not part of its engine and the NHS is not part of a hospital. This means that *the parts are always at a lower level to the whole*, and there is an immutable upwards arrow that establishes a difference in levels.

It is common to call the highest level of a system its macrolevel, the lowest level its microlevel and the intermediate level its mesolevel. The problem with this is that many systems have more than one intermediate level, as illustrated in Figure 18.5. One way to discriminate the intermediate levels is to number them with Level 1 at the microlevel, Level 2 the first mesolevel, Level 3 the next intermediate level, Level 4 the next intermediate level, and Level 5 the macrolevel of the whole system, as shown at the right of Figure 18.5.

Since building a multilevel vocabulary often involves adding or removing levels between existing levels, there is no absolute interpretation for the number representing a level. This relatively can be emphasised by the '*Level N+k*' notation, where N is an arbitrary level, *Level N+1* is the level above it and *Level N-1* is the level below.

Defining part–whole structures has to be done with care, as illustrated by the following example taken from Winston, Chaffin and Herrmann (1987):

Simpson's finger is part of Simpson,
Simpson is part of the Philosophy Department,
Simpson's finger is part of the Philosophy Department.

The university's rules and statutes determine the assembly of Simpson and other academics into the Philosophy Department. The statutes establish which people are members (the parts) and the part–whole relationship of interaction between those people to form the higher-level structure, for example, hold weekly meetings with minutes. By itself,

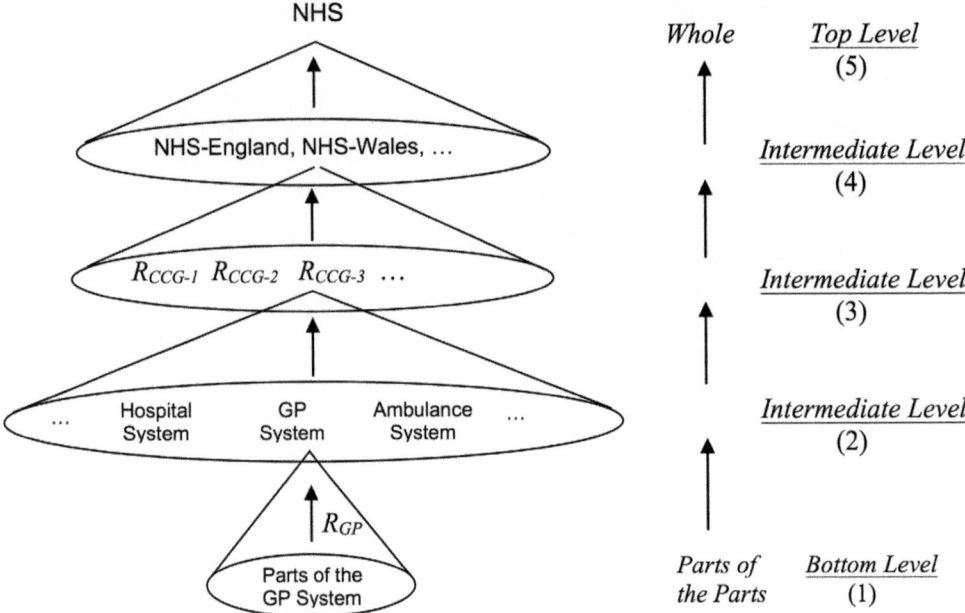

Figure 18.5   There may be many intermediate levels between the top and bottom levels

Simpson's finger is not eligible to be a member of the Philosophy Department and the apparent conundrum is artificial.

More generally, when describing the structures of multilevel systems it is necessary to know explicitly the relationships that makes the parts into the whole. To summarise this section, the key features of part–whole systems are that:

- The components of a part–whole structure exist at a lower level than the whole;
- It is necessary to make explicit how the parts fit together to make the whole;
- A structure at one level may become a component in a higher level structure;
- There can be any number of levels between the lowest and highest levels.

18.4   MULTILEVEL TAXONOMIC AGGREGATION

Taxonomic aggregation plays an important role in policy and management. For example, the *Statistical Classification of Economic Activities in the European Community*, known as NACE (*Nomenclature statistique des Activités économiques dans la Communauté Européenne*), has twenty-one highest-level classes:

A –   Agriculture, forestry and fishing
B –   Mining and quarrying
C –   Manufacturing
D –   Electricity, gas, steam and air conditioning supply

E – Water supply; sewerage; waste management and remediation activities
F – Construction
G – Wholesale and retail trade; repair of motor vehicles and motorcycles
H – Transporting and storage
I – Accommodation and food service activities
J – Information and communication
K – Financial and insurance activities
L – Real estate activities
M – Professional, scientific and technical activities
N – Administrative and support service activities
O – Public administration and defence; compulsory social security
P – Education
Q – Human health and social work activities
R – Arts, entertainment and recreation
S – Other services activities
T – Activities of households as employers; undifferentiated goods – and services – producing activities of households for own use
U – Activities of extraterritorial organisations and bodies

Each of the classes has a variety of subclasses, as illustrated for Construction in Table 18.1. The taxonomic aggregations can be drawn in the form of trees, as illustrated in Figure 18.6.

Usually taxonomies are created for a purpose, such as the collection of statistics. For example, Table 18.2 shows the added value of a company's activities classified as Construction (C), Wholesale and retail trade; repair of motor vehicles and motorcycles (G), and Professional, scientific and technical activities (M). In principle, firms can assign the added value of their activities to the different classes, and this can be summed over all firms to give national statistics of the activities at the various levels of aggregation. Typically such figures are used for economic planning by national government and the European Commission.

18.5   NAMING ELEMENTS AND CLASSES BY INTENSION AND EXTENSION

There are two ways of defining classes. One way involves listing all the things referred to by a word or phrase. For example, at the time of writing the class of 'twenty-first century British Prime Ministers' has members Blair, Brown, Cameron and May. This listing is called a definition of 'twenty-first century British Prime Ministers' by *extension*.

In contrast to defining terms by extension, they can be defined by *intension*, which usually means establishing a rule defining class membership. For example, according to the Oxford Dictionary, a car is "a road vehicle, typically with four wheels, powered by an internal-combustion engine and able to carry a small number of people". Thus my desk is not a car, but the machine I drove to work this morning is a car. Intentional definitions can be difficult to apply in practice because they require interpretation. For example, some 'cars' can carry nine people, so may not qualify as cars. Also as shown in Figure 18.7b,

Table 18.1   The NACE classification for construction

Division	Group	Class	Section F– constuction	ISIC Rev. 4
41			Construction of buildings	
	41.1		Development of building projects	
		41.10	Development of building projects	4,100*
	41.2		Construction of residential and non-residential buildings	
		41.20	Construction of residential and non-residential buildings	4,100*
42			Civil engineering	
	42.1		Construction of roads and railways	
		42.11	Construction of roads and motorways	4,210*
		42.12	Construction of railways and underground railways	4,210*
		42.13	Construction of bridges and tunnels	4,210*
	42.2		Construction of utility projects	
		42.21	Construction of utility projects for fluids	4,220*
		42,22	Construction of utility projects for electricity and telecommunications	4,220*
	42.9		Construction of other civil engineering projects	
		42.91	Construction of water projects	4,220*
		42.99	Construction of other civil engineering projects n.e.c.	4,290*
43			Specialised construction activities	
	43.1		Demolition and site preparation	
		43.11	Demolition	4,311
		43.12	Site preparation	4,312*
		43.13	Test drilling and boring	4,312*
	43.2		Electrical, plumbing and other construction installation activities	
		43.21	Electrical Installation	4,321
		43.22	Plumbing, heat and air conditioning installation	4,322
		43.29	Other construction installation	4,329
	43.3		Building completion and finishing	
		43.31	Plastering	4,330*
		43.32	Joinery installation	4,330*
		43.33	Floor and wall covering	4,330*
		43.34	Painting and glazing	4,330*
		43.39	Other building completion and finishing	4,330*
	43.9		Other specialised construction activities	
		43.91	Roofing activities	4,390*
		43.99	Other specialised construction activities n.e.c	4,390*

Source: From: http://ec.europa.eu/eurostat/documents/3859598/5902521/KS-RA-07-015-EN.PDF, page 72.

the definition of car may need updating to include things excluded by the intentional definition but desired in the extension such as being powered by electricity, as with the BMW i3 electric car.

When building structured vocabularies inconsistencies can arise between intension

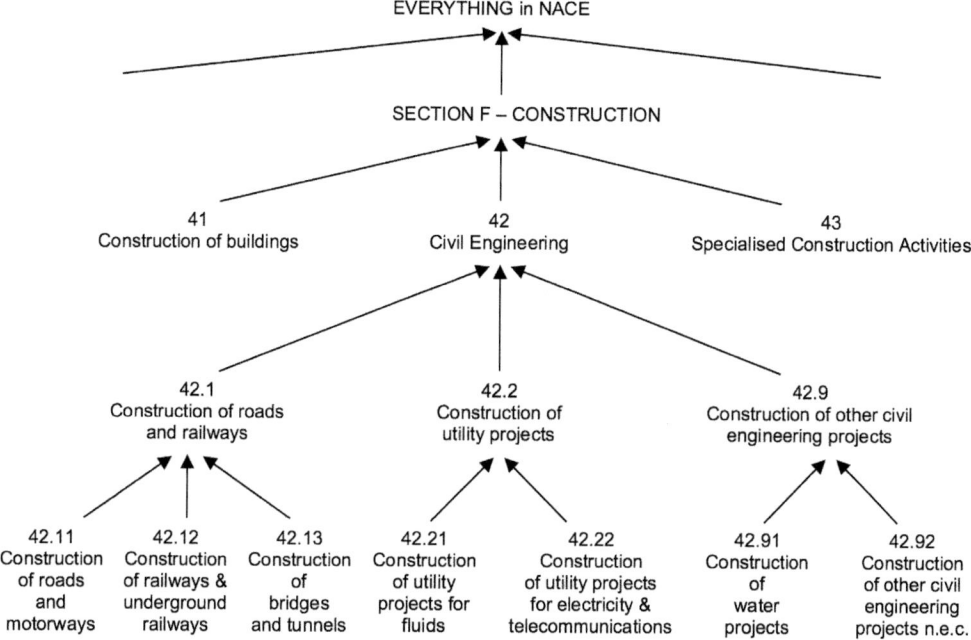

Figure 18.6   Part of the NACE taxonomic aggregation drawn as a tree

Table 18.2   An example of using the NACE classification to record a company's added value

Section	Division	Group	Class	Description of the class	Share
C	25	25.9	25.91	Manufacture of steel drums and similar containers	10%
	28	28.1	28.11	Manufacture of engines and turbines, except aircraft, vehicle & cycle engines	6%
		28.2	28.24	Manufacture of power-driven hand tools	5%
		28.9	28.93	Manufacture of machinery for food, beverages and tobacco processing	23%
			28.95	Manufacture of machinery for paper and paperboard production	8%
G	46	46.1	46.14	Agents involved in the sale of machinery, industrial equipment, ships & aircraft	7%
		46.6	46.61	Wholesale of agricultural machinery, equipment and supplies	28%
M	71	71.1	71.12	Engineering activities and related technical consultancy	13%

Source: http://ec.europa.eu/eurostat/documents/3859598/5902521/KS-RA-07-015-EN.PDF.

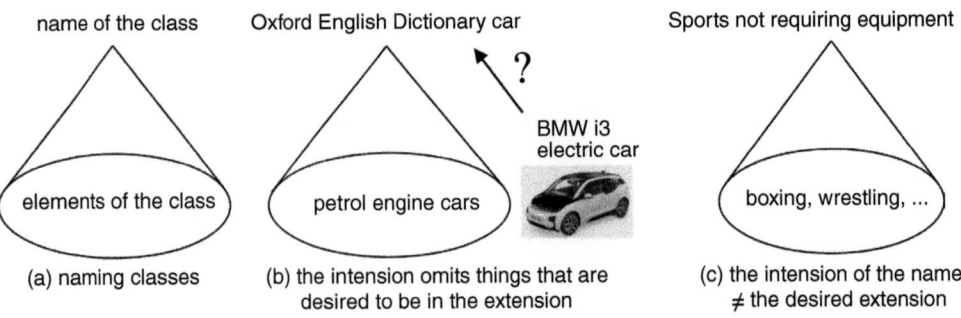

Figure 18.7   Care is needed when naming classes and defining them by extension and intension

and extension. For example, a vocabulary to classify television programmes had a class 'Sports Not Requiring Equipment' illustrated extensionally with the examples wresting and boxing (Figure 18.7c (Gould, Johnson and Chapman 1984). However boxing requires equipment such as gloves, gum shields, and a ring and is not a 'sport not requiring equipment'. The problem here is that the *name* of the class itself has meaning, and in this case the intension contained in the name does not match the desired extension.

Note that taxonomic cones are different to part–whole cones. An element in the base of a cone defined by extension means 'belongs to this list', while intension means 'the element obeys this rule'. In contrast, part–whole aggregation requires an assembly relation saying 'the elements are put together this way'. Part–whole assemblies create new things.

A definition by extension corresponds to using a computer lookup table with all the elements listed as data. A definition by intension corresponds to formulating a computer function that accepts data and decides if the required pattern holds between them or not.

18.6   GROUNDING AND TRANSITIVITY IN TAXONOMIC AGGREGATION

Taxonomic aggregation does not create new entities, it simply provides a way of defining subsets of entities. For example, Figure 18.8a shows a taxonomy of scientists with a subclass of physicists. It also shows the chemist Lavoisier. All of these scientists existed irrespective of taxonomic definitions such as 'physicist' and 'chemist' which give just one of many ways these they could be grouped, for example, they could be grouped as women and men (Figure 18(b)). Figure 18.8 also illustrates how words can be *grounded* in real things. Here Einstein, Curie and Lavoisier are *grounded elements*, for example the name 'Curie' here uniquely identifies the real person Marie Curie. In contrast the word 'physicist' does not uniquely name any individual, but names a group of individuals. The extension of the word 'physicist' uncontroversially includes Einstein and Curie. Whether or not it should include Lavoisier is a matter of debate and agreement between those defining and using the taxonomy.

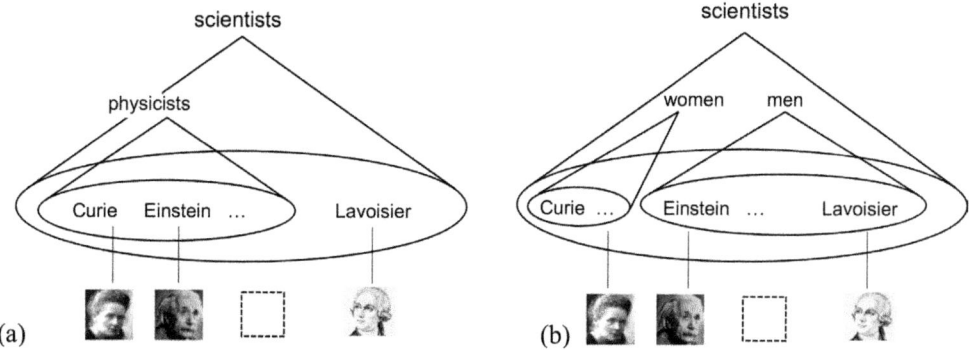

Figure 18.8   The words Einstein, Curie and Lavoisier are grounded elements, not classes

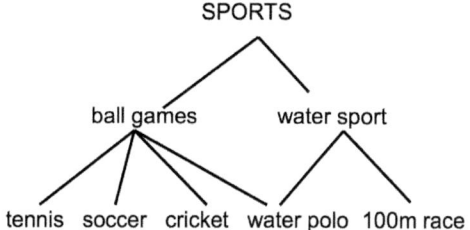

Figure 18.9   A lattice taxonomy allows elements to belong to more than one class

Taxonomies are usually *transitive*. If a grounded element belongs to a lower level class, and that lower level class aggregates into a higher-level class, then the grounded element also aggregates into the higher-level class. This is because the classes of the taxonomy define *sets* of grounded elements, and a subset of a subset of a set is also a subset of that set.

Taxonomic aggregations are usually tree-like partitions with disjoint classes for any level. This can cause distortion and is not essential. For example, in a taxonomy to classify sports, water polo belongs to both the classes 'ball games' and 'water sports'. When counting the number of 'ball game' and 'water sports' teams, a water polo team counts for both classes. However, there is no need to double count the water polo team when counting the number of SPORTS teams at a more aggregate level.

18.7   INTERLEAVED MULTILEVEL PART–WHOLE AND TAXONOMIC AGGREGATION

Words and phrases are used to denote objects, relationships between objects to form systems, the systems formed, and classes of objects and subsystems. This vocabulary has a structure of levels reflecting the things identified. This can be signified by drawing an upward arrow between them. Figure 18.10a shows a three level taxonomic aggregation while Figure 18.10b shows a three level part–whole aggregation.

Although they are different, part–whole aggregations and taxonomic aggregations

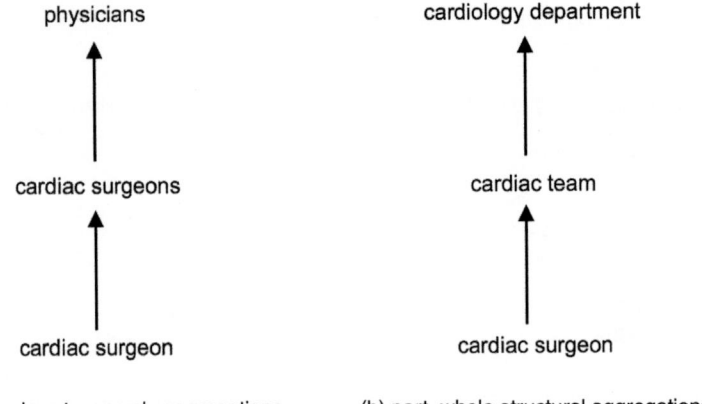

(a) element–class taxonomic aggregations   (b) part–whole structural aggregations

Figure 18.10   Taxonomic versus structural aggregation

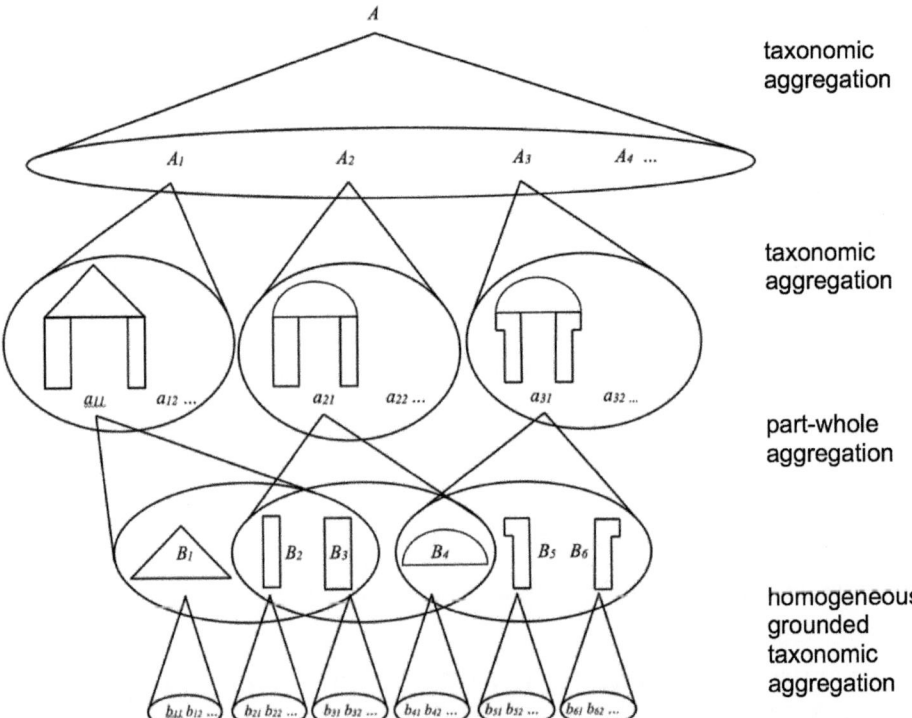

Figure 18.11   Interleaved taxonomic and part–whole aggregations

work together in multilevel systems. To illustrate this consider the arch construction in Figure 18.7.

In this multilevel system blocks are assembled to form three types of arches. Each individual arch is assembled into a class of arches of that type. At the lowest level, the

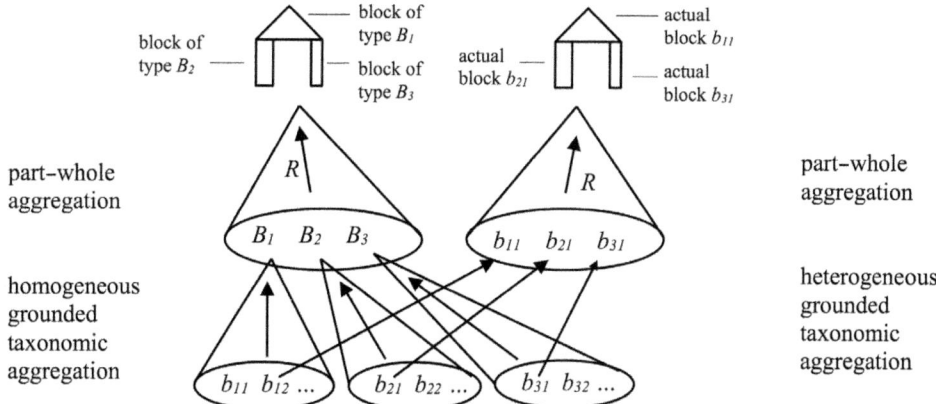

Figure 18.12   *The interleaved taxonomic and part–whole aggregation in action*

individual blocks are assembled into sets of their type. Thus the part–whole aggregation is sandwiched between two taxonomic aggregations. At the top level the different classes of arches are brought together under a single taxonomic class of arches, denoted *A*. In this example there are three taxonomic aggregations and one part–whole aggregation.

Figure 18.12 illustrates how such a scheme might be used and the subtle distinction between an organisational description of the objects in a multilevel system, the objects themselves, and the part–whole structure of the objects. On the left is a multilevel description of the arch in terms of the types of blocks that are assembled to make it. This is different to an actual instance of the block, as shown on the right. The multilevel description of the arch means that one can work top-down on the left to find out the components are necessary to build an instance of the arch. This information enables the set of parts to be assembled, as with a furniture flat pack. The relation *R* then says how those parts should be assembled, as with the instructions in a flat pack.

Bottom-up, the assembly of the actual parts into sets of parts of the same kind is called a *homogeneous grounded taxonomic aggregation* at the bottom left of Figure 18.12. This aggregation is like putting components of the same kind into a bin container.

The bottom-up assembly of the set of components needed to make the arch is called a *heterogeneous grounded taxonomic aggregation* at the bottom right of Figure 18.12. This is like taking the parts from bins of homogeneous components to make the set of heterogeneous components, ready for the bottom up part–whole arch assembly. This process creates a real object, *grounded* in the real components at the lowest level.

Although this illustrative example involves physical objects, these ideas can equally well be applied to social systems. For example when building a team of people, the types of people will be specified and so will the way they must work together to achieve the team objective. Then individual people will be appointed as part of a heterogeneous grounded taxonomic aggregation. Following this, part–whole aggregation will be needed to meld the individuals into a well-working team. This might involve training for individuals and the whole group.

18.8　TRAFFIC ON THE MULTILEVEL BACKCLOTH

In an early attempt to formulate a mathematical theory of multilevel systems, R.H. Atkin (1974, 1977) made a distinction between relational structure and patterns of numbers distributed over relational objects called *simplices*. The distinction is clear to see in network theory. Networks are formed from *vertices* and *edges*, also called *nodes* and *links* in social analysis. Whereas a network such as a road system may not change over a given period of time, the speed and number of vehicles may change considerably during that time. Atkin called the relatively static relational structure the *backcloth* and he called the relatively dynamic patterns of numbers the *traffic*.

　　Backcloth structure at the microlevel could include the relational structure of a family. The traffic on this structure could include the number of family holidays taken and the money spent on them. At higher levels, institutions such as businesses form relatively fixed relational structures supporting patterns of numbers such as their costs and incomes. Another example is the relational structure of a hospital supporting a traffic of patients treated and money spent.

　　For decision-making purposes, the patterns of numbers over the multilevel structure are of great importance. In particular it is important to understand how numbers at the lower level aggregate into number at the higher level, and how higher level quantities can be distributed top-down. Figures 18.13 and 18.14 illustrate the bottom up aggregation of backcloth and traffic.

　　Figure 18.14 shows disaggregate parts mapped to disaggregate numbers at the lower level mapped to aggregate numbers at the higher level. An important feature for these mappings is that they are coherent as the multilevel traffic aggregates over the multilevel backcloth.

　　The general question is whether or not the *Level N+1* numbers can be reconstructed from the *Level N* number? The answer is that it can if the mappings have appropriate aggregation properties.

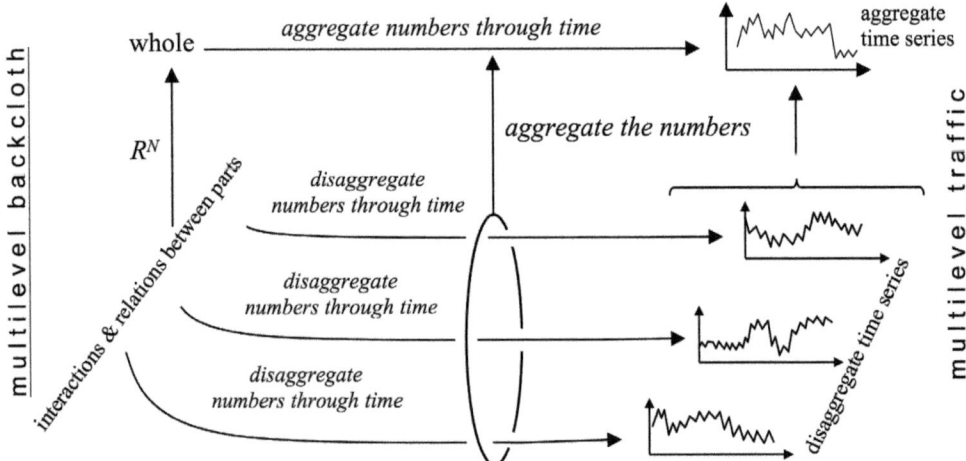

Figure 18.13　Aggregation of numbers over the multilevel backcloth

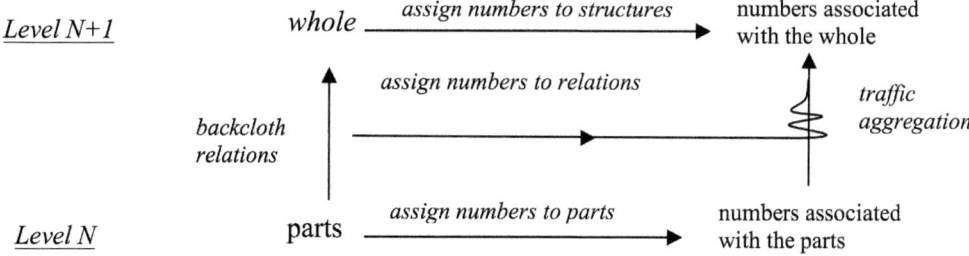

Figure 18.14   Aggregating backcloth and traffic from Level N to Level N+1

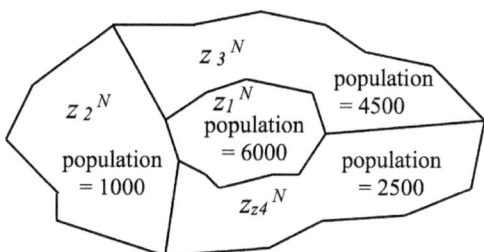

Figure 18.15   A city assembled from four zones

As an example of aggregation of lower level statistics consider a city denoted by the symbol z^{N+1} at *Level N+1*. Suppose it is made up of four *Level N* zones, z_1^N, z_2^N, z_3^N, and z_4^N. The population of the city at *Level N+1* is simply the sum of the populations of the zones at *Level N*. It will be called a *linear* aggregation. The widespread use of spreadsheets for managing social systems shows that linear aggregation is very common. However, complex multilevel systems have many non-linear structures and, by themselves, spreadsheets are not powerful enough to represent their dynamics.

18.9   INFORMATION CLOSURE AND THE LOWEST LEVEL OF REPRESENTATION

What is the lowest level necessary to understand the dynamics of multilevel systems? For example, when trying to understand the behaviour of the people in a social system is it necessary to know their genetic makeup? The answer is that it depends on the purpose of the analysis. Genetic makeup is central in personalised medicine (Royal Society 2005), but currently it is not considered relevant to the majority of social systems in the public and private sectors.

In a different context and using a different notation, Pfante et al. (2014) suggest the following properties for multilevel systems:

I. Information closure: The higher level process is information closed, i.e., there is no information flow from the lower to the higher level. Knowledge of the microstate will not improve predictions of the macrostate.

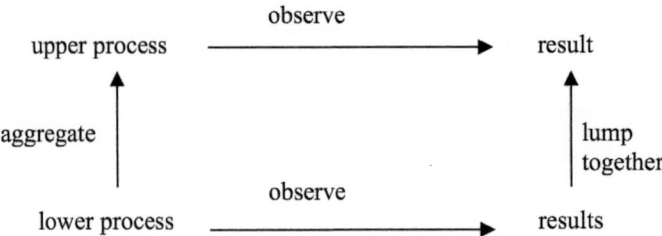

Figure 18.16   Observational commutativity

II. Observational commutativity: It makes no difference, whether we perform the aggregation first, and then observe the upper process, or we observe the process on the microstate level, and then lump together the states.

These properties are illustrated in Figure 18.16. The first says that the upper process is information closed if knowledge from the lower process adds nothing new to knowledge of the higher process. For example, Hook's law states that the extension of a spring is proportion to the mass attached to it. This could be modelled at a lower level using Finite Element methods but for most purposes this would add nothing useful to Hook's Law. Observational commutativity means that going the round diagram either way makes no difference.

18.10   LOWEST LEVEL TRAFFIC DISAGGREGATION – SYNTHETIC MICROPOPULATIONS

In complex systems science and its applications to decision-making, computer simulation is a major tool for investigating the possible consequences of policy actions. Moeckel, Spiekermann and Wegener (2003) write:

> Microsimulation models require micro data. However, the collection of individual micro data, i.e. data that can be associated with single buildings, or the retrieval of individual micro data from administrative registers is neither allowed in most countries nor desirable for privacy reasons. Therefore these models work with synthetic micro data that can be retrieved from general accessible aggregate data. A synthetic population has to be generated that represents individual actors in the form of households and household members. A synthetic population is statistically equivalent to a real population. For each household characteristic such as household size, income, number of cars and address are generated. Each person is described by characteristics such as age, sex, religion, and work location. For creating addresses for the synthetic population, land-use data are disaggregated to raster data by GIS techniques.

The first large-scale applications of synthetic micropopulations for micro simulation was the TRANSIMS systems developed at Los Alamos National Laboratory for road traffic modelling in the 1990s (Barrett et al. 1999).

Figure 18.17 shows the generality of creating synthetic micropopulations. On the left, at *Level N* each sampled individual has some characteristic. These values are aggregated to give the population statistics for the real population. Implicitly there is a population to which these statistics can be extrapolated. For example, if data were collected on the

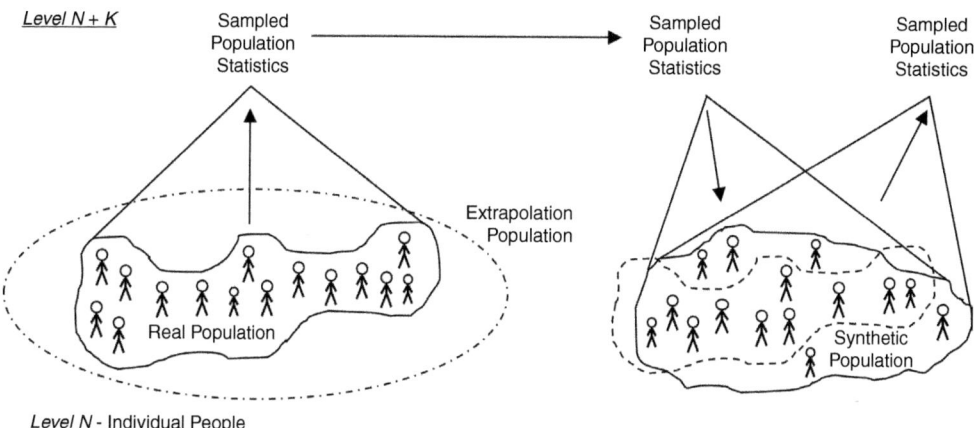

Figure 18.17   *Synthetic micropopulations*

incomes of people in various British cities, the *extrapolation population* could be Britain, but would not include Spain or Germany, and certainly not India, China or Brazil.

Higher-level statistical values can be distributed across a new 100 per cent synthetic population at *Level N*, that is, every member of the new population is assigned a value. This is done by a process of *disaggregation* (the down arrow on the right) which typically uses Monte Carlo methods to assign values to the members of the new population. Synthetic micropopulations are very important for modelling multilevel systems in a policy context. Many social systems are information complete at the lowest level of individual people. To explore their multilevel dynamics it is essential to be able to run computer models at the lowest level of individual people. Synthetic micropopulations enable this.

18.11   THE FORMAL MODEL OF MULTILEVEL SYSTEMS

Figure 18.18 shows the *Formal System Model* (FSM) (Fortune 1993). It is adapted from Checkland (1981) who in turn drew on the work of Churchman (1971) and Jenkins (1969). It has been widely used to identify system risk and causes of system failure. In the context of multilevel systems, this model explicitly considers three levels: the system, its subsystems, and the wider system in which it is embedded. Of course, a subsystem of one FSM could itself be modelled as a system with its own subsystems and wider system in a different FSM, and similarly, the wider system could be represented as a system in a different FSM with its own wider system but in this section it will be shown how a single model can be extended to any number of levels of representation.

This multilevel model suggests a number of important questions for those analysing systems for policy purposes. The questions, adapted from Fortune and Peters (1995), are:

- What is the continuous purpose or mission of the system and its subsystems?
  - What are the system and subsystems supposed to do? Who set the purpose/

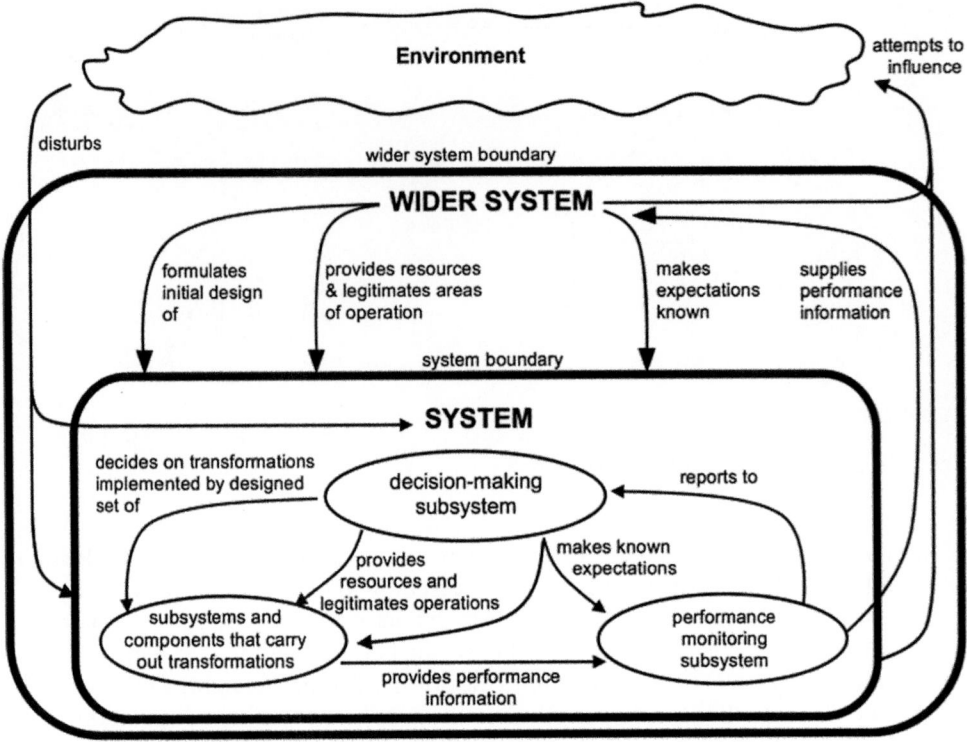

Figure 18.18   The Formal System Model (adapted from Fortune 1993)

mission? Does everyone agree at every level? Have the expectations been made known within the system? Are they consistent between levels?

- What is the structure of the system?
  – What are the components of the system and its subsystems? What are the interactions between the parts? At each level, how do the system and subsystems bring about the transformations to convert inputs into outputs?
- What is the nature of the decision-making sub-system at each level?
  – How is the system managed? Who is responsible for deciding how the purposes of the system are achieved at each level? Who is responsible for providing the resources to enable this to happen?
- What is the nature of the performance-monitoring sub-systems?
  – Are the transformation processes being monitored at each level? Are deviations from the expectations being reported to the decision-making subsystem so that corrective actions can be initiated where necessary?
- What is the degree of connectivity between the multilevel components?
  – At each level, what are the essential interactions for the system to perform without failure? Where is there feedback between the components of the system? What types of influences link the components within and between levels?
- What is the environment that interacts with the system?

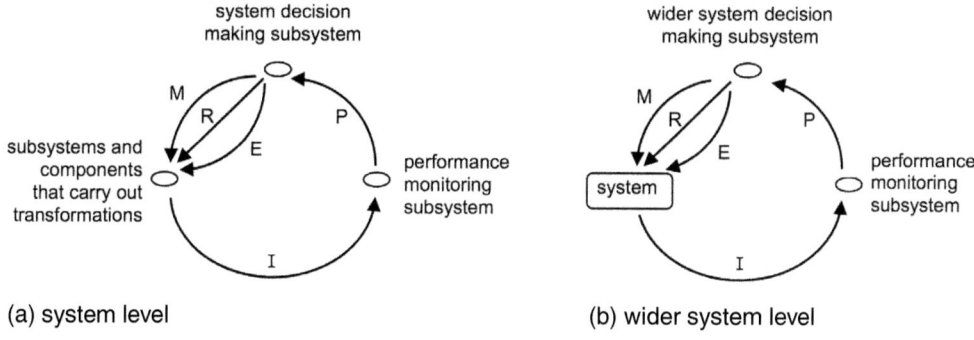

Key to meaning of the arrows

P – Performance: reports performance information to
R – Resources: provides resources and legitimates operations
I – Information: provides information on
M – Mission – decides on transformations implements by designed set of subsystems and components
E – Expectation: makes known expectations.

Figure 18.19   The systems – decision-making – monitoring motif of multilevel systems

- What are the components of the multilevel environment? Where are the boundaries between the system, its subsystems, and their environments?
● How are resources distributed across the system?
  - At each level, does the decision-making subsystem control the resources? Have sufficient resources (in terms of quality and quantity) been allocated to carry out the transformations that are required within and between levels?
● How does the system maintain continuity and adapt to change?
  - At each level, how does the system monitor its changing environment? What is the capability of the system and its subsystems to adapt to change? How effective are the system and its subsystem's attempts to influence the environment? Are they well-coordinated between levels?

In network science, a repeated or noteworthy configuration of nodes and arrows is called a *motif* (Johnson, Fortune and Bromley 2017). Inspection of the Formal System Model shows the motif illustrated in Figure 18.19 at both the system and wider system levels.

This kind of 'fractal' structure is exactly what is required for modelling multilevel systems for policy applications. In principle *every* managed social system has subsystems and components, and the managers' responsibilities include making clear the mission and the expectations at each level and allocating the resources necessary to enable the desired transformations to be made. The managers' responsibilities also include monitoring the performance of the system, which in turn requires that the necessary information be made available. The configuration of components and arrows in Figure 18.19 will be called the *multilevel systems motif*.

This motif structure gives a general architecture for the multilevel management of multilevel systems, as illustrated in Figure 18.20.

As an example, consider the so-called bed-blocking problem which the National Audit Office (2016) summarises thus:

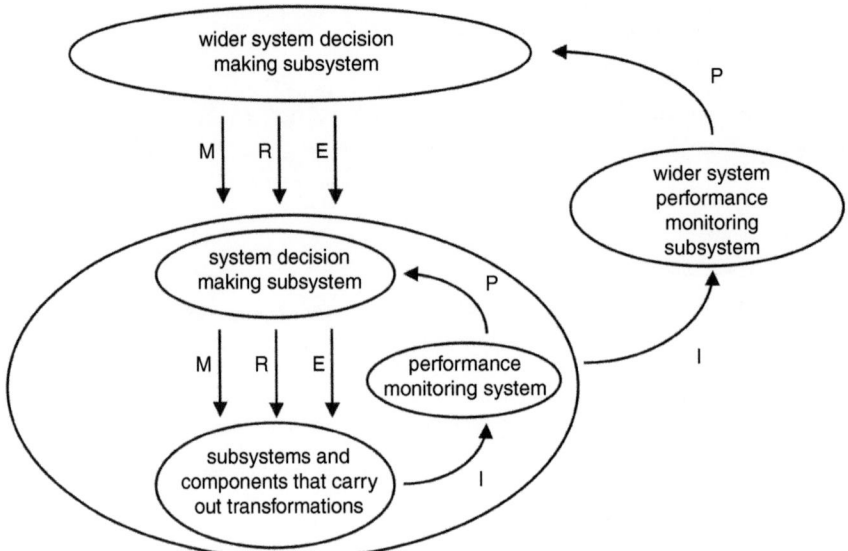

(a) the combined system– and wider system– decision-making – monitoring motifs

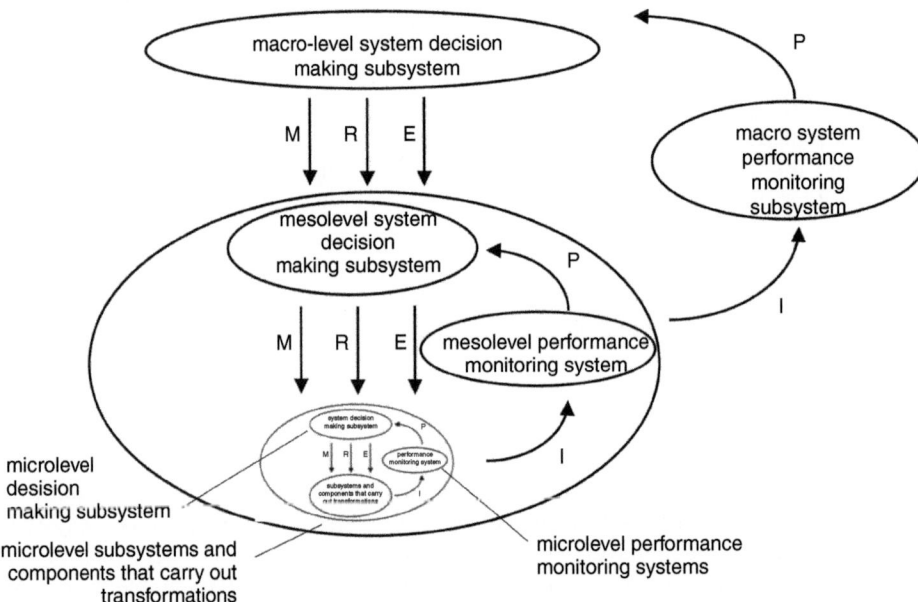

(b) the system – decision-making – monitoring structures at micro-, meso- and macro-levels

Figure 18.20   Nested system– decision-making – monitoring motifs

Unnecessary delay in discharging older patients . . . from hospital is a known and long-standing issue. For older people [this] can lead to worse health outcomes and can increase their long-term care needs [and] is an additional and avoidable pressure on the financial sustainability of the NHS and local government. . . . Older people are cared for in hospital by the NHS, but once discharged some may need short- or long-term support from their local authority or community

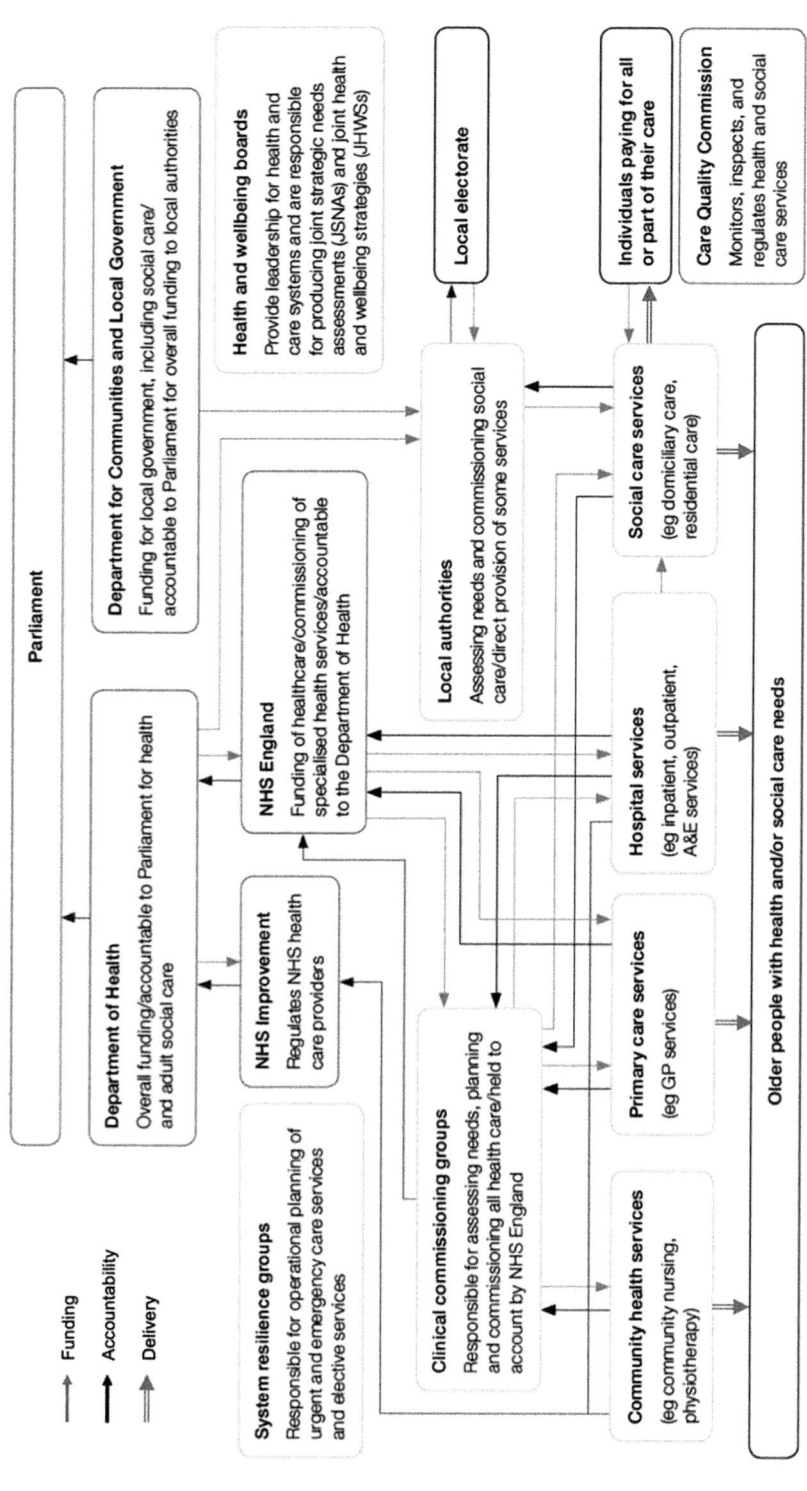

Figure 18.21   The health and social care accountability and funding structure (National Audit Office 2016)

health services. This may involve living at home with some support or living in a care home. NHS community healthcare and short-term care to increase people's independence provided by local authorities are free. Local authorities have to apply a financial means test and an eligibility test based on levels of need for other types of care.

The number of recorded delayed transfers of care has increased substantially over the past two years [a 31 per cent increase between 2013 and 2015]. The main drivers for this increase are the number of days spent waiting for a package of home care (which more than doubled between 2013 and 2015, from 89,000 to 182,000) and waiting for a nursing home placement or availability (which increased by 63%) . . . The delayed transfers of care data substantially underestimate the range of delays that patients experience . . . we estimate that the number of older patients in hospital who are no longer benefiting from acute care to be . . . about 2.7 million bed days a year. (National Audit Office, 2016, pp. 5–7, Sections 1, 2, 9 and 10)

The analysis of this problem by the National Audit Office includes a representation of the health and social care accountability and funding structure reproduced here as Figure 18.21. The left and right of the diagram can be separated to give the formal multilevel system model sketched in Figure 18.22.

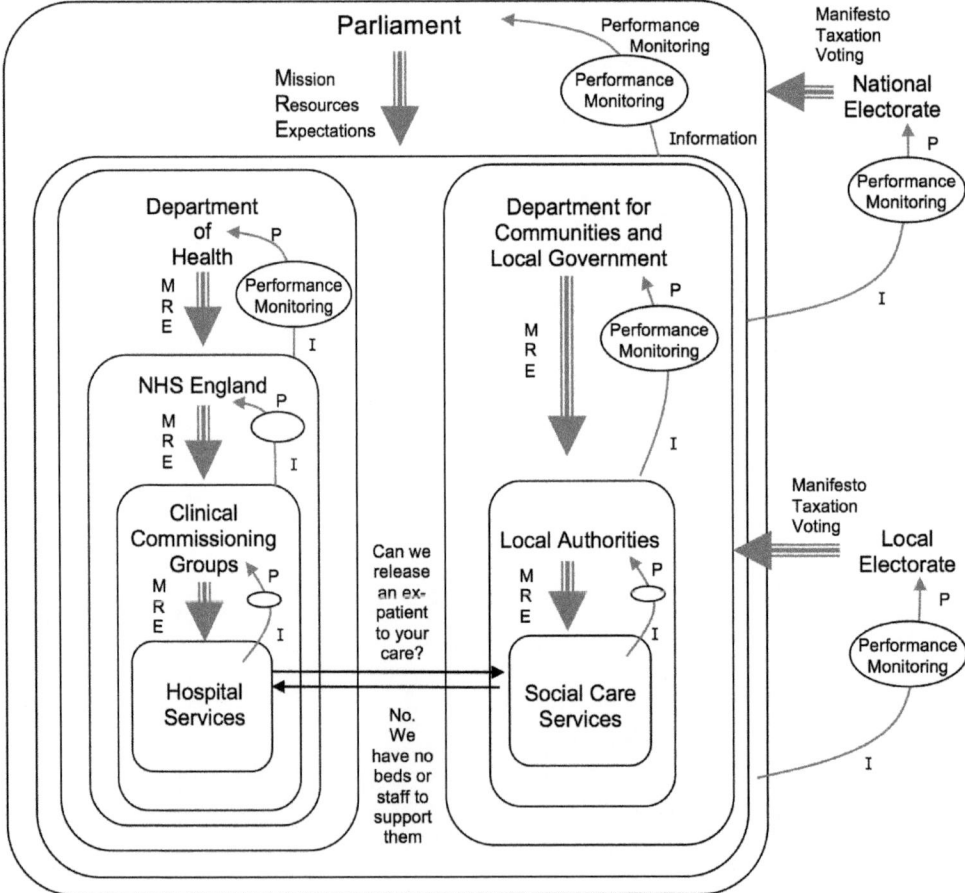

Figure 18.22   An abbreviated formal multilevel system model of the bed-blocking system

A complete version of the abbreviated model shown in Figure 18.22 would require the multilevel structure of the Hospital Services system and Social Care Services system to be made explicit But even without this the model illustrates that the problem of a hospital administrator not being able to discharge a patient is due to a local authority administrator being unable to put together a suitable care package at the microlevel.

The NAO analysis includes the possibility of the Local Electorate monitoring and improving the performance of local authorities in providing the social care packages required to alleviate the bed-blocking problem. However, the range of services provided by local authorities is much wider than those provided by the NHS. The resources allocated by local authorities to social care compete with those allocated to education, transport, planning, parks, leisure, food standards, waste disposal, and much else. Voters may be dissatisfied with local politicians for their performance in any of these areas. Furthermore in the UK, local elections are often influenced by national issues as much as they are by local performance. This suggests that electoral pressure on local councils is not an effective performance monitoring tool.

There are major problems underlying the UK's bed blocking crisis. The broad political consensus is that this care should be provided by the state, and all the major parties are committed to this principle but the level of social care required for a growing aging population would be very expensive and thus require significant tax increases. However, there is no consensus on how the extra money should be raised, and what would be the best way to spend it. There are many suggestions reflecting individual and collective norms and values, but no practical way to choose between them.

On 6 January 2016 Norman Lamb MP proposed setting up a cross party commission to address the problem, requesting that "leave be given to bring in a Bill to establish an independent commission to examine the future of the National Health Service and the social care system; to take evidence; to report its conclusions to Parliament; and for connected purposes." He said

> The Government face a choice – either the system will drift into a state of crisis or we confront the existential challenge now. This transcends narrow party politics. We have to decide as a country how much we want to spend on our NHS and care system. What can we do differently to make better use of the resources available? Should we consider, as I have proposed, a dedicated NHS and care tax, and give local areas the ability to vary it? Should we end the artificial divide between the NHS and social care? (Lamb, 2016a, 2016b)

If the request for an independent commission is granted how might the science of complex systems be useful? The question of how much a country wants to spend on its health and welfare system is political. The question of what can be done differently to make better use of the resources available is technical, and this is where science can help. Generating and evaluating different ways of doing things means *designing* and testing new structures and procedures. This involves making explicit the requirements of the system and the constraints that must be satisfied. The health and welfare system for an aging population is a complex multilevel system and the methods of complex systems science can be used to investigate the possible outcomes of any suggested policy attempting to solve this long standing problem. A more precise prescription for combining policy questions, complexity science, policy informatics and citizen engagement as *Global Systems Science* is given in Dum and Johnson (2017).

The purpose of the multilevel system model presented here is to support computer analysis of the possible outcomes of policy. There are many actors and agents in this system, from national politicians to those delivering services at the local microlevel, including individual nurses, doctors, administrators, porters, and so on. The multilevel structures sketched here can support computer models of these agents and their interactions at all system levels, allowing any policy options to be tested and different policies to be compared. Without computational assistance analysts may get lost or miss something important at one of the levels.

Computer simulations cannot say definitively that a policy will work but they can investigate the range of likely outcomes as an input to the policy process. Simulations can sometimes give strong evidence that a policy will not work, for example, by showing that the policy never gives the outcome expected, no matter how the assumptions and associated parameters are varied.

Multilevel social systems are often information-complete at the lowest level of the individual agent. What people do at the microlevel can impact on all higher levels and cause a system to fail. The theory of synthetic micropopulations provides the data necessary for simulation at all levels including the lowest.

18.12   CONCLUSIONS

This chapter has outlined a method of representing multilevel systems for policy purposes. Making clear the multilevel structure is particularly important when the impact of policy is to be investigated by computer simulation. However, even for less formal analyses of policy awareness of the subtleties of multilevel systems can add greater precision and make successful policy more likely. The main ideas covered include:

- Socio-economic systems have two kinds of multilevel structure.
  - Part–whole systems play a major role in multilevel systems, where intermediate level wholes may become parts in higher-level structure.
  - Taxonomic aggregation plays another major role in multilevel systems but, taxonomic aggregation is very different to part–whole aggregation.
- Part–whole and taxonomic aggregations are interleaved in multilevel systems.
- Multilevel classes can be defined by extension (listing the members) or intension (giving a rule for membership). Care is required to ensure that class names with intentional meaning are consistent with the intended extensions.
- Systems thinking provides fundamental ideas and motivation for analysing systems, including providing a multilevel representation by generalising the Formal Systems Model to include all levels of decision-making.
- A theory of multilevel systems must be grounded in real things. This includes part–whole aggregations where the part-of relation may or may not be transitive.
- Taxonomic aggregations need not force observers to assign things to just one class.
- Confusing anomalies can appear in multilevel systems, for example, part–whole aggregation is not transitive.
- Objects in multilevel systems can be grounded and information-closed at intermediate levels so that analysing them at lower levels gives no extra information.
- Patterns of numbers are distributed over the multilevel structure which acts as a kind of backcloth for the system traffic.

- Data can be disaggregated top-down to form synthetic micropopulations to enable massive multilevel computer simulations to support policy.

In an increasingly complex multilevel world the risks inherent in ignoring or not knowing the dependencies between subsystems are increasing. An operational theory of multilevel systems will not prevent the willful implementation of bad policies, but can support the design and implementation of policies that achieve their objectives without unintended consequences.

BIBLIOGRAPHY

Atkin, R.H. (1974) *Mathematical Structure in Human Affairs*, London: Heinemann Educational Books.

Atkin, R.H. (1977) *Combinatorial Structure in Social Systems*, Berlin: Birkhäuser Verlag.

Barrett, C.L. et al. (1999) TRansportation ANalysis SIMulation System (TRANSIMS). Version TRANSIMS-LANL-1.0. Volume 0: Overview. LA-UR 99-1658. Los Alomos, NM: Los Alamos National Laboratory.

Bignell, V. and Fortune, J. (1984) *Understanding Systems Failures*, Manchester: Manchester University Press.

Checkland, P.B. (1981) *Systems Thinking, Systems Practice*, Chichester: Wiley.

Churchman, C.W. (1971) *The Design of Inquiring Systems*, New York: Basic Books.

Dum, R. and Johnson, J. (2017) 'Global systems science and policy', in *Non-equilibrium Social Science and Policy*, J.H. Johnson et al. (eds), Berlin: Springer, pp. 209–225.

Fortune, J. (1993) *Systems Paradigms*, Milton Keynes: Open University Press.

Fortune, J. and Peters, G. (1995) *Learning from Failure*, Chichester: Wiley.

Gould, P., Johnson, J. and Chapman, G. (1984) *The Structure of Television*, London: PION.

Jenkins, G.M. (1969) 'The systems approach', *Journal of Systems Engineering*, 1, 3–49.

Johnson, J.H. (2006) 'Hypernetworks for reconstructing the dynamics of multilevel systems, in *European Conference on Complex Systems*, 25–29 September 2006. Oxford. Accessed 30 December 2016 at http://oro.open.ac.uk/4628/1/ECCS06-Johnson-R.pdf.

Johnson, J.H. (2014) *Hypernetworks in the Science of Complex Systems*, London: Imperial College Press.

Johnson, J.H., Fortune, J. and Bromley, J.M. (2017) 'Systems, networks, and policy', in *Non-equilibrium Social Science and Policy*, J.H. Johnson et al. (eds), Berlin: Springer, pp. 111–134.

Lamb, N. (2016a) *Motion for leave to bring in a Bill (Standing Order No. 23) National Health Care and Social Care (Commission)*, Daily Hansard Debate, 6 January 2016: Columns 287–289. Accessed 29 December 2016 at http://www.publications.parliament.uk/pa/cm201516/cmhansrd/cm160106/debtext/160106-0001.htm#16010634000002.

Lamb, N. (2016b) *NHS and Social Care Commission*. House of Commons Hansard, Volume 605, 28 January 2016. Accessed 29 December 2016 at https://hansard.parliament.uk/Commons/2016-01-28/debates/16012841000002/NHSAndSocialCareCommission.

Moeckel, R., Spiekermann, K. and Wegener, M. (2003) 'Creating a synthetic population', Paper presented at the 8th International Conference on Computers in Urban Planning and Urban Management (CUPUM), Sendai, Japan, May 2003. Accessed 30 December 2016 at http://spiekermann-wegener.de/pub/pdf/CUPUM_2003_Synpop.pdf.

NACE (2008) *Nomenclature statistique des activités économiques dans la Communauté européenne*. Accessed 29 December 2016 at http://ec.europa.eu/eurostat/documents/3859598/5902521/KS-RA-07-015-EN.PDF.

National Audit Office (2016) *Discharging older patients from hospital*, National Audit Office, 26 May 2016. Accessed 29 December 2016 at https://www.nao.org.uk/wp-content/uploads/2015/12/Discharging-older-patients-from-hospital.pdf.

NHS (2016), *The NHS in England*, accessed 28 December 2016 at http://www.nhs.uk/nhsengland/thenhs/about/pages/nhsstructure.aspx.

Pfante, O., Bertschinger, N., Olbrich, E., Ay, N. and Jost, J. (2014) 'Comparison between different methods of level identification', *Advances in Complex Systems*, 17(2).

RioTinto (2016) *Our business*, accessed 28 December 2016 at http://www.riotinto.com/our-business-75.aspx.

Royal Society (2005) *Personalised Medicines: Hopes and Realities*, Report of Royal Society working group on pharmacogenetics, chaired by David Weatherall, accessed 30 August 2015 at https://royalsociety.org/~/media/Royal_Society_Content/policy/publications/2005/9631.pdf.

Winston, M., Chaffin, R. and Herrmann, D. (1978) 'A taxonomy of part–whole relations, *Cognitive Science*, 11(2), 417–444.

19. Complex scenarios in socio-economic data: a comprehensive analytical study

Assistant Professor Sanjay Kumar Palit, Calcutta Institute of Engineering and Management, Associate Professor Santo Banerjee, University Putra Malaysia and Assistant Professor Sayan Mukherjee, Sivanath Sastri College

19.1 INTRODUCTION

Economy is a manifestation of complex social systems. Complex systems are those which are composed of many particles, or objects, or elements of same or different kinds. The elements may interact with each other in a more or less complicated manner by various nonlinear couplings. The global human society, especially the economy, with its numerous participants – managers, employers and consumers, its financial systems, its capital goods, its natural resources, and its traffic probably form the most complex system in our world. Economic systems are embedded in the more comprehensive human societies, with different human activities and different political, ideological, ethical, cultural, or communicative habits.

According to Herbert Simon, "Economics and the social sciences are, in fact, the hard sciences, because the complexity of the problems dealt with cannot simply be reduced to analytically solvable models or decomposed into separate sub processes" (in Schweitzer 2002). However, in the recent few years, the emerging interdisciplinary sciences of complexity have provided some new methods and tools for dealing with these problems, ranging from complex data analysis to sophisticated computer simulations. In fact, the researchers throughout the globe have recently started to apply different advanced methods developed for the natural sciences to social and economic problems.

In this chapter, our design of the experiment is scheduled on two types of complex socio-economic data – Country-level population and Country-level Gross Domestic Product (GDP), which are downscaled projected based on Special Report on Emissions Scenarios (SRES) with A1, A2, B1, B2 marker scenarios. The data A1, A2 represent the projected economic growth rates with very high for the A1 family, medium for the A2 family, while the data B1, B2 represent population with high for the B1 family and medium for the B2 family. SRES regional GDP growth rates and population data were calculated from 1990 to 2100 based on the SRES marker model regional data and applied uniformly to each country that fell within the SRES-defined regions. The whole data and information are collected from a Data Centre in NASA's Earth Observing System Data and Information System (EOSDIS) – hosted by Center for International Earth Science Information Network (CIESIN) at Columbia University.

GDP is nothing but the market value of all officially recognized final products and services within a country in a year or over a given period of time and so it is a strong indicator of a country's material standard of living. On the other hand, worldwide population is another major issue as it greatly influences world economy. Various studies have already been done regarding these two prime factors and several standard conclusions have been derived. However, all of those conclusions are based on general theories of Economics. Thus it is necessary to implement some popular nonlinear tools used in natural sciences and verify whether the conclusions derived by these tools match with the aforesaid standard conclusions. If these validations are found to be successful, then the respective nonlinear tools can be used as an alternative way to study different socio-economic data. Being motivated with this thought, we have made use of three popular nonlinear tools – τ – recurrence rate (Marwan et al. 2007), Mean conditional recurrence (MCR) (Romano et al. 2007) and Complex networks (CN) in analyzing the country's GDP and populations. τ – Recurrence rate is used to show how two non-identical systems get synchronized through their phase spaces. MCR represents another much stronger type of synchronization, called generalized synchronization (GS) (Rulkov et al. 1995; Kocarev and Parlitz 1996). It can also detect the driver and response system in unidirectional coupling occurred in GS. On the other hand, CN reflects the overall scenarios of the complex systems by its various statistical measures.

The word synchronization (Pikovsky, Rosenblum and Kurths 2001) means 'to share the common time'. Formally, synchronization is defined as a general process wherein two (or many) dynamical systems (either equivalent or non-equivalent) are coupled or forced (periodically or noisy), in order to realize a collective or synchronous behavior. There exist different types of synchronization in coupled complex systems of which the mostly important are Complete Synchronization (CS) (Fujisaka and Yamada 1984; Afraimovich, Verichev and Rabinovich 1986; Pecora and Carroll 1990), Generalized Synchronization (GS) (Rulkov et al. 1995, Kocarev and Parlitz 1996) and Phase Synchronization (PS) (Rosenblum, Pikovsky and Kurths 1996; Rosa, Ott and Hess 1998). Complete synchronization (CS) is the most natural synchronization state that emerges in case of identical systems, which corresponds to the equality of the state variables of the two systems while they evolve in time. On the other hand, generalized synchronization (GS) phenomena occurs between non-identical systems for a sufficiently large coupling strength, which eventually leads to the emergence of specific functional relationships between the outputs of the coupled systems. If the phases of two systems adapt to each other, that is, if they evolve in the same manner, then the synchronization is called phase synchronization (PS). CS is mainly studied only for the coupled identical systems, which is very rare in experimental cases. Thus there is a basic need to study synchronization between two non-identical systems. This is done mainly through PS and GS. PS is basically a weak degree of synchronization in which phases and frequencies of the systems become locked though their amplitudes often remain almost uncorrelated. On the other hand, in case of GS there exists a specific functional relationship between the outputs of the coupled systems. GS in most of the cases is intimately connected with a unidirectional coupling scheme, which has one drive and one response system. Both of PS and GS can be described by the recurrences and joint recurrences between the trajectories of different systems. This is because if two systems are synchronized then obviously their recurrences are dependent on each other.

Recurrence plot (RP) (Marwan et al. 2007) is a 2D graphical tool, which describes movements of the trajectories by only decoding the phase spaces points into binary image. The basic idea behind the construction of RP is that if two points in a phase space are recurrent in the sense that they fall in a neighborhood of each other, then the pair of points are represented by the number '1',otherwise they are represented by '0'. In this manner, all the points in the phase space are mapped into binary matrix, that is, a matrix whose elements are binary numbers (0 and 1). The image of that matrix reveals a 2D diagram consists of black and white dots, where black dots indicate recurrent points while white dots represent non-recurrent points of the phase space trajectories. In RP, a diagonal line explains how two trajectories move parallel in the phase space. In other words, the diagonal lines in RP indicate the periodic trajectories. In fact, for a periodic trajectory, RP shows some uninterrupted diagonal lines which are equally spaced. The distances between these diagonal lines are the periods of the trajectory. Hence, there is a relation between the distances of diagonals and the time scales of the system. Horizontal or vertical lines on the other hand describe how the trajectories are trapped in a position. There are several measures for quantifying the phase space, which are based on these diagonal, horizontal and vertical lines. However, not all of them are relevant in the present context. For the present purpose, the only measure that is very much useful is the recurrence rate (RR) (Marwan et al. 2007). RR describes how many points are recurrent in a phase space that means it calculates the recurrence density of a phase space. RR is computed for each diagonal line that is parallel to the line of identity (LOI). τ − recurrence rate is a special variety of RR, which is calculated over all diagonal lines that are parallel to the diagonal line located at a distance τ from LOI. This τ − recurrence rate can quantify the degree of PS, which helps us to analyze PS by means of τ-recurrence of the corresponding reconstructed phase spaces of two non identical systems, where phases or time scales are locked.

Joint recurrence is another way to describe recurrences between the trajectories of different systems, which measure the time indices when both of them recur simultaneously. The corresponding plot is known as Joint Recurrence Plot (JRP) (Marwan et al. 2007). In this approach, the basic features of the individual phase spaces are preserved. JRP can detect GS, which cannot be understood by RP. However, JRP still cannot detect the driver and response system in a unidirectional coupling occurred in GS. The detection of driver − response systems is normally done by mean conditional recurrence (MCR). MCR of the systems X and Y is the mean conditional probabilities of recurrence between them. Thus MCR is calculated from both of JRP and RP.

All of the aforesaid measures including RP and JRP are based on the phase space of the dynamical system. However, it is not always possible to get dynamical models for every real life phenomena. Rather, in most of the cases, only the data in the form of time series are given. Thus for time series data, construction of RP and JRP is possible only if the phase space can be reconstructed from this data. Thanks to the celebrated Takens theorem (Takens 1981) ensuring that such reconstruction is possible from one single observation of the real data.

Phase space reflects the long-term dynamics of a system. By virtue of the Takens theorem a phase space can be reconstructed from real data, which requires suitable time-delay and proper embedding dimension. For a nonlinear time series data, the suitable time-delay is generally calculated by Average Mutual Information (AMI) and the proper embedding dimension is calculated by the False Nearest Neighborhood (FNN) method (Kennel, Brown and Abarbanel 1992). Nonlinearity is normally verified by surrogate data test (Theiler 1992). The joining path of the successive embedding points in the phase spaces is called trajectory. The different natures of the trajectories in the phase spaces reveal a different kind of dynamics. For example, whenever trajectory moves in a bounded region with an erratic way, the system is called chaotic. One of the promising tools that provide evidence consistent with chaos in the reconstructed phase spaces is the Lyapunov Exponent (LE) (Strogatz 1994).

The words complexity and networks, originated in different contexts are getting popular day by day even in our everyday life. Nowadays, they are found to be jointly associated with characterizing the so-called complex networked systems in different areas of our society that range from natural phenomena to technological models. A common example is the Internet network, which is accessed by at least one-third of the world population and is still expanding continuously. Networks are basically graphs consisting of nodes and edges connect the nodes. Complex networks (CN) are large and irregular networks that are 'interesting' in the sense that although they are irregular they do convey certain statistical regularity. CN are involved in a number of different fields that include physics, chemistry, computer science, sociology, economics, finance, biology, epidemiology, and so on. In the present arena of research, CN is used as an alternative tool to understand the dynamical pattern of different real time systems (Albert and Barabasi 2000; Dorogovtsev and Mendes 2002; Ghoshal and Newman 2007; Newman 2003; Newman, Barabasi and Watts 2006). CN can be constructed from time series and so it is also used in time series analysis. Once a CN is constructed from a time series, different statistical measures – degree distribution of nodes (Ghoshal and Newman 2007), small-world property (Newman and Watts 1999; Watts and Strogatz 1998), clustering coefficients (Fronczak et al. 2002), Assortativity coefficient (Newman 2003), degree centrality, eigenvector centrality (Newman 2003; Newman, Barabasi and Watts 2006) and such others are used to properly characterize the system generating the time series data. In the past few years, several researches regarding the application of CN to the study of time series have been done throughout the globe (Nicolis, Garcia-Cantu and Nicolis 2005; Zhang and Small 2006; Small, Zhang and Xu 2009; Donner et al. 2010; Xu, Zhang and Small 2008; Lacasa et al. 2008; Li, Cao and Tan 2011). The network formed from the chaotic time series was found to possess small-world and scale-free characteristics, while the noisy time series correspond to the characteristics of a random graph (Zhang and Small 2006; Small, Zhang and Xu 2009). The formation of CN has also been done from the financial time series by many researchers (Mantegna 1999; Bonanno, Lillo and Mantegna 2001; Bonanno et al. 2004; Kim et al. 2002; Caldarelli 2007; Garlaschelli et al. 2005; Tse, Liu and Lau 2010). These types of CNs have the capability to explain different characteristics of the financial data that includes stock returns, global stock markets and securities markets data.

In the present context, the synchronization tools have been applied to analyze the worldwide population data so as to verify whether the population data of a particular year is synchronized with that of the previous years. On the other hand, CN and its associated statistical measures have been applied to the worldwide GDP data to discuss and compare the economic scenarios of different countries during two different time frames – 1990–2015 and 2020–2100. The whole chapter is organized as follows: Section 2 describes the data protocol and data processing, section 3.1 deals with the synchronization analysis of the country-level population data and section 3.2 performs the study of CN and its associated measures for the analysis of the country-level GDP data. The whole chapter is summarized in the conclusion section.

19.2   DATA PROTOCOL AND DATA PROCESSING

For the purpose of analyzing socio-economic data, we have considered only the country level population data and GDP data based on Special Report on Emissions Scenarios (SRES). The entire data and information are collected from a Data Centre in NASA's Earth Observing System Data and Information System (EOSDIS), hosted by the Center for International Earth Science Information Network (CIESIN) at Columbia University. The GDP data and country level populations are given for 1990–2100. For each case downscaled projections are predicted by SRES A1, A2, B1 and B2 Markers.

Population growth and economic growth rates are two exogenous assumptions incorporated within the four IPCC SRES scenario families A1, A2, B1 and B2. According to the SRES report economic growth rates were assumed to be 'very high' for the A1 family, 'medium' for the A2 family, 'high' for the B1 family and 'medium' for the B2 family. Quantitatively these assumptions translated into World GDP for the year 2100 of approximately 525–550 trillion US1990$ (market exchange rates)/year for the A1 family, 243 trillion US1990$/year for the A2 family, 328 trillion US1990$/year for the B1 family and 235 trillion US1990$/year for the B2 family. The corresponding per capita GDP growth rates depend on the corresponding regional population data used in the SRES report. The present SRES GDP projections for individual countries are downscaled out to 2100 by using the regional growth rate method, where the regional GDP growth rates were calculated from the marker model regional data and applied uniformly to each country that fell within the SRES's defined regions.

The population projections for the IPCC SRES emissions scenarios (A1, A2, B1 and B2) are taken from both the United Nations (UN) and the International Institute for Applied Systems Analysis (IIASA). The SRES A1–B1 and A2 population scenarios for world regions were adopted in 2000 from population projections realized at IIASA in 1996 and published in Lutz (1996). Both of the IPCC SRES A1 and B1 scenarios used the same IIASA 'rapid' fertility transition projection, which assumes low fertility and low mortality rates. The SRES A2 scenario used a corresponding IIASA slow fertility transition projection (high fertility and high mortality rates). The downscaling from region to country level of the IIASA scenarios is based on the calculation of the fractional shares of each country into regions according to the latest country population estimates and projections for 1990–2050, from the United Nations 2000 Revision. For each SRES population scenario, the UN variant that was the closest to the SRES scenario was chosen as the starting point for the population downscaling. Thus, UN medium variant was chosen for the A1 and B1 scenarios, while United Nations 2000 high variant was used for the A2 scenario. On the other hand, the SRES B2 population scenario was based on the UN 1998 Medium Long Range Projection for the years 1995–2100. The official version projects population are available for eight regions of the world: Africa, Asia (excluding India and China), India, China, Europe, Latin America, Northern America and Oceania. The UN 1998 Long Range Projections are a regional extension of the 1996 UN 'Revision' that projects population by country between 1995 and 2050. Projections are done for three variants: high, medium and low. Countries with populations less than 150,000 have not been included in the present database because these were not readily available from UN data sources in electronic form. The 1990 population estimates is also available in the database as a base year, which is collected from the UN Common Statistics Database (UNSTATS). The data were accessed at: http://unstats.un.org/ in April 2002. However, the present downs-caling procedure is not free from artifact. This is because of the post-2050 transition to the uniform growth rate method. If a country is projected by the UN 1996 revision to have a declining (or growing) population at 2050 but falls within a larger region that has a growing (or declining) population after 2050, a discontinuity will occur. For example, Cuba and Barbados are problematic in this regard and should not be used. Other coun-tries may have a slower or faster projected growth rate at 2050 than the regional projec-tion. In these cases, the population slope for such countries will show a discontinuity in post-2050. If an attempt is made to remove these discontinuities on a case-by-case basis, such as by using additional country-specific information, or even deleting them from the database altogether, then the regional totals will likely develop discrepancies with those in the SRES report. Artifacts are also associated with the simplicity of the linear downscal-ing approach, and these can only be removed with more sophisticated treatments, or by relaxing constraints on consistency with the SRES report.

The key difference in the downscaling procedure to GDP and population lies in the fact that uniform GDP growth rates were applied starting in the base year of 1990 but for population, uniform growth rates were applied only after 2050 (prior to the UN 1996 Revision population data were available to simulate growth rates changes over time). Therefore, the GDP downscaling introduces inaccurate national GDP growth rates in the near-term, when compared to actual near-term data for countries, because national GDP growth rates are obviously not uniform within regions. These are just the limitations of the GDP and Population databases we have considered for the present study of complexity analysis and in no way constrain the study on such data.

For the purpose of our study, we have subdivided both of GDP datasets into two groups. One is for the years 1990–2015 and another is for the years 2020–2100.

19.3   COMPLEXITY ANALYSIS OF SOCIO-ECONOMIC DATA

Complex dynamics and their interactions reveal the most important scenarios of a complex system. The complex interactions can be described by establishing the relations of the coupled systems and the measures of their complex network. There exist several standard complex systems which are predicted by the measures of complex coupling and complex networks. Studies of synchronization and complex networks are two essential mathematical tools, which have immense capability to describe the nature of complex coupling and the complex interactions respectively. Thus, in order to analyze the complex-ity of the socio-economic conditions (based on GDP and population data), we described some standard and well-known measures related to synchronization and complex net-works and then applied those measures to study GDP and population data.

19.3.1   Complex Coupling – Synchronization Analysis

Synchronization means an adjustment of rhythms of oscillating objects due to their weak interactions (Pikovsky, Rosenblum and Kurths 2001). Christiaan Huygens was the first scientist, who experimentally observed that – two pendulum clocks hanging from a common support, initially moving in opposite directions can be ultimately moved with same oscillation. It was the first experimental observation of synchronization phenomena before 1658 (Pikovsky, Rosenblum and Kurths 2001). But Huygens called it 'sympathy of two clocks'. He did not recognize this phenomenon as synchronization. Through a long process the study of synchronization has been made more sophisticated and still research is going on. Now to recognize whether a phenomenon is synchronized or not, the following criteria are verified:

- Systems must produce self-sustained oscillation, that is, systems can generate their own rhythms;
- System can adjust two oscillations with weak interactions;
- There exists a certain range in which the oscillations are mismatched.

Thus, synchronization is defined as a general process wherein two (or many) dynamical systems (equivalent or non-equivalent) are coupled or forced (periodically or noisily), in order to realize a collective or synchronous behavior. However, for nonlinear chaotic or complex dynamical systems arousal of such collective or synchronized behavior is, in general, not obvious. Chaos refers to the evolution that depends crucially on the initial conditions. Under the influence of chaos, even if two identical (but separated or uncoupled) systems evolve from slightly differing initial states, they give rise to two exponentially separated trajectories. In other words, the trajectories evolve in a non-synchronized manner. Thus, it seems that chaotic systems defy synchronization and this motivated the researchers to study how a synchronized behavior can be set in coupled chaotic phenomena.

Classification of different behavior in a synchronized coupled chaotic system can be done from the following three different viewpoints:

The first is to classify the observed collective states depending on the nature of the synchronization phenomena. Among several such synchronization states, complete or identical synchronization (CS) (Fujisaka and Yamada 1984; Afraimovich, Verichev and Rabinovich 1986; Pecora and Carroll 1990), phase synchronization (PS) (Rosenblum, Pikovsky and Kurths 1996; Rosa, Ott and Hess 1998), lag synchronization (LS) (Rosenblum, Pikovsky and Kurths 1997) and generalized synchronization (GS) (Rulkov et al. 1995; Kocarev and Parlitz 1996) have received tremendous attention from the researchers.

The second is to distinguish different synchronization phenomena as a function of the nature of the coupling. Sometimes a major difference is observed in the process leading to synchronized states depending upon whether the coupling is symmetrical or asymmetrical. This leads to a bidirectional and the unidirectional coupling configuration. The nature of such coupling is not always linear. The coupling may be nonlinear and even impulsive (that is, a coupling that acts only at discrete time intervals). In unidirectional coupling, there is associated a drive (master) system, output of which controls the behavior of a response (slave) system. This further means that the master system evolves uncoupled and drives evolution of the slave system. The situation is very different for bidirectional coupling where both subsystems are coupled to each other, and an adjustment of the dynamics onto a common synchronized manifold is induced by the coupling factors.

The third is to distinguish the different cases of synchronization as a function of the nature of the coupled systems. This is a comparatively new approach and we are not interested in discussing this point in the present context. Synchronization can be made possible for both identical systems and non-identical systems. However, synchronization of non-identical systems needs more attention as the experimental and real systems are never identical. Since in the present case we are dealing with some real data, it is worthy studying synchronization only for non-identical systems. Worldwide population depends on several factors and in this concern one of the major issues is whether or not the population data of a particular year is synchronized with that of the earlier years. This can only be verified by finding the existence of coupling between them. Finding the direction of such coupling is another interesting issue in this context. The respective methodologies are sequentially described in the following subsections.

Concept of phase space and phase space reconstruction
Since synchronization of two time series is based on their reconstructed phase spaces, the idea of phase space and reconstruction of phase space must be known in detail. Let us consider n- nonlinear autonomous simultaneous differential equations:

$$\frac{dy_k}{dt} = f_k(y_1, y_2, y_3, ..., y_n) \ k = 1, 2, 3, ..., n \tag{1}$$

Then, flow or solution space or phase space of the system (1) is given by $(y_1(t), y_2(t), ..., y_n(t)) \in \mathbb{R}^n$. In this context, a solution with an initial condition of the system (1) is like a path of a fixed point which is dropped in that flow $(y_1(t), y_2(t), ..., y_n(t)) \in \mathbb{R}^n$. The path is called trajectory in the phase space. There are several linear mathematical tools, by which we can extract some linear information, viz., equilibrium with stability, instability or more precisely saddle, node and focus, saddle-node bifurcation, saddle-focus for the linear version of the nonlinear system of two or more non-autonomous differential equations (Strogatz 1994). But according to the famous Hartman–Grobman theorem (Hartman 1960), all the nonlinear information (Hopf bifurcation: subcritical limit cycle, supercritical limit cycle) regarding the solution in the phase space cannot be extracted without using nonlinear mathematical tools (Strogatz 1994). Chaos (Strogatz 1994) is also a nonlinear phenomenon, which may occur for a system of non-autonomous differential equations, whenever the number of equations is at least three. Whenever the solution $(y_1(t), y_2(t), ..., y_n(t)) \in \mathbb{R}^n$ of (1) with some initial conditions $(y_1(a_1), y_2(a_2), ..., y_n(a_n))$, goes from the initial point to another position and moves on a closed curve with the change of t in the phase space, we say that Hopf bifurcation occurs in that system. The closed curve is called limit cycle. It has been observed that this occurs due to a small perturbation of some parameters involved in the system (1). If the limit cycle is formed by the trajectories of the system (1) with initial point $(y_1(a_1), y_2(a_2), ..., y_n(a_n))$ lying outside the limit cycle, then it is called a subcritical limit cycle. If the limit cycle is formed by the trajectories with initial point $(y_1(a_1), y_2(a_2), ..., y_n(a_n))$ lying inside the limit cycle, then it is called a supercritical limit cycle. The subcritical or supercritical limit cycles are called regular attractors and they are also stable for a particular value of the parameters involved in the system (1). Whatever the form of limit cycles, it has been seen that the path of the phase spaces always moves in a closed path. This implies that the trajectories are always attracted by a periodic phase space and hence it is known as periodic attractors. However, if the solution $(y_1(t), y_2(t), ..., y_n(t)) \in \mathbb{R}^n$ of (1) with some initial conditions, goes from the point $(y_1(a_1), y_2(a_2), ..., y_n(a_n))$ to another position and moves in an erratic manner with the change of t in the phase space, we say that chaos may occur for the system (1). In this case, trajectories of the system move in a bounded region and formed a dense orbit, called a chaotic attractor.

To explain all these nonlinear features, let us consider a standard Food web model (Mandal et al. 2010), which is given by:

$$\frac{dx}{dt} = \alpha x\left(1 - \frac{x}{k}\right) - \frac{a_1 x}{1 + b_1 x} y$$

$$\frac{dy}{dt} = h_1 \frac{a_1 x}{1 + b_1 x} y - \frac{a_2 y}{1 + b_2 y} z - d_1 y \qquad (2)$$

$$\frac{dz}{dt} = h_2 \frac{a_2 y}{1 + b_2 y} z - d_2 z$$

where x, y, z are variables and $\alpha, k, a_1, b_1, h_1, a_2, b_2, h_2, d_1, d_2$ are parameters with the initial condition $x(0) = 0.7; y(0) = 0.11; z(0) = 9.1$.

Figure 19.1 shows point, regular and chaotic attractors of the system (2). There are several illustrations of mathematical models, where the aforesaid nonlinear qualitative behaviors are observed. There are some standard quantifying measures such as Lyapunov exponent (Strogatz 1994), sensitivity analysis (Strogatz 1994), Bifurcation analysis (Strogatz 1994) by which these qualitative behaviors can be recognized. In fact, constructing a proper model for a complex phenomenon and observing their nonlinear behavior is one of the efficient ways to predict the long-term dynamics of the system. However, if the parameters and variables of the system are not properly taken, the mathematical model does not reflect the real scenario. Thus reduction of the dimension of the dynamics (known as embedding dimension) and improper choice of the parameters reveals huge error in the prediction of the dynamics. So, a general question arises: 'Is there any way to reconstruct the dynamics from the time series observation of a dynamical system?' In the famous Takens theorem, Takens first proved that it is possible to reconstruct the attractor of a given continuous dynamical system from one of the components of its solution vector and that the reconstructed attractor is topologically equivalent to the given attractor.

The method of attractor reconstruction or phase space reconstruction needs two important parameters, viz. suitable time-delay and proper embedding dimensions. There are several methods for finding the suitable time-delay and embedding dimension respectively. To calculate suitable time delay, methods of autocorrelation (AC) (Williams 1997), Average Mutual Information (AMI) (Fraser and Swinney 1986) are very useful. The reconstruction of phase space can also be done with different time-delays. In this case, suitable time-delays are calculated by AMI of several variables (Simon and Verleysen 2007), Cross auto-correlation (Palit, Mukherjee and Bhattacharya 2013). On the other hand, embedding dimension can be best approximated by the False Nearest Neighborhood (FNN) method (Kennel, Brown and Abarbanel 1992). Suitable time-delay and embedding dimension are known as the reconstruction parameter and it is denoted as (τ, m) (for example, $(\tau, m) = (13, 3)$ for the system (2), in Figure 19.2). For a signal $\{x(t)\}$ with reconstruction parameter (τ, m), the reconstructed attractor is given by

$$\{x(t), x(t + \tau), x(t + 2\tau), \ldots, x(t + \overline{m - 1}\tau)\} \qquad (3)$$

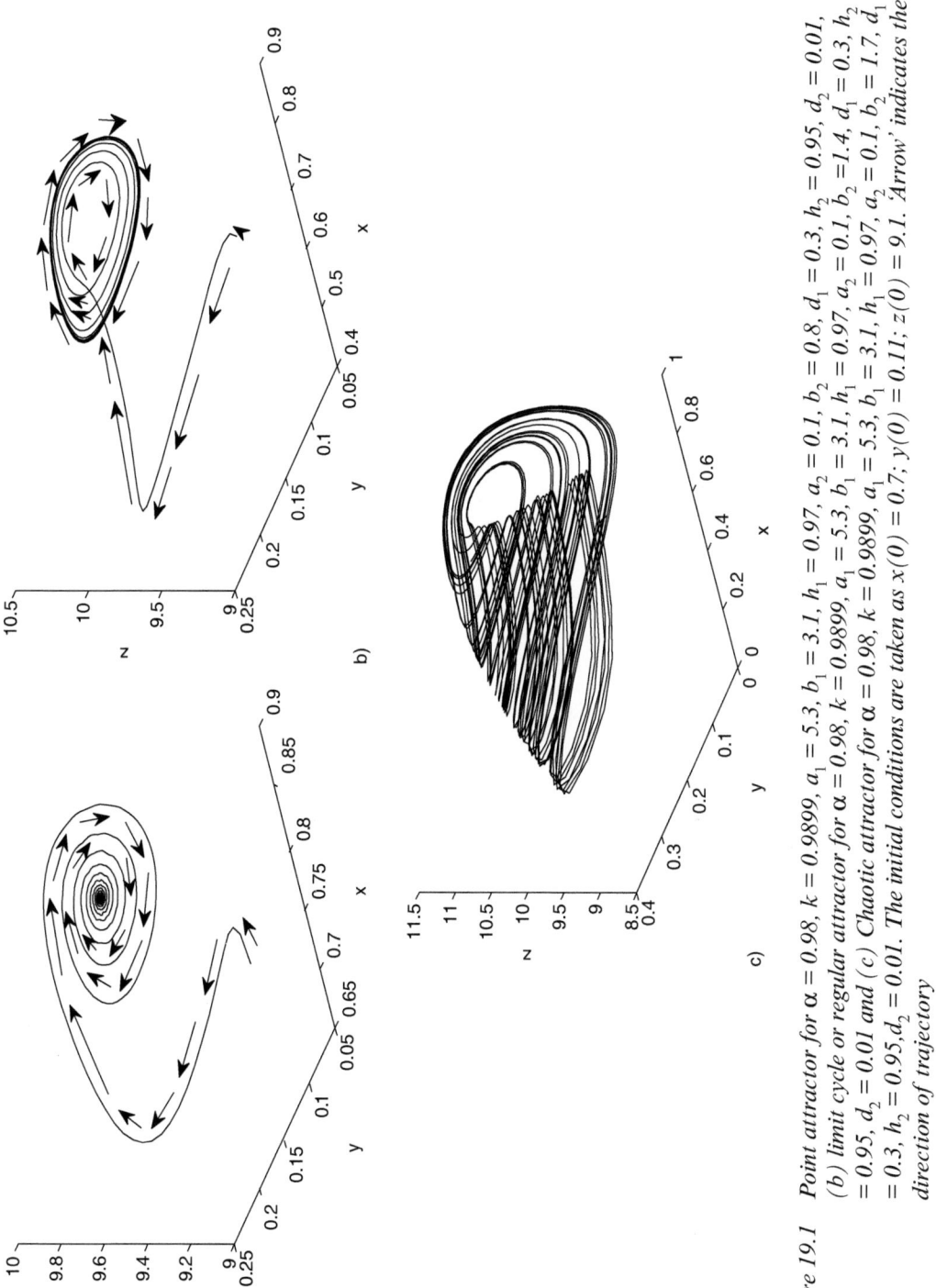

Figure 19.1 Point attractor for $\alpha = 0.98$, $k = 0.9899$, $a_1 = 5.3$, $b_1 = 3.1$, $h_1 = 0.97$, $a_2 = 0.1$, $b_2 = 0.8$, $d_1 = 0.3$, $h_2 = 0.95$, $d_2 = 0.01$, (b) limit cycle or regular attractor for $\alpha = 0.98$, $k = 0.9899$, $a_1 = 5.3$, $b_1 = 3.1$, $h_1 = 0.97$, $a_2 = 0.1$, $b_2 =1.4$, $d_1 = 0.3$, $h_2 = 0.95$, $d_2 = 0.01$ and (c) Chaotic attractor for $\alpha = 0.98$, $k = 0.9899$, $a_1 = 5.3$, $b_1 = 3.1$, $h_1 = 0.97$, $a_2 = 0.1$, $b_2 = 1.7$, $d_1 = 0.3$, $h_2 = 0.95$, $d_2 = 0.01$. The initial conditions are taken as $x(0) = 0.7$; $y(0) = 0.11$; $z(0) = 9.1$. 'Arrow' indicates the direction of trajectory

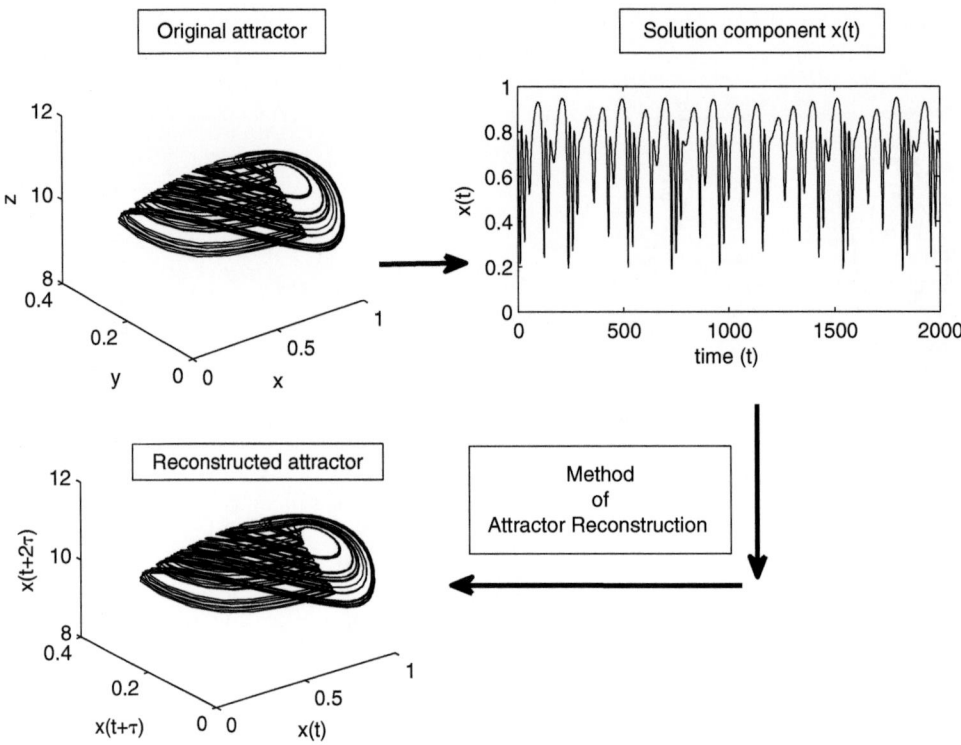

Figure 19.2   *Original attractor of system (2). AMI reveals* $\tau = 13$. *With* $\tau = 13$, *the 3D attractor is reconstructed, which is topologically equivalent to the original attractor*

However, it is not always possible to explain the movements of the phase space trajectories by the aforesaid measures, especially for the phase spaces with higher embedding dimension. Recurrence plot (RP) (Eckmann, Kamphorst and Ruelle 1987) and its related measures (Marwan et al. 2007) stand as a most promising tool in this context. In fact, RP can be used in the detection of synchronization and also in finding the coupling directions.

Recurrence plot
Let $x_i \in \mathbb{R}^n$ be any point on the trajectory in n-dimensional phase space. Then, recurrence between any two points $x_i, x_j \in \mathbb{R}^n$ means how much they are closed to each other in that phase space. Mathematically, $x_i, x_j \in \mathbb{R}^n$ have a recurrence if $x_i \in N_\varepsilon(x_j) = \{x \in \mathbb{R}^n : \|x - x_j\| < \varepsilon\}$, where ε is a proper threshold. Recurrence plot is a two dimensional (2D) diagram constructed from a matrix $(R_{ij})_{n \times n}$, where $R_{ij} = \Theta(\|x_i - x_j\| - \varepsilon)$, $\|\cdot\|$ is the Euclidean norm in $\mathbb{R}^n$ and Θ is Heaviside function. In other words,

$$R_{ij} = \begin{cases} 1, \text{ if } \|x_i - x_j\| < \varepsilon \\ 0, \text{ otherwise} \end{cases} \tag{4}$$

In the pictorial representation, '1' is considered as a 'black' dot and '0' as 'white' dot. Therefore, recurrence of any two points in n-dimensional phase space is visualized as a 'black' spot in the two-dimensional diagram. Thus, a complete picture of recurrence of all points in phase space can be found out only in the 2D diagram, which is known as recurrence plot (RP). Recurrence is a fundamental property of a dissipative dynamical system. The n-dimensional phase space trajectories can be investigated through 2D representation of their recurrences (Marwan et al. 2007; Eckmann, Kamphorst and Ruelle 1987). Initially RP was used for the visual inspection of higher dimensional trajectories. In RP, periodic orbits are recognized by a rectangular region made by the diagonal lines. The occurrence of more than two rectangular regions indicates the presence of chaos due to unstable periodic orbits. In RP, the diagonal lines represent the parallel movements of the trajectories, while the vertical/horizontal lines represents the trapping time of the trajectories. In this concern, recurrence rate (RR) (Marwan et al. 2007) is a very useful measure. This is defined by

$$RR(\varepsilon) = \frac{1}{N^2} \sum_{i,j=1}^{N} R_{i,j}(\varepsilon) \tag{5}$$

The measure is computed for each diagonal line which is parallel to the line of identity (LOI). A particular variety of recurrence rate is $\tau-$ recurrence rate (Marwan et al. 2007), which is calculated over all diagonal lines which are parallel to the diagonal line located at a distance τ from LOI. Mathematically, the $\tau-$ recurrence rate (Marwan et al. 2007) is defined by:

$$RR_\tau = \frac{1}{N-\tau} \sum_{i=1}^{N-\tau} R_{i,i+\tau} \tag{6}$$

Let us now first illustrate RP by means of the famous example of the Lorenz system (Strogatz 1994) used to model the atmospheric systems. This is given here:

$$\frac{dx}{dt} = s(y - x)$$

$$\frac{dy}{dt} = rx - y - xz \tag{7}$$

$$\frac{dz}{dt} = -bz + xy$$

The initial condition is taken as $x(1) = 8, y(1) = 9, z(1) = 25$.

The chaotic attractor is obtained for the parameter values $s = 10, r = 28, b = 8/3$, which is shown in Figure 19.3a. The corresponding RP is given by Figure 19.3b.

The same technique is now applied on the population data (Figure 19.4). Since there does not exist any mathematical model for these data, the phase spaces are first reconstructed from the data by using equation (3) with $(\tau, m) = (1, 2)$, where τ is the delay and m is the embedding dimension. Finally, the corresponding RP's are obtained from that reconstructed phase spaces. As a sample illustration, the time series plot and the corresponding RPs of the population data only for the year 1990 and 2020 under A1B1A2 and B2 marker scenarios are presented in Figure 19.4.

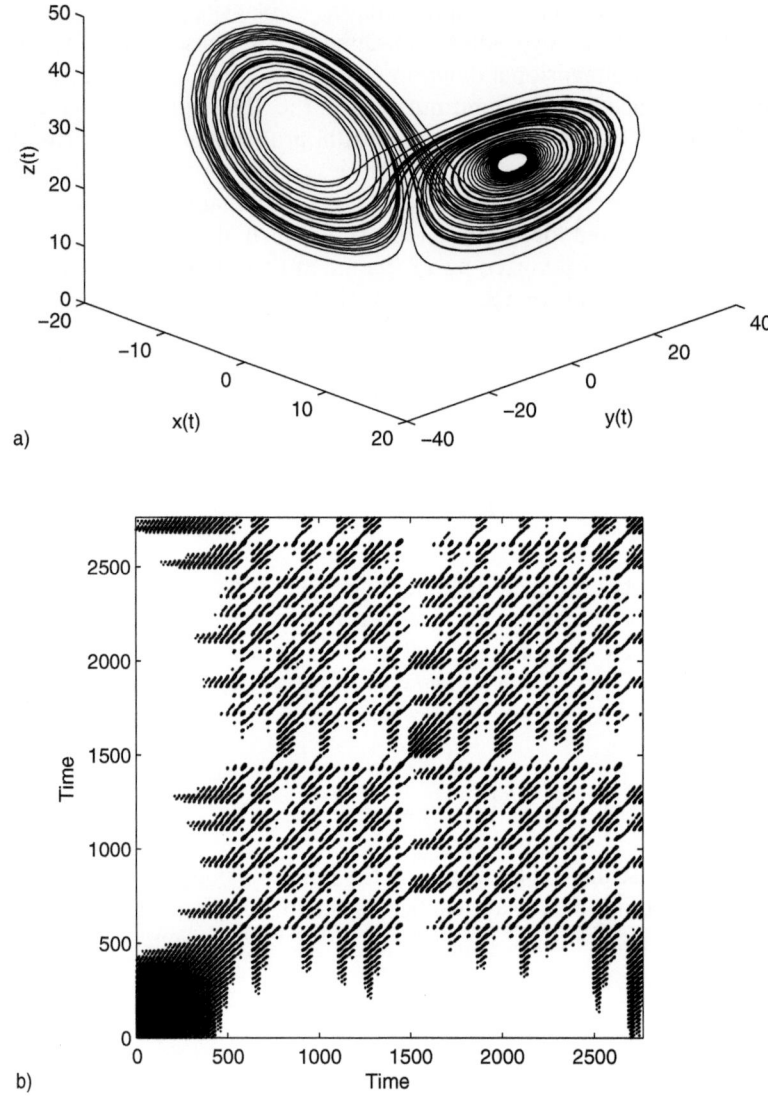

a)

b)

Figure 19.3a   Chaotic attractor of the Lorenz system given by (7). Chaotic solution
is found out by solving eq. (7) by integrating the differential equation.
As system parameters control the dynamical behavior we have taken
s = 10, r = 28, b = 8/3 for getting chaotic attractor. The time delay is taken as
dt = 0.01. (b) Recurrence plot for the chaotic attractor of Lorenz system with
ε = 5

Phase synchronization (PS) by means of τ-recurrences
Two non-identical systems meets in PS (Boccaletti et al. 2002; Marwan et al. 2007;
Mukherjee et al. 2014) means that the distances between their diagonal lines in RP are
coinciding as their time scales adapt to each other. Now, if probability of first system

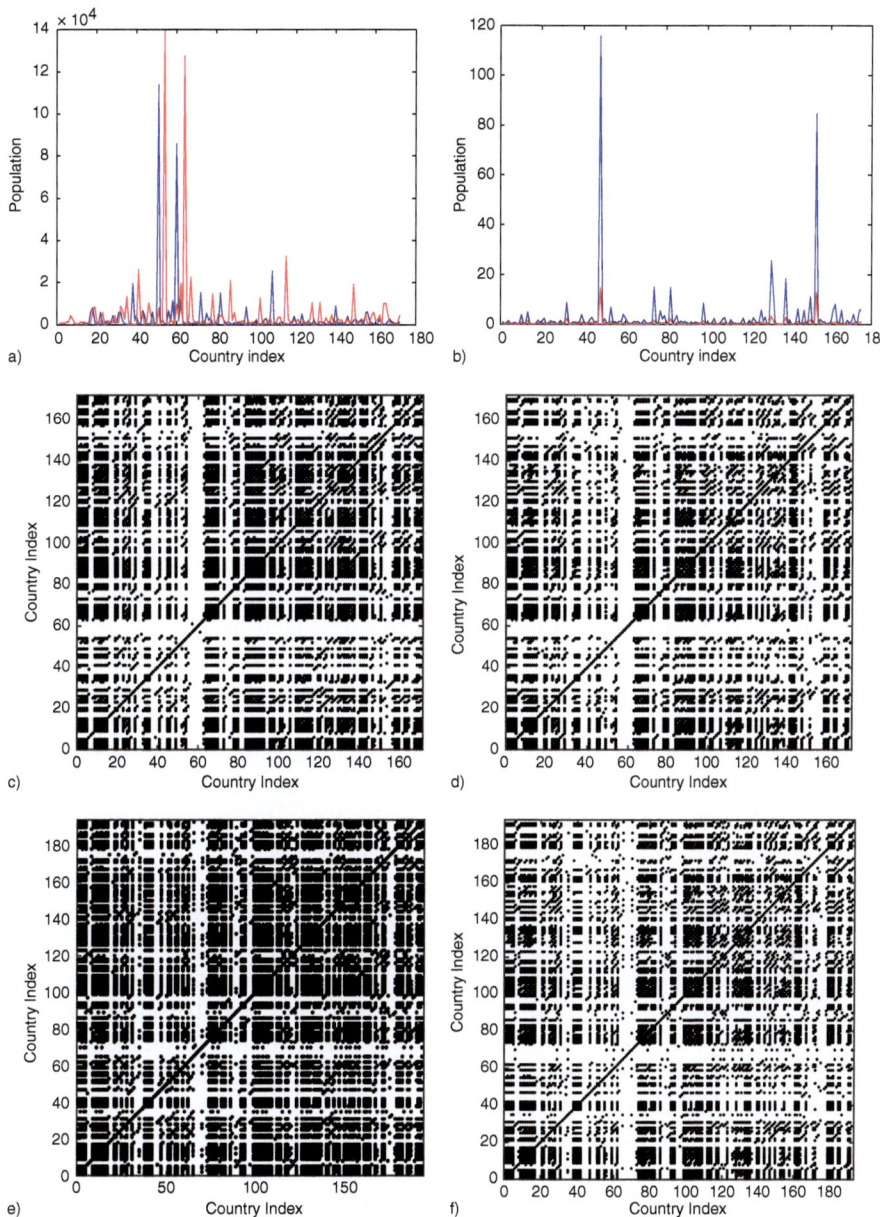

Figure 19.4   *(a) and (b) represents population data for A1B1A2 and B2 marker scenarios respectively. The data are visualized after downscaling it by 10^{-11}. The data are plotted only for the year 1990 (blue color) and 2020 (red color) under A1B1A2 and B2 marker scenarios. (c) and (d) represent RP of the population data for the year 1990 and 2020 respectively under A1B1A2 marker scenarios. (e) and (f) represent RP of the population data for the year 1990 and 2020 respectively under B2 marker scenarios. In each case, threshold is taken as $\varepsilon = 0.1\sigma(x)$, $\sigma(x)$ is the standard deviation of the population data $x(t)$*

recurs after τ time steps is large, then same for the second system recurs after the same time interval will be also large. Let $p(\varepsilon,\tau)$ be the probability that a system recurs to ε-neighborhood of a point x_i of the trajectory after τ time steps. Then by comparing probability $p(\varepsilon,\tau)$ of the two systems, it is possible to quantify PS (Marwan et al. 2007) clearly. The probability $p(\varepsilon,\tau)$ can be estimated directly from the RP by:

$$p(\varepsilon,\tau) = RR_\tau(\varepsilon) = \frac{1}{N-\tau}\sum_{i=1}^{N-\tau}R_{i,i+\tau} \tag{8}$$

which is basically the $\tau-$ recurrence rate as defined earlier by equation (6).

Let $RR_\tau^X = \frac{1}{N-\tau}\sum_{i=1}^{N-\tau}R_{i,i+\tau}^X$ and $RR_\tau^y = \frac{1}{N-\tau}\sum_{i=1}^{N-\tau}R_{i,i+\tau}^y$ be the $\tau-$ recurrence rates for the two phase spaces X and Y. Basically, $\overline{RR_\tau^X}$ and $\overline{RR_\tau^Y}$ are the probabilities normalized to zero mean and standard deviation one. The correlation coefficient CPR between $\overline{RR_\tau^X}$ and $\overline{RR_\tau^Y}$ is defined as:

$$CPR = \langle \overline{RR_\tau^X}, \overline{RR_\tau^Y} \rangle \tag{9}$$

where $\langle X, Y \rangle$ denotes the inner product of X, Y. If $CPR \approx 1$, then both systems are in PS in the sense that the probability of recurrence will be maximal at the same time (Marwan et al. 2007). The value of CPR also indicates the degree of synchronization.

Since our goal is to check whether or not the population data of different years (1990–2100 with a gap of five years) are synchronized with each other, we calculate the $\tau-$ recurrence rate for all pairs of the given population data for the year 1995 up to 2100. To determine their degree of synchronization, we also calculate the CPR values for all such pairs. This actually helps to detect whole scenarios of the phase synchronization for all pairs of population data. Thus, we calculate $RR_\tau^{X^{(i)}}, RR_\tau^{X^{(j)}}$ for each pair of population data $(X^{(i)}, X^{(j)}), i, j = 1, 2, 3, ..., 23$ (1 denotes the year 1990, 2 denotes 1995 and so on) and then obtain $CPR(i,j) = \langle \overline{RR_\tau^{X^{(i)}}}, \overline{RR_\tau^{X^{(j)}}} \rangle$ for each i, j. The contour plots of $[CPR(i, j)]$ for the population data under A1B1A2 and B2 marker scenarios are given by Figures 19.5a and 19.5b respectively.

As both the plots are symmetric, it is sufficient to look after any half of the contour plots to draw any conclusions. Both the plots reveal that all pairs of population data are phase synchronized and degree of PS for both the marker scenarios show an almost decreasing trend. For better understanding, the CPR values of the year indices $(1,j), j = 1, 2, ..., 23$ are presented in Figure 19.6 as a sample illustration.

Practically this type of trend indicates that the underlying dynamics of population data of past present and future are synchronized but as the time grows, the degree of synchronization gradually decreases. This is quite obvious; as the time grows different new parameters come into existence, which greatly influence the underlying dynamics of the population data.

From the above analysis, behavior of the phase synchronization of two non-identical systems can easily be understood. In other words, phase dependency of two systems can be described by $\tau-$ recurrence rate. Correlation CPR measures degree of phase synchronization. However, PS is a weaker type of synchronization in the sense that in PS, only the long-term reconstructed dynamics (phase space) of two non-identical systems get correlated. Moreover, in PS, when the dynamics are reconstructed from two different time

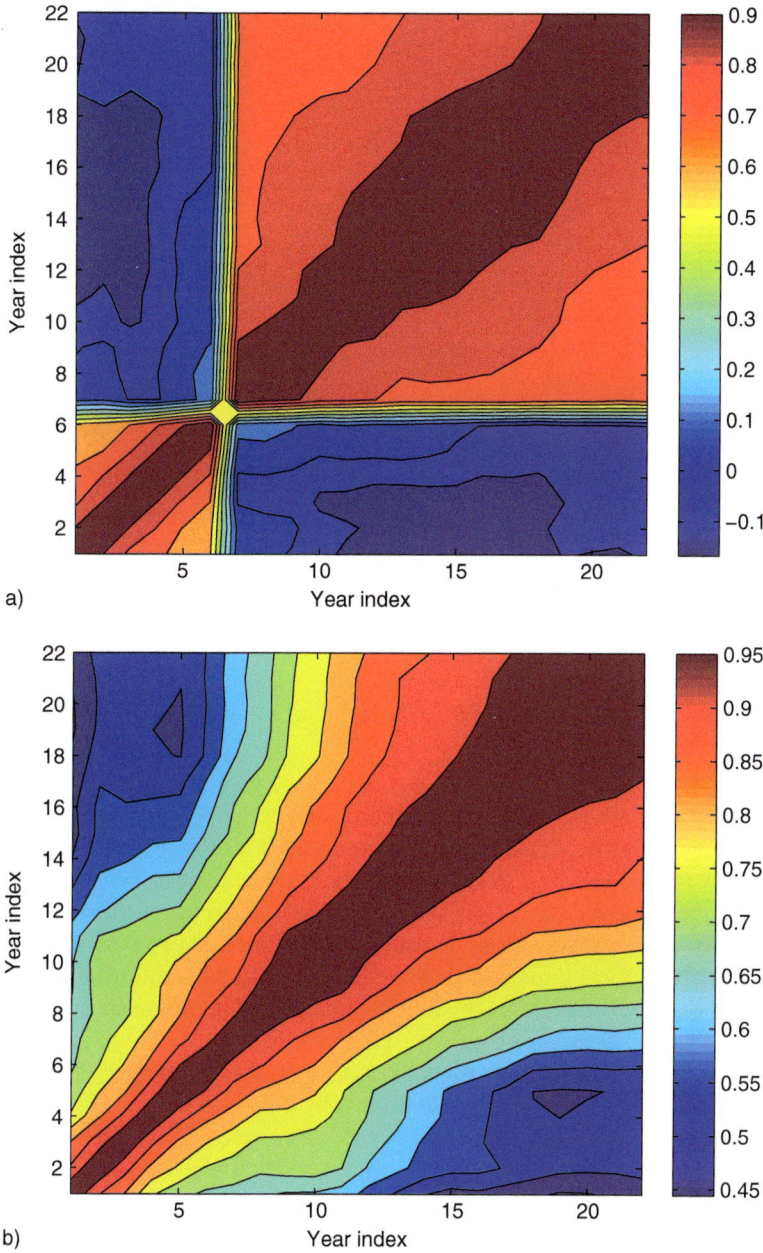

a)

b)

Figure 19.5   *Contour plots constructed with the values of [CPR(i, j)] for the Year index(i,*
j = 1, 2,.., 23). In each case, (τ, m) for the phase space reconstruction and ε for
RP construction are taken as (τ, m) = (1, 2) and ε = 0.5σ(x) σ(x) being the
standard deviation of the population data x). Color bar denotes the values of
[CPR(i, j)]. (a) represents contour plot of the population data for A1B1A2
marker, while (b) represents the same for B2 marker

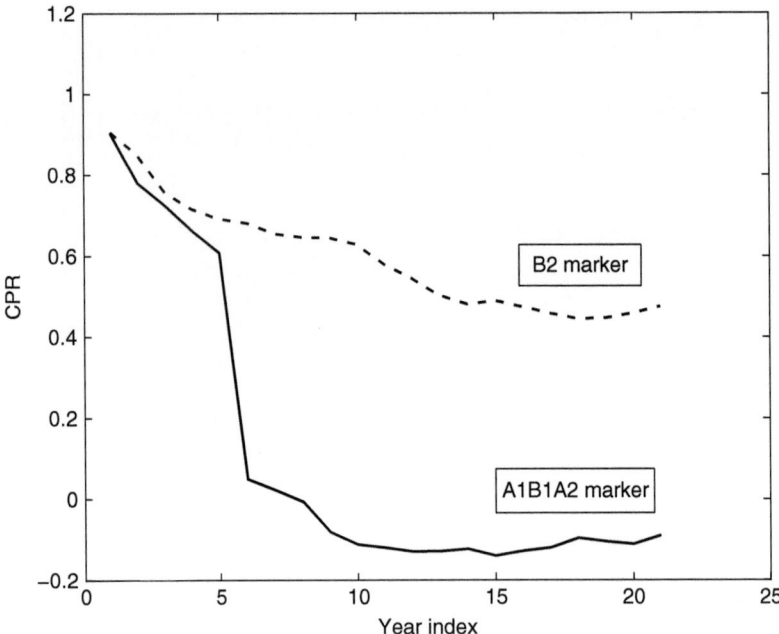

Figure 19.6   CPR values plotted with respect to Year index (1,j), j = 1, 2,.., 23. Solid and dotted lines represent CPR of the A1B1A2 marker and B2 marker respectively

series the delay and embedding dimensions are considered as the same. So we must look for a comparatively stronger version of synchronization, which is no longer possible by RP. In this concern, another concept, Cross Recurrence Plot (CRP) was developed, which is simply a bivariate extension of RP. CRP measures dependencies between two different systems by comparing their states (Marwan et al. 2007). However, the CRP concept is not workable for two different time series. In fact, the difference of two different state vectors with two different physical units reveals completely meaningless information. Moreover, different phase space with different embedding dimension does not make any sense in CRP (Marwan et al. 2007). There is another way to describe recurrences between the trajectories of different systems in which phase spaces are reconstructed separately and measure the time indices when both of them recur simultaneously. This is known as Joint recurrence and the corresponding matrix plot is called Joint Recurrence Plot (JRP) (Marwan et al. 2007). In this approach, the basic features of the individual phase spaces are preserved.

Joint recurrence plot and mean conditional recurrence
Let the embedding dimensions of the reconstructed phase spaces of two systems X and Y be d_X and d_Y respectively and $\varepsilon^X, \varepsilon^Y$ be the respective thresholds of JRP. Then, JRP is given by a joint recurrence matrix $JRP_{i,j}^{X,Y}$ is given by:

$$JRP_{i,j}^{X,Y} = \Theta(\varepsilon^X - \|X_i - X_j\|)\Theta(\varepsilon^Y - \|Y_i - Y_j\|), \; i,j = 1,2,...,N, \tag{10}$$

where $\Theta(.)$ is the Heaviside function.

From the above definition, it implies that the system X and Y recurs in its own phase space simultaneously. Using JRP, different kinds of complex coupling can be measured. In fact, generalized synchronization (GS), a much stronger version of synchronization can be detected by the JRP. As far as direction of coupling is concerned, JRP alone cannot recognize the driver and response in a synchronized phenomenon. Thus, in order to recognize driver and response, Romano et al. first proposed a method of Mean Condition Recurrence (MCR) (Romano et al. 2007) based on JRP and RP. MCR is nothing but the mean conditional probabilities of recurrence between the systems X and Y. This is defined as follows:

$$MCR(Y/X) = \frac{1}{N}\sum_{i=1}^{N}\frac{\sum_{i=1}^{N}JRP_{i,j}^{X,Y}}{\sum_{i=1}^{N}R_{i,j}^{X}}, \tag{11}$$

$$MCR(X/Y) = \frac{1}{N}\sum_{i=1}^{N}\frac{\sum_{i=1}^{N}JRP_{i,j}^{X,Y}}{\sum_{i=1}^{N}R_{i,j}^{Y}}, \tag{12}$$

where $R_{i,j}^{X} = \Theta(\varepsilon^{X} - \|X_i - X_j\|)$ and $R_{i,j}^{Y} = \Theta(\varepsilon^{Y} - \|Y_i - Y_j\|)$.

The criterion for detecting the asymmetry of the coupling is given as follows:

If X drives Y, $MCR(Y/X) < MCR(X/Y)$,
If Y drives X, $MCR(X/Y) < MCR(Y/X)$.

To compute MCR, it is necessary to fix four parameters: the embedding parameter (τ, m) for the reconstruction of the phase space, and the thresholds $\varepsilon^{X}, \varepsilon^{Y}$. But it has been established that (τ, m) does not influence the results of MCR (Romano et al. 2007). It has also been verified that MCR does not depend on the choice of $\varepsilon^{X}, \varepsilon^{Y}$ (Romano et al. 2007). In order to detect direction of the coupling, a broad range of the values of $\varepsilon^{X}, \varepsilon^{Y}$ reveals correct estimation of (11) and (12) (Romano et al. 2007).

For our present analysis of downscaled population data through JRP and MCR, (τ, m) is taken as $(1, 2)$ for each pair of dataset x, y. Using $(\tau, m) = (1, 2)$, joint recurrence matrices are calculated with $\varepsilon^{X}, \varepsilon^{Y} = 0.5\sigma(x)$ ($\sigma(x)$ being the standard deviation of the population data x). As sample illustrations, some of them are visualized by their corresponding JRP given by Figure 19.7.

In Figure 19.7, a large number of rectangular and square regions are found in JRP. Also there exist few diagonal, horizontal/vertical lines and single isolated points. Rectangular or square regions and long diagonal, horizontal and vertical lines indicate strong correlations between two systems (here two datasets). Thus by joint recurrence, functional dependencies between two sets of population data is established. In fact, JRP ensures the existence of functional correlations between each pair of population data. But, it is impossible to recognize driver and response systems from JRP. This is because JRP can detect only recurrence of the trajectories of the two phase spaces.

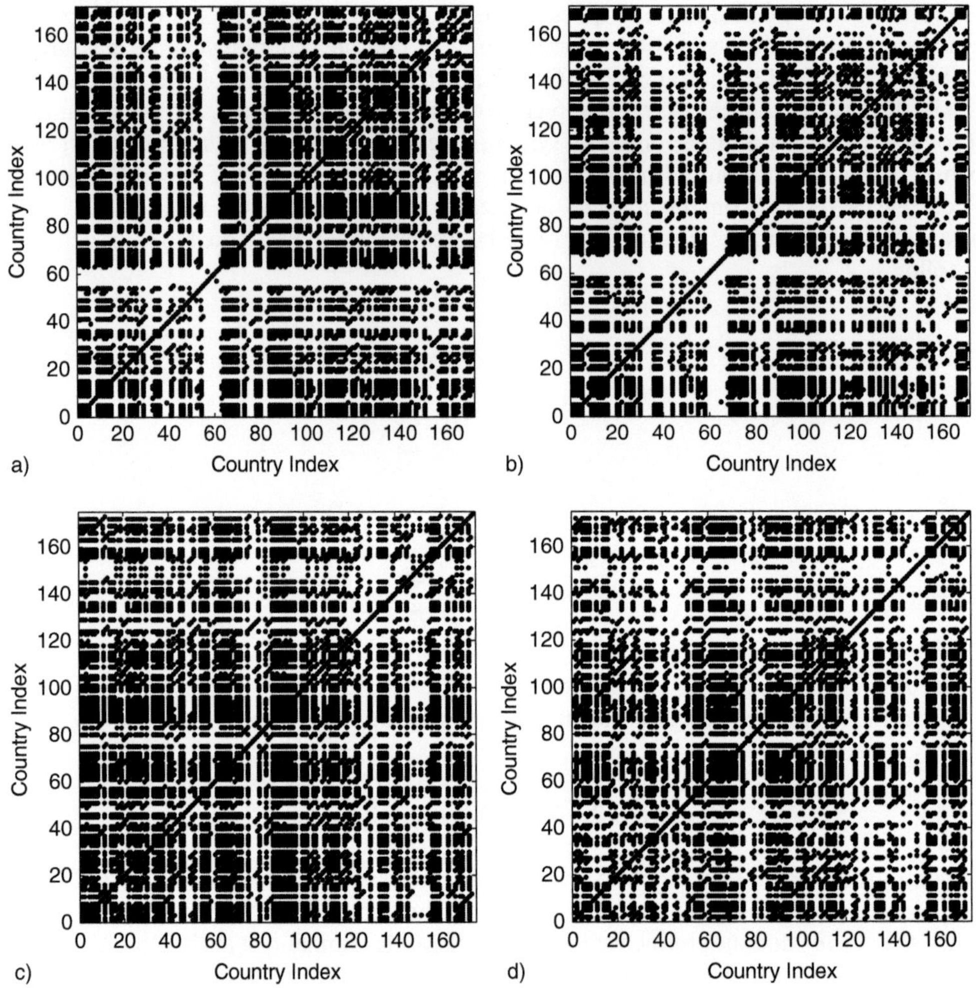

Figure 19.7   *JRP of the population data for (a) 1990 and 1995 of A1B1A2 marker, (b) 1990 and 2020 of A1B1A2 marker, (c) 1990 and 1995 of B2 marker and (d) 1990 and 2015 of B2 marker. In each case, threshold is taken as $\varepsilon^x = \varepsilon^y = 0.5\sigma(x)$ and delay embedding $(\tau, m) = (1.2)$*

To detect driver and response, we apply MCR analysis on each pair of data in both forward and backward directions. Thus, we measure both $MCR(Y/X)$ and $MCR(X/Y)$ for each pair of population data and calculate the difference $\Delta(MCR) = MCR(X/Y) - MCR(Y/X)$ between them. These differences are stored in a matrix $\Delta(MCR)_{i,j}$. The matrix plots of the matrix $\Delta(MCR)_{i,j}$ for A1B1A2 and B2 marker scenarios are respectively given by Figure 19.8a 19.8b.

Figures 19.8a and 19.8b reveal that the matrices $\Delta(MCR)_{i,j}$ for both of A1B1A2 and B2 marker are asymmetric. This means that the values of $\Delta(MCR) = MCR(X/Y) - MCR(Y/X)$ are not same for (i,j) and (j,i) indices. So $[\Delta(MCR)]_{i,j} \neq [\Delta(MCR)]_{j,i}$. Physically, it implies that

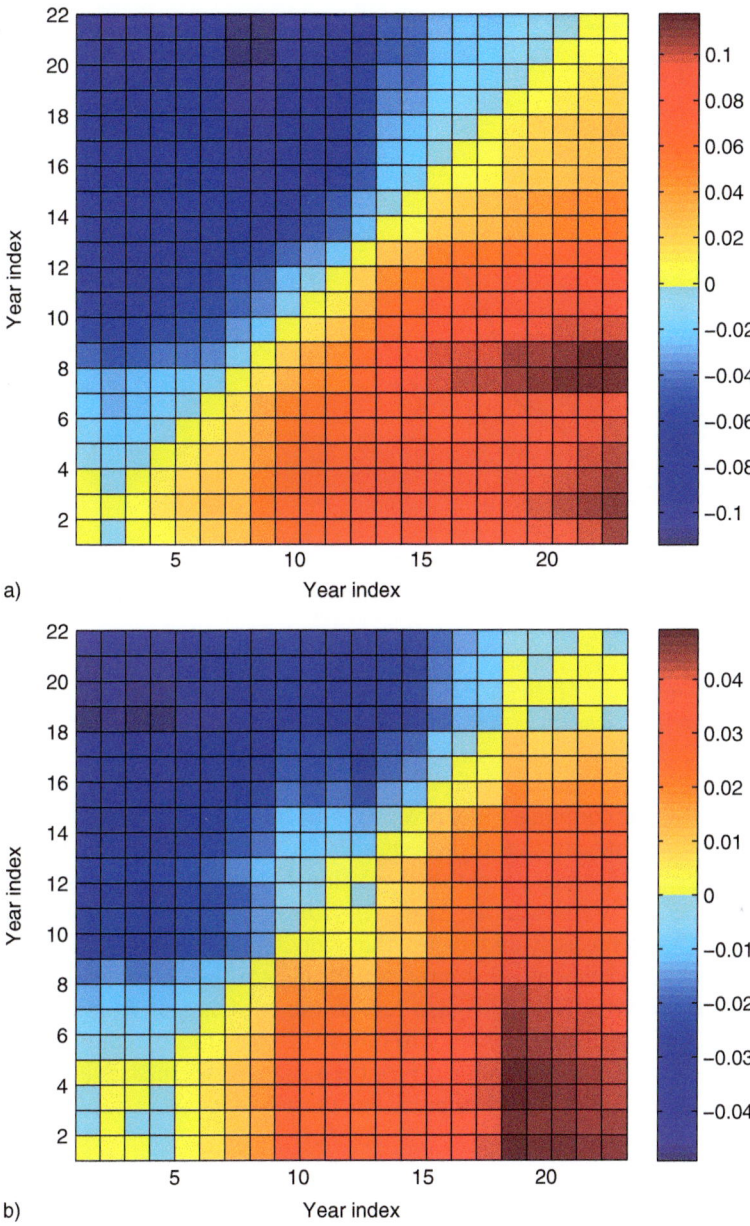

Figure 19.8 *Matrix plots constructed with the values of $[\Delta(MCR)]_{i,j}$ for the Year index $(i, j = 1, 2,..., 23)$. In each case, (τ, m) and $(\varepsilon^X, \varepsilon^Y)$ are taken as $(\tau, m) = (1, 2)$ and $\varepsilon^X, \varepsilon^Y = 0.5\sigma(x)$ $(\sigma(x)$ being the standard deviation of the population data $x)$. Color bar denotes the values of $\Delta(MCR)$. (a) and (b) represents matrix plot of the population data for A1B1A2 and B2 marker respectively*

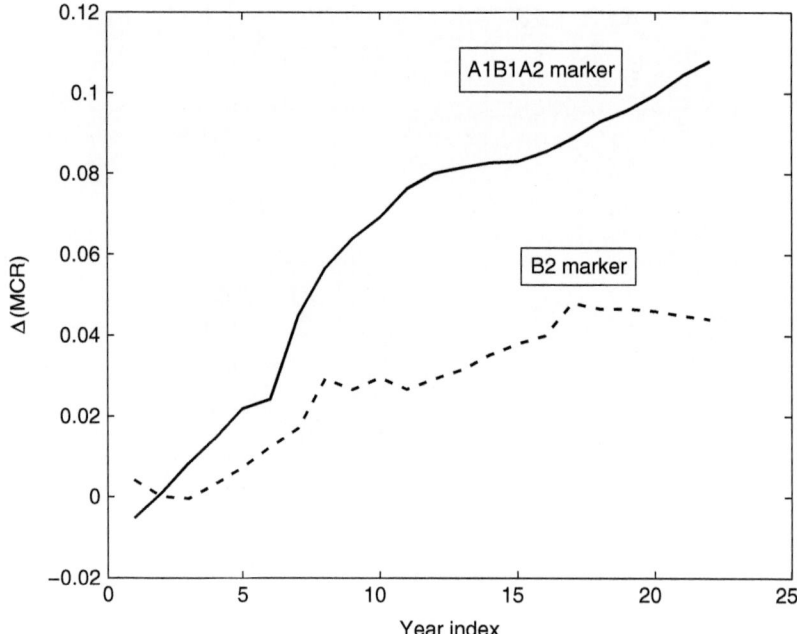

Figure 19.9   Values of Δ(MCR) = MCR(Y/X)− MCR(X/Y) plotted with respect to Year index (1, j), j = 1, 2,..., 23. Solid and dotted line represents Δ(MCR) of the A1B1A2 marker and B2 marker respectively

there exists asymmetric coupling. Since positive value indicates $MCR(Y/X) < MCR(X/Y)$, that is, X drives Y and negative value indicates $MCR(Y/X) < MCR(X/Y)$, that is, Y drives X, both the matrix plots indicate that the population data of ith year drives the same of jth year only when $i < j$. As an example, $\Delta(MCR)$ values computed for $(1, j), j = 1, 2, ..., 23$ is given in Figure 19.9 to precisely understand the trend.

From the practical viewpoint, this implies that the worldwide population data can be synchronized by means of GS. This also reveals the existence of coupling between the population data of any two different years and the direction of coupling. It is found that the population data of any year drives the same of the subsequent years. This is quite expected as present and past populations must have some influence on the future populations.

19.3.2   Analysis by Complex Network

The presence of network is everywhere in the real world starting from the biological systems, communication systems, traffic systems, Internet to social and economic systems. However, we do not often realize the presence of these networks in our daily activities unless failures occur in these systems. Study of networks is not a new discipline in research, yet the extent to which technology allows these networks to take on new roles is a new phenomenon. A network can be essentially represented by graph, an abstract mathematical structure consists of a set of generically called nodes (vertices) connected by links (edges) representing some binary relationship. In the case of the World Wide Web, the websites are the nodes and the hyperlinks between the websites act as links. In a social network the nodes are the people and links exist in the form of friendship, professional collaboration, political alliances and so on. In the Internet, the nodes are the routers and the links represent electrical connections between the nodes. Another fascinating example is the network of molecular events of living cells, which allow cells to multiply and hence the organisms to grow (metabolic pathways and gene expression networks), that allow inter and intra cell communication. Organisms themselves form networks of predator and prey in food webs that have been studied by biologists for a long time. A complex network is one that comprises of a large number of nodes which are linked to each other according to some specific connection topologies. In other words a complex network is a graph comprising of a large number of nodes. A Graph $G = <V, E>$ consists of a set of nodes (vertices) V and a set of edges E. If V contains n elements then the graph is said to be of order n. If there exists an edge e between two vertices u and v, then u and v are said to be incident on e. In this case, u and v are said to be adjacent to each other. A graph of order n can have no more than $\frac{n(n-1)}{2}$ edges. When all these possible edges are present the graph is called a complete graph and it is denoted by K^n. If a real number is associated to the existing edges, then the graph is called a weighted graph. If direction is associated with the existing edges, the graph is called a directed graph, otherwise it is known as an undirected graph. Since our main intention is to study complex networks, we are not entering into the details of graph theory.

Complex networks of socio-economic GDP data

Let us now form a complex network from the socio-economic GDP data. For this purpose, we propose a methodology based on the mapping algorithm given in Li, Cao and Tan (2011). As GDP data is given in respect of different countries of the globe, we considered the nodes as the country. Data is given year wise up to the year 2100 starting from 1990 with a gap of five years.

Let $\{x_{i1}, x_{i2}, x_{i3}, ..., x_{iN}\}$ be the GDP data for the ith country for N consecutive years (with a gap of five years). Define the distance function on the GDP values of any two countries i, j as:

$$d_{ij} = \sqrt{\sum_{k=1}^{N} |x_{ik} - x_{jk}|^2}, (i, j = 1, 2, 3, ..., n), \tag{13}$$

where x_{ik} denotes the GDP values of the ith countries in kth year and n denotes the number of such countries. The connection between the nodes is defined as follows:

Let d_{max} be the maximum difference in GDP values, that is, $d_{max} = Max(d_{ij})$. The judgment value to decide the connection is then defined as $\Delta = \frac{d_{max}}{k-1}$. Finally the adjacency matrix $(a_{ij})_{k \times k}$ of the network is formed as

$$a_{ij} = \begin{cases} 1, \text{ if } d_{ij} \leq \Delta \\ 0, \text{ otherwise} \end{cases}.$$

Since GDP is one of the strong indicators of a country's economic growth, these complex networks are expected to explore some important features of the country's socio-economic condition.

As stated earlier, there are four types of country-level Gross Domestic Product (GDP) data downscaled projected based on Special Report on Emissions Scenarios (SRES) with A1, A2, B1 and B2 marker scenarios. In A1 data, the projection is based on very high economic growth rates, while it is based on medium economic growth rates for A2 data. On the other hand, the basis of projection is high population for the B1 family and medium population for the B2 family. SRES regional GDP growth rates and population data were calculated from 1990 to 2100 based on the SRES marker model regional data and applied uniformly to each country that fell within the SRES-defined regions. Thus we form complex networks for each of the markers by partitioning the data into two groups. The first group contains GDP data from 1990 to 2015 and the second group contains data for the years 2020 to 2100. The complex networks of both the groups for A1, A2, B1 and B2 are given by Figure 19.10.

Characterization of the socio-economic GDP network

Now since the complex networks constructed for the GDP data comprising of large number of nodes, the use of statistical analysis is one of the proper tools for a useful mathematical characterization. In this concern, three major properties are mainly studied. These are: degree distribution, clustering and small-world properties.

A. Degree distribution The most basic statistical characterization of a graph is the degree distribution $P(k)$ (Ghoshal and Newman 2007; Newman 2003). For an undirected graph this is defined as the probability that any randomly chosen vertex is of degree k, while for the directed graphs, we need to consider two different degree distributions, one is the probability that a randomly chosen vertex has in-degree k_{in}, called the in-degree distribution $P(k_{in})$ and the other is the probability that a randomly chosen vertex has out-degree k_{out}, called the out-degree distribution $P(k_{out})$. The real networks can be broadly classified into two classes based on the functional form of this degree distribution. The first one refers to the so-called homogeneous networks, which corresponds to the light tailed Poisson's or Gaussian distributions. The second class is the networks with heterogeneous connectivity pattern usually corresponding to heavy tailed degree distribution. The homogeneous and heterogeneous networks can be discriminated by looking at the first two moments of the degree distribution. For homogeneous networks, both the moments are bounded, whereas for the heterogeneous network, at least the second moment must be unbounded. In the distribution with a heavy tail there is a finite probability of finding vertices with degree much larger than the average degree <k>. In most of the times vertices possess a small degree, but there is an appreciable probability of getting vertices of large degree. In this type of degree distribution, all intermediate values are probable and the average degree does not represent any special value for the distribution. On the contrary, in the light tailed bell-shaped distributions with fast decaying tails, the average value is very close to the maximum value of the distribution and represents the most probable value in the system. Thus, a very large level of degree fluctuations occurs in the case of heterogeneous networks, which is not there in homogeneous networks.

In the power-law degree distribution (Adamic and Huberman 2001) given by $P(k) = Ak^{-\gamma}$ with $2 \leq \gamma \leq 3$, which is basically a heavy tailed distribution, the degree fluctuation is, therefore, unbounded and depends only on the system size. Thus any intrinsic scale for the fluctuations is absent and so the average value is not a characteristic scale for the system.

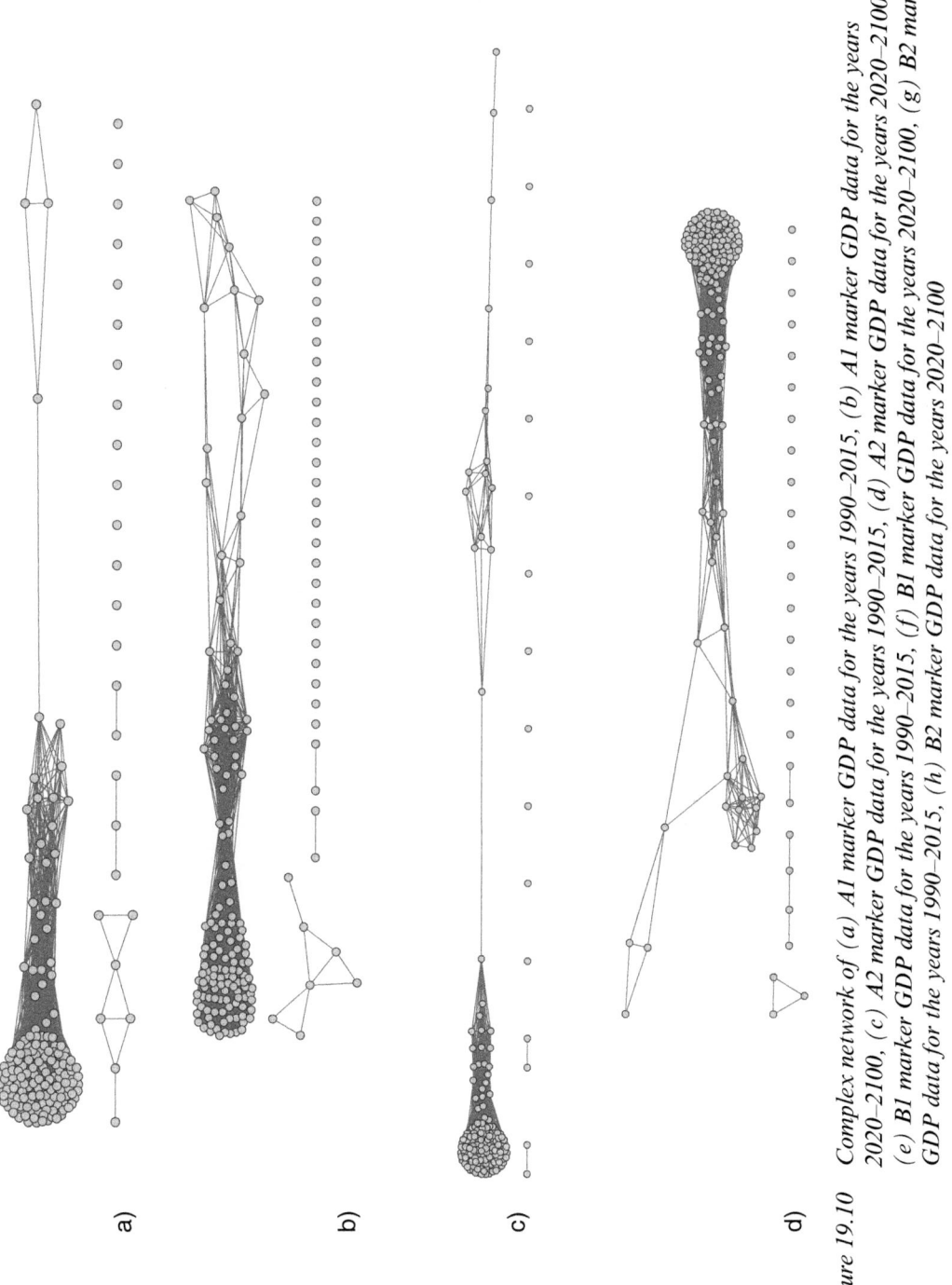

Figure 19.10 Complex network of (a) A1 marker GDP data for the years 1990–2015, (b) A1 marker GDP data for the years 2020–2100, (c) A2 marker GDP data for the years 1990–2015, (d) A2 marker GDP data for the years 2020–2100, (e) B1 marker GDP data for the years 1990–2015, (f) B1 marker GDP data for the years 2020–2100, (g) B2 marker GDP data for the years 1990–2015, (h) B2 marker GDP data for the years 2020–2100

411

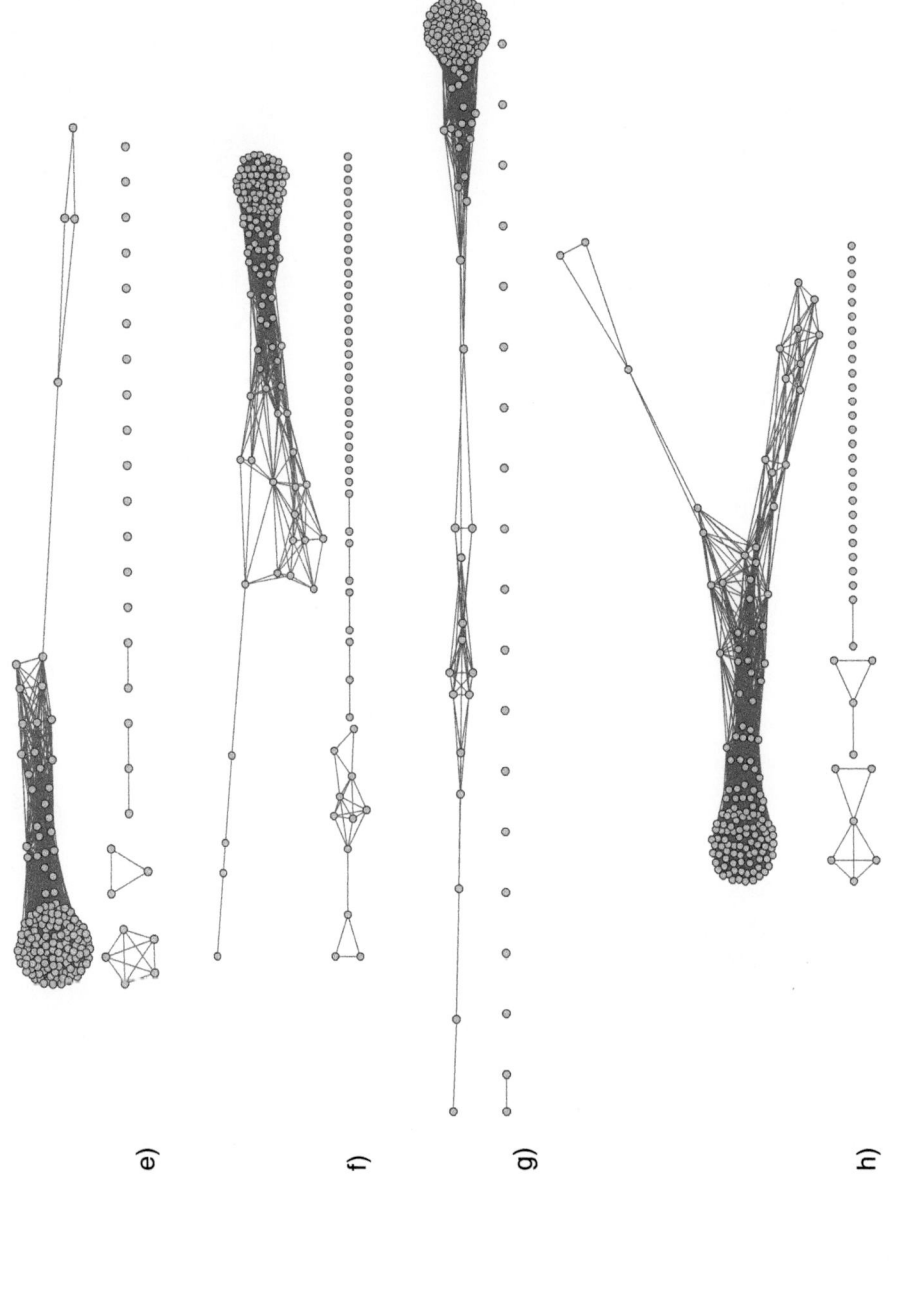

e)

f)

g)

h)

Figure 19.10 (continued)

In other words, the network is scale free (Caldarelli 2007). This idea can also be extended for $\gamma < 2$ as in this case even the first moment is unbounded. Therefore, the power-law behavior and the relative exponent are good quantitative measures of heterogeneity of the network degree. However, there may also exist some other types of heterogeneous networks. The man-made networks (Mathematical models) like regular lattices and trees always have the lowest heterogeneity and lowest randomness in the sense that the probability of any two randomly chosen nodes being adjacent to each other is very low or zero. On the other hand, the real world networks lie between homogeneous and heterogeneous conditions.

Let us now determine the nature of the complex networks we have formed for the GDP data of past and future scenarios for each of the marker A1, A2, B1 and B2 by finding the degree distribution in each of the cases. Although, degree distribution clearly captures only a small amount of information of the network, it can still provide some important clues regarding the structure of the network. The frequency distribution of the degree of the vertices with respective to the bins (bins are counted among the vertices) is given Figure 19.11.

It is evident from Figure 19.11 that most of the nodes (countries) have a low degree, while very few of them have large degree for each of the four marker scenarios. Such types of degree distribution are said to possess a long tail, which indicates that the networks are scale free. Thus, all of the above complex networks are scale free. Since our network is formed on the basis of the GDP difference of the countries, the degree distribution also indicates that there are very few countries whose GDP is close to the GDP of many other countries. For most of the countries, the GDP is similar to the GDP of only few countries and for some countries, the GDP is completely different, that is, they are isolated nodes. This is quite obvious as the economic policy for setting national and foreign companies in a country normally differs widely from country to country in most of the cases.

B. Small-world properties In many large-scale networks, sometimes it is found that the average distance between nodes is very small compared to the size of the graphs. The distance between two nodes in a graph is defined as the shortest path length $<l>$ among them. The average shortest path length among all possible pairs of nodes in the network is known as average distance between vertices. Such large-scale networks are said to have small-world properties (Newman and Watts 1999; Watts and Strogatz 1998). Small-world actually means that it is possible to reach one node from any other in the network through a very small number of intermediate nodes. Thus, small-world property exists in a network if the average distance $<l>$ scales logarithmically with the number of nodes. Most of the sociological networks possess this small-world property, where it is often called 'six degrees of separation'. It means that a small number of acquaintances (on an average six) is enough to create a connection between any two people chosen at random.

For each of our GDP networks, we compute the average distance between nodes (countries) and compare with the size of the network to check whether the GDP networks possess Small-world properties. The result is given in Table 19.1.

It follows from Table 19.1 that all of the GDP networks possess small-world properties, as average distance between nodes is very small compared to the size of the network for each of the four marker scenarios during 1990–2015 and 2020–2100.

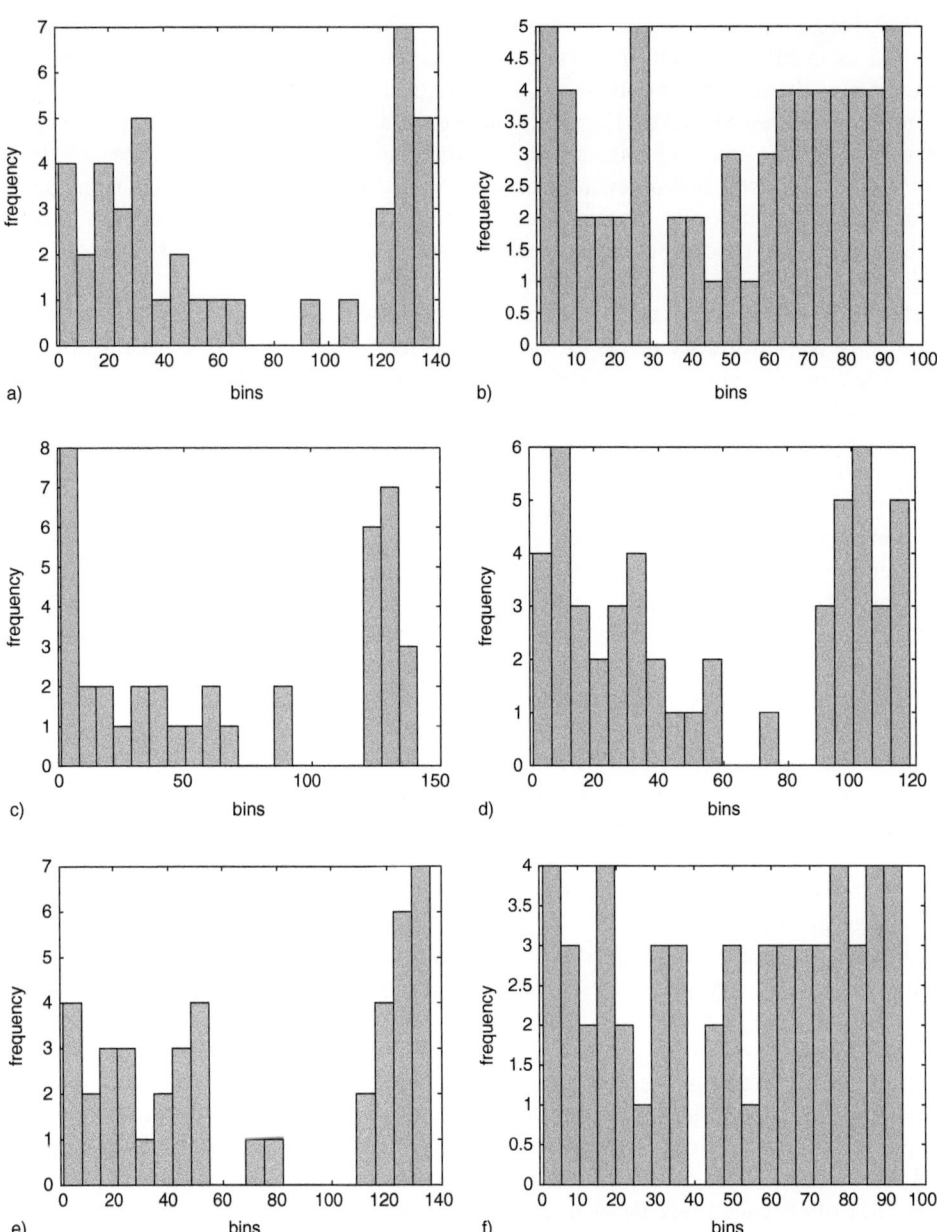

Figure 19.11   *Frequency distributions of degree of the nodes in the complex networks of (a) A1 marker GDP data for the years 1990–2015, (b) A1 marker GDP data for the years 2020–2100, (c) A2 marker GDP data for the years 1990–2015, (d) A2 marker GDP data for the years 2020–2100, (e) B1 marker GDP data for the years 1990–2015, (f) B1 marker GDP data for the years 2020–2100, (g) B2 marker GDP data for the years 1990–2015, (h) B2 marker GDP data for the years 2020–2100*

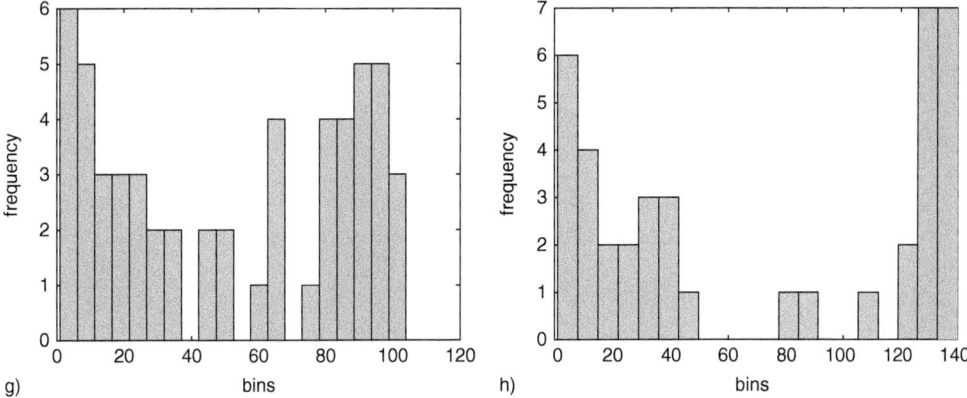

Figure 19.11   (continued)

Table 19.1   *Average distance between nodes (countries) and size of the complex networks of GDP data for each of the four marker scenarios during 1990–2015 and 2020–2100*

Scenarios	1990–2015		2020–2100	
	Average distance	Network size	Average distance	Network size
A1	1.60049	183	2.40771	179
A2	2.12298	179	2.24771	179
B1	1.51522	179	2.21815	179
B2	1.84833	179	2.09029	179

C.   Transitivity or clustering   Transitivity or Clustering of a graph (Fronczak et al. 2002) refers to the tendency of forming cliques (triangles) in the neighborhood of any given vertex. Clique in an undirected graph is a subset of the vertices, where every two distinct vertices are adjacent. Small-world property cannot fully characterize a network alone. In many sociological and technological networks, small-world property goes with a high level of clustering. In high clustering networks, if the vertex i is connected to j, and j is connected to k, then very likely i is also connected to k. For an undirected graph, clustering is measured by means of a clustering coefficient. Let k_i denote the degree of the vertex i and e_i be the number of edges existing between the k_i neighbors of i. The clustering coefficient C_i of the vertex i is defined as the ratio of e_i and the maximum possible edges $\frac{k_i(k_i - 1)}{2}$ among its neighbors, that is:

$$C_i = \frac{2e_i}{k_i(k_i - 1)}. \tag{14}$$

The coefficient C_i represents the average probability in which two neighbors of the vertex i are connected to each other. Naturally, C_i is defined only for $k_i > 1$. For $k_i \leq 1$, $C_i = 0$.

The clustering coefficient clearly distinguishes the small-world effect from random graphs and regular grids in the sense that the random graphs feature the small-world effect but are not clustered (that is, the clustering coefficients of the vertices are too small), while the regular grids are well clustered but do not possess small-world properties. This clustering coefficient is often called the local clustering coefficient. The global clustering coefficient is basically the mean over all local clustering coefficients of vertices of the network. Naturally, it represents the overall scenario of clustering in a network.

The local clustering coefficients for each node (country) of the networks for each of the two groups of A1, A2, B1 and B2 marker data are given by the contour plot in Figure 19.12.

It is observed from Figure 19.12 that for each of the four marker scenarios, the clustering coefficient is very high (close to 1) for the first 120 countries (nodes) during 1990–2015 with some exception, which dramatically decreases during 2020–2100. However, for the countries with lower clustering values during 1990–2015, practically there is no change during 2020–2100. In the present context of GDP networks, this actually indicates that the tendency of forming cliques (triangles) was very high during 1990–2015 for most of the countries. Here clique at ith country means, if GDP of ith country is sufficiently close to the jth country and the same of jth country is sufficiently close to that of kth country, then GDP of ith country must be sufficiently close to kth country. This further means that economical policies, political and sociological environment, climate of these three countries i, j, k have some similarities for which the GDP of these countries are close to each other. Possibly there will be some changes in these parameters for which the tendency of forming cliques gets decreased for most of the countries during 2020–2100. To find the overall scenarios of the GDP networks for each of the markers A1, A2, B1 and B2 during 1990–2015 and 2020–2100, the global clustering coefficients are calculated, which are given in Table 19.2.

It is evident from Table 19.2 that for all the markers the value of the global clustering coefficient decreases for the years 2020–2100. This actually strengthens the interpretation already made from the local clustering coefficient for the individual countries.

Besides the above three statistical properties, the following measures are often useful to characterize complex networks:

D.   Assortativity   For some types of complex networks, the natural question is whether nodes of a certain type connect only to nodes of similar type, or to unlike types, or do the links exist between multiple types? For example, in an academic acquaintance network, two mathematicians are more likely to know each other than someone who is a biologist. In the food web, a number of links are there that connect plants and herbivores, many more that connect herbivores and carnivores, but not too many that link a plant to another plant, or a herbivore to a herbivore. This type of selective linking is conventionally referred to as assortative mixing or homophily (Newman 2002). In most of the real networks, for example, Technological networks, Biological networks and Social networks, assortative mixing is a very common feature. The assortative mixing is quantified by means of assortative coefficient. Let E_{ij} be the number of links (edges) in the network that connects a node of type i to a node of type j, with $i, j = 1, \ldots, n$. These edges are represented in form of an edge incident matrix E. Thus, $E = (E_{ij})$. A normalized mixing matrix $e = (e_{ij})$ is then defined as $e = \frac{E}{\|E\|}$, where $\|E\|$ the sum of the elements of the matrix E and e_{ij} represents the fraction of edges that connects nodes of types i and j, and satisfies the normalization condition $\sum_{i,j} e_{ij} = 1$. The Assortativity coefficient r is finally defined as:

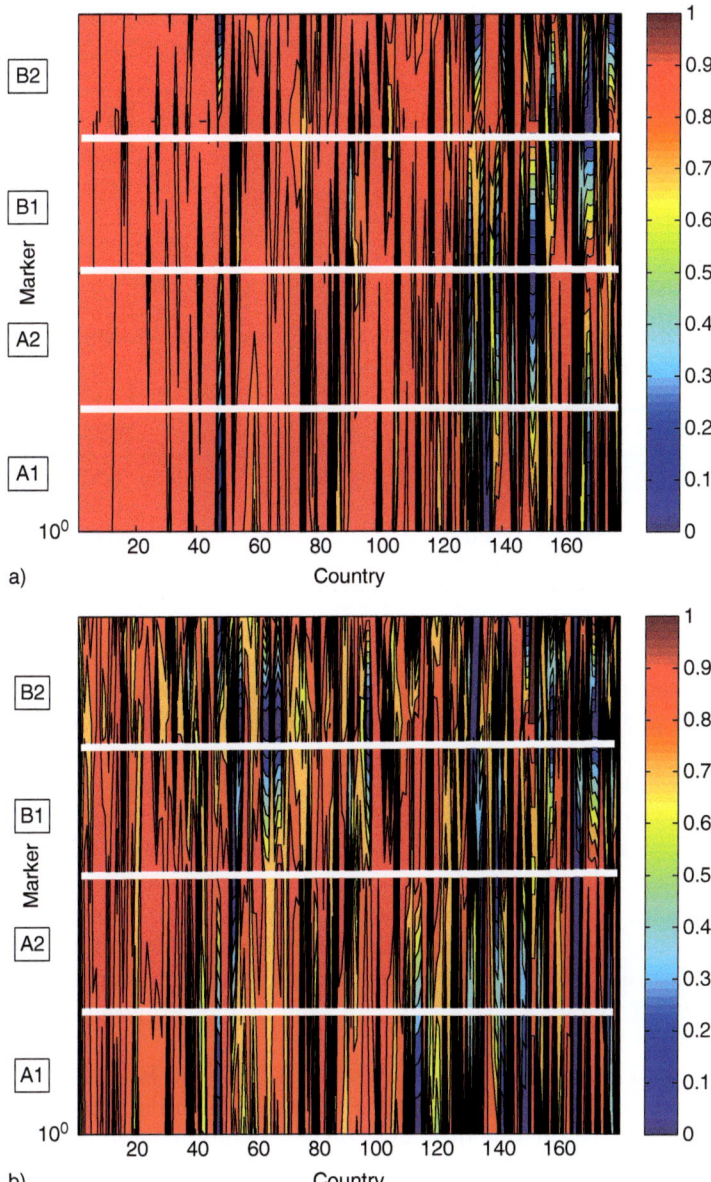

Figure 19.12   *Contour plot of local clustering coefficient of the A1, A2, B1 and B2*
marker GDP data for the years (a) 1990–2015, (b) 2020–2100

$$r = \frac{Tr(e) - \|e\|^2}{1 - \|e\|^2},$$ (15)

where $Tr(e)$ is the standard matrix trace – the sum of the diagonal elements e_{ii}. The value
of r lies in the range $[-1, 1]$, where 1 represents a perfectly assortative network and -1

Table 19.2   Global clustering coefficient of the complex networks of GDP data for each of the four marker scenarios during 1990–2015 and 2020–2100

Scenarios	1990–2015	2020–2100
A1	0.817975	0.679803
A2	0.821379	0.770558
B1	0.830422	0.637946
B2	0.819891	0.715951

Table 19.3   Assortativity (Degree correlation) coefficient of the complex networks of GDP data for each of the marker scenarios

Scenarios	1990–2015	2020–2100
A1	0.63364	0.67222
A2	0.58047	0.70976
B1	0.60286	0.62269
B2	0.52275	0.68658

indicates a perfectly disassortative network, while 0 represents a randomly mixed network. The assortative coefficient is closely related to the Pearson correlation in statistics.

A special case of assortative mixing based on the degrees of vertices is referred to as degree correlation. Thus, the tendency among the higher (lower) degree nodes to get connected with the higher (lower) degree nodes or the tendency of the higher degree nodes to get associated with lower degree ones, is known as degree correlation. The degree correlation is obtained similarly as assortative coefficient by using equation (15), but in this case the elements e_{ij} of the matrix e represent the fraction of edges that connect a node of degree i to that with degree j. Since the degree is an important topological measure, degree correlations play a significant role to understand the complicated network structural effects. The assortativity of the complex networks calculated for each of the four marker scenarios and for each of the two groups of data is summarized in Table 19.3.

It follows from Table 19.3 that for each of A1, A2, B1 and B2 marker scenarios the Assortativity coefficient has increased for the projected data for the years 2020–2100. This indicates that the tendency among the higher (lower) degree nodes to get associated with the higher (lower) degree nodes increases during 2020–2100. In the GDP network, high degree of a node means that the corresponding country's GDP is close to the GDP of many other countries. On the contrary, low degree means that the corresponding country's GDP is close to the GDP of very few countries.

Therefore, if the Assortativity coefficient is close to 1, then high (low) GDP countries show a tendency to associate themselves with high (low) GDP countries. In the present GDP networks, we observe that the Assortativity coefficient is neither too large nor too small for each of the marker scenarios. This indicates that there is a mixed trend in the GDP networks in the sense that some (50 percent to 65 percent) of the high (low) GDP countries are connected with high (low) GDP countries and some other high (low) GDP countries are connected with low (high) GDP countries. This tendency is expected to increase for each of the four marker scenarios during 2020–2100. The rate of change is found to be remarkable for A2 and B2 marker scenarios.

E.   Centrality   Another question of prime importance regarding a CN is to determine the relative importance of a node within the network. For example, in a social network, this is similar to finding the most important person, in the World Wide Web; this is similar to identifying the most popular web-page. The degree of importance of a particular node in a network is referred to as centrality. Centrality can be defined in different ways depending on the type of the network. The simplest measure of centrality is the degree centrality. The degree centrality (Newman 2003; Newman, Barabasi and Watts 2006; Ghoshal and Newman 2007) is defined as the number of links incident on a node. Thus, the higher the degree of the node, the more central is the node. Mathematically, degree centrality of the node i is given by:

$$c_d(i) = \frac{k_i}{n-1},\tag{16}$$

where k_i is the degree of vertex i and n is the number of nodes in the network. However, the large number of connections to other nodes is not always the only criteria for a node to be central. Rather, if a node is connected to highly central neighbors, then the node is itself more central. This type of centrality is referred to as eigenvector centrality (Newman 2003; Newman, Barabasi and Watts; Ghoshal and Newman 2007). In this case, the centrality of the node i (x_i, say) is given by the equation:

$$x_i = \lambda^{-1} \sum_{j=1}^{n} A_{ij} x_j,\tag{17}$$

where A_{ij} is the adjacency matrix of the graph (network) and λ is some constant. The above equation can be equivalently written as:

$$A\vec{x} = \lambda\vec{x},\tag{18}$$

which is basically the eigenvector equation with $\vec{x}$ be an eigenvector of $A = (A_{ij})$. According to Perron–Frobenius theorem in linear algebra, the requirement that all entries of the eigenvector $\vec{x}$ be positive implies that the eigenvector corresponds to the largest eigenvalue λ gives the correct centrality measure. The centrality, in fact the eigenvector centrality of the vertex i is given by the ith component of the eigenvector, thus obtained. The degree centralities of each country for each of the four marker scenarios during 1990–2015 and 2020–2100 are given by the matrix plot in Figure 19.13.

It is observed from Figure 19.13 that degree centrality is very high for the first 100 countries with few exceptions for each of the four marker scenarios during 1990–2015. However, the scenarios have been changed completely during 2020–2100 except for the A2 marker. In fact, almost all of the countries with high degree centrality during 1990–2015 have been changed to low degree centrality during 2020–2100 under all marker scenarios except A2. For A2 marker scenarios, the situation is almost the same in both 1990–2015 and 2020–2100 sessions. Another remarkable thing to notice is that for all four marker scenarios, the status of almost all countries with low degree centrality remains unchanged

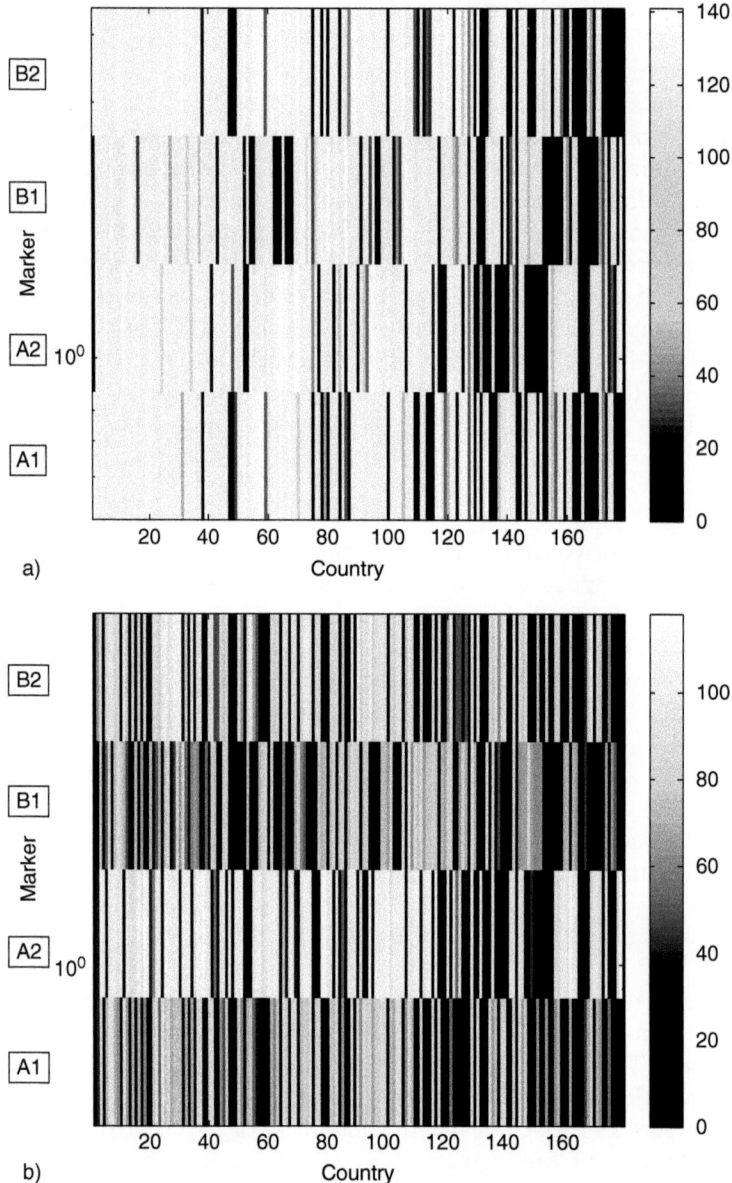

*Figure 19.13   Matrix plot of degree centrality of the A1, A2, B1 and B2 marker GDP
data for the years (a) 1990–2015, (b) 2020–2100*

in both the sessions. From the viewpoint of economics, this indicates that most of the
countries with almost the same GDP values (GDP difference lies within a threshold Δ)
during 1990–2015 become countries with different GDPs (GDP difference lies outside the
threshold Δ) during 2020–2100 except under A2 marker scenarios. Possibly this is due to
the change in the government policy of the individual countries.

Now, to check whether the highly central nodes (countries) are connected with highly central nodes (countries), we compute the eigenvector centrality of each node for each of the marker scenarios during 1990–2015 and 2020–2100. This is given by Figure 19.14. If this eigenvector centrality of the highly central node is high, then the corresponding node is connected with more highly central nodes. From the viewpoint of economics, this indicates that the country's GDP is sufficiently close to many countries and most of them also have a sufficiently close GDP with many other countries. On the other hand, if the eigenvector centrality is low for a highly central country (node), then the country's GDP is sufficiently close to many countries, GDP of which are sufficiently close to very few countries.

It is evident from Figure 19.14 that eigenvector centrality is very low for almost all of the highly central countries for each of the four marker scenarios during 1990–2015. During 2020–2100, the situation remains almost same for most of the highly central countries and for some other highly central countries it is even worse than 1990–2015.

19.4 CONCLUSIONS

The study introduced in this chapter is new of its kind. Two types of socio-economic data (worldwide population data and worldwide GDP data) under two different marker scenarios A1B1A2 and B2 have been studied by means of Complex coupling (synchronization analysis) and Complex network (CN). Nature of complex coupling has been quantified by phase synchronization (PS) and mean conditional recurrence (MCR). It is observed that the worldwide population data of every two different years (1990, 1995, 2000,. . ., 2100) are in PS. In this case, CPR quantifies degree of PS between each pair of population data, which shows a decreasing trend. From the practical viewpoint, this type of trend indicates that the underlying dynamics of population data of past, present and future are phase synchronized but as the time grows, the degree of synchronization gradually decreases. This is quite obvious; as the time grows different new parameters come into existence, which greatly influence the underlying dynamics of the population data. Furthermore, decreasing trend of CPR has been observed for each year of population data with respect to the base year 1990. Thus PS analysis shows a continuous degradation in the degree of PS from its base year. However, PS is a weaker type of synchronization. So we have made an attempt to establish a stronger variety of it viz. GS. This reveals that the worldwide population data of every two different years are also in GS and hence there exist coupling between the populations data of any two different years. In this context, the direction of coupling is detected by MCR. It is found from the MCR values that the population data of any year drives the same of the subsequent years. This is at par with the practical viewpoint as present and past populations must have some influences on the future populations. On the other hand, the analysis of the worldwide GDP data for four different marker scenarios – A1, A2, B1 and B2 have been done by forming their CNs. All the networks have been formed on the basis of the GDP difference of the countries. For a comparative discussion, the GDP data is subdivided into two groups – 1990–2015 and 2020–2100. For each of the CN's, some relevant measures viz. degree distribution, small-world property, local and global clustering coefficients, Assortativity coefficient, degree centrality and eigenvector centrality have been computed.

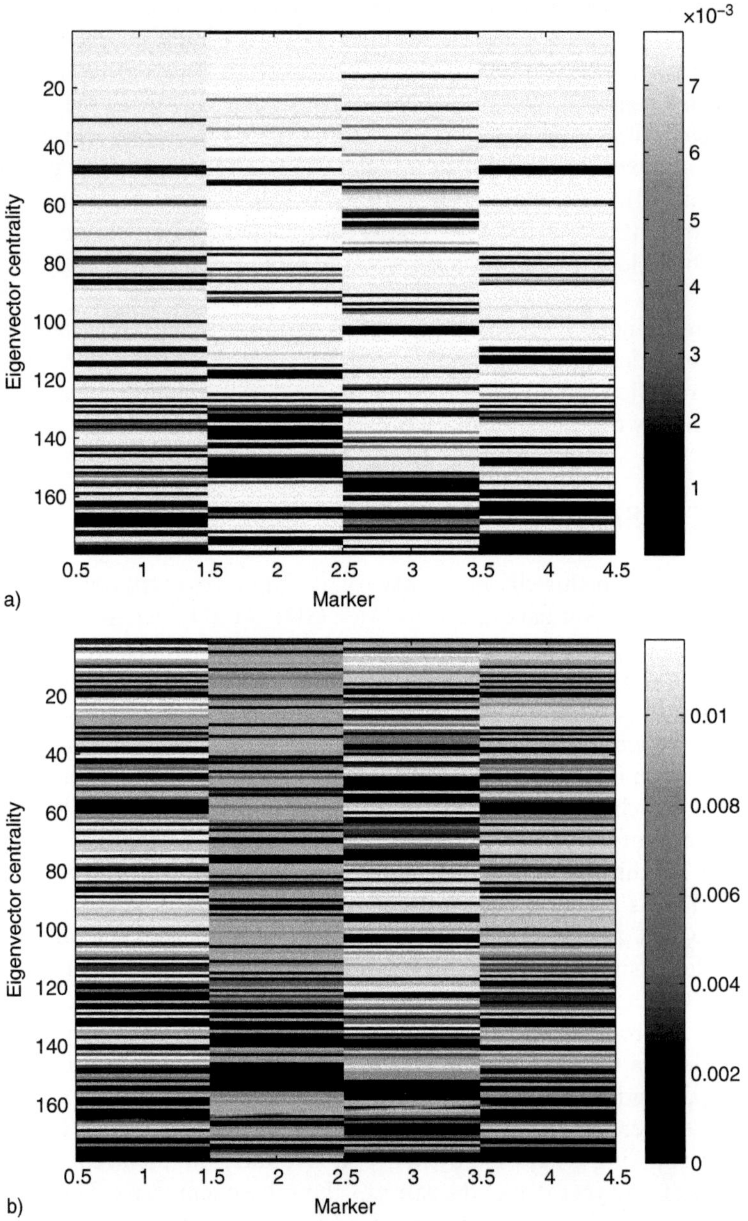

Figure 19.14   Matrix plot of eigenvector centrality of the A1, A2, B1 and B2 marker GDP data for the year (a) 1990–2015, (b) 2020–2100

Degree distribution reveals that all the networks are scale free in the sense that there are very few countries whose GDP is close to the GDP of many other countries. For most of the countries, the GDP is similar to the GDP of only few countries and for some others, the GDP is completely different. This is quite obvious as in most of the cases; the economic policy of a country for setting national and foreign companies normally differs widely from country to country. It is also revealed that all of the CNs thus constructed for the GDP data are scale free. They also possess small-world property.

Local clustering coefficients computed for each country (node) indicates the tendency of forming triangles was very high during 1990–2015 for most of the countries, which gets decreased during 2020–2100. Here existence of triangles practically means that the economical policies, political and sociological environment, climate of the three countries (nodes) forming the triangles have some similarities for which the GDP of these countries are close to each other. Possibly there will be some changes in these parameters for which the tendency of forming triangles gets decreased for most of the countries during 2020–2100. Global clustering coefficients also support these observations in each and every case.

Further, it is found that for each of A1, A2, B1 and B2 marker scenarios the Assortativity coefficient has increased for the projected data for the years 2020–2100. This indicates that the tendency among the higher (lower) GDP countries to get associated with the higher (lower) GDP countries increases during 2020–2100. Since the Assortativity coefficient is neither too large nor too small for each of the marker scenarios, there is a mixed trend in the GDP networks. This tendency is expected to increase for each of the four marker scenarios during 2020–2100. The rate of change is found to be remarkable for A2 and B2 marker scenarios.

From the viewpoint of economics, degree centrality indicates that most of the countries with almost the same GDP values (GDP difference lies within a threshold Δ) during 1990–2015 become countries with different GDPs (GDP difference lies outside the threshold Δ) during 2020–2100 except under A2 marker scenarios. Possibly this is due to changes in the government policies of the individual countries.

Finally, eigenvector centrality is found to be very low for almost all of the highly central countries for each of the four marker scenarios during 1990–2015. During 2020–2100, the situation remains almost the same for most of the highly central countries and for some other highly central countries it is even worse than 1990–2015. From the viewpoint of economics, eigenvector centrality of the highly central node is high indicating that the country's GDP is sufficiently close to many countries and most of them also have a sufficiently close GDP with many other countries. On the other hand, if the eigenvector centrality is low for a highly central country (node), then the country's GDP is sufficiently close to many countries, GDP of which are sufficiently close to very few countries.

This chapter gives an essence of complexity analysis by means of complex coupling and CN in socio-economic scenarios and also provides an insight into a new type of analysis in this context. The chapter highlights the use of RP, JRP, and MCR for the synchronization of complex socio-economic data and describes the effectiveness of forming CN in the socio-economic context. It also shows one of the methodologies of forming the CNs from socio-economic data. One can make some newer and newer sophisticated analysis of the socio-economic data, just by changing the rule of forming the networks as and when required.

REFERENCES

Adamic, L.A. and B.A. Huberman (2001), 'Power-law distributions of the world wide web', *Science*, **287**, 2115a.

Afraimovich, V.S., N.N. Verichev and M.I. Rabinovich (1986), 'Stochastic synchronization of oscillations in dissipative systems', *Radiofizika*, **29** (9), 1050–1060.

Albert, R. and A.L. Barabasi (2000), 'Topology of evolving networks: Local events and universality', *Physical Review Letter*, **85**, 5234.

Boccaletti, S., J. Kurths, G. Osipovd, D.L. Valladares and C.S. Zhouc (2002), 'The synchronization of chaotic systems', *Physics Reports*, **366** (1), 1–101.

Bonanno, G., F. Lillo and R. Mantegna (2001), 'High-frequency crosscorrelation in a set of stocks', *Quantum Finance*, **1**, 96–104.

Bonanno, G., G. Caldarelli, F. Lillo, S. Miccieche, N. Vandewalle and R. Mantegna (2004), 'Networks of equities in financial markets', *European Physics Journal B*, **38**, 363–371.

Caldarelli, G. (ed.) (2007), *Scale-free Networks*, Oxford: Oxford University Press.

Donner, R.V., Y. Zou, J.F. Donges, N. Marwan and J. Kurths (2010), 'Recurrence networks: A novel paradigm for nonlinear time series analysis', *New Journal of Physics*, **12**, 033025.

Dorogovtsev, S.N. and J.F.F. Mendes (2002), 'Evolution of networks', *Advance Physics*, **51**, 1079.

Eckmann, J.P., S.O. Kamphorst and D. Ruelle (1987), 'Recurrence plot of dynamical systems', *Europhysics Letter*, **5**, 973–977.

Fraser, A.M. and H.L. Swinney (1986), 'Independent coordinates for strange attractors from mutual information', *Physical Review A*, **33**, 1134–1140.

Fronczak, A., J.A. Holyst, M. Jedynak and J. Sienkiewiecz (2002), 'Higher order clustering coefficients in Barabasi–Albert networks', *Physica A*, **316**, 688.

Fujisaka, H. and T. Yamada (1984), 'Stability theory of synchronized motion in coupled-oscillator systems', *Progress of Theoretical Physics*, **72** (5), 885–894.

Garlaschelli, D., S. Battiston, M. Castri, V. Servedio and G. Caldarelli (2005), 'The scale-free topology of market investments', *Physica A*, **350**, 491–499.

Ghoshal, G. and M.E.J. Newman (2007), 'Growing distributed networks with arbitrary degree distributions', *European Physics Journal B*, **58**, 175.

Hartman, Philip (1960), 'A lemma in the theory of structural stability of differential equations', *Proc. A.M.S.* **11** (4), 610–620.

Kennel, M.B., R. Brown and H.D.I. Abarbanel (1992), 'Determining embedding dimension for phase-space reconstruction using a geometrical construction', *Physical Review A*, **45** (6), 3403–3411.

Kim, H.-J., Y. Lee, B. Kahng and I. Kim (2002), 'Scale-free network in financial correlations', *Journal of the Physical Society of Japan*, **71** (9), 2133–2136.

Kocarev, L. and U. Parlitz (1996), 'Generalized synchronization, predictability, and equivalence of unidirectionally coupled dynamical systems', *Physical Review Letters*, **76**, 1816–1819.

Lacasa, L., B. Luque, F. Ballesteros, J. Luque and J.C. Nuno (2008), 'From time series to complex networks: The visibility graph', *Proc. Natl. Acad. Sci. USA*, **105**, 4972–4975.

Li, Y., H.D. Cao and Y. Tan (2011), 'A novel method of identifying time series based on network graph', *Complexity*, **17**, 13–34.

Lutz, W (ed.) (1996), *The Future Population of the World: What Can We Assume Today?* London: Earthscan Publications.

Mandal, S., M.M. Panja, S. Ray and S.K. Roy (2010), 'Qualitative behavior of three species food chain around', *Nonlinear Analysis: Modelling and Control*, **15** (4), 459–472.

Mantegna, R.N. (1999), 'Hierarchical structure in financial markets', *European Physics Journal B*, **11**, 193–197.

Marwan, N., M.C. Romano, M. Thiel and J. Kurths (2007), 'Recurrence plots for the analysis of complex systems', *Physics Reports*, **438** (5–6), 237–329.

Mukherjee, S., S.K. Palit, S. Banerjee, M.R.K. Ariffin and D.K. Bhattacharya (2014), 'Phase synchronization of instrumental music signals', *The European Physical Journal Special Topics*, **223** (8), 1561–1577.

Newman, M.E.J. (2002), 'Assortative mixing in networks', *Physical Review Letters*, **89**, 208701.

Newman, M.E.J. (2003), 'The structure and function of complex networks', *SIAM Review*, **45**, 167.

Newman, M.E.J. (2005), 'Power laws, Pareto distributions and Zipf's law', *Contemporary Physics*, **46**, 323.

Newman, M.E.J. and D.J. Watts (1999), 'Scaling and percolation in the small-world network', *Physical Review E*, **60**, 7332.

Newman, M.E.J., A.L. Barabasi and D.J. Watts (eds) (2006), *The Structure and Dynamics of Net-Works*, Princeton, NJ: Princeton University Press.

Nicolis, G., A. Garcia-Cantu and C. Nicolis (2005), 'Dynamical aspects of interaction networks', *International Journal of Bifurcation and Chaos*, **15**, 3467–3480.

Palit, S.K., S. Mukherjee and D.K. Bhattacharya (2013), 'A high dimensional delay selection for the reconstruction of proper phase space with cross auto-correlation', *Neurocomputing*, **113** (3), 49–57.

Pecora, L.M. and T.L. Carroll (1990), 'Synchronization in chaotic systems', *Physical Review Letters*, **64**, 821–825.

Pikovsky, A., M. Rosenblum and J. Kurths (2001), *Synchronization A Universal Concept in Nonlinear Sciences*, New York: United States of America by Cambridge University Press.

Romano, M.C., M. Thiel, J. Kurths and C. Grebogi (2007), 'Estimation of the direction of the coupling by conditional probabilities of recurrence', *Physical Review E*, **76**, 036211.

Rosa, E.R., E. Ott and M.H. Hess (1998), 'Transition to phase synchronization of chaos', *Physical Review Letters*, **80**, 1642–1645.

Rosenblum, M.G., A.S. Pikovsky and J. Kurths (1996), 'Phase synchronization of chaotic oscillators', *Physical Review Letters*, **76**, 1804–1807.

Rosenblum, M.G., A.S. Pikovsky and J. Kurths (1997), 'From phase to lag synchronization in coupled chaotic oscillators', *Physical Review Letters*, **78**, 4193–4196.

Rulkov, N.F., M.M. Sushchik, L.S. Tsimring and H.D.I. Abarbanel (1995), 'Generalized synchronization of chaos in directionally coupled chaotic systems', *Physical Review E*, **51**, 980–994.

Schweitzer, F. (2002), *Modeling Complexity in Economic and Social Systems*, London: World Scientific Publishing.

Simon, G. and M. Verleysen (2007), 'High-dimensional delay selection for regression models with mutual information and distance-to-diagonal criteria', *Neurocomputing*, **70**, 1265–1275.

Small, M., J. Zhang and X.K. Xu (2009), 'Transforming time series into complex networks', *Complex Sciences*, **5**, 2078–2089.

Strogatz, S.H. (1994), *Nonlinear Dynamics and Chaos: With Applications to Physics, Biology, Chemistry, and Engineering*, New York: Perseus Books.

Takens, F. (1981), 'Detecting strange attractors in turbulence in dynamical systems and turbulence', *Lecture Notes in Mathematics*, **898**, 366–381.

Theiler, J., S. Eubank, A. Longtin, B. Galdrikian and J. Farmer (1992), 'Testing for nonlinearity in time series: The method of surrogate data', *Physica D*, **58**, 77–94.

Tse, C.K., J. Liu and F.C.M. Lau (2010), 'A network perspective of the stock market', *Journal of Empirical Finance*, **17**, 659–667.

Watts, D.J. and S.H. Strogatz (1998), 'Collective dynamics of "small-world" networks', *Nature*, **393**, 440.

Williams, G.P. (1997), *Chaos Theory Tamed*, Washington, DC: Joseph Henry Press.

Xu, X., J. Zhang and M. Small (2008), 'Superfamily phenomena and motifs of networks induced from time series', *Proc. Natl. Acad. Sci. USA*, **105**, 19601–19605.

Zhang, J. and M. Small (2006), 'Complex network from pseudo periodic time series: Topology vs dynamics', *Physical Review Letter*, **96**, 238701.

20. Employment of tools and models appropriate to complex, real-world situations
Patrick Beautement, The abaci Partnership LLP

20.1 INTRODUCTION

The Principles of Complexity described by Mitleton-Kelly (2003) are enduring and the manner in which they are manifested as patterns, influences and other phenomena depends on the situation in which practice is carried out. At its core, practice involves dealing with three aspects of change: the perceptions of change; the characteristics of what can be understood or predicted about change; and the approaches that actors wish to take to influence change. A 'space of possibilities' emerges from the intersections between, and exclusions from, these three aspects. A major challenge for practitioners is selecting, matching and employing tools, models and methods appropriately when their understanding of some situation, that is, their context (and the intentions and the desired outcomes that follow) motivates them to intervene.

This chapter addresses this challenge by introducing a model of practice for complex situations based on the three Aspects of Practice mentioned above – where the aim of practice is to shape and influence purposefully and effectively, as far as the situation permits. The chapter then examines the issues arising of change and adaptation and develops an assessment framework. The Framework enables practitioners to judge the degree to which candidate tools and methods match appropriately both to the imperatives of the real-world situation and to the wider contexts (that is, mental models) of the actors in the situation. There are two parts to this matching: one is having good practice for dealing with a range of complex mental models; and the other is for making sure that one's mental models of complex situations are adequate to deal with the changing situation. Key to this is to be continually expanding one's range of options to anticipate changing need. To illustrate its use, the Framework is then employed in assessing the capabilities of autonomous agents and multi-agent systems (used for exploring intervention options). The chapter then shows how practitioners might act upon those options to bring about change through the use of multi-modal influence networks. Throughout the chapter the discussion is illustrated with techniques which have been found to be useful when dealing with complex situations. As a summary, the chapter concludes by listing eleven principles of practice derived from recent real-world experiences.

20.2 A REVIEW OF THE STATE OF PRACTICE

Practitioners, in the sense used in this chapter, are those people who bring about real-world change in their day-to-day work. This definition includes, for example, doctors and nurses, teachers, community policy and decision makers, social workers, humanitarian aid and emergency service workers. In other words, those who take a lead in acting in complex situations where it is not always possible to determine in advance, unequivocally, what is exactly the right thing to do and who should do it – ambiguity is taken for granted. It does not really include those who follow set processes and procedures (which have been designed to enable repeatable and dependable outcomes to be achieved) such as call-centre workers, or people employed on industrial production lines or in supermarket logistics. It also does not *usually* include academics, managers of business administration, or 'back-office' personnel who work with the abstractions, rather than with the direct consequences, of a situation.

In reviewing the state of practice, humanitarian and development aid has been carefully documented by Chambers (2008) and a range of tools and techniques evaluated, especially those designed to enable development enquiry to be used understand the needs on the ground. Chambers concludes that whilst many excellent tools exist, there is a great deal of duplication among them and a lack of a mechanism for matching tools to requirement. Ramalingam and Jones (2008) approached the practitioners' task from a complexity science perspective and used this lens to expose many of the weaknesses in the ways-of-working that are often employed. Primarily these include: unrealistic expectation on behalf of donors and tasking agencies, the mandating of approaches that are inappropriate to the situation in the field, the setting of arbitrarily short project timescales, and the use of monitoring and evaluation criteria which are not relevant given the behaviours and phenomena observed on the ground. In 2009, a workshop was held at Warwick University in the UK where academics, practitioners and policy makers analyzed and documented the factors to be addressed if insights from complexity science were to be put into practice effectively (Beautement and Broenner 2009a). The workshop highlighted the importance of the cross-disciplinary, and trans-disciplinary collaboration needed to deal with complex situations. As Hoffman and Klein (2011) explain, alternative and challenging viewpoints are essential and need to be aligned, but not homogenized.

Conn (2010) has also approached practice with an eye to complexity science and has examined the patterns of practice in local communities compared to that of governments. Conn has noted the mismatch between the various organizational forms involved and has indicated the pathologies of behaviour that result. She points to the need for more effective selection and employment of tools and techniques depending on the mix of actors involved. In considering the challenges of climate change adaptation, Levine, Ludi and Jones (2011) have noted the need for a suitable high-level framework through which alternative approaches (such as organizational transitions, knowledge use, institutions and entitlement, skill development, and potential innovations) can be compared systematically. In medicine, Sturmberg and Martin (2012) have highlighted the mismatches between the management of health institutions and the professional practices of doctors and nurses. Particularly, Sturmberg has indicated that fixed, linear processes are being required by managers – but that these reduce the flexibility of ward staff and make it more difficult for them to deal with the unexpected.

A number of common themes are highlighted by this research – not least among them is that lessons are not being learned – and include that:

- a variety of alternative perceptions among the actors are necessary and inevitable – but that practitioners must be able to align them appropriately;

- the variety of organization forms, ways-of-working and approaches to situations employed in practice can themselves be part of the problem – unless their strengths and weaknesses are acknowledged, challenged and accommodated;
- and that practitioners must be realistic about the degree to which any situation can be understood, information gathered and insight gained – as depth of these depends on the circumstances of that practice.

The interplay among these themes is to be examined further in this chapter – mostly from the practitioners' viewpoint (as they will be the main users of the Framework described below).

20.3   A MODEL OF PRACTICE

This section offers a Model of Practice which characterizes the nature of practice in complex situations. The Model, Figure 20.1, provides a simple way of appreciating the factors affecting practice – it also illustrates where three Aspects of Practice discussed below sit with respect to one other.

Practice, and hence the Model of Practice, is concerned with the relationship between the contexts, that is, mental models, of the practitioner(s) who are active in some situation. Each context of each practitioner(s) will have their own flow of change over time. They will remain relatively independent until they become entangled in some way. The nature

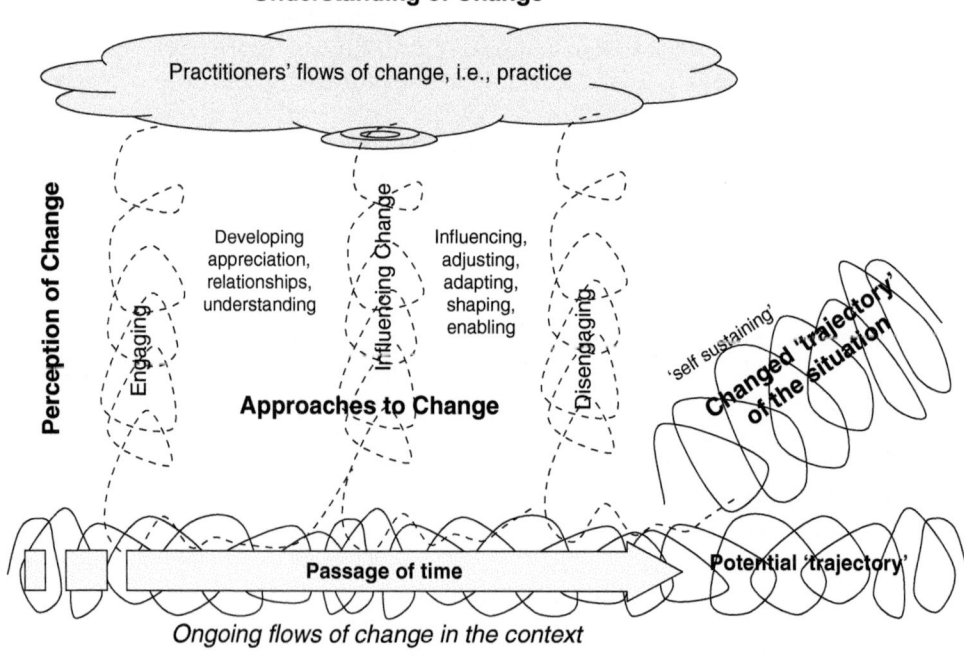

Figure 20.1   A model of practice for complex situations

of the resulting inter-connectivity is then key to the phenomena that will arise. However, the degree of coupling will not be by chance, but will depend on the intentions of the practitioner(s). When intentions are put into practice then this causes engagement with and between the contexts and a space of possibilities arises. In this space, practitioner(s) now co-evolve dynamically because of their interdependence and the emergence of novelty in the changing situation. Different practitioners will perceive these changes to themselves in different ways depending on their mental models (contexts) and their degree of familiarity and experience of such situations – this is Aspect One. In addition, because practitioner(s) will reason about the changes in different ways, depending on their capabilities, this will alter what they (can) find out and know and hence alter their perceptions of those changes. Collaboration can be an important tool in aligning otherwise separate viewpoints into an holistic appreciation – and also in challenging and ameliorating entrenched, inward-looking and dysfunctional perceptions.

The extent to which change can be understood is affected by the nature of a situation which determines what practitioners can predict about it – this is Aspect Two. An important concept here is that of the 'prediction horizon' – the area beyond in which it is not possible to detect future events based on past occurrences. Changes in the situation, because they are in the real world, will be far-from-equilibrium and therefore not easily predictable (Marsay 2006 shows the relevance of using the fluid metaphor for thinking about change). The nearness of the prediction horizon varies depending on what it is that practitioners are trying to predict (and their depth of understanding of it).[1]

The nature of the ongoing engagement then depends on the practitioner(s) attitudes to change and their stance on adaptation – this is Aspect Three. For example, they may be reactive (and so driven by the change) or deliberative (and try to plan for known change), or anticipatory (and set about envisioning and bringing about purposeful and novel changes to the future). They may, of course, try to disengage – or try to simplify the situation to make it tractable (given their, maybe limited, capabilities).

In any case, effective practitioners will monitor what is happening, reflect on it and alter the nature of their engagement as change occurs. The iterative practice that results has to adapt over time and this depends on the abilities of practitioner(s) and the means at their disposal to influence change. Any interventions, always involving other actors, will have to be made at the relevant level(s) and scales, when conditions are favourable, and over the necessary timescales. Structures (within and between actors and entities) may self-organize because of the coupling and the inevitable feedback. These changing structures need to be sensed if practitioner(s) are to maintain any kind of mature awareness of self, of others and of the situation. Appreciating the appropriate indicators to use is very important. Over time there will be historicity and path-dependence meaning that certain courses of action have consequences which limit the future – even though the future will never repeat the past exactly. This is why there will be a need for the continuous iterative activity which is so characteristic of, and necessary in, the real-world situations discussed in the rest of this chapter.

As a summary of this section, Figure 20.2 (adapted from Dodd, Prins and Stamp 2007) shows the behaviours, good and mal-adapted, that can result from practitioner(s) having various degrees of competence. This Figure can be used as a simple diagnostic tool to assess the degree to which practitioners are 'complexity-worthy' (in the way we expect

Aspects of Change

	1. Perceptions of Change - sense appropriately	2. Understanding Change - reason appropriately	3. Approach to Change - influence appropriately	
Effective practice	Yes	Yes	Yes	Well-placed
Inhibited practice	Yes	Yes, but not how	No, so ineffective	Aware, well meaning, but inhibited
Directed practice	Yes, but forced to	No, 'empty-headed'	Yes, possibly inappropriate	Outside Intent provided. 'Dysfunctional'
Watcher/ 'lurker'	Yes, 'voyeur'	No	No	Aware, not interested in opportunity
Ill-informed volunteer	No, so 'blind'	Yes, based on own doing	Yes, but ill-informed	Could do it, can't detect what or when
'Arm chair' volunteer	No	Yes, hypothetically	No	Has visions, dreams about change
Interfering volunteer	No	No	Yes, impulsively	'Loose-cannon' - miss aligned
Entrenched institution	No	No, in 'world of their own'	No	Detached, indifferent

Example Practitioner 'Caricatures'   Consequences in Real-World Terms

NB: Assume the natural complexity is a similar for all cases

Figure 20.2   Pathologies of adapted and mis-aligned practice

things to be sea-worthy) – that is, to assess the extent of their ability to deal with complex situations).

20.4   PRACTICE AND ACTORS' INTENTIONS

Without engagement there is no practice, and engagement only occurs when there is an intention to become involved. This practice may be as simple as deciding to visit somewhere (where, for example, entanglement will inevitably follow whether intended or not) or may be as complicated as deploying to intervene in an humanitarian crisis.

An important part of practice therefore is understanding why 'we' have become involved and to what degree. This is guided by addressing the following questions:

● What is motivating us to intervene in this complex situation – what is the problem, or opportunity, we perceive? What discomforted us enough to trigger our involvement?
● What are our implicit and explicit intentions? We are intervening in order to . . . do what? Have we surfaced our hidden assumptions about our purpose, are they being challenged? Given what is knowable about the situation, would our interventions make sense – and how would we know if they did?

- What is the nature of the current situation? What has caused the phenomena we perceive, and those we can't perceive (yet are there), to come about? Have we all faced up to the givens and realities of the changing situation?
- How well do we understand how this situation 'works' and what the viewpoints, contexts, motivations and influences of others are?
- What is the nature of the change we are trying to bring about? Do we have the requisite variety to enact them? Which changes are already underway and could we engage with them to shape and influence the phenomena purposefully? What are our options for bringing about self-sustaining transitions which exploit the existing flows?
- Which approaches and capabilities are suitable for use in these complex environments? How do we work out if they are appropriate to the changes occurring/desired over time? What do we do if we can't?
- How do we go about making decisions about possible futures beyond the prediction horizon' Do we need to?
- Who is best placed to bring about the changes? Do we (if we are the change agents) have the mindset, capabilities, alliances and tools available to influence appropriately?
- In any case, to what degree are our current capabilities complexity-worthy? And how would I know that they are – how do I judge 'appropriateness'?

All actors could use these as a diagnostic tool and ask themselves the questions above before engagement. If they cannot be adequately answered then it is probably best not interfere (as Figure 20.2 has illustrated) – because the practice would then be poorly matched to the situation and so mal-adapted outcomes, such as those in the Figure, could be inevitable.

20.5   ASPECTS OF PRACTICE

Once it has been decided to become engaged then the space of possibilities that follows is inevitably complex. The degree of complexity can be characterized by plotting the three Aspects of Practice on the simple graph shown in Figure 20.3. The right-hand axis is Aspect One – the Perceptions of Change, and the left-hand side shows Aspect Two – the Understanding of Change. The lower axis shows the Aspect Three – Approach to Change. A number of types of situations involving different practitioners have been plotted on the curve. To the lower left of the graph, logistics systems easily deal with a situation which they find routine and where data is familiar and self-evident – their way of working is largely deliberative and rational. Moving up the curve, situations become increasingly challenging, the changing information more ambiguous and hard to understand, and the ways-of-working required have to be much more adaptive.

A key point illustrated by the graph is that transitions are not always smooth – there can be dislocations, as happened on the Deep Water horizon oil rig in 2004 (discussed later in the chapter). Such transformations are always a shock (largely because of the lack of time to understand and deal with an unfamiliar experience), and will be especially difficult to deal with if systems are optimized to a status quo and practice has not explored the consequences of such shocks. Such shock situations are low-frequency, though high

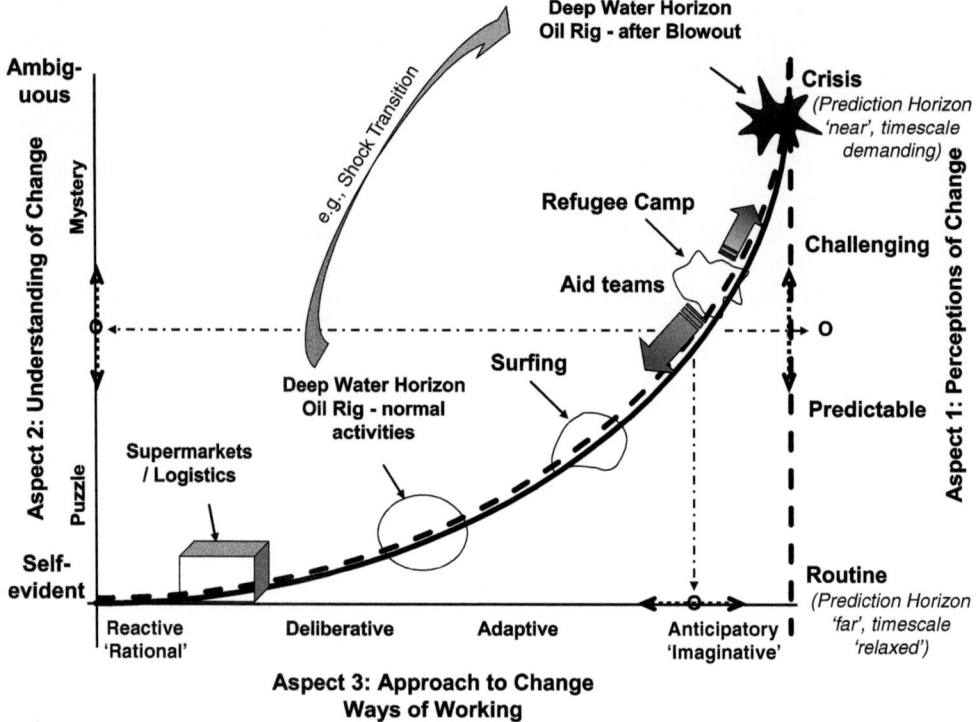

Figure 20.3   The three aspects of practice and the space of possibilities

impact, events – adaptive practice has effective risk-mitigation and contingency capability build-in on purpose to deal with them.

The Aspects of Practice, which are the axes of Figure 20.3, are also those of the Assessment Framework presented in this chapter. The Aspects will each now be examined in more detail. Note that the order in which they are described *does not* imply that they are steps 1, 2 and 3.

20.6   ASPECTS OF PRACTICE 1: PERCEPTIONS OF CHANGE

When dealing with real-world practice one is inevitably dealing with a degree of subjective perception of the situation – absolute objectivity is only theoretically, maybe never, possible. This is because underlying everything that happens in the world is what may be termed 'natural complexity' – which can largely be taken as a given. It is just how our world happens to be. Then, the scientist's analytical view of that natural complexity could be called 'academic complexity' – as its abstract viewpoint aims to eradicate subjective perceptions. However, the terms used by scientists to describe observed phenomena are themselves largely abstractions and may not be meaningful to practitioners. However, for practitioners, depending on their perceptions of the 'complexity' of the situation and their capabilities, then their behaviours and the resulting outcomes may differ, as Figure 20.4

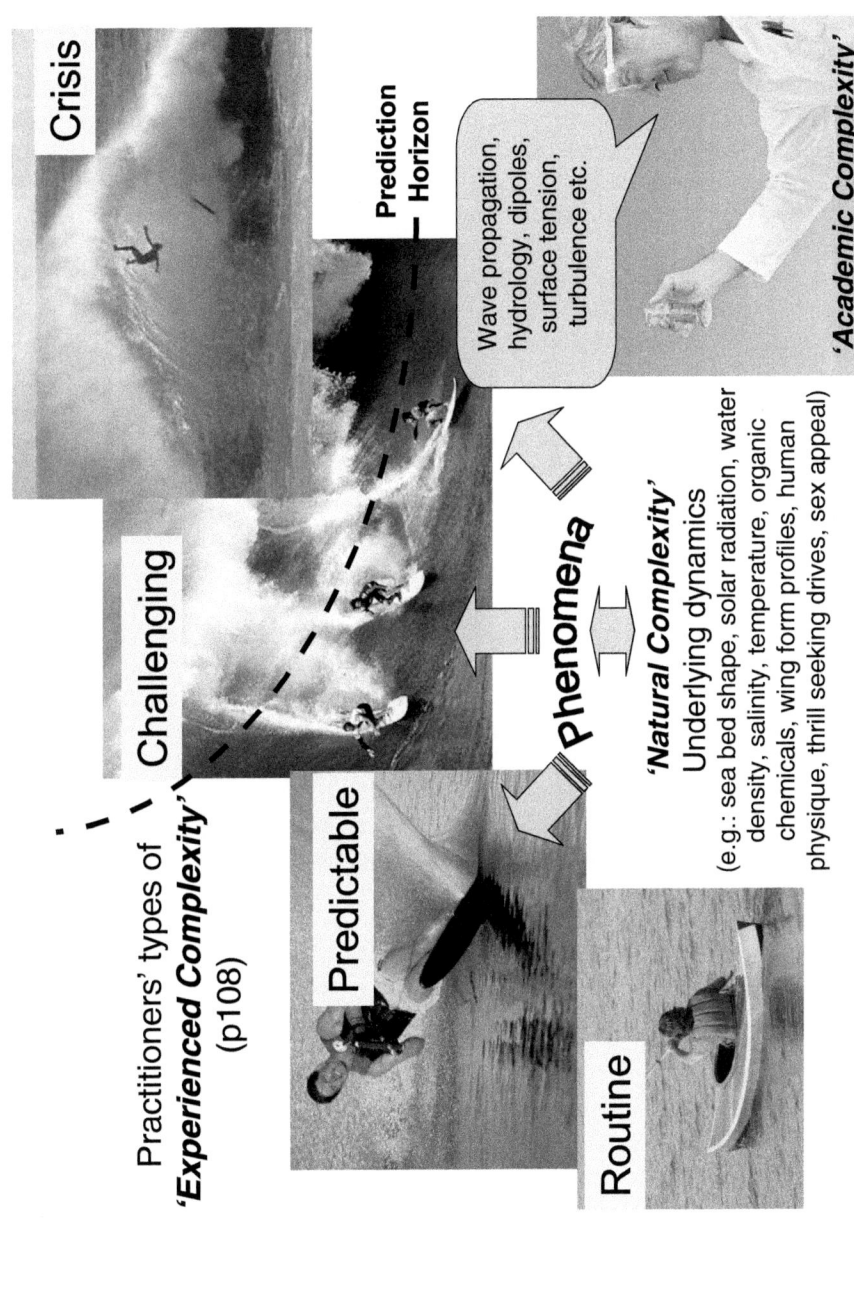

Figure 20.4   Perceptions of change – natural, academic and experienced complexities

illustrates.[2] Hence there will always be a range of contexts (mental models) which may have to be aligned when people work together – and these may change at any time (in the Figure, maybe a tsunami is coming!).

In this chapter the focus is on practitioners and their 'experienced complexities'. Research and workshop studies[3] have shown that people tend to describe their contexts, their perceptions of change, in four main ways: as routine, predictable, challenging, or as a crisis. What discriminates between the perceptions is the degree to which the situation can be meaningfully understood within the time available (which may be seconds, days, weeks or months depending on the prediction horizon) – set against their approach to change. In other words, the degree to which those involved have the capability to cope with the experienced change, that is, their overall complexity-worthiness at that moment in time, which may change when the situation changes.

A routine context can be easily understood and dealt with. Whereas the challenging context shown in Figure 20.4 would defeat a novice but is mostly manageable by an expert. However, beyond the prediction horizon lurks the unexpected, ready to catch out all but the most exceptional practitioners. For many scientists, from the academic complexity viewpoint, all these contexts are essentially equivalent situations – the nature and properties of water are invariant, rates of change may be different but, overall, one set of models applies throughout.

One important point here must be made about academic complexity. The impression that practitioners have articulated in workshops, erroneously, is that complexity on complexity begets only more complexity. Yet, if this were so, the world would be unmanageable. Instead, as Cohen and Stewart (2000) indicate, what practitioners work with are the features (of the situation) which they perceive to be invariant enough, at their level of emergence, to be engaged with and influenced. A simple example illustrates this. A child is pouring out milk for a cat and does this perfectly competently because, for the child (and the cat), the discernable features (that is, their mental models, that Cohen and Stewart call 'simplicities') in this natural situation are self-evident. Yet for scientists, with their different mental models, there is turbulent flow in the milk and massive underlying complexity in the bodies of the child and the cat and a myriad of interactions with bacteria in their environment and so on. In short, practitioners are really only concerned with experienced complexity, though they will wish for scientists to partner with them to advise on the underlying natural complexities, that is, on the phenomena studied as part of academic complexity. It is this partnering, and the use of appropriate tools and methods, that is one of the concerns of this chapter.

20.7   ASPECTS OF PRACTICE 2: UNDERSTANDING OF CHANGE

Though US Defense Secretary Rumsfeld was often mocked for his 'known knowns, unknown unknowns' speech[4] he was expressing an important truth – that one must understand what one does know, doesn't know, could know and can't know. With complex contexts this is especially important for, as this chapter will show later, 'we' (whoever we are) may not be in the best position to observe the significant indicators of the situation. Treverton (2003) has simplified Rumsfeld's statement in terms of what he calls 'puzzles' and 'mysteries', where:

- Puzzles (knowns and knowables – of existing mental models) have the following characteristics. They can be a solved by following a procedure or process. The puzzler knows in advance what the puzzle is and so can bound the problem. When something is missing, it is easy to classify the missing item, describe it in 'fact-like' terms (a red piece with a face on it) and then search for or collect the missing items. Once candidate pieces have been found, it is possible to compare them (because a prior model exists) or search results or items collected and then match a candidate item as being the missing one. It is then possible to fit the new fact into the puzzle (as the puzzle is 'static') and confirm it is the 'right' piece. Puzzles are problems which have solutions, for example, "How many people bought discounted items in our supermarket today?" So-called 'big data' solves puzzles.
- Mysteries (unknowns and unknowables) are different – for example, real situations with no satisfactory model, or for which there is no satisfactory model of a type currently considered legitimate. Exploring mysteries involves imagination, creativity and what-if-ing (not process-following). The thinker starts with little or no knowledge of the nature of the situation and builds theories/hypotheses or 'fantasies' of what might be going on and may project these into the future or 'backcast' them from possible ones. As Smith (2005) explains, potential indicators/weights of evidence that might exist/be required to support or refute hypotheses are looked for to see if they exist in the perceived (or are implied) in the hidden world. This looking is purposefully directed, not looking and then 'sensemaking' afterwards. The thinker may 'shake the tree' first (known as enacting) to throw up items which may prove to be of significance to support or refute hypotheses. There is no final, correct answer, instead, judgement, re-assessment, and continued purposeful probing are required.

As a result, the nature of the general discourse in some situation can be characterized in terms of the type of information being employed. For example, if the practitioner is dealing with the activities of a supermarket chain then the information exchanges are largely factual and rational, that is, puzzles. Whereas, the situation in Syria and Iraq, for example, is usually of a different type (Beautement 2011) – as much of the information exchanged is non-verbal and so is very hard to observe or capture (Dack et al. 2009) – in other words, it is of mysteries.

The differences between puzzles and mysteries are made clear when one examines, say, prediction. Consider an accident black spot at a cross roads with a bar nearby where alcohol is sold. To predict, in general terms, at which times of day accidents are more probable is merely a puzzle and routine if good historical data is available. If little data is available then complexity science can be employed to identify likely patterns – which can then be searched for, as Erdi (2008) explains. However, it would be very challenging to predict a single specific accident before it happens – and say exactly which person, in which car, at which time, and on which day they will crash – as it's a mystery, something beyond the prediction horizon.

The systematic assessments of practitioners' experiences already mentioned show that the four types of Perceptions of Change correspond to four general strategies used for understanding: observation (routine) and deduction (prediction) for puzzles; and induction (challenging) and abduction (crisis) for mysteries. Figure 20.5 provides a simple tool for matching appropriate ways-of-working to the level of understanding required.

The significance of Figure 20.5 becomes apparent when one starts to think about the use of information technology to support and augment human decision-making. Look at each row in the Figure and consider which of them are tractable with the computing tools we have now. Even the performance of sophisticated agent-based techniques struggle to work effectively above Level 2 – the reasons for which will be explained later in the chapter.

20.8   ASPECTS OF PRACTICE 3: APPROACHES TO CHANGE

Regardless of whether practitioners are acting as individuals, or as members of a community, or as part of some organization they will find that there are four main approaches to dealing with change – as Adger et al. (2007) note – one can be reactive, deliberative, adaptive, or anticipatory.

Unewisse and Grisogono (2007) extend this by introducing the idea of having an 'adaptive stance' towards change – where one acts, observes, decides and adapts iteratively – in other words, engages in learning by doing. Thereby they show the link between an organization's approach to change, its levels of learning and the capabilities it requires, where being:

- reactive: rarely learns from experience, even after the fact;
- deliberative: is about (so-called single-loop) learning how to learn from past experience;
- adaptive: is (double-loop) learning ways-of-working that provide an adaptive stance; and
- anticipatory: involves adapting how to anticipate, influence and exploit change adaptively (triple-loop learning), *and* how to adapt capabilities (quadruple-loop).[5]

Hence, to be fully adaptive we might, for example, develop social skills that enable us to forge new relationships and adapt. To be anticipatory we might socialize widely in advance of any manifest need and would continually be checking our foundations and (hidden) assumptions. Epistemologically, we might develop a new range of skills relevant to the changing situation.

These insights were extended into a diagnostic tool by Jones et al. (2014), a version of which is shown at Figure 20.6. Each column corresponds to one of the four approaches to change, with the rows representing the key features of capability making up what they term 'adaptive capacity'. The bottom row of the Figure summarizes the type of outcome that each approach can achieve in a complex situation.

The approaches in the first two columns are essentially backward-looking, in that they expect the past to be a guide to the future and rely on probability data for prediction. Most risk-management is essentially backward-looking for this reason and so is often surprised by the unexpected (Lissack 2011). The approaches in columns three and four are forward-looking in that their practitioners see the need to conceive novel, possible futures and to position themselves to take advantage of them. Therefore, complex situations require practitioners to adopt forward-looking approaches because of the endless novelty that such situations generate.

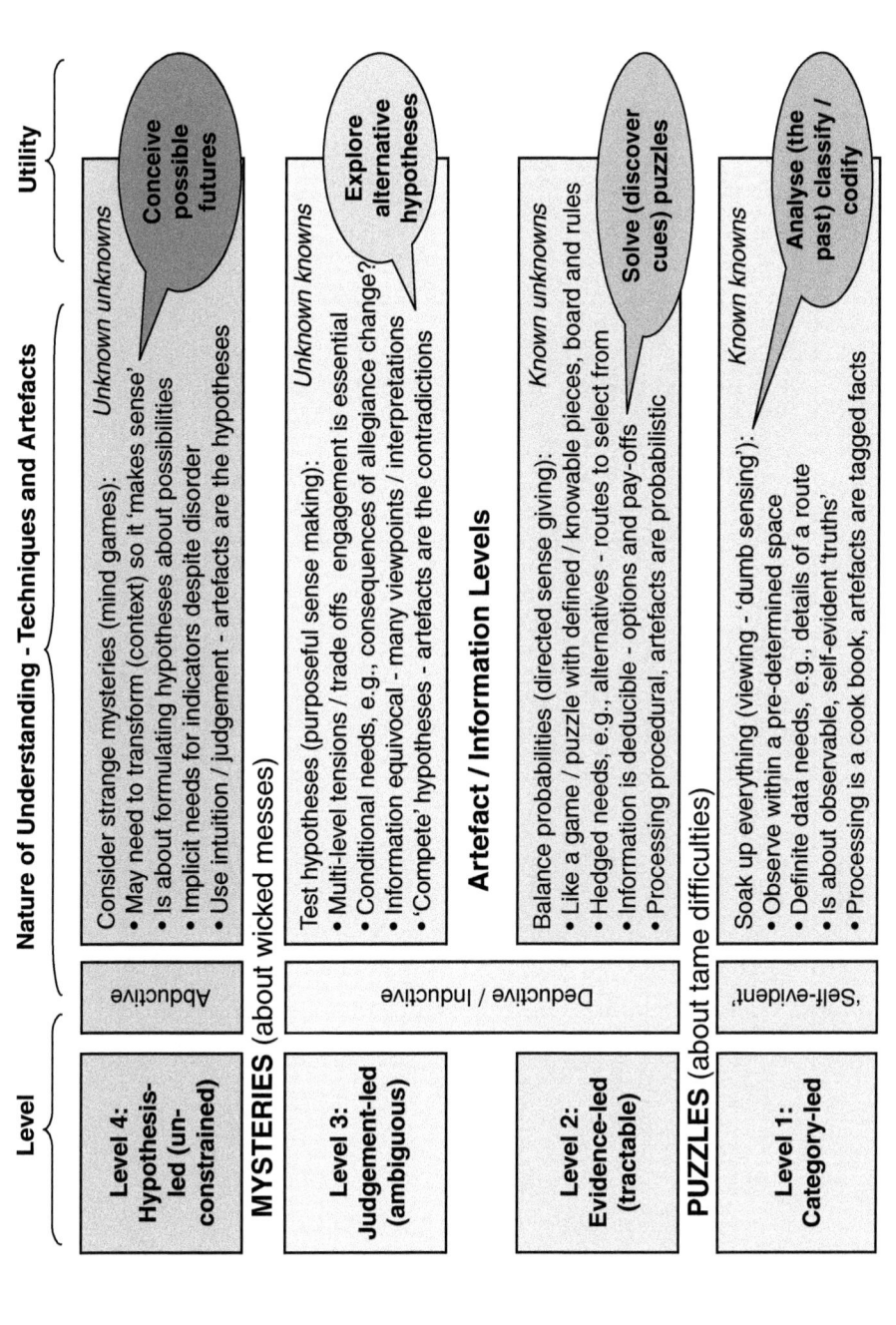

Figure 20.5   Levels of understanding matched to ways-of-working

437

Approach to Change - Ways of Working

Adaptive Capacity Characteristics	Reactive [Backward-looking]	Deliberative [Backward-looking]	Adaptive [Forward-looking]	Anticipatory [Forward-looking]
Governance and Policy	- Absent - [Have to deal with the now]	Deducing policy / plans based on hindsight - minimise 'risk'	Trade off possible futures and consequences - given uncertainty	Open mindset able to work with ongoing change / uncertainty
Innovation - exploration	Minimal / habits, but necessity the 'mother of invention'	Converging on 'best practice' – test wiggle-room	Explore / experiment. Discover / exploit degrees-of-freedom / opportunities	'Imagineering' e.g. forecasting, backcasting, horizon scanning
Knowledge and Information	Use of traditional wisdom / lore - as seems fit	Generating 'facts' from historical data / 'lessons identified'	Formulate / iterate possibilities and establish 'measures' of change	Able to work with plausible narratives and relevant indicators / warnings
Institutions and Entitlements	'Fixed', based on historical / tribal precedents	Organizing in line with plan delivery	Devolve / adapt for change. Support initiative / proven resilience	Flexible mix of formal and ad-hoc communities of interest
Asset base - availability	Indigenous - use what's on hand, current skills	Matching assets (usually centrally held) to plans	Accessing what's relevant / best placed given ongoing needs	(Re)distributable for contingency - plus strategic / central for crises
Nature of Outcomes	Driven by events, people 'passive'	Events dealt with only if planned for, otherwise 'crisis'	Events mitigated - as possible change is anticipated and adaptive capacity in place	

Figure 20.6 Approaches to change compared – from reactive to anticipatory

20.9 JUDGING APPROPRIATENESS

20.9.1 Practice as Synthesis

Clearly the three Aspects of Practice do not exist in isolation – practice is a synthesis of them. As Figure 20.3 above shows, there is continuous dynamic interaction between each of the Aspects which is highly situation dependent. Practitioners must continually match, employ and assess the degree to which their overall ways-of-working are appropriate to the changing situation.

So, when practitioners are considering how to influence change they have to think about whether they are going to work in a backward- and/or forward-looking manner. If backward looking, then Figure 20.3 can be used as a basic diagnostic tool to see if they are appropriately stanced for the current realities. If forward-looking, then the Figure can be used to characterize the expected situation – then, from this, the necessary approach to, and understanding of, change can be read off the graph and adapted over time. Best of all is a mixture of the two.

A most important part of this synthesis of practice is how to go about engaging with a situation (or changing the nature of the existing engagement, or even disengaging from it). This can be thought of as three interrelated phases of engagement, as Figure 20.7 shows (adapted from Alston et al. 2006, who term them the design, assemble and run-time,

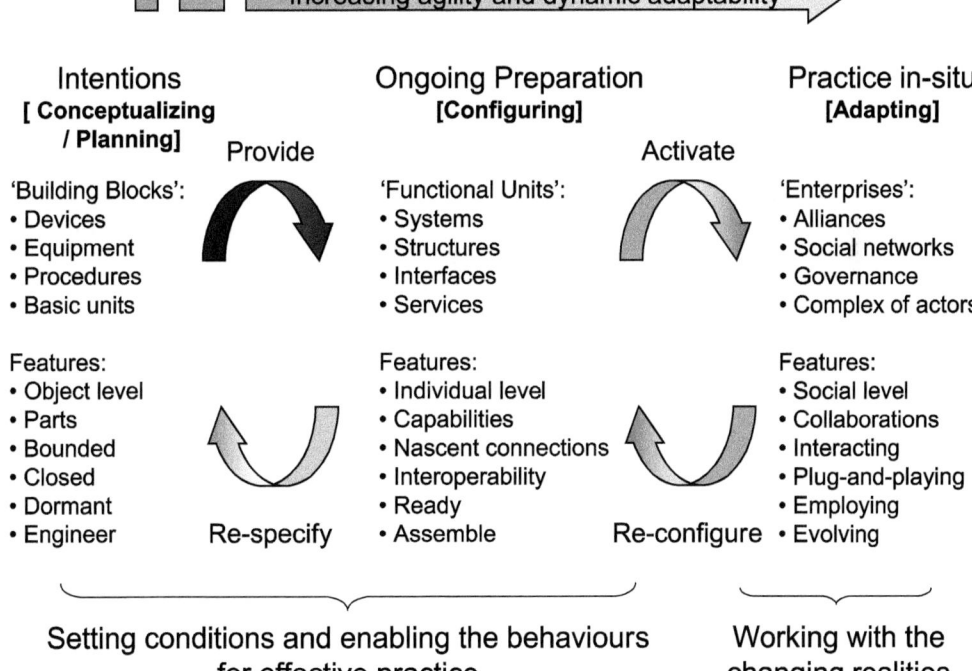

Figure 20.7 Phases of engagement

DART, phases). In every phase, some kind of transition or transformation needs to occur – where practice morphs and flows back and forth between them as the circumstances change (this notion of practice as intersecting loops has been called 'dancing with ambiguity', Leifer and Steinert 2011).

To the left is the intentions phase where conceptualizing and planning takes place. The middle, configuring, phase is where the necessary preparations for, or adjustment to, capability are made. To the right is practice, where the ongoing engagement and adaptation occur. For practitioners, appreciating clearly which tools and methods are appropriate to use in each phase is an important part of the competencies that they have.

Linking the phases are the transition stages – often the most challenging part of practice as they involve changes in, for example, form and structure, in authorities and responsibilities and, most importantly, in the nature of the phenomena being experienced. When these transitions go wrong, or are poorly understood, then the results can be very public. For example, in 2008 in the UK, when London's Heathrow Airport Terminal 5's baggage handling spectacularly failed to operate 'as expected' when first opened.[6] Part of the reason for this was that the phases were treated as if they were isolated and the nature of the dynamic transitions was not fully appreciated. In practice the phases are always 'intertwingled' – and it is this inter-connnectedness that generated the complexity of the situation that was not anticipated.

20.9.2   Appropriateness Framework

To judge appropriateness effectively, given the richness of practice, a rigorous yet pragmatic tool is needed. In the Millennium Project, Glenn (2009) provides a taxonomy of research methods which are characterized under the headings of: quantitative, qualitative, normative, and exploratory (corresponding roughly to the Approaches to Change used here). In addition, Aaltonen (2009) offers a grid, whose axes ('understanding of systems' and 'means of controlling' relate to the left-hand and lower axes of Figure 20.3 respectively) represent a landscape on which tools and methods are positioned – depending on their suitability to inform practitioners' contexts. This chapter adds practice, the Perceptions of Change axis, as a third one. Overall, these three Aspects define the Appropriateness Framework shown in Figure 20.8, on which a number of methods have been plotted as examples.

The Framework can be used in two main ways – either to select tools and methods suitable for use, given a context; or, to identify for which type of contexts a candidate set of tools would be appropriate. In the Figure, eight regions are shown along with the kinds of reactions to the situation that might be typical in those regions.

In the first way of using the Framework, the nature of the context would need to be drawn as an 'envelope' within the cube space. This envelope characterizes the set of functions or behaviours that must be addressed by the tools and methods selected. For each tool, its envelope of capabilities can be plotted on the cube. Where there is overlap, there is a degree of suitability; where disjoint, none; and where complete the tool or method would appear to be suitable. The set of contenders so produced need to be subjected to a further level of judgement-based practitioner analysis (supported by expert advice from scientists, which itself must be appropriate). A matrix such as the one

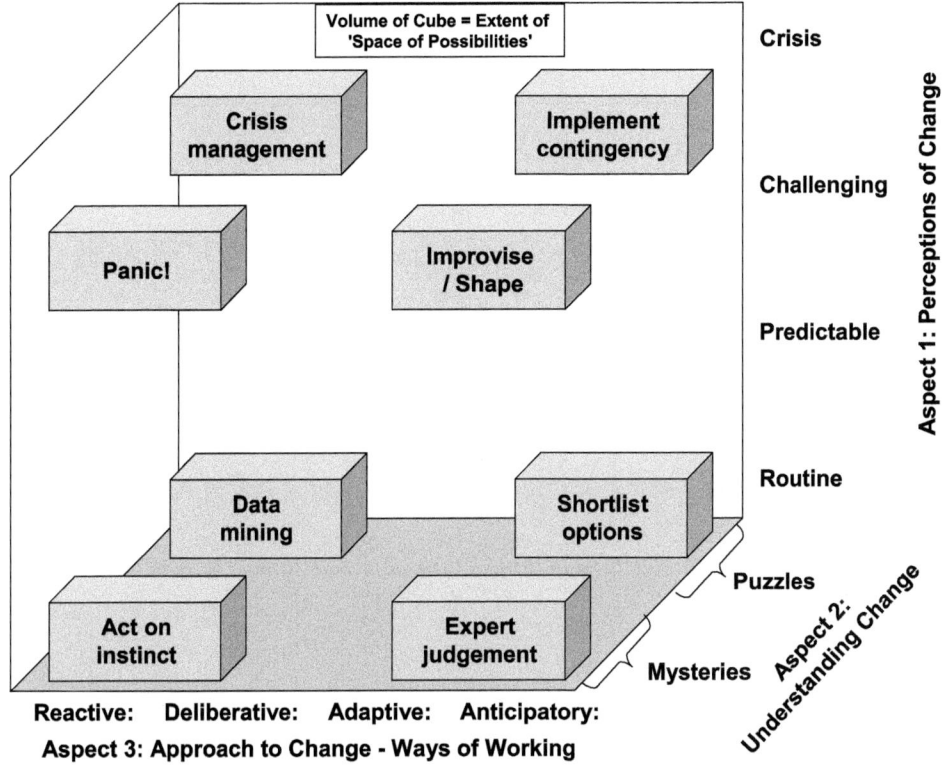

Figure 20.8   An appropriateness framework with examples

in Figure 20.9 might be used to do this. For a particular task and candidate technique is selected (in this example, social network analysis). The elements of the task are listed down the left-hand side. Across the top (in the centre column) are four factors which are used to define the characteristics of the task (for example, how easy and critical is it to the overall endeavour). To the right are the criteria for assessing the appropriateness of the candidate method (is it well matched, mature enough and so on). The method is scored against each sub-task and an overall assessment of suitability made. This would be followed by the use of expert judgement to compare methods.

In the second case, to identify for which type of context tools and methods are suitable, the envelope of capability for the tools and methods available would be plotted. Then, by reading from the axes of the Framework, the nature of the context that the tools would address can be read off. Various combinations of tools can be tried to refine or expand coverage accordingly. Such an analysis could be very useful in informal federations or in coalitions of communities where different organizations are providing different capabilities. The analysis could show up duplications, omissions, conflicts and so aid in building overall effectiveness and efficiency. For example, as Figure 20.2 indicates, the consequences of a mismatch of tools to contexts can lead to confusion, making dysfunctional interventions and so on – even panic.

List of Required sub-Tasks:	Characteristics of Task:					Candidate Method / Technique to employ on the task: (e.g., Social network analysis)			
	Degree to which we know how to do this?	How easy to provide this ability?	How critical is this task to overall intention?	Overall Task Score		How appropriate is the Technique to the task?	Mature Technique? Does it need development?	Overall assessment of Task's Capability	
T001: Ability to analyse others' Intentions	Medium	Medium	Critical	Issue		Weak	Mature	Issue	
T002: Ability to analyse the demographics of the human environment	Easy	Easy	Medium	Routine		Supports	Mature	Trivial	
T003: Ability to align or coordinate with coalition partners	Medium	Medium	Critical	Issue		Supports	Medium	Challenge	
etc.									

Figure 20.9   Example of a detailed Assessment Matrix

20.9.3 Assessment Example: Autonomous Agents and Multi-agent Systems

The field of autonomous agents and multi-agent systems (AAMAS) is broad and has been well researched and described (see, for example, the journal and conference series of that name, Jennings, Sycara and Wooldridge 1998; Dignum 2009 and Trajkovski 2009). Though major agent-based experiments have been successfully carried out (Kirton et al. 2003) there are many general issues and challenges still to address if the use of agents is to be 'industrial-weight', and these have been clearly stated (Allsopp et al. 2006). However, for the purpose of this chapter we can identify three main strands of AAMAS, where agents are:

- embodied (for example, robots), or disembodied, and emulate or simulate human behaviour in a manner such that the agents work as colleagues or partners in the real world – with so-called adjustable autonomy in 'human-machine teaming' (Bradshaw et al. 2004);
- software entities (for example, avatars) in a game or virtual world where they are programmed to apparently behave like human beings – such that the simulated behaviour is plausible to people immersed in the virtual world – that is, agent-based gaming;
- software entities in a model or simulation whose behaviours are representative of what humans would do – such that the outcomes in the model are analogous enough to those in the real world to inform policy and decision-making – that is, agent-based modelling.

The example in this chapter will focus on third type, agent-based modelling (ABM). ABM has been in general use for some time, but has been hampered by the lack of a clear definition of 'agent'. If agents are merely code blocks in a simulation tool then they do what they are programmed to do (Jennings 2001). If they are reflexive, see Mathieson (2005), and self-modifying (as they should be to emulate simple human behaviour), then modelling runs are not repeatable in the manner required for operational analysis, for example. Applying the Appropriateness Framework to ABM, see Figure 20.10, throws up some interesting patterns.

An obvious one is the appropriateness of ABM for exploring deliberative ways-of-working in predictable situations. The use of ABM here is suitable if, say, the activities of a logistics enterprise are being modelled (as this is essentially a puzzle) and so ABM scores highly. Whereas, when dealing with crisis situations where it is hard to anticipate (such as those involving the so-called 'Islamic State' in Iraq in 2014/15), then ABM scores poorly because the meaning and significance of data to be used, if available, changes too rapidly to be fed into a model (if building a model is indeed possible). Hence the distinct fall-off in applicability.

Another pattern is the step down in capability of ABM to deal with mysteries. Why is this? One factor is that in most situations it is almost impossible to capture the data on the intentions of actors in a manner that allows combinations of interactions and effects to be modelled, as at many levels of conflict they are mysteries. This is because much of what people would call their intentions is concealed, indeed, even from themselves – and therefore largely uncollectable. Yet intentions and motivations drive the dynamics of

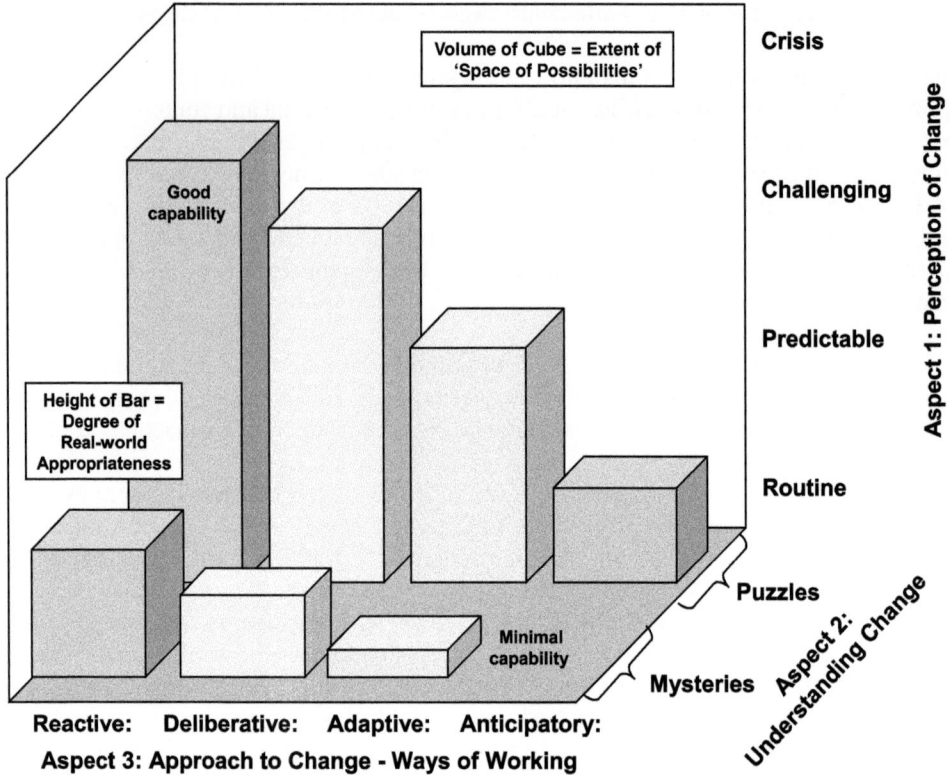

Figure 20.10   Assessment framework applied to agent-based modelling

situations almost more than almost anything else. So models without this information would be of little value, as Feltovich et al. (2004) note. Kurtz and Snowden (2003) offer a partial solution to this in the use of abduction to gather micro-narratives which can then be analyzed to reveal otherwise hidden motivations and concerns.

Nevertheless, ABM is a useful tool which is appropriate to some types of contexts, especially when used in an iterative manner as part of human decision-making. Where contexts are multi-modal (involving heterogeneous agents working in different ways) and multi-level (from local to global, micro to macro) then ABM can be used with care. For example, the Santa Fe Institute's Steve Lansing (2003) has been dealing with issues such as the tensions between government and local communities over rice growing in Bali. Lansing has used ABM to show that the strategies (based on social and religious controls) used by local farmers to determine planting times were robust and appropriate. The data was used, not to tell the farmers something they already knew but, to 'prove' to government officials that their centrally-imposed scheme for fertilizer and pesticide application was causing damage, and this persuaded the government to withdraw the scheme.

20.9.4 Conclusions on the Real-world Readiness of AAMAS

Real-world-readiness can be defined as the degree to which a method or tool can have traction in the context such that the insights it generates can add value (in the widest sense of value) for the practitioners who use it. This is clearly subjective, based on the nature of the context of interest. Many complexity scientists would be uncomfortable with what they would consider a lack of rigour in these assertions but, for practitioners, these are the givens and realities that they have to resolve – and pragmatism has to win out in the end. The Assessment Framework assists in this debate by providing a *structured way for practitioners and scientists to work together to the evaluate real-world readiness of tools and methods*. Such judgement-based professional approaches are often the best ones to use when assessing appropriateness in practice (Jones et al. 2014).

20.10   EMPLOYING APPROPRIATELY

20.10.1   Appropriate Practice Over Time

Once practitioners have decided to engage and have assembled the capability required for that engagement, they then need to sustain their engagement appropriately and effectively over time. How should this be done and how can insights from complexity science assist? To answer this it is necessary to look a little closer at the natural complexity, the underlying dynamics that drive change (regardless of perceptions) that was mentioned at the start of this chapter. Natural complexity operates at a number of levels and scales which, for example, Morowitz (2002) started to classify. The significance of this insight is that practitioners cannot act or intervene at only one level – they may have to work through a number of levels simultaneously, or in different places or at different times or with a variety of partners and intermediaries, maybe over years. In addition, the level at which practitioners work may not be the one where sense can best be made of the situation – it may only be clear at lower or higher levels of emergence from the one that they occupy.

Some of these levels of emergence are illustrated in the hypothetical example shown in Figure 20.11. Here, government policy makers are the practitioners (no particular government is suggested or implied). They decide to introduce a radical change that cuts the state benefits given to those who have previously depended on them (such as happened in Ireland/ Eire after the financial crash of 2008 when the Financial Emergency Measures in the Public Interest Act 2009 was introduced). Over time, microbes exploit the opportunity to multiply offered by the bodies of those weakened by poor nutrition. This might eventually lead to more demand on the health service, the increased cost of which largely negates the savings made by cutting health benefits.[7] The media and popular opinion criticize the cuts. In this example, the interplay between levels is commonplace in practice – and is vastly simplified in the Figure. Research underway by the author, and not yet published, suggests that there may be as many as sixty levels of emergent transition from the sub-atomic to the cosmic and through to cyber – which would explain why natural complexity generates so much of the rich interaction in the real world that can be such a challenge to deal with in practice.

If one is to put this natural complexity to work then one has to appreciate how to affect the underlying generators of the phenomena that practitioners perceive in the real-world.

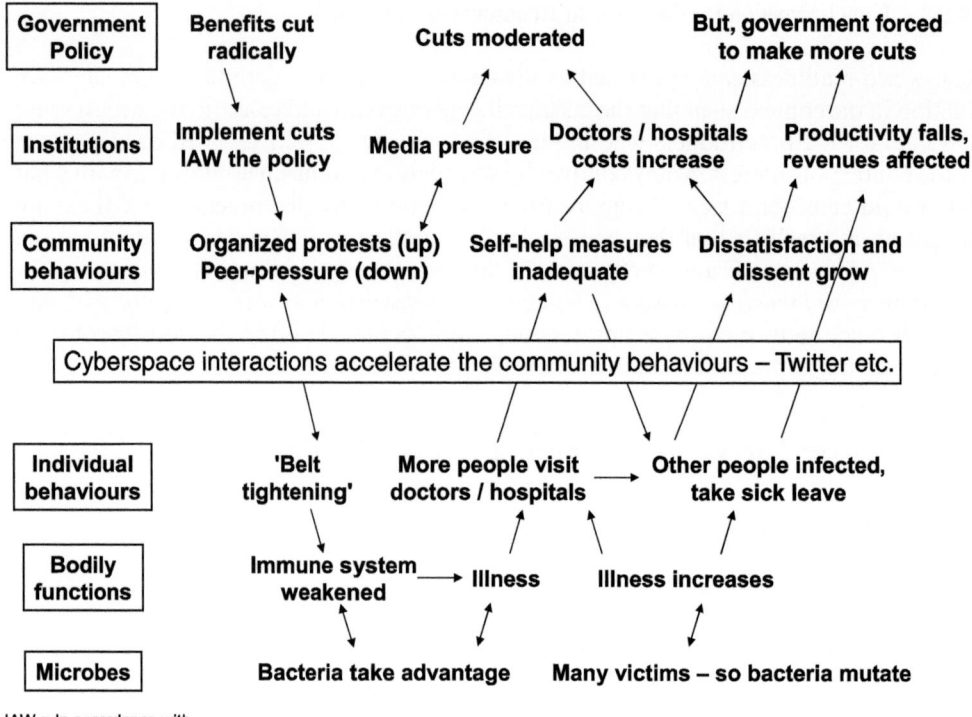

Figure 20.11   Interplay of levels and scales in a multi-model, multi-level manner (simplified)

In practice, there are plenty of degrees of freedom and options available for, selectively and pragmatically, shaping and influencing change in a situation. Indeed, everyday experience tells us that there are least six main avenues available for triggering change in practice over time.

These six are by:

- making changes in the environment: for example, 'seeding', setting the conditions so that preferred phenomena are more likely to come about (such as providing infrastructure that strengthens livelihoods) or by directly altering the geography (for example, tree-planting schemes);
- changing the nature of top-down influences: for example, via orders, policy directives, direct interventions, incentives, rules and permissions. This includes the ways that accessibility rights or 'boundaries' might be changed (say, in permissions to use land), or that behaviours might be encouraged or clamped (for example, by offering or withdrawing rewards), or that connectedness and interactions might be tuned (for example, through imposing safety rules, or providing funding incentives to groups that agree to collaborate);
- changing or manipulating self-organization and/or self-regulation (self-) 'mechanisms': for example, by influencing social drivers (such as ethos, behavioural norms

and the formation/membership of community groups so certain modes of behaviour become more or less acceptable). In some situations, tensions and synergies between groups may change the dynamics (for example, local residents may object to commercial development or they may partner with developers to help conserve the surroundings);

- changing the nature of bottom-up influences: for example, via 'the people' who can think/act locally but cause effects with potentially broad impact (for example, 'the public mood', blogs, individual initiative leading to self-help campaigns). Many phenomena are enablers for change at 'higher' levels and so have a bottom-up influence, for example, waves enable surfing which enables people to have fun – these sort of 'cascades' of emergence will, of course, change depending on the situation;
- influencing cyberspace: because of the role it plays in shaping human behaviour. Given the importance (in any situation) of the 'environmental' effects of co-evolution, the mappings between cyberspace and 'real space' are profound drivers which have been discussed elsewhere (for example, by Levy 1993). Phenomena in cyberspace can be affected using the same top-down, bottom up and self-organizing strategies as those discussed above – indeed, the Internet's self-repairing (autonomic) capabilities are already actively influenced by policy makers, bloggers, social-networkers and system administrators;
- also 'doing nothing', which may involve active disengagement or just 'letting things follow their course'. Given that 'timeliness' and timing are key factors in perceptions of situations, someone who apparently 'does nothing', may in fact be waiting for opportunities with timescales over years or decades (as Machiavelli did in his political scheming).

In terms of influence, at the human level there is also a great deal of diversity. Thompson (2008) identifies five groupings of people which he terms 'solidarities' (hierarchy, individualism, fatalism, egalitarianism and autonomy). Thompson makes the important point that solidarities do not partition society, they are grouping that people move between – indeed, an individual may simultaneously belong to several of them at once. Figure 20.12 amplifies these groupings and adds a further twist – those contexts of the 'dark-solidarities' of the underworld (that is, criminal, dysfunctional and/or deliberately malicious groupings).

The importance of this underworld is seen in practice all the time. Think of any of your favourite famous detectives and crime programmes. On how many occasions have you seen them 'break the law' to achieve justice (for example by working with criminals to reveal evidence and so on)? The notion of good criminals, Felbab-Brown (2013), is now well enough established to part of the US President's Task Force on 21st Century Policing.[8] Also, in politics and diplomacy, partnerships have sometimes to be made with undesirables to bring about the conditions for change, such as occurred in Northern Ireland – unpalatable though this may be. However, in commerce and economics, as the Enron scandal and the crash of 2008 and its aftermath have shown, many involved worked illegally in the space of the underworld solidarities on a routine basis. Practitioners, such as those working with humanitarian aid, may face the dilemma of deciding whether to stick to their principles and maybe delay beneficial outcomes, or to be pragmatic and get effective results quickly. Engaging in appropriate practice

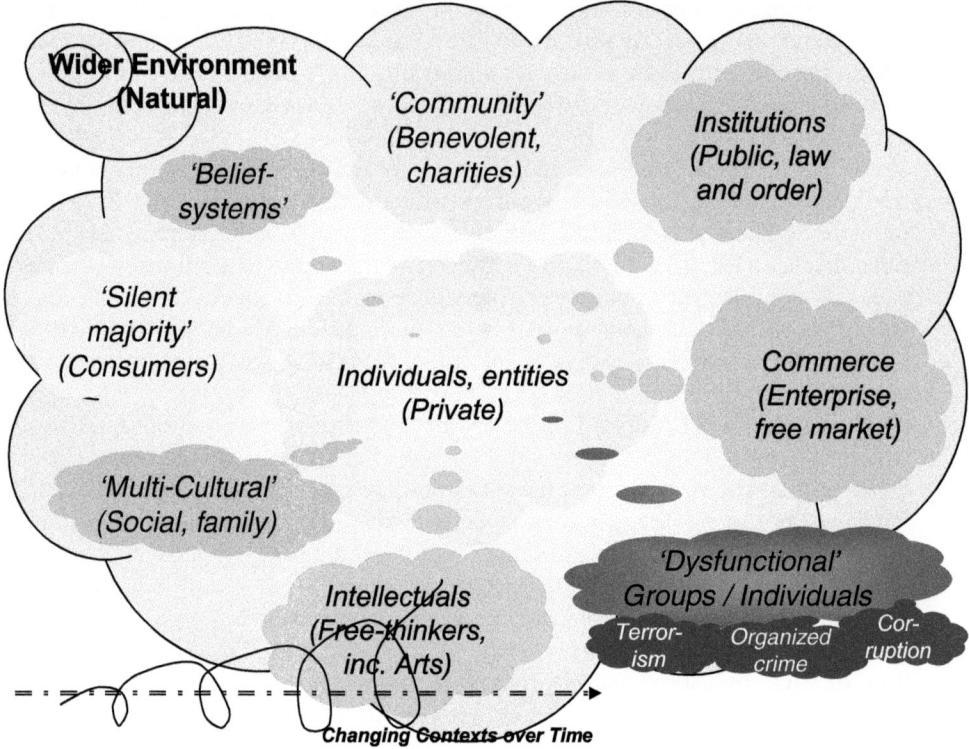

Figure 20.12   The wider contexts of practice

over time can, therefore, require hard choices – and complexity scientists who assist would have to be open to dealing with unpleasant truths if their contributions are to be relevant.

20.10.2   Multi-modal Networks for Influencing Change

A complexity insight for practice is that not only are the 'nodes' in influence networks heterogeneous and complex in themselves, but also that the 'links' are too. Therefore, influence networks are multidimensional – for example, in the 'levels' of their granularity (micro to macro) and scale (nano-second to years) and in terms of the ways that influence is mediated and propagated through networks.

Also, influence is multi-modal – for example, influence may propagate via changes in the environment, by direct contact, or via emissions of mediating energy. So, depending on the sensing and perceiving techniques of the observer, influence may seem to become 'invisible' (as it propagates via another modality) – which may lead the observer to substitute false influence mechanisms and even, in the human case, develop conspiracy theories.

It is not possible to understand these complex behaviours by trying to break them down in a reductionist manner. Most influence networks are depicted as nodes and links operating in the manner of a feed-forward neural network where both nodes and links appear to be homogenous in their nature. The assumption is that each link represents a direct 'cause–effect' mapping. Smuts (1926),[9] recognized the limitations of this everyday cause–effect thinking and that usually the notions of 'cause' and 'effect' are hopelessly simplistic and too tightly bounded. Models, such as simple agent-based ones, based on such ideas fail to deal with reality, both because of their naivety and also because they easily fall prey to Gödel's Incompleteness Theorem (for example, see Hofstadter 1983).

At the heart of appropriate practice, especially in partnerships, coalitions of the willing and in loose federations, is understanding the modes and degrees of freedom of the partners. Especially how they overlap and intersect in an enabling manner and/or how they interfere and inhibit in a way that undermines desired change. A full examination of these issues, and a tool to employ in practice, can be found in Beautement and Broenner (2011, pp.137–140). The factors that need to be considered include, depending on circumstances, the extent of:

- the possible actions that partners can perform in theory (no restrictions);
- the space of actions that partners together can achieve in theory (no restrictions);
- the actions that are available and achievable alone and together (in this situation);
- actions that partners are permitted or authorized to do (external constraints);
- actions that people are obliged to do or expected to do together (social, personal norms).

The totality of interactions between these factors defines one of the spaces of practice that would need to be considered, often called the 'wiggle-room'. Intersections between this space and the spaces of other actors in the situation reveals where collaboration, confrontation or conflict may occur. Practitioners can 'wiggle the wiggle room' to see where alternative arrangements may offer opportunities and/or release constraints.

A technique for using these multi-modal, multi-level influence networks has been partly adapted from Howard, Bennett and Bryant (2001, whose work involved peace negotiations in Northern Ireland and the Balkans) and involves taking the perceptions of actors one at a time (in an order determined by the spirit of the exploration). One then works with, or on behalf of, the individuals and organizations concerned to identify the factors and influences arising as they see them. This includes analysis of the 'what do I think' and 'what do I think they think' type. Viewpoints within and across perceptions can be compared to expose wiggle room, contradictions, alternatives, 'invisibles' and so on. The added value of this approach is that, by examining a rich variety of levels and modalities, novel influences, effects and linkages can be identified which would normally go unnoticed. This analysis has proved to be a powerful technique for triggering insight and debate, as Beautement and Broenner (2009b) discuss, and as the examples below illustrate.

20.10.3   Example 1: Climate Change Adaptation in Africa, 2014

This example examines how a number of different perceptions of change (that is, Aspect One) are being aligned to improve adaptive practice. It concerns climate change adaptation in Africa, where policy makers are often faced with the difficult task of making decisions in the face of an uncertain future outlook. Despite this need, as Jones et al. (2013)

explain, the development and humanitarian sectors continue to face criticisms over their relative rigidity and short-termism with regards to project funding and delivery. Recent emphasis on promoting a 'resilience approach' has resulted in calls for more longer-term objectives and deliverables, greater flexibility in planning processes, as well as better collaboration and coordination amongst key actors. At the core of this example are misaligned perceptions of change.

In Phase 2 of the Africa Climate Change Resilience Alliance's (ACCRA) work, posters were used to express in detail, in a way that would make sense at the community level, the practical behaviours to adopt in order to realize appropriate adaptation. Figure 20.13 shows the top-level view of six posters, each with its own detailed explanation. It became apparent during these activities that the content of the posters[10] *actually provide a diagnostic tool* which could be used at all levels to examine and evaluate:

- current policies, structures, behaviours and incentives so that groups could assess their current state of readiness for coping with adaptive change;
- indicators of change, and of the factors and conditions that might be preventing change – or that were available but not exploited;
- learning, capacity-building and educational initiatives – to inform the development of curricula material and to see how well they were progressing; and
- ongoing performance and achievements in real-world adaptive change – as it happened.

In addition, a participatory game was produced for use by district-level decision makers which enabled them to explore what constitutes the relevant elements of adaptation in each situation. They used it to identify the wiggle room that they potentially had, to explore it and expand it by removing the barriers to more effective practice by aligning contexts. Both the local and district-level activities are required to support the overall transitions – including that of influencing the government policy makers above them, not an easy task as Crook and Booth (2011) discuss. The activities, started in 2013 and ongoing to date, show that a change in ways-of-working is needed, is theoretically possible, of potential value, and is actually and practically achievable through capacity building at all levels.

20.10.4　Example 2: Deep Water Horizon Oil Rig Disaster, 2004

This example examines how a poor understanding of the nature of change (that is, Aspect Two) were not able to deal with the sort of shock transition that was described at the start of the chapter. It concerns the case of the 'Deep Water Horizon' oil rig that exploded in the Gulf of Mexico in 2010 while it was drilling the Macondo well. According to the Final Report by the National Commission on the BP Deepwater Horizon Oil Spill and Offshore Drilling (2011):

> Macondo was not the first well to earn that nickname "the well from hell"; like many deepwater wells, it had proved complicated and challenging. As they drilled, the engineers had to modify plans in response to their increasing knowledge of the precise features of the geologic formations thousands of feet below. Deepwater drilling is an unavoidably tough, demanding job, requiring tremendous engineering expertise. (p.2)

Figure 20.13   Criteria for assessing climate change adaptation capability

The oil industry is used to dynamic transitions, they occur every day – such as down-hole tool failures and surface equipment down-time (cables break, pipes burst, machinery goes wrong, power fails) – these things 'just happen'. As they are inevitable, and in a way are expected, their effects can, largely, be mitigated. These events may be sudden but it is still possible to make sense of them and act accordingly. The workers are backed up by manuals and guides covering the operation of the rig, policies and rules concerning safety and environmental protection, and there are instructions that document 'best practice' in check-lists provided by the companies involved. Normally then, relationships between workers and systems (the oil rig) are defined by these procedures through an established 'chain-of-command'. The normal zone of human influence extends across the puzzle space, up to the predictable level and mostly in the backward-looking region. This zone is usually well matched to the function and behaviour of the rig when it is working normally.

The blow-out changed all this by triggering a whole series of shock transitions where previously-held assumptions became invalid in the face of the ever-new, dynamically-changing contexts. The nature of the rig, that is, the situation, had changed radically because of the explosions. Automatic systems were destroyed (or lacked power) and so could no longer assist people's capabilities – the rig was no longer functioning with the workers as a 'human-machine team'. The Report gives a clear sense of there being fewer and fewer options for the workers and a degradation of influence – from being able to manoeuvre and control the whole rig as a unit, reduced to just parts of it, then to only single machines and finally to where the lack of a single knife was a life-threatening issue. This reduction in this envelope of influence fundamentally changed people's degrees of freedom and wiggle room. The post-blow out envelope of required behaviours (mostly in the mystery space) are a mismatch with the usual established procedures. The unrecognized, unchallenged and unresolved presence of this mismatch led to the final breakdown of control and to complete disaster.

The survivors were now called upon to operate the crippled rig systems in manual mode. In effect, what had once been a well-mannered colleague was now like a dangerous alligator. People's ways-of working were forced to change radically as normal procedures were no longer available actions for the control-room staff. Instead, workers needed to be 'allowed' to improvise to deal with the situation, but regulations prohibited this. Yet it was essential that people self-organized, to collaborate locally and ad-hoc, to try and recover control – but, because of the damage, the workers could not even come together as previously exercised to form the usual fire and rescue teams. Instead, they had to work with whomever and whatever they could find and develop what seemed to be the best purpose/intent at that time.

After the incident the inquest took place and the Report mentioned was produced. Yet, almost without exception, the recommendations concern top-down policy and regulative frameworks – of which there already seems to be too much. Even the Safety section on page 231 of the Report focuses on procedures during normal operations rather than on worst case transitions. The lessons from complexity science have not been learned and, more importantly, not implemented. Deep Water Horizon could happen again – almost exactly as before – and BP would be faced once more with having to make crippling compensation payments.

20.10.5   Example 3: Fuel Protests in the UK, 2000

This example is about the mismatch between two different approaches to change (that is, Aspect Three). In 2000 there were protests in the UK over the rising price of fuel for vehicles. The most vocal protesters were the lorry (truck) drivers who blockaded city centres, oil depots and caused traffic jams on motorways by driving very slowly. Fuel shortages led to high-demand for petrol – which was serious for the UK government who had, as part of cost-saving measures, abolished many of their fuel storage facilities. This meant that emergency personnel, and some government workers, had to get their fuel from commercial petrol stations. As these ran out of supplies the government faced the prospect of losing control of the situation (when, for example, police cars ran out of petrol).

The fuel protesters were individuals with a common interest to see tax on fuel reduced – they had no formal leadership and did not belong to an established group of any sort and this made it difficult for the government to find a way of engaging with them or of exerting influence. An appeal on television to stop the protests by the then Prime Minister (Tony Blair) received a derisive response from the general public who were, generally, sympathetic to the protests. Mr Blair ordered the formation of a task force including oil executives and police officers to work out how to deal with future demonstrations. Home Secretary Jack Straw, who headed the group at that time, admitted that the government had not been fully prepared for the events. "We were not caught on the hop about the concerns – it is certainly the case, however, that the scale and escalation of the protest was something which was unexpected," he said.[11] But one truck driver, encouraged by the widespread public support of this protest, promised more chaos if fuel prices did not fall. "We could have brought this country down – we've got them on the rack, let's keep them on the rack", he told colleagues.

There was clearly a profound mismatch on many levels here. The protesters and general public together acted like an undefined 'swarm' of millions of people generating transient, yet nationally significant, effects. They were using mobile phones and social media, new phenomena, to gather support and to coordinate their activities on a single issue across space, and then transition in a short time. The UK government, in contrast, is a federation of institutions with identities, structures and capabilities formed over many centuries. This is intended to provide certainty and assurance over time. However, its stability undermines agility. As Figure 20.14 indicates, there is a mismatch between what Conn (2010) calls the government's formal 'teeth' structures and the informal dynamics of community activity.

The Figure shows a continuum of organizational forms based on Thompson (2008), their origins, dynamics and ways-of-working. To the right are where the protesters started out – as individual activists – to the left are institutional groupings, such as governments. The government's organizational form cannot easily become 'unstructured', whereas the fuel protesters and general public were an unstructured swarm – some properties of which are partially understood by current views on complexity science. The government was advised to introduce a trigger that would cause the swarm to 'crystallize out' and so become a cohesive group – which was more tractable to government methods.

How was this done? The challenge was information. Much of that needed concerning the protestors was entirely in the mystery space, whilst the government worked mostly in the puzzle space and struggled to find out what was going on. As described in an article

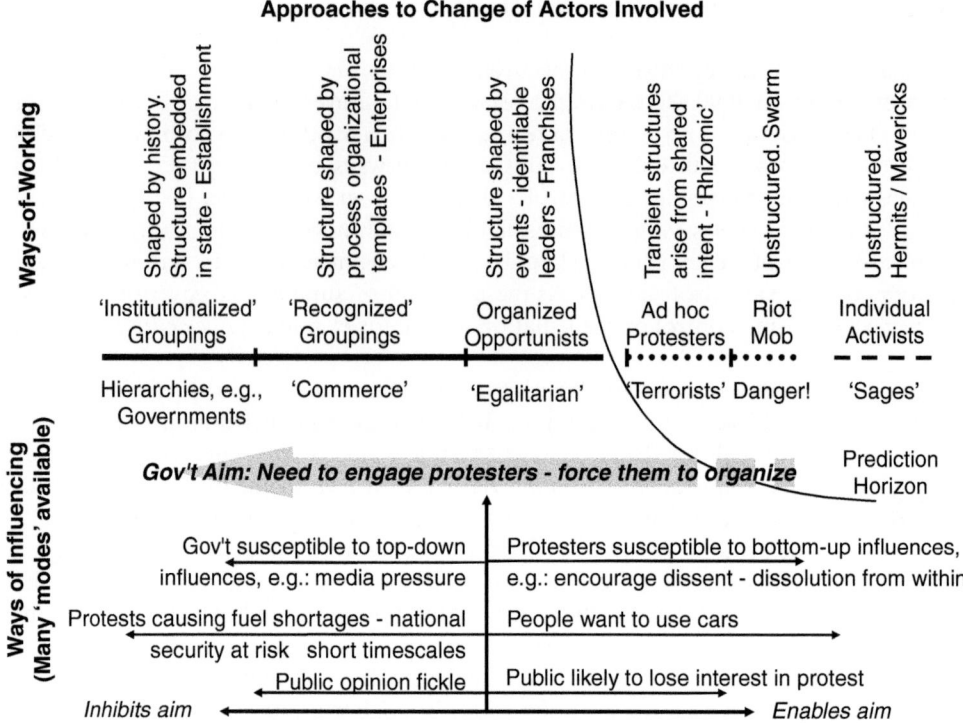

Figure 20.14   The situation of the 2000 UK fuel protests

from 14 September 2000 in *The Telegraph*, the UK government agreed to negotiate with the protesters if they 'elected a representative'. The People's Fuel Lobby was formed and a spokesman elected. Other groups also joined because of their interest in fuel issues.

The government's strategy worked – formal structures were appearing which could be reasoned with and influenced. The general public lost interest at this point, fuel flows were restored and the issue faded from the headlines. If the fuel protestors had fully understood how the power of their multi-modal influence worked then the crisis might have continued for a far longer time. In more recent times the 'Occupy' anti-capitalist movements have suffered similar fates because they have not managed to form resilient and durable structures.

20.11   PRINCIPLES FOR EFFECTIVE CHANGE IN PRACTICE

Eleven 'Principles of Practice' have been derived from these experiences. They reflect the integrative and iterative manner in which effective change can be maximized in practice:

- *Principle 1: Dynamic, ongoing change can be influenced purposefully.* This is because: (a) the underlying 'drivers of complexity' have been identified, and (b) practical techniques are available to purposefully engage with and shape the underlying drivers.

- *Principle 2: Context understanding and perceptions are diverse – there is no 'single view of the truth'*. The experienced complexities of stakeholders are necessarily different, as are their knowledge and information needs.
- *Principle 3: Change is ongoing, dynamic and multi-level – there may be no end*. This means that trends, flows, gradients, potentials and other 'energy metrics' are appropriate dynamic indicators of the progress of practice.
- *Principle 4: There are many qualities of power and influence to accommodate*. These affect people's ability to adapt and may arise from individuals, their beliefs and vulnerabilities, or from community values, from gender issues, institutional structures and the political economy, and from the changing environment and so on.
- *Principle 5: It is necessary to appreciate who is/what are best placed to bring about change*. Given the inevitable natural complexity of practice, those tasked with achieving change may not be the ones best placed in a situation to be drivers of change. Given the time horizons, practitioners should work adaptively through those who are best placed.
- *Principle 6: Interventions must have the necessary requisite variety, that is, have appropriate complexity-worthiness given the desired changes*. This insight arises from Ashby's 'Law of Requisite Variety' (1957) which states, in essence, that to influence something your practical behaviours must be equivalent to, and preferably exceed, the repertoire of behaviours of that which you are trying to influence – that is, to deal with innovation by innovating.
- *Principle 7: Practice is not just about adapting, but is also about being able to adapt the adapting and learn*. This is because people are in continual co-evolution with the environment and, as there will never be a 'steady-state' balance or equilibrium, anticipatory innovation will always be required. One cannot adapt once and then stop.
- *Principle 8: Different decision-making and problem-solving styles are required for different situations*. Because practice involves inevitable novelty and change over time, there can never be a 'one-size-fits-all' solution, nor can 'optimized' processes be used as 'best practice' in all situations, see Beautement (2011).
- *Principle 9: When reasoning about change, past evidence does not guarantee future prediction*. This means that, though we may have evidence of a past train of events, there is no guarantee that we can extrapolate a reliable prediction from this into the future. As there are limits to what we can know and observe, there will always be uncertainties and unknowns, and that we must accept this as a given. A key skill for practice and risk management is envisioning, and being prepared to act on possible, not probable, futures.
- *Principle 10: When innovating, transition to new forms may be the only valid option*. Because of the inevitable novelty already mentioned, the transformation from non-adaptive capabilities to being appropriately complexity-worthy will require purposeful, ongoing, innovation and adaptation. Gradual, superficial, incremental transition is just not an option in some unsettling circumstances.
- *Principle 11: Change will be impeded unless appropriate degrees of freedom and 'wiggle room' are available*. Being open to change means appreciating where the 'spaces of possibilities' are, and how to maximize and exploit them. A misplaced drive for control, repeatability and certainty may clamp down on the very space that is needed for adaptive behaviour to flourish.

20.12 SUMMARY

This chapter has offered a way of characterizing complex situations in a manner suitable for use when practitioners wish to select suitable tools and techniques. The complementary nature of the roles of complexity scientists and that of the practitioners to work together as a partnership has been highlighted. A key challenge to address is that of appropriateness – in selecting, matching and employing, intelligently, tools and methods which will generate useful insights. An Assessment Framework has been described which can be used to aid that dialogue – such that the outputs are more likely to be applicable to the real-world. Engaging in agent-based modelling to generate options for further consideration is not a trivial task and may require significant investment over time, especially if major questions are being addressed, Bishop et al. (2013). However, we should be appropriately cautious in our use of agent-based modelling, as Professor Michael Batty from UCL said in at the ECCS conference in 2009:

> We do not have any idea how the people in our models will adapt to change and this is not new. The very fact [that] a generation ago we thought [that] we could treat cities [as if they are] in equilibrium is testament to the limits of our knowledge. But I believe that what this is showing is that we need new forms of intelligence system to deal with the future where we will have many different models running in parallel, mediated by a context that seeks to 'inform' rather than 'predict'. The quest is to find the appropriate milieu in which to act this way.

This chapter aims to help analysts and practitioners identify that milieu as it relates to their contexts such that their work acts for the greater good in practice, as Mortenson and Relin (2007) describe, not only for academic interest in general. To emphasize this, eleven Principles of Practice have been listed – drawn from the real-world experiences of practitioners.

NOTES

1. For example, for the movement of continental plates the prediction horizon is far away in time; for the next action of a terrorist the horizon is near in time, and may be so near as to not even exist at all.
2. For readers who wish to delve further into issues of perceptions of complexity, then the page number on Figure 20.4 corresponds to that in the book by Beautement and Broenner (2011).
3. For example, Panos dialogue in London, July 2009: http://aidontheedge.info/2009/10/27/complexity-theory-and-evaluation-july-2009-meeting-report-now-available/ accessed 3 April 2015.
4. See summary here: http://www.theatlantic.com/politics/archive/2014/03/rumsfelds-knowns-and-unknowns-the-intellectual-history-of-a-quip/359719/ accessed 31 March 2015.
5. See https://complexitydemystified.wordpress.com/2012/05/31/effective-adaptation-involves-quadruple-loop-learning/#comment-25 for a further discussion of these loops. Accessed 31 March 2015.
6. See 12th Report of the UK's Transport Committee 2008: '*The Opening of Heathrow Terminal 5*'. Accessed March 2015: http://www.publications.parliament.uk/pa/cm200708/cmselect/cmtran/543/543.pdf.
7. See: http://www.thejournal.ie/general-practice-in-ireland-at-breaking-point-says-imo-748211-Jan2013/ and http://www.stpatricks.ie/blog/mental-health-and-economic-recession. Accessed 1 May 2015.
8. Building Trust and Legitimacy. President's Task Force on 21st Century Policing (2015). See: http://www.cops.usdoj.gov/default.asp?Item=2761. Accessed 2 April 2015.
9. Smuts is often ignored because of his racist views, but his work on holism is of merit nevertheless.
10. Copies of the full posters are available from: http://www.beautement.com/Library/abaci/accra-cca-post ers-2013.ppt [Accessed 02 October 2017]. These posters amplify the capabilities listed on the left-hand side of Figure 20.6.
11. See the BBC's website 'On this Day' entry for a summary (accessed 2 April 2015): http://news.bbc.co.uk/onthisday/hi/dates/stories/september/15/newsid_2518000/2518707.stm.

REFERENCES

Aaltonon, M. (2009) *'Evaluation and Organization of Futures Research Methodology'*. In Millennium Project's 'Futures Research Methodology – Version 3.0' report.

Adger, W.N., Agrawala, S., Mirza, M.M.Q., Conde, C., O'Brien, K., Pulhin, J., Pulwarty, R., Smit, B. and Takahashi, K. (2007) 'Assessment of Adaptation Practices, Options, Constraints and Capacity'. In M.L. Parry, O.F. Canziani, J.P. Palutikof, P.J. van der Linden and C.E. Hanson (eds), *Climate Change 2007: Impacts, Adaptation and Vulnerability. Contribution of Working Group II to the Fourth Assessment Report of the Intergovernmental Panel on Climate Change*. Cambridge University Press, Cambridge, pp. 717–743.

Allsopp, D. Greaves, M., Goldsmith, S., Spires, S. Thompson, S. and Janicke, H. (2006) *'Autonomous Agents and Multi-agent Systems (AAMAS) for the Military – Issues and Challenges'*. Published in "Lecture Notes in Computer Science". Revised and Invited Papers Editors: Simon G. Thompson, Robert Ghanea-Hercock, 2006. Springer, Berlin Heidelberg.

Alston, A.J., Beautement, P., Dodd, L. and Lloyd, M.A. (2006) *'Agile and Adaptive Coalition Operations – Leveraging the Power of Complex Environments'*. At the 11th ICCRTS Symposium.

Batty, M. (2009) *'Evolution, Cities and Planning: Exemplars from the Sciences of Complexity'*. Keynote Speaker at Plenary Session 3 (MS.02) of the European Conference on Complex Systems, Warwick, UK.

Beautement, P. (2011) 'Making Complexity Work in Practice'. *Information, Knowledge, Systems Management* (IKSM), 10(1–4), 345–372.

Beautement, P. and Broenner, C. (2009a) *'Putting Complexity to Work, Supporting the Practitioners'*. A White Paper describing in full the conduct of the Workshop. [Online] Available from: http://www.beautement.com/Library/abaci/eccs09_pctw_white-paper_v1-1.pdf [Accessed 02 October 2017] .

Beautement, P. and Broenner, C. (2009b) *'Complex Multi-modal Multi-level Influence Networks – Affordable Housing Case Study'*. In Jie Zhou (ed.), First International Conference, Complex 2009, Shanghai, China. Revised Papers, Part 2. Springer, Berlin Heidelberg.

Beautement, P. and Broenner, C. (2011) *Complexity Demystified – a Guide for Practitioners*, Triarchy Press, Axminster.

Bishop, S. et al. (2013) *'Global System Dynamics and Policy: Best Practice Guidelines Towards a Science of Global Systems'*. [Online] Available from: http://www.gsdp.eu/nc/news/news/date/2013/10/06/global-system-dynamics-and-policy-best-practice-guidelines-towards-a-science-of-global-systems/ [Accessed 20 March 2015].

Bradshaw, J.M., Beautement, P., Breedy, M.R., Bunch, L., Drakunov, S.V., Feltovich, P.J., Foggman, R.F., Jeffers, R., Johnson, M., Kulkarni, S., Lott, J., Raj, A.K., Suri, N. and Uszok, A. (2004) *'Making Agents Acceptable to People'*, in N. Zhong and J. Liu (eds), *Intelligent Technologies for Information Analysis*. Springer, Berlin Heidelberg, pp.361–398.

Chambers, R. (2008) *Revolutions in Development Inquiry*. Earthscan, London.

Cohen, J. and Stewart, I. (2000) *The Collapse of Chaos. Discovering Simplicity in a Complex World*. Penguin Books, London.

Conn, E. (2010) *'Community Engagement in The Social Eco-System Dance – Tools for Practitioners'*. International Workshop on Complexity and Real World applications. Southampton, UK, 21–23 July 2010. Reference with permission of the author.

Crook, R. and Booth, D. (2011) 'Working with the Grain? Rethinking African Governance', *IDS Bulletin*, 42(2).

Dack, L., Sanderson, J., Allen, J., Schranz, N. and Beautement, P. (2009) *'Engaging with Complexity – Human Behaviour Representation'*. At RTO-MP-MSG-069 Symposium on the Use of modelling and Simulation in Support to Operations.

Dignum, V. (2009) *Handbook of Research on Multi-Agent Systems: Semantics and Dynamics of Organizational Models*. Hershey, PA, USA.

Dodd, L., Prins, G. and Stamp, G. (2007) *'Going From Closed To Open: How May We Help To Make It Bearable?'* [Online], International Conference on Complex Systems 2007, 28 October to 2 November, Boston, MA.

Erdi, P. (2008) *Complexity Explained*. Springer Verlag, Berlin, pp.328–351.

Felbab-Brown, V. (2013) *'The Purpose of Law Enforcement is to Make Good Criminals? How to Effectively Respond to the Crime-Terrorism Nexus'*. http://www.brookings.edu/research/presentations/2013/11/21-how-effectively-respond-crime-terrorism-nexus-felbabbrown [Online] Accessed 2 April 2015. At the Potomac Institute for Policy Studies seminar entitled, "Convergence of Crime and Terrorism?" held on 21 November 2013.

Feltovich, P.J., Bradshaw, J.M., Jeffers, R., Suri, N. and Uszok, A. (2004) *'Social Order and Adaptability in Animal and Human Cultures as an Analogue for Agent Communities: Toward a Policy-Based Approach'*. In A. Omicini, P. Petta and J. Pitt (eds). *Engineering Societies in the Agents World IV*. SpringerLink, London, pp. 21–48.

Glenn, J.C. (2009) *'Introduction to the Futures Research Methods Series'*. Millennium Project, Futures Research Methodology (version 3).

Hoffman, R. and Klein, G. (2011) 'Naturalistic Investigations and Models of Indeterminate Causal Reasoning: Things Aristotle and Hume Never Told You'. *Information, Knowledge, Systems Management*, 10(1–4).

Hofstadter, D.R. (1983) *Gödel, Escher, Bach: an Eternal Golden Braid*. Penguin Books, London.

Howard, N., Bennett, P. and Bryant, J. (2001) '*Drama Theory and Confrontation Analysis*'. In J.V. Rosenhead and J. Mingers (eds). *Rational Analysis for a Problematic World Revisited: Problem Structuring Methods For Complexity, Uncertainty And Conflict*. Wiley, London, pp. 225–248.

Jennings, N.R. (2001) 'An Agent-based Approach for Building Complex Software Systems'. *Communications of the ACM*, 44(4), 35–41.

Jennings, N., Sycara, K., Wooldridge, M. (1998) 'Roadmap of Agent Research and Development'. *Journal of Autonomous Agents and Multi-Agent Systems*, 1(7–38). Kluwer Academic Publishers, Boston.

Jones, L., Ludi, E., Carabine, E. and Grist, N. (2014) '*Planning for an Uncertain Future: Promoting Adaptation to Climate Change through Flexible and Forward-looking Decision Making*'. Overseas Development Institute.

Jones, L., Ludi, E., Beautement, P., Broenner, C. and Bachofen, C. (2013) '*New Approaches to Promoting Flexible and Forward-Looking Decision Making: Insights from Complexity Science, Climate Change Adaptation and "Serious Gaming"*'. Overseas Development Institute.

Kirton, M. et al. (2003) 'Coalition Agents Experiment: Network-enabled Coalition Operations'. *Journal of Defence Science* (Special edition), 8(3).

Kurtz, C.F. and Snowden, D. (2003) 'The New Dynamics Of Strategy: Sense-making in a Complex World'. *IBM Systems Journal*, 42(3), 462–483 [Cynefin framework].

Lansing, S. (2003) *Social Science Models and Tropical Disasters: When Emergence Really Counts*. Santa Fe Institute, Santa Fe, USA.

Leifer, J.J. and Steinert, M. (2011) 'Dancing with Ambiguity: Causality Behavior, Design Thinking, and Triple-loop-learning'. *Information, Knowledge, Systems Management*, 10(1–4).

Levine, S., Ludi, E. and Jones, L. (2011) '*Rethinking Support for Adaptive Capacity to Climate Change: The Role for development Interventions*'. Report for the Africa Climate Change Resilience Alliance (ACCRA) by the Overseas Development Institute (ODI), London.

Levy, S. (1993) *Artificial Life*. Vintage Books, New York.

Lissack, M. (2011) 'Miracles and Nasty Surprises'. In H. Letiche, M. Lissack and R. Schultz (eds), *Coherence in the Midst of Complexity: Advances in Social Complexity Theory*. Palgrave Macmillan, New York, pp. 1–10.

Marsay, D.J. (2006) '*ISTAR and C2 – A Holistic View*'. At the 11th International Command and Control Research Technology Symposium.

Mathieson, G. (2005) '*Complexity and Managing to Survive it: Complex Adaptive Reflexive Systems (CARS)*'. DSTL. Institute for the Study of Coherence and Emergence. [Online] Available from: http://ismor.cds.cranfield.ac.uk/collected-papers-of-graham-mathieson/complexity-and-managing-to-survive-it-playing-with-cars/@@download/paper/cars.pdf [Accessed 30 March 2015].

Mitleton-Kelly, E. (2003) 'Ten Principles of Complexity and Enabling Infrastructures' in *Complex Systems and Evolutionary Perspectives on Organisations: The Application of Complexity Theory to Organisations*, Chapter 2. Elsevier, London.

Morowitz, H.J. (2002) *The Emergence of Everything: How the World became Complex*. Oxford University Press, Oxford.

Mortenson, G. and Relin, D.O. (2007) *Three Cups of Tea: One Man's Mission to Promote Peace One School at a Time*. Penguin Books, London.

National Commission on the BP Deepwater Horizon Oil Spill and Offshore Drilling. '*Final Report*', 2011 [Online]. Available from: http://www.gpo.gov/fdsys/pkg/GPO-OILCOMMISSION/pdf/GPO-OILCOMMISSION.pdf [Accessed 2 April 2015].

Ramalingam, B. and Jones, H. (2008) '*Exploring the Science of Complexity: Ideas and Implications for Development and Humanitarian Efforts*'. Working Paper 285, Overseas Development Institute, London.

Smith, R. (2005) *The Utility of Force*. Allen Lane, London, pp.323–331.

Smuts, J.C. (1926) *Holism and Evolution*. Macmillan, London, pp.17–18.

Sturmberg, J.P. and Martin, C. (eds) (2012) *Handbook of Systems and Complexity in Health*. Springer Publications, Berlin Heidelberg.

Thompson, M. (2008) *Organising and Disorganising*. Triarchy Press, Axminster.

Trajkovski, G. (2009) *Handbook of Research on Agent-Based Societies: Social and Cultural Interactions*. IGI Global, London.

Treverton, G.F. (2003) *Reshaping National Intelligence for an Age of Information*. Cambridge University Press, New York.

Unewisse, M. and Grisogono, A. (2007) '*Adaptivity Led Networked Force Capability*'. At the 12th International Command and Control Research Technology Symposium.

21. Leadership network structure and influence dynamics

Dr Michael Gabbay, University of Washington

21.1 INTRODUCTION

Pivotal policy decisions in states or organizations like militant movements are often made by a small group of top leaders (Hermann 2001). This speaks to the importance of developing systematic methods for improving the ability to understand and anticipate the dynamics of leadership groups. This chapter describes a quantitative methodology for the analysis and modeling of leadership networks which leverages research in complex systems, in particular nonlinear dynamical systems theory (Strogatz 1994) and network science (Newman 2010). The nonlinear systems element is the model of social influence dynamics which can exhibit complex phenomena such as large, discontinuous transitions (bifurcations) as a parameter is varied and non-trivial interactions with network structure. Factional and other divisions within leadership networks can induce meaningful structure in them; algorithms developed in complex networks research for analyzing community structure can probe this factional structure and, crucially, relate that structure to policy divisions. Investigation of both the network and issue space, as well as their integration, is a core focus of the methodology and is accomplished statically via structural analysis and dynamically via the nonlinear social influence model which evolves leader positions on issues in response to their mutual influence over their network of ties.

This chapter introduces a recently developed prototype software package, PORTEND, that provides a user interface for the analysis and simulation methods. PORTEND's analytical capabilities are illustrated for an application to Iranian leadership elites regarding seven major issues with a particular focus on whether their nuclear technology capabilities should or should not be constrained and subject to international monitoring. Previous applications of the methodology to Russian and Afghan leadership networks have been reported elsewhere (Gabbay 2007a; Gabbay 2013). The factional structure of the Iranian leadership group is analyzed first based on their positions on the issues, then with respect to the network of inter-actor influence relationships, and finally by a synthesis of the issue and network data. Moving from structural analysis to simulation, a qualitative description of the nonlinear social influence model is presented followed by application of the simulation to the nuclear issue and discussion of its implications with respect to Iranian decision-making concerning the nuclear negotiations that took place from 2013 to 2015.

21.2 PORTEND SOFTWARE

PORTEND (Political Outcomes Research Tool for Elite Network Dynamics) integrates quantitative techniques from nonlinear systems theory and network science to aid the analysis of policy and factional outcomes with respect to the internal dynamics of a system of political actors. The political actors may be individual leaders or organizations within a government or movement. The outcomes of concern may be policy decisions, winning and losing factions, the positions of individuals, or the potential for issues to cause dissension or factional realignment. Political actors are represented mathematically with respect to their preferences on one or more issues, the saliences of those issues, the network of inter-actor influence, and actor power and susceptibility to influence. The data from which these quantities are calculated is obtained from surveys given to expert analysts. PORTEND imports these surveys and aggregates them to form a composite analyst if desired. It then allows for structural analysis regarding issues and the inter-actor network and for the simulation of social influence and group decision-making outcomes. The analyses can be performed for the composite analyst or separately for the individual analysts. An overview of the methodology is shown in *Figure 21.1*. PORTEND is currently in a prototype stage of development and is implemented in Matlab.

21.3   IRAN APPLICATION

This section introduces the Iranian leadership case study which will be used to illustrate the capabilities of the methodology implemented within PORTEND in this chapter. The case study, which was initiated in 2013, considered fifteen top members of the Iranian leadership, as identified by analysts of Iranian politics *Table 21.1*). A survey was developed and then completed by two Iran experts in the autumn of 2013. The elements of the survey will be discussed in the next section. While a major concern of the study involved the Iranian nuclear program, the broader context of Iranian elite politics was also of interest and so the survey included the seven issues below (abbreviations in parentheses):

- Liberalism (LIB): The proper role for Western culture, Islam, media sources, and democratic institutions.
- Economic Reform (ECON): Whether economic policies should benefit the current elites or a wider set of interests.
- Arab States (ARAB): Whether Iran's peers in the Arab world are potential allies or enemies.
- Syrian Regime (SYR): Whether the Assad regime in Syria should be supported.
- US/Israel (USISR): The extent to which Iran should confront the US and Israel.
- Nuclear Issues (NUKE): The extent to which Iran should develop nuclear technology.
- IRGC Influence (IRGC): The appropriate role for the Islamic Revolutionary Guard Corps (IRGC).

The analytical questions of interest included:

- Will Iran agree to a nuclear deal that places strong restrictions on enrichment?
- Who might dissent from a nuclear deal and who are possible swing players?

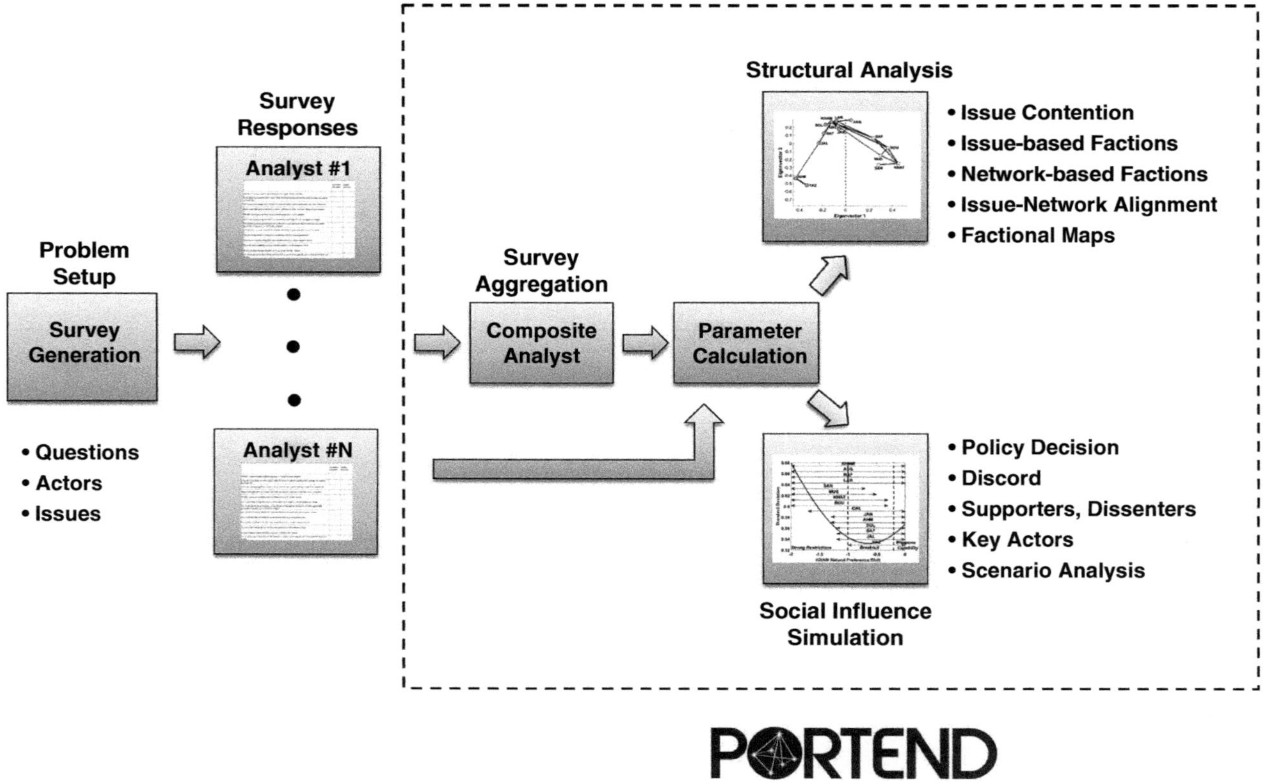

Figure 21.1   *Methodology overview*

461

Table 21.1   Iranian elites in case study. The abbreviations used in plots are shown in parentheses. Information on roles is as of late 2013

Actor (Abbr.)	Role/Notes
Ali Hoseini Khamenei (KHAM)	The supreme leader, the highest political and religious authority in the Islamic Republic of Iran.
Qasem Soleimani (SOL)	Commander of the Quds Force, a unit of the Islamic Revolutionary Guard Corps (IRGC).
Mir Hossein Musavi (MUS)	Prime Minister of Iran from 1981 to 1989. In 2009 he was the reform candidate for president, around whom the Green Movement coalesced. He has been under house arrest since February 2011.
Mohammad Taqi Mesbah Yazdi (YAZ)	A hardline cleric and politician. He is a member of Iran's Assembly of Experts and is seen as the most conservative cleric in Iran.
Ahmad Janati (JAN)	A hardline cleric and chairman of the Guardian Council.
Asadollah Asgaroladi (ASG)	An important businessman with interests in exports, banking, real estate and healthcare. President of several of Iran's international Chambers of Commerce.
Ali Akbar Hashemi-Rafsanjani (RAF)	Served as president of Iran from 1989 to 1997 and chairman of the Expediency Council.
Ali Ardeshir Larijani (LAR)	Current chairman of the Iranian Parliament and former secretary of Iran's Supreme National Security Council.
Yousef Sanei (SAN)	An Iranian scholar and Islamic theologian and philosopher. He serves as a Grand Marja of Shia Islam.
Mohammad Baqr Qalibaf (QAL)	The current mayor of Tehran.
Yahya Rahim Safavi (SAF)	An Iranian military commander and former Chief Commander of the IRGC.
Mahmud Ahmadinejad (AHM)	The former president of Iran.
Seyyed Mohammad Khatami (KHAT)	President of Iran from 1997 to 2005. One of Iran's most prominent reformers.
Saeed Jalili (JAL)	Secretary of Iran's Supreme National Security Council, the equivalent of the US National Security Council.
Hassan Rouhani (ROU)	The current president of Iran.

- What are the most controversial issues? Which actor inter-relationships do they stress?
- What issues have the potential to lead to factional realignments?

In November 2013, after the survey had been developed, an interim nuclear deal was announced between Iran and its negotiating counterpart, the P5+1 countries, consisting of the five permanent members of the UN Security Council (China, France, Russia, US, UK) and Germany. This spawned an additional question as to what may have caused the shift in Iran's posture toward nuclear negotiations which will be discussed in the section on simulation results. Space does not allow background on Iranian politics to be provided here – a good discussion of Iranian factional politics can be found in Rieffer-Flanagan (2013).

21.4 ANALYST SURVEY

The analyst survey elicits expert judgment on the leadership group under study. The use of a survey methodology allows analysts to complete the survey at their convenience and avoids potential groupthink effects associated with oral elicitation of a group of analysts at one sitting. Only the Actor Opinions and Influence Network components of the survey are discussed here as they are the ones most essential for understanding the results presented below (other components are described in Gabbay 2013). The surveys can be averaged to form a composite assessment or analyzed individually in order to bring out differences in analyst perspectives.

The Actor Opinions survey section contains a list of statements designed to assess the attitudes of the group members relevant to the policy issues of concern. For each member, analysts are asked to estimate the member's level of agreement/disagreement with a series of statements covering a range of issues, goals, identities, and specific policies. Examples include 'The production, stockpiling, and use of nuclear weapons are all forbidden in Islam' and 'The IRGC should play a guiding role in maintaining Iran as an Islamic republic.' The instructions direct analysts to score the statements on the basis of the private beliefs of the members if thought to be at odds with their public rhetoric. The Actor Opinions section is used to calculate member issue positions known as 'natural preferences', a key parameter in both structural analysis and the simulation.

The Influence Network section contains a matrix in which analysts estimate the strength of each actor's direct influence upon each of the other members in the group and vice versa. This (directional) dyadic influence strength depends on factors such as the frequency of communications, status within the group, common or rival factional membership, and personal relationships of friendship or animosity. The influence network is used directly in structural analysis and to calculate the 'coupling strengths' which scale the persuasive force of one member on another in the social influence dynamics simulation.

21.5 STRUCTURAL ANALYSIS

Structural analysis involves quantitatively and visually probing the factional composition of the group as a whole and how individuals are situated within the group. Analyst judgments on discrete elements concerning individual actors and actor dyads are synthesized to enable the discovery of broader features and patterns in the group. In addition to being illuminating in its own right, structural analysis can help focus the simulation effort on particular issues such as those which are most polarizing or have the potential to result in new alignments of actor subgroups distinct from the dominant factional configuration. It also allows for insight into dynamics not encompassed by the simulation such as interactions between multiple issues, alliance formation, and succession considerations.

21.5.1   Issue Analysis

The methods for issue analysis utilize only the group member issue positions (natural preferences) calculated from the actor opinions. The analyses can address how contentious an issue is, how similar actor positions are for any given pair of issues, and patterns of actor alignment across the whole set of issues. This section presents examples of these analyses for the Iran case.

The most fundamental element of issue analysis is simply the actor natural preferences themselves as is shown in the plots of *Figure 21.2*. The positive end of the scale indicates support or a favorable attitude with respect to the issue and has a maximum value of 2. Similarly, the negative axis signifies opposition or an unfavorable attitude. These plots are useful for visual inspection of individual actor positions and their distribution within an issue as well as examining clustering across issues. To better highlight clustering patterns and deviations from them, conservatives are identified as those actors having negative scores on the Liberalism plot and marked by solid gray circles; reformers have positive Liberalism scores and are marked by open squares. The Liberalism plot shows a bloc consisting of KHAM (the Supreme Leader), SOL, SAF, JAN, YAZ and JAL at the far negative end of the axis indicating strong opposition to political and cultural liberalization whereas ROU (the president), KHAT, MUS and SAN are found oppositely at the pro-liberalization side. This pattern of opposed clustering is repeated for other issues as well thereby leading to the interpretation of the former subgroup as a core conservative or hardline faction and the latter one as a core reformist or moderate (from a US/Western viewpoint) faction. Note that RAF is usually aligned with the reformists except on the Economic Reform issue towards which he is most opposed. A subgroup composed of LAR, QAL and ASG typically forms a conservative-leaning centrist bloc with Economic Reform again a notable exception. The level of disagreement over an issue is indicated by the amount of spread in the actor positions as can be quantified by standard deviation (see *Table 21.2*). Nuclear Weapons, in which the actor positions appear most compressed, is the least contentious issue by this measure.

To get a sense of the relationship between issues, *Figure 21.3* shows plots of actor natural preferences on two pairs of issues. Observe in *Figure 21.3*a that the actor positions in the joint US/Israel and Nuclear Weapons space fall essentially on a line as is indicated by the almost perfect (anti-) correlation of −0.98. This implies that, although the issues are plotted on a two-dimensional plane, the system is essentially one-dimensional in the sense that if given the actor positions on one issue, then their positions on the second can be inferred with high accuracy. In *Figure 21.3*b, Economic Reform appears on the vertical axis: there is now more scatter of actor positions and the correlation is lower in magnitude (although still highly statistically significant) indicating a less one-dimensional aspect. The core conservative and reform factions are still effectively at the opposite ends of the main axis (PC 1) but RAF and AHM are significantly off axis as are, to a lesser extent, ASG and LAR. The two plots have different implications with respect to potential coalitions if the two issues interact so that changing position on one issue affects an actor's position on the other. In *Figure 21.3*b, RAF is nearer the conservatives and could side with them increasing his support for a more robust nuclear capability and bolstering their opposition to economic reforms. An analogous implication holds for AHM with respect to the reformist faction. Such realignment would not be possible if the two issues

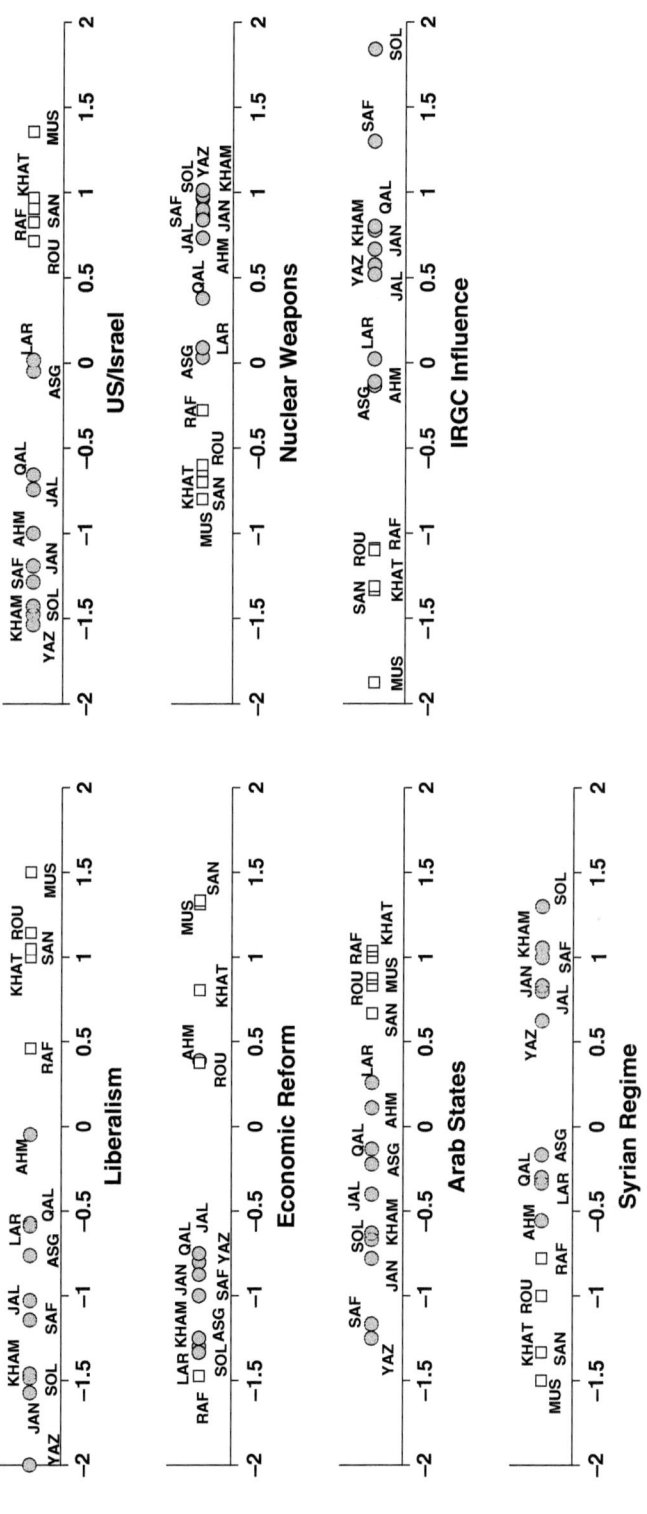

Figure 21.2   Actor natural preferences for the seven issues. Conservatives are gray circles, reformers are open squares

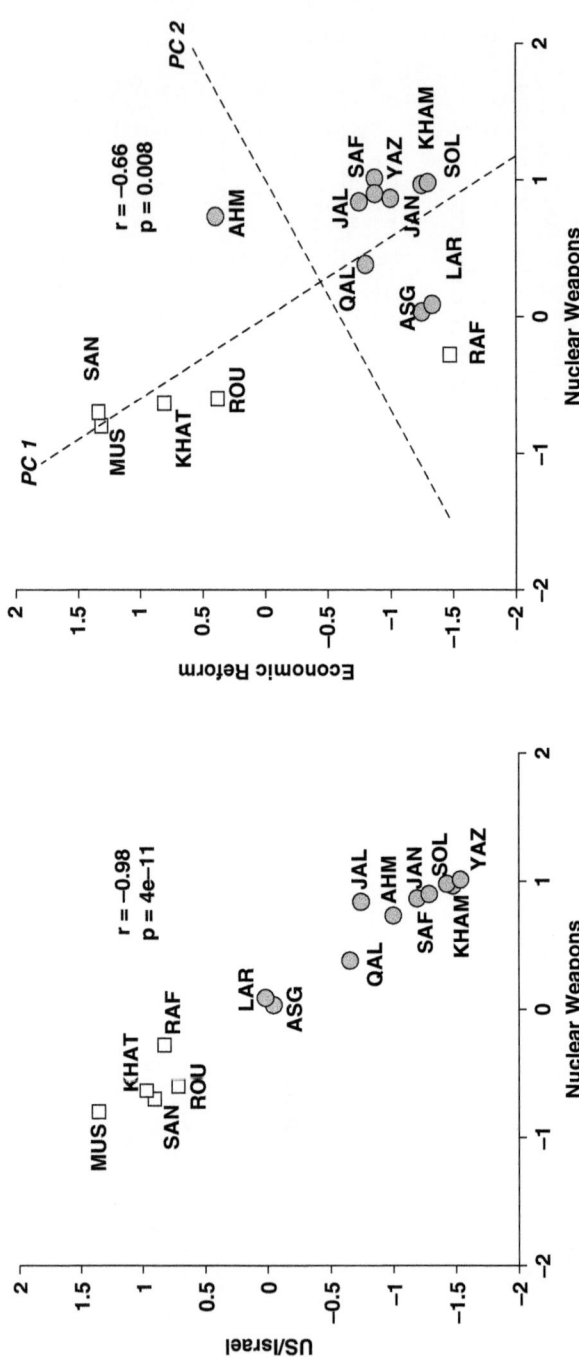

Figure 21.3 Two-dimensional issue plots: (a) US/Israel and (b) Economic Reform plotted versus Nuclear Weapons. The numbers in the upper right-hand corner are: the correlation between the actor positions on the two issues (r); and the p-value measure of statistical significance (p) which indicates the probability that the observed correlation could have occurred by chance given no underlying relationship – lower p-values imply stronger statistical significance. The dashed lines in (b) correspond to principal component axes

466

in play were US/Israel and Nuclear Weapons as in *Figure 21.3*a: RAF would remain close to the reformers and AHM to the conservatives. However, it could be possible for ASG and LAR to be forced to side with one of the main factions if maintaining their centrist positions were to become untenable.

While the discussion of factional alignments so far has involved visual inspection across issues, numerical methods exist for automatically revealing patterns of alignment. One such technique is Principal Components Analysis (PCA) which seeks to represent a data matrix by a series of coordinate vectors, known as principal components (PCs), each of which corresponds to a pattern of covariation in the data (Webb and Copsey 2011). The PCs are ranked in descending order of importance as determined by how much of the variance (the data scatter around the mean) they carry which is given by their 'eigenvalues'. Each PC is uncorrelated with the others so that they run as perpendicular directions through the data; in fact, they correspond to an alternative set of coordinate axes to the direct data variables.

For example, we can interpret *Figure 21.3*b as measurements of the two issue variables, Nuclear Weapons and Economic Reform, with each actor's natural preference pair as a data point. The first PC then points in a direction along the dashed line running from upper left to lower right and the second is the line perpendicular to that. In essence, PCA has rotated the standard coordinate system, wherein each axis corresponds to one issue, to the dashed system where each PC is a weighted combination of the two issues (the weights can be negative). The origin is the intersection of the two PCs located at the point given by the mean along each issue. An actor's coordinate on each PC is the (signed) distance between this origin and where he falls on the PC axis (the nearest point on that axis to him). The variance in the actor coordinates on PC 1 is given by its eigenvalue of 4.22 whereas PC 2's eigenvalue is 1.75 so we see that PC 1's share of the total variance (71 percent) is much larger than that of PC 2 (29 percent) indicating that PC 1 is more important in approximating the data. (The disparity between the two PCs would be even greater for *Figure 21.3*a given that it is much more one-dimensional.)

Turning to the complete set of issues, *Figure 21.4* shows the first two (out of seven) principal components obtained from the data matrix formed by the natural preferences of each of the fifteen actors on all seven issues. The top plot on the left side shows the actor coordinates for the first principal component. This corresponds well to the dominant factional alignment identified in our discussion of the issue plots of *Figure 21.2*. The core conservative bloc of KHAM, SOL, SAF, JAN, YAZ and JAL is on the extreme negative end; the conservative-leaning centrists QAL, LAR and ASG are just left of zero and the core reform bloc of ROU, KHAT, MUS and SAN is on the far positive side. Rafsanjani is aligned with the reformers on PC 1 as is the case on six of the issue plots. Ahmadinejad's location as a centrist may be surprising given his international reputation as a hardliner during his presidency but is supported by his position near the center or on the reform side for four of the issues. The eigenvalues in the corresponding table show that PC 1 carries 57 percent of the total variance, much larger than PC 2's 17 percent share. This supports the interpretation of PC 1 as the dominant factional alignment. The PC 1 Issue Value column shows that there is no single primary issue whose magnitude is much larger than the others, again suggesting that PC 1 represents the most common pattern across the set of issues. This is not the case, however, for PC 2 where the Economic Reform component of 0.89 is by far the strongest. The plot of the PC 2 coordinates shows RAF

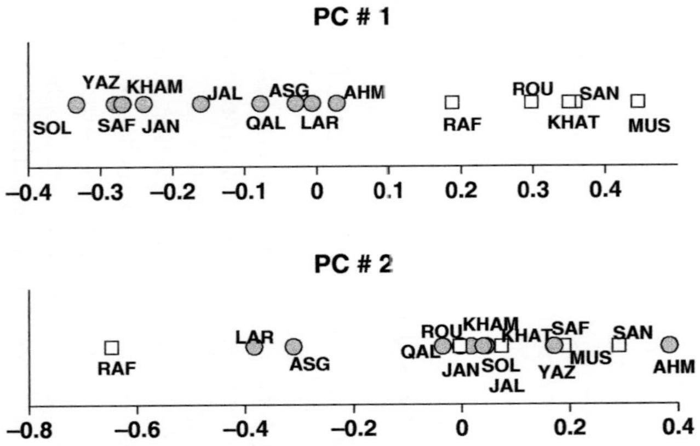

PC # 1

PC # 2

PC 1 Eigen-value	PC 1 Issue	PC 1 Issue Value	PC 2 Eigen-value	PC 2 Issue	PC 2 Issue Value
0.57	LIB	0.46	0.17	ECON	0.89
	IRGC	−0.44		USISR	−0.34
	USISR	0.41		ARAB	−0.26
	SYR	−0.39		NUKE	0.16
	ECON	0.32		IRGC	0.09
	ARAB	0.30		LIB	0.06
	NUKE	−0.28		SYR	0.03

Figure 21.4 First two principal components of actor natural preferences. Left: Actor coordinates. Right: Eigenvalues and issue values. Eigenvalues are expressed as the fraction of the total sum of eigenvalues. Issue values are listed in descending magnitude

and AHM at opposite ends reflecting the fact that, while the majority of the actors pre-serve the standard factional composition for Economic Reform, RAF and AHM make large against-the-grain shifts in the conservative and reformist directions respectively as observed in *Figure 21.2* (RAF and AHM also appear at opposite ends of the second PC for the two-issue example of *Figure 21.3*b).

21.5.2   Network Analysis

Parallel to the investigation of issue-based factions described above, the factional structure which arises from the network of inter-actor influence relationships is also of concern. Network science has developed many algorithms for detecting community struc-ture in networks. Intuitively, the goal is to find subgroups of nodes which have more links among them than they do with other subgroups. Community structure may reflect simi-larities in preferences among network members via: the homophily principle (also known as assortative mixing), a formal construct for the commonplace that 'birds of a feather flock together' (Newman, 2010); or the mechanism of social influence which assumes that people who interact more often tend to become more similar (Friedkin and Johnsen 2011). This section presents the application of a community structure algorithm which is then extended to illustrate how community structure and actor natural preferences can be integrated to address joint issue-network alignment.

The algorithm employed in PORTEND seeks to divide a network into two communities so that the network 'modularity' is maximized (Newman 2006). The contribution to the total modularity from a given pair of nodes is proportional to the difference between their observed tie strength and that which would be expected if their interactions were solely due to chance; these contributions for all the dyads form the elements of the modular-ity matrix. The total modularity expresses the extent to which a putative division of the network into two communities exhibits a level of intra-community linking exceeding the level expected if the division were, in fact, arbitrary with no correspondence to behavio-rally meaningful subgroups. The maximization is done in an approximate but efficient way by calculating the first eigenvector of the modularity matrix (eigenvectors are ranked in order of descending eigenvalue) and then assigning all nodes whose components in the first eigenvector are positive to one community and the nodes with negative components to the other. As an example, Newman (2006) presents an application to a network of 62 dolphins and finds that the two communities identified by the first eigenvector matched to high accuracy the two groups into which the network actually split after a key dolphin died (only three dolphins were misclassified).

The application of the community detection algorithm to the Iranian influence network is shown in *Figure 21.5* which plots the actor coordinates obtained from the first two eigenvectors of the modularity matrix (using the symmetrized network in which tie strengths are the same in both directions in a dyad). We refer to the eigen-vectors as 'factional dimensions'. The initial discussion of *Figure 21.5* will center on the meaning of Factional Dimension (FD) 1 but, as will be seen below, FD 2 also has a significant interpretation regarding the Economic Reform issue. The dashed line corresponds to the division formed by separately grouping nodes with positive and negative signs in eigenvector 1. The left and right sides correspond to conservative and reformer classifications respectively. The correspondence with the issue-based factions

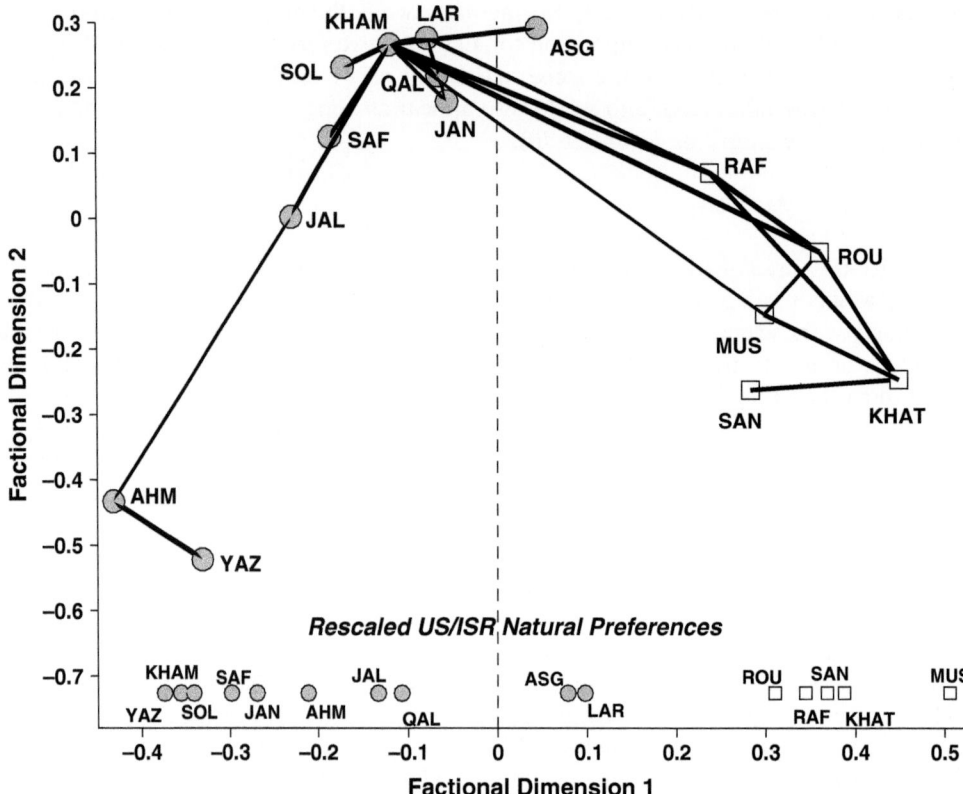

Figure 21.5   *Community structure in the Iran influence network. Dashed line partitions network into conservative (left) and reformist (right) communities. Link thickness between actors is proportional to relationship strength (weak links have been thresholded). Points at bottom of plots are actor natural preferences for the US/Israel issue rescaled to fit inside the horizontal axis*

is immediately apparent because, as in PC 1 in *Figure 21.4*, all the gray circles are on one side and the white squares on the other. All members of the core conservative and reform blocs as identified by the issue analysis above are correctly classified. Only ASG can be considered to be misclassified as a reformer, perhaps understandable given that he is more of a centrist than a hardline conservative (and in fact he appears in the middle of FD 1).

It is also possible to combine issue and network data for purposes of addressing polarization and factional realignment. As used here, polarization refers to the extent to which disagreement over an issue reinforces divisions present in the network. Hence, polarization is not simply the level of disagreement over an issue as might be gauged from the standard deviation of actor issue positions. Quantitatively, the contribution of an actor dyad to the polarization for a given issue is found by multiplying the corresponding modularity matrix element – network-derived data – by the product of the two actor natural preferences – issue data (which makes the polarization equivalent to the

Table 21.2 *Integrated issue-network analysis metrics. Issues are listed in descending order of polarization. Statistical significance levels of correlations: * p<.01, ** p<.001, *** p<.0001. The last column shows the standard deviation of the actor natural preferences (rank in parentheses)*

Issue	Polarization	Aligned Dimension	Correlation Magnitude	Standard Deviation
US/Israel	0.395	1	0.885***	1.036 (3)
IRGC Influence	0.393	1	0.779**	1.095 (2)
Liberalism	0.358	1	0.799**	1.142 (1)
Syrian Regime	0.248	1	0.731*	0.974 (5)
Economic Reform	0.198	2	0.645*	0.998 (4)
Nuclear Weapons	0.195	1	0.907***	0.701 (7)
Arab States	0.158	1	0.796**	0.786 (6)

covariance between natural preferences over all the ties in the network (Newman 2010)). The polarization value for each issue is shown in *Table 21.2*. For comparison purposes, the standard deviation of actor natural preferences is shown in the last column. The US/Israel issue is most polarizing even though Liberalism has the highest standard deviation. Nuclear Weapons and Economic Reform have very nearly the same polarization whereas the latter has a substantially larger standard deviation. Consequently, we see that the integration of network and issue data gives a different and perhaps more significant picture of issue divisiveness than issue data alone.

The Aligned Dimension column in *Table 21.2* is the number of the factional dimension with which the issue has the highest magnitude correlation. The larger value of polarization of US/Israel as compared with Liberalism, despite the latter's higher standard deviation, is a reflection of the greater alignment that US/Israel has with FD 1 as seen by its better correlation in *Table 21.2*. A visual sense of this alignment can be gleaned from *Figure 21.5* by comparing the actor network positions along the horizontal axis with their (rescaled) US/Israel natural preferences at the bottom. Whether or not the correlation represents a genuine relationship between the eigenvector and the actor natural preferences can be assessed from the p-value column with lower values indicating greater significance. All of the correlations in *Table 21.2* are highly significant. Six of the seven issues best correlate with FD 1 reinforcing the conclusion that it represents the dominant factional division in the network. Consequently, these issues stress the major faultline in the group but are not likely to cause a fundamental factional realignment (although centrists may be forced to side with one camp or another as noted in connection with *Figure 21.3*a). However, Economic Reform is seen to align best with FD 2 (the vertical axis in *Figure 21.5*) and, therefore, if it were to become more salient, a factional realignment could be induced in which RAF allies more strongly with the conservatives and AHM does likewise with the reformers.

21.6   NONLINEAR SOCIAL INFLUENCE SIMULATION

21.6.1   Model Description

The nonlinear model of social influence simulates the evolution of group member positions along the policy axis due to their mutual interactions. The social science underpinnings of the model derive primarily from social psychology theories of attitude change and small group dynamics and theories of foreign policy decision-making (Eagly and Chaiken 1993; Hermann et al. 2001). A brief summary of the model is presented in this section; fuller descriptions can be found in Gabbay (2007c, 2007b). It should be noted that since the model is focused on group dynamics, it does not involve a representation of the decision-making calculus associated with particular policy choices (see Davis and O'Mahony (2013) for an example of a computational model that does so in the context of insurgent groups). With respect to other models of group dynamics, on a mathematical level, the nonlinear model is most similar to that of social influence network theory (Friedkin and Johnsen, 2011) to which the model can be made equivalent in the (linear) limit of low disagreement. The most prominent formal model of decision-making applied to real-world political contexts is that of Bueno de Mesquita (1997, 2009) which however has received some criticism regarding lack of transparency (Scholz, Calbert and Smith 2011). While Bueno de Mesquita's model uses analyst input and a one-dimensional issue axis as does the present model, it is based on expected utility theory whereas PORTEND is rooted in nonlinear dynamical systems theory and network science, cornerstones of complex systems research.

In the model, an actor's position changes under the influence of two separate forces: the 'self-bias force' and the 'group influence force.' Considering the self-bias force first, each actor is assumed to come to the debate with an initial issue position given by his natural preference (also called the natural bias) which reflects the actor's underlying beliefs, attitudes, and worldview pertinent to the issue. If an actor's position is shifted from his natural preference due to group pressures, he will experience a cognitive dissonance that resists this change and strives to move the actor's position back toward the natural preference.

The group influence force is the total force acting to change an actor's position due to the persuasive efforts of the other actors in the group. It is assumed to operate in a pairwise manner so that an actor – the message receiver – experiences a persuasive 'coupling force' from another actor – the message sender – to whom he is connected (and vice versa). The functional form of the coupling force is nonlinear in the difference between the sender and receiver positions: if the difference is small, the force increases roughly linearly; the force then reaches a peak at a difference known as the 'latitude of acceptance,' beyond which it begins to wane towards zero. This form is motivated by social judgment theory which posits that the amount of opinion change in a person receiving a persuasive message follows an inverted U-curve as a function of the difference between the opinion advocated in the message and that of the receiver (Eagly and Chaiken 1993) (however, the coupling force in the model has a long tail rather than ending abruptly as in an inverted-U). The coupling force that actor j exerts on actor i also depends on the 'coupling strength' from j to i, which is obtained from the influence network. The 'coupling scale' is the mean of the incoming coupling strengths (in-degree).

The model description above governs how actors change their positions under their mutual influence but does not yield the decision itself. In order to do so, the appropriate decision rule – leader choice, weighted majority, or consensus – must be applied. Typically, this is done after the simulation reaches equilibrium so that the actor positions reach steady-state values that no longer change perceptibly. For purposes of determining whether an actor supports or dissents from a policy decision, an actor is considered to support a policy if it lies within a specified maximum distance, usually taken to be the latitude of acceptance, from the actor's final position. Similarly, actors are taken to dissent from a policy if it lies beyond this distance.

Complexity enters into the model via the nonlinear form of the influence between actors and its interaction with the network formed by the inter-actor coupling strengths. The model can be considered to have two regimes of behavior: a 'linear' one, in which the behaviors typically correspond to initial intuition, and a 'nonlinear' regime corresponding to high disagreement (roughly, position differences exceeding twice the latitude of acceptance) in which behaviors can run counter to initial intuition. The linear regime is always characterized by gradual changes in outcomes as parameters such as the level of disagreement or coupling scale are varied whereas the nonlinear regime can exhibit discontinuous transitions, referred to as bifurcations, between states such as deadlock, majority rule, and consensus (Gabbay and Das 2014). With respect to the interaction of nonlinearity and network structure, at high disagreement levels networks with lower tie density (for example, a chain) can be more effective at reducing group discord and yielding consensus than ones with higher density (for example, a complete network) in contrast with the 'linear' expectation that a higher number of ties is better for consensus formation (Gabbay 2007b).

21.6.2   Simulation Results

All seven issues were simulated. Here only the Nuclear Weapons results are discussed as that issue was of primary analytical concern. The simulation using the set of parameter values calculated directly from the composite analyst is shown in *Figure 21.6*a. The latitude of acceptance is taken to be one unit along the issue axis as that corresponds to a step along the attitude survey scale, say from 'neutral' to 'weak agreement' or from 'weak agreement' to 'strong agreement'. Actors start out at their natural preferences and the time units are essentially arbitrary given that the equilibrium is of concern.

The policy labels and corresponding intervals in *Figure 21.6*a are calculated from the Actor Opinions section of the survey (they can also be set manually) and are intended to be rough guides to assist in interpretation of simulation results rather than hard and fast boundaries. The Weapons Capability policy corresponds to an actor believing that a nuclear weapons capability is critical to ensuring the survival of the Iranian regime. Breakout signifies that the actor prefers that Iran should have the ability to develop nuclear weapons without building or testing them. Strong Restrictions signifies that the actor is willing to accept more forceful constraints on Iran's nuclear enrichment program such as intrusive monitoring of nuclear facilities in exchange for the removal of economic sanctions (a fourth policy of No Enrichment was not preferred by any actor).

The decision rule is leader choice and the open diamond indicates the final policy, coincident with KHAM's final position. We see that the policy choice is located in the

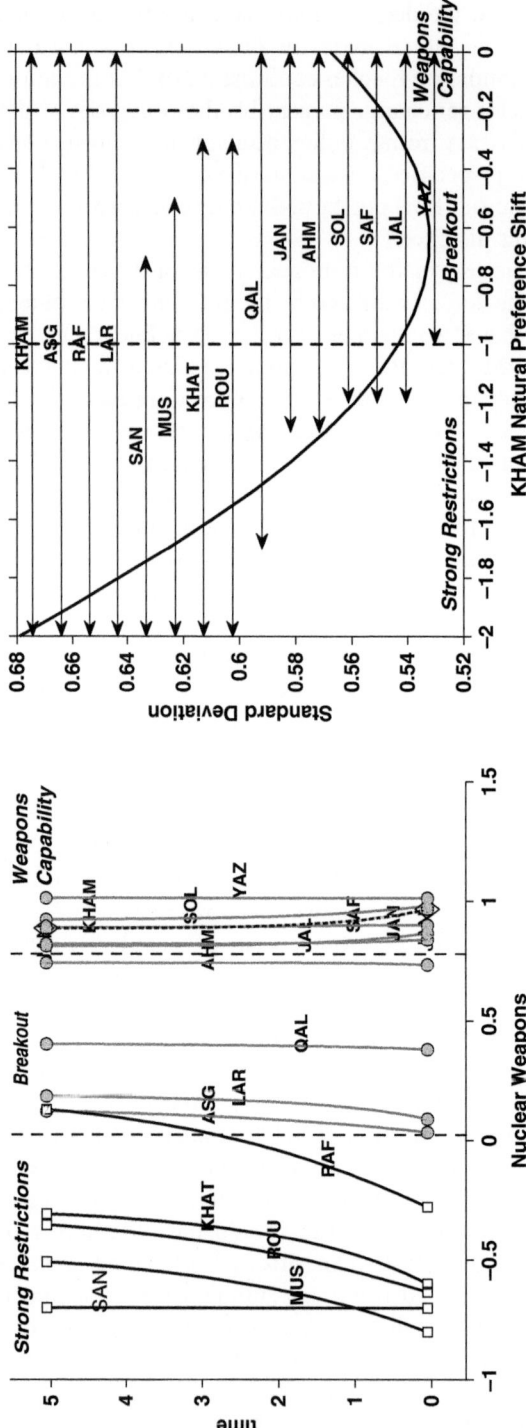

Figure 21.6 Nuclear Weapons issue simulation. (a) Actor trajectories using composite analyst values (first letter of actor abbreviation intersects with trajectory curve). Dashed lines demarcate boundaries between different policy labels. Dotted line is policy value (same as KHAM trajectory due to leader choice rule) and open diamond at top is policy decision. (b) Effect of Khamenei softening his natural preference: standard deviation of actor final positions (solid curve) and actor concurrence intervals (double-headed arrows, actor listed above). Horizontal axis is shift from KHAM original natural preference from composite analyst

Weapons Capability zone, just slightly less hardline than KHAM's initial natural preference. This is not surprising given the outsize influence that KHAM has on the group; his network out-degree – the sum of all his outgoing influence network values on the rest of the group is more than three times the second highest actor. Rafsanjani does move sufficiently towards a harder line so that he can support the policy. However, the core reformers, most notably ROU, dissent as they end up greater than one unit (the latitude of acceptance) from the policy.

The above result, however, is inconsistent with the more conciliatory posture that Iran took in reaching the interim nuclear agreement in November 2013. It is not tenable that the Iranian president Rouhani, a savvy political insider, would have been vigorously pursuing a nuclear deal with the United States completely at odds with the Supreme Leader's policy, thereby setting himself up for failure. This leads to the inference that the Iranian policy may have shifted to a softer line than represented in the original analyst data. Possible scenarios underlying this shift can be investigated by changing the simulation parameters. Simulations of scenarios involving increased group cohesion or increased reformer status due to Rouhani's election in June 2013 could not produce a significant enough policy shift. But another potential explanation is that Khamenei himself softened his position, which can be modeled by shifting his natural preference in the negative direction along the Nuclear Weapons issue axis. This can indeed account for the softer line policy: given the leader choice decision rule and his great influence, the policy essentially follows his natural preference; a shift of −0.2 brings the policy into the Breakout range and a shift of −1 moves it into the Strong Restrictions range.

While it is clear that Khamenei can shift the policy if desired, considerations of minimizing discord within the leadership as a whole and, in particular, maintaining the support of key hardliners – the IRGC members Soleimani and Safavi, and Janati, the chairman of the Guardian Council – are doubtlessly important in his decision-making calculus. These factors can be assessed using *Figure 21.6b* which plots the standard deviation of the final positions and the concurrence interval for each actor – the range of KHAM's natural preference shift over which the actor supports the policy decision. Observe that there is a minimum in the standard deviation at a shift of about −0.6 approximately in the middle of the Breakout interval. Furthermore, there is a range from about −0.7 to −1 for which all actors concur with a policy in the Breakout zone. These observations imply that a Breakout policy would minimize discord within the group. Indeed, as KHAM moves into the Strong Restrictions zone, he rapidly begins to lose conservative support: first YAZ, then crucially at −1.2 the IRGC members SOL and SAF, followed shortly thereafter by JAN.

The above analysis leads to the conclusion that the Khamenei softening scenario is a plausible explanation for Iran's shift to a posture more amenable to reaching a deal on the nuclear issue; he can maintain consensus while pursuing a Breakout policy, which is consistent with trying to reach a nuclear agreement, albeit one which would be very weak from the perspective of the United States. The fact that there were secret meetings between US and Iranian officials on the nuclear issue starting in 2012, a year prior to Rouhani's election (Associated Press 2014), suggests that Khamenei may very well have shifted towards a more flexible position than the original hardline Weapons Capability ascribed to him from the analyst surveys. With respect to the prospects of reaching a final deal, his original analyst-derived position would imply that a deal would be extremely unlikely. The analysis of the softening scenario indicates that Khamenei's room for maneuver is limited so that he can only move a small amount into the Strong Restrictions zone before losing the support of key conservatives. This suggests that a deal which provides robust provisions against an Iranian breakout capability – in particular, the US stated that it sought a minimum breakout time of one year – would indeed be possible but very difficult to reach. A deal between Iran and the P5+1 was in fact announced in July 2015. An assessment as to the strength of the deal from the P5+1 perspective – whether the monitoring and other restrictions on Iranian nuclear activities are sufficiently robust as to prevent a rapid breakout capability or a covert program – cannot be made here. However, the fact that the negotiations took twenty months from when the interim deal was announced to reach a final agreement, including two six-month extensions of the interim deal, attests to the difficulty in consummating the negotiations.

21.7   CONCLUSION

Relationships among leadership elites and their preferences on important issues are essential elements in determining the outcomes of policy debates. This chapter has presented a methodology, implemented with the PORTEND software package, for the analysis of the factional structure of leadership elites and the simulation of their group decision-making via the nonlinear social influence model. Methods for investigating factional structure based on issue data alone range from simple standard deviations and plots of actor natural preferences to more sophisticated pattern extraction using Principal Components Analysis, which revealed meaningful dominant and subordinate factional alignments among the set of Iranian leaders; the first corresponding to the primary conservative–reformer divide over most of the issues and the second reflecting key departures from this alignment with respect to economic reform. Complementary to the issue analysis, the application of a community structure algorithm to the inter-actor influence network also yielded similar dominant and subordinate structures via the first two eigenvectors. The uncovering of the parallel structure in the issue and network data illustrates the power of applying methods from research in complex networks. This research also forms the basis for the polarization metric which quantifies the extent to which differences in actor issue positions also stress network faultlines, thereby providing an integrated measure of how divisive an issue is.

   The social influence model entails complexity via its nonlinear coupling of actors over their influence network and was applied to Iranian nuclear decision-making. Simulation of the original analyst values yielded a policy decision that was so hardline as to be inconsistent with apparent Iranian moves towards more negotiating flexibility in late 2013. The model's capability for scenario analysis was illustrated to address this inconsistency. Khamenei's shift towards a more moderate natural preference was found to be the most plausible explanation. Sweeping over his natural preference shift, the simulation indicated that he had sufficient room to maneuver before losing the support of key hardliners so as to make negotiations tenable. However, his ability to enter the Strong Restrictions zone, which presumably would have considerable overlap with the goals of the P5+1 countries, was found to be quite limited implying that achieving an agreement would be quite difficult – a conclusion perhaps supported by the long period of time required to reach a deal.

Finally, addressing further research, one area could involve the investigation of whether automated content analysis of actor rhetoric could be a viable input source for either the structural analysis or the simulation. Another area could be extending the social influence model to a multidimensional issue space in order to allow issues to trade off against each other. Additionally, complexity research on adaptive networks could be used to develop an issue-network coevolution model in which both issue positions and network ties would interact and change dynamically, thereby explicitly modeling alliance formation processes, a capability not present in the current model.

ACKNOWLEDGMENTS

This work was supported by the Defense Threat Reduction Agency and the Office of Naval Research under grant HDTRA 1-10-1-0075.

REFERENCES

Associated Press. 2014. Iran's Supreme Leader Dismisses Direct US Talks. *New York Times*, August 13.

Bueno de Mesquita, B. 1997. A Decision Making Model: Its Structure and Form. *International Interactions*, 23, 235–266.

Bueno de Mesquita, B. 2009. *The Predictioneer's Game*, New York: Random House.

Davis, P.K. and O'Mahony, A. 2013. *A Computational Model of Public Support for Insurgency and Terrorism*, Santa Monica, CA: RAND.

Eagly, A.H. and Chaiken, S. 1993. *The Psychology of Attitudes*, Fort Worth, TX: Harcourt College Publishers.

Friedkin, N.E. and Johnsen, E.C. 2011. *Social Influence Network Theory: A Sociological Examination of Small Group Dynamics*, Cambridge, UK: Cambridge University Press.

Gabbay, M. 2007a. Application of a Social Network Model of Elite Decision Making. *Annual Meeting of the International Studies Association*. Chicago, IL.

Gabbay, M. 2007b. A Dynamical Systems Model of Small Group Decision Making. In: Avenhaus, R. and Zartman, I.W. (eds) *Diplomacy Games: Formal Models and International Negotiations*, Berlin: Springer, pp. 99–121.

Gabbay, M. 2007c. The Effects of Nonlinear Interactions and Network Structure in Small Group Opinion Dynamics. *Physica A*, 378, 118–126.

Gabbay, M. 2013. Modeling Decision-Making Outcomes in Political Elite Networks. In: Glass, K., Colbaugh, R., Ormerod, P. and Tsao, J. (eds) *Complex Sciences*, Berlin: Springer International Publishing, pp. 95–110.

Gabbay, M. and Das, A.K. 2014. Majority Rule in Nonlinear Opinion Dynamics. In: In, V., Palacios, A. and Longhini, P. (eds) *International Conference on Theory and Application in Nonlinear Dynamics (ICAND 2012)*, Berlin: Springer International Publishing, pp. 167–180.

Hermann, C.F., Stein, J.G., Sundelius, B. and Walker, S.G. 2001. Resolve, Accept, or Avoid: Effects of Group Conflict on Foreign Policy Decisions. *International Studies Review*, 3, 133–168.

Hermann, M.G. 2001. How Decision Units Shape Foreign Policy: A Theoretical Framework. *International Studies Review*, 3, 47–81.

Newman, M.E.J. 2006. Finding Community Structure in Networks Using the Eigenvectors of Matrices. *Physical Review E*, 74, 036104.

Newman, M.E.J. 2010. *Networks: An Introduction*, Oxford, UK: Oxford University Press.

Rieffer-Flanagan 2013. *Evolving Iran: An Introduction to Politics and Problems in the Islamic Republic*, Washington, DC: George town University Press.

Scholz, J.B., Calbert, G.J. and Smith, G.A. 2011. Unravelling Bueno De Mesquita's Group Decision Model. *Journal of Theoretical Politics*, 23, 510–531.

Strogatz, S.H. 1994. *Nonlinear Dynamics and Chaos*, Reading, MA: Perseus Books.

Webb, A.D. and Copsey, K.D. 2011. *Statistical Pattern Recognition*, West Sussex, UK: Wiley.

PART VI

MIXED METHODS AND COMPLEX ANALOGIES

Associate Professor Benyamin Lichtenstein

The final section of this book puts emphasis on combinations of methods—synthetic analysis—rather than focusing on any given approach on its own. The five chapters here explore complex phenomena through a range of methods, each complementing the others in a holistic examination of the system/phenomenon.

I begin with Goldstein's work, based on his long-standing influence in creating a discipline of complexity (Goldstein 1988, 1999, 2000; Goldstein, Hazy and Silberstang 2010). His work is also the most theoretically broad in this section, as reflected in his main research question: 'How can we best conceptualize, and therefore measure, the nature of emergence itself, in our world?' His primary claim is that we often see emergence in the wrong way, as an outcome derived ('self-organized') by a recombination of existing elements. Instead, emergence should be seen as the creation of something 'radically novel.' Whereas in self-organization the components stay the same but are re-ordered, in radical novelty the micro elements themselves transform, leading potentially to new types and levels of macro order, and sometimes to new laws and processes within the system.

To study this emergence requires first considering the many different ways these macro-phenomena may appear in emergence, and then seeking to identify at least one data collection/analysis method for each. These may be quantitative—he talks about dissipative structures, cybernetics, physics and mathematics (Cantor sets). They may be qualitative—he includes Goethe's *alternative qualitative methods*, and uses case studies as examples. They may be actionable, as he illustrates through the creative arts, and his own approaches of *difference questioning* and *purpose contrasting* for enacting social change. Mostly, these need to be synthesized, brought together into a shared view (story) about the complex phenomenon. Overall, Goldstein's emphasis is in exploring methods that can identify the radical nature of emergent, 'self-transforming constructions.'

Varga's chapter pushes this theme in a more operational direction, identifying various complexity methods and approaches that can be used to "integrate across scales and boundaries." Here again, the goal is to synthesize information, to gain the clearest understanding of the complex systems being studied. But she reminds us that integration is challenging, requiring carefully designed strategies of data collection and analysis.

The core of her chapter is a detailed introduction of ten specific methods/designs and seven paradigmatic stances that researchers can use to explore complex systems. Each of these approaches is valuable for seeing or enacting a specific quality of emergence. In addition, she provides remarkable detail on a panoply of data collection techniques and research methods, which span qualitative/quantitative, computational, historical, and more. Likewise, she provides strategies for dealing with the inevitable challenges that such complicated research designs can face. Finally, she presents a 3 x 3 matrix of data collection and analysis techniques, which reveal the potential of multi-method designs in complexity studies. In sum, Varga offers a kind of complexity methods road map, with strategic advice for how best to collect and analyze wide-ranging data on complex systems.

Drawing on the theme of complexity as a set of methods, Lichtenstein's chapter argues that complexity is really a combination of 15 sciences, each one a distinct disciplinary approach for identifying and understanding complex systems. Examples include NK landscapes, dissipative structures, cybernetics, power laws, fractals, multi-agent models, and ecosystem resilience. Each of these examines emergence in a unique way, by invoking a certain kind of research question, collecting a distinct type of data, and utilizing a novel analytic method to pursue the research. The bulk of the chapter is a brief introduction into each of these sciences, and the stream of studies that have used them.

The chapter also examines three main avenues for pursuing complexity research on emergence. The first, computational and mathematical models, is well-known through the work of scientists at the Santa Fe Institute (for example, Kauffman 1993; Holland 1998) and many parallel centers of simulation-based complexity studies. Unfortunately, some people think that simulations *are* complexity science, as opposed to one of its categories. However, although computational methods are good at answering certain types of questions, they are limited in exploring social systems in situ. Second are ideographic analogies to specific sciences (Tsoukas 1989; 1993), including dissipative structures, autocatalysis, and ecological resilience. Here, a careful mapping between a theory of natural systems and a particular social situation can reveal how causal processes in the former may be reflected in the unfolding of the latter. Finally, a third category is narrative and multi-method designs, which are best used to show instances of emergence as they are happening. Garud and his colleagues (Garud and Guiliani 2013; Garud, Gehman, and Giuliani 2014; Garud et al. 2015) have mastered these approaches, generating great insight into the nature and dynamics of emergence. In sum, Lichtenstein's chapter furthers the road map idea, presenting a diverse set of methods that are used by complexity scientists.

Andriani and Carignani focus on methods that are based on strong ideographic analogies, mainly to the evolutionary concept of *modular exaptation*. Exaptation occurs when a trait or technology is applied in a different context to accomplish a new function or generate new modalities for which it was not designed. Often this yields radical improvements in the new system/context. A bulk of their chapter examines how to make ideographic analogies valid, avoiding the problem of sloppy metaphors. Specifically, they clarify the connections between the base theory of exaptation and the target of technology innovation, and show how to make convincing analogies through the method of structure-mapping (Gentner 1983).

Their chapter offers two contributions to complexity methods. First, the authors identify modular exaptation as a powerful driver of innovation, and show how it can be used to explain some of the most radical innovations in our time. In particular it allows us to track the emergence of entirely new product categories, in contrast to the incremental improvements shown in traditional studies of organizational evolution. Second, they introduce *ideographic analogy* as a powerful analytic tool for exploring complex systems. The detail they provide is extremely useful for any complexity researcher who draws on scientific theories to explain complex social systems.

The last chapter in this section, by Shapiro and Scott, uses ideographic analogy as the basis for a new method called Dynamical Systems Therapy, which is dedicated to the practice of psychotherapy rather than research on complex systems per se. DST, the psychotherapeutic relationship is seen as a *complex adaptive system* that incorporates multiple interactive levels, including neurons and hormones, subjectivity (perception, feeling, intentionality), and inter-subjectivity. In contrast to the traditional psychiatric view of patients with diseases that needs to be fixed, DST views a patient's symptoms as 'breakdowns of adaptation,' and the therapy as a means for bringing more control to patients themselves. This psychobiological model expands the range of interventions to include psychopharmacology, psychodynamics, and cognitive-behavioral changes, all of which improve the chances of transforming the current system into one that's more adaptive and appropriate.

Of all the chapters in this section, Shapiro and Scott's is the most applied, focusing specifically on the practice of psychotherapy. The chapter exemplifies its key points through excerpts from actual therapy sessions. These reveal the skillful application of DST that clearly open up new cognitive territory for patients. The method is also explained through an analogy to *adaptive landscapes*, which shows how individuals can move away from rigid habits into new ways of feeling, thinking and acting. Likewise, they show how therapists can expand their understanding of the process, especially in how "the therapist's and the patient's systems interpenetrate and co-evolve together. . ." In this way it is truly a 'mixed' method, as it incorporates biochemical, emotional, behavioral and interpersonal realms, while bringing the practitioner directly into the focal system.

As a group, these chapters present complexity science in an expanded way, including a spectrum of approaches that include and transcend most common notions of complex systems. Their phenomenological foundations make them especially useful in practice, in understanding, as well as creating/enhancing emerging situations.

REFERENCES

Garud, R., and Guiliani, A.P. 2013. A narrative perspective on entrepreneurial opportunities. *Academy of Management Review*, 38(1): 157–160.

Garud, R., Gehman, J., and Giuliani, A.P. 2014. Contextualizing entrepreneurial innovation: A narrative perspective. *Research Policy*, 43: 1177–1188.

Garud, R., Simpson, B., Langley, A., and Tsoukas, H. 2015. How does novelty emerge?.In R. Garud, B. Simpson, A. Langley, and H. Tsoukas (eds), *The Emergence of Novelty in Organizations*: 1–26. Oxford: Oxford University Press.

Gentner, D. 1983. Structure-mapping: A theoretical framework for analogy. *Cognitive Science*, 7(2): 155–170.

Goldstein, J. 1988. A far-from-equilibrium systems approach to resistance to change. *Organizational Dynamics*, 15(1): 5–20.

Goldstein, J. 1999. Emergence as a construct: History and Issues. *Emergence*, 1(1): 49–72.

Goldstein, J. 2000. Emergence: A concept amid a thicket of conceptual snares. *Emergence*, 2(1): 5–22.

Goldstein, J., Hazy, J., and Silberstang, J. 2010. A complexity science model of social innovation in social enterprise. *Journal of Social Entrepreneurship*, 1(1): 101–125.

Holland, J. 1998. *Emergence: From Chaos to Order*. Cambridge, MA: Perseus Books.

Kauffman, S. 1993. *The Origins of Order*. New York: Oxford University Press.
Tsoukas, H. 1989. The validity of idiographic research explanations. *Academy of Management Review*, 14(4): 551–561.
Tsoukas, H. 1993. Analogical reasoning and knowledge generation in organization theory. *Organization Studies*, 14(3): 323–346.

22. Complex analogy and modular exaptation: some definitional clarifications

Professor Pierpaolo Andriani, Kedge Business School
and Professor Giuseppe Carignani, Technical High School
Malignani and University of Udine

22.1 INTRODUCTION

Charles Darwin took very seriously the problem of the origin of *organs of extreme perfection and complication.*[1] It is a major difficulty of his gradualist evolutionary theory of 'descent with modification'. The objection, raised in particular by St. George Mivart (an English Catholic biologist) is based on the fact that natural selection can hardly explain the incipient stages of organs because they can improve the fitness of an organism only when they are completely developed and functional (Mivart 1871).[2]

In order to tackle the problem, Darwin cautiously advanced a hypothesis that was actually a brilliant insight into later research: he suggested the possibility of functional shift, an evolutionary process later redefined 'exaptation'. This first Darwinian example of functional shift is quite clear:

> The illustration of the swimbladder in fishes is a good one, because it shows us clearly that an organ originally constructed for one purpose, namely flotation, may be converted into one for a wholly different purpose, namely respiration . . . The swimbladder is homologous, or ideally similar, in position and structure with the lungs of the higher vertebrate animals: hence there seems to me to be no great difficulty in believing that natural selection has actually converted a swimbladder into a lung, an organ used exclusively for respiration. (Darwin 1859, p. 190)

The same process described by Darwin can be recognized in technological change: for example, in early agricultural tractors equipped with internal combustion propulsion plants the engine/gearbox case was "so huge and so rigid" (Dew et al. 2004, p. 78) that it could be 'exapted' into a chassis (the structural frame of the vehicle). The engine/gearbox assumed the structural function, reducing the tractor's overall weight and cost and ultimately contributing to its success (Dew et al. 2004). Clearly, the breakthrough was enabled by the fact that the engine was 'ideally similar in position and structure' to the necessary structural component: the very words used by Darwin apply perfectly to technological innovation. The case is not unique: several scholars have discussed dozens of significant cases of 'functional shift' in technological change (Dew et al. 2004; Cattani 2005, 2006; Andriani and Carignani 2014).

We can easily recognize these cases of 'functional shift' as a similar process shared by two ostensibly very different complex systems (biological evolution and technological change). The similarities between the technological and the biological domains, however, are not limited to exaptation: Wagner and Rosen propose nine 'commonalities' between 'nature' and 'technology and science', in order to show that technological innovation "reflects almost everything we have learned about biological evolution in the two centuries since Darwin", so that "innovations in biological evolution and in technology have many common features" (Wagner and Rosen 2014, p. 1).

One may wonder why two complex systems as different as technological change and biological evolution share such an impressive number of similar processes. This chapter argues that the similarity between these two domains is not just a nice coincidence, but is based on structural reasons: it is actually a 'complex analogy' that legitimizes the transfer of concepts across the two domains. This is clearly an interesting methodological approach to the study of complex systems, because it suggests that the knowledge acquired in one of the systems (the 'base' domain) can be applied to the other (the 'target' domain). But under what conditions can concepts developed in one discipline, and resting on an accepted and historically well-tested base of related concepts, definitions and operational methods be transferred to a new domain?

In particular, can the concept of exaptation, steeped in evolutionary biology, and anchored on the related evolutionary concepts of functions, selection and fitness, be legitimately used to analyse technological change? This is the question we tackle in this chapter. Building on the concept of *complex analogy* (Gentner 1983) we assert is that the transfer of concepts across domains is possible if the domains share not just generic similarities, but a coherent set of *relations* between analogous concepts (structure mapping).

The chapter is organized as follows: after discussing some properties of complex systems, section 22.2 introduces the concept of exaptation, briefly describes the literature associated with it and highlights a few critical issues related to the validity of the exaptation concept in innovation studies. Section 22.3 deals with the foundational concepts underlying exaptation: in particular, we discuss function and functional modules in biology and technology, and derive the concept of modular exaptation on which this chapter is focused. In section 22.4, we present an overview discussing the role of analogy in scientific (and philosophic) thinking and introduce the concepts of *structure mapping* and *complex analogy* (Gentner 1983). We show that analogic thinking across the coupled field of evolutionary biology and technological change is legitimate provided that a certain number of conditions are in place. We apply this approach to the case of exaptive evolution in technological change, thereby providing one of the first building blocks in methodological terms to the effective study of exaptation. At the core of the theoretical argument there is an artefact-centred technological-biological evolutionary analogy. In section 22.5 we summarize our methodological contributions and discuss implications and possible avenues of research.

22.2   EXAPTATION IN COMPLEX SYSTEMS

Complex systems science focus on the study of emergent properties in systems composed of multiple heterogeneous agents interacting over mostly non-regular and time-dependent network configurations often nested within one another. The astronomical number of configurations of complex systems' agents occasionally gives rise to emergent properties that are evolutionary novel, thereby increasing the complexity of the system and in some cases conferring some kind of evolutionary advantage to a system that expresses such properties.

Complex systems theory stresses the ontological unpredictability of such properties. Unpredictability goes beyond quantum mechanics unpredictability (Prigogine and Stengers 1997). It is steeped in the multiple phenomena related to:

(a) The huge number of configurations of many complex systems that will eschew any type of information processability even by the most powerful information-processing machines designable in the foreseeable future. Wagner (2011) has calculated that for typical metabolic systems, protein and genetic systems (both at the genotypic and phenotypic level) the number of configurations of even simple systems extends to numbers of the order of (or bigger of) 10^{100}. Similar numbers characterize technological systems as well; and
(b) 'Butterfly-effect'[3] dynamic that may activate self-sustaining information cascades that may lead to the emergence of radical different behaviour in existing systems. As bifurcations between dynamical trajectories may be triggered by noise or apparently inconsequential events, predictability is inherently impossible.

Complexity is both interdisciplinary and cross-disciplinary. Interdisciplinarity derives from the fact that solutions of complex problems demand high cognitive diversity, whereas cross-disciplinarity refers to the fact that solutions or frameworks are often transferrable across disciplines.

22.2.1 Exaptation

One emergent property of complex systems is exaptation. Exaptation, the co-opting of a technology or biological trait or module for a function for which it was not designed or selected, is a major but overlooked evolutionary force (Gould and Vrba 1982; Dew et al. 2004; Kauffman 2000; Cattani 2006; Wagner 2011; Andriani and Carignani 2014). Exaptations result from novel associations between artefacts and domains of applications. The association is mediated through an historical context. For instance, the emergence of the modern trumpet and horn is due to the exaptation (and subsequent adaptation) of mechanical valves,[4] originally used to control airflow in blast furnaces. Friedrich Blühmel, a horn player who worked in a mining company (context), transferred a technology well known in the mining industry to the music industry (Tappi 2005).

'The wonderful metamorphoses in function' hypothesized by Darwin (1859, summary of Chapter 6) was later reframed and defined 'preadaptation' by the French naturalist Lucien Cuénot (1914). Cuénot's 'preadaptation' refers to traits whose functional shift increases the fitness of an organism. The concept was reframed by Gould and Vrba (1982), in which they coined the term 'exaptation' to describe the process (rather than the effect) of functional shift. Gould and Vrba contrasted exaptation, or the discovery of a new function for an existing trait, with adaptation, defined as the improvement of a trait through natural selection. Exaptation is intrinsically non-adaptive. In an evolutionary fitness landscape, adaptation is akin to peak-climbing, driven by increase in a well-defined fitness function; as exaptation (especially the most radical ones) is inherently about the emergence of new functions, that is, exaptation entails a structural modification of the fitness landscape and corresponds to the introduction of a new peak.

To qualify as an exaptation, a feature cannot have arisen "as an adaptation for its present role" (Gould 1991, p. 43) but instead must have been coopted for its current function and enhance fitness (Gould 1991). Exaptations appear ubiquitous in the history of technology and markets. Mokyr claims (2000) that exaptation is frequently encountered in the history of technology. Several authors have shown the ubiquity of exaptation in technological evolution (Lane 2011; Kauffman 2000; Andriani and Carignani 2014; Cattani 2006; Dew et al. 2004), in biological evolution (Barve and Wagner 2013; Gould 1985; Andriani and Cohen 2013; Wagner 2011) and in other disciplines, such as anthropology, psychology, and language (Buss et al. 1998; Brown and Feldman 2009; Fitch 2005). Exaptations can be conceptualized as *jumps* of technologies across different application domains.

Transferring technologies across domains is one way to reduce the cognitive complexity associated with radical innovation, which involves the parallel creation of technologies and markets. Another way of dealing with the burden of cognitive complexity is to transfer ideas from one domain (or discipline) to another. For instance, the origin of the events that would lead more than a century later to the discovery of the most successful drug in history, that is, the aspirin, lie in a curious event that happened in England in 1758. The Reverend Stone tells the story in a letter to the Royal Society:

> [T]here is a bark of an English tree, which I have found by experience to be a powerful astringent, and very efficacious in curing agues and intermitting disorders.[5] About six years ago, I accidentally tasted it, and was surprised at its extraordinary bitterness; which immediately raised in me a suspicion of its having the properties of the Peruvian bark.[6] As this tree delights in a moist or wet soil, where agues chiefly abound, the general maxim, that many natural remedies lie not far from their causes, was so very opposite to this particular case, that I could not help applying it; and that this might be the intention of Providence here I must own I had little weight on me. (Jeffreys 2008, p. 17)

Reverend Stone discovered that the bark of the willow was an effective treatment for a symptom of malaria, that is, high fever. The letter contains several interesting points for the historian of technology but one in particular is important for this chapter. The first concerns the logic used by Stone. Stone assumes that as the bitterness of the willow reminded him of the Chinchona tree, then, the therapeutic action of the Chinchona may be shared by the willow. In other terms, a shared attribute stands as an indicator for a shared causal relationship between the attribute and the action on the organism. Although Reverend Stone's hypothesis was revealed to be entirely wrong, Stone's analogical thinking led via unintended routes to a major technological revolution. Stone's hypothesis was rooted in a popular folk theory, promoted by Paracelsus, known as *Doctrine of Signatures*. This doctrine predicated an early form of analogic thinking based on shared attributes, rather than shared relations. The doctrine asserted that Nature gave away signs about the functions of natural things. So a flower shaped as an eye could cure eye problems. The bitterness of the willow bark thriving in stagnant water, hence related to the then reputed cause of malaria ('bad air'), was a sign that revealed its healing property, that is, an unexpected function.

22.2.2   Source of Selectable Variation in the Technosphere

According to a generalized Campbellian *variation-retention-selection* epistemology (Bickhard and Campbell 2003), the expansion of the technosphere requires a supply of variations, on which selection can act. Along with traditional channels of production for variations, such as random mutations and purposeful design, an additional channel is provided by exaptation. The emergence of exaptive variations reflects a fundamental property of technologies, that is, the non-prestateability of technological applications (Kauffman 2000). The term non-prestateability indicates that the full set functions (current and potential) of any technology cannot be listed at any moment in time. This is due the fact that the number of applications of any technology depends on the interaction technology-context. Because such interaction is inherently unlimited and subject to combinatorial explosion, the number of potential applications of existing artefacts must inherently be larger than what designers or inventors can conceive. The exaptive channels thus feeds on the heterogeneity of the economy (Jacobs 1969; Mokyr 2002, 1990). Drawing on Gibson's (1950) idea of artefacts as affordances, Tuomi (2002) also extends the non-prestateability argument by noting:

> [T]echnologies and technical products have 'Interpretative flexibility'. . . Different user groups and stakeholders impute different meaning to a given technological artefact . . . Instead of being a 'well defined' objective artefact, with characteristics that could be described without reference to social practice, the artefact in question has many, and possibly incompatible, articulations. These 'meaningful products' may develop independently of each other, and one technological artefact can embed several meaningful products simultaneously.

Hence exaptations provide an inexhaustible source of variations in the economy, complementing random and purpose-driven inventions and innovations.

Non-prestateability extends from artefacts to capabilities. Whence do capabilities that sustain competitive advantage come? The traditional answer is that they are designed to fill an identified market opportunity. Yet Cattani (2005, 2006) shows that Corning's supremacy in the fibre optic industry is due to capabilities it developed prior to the emergence of the fibre optic industry. Cattani thus extends the discussion of exaptation[7] from artefacts to capabilities. Using the notion of *transformative capability* (Garud and Nayyar 1993), Cattani also shows that the value of proprietary capabilities is similar to that of shadow options, so the role of foresight depends on the redeployment of capabilities in new domains.

22.2.3 Radical Innovation

Exaptation is linked to radical innovation via the concept of speciation. Levinthal (1998)[8] explains the sudden rise of market-changing innovations in terms of speciation. He notes that in several instances, technological change associated with the emergence of new market-defining applications is minimal or non-existent. A new market speciation results from technology-domain changes (Adner and Levinthal 2002), not a Schumpeterian technology–technology combination. A speciation process thus leads to branching of industries and the emergence of new market lineages (Dew et al. 2008).

Another line of inquiry focuses on the interaction between exaptive and adaptive processes. Exaptation as a fundamental micro-process of innovation (Cattani 2006) that usually triggers adaptive technology market responses, which, as Jacobs (Jacobs 1969, 1985, 2000) shows, may become self-reinforcing and assume the character of avalanches. Levinthal (1998) theorizes about an exaptive–adaptive cycle in the context of the wireless market; Lane (2011) elaborates an adaptive–exaptive model of innovation based on technology adoption that includes technological, organizational, and societal considerations, which he calls *exaptive bootstrapping*.

22.2.4   Strategy and Entrepreneurship

A behavioural theory of entrepreneurial firms, effectuation theory (Dew et al. 2008) moves from the observation that entrepreneurship involves the creations of new market niches by firms and markets in the absence of well-defined goals. In a state of Knightian (1921) uncertainty (Dew et al. 2004), entrepreneurs make choices using a logic of design, rather than a logic of decision, such that they "make decisions now in terms of goals that will only be knowable later" (March 1976, p. 75). Accordingly, effective firms act not only within market environments, but also upon them, and in some cases end up creating new markets not predictable *ex ante* even by the very stakeholders involved in the negotiation process. Not taking the environment as given and/or predictable also implies that adaptive or other types of reactive strategies are inadequate and even inappropriate in the effectual process. Instead, entrepreneurial firms that use an effectual logic tend to develop exaptive strategies[9] (Dew et al. 2008).

22.2.5   R&D Management

Exaptation implies a change of perspective on R&D and inventions. If a significant fraction of innovations are fuelled by the dormant potential of existing technologies (to generate new functions), then innovation-by-exaptation complements a market-pull or technology-push logic and provides a strong rationale for the management of serendipity (Merton 2004). Grandori (2007) suggests looking for the *performance potential* implicit in existing technologies rather than the *performance gaps* that may drive the development of new technologies. With this perspective, the exaptive search focuses on potential applications of endogenous knowledge and technology. The *unshelving* technique rooted in the transformative capacity idea (Garud and Nayyar 1993) offers a starting point. Similarly, Dew et al. (2008) suggests that slack is a crucial factor for exaptive discoveries. By reducing slack, efficiency-driven management approaches actually put sources of organizational creativity at risk.

This brief review of relevant literature hints at the breadth of issues and range of frameworks that exaptation has inspired. However, significant ambiguities, inconsistencies, and problems persist with regard to meaning of exaptation and its application in social sciences. These difficulties stem from a double source: first, as Kauffman argues in several publications, exaptation is a dynamic process that makes sense only in an open *non-ergodic* universe, that is, a universe in which the possibilities of evolution are ontologically, and not only experimentally, unknown and unknowable. The lack of foreseeability of the evolution of complex systems extends from the macro to the micro picture: technologies also show an ontological lack of foreseeability regarding their range of applications. Exaptation is an attempt to frame such indeterminacy within an evolutionary framework. However, the multifarious role of technologies generates difficulties as to how to exactly define the parameters of the evolutionary frameworks. This difficulty is reflected in the way the literature describes the change of applications that are at the core of exaptation. Terms such as 'function', 'functionality', 'applications', and 'purpose' get used interchangeably in extant literature – often in the same text – to broadly indicate the emergent functional shift (often including the emergence of a new economic entity) typical of exaptation. In reality, these terms imply significantly different aspects, such as what a module does, what a module is for, what it has been selected for, or what it has been designed for. Scholars rarely specify which meaning their terminology prioritizes.

Second, the use of concepts imported from evolutionary biology requires a careful discussion about the validity and limit of the evolutionary biology–technological evolution analogy. The fundamental evolutionary concepts of function and selection are critical to understand and define exaptive processes, but without a critical discussion about the technological context, they risk becoming metaphors without analytical power. Although the evolution of species and technologies share some fundamental dynamics, their differences must be identified to ensure the validity of analogical thinking. In particular, the concepts of modularity and function, common to both technological and biological systems and crucial to understand exaptation, are characterized by specific features that are context-dependent and as such need to be carefully evaluated in the specific context of operation.

22.3 FUNCTION, FUNCTIONAL MODULES, AND MODULAR EXAPTATION

A key concept in Joseph Schumpeter's theory of economic development is the necessity of a dynamic process of innovation (that he famously defined 'creative destruction') as a fundamental complement to the static 'equilibrium flow' (Schumpeter 1934). By means of relentless 'creative destruction' the capitalist economy incessantly changes itself from within. Schumpeter sometimes used evolutionary terms, and the general features of an evolutionary theory show up in his works, but was very critical about the usage of theoretical concepts or techniques borrowed from biology in the economic domain (Fagerberg 2003): in fact, his interest was to create a radical, dynamic general theory of economic development rather than to propose an analogy between economic and biological microprocesses. However, Schumpeter's work is all-important in justifying evolutionary economics because by exposing the limits of the equilibrium vision of economic development he postulated the necessity of models able to analyse complex dynamic systems.

After Schumpeter, the usefulness of an evolutionary theory of technological change (and more broadly of evolutionary theories in economics studies in general, for example, Penrose (1952) has been critically debated. Unfortunately, a number of evident disanalogies (for example, Ziman 2000, p. 5) create problems in the direct application of evolutionary models to the technological domain, suggesting to some scholars the necessity of invoking Lamarkian features, which are very controversial in biology (Nelson and Winter 1982). Recent discoveries in evolutionary biology, related in particular to bacterial evolution, have reignited the interest in the analogy between technology and biology (for example, Hartwell et al. 1999) and confirmed the role of exaptation (for example, Ganfornina and Sanchez 1999), outlining also a new, broader perspective of evolution, in which a multiplicity of patterns can be legitimately adopted (Doolittle and Bapteste 2007).

Building on these results we explore the viability of a rigorous biological-technological analogy, choosing *the artefact* as the evolving unit in our evolutionary theorizing: although the choice of the artefact as the primary unit for the study of technological change may seem naïve, a number of scholars endorse it. According to technology historian George Basalla:

> The artefact – not scientific knowledge, nor the technical community, not social and economic factors – is central to technology and technological change. (Basalla 1988, p. 30)

22.3.1   Modularity

Our departure point from Basalla's and other extant theories of technological change is the explicit inclusion of the modularity of the artefact. Indeed, in our conceptualization the artefact is modular, that is, an artefact can be modelled as composed of separate elements called modules, each of which is recursively[10] composed of lower-level modules. An artefact can be represented by a multi-level tree in which at each level the modular set represents a partition of the whole artefact: this is a simple conceptualization often used in the technological domain (Figure 22.1a). A similar representation is accepted in biology for describing – for example, modularity at sub-cellular level (Figure 21.1b).

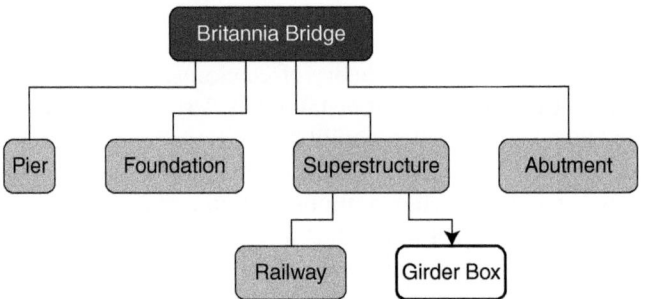

Note: Britannia Bridge: the 'Girder Box' functional module is evidenced.

Figure 22.1(a)   Modular tree of a technological artefact

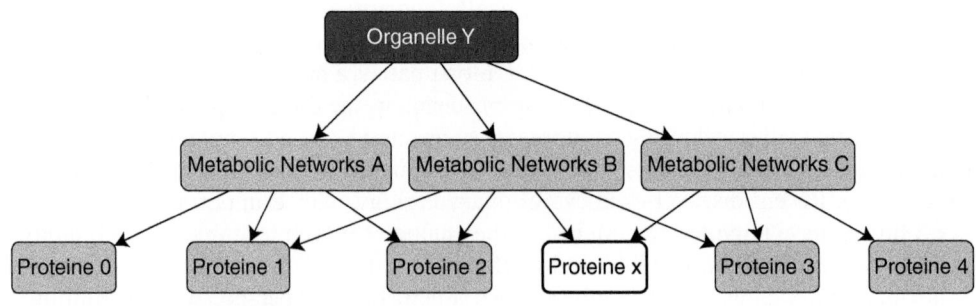

Source: Reproduced and modified from Zhu et al. (2012).

Figure 22.1(b)   Modular tree of a microbial cell

The general concept of modularity, however, is much more complex than the stylized one described before, and central in understanding complex systems. Herbert Simon (Simon 1962) proposed modularity as a systemic concept, describing how a complex system can be decoupled into subsystems (modules) that perform nearly independently of each other. Simon describes near-decomposability as characterized by the following properties:

1. "the short-run behaviour of each of the component subsystems is approximately independent of the short-run behaviour of the other component"
2. "in the long run, the behaviour of any one of the components depends in only an aggregate way on the behaviour of the other components" (Simon 1962, p. 474).

Near-decomposability applies to various levels of a complex system, because it "is composed of interrelated subsystems, each of the latter being, in turn, hierarchic in structure" (Simon 1962, p. 468). Simon argues that near-decomposability is common in complex systems, as exemplified in the domains of physical, chemical, biological, technological, and social systems.

22.3.2 Function

The analogy between biological and technological modules, however, is problematic, because of the presence of evident disanalogies (for example, the intentionality of technological design). It is all the more surprising, therefore, that recent biological research has re-proposed modularity as a key concept shared by technological and biological systems, constituting a powerful link between the biological and technological domain. Hartwell et al. (1999) for example, write that "the notion of function and functional properties separates biology from other natural sciences and links it to synthetic disciplines such as computer science and engineering". In good accordance with the technological concept, they define a functional module as "a discrete entity whose function is separable from those of other modules". Moreover, they argue that functional modules constitute "a critical level of biological organization" (Hartwell et al. 1999, pp.C47–C48). It is important to notice that Hartwell and colleagues do not refer just to 'modules', but to 'functional modules'. Function constitutes a new definitional concept that enters the picture. Unfortunately, the concept of function is difficult and problematic in technology, thus potentially capable of disrupting the biological-technological analogy.

Indeed, while the term 'function' and the related concept of contextually driven 'functional shift' emerge from all the definitions of exaptation, a variety of different terms holding ambiguous and sometimes overlapping meaning are used. This aspect is crucial: in fact, depending on how broadly or narrowly we define a function of a trait/module, we end up with different results regarding the occurrence or not of exaptation.

The ambiguity is borne out of the fact that exaptation implies functional discontinuity but technological continuity. As a function depends on what the object does and/or what the object has been designed for, and is linked to the form and phenomena that the underlying technology exploits, then assessment of functional shift depends on the definition of function itself. For instance, solid gripping of a tennis racquet or a pen constitutes only a subset of the possible uses of the human hand. What is the function for which the human hand was selected? There are two options; the former encompasses a broad definition of function for the human hand, that is, the hand's function is tool-using. In this case holding a pen, tennis racquet or Palaeolithic sharpened stone would not qualify as exaptations. They are a subset of the generic functions of the human hand. The latter option is based on the observation that in evolution selection pressures are specific. Organisms are selected on the basis of fitness-increase deriving from specific capabilities. For the human hand, that was the modification of available naturifacts to produce weapons. From this standpoint tennis playing and writing become exaptations of the human hand (Linde-Medina 2011).

This problem becomes even more acute in the technology domain. The classical exaptation example of the microwave oven is instructive. In 1945 working on a magnetron (a subsystem of a radar system that emits microwave radiation) engineer Percy Spencer noticed the effect of magnetron radiation on food. Spencer studied the novel phenomenon and applied for a patent (Osepchuk 1984). Two years later Raytheon commercialized the first microwave oven.

Indeed, the question whether the microwave oven is an exaptation of the magnetron hinges on the definition of function. If we base it on the underlying scientific/technological phenomenon, that is, conversion of electrical energy into electromagnetic radiation, then *strictu sensu* there is no exaptation. Both magnetron and microwave oven convert electricity into microwave radiation. If instead, as we suggest, we adopt a contextual and historical definition (what the object was selected for), according to which the magnetron function is to supply power to radar systems and the microwave oven function is to cook food, then there no doubt about the magnetron exapting into a microwave oven. Clearly we need an operational definition of function to support a theory of exaptation in the technological domain.

When we adopt an evolutionary analogy in studying technological change and innovation we deliberately embrace an etiologic concept of function: "The explanation to why something exists intimately rests on how it became what it is" (Dosi 1997, p. 1531). Neander (1991) defends the etiological concept defining the proper function of an item as a 'selected effect'. Millikan (1999) identifies the direct proper function of items with the reproduced physical dispositions that *causally* contribute to the existence of the items. In other words, an etiological definition of function responds to functional *backward-looking questions* of the kind: 'Why is X there?' (De Winter 2010).

The term function, widely used in papers on Darwinian evolution, starting with Darwin himself (especially when discussing *organs*), is therefore to be intended in this etiological sense, albeit used in a more liberal way in biological practice (De Winter 2010). The standard etiological theories of biological function (for example, Neander 1991) are based on the causal history of biological items.

The etiological concept on which biological function is founded seems to pose a critical threat to the biological-technological artefact-centred analogy we are discussing. Indeed, technological artefacts (artefacts intended for a practical purpose) are intentionality created by a designer and a manufacturer: intentionality seems therefore to play a constitutive role (Krohs and Kroes 2009). Technological artefacts ostensibly exist (that is, were (re)produced) deliberately for performing the function(s) intended by the designer and manufacturer. The connection seems so obvious that the intended function often gives artefacts their proper name: (a nutcracker, a dumper, Wright's 'Flyer' and so on). In De

Winter's formulation the intended function responds to *'forward-looking questions* of the type: *Why will x be (re)produced? Why will x be maintained? Why will x be integrated in system s?'* (De Winter 2010, p. 5).

This leads to an apparent contradiction between the etiological concept necessary to biology ('the effect for which x was selected') and the intentional one necessary to technology ('the effect for which x will be reproduced'). Indeed, some scholars (for example, Vermaas and Houkes 2006) argue that an etiological concept of artefacts' function is untenable and therefore any evolutionary analogy of technological innovation is doomed from its very beginning. Countering this objection is therefore necessary to our conceptualization: we believe it can be countered in two ways:

First, in a narrow way, our focus is on inventions, prototypes and in general technologies in the fluid state of evolutionary development where the role of historical accidents is more pronounced than in incremental innovations that follow the so-called 'linear innovation model'.

Second, a more general objection is that emergence of technologies is a collective co-evolutionary process where agency and intentionality play a role alongside multiple other causal and contextual factors, making technological change much more akin to a generalized evolutionary process (Arthur 2009; Jacobs 2000; Bickhard and Campbell 2003) than the simple consideration of intentionality would lead us to believe. It follows that the use of an etiological definition of function is justified in the technological arena, especially in the cases in which serendipitous discovery (Dew 2009; Merton 2004) plays a role.

The use of an etiological definition of function clarifies another ambiguous aspect of exaptation. The literature on exaptation presents two broad classes of examples. The former, the most extreme case, concerns functional shift based on the exploitation of a different scientific/technological phenomenon. For instance, a table lamp converts electricity into light, and at the same time converts electricity into heat. According to this kind of example, using a table light to keep food warm qualifies unquestionably as exaptation.

The latter concerns functional shift absent phenomenon change. These cases are less clear-cut. The use of MP3 technology (Dew et al. 2004) for music storage and reproduction is one of such cases. MP3 was designed as a digital compression technology originally for legal documents. The extension to music doesn't change the underlying phenomenon but represents an unexpected market change. From an etiological point, selection is driven by features that enhance fitness (or market appealing, in the technology case) and hence both types qualify as exaptation.

On the basis of these considerations we adopt an etiological concept of function that is in our opinion tenable in both the biological and the technological domain and can therefore be used to found a well-formed complex analogy between biological and technological exaptation. We argue that our concept of function applies to all artefacts, but for most 'mature' artefacts it is less significant: in fact, for most artefacts the intentional concept of function and the etiologic concept of 'proper' function coincide, because the function(s) for which the artefact (and indirectly its modules) were selected by the user are exactly those for which the designer designed them and the manufacturer manufactured them.

On the contrary, the intentional and etiological concept of function are different for novel artefacts, especially radical innovations: they are often modular and are assembled from existing modules/subsystems some of which were originally designed and manufactured for different functions.

While certainly belonging to the world of technical artefacts, radical inventions are possibly less studied from the point of view of the philosophy of technology, and maybe less understood. Actually Preston (2009, p. 48) states that "Novel artefacts, in short, have no proper function." While this could be true for the whole novel artefact, that has no history, it is not true for its modular components, often horizontally transferred,[11] from other industries or technological lineages (Utterback 1996; Hargadon 2003; Arthur 2009).

In this regard it is interesting to reconsider a famous historical case of innovation, the Britannia Bridge, a Victorian engineering marvel built in 1845–1850 across the Menai Strait. This railway bridge has been regarded as a discontinuity in structural engineering, which at first sight confirms the untenability of the etiological concept of function in technology (Vermaas and Houkes 2003). The bridge featured a unique engineering solution: "On this bridge, regarded as highly innovative in its days, trains travelled through a horizontal, hollow iron tube, upheld by piers" (Vermaas and Houkes 2003, p. 278). Vermaas and Houkes cite this case to demonstrate why the etiological definition of technological functions may be untenable in the technology domain, in that "surely the function of the railway bridge was to make possible railway traffic across a body of water, not to withstand water pressure while keeping the ship afloat" (p. 280).

But a closer historical scrutiny that takes into account the modularity of the artefact (Figure 22.1a) reveals the true origin of the 'new' technological function (Vincenti 2000). The function of the whole bridge was indeed "making possible railway traffic", as Vermaas and Houkes (2003, p. 279) write, but its tubular beam – the hollow iron tube – (a modular component, referred to as a box girder in both bridge and ship construction) was designed to resist flexural and torsional structural challenges (not only to withstand water pressure) and was borrowed from naval engineering. In fact, Sir William Fairbairn, an experienced naval engineer, was involved in the Britannia Bridge design. This case thus reveals that what at a higher modular level seems a new function (a bridge enabling railway traffic) reveals itself at component level as an existing function transferred from a different technological lineage (tubular beams designed to withstand structural stresses). The transfer of box girder technology form shipbuilding to civil engineering – is therefore a confirmation rather than a critique to the etiological concept of function we propose.

The Britannia Bridge case is not unique in the history of technology: most radical, seemingly discontinuous inventions from the bow and arrow (Carignani 2016; Lombard and Phillipson 2010), to the turbojet (Giffard 2016), reveal historical continuity at modular lower levels. We can therefore adopt the etiological concept of function in technology.

So far, we posit the centrality of the concept of functional module recognized by recent biological research as 'a critical level of biological organization', as a key concept in founding a robust biological-technological analogy. We also show that the etiological concept of function is tenable in technology when applied to modules rather than to whole artefacts. Under this assumption, the concept of technological functional module closely resembles the analogous biological concept.

Since exaptation is by definition associated to functions, and functions are associated to modules (functional modules), the concept of *modular exaptation* follows a straightforward consequence of this reasoning.

22.3.3 Modular Exaptation

We define *modular exaptation* the process in which a module previously designed and manufactured for a certain function is later selected (that is, replicated) for a different function. A few examples are instructive in understanding why the concept is important.

We observed previously that the behaviour of any artefact or module is a larger set than that imagined by its designer and manufacturer, which we defined its purpose. In most cases, purpose and function coincide, and users select artefacts for their designed function. But this coincidence does not have to exist. Users (and manufacturers or inventors) co-opt artefacts for uses for which they were not designed. Co-opted usages may be trivial: "John may want to stand on a folding chair to clean the shelves of his kitchen" (Vermaas and Houkes 2003, p. 264),[12] or very significant, such as if the user's selection triggers a purpose–function bifurcation that prompts the emergence of new artefacts and markets. For example:

- The CD-ROM was originally designed and patented as a digital-to-optical recording and playback system. Later it was selected by the users for a different, unexpected function, namely, as a data storage medium for computers (Dew et al. 2004).
- The engine of agricultural tractors was strong and stiff enough to provide a structural function for the whole tractor, substituting the chassis and enabling the emergence of the modern internal combustion engine tractor (Dew et al. 2004).
- The magnetron (a component of a radar system) was used as the core of a new technology (microwave oven) to convey thermal energy into food (Andriani and Carignani 2014).

As these examples show, the term 'exaptation' describes different processes, sharing a common selective mechanism. In each of these cases, the user or the inventor discovered that a subset of the artefact's behaviour (or of one of its modules), already in existence but not coincident with the one for which the artefact or module was designed and built, could be exploited to support a new function. After the exaptation process, the artefact was reproduced to support the new function, which became the purpose of the new artefact.

A cursory look at the examples shows that for the CD-ROM, the entire artefact was exapted, and a new product – memory storage – appeared in perfect technological continuity in the market space; in the tractor case, an internal component of an existing product was exapted and generated differentiation in the (existing) tractor market, without however altering the artefact's purpose. The magnetron case differs from the previous two, insofar an internal component of the radar system was exapted and, through a process of technology–technology combination, created a new product (microwave oven) and a new market. Moreover, in the first case, selection operated on the same level, whereas in the second and third examples, it applied to both the final product level and the exapted component level. The process of selection can take place at the same or at *a higher modular level* from that of the exapted module. Indeed, the exaptation of a biological module (for example, an organ) improves the fitness and therefore triggers the selection of the organism, at a higher level, while in technology the exaptation of a module can have affect at a higher level (the microwave oven case) or at the same level (the Viagra case).

These examples show that a non-ambiguous definition of exaptation should include first, the artefact's modularity, to integrate the possible multi-level effects, second, the distinction between function and purpose and third, a selection-based use of function.

22.4   MODULAR EXAPTATION AND COMPLEX ANALOGY

The choice of the artefact as the evolving unit, based on Basalla's and other scholars' arguments (for example, Arthur 2009), is reinforced by a number of recent discoveries in biological evolution, proposing in particular the centrality of functional modules. However, borrowing such a concept from biology requires solving the apparent contradiction between 'intentional' function (purpose) and biological 'proper' function, an etiological concept. Our discussion above leads to the conclusion that the etiological concept of function is tenable also in the technological domain, suggesting that the resemblance between the biological and technological domain can be formalized into a complex analogy. The main advantage of founding a 'complex analogy' is that we can reliably base scientific reasoning on it. This section is aimed at discussing how to clarify such an analogy.

22.4.1   Analogy: A Very Short Background

Analogic reasoning is a fundamental cognitive competence used by humans for understanding new domains on the basis of known ones. It is a powerful tool for several diverse human activities, including management (Gavetti et al. 2005). Even more, analogy is, according to Hofstadter "the very blue that fills the whole sky of cognition" (Hofstadter 2001, p. 499). Indeed, the power of analogy in science has been emphasized over the centuries, from Aristotle ("Hidden nature, is known only through analogy" (Physics, Book One) to Konrad Lorenz in his Nobel Prize acceptance lecture (Lorenz 1974).

The core idea of analogical scientific reasoning is that two *domains* can be described, at a certain level of abstraction, by the same cognitive structure. The knowledge we already possess about one of the domains (the *base* domain) can therefore be useful in understanding the second one (the *target* domain). Albeit straightforward in principle, setting a 'deep' analogy in scientific research needs a rigorous methodological approach. In fact, a number of authoritative scholars have warned about the traps associated with the superficial usage of analogy in interdisciplinary research, often leading not to useful analogies but rather to *sloppy metaphors* (Lass 1990, p. 79). Gavetti et al. (2005), more specifically, warn against faulty analogical reasoning in strategic management, but at the same time posit that sound analogies are critical to strategy, especially in presence of radical change. The novel construct 'modular analogy' (Andriani and Carignani 2014) that we have already introduced and that we are going to better formalize in this section has therefore the potential of contributing to strategic thinking.

22.4.2   Gentner's Complex Analogy

According to Gentner's *structure-mapping theory* (Gentner 1983) a *complex analogy* is a comparison in which (most) relations, but few or no objects' attributes can be mapped from base domain to target domain. In a complex analogy:

- *Attributes* tend not to be transferred;
- *Relations* are transferred;
- Coherent sets of relations are preferred (*systematicity principle*).

Other principles coherently complete the original definition (Gentner and Jeziorski 1993, p. 450):

- No extraneous associations: only commonalities strengthen the analogy;
- No mixed analogies: the relational network to be mapped should be entirely contained within one base domain.

Under these conditions, the two domains share a similar *inherent relational structure*: "An analogy is an assertion that the structure that usually applies to a domain can be applied to another domain" (Gentner 1983, p. 156).

A beautiful example drawn from the history of science (Gentner and Markman 1997) is Johannes Kepler's analogous thinking in *Astronomia Nova*. In order to demonstrate the possibility of an action at a distance – an abhorrent notion to natural philosophers of the time – Kepler posited the analogy between motive power (causing the motion of planets around the Sun) and light, transmitted similarly through ethereal empty space, but not visible in the ethereal empty space.

The history of science and technology is replete with similar successful cases of analogical thinking. Indeed, in ideal continuity with Kepler's scientific thinking, Rutherford's famous analogy between the solar system and the atom exemplifies the key concept that relations, rather than attributes, are the foundational link between different domains in scientific *complex analogies*. The atomic nucleus hardly shares any *attribute* with the Sun (it is not hot, nor massive, nor yellow), nor planets share any *attribute* with electrons, but the *relation*s between Sun and Planet ('attracts'; 'revolves around') led eventually to the common abstraction of the 'central force' (Gentner 1983, p. 160).

In synthesis, in order to found a *complex analogy* "common relations (between the base and target domain) are essential, common objects are not" (Gentner and Markman 1997, p. 46). This is promising, because, when considering a biological-technological analogy, it is apparent even at a cursory glance that artefact and organism are hardly similar in terms of shared attributes. But the relations the two domains share support the building of the complex analogy: "a clever, sophisticated process that can be used in creative discovery" (Gentner and Markman, 1997, p. 45).

22.4.3 Definitions

This section is aimed at framing the concept on Modular Exaptation into an artefact-centred technological-biological complex analogy conforming to Gentner's requirements. In order to do so, we formalize the discussion we proposed in the preceding sections by defining key definitions and constructs.

Artefact
The standard definition of artefact is associated to purpose: for example, "An artefact may be defined as an object that has been intentionally made or produced for a certain

purpose" (Hilpinen, Artifact and Zalta 2011). In this chapter we adopt an extended concept of artefact, defined as follows: An artefact is an object that has been intentionally made or produced or selected (and replicated) for a certain purpose.

Functional module

A functional module is a component of a modular artefact that is:

(a)   internally cohesive;
(b)   clearly separated from the other modules by a boundary while connected to them through an interface;
(c)   associated to a single *function* or a discrete set of *functions*.[13]

Modular artefact

We define modular artefact an artefact composed of functional modules, each of them an artefact.

Function and behaviour

According to the definition, each functional module is associated to a function (or a set of functions). It is useful to encase the concept of function into a broader frame: we therefore define the following concepts:

Behaviour[14] (B-set): the complete set of effects, included all the possible actions, processes and operations that a module can perform.

Function (F-set): the subset of Behaviour for which a module (or the artefact of which the module is a component) was selected for replication.

We notice that Function may or may not coincide with the intentional function (purpose) intended by the functional module's designer.

22.4.4   Making Explicit the Complex Analogy

Biology is the base domain in our complex analogy. In particular, the elements (objects) composing the base domain at our chosen level of organization are living entities, usually defined *organisms,* while those of the target domain are *artefacts,* as defined before. According to Gentner (1983, p. 156) "Objects may be clear entities (for example, 'rabbits'), components parts of a larger object (for example, rabbit's ear) or even combinations of smaller units, for example, 'herd of rabbits'); the important point is that they function as whole at certain level of organization".

Our chosen level of organization at the biological level is that of the organism, which is generally considered the preferred object of selection in biological evolution (Mayr 2001).

Attributes and relations

The distinction between attributes and relations is important: "*Attributes* are predicates taking one argument and *relations* are predicates taking two or more arguments. For example COLLIDE (x,y) is a relation, while LARGE (X) is an attribute" (Gentner 1983, p. 157).

In setting up a complex analogy between the biological and technological domain, therefore, the fact that the attributes of an artefact (say, a turbojet engine) are ostensibly different from an organism (say, a bacterium) is not a fatal critique to the viability of the analogy. What is important is that a number of relations can be coherently mapped across the domains. Indeed, recalling the concepts presented in the previous sections of this chapter we can easily recognize the resulting predicates as relations rather than attributes: artefacts *are made of* modules; organisms *are made of* modules; each module (organic, biological) is *separated from* other modules by a boundary; each module *is associated to* a Behaviour (a subset of which is called *Function*).

Evolution and systematicity

Evolution is made possible by two[15] fundamental processes: *variation* and *selection* (Bell 2008). Variation provides the super-production of diverse evolving phenotypic objects[16] to the evolutionary processes. Selection is the key mechanism through which only a limited number of the evolving objects are given the resources for further replication.

Variation and *selection* (and the third process '*retention*', through which organisms – and artefacts – are replicated and transmitted to the next generation) are the epistemological foundation of a generalized evolutionary system (Bickhard and Campbell 2003). What we propose in this chapter is founded on the idea that technological change is analogous to biological evolution, in that both domains are generalized evolutionary processes (Arthur 2009; Jacobs 2000; Bickhard and Campbell 2003). Gentner's theory of complex analogy supports the legitimacy of this analogy across two different complex systems.

In order to complete the verification of the theory we discuss the *sistematicity principle*. The *sistematicity principle* states that not all the relations that can be selected in the base and target domain are equally significant. According to Gentner, a complex analogy should be based on "a system of connected knowledge, not a mere assortment of independent facts. Such a system can be represented by an interconnected predicate structure in which higher-order predicates enforce connections among lower-order predicates" (Gentner 1983, p. 162).

The *high-order predicates* required are therefore high-level relations that hold and maintain their meaning in both systems. It seems natural to associate them to the foundations of generalized evolutionary systems. We think that the Campbellian epistemological approach we build upon provides a strong platform as a requisite 'system of connected knowledge'. This approach allows for overcoming a critical issue in applying the sistematicity principle, which concerns the difficulty of defining ex ante what 'connected knowledge' may mean.

Indeed, as a consequence of our previous discussions, it becomes straightforward to describe the fundamental processes of *variation* and *selection* in generalized terms maintaining their meaning in both the biological (base) and the technological (or target) system.

Variation. By definition, each functional module is associated to a *Behaviour*, which can be seen as a discrete set of effects: all the possible actions, processes and operations that the module can perform given the current state of its development. Function is a subset of Behaviour. Due to adaptation and to other evolutionary processes (for example, drift, self-organization), behaviour changes over time, gradually generating variations that spread across the population, so that there are objects (functional modules) characterized by different behaviours (Figure 22.2).

Selection. When certain boundary conditions are fulfilled exaptation takes place. The

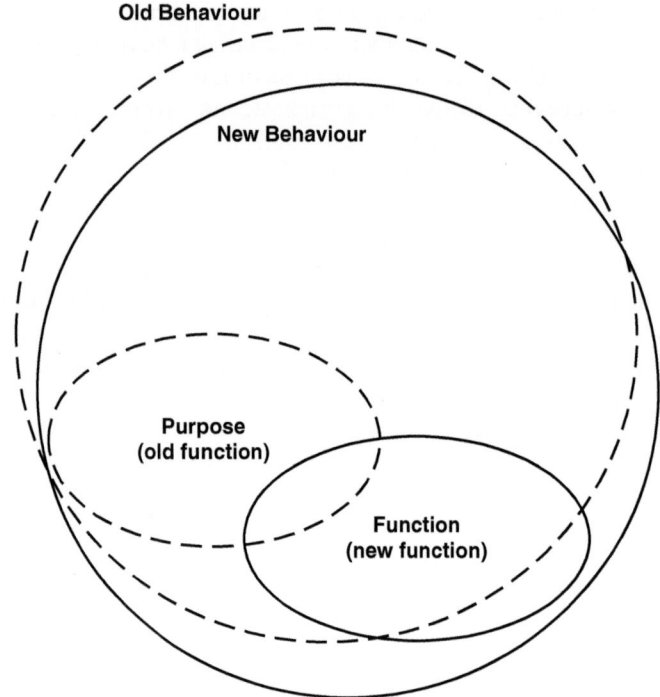

Note: Exaptation as modular functional shift, from an existing function P (that could coincide with the purpose of the) to a new function F. In general the new function may overlap the old function (as shown in figure). A Variation of Behaviour (in figure, from Old Behavior to New Behaviour) may enable exaptation.

Figure 22.2   Exaptation as modular functional shift

event is simple: a subset of Behaviour, different from the previous Function (Purpose), is selected as the new function (Figure 22.2). The novel function is selected either because it gives a competitive advantage to the artefact in its present niche, or because it gives it the opportunity of colonizing or creating a new niche. In the most radical cases, exaptation gives origin to a new technological lineage, thus marked by a bifurcation in the technological trajectory, that gradually drifts away from the original lineage.

As we can see, exaptation in an emergent phenomenon that is analogous in both the biological and the technological system, because they share the coherent set of relations that are the foundation of a generalized evolutionary system. Identifying the systematicity principle as the fundamental cognitive structure of a generalized evolutionary system completes the application of Gentner's 'complex analogy' theory thereby confirming the legitimacy of analogy across two different complex systems. It also contributes to ascertaining under what conditions transferability across complex domains is legitimate.

Empirical support
While not required by the original theory of *structure mapping*, the necessity of empirical validation of the *complex analogy* has emerged in successive research. Modular exapta-

tion describes at a detailed, artefact-specific level how the process of innovation driven by the emergence of functional shift takes place. Since the process includes morphological changes, it is possible to identify in the historical records the events marking the exaptive phenomena, finding actual instantiations of the stylized process that can described as follows:

- Competition and other social and technical phenomena incessantly expand the set of available functions, by changing the way in which the intentional *function* (*purpose*) of each module is obtained and by generating new intentional functions. New materials are developed, production systems improved, and ultimately the *behaviour* of each functional module changes over time. While *intentionally* improving the purpose, this restless activity *blindly* changes the behaviour of each functional module (*variation*), generating a growing set (in evolutionary terms a *population*) of diverse functional modules.
- In other words, competition changes the functional modules' 'adjacent possible' (Kauffman 2000), that is, the set of new functions that become theoretically achievable given the state of development of the modules. Since this happens in each module and at each modular level, the possibilities of functional shift and successful recombination are consequently multiplied. Exaptation is therefore a process in which emergence of new functions is reinforced by a combinatorial dynamics, whose space is expanded by modularity. Consequently, we posit that the emergence concept of *modular exaptation* is central for a complete understanding of technological change.

By grounding modular exaptation into the empirical historical record we also show how the process can be classified into different evolutionary phenomena (internal exaptation, external exaptation, radical exaptation). The two-way exchange of knowledge supported by the complex analogy is confirmed when we notice that in the technological domain each functional module can be transferred to a different industry (across different technological lineages), and the analogous process (Horizontal Gene Transfer) is documented in biological evolution.[17]

22.5   CONCLUSIONS: THE POWER OF ANALOGY

In this final section we summarize the contribution and discuss its methodological value. The application of the complex analogy framework to the problem of exaptation is the general methodological contribution to the study of complex systems we propose in this chapter. The opportunity of rethinking the biological-technological analogy stems from recent discoveries in evolutionary biology, and in particular in bacterial evolution. The recognition of the functional module as "a critical level of biological organization" (Hartwell et al., 1999, C47–C48), suggests the possibility of reframing the artefact-organism analogy already proposed by several scholars (for example, Basalla 1988), explicitly including the modular nature of the artefacts. However, we explicitly observe that the viability of the concept of 'functional module' in the technological domain is disputed, hence potentially disrupting the foundation of our analogy. This is due to the fact that technological function is tightly, though not exclusively, coupled with human intentionality, which instead plays a limited role in biological evolutionary systems.

The first contribution of this chapter is, therefore, to indicate the tenability of etiological concept of biological function ('proper function') in the technological domain. A consequence of this deceptively subtle idea is that technological function becomes intrinsically (by definition) associated to selection (an evolutionary concept) rather than to intentionality. Framed this way (this implies an extended definition of artefact), the biological concept of functional module is therefore a valid analogue of the technological one.

This is the gateway towards a formal *complex analogy* between the biological and the technological domain, our second contribution. Following this thread, we formalized the analogy between the two domains evidencing their common *structure mapping*. We do this by showing that *relations*, rather than *attributes*, are shared by the biological and technological domain, and that the epistemological foundations of generalized evolutionary systems (based on variation, selection, retention[18]) can fulfil the requirement of systematicity.

Finally, we outlined a stylized process describing in evolutionary terms how technological innovation can have an exaptive origin. Many historically documented innovations seem to confirm this framework.

The *complex analogy* we outline can in fact be considered as a tool for studying complex systems, taking advantage of the foundational property of the complex analogy: the core idea of analogical reasoning is that two *domains* can be described, at a certain level of abstraction, by the same *relational structure*. The knowledge we already possess about one of the domains (the *base* domain) can therefore be useful in understanding the second one (the *target* domain). The flow of concepts between the biological and the technological domain can therefore be reliably (even if carefully) explored, considering the knowledge accumulated in the base domain for usage in the target domain – and vice-versa.

We elucidate the usefulness of the complex analogy by showing how the concept of functional module based on the etiological vision of function naturally led to the novel concept of *modular exaptation*. Modular exaptation provides a potentially useful concept based on analogy. Even more, the concept suggests a different way of thinking about innovation, in which concepts drawn from evolution coexist with these developed in innovation science.

NOTES

1. *'Difficulties on theory'* is the title of Chapter VI of Origin (Darwin 1859), while 'Organs of extreme complexity and perfection' is the paragraph in which the swimbladder case is presented.
2. "Natural Selection, simply and by itself, is potent to explain the maintenance or the further extension and development of favourable variations, which are at once sufficiently considerable to be useful from the first to the individual possessing them. But Natural Selection utterly fails to account for the conservation and development of the minute and rudimentary beginnings, the slight and infinitesimal commencements of structures, however useful those structures may afterwards become" (Mivart, 1871, p. 24).
3. The Buttefly Effect indicates a particular dynamic in which a system becomes extremely sensitive to initial conditions. The sensitivity may cause the onset of a chaotic behaviour whereby tiny inconsequential factors trigger an explosive divergence from the previous state of the system.
4. "The numerous uses of the mechanical forces, which I had an opportunity of seeing during my presence in

Upper Silesia, particularly the various air pipes used in the blast apparatus of the high and low furnaces which always led me back to the basic idea of executing an improvement on these instruments, I believe I could use to reach my goal and therefore sought the company of the keepers of the machines and other experts in order to comprehend the closing and opening of the wind pipes, whilst I started out with the idea of which way the air must pass through the tubes of the instrument, to lengthen or shorten according to certain dimensions, in order to make up the missing notes of the compass" (Heyde 1978, p. 21).

5. Agues stands for malaria. Malaria was at the time found all over Europe including the South of England and the Netherlands.
6. The Peruvian bark was so well known as the only treatment and prophylaxis against malaria, that the author doesn't find necessary to use its proper name, the Chinchona tree from which bark quinine was extracted.
7. Cattani prefers the term preadaptation. We consider preadaptation and exaptation synonyms.
8. Levinthal (1998) does not use the term exaptation, but his definition of speciation depends on the cooption of technologies mechanism of exaptation described above.
9. Several examples of effectual entrepreneurship can be found in (Read et al. 2010).
10. A module is composed of other modules, at each level of the modular tree. The hierarchical decomposition of a module is therefore recursive.
11. Horizontal Genetic Trasfer (HGT) is the transfer of genetic materials codifying functional modules across different lineages. It is recognized in biological evolution as an important evolutionary mechanism, especially in bacterial evolution (Hartwell et al. 1999). By analogy, in this chapter we define horizontal transfer the transfer of technological functional modules across different technological lineages.
12. This case is inconsequential in our selectionist approach because it is unrelated to the selection of the chair.
13. In principle, one or more function can be associated to each module. The (functional) modularity of a system is not a discrete attribute (modular/non-modular). In a technological system (artefact) there may be different degrees of modularity between a perfectly modular system in which functions are mapped to modules one-to-one (each module performs one function) and a perfectly integral system in which all modules perform all functions (Ulrich 1995). In a biological system (microorganism) the situation could be more complex, because a functional module "may belong to different systems at different times" (Hartwell et al. 1999, p.C48). The key point, however, is that in microbiology: "the notion of a module is useful only if it involves a small fraction of the cell components in accomplishing a relatively autonomous function"(Hartwell et al. 1999, p.C48). Hartwell's note reinforces the analogy between the biological and the technological concept of function, in that that technological functions are usually autonomous.
14. We borrow the concept of behaviour from John Gero's FBS (Function Behavior Structure) conceptual framework developed for engineering design (Gero 1990).
15. Biologists (and complexity scientists) such as Kauffman (1995, 2000) and Wagner (2011) add a further process, self-organization, to the triad variation-selection-retention.
16. Organisms in the biological domain, artefacts in the biological domain. The term 'phenotypic' refers to the fact that both organisms and artefacts are 'tangible' objects that can interact with the environment.
17. Most organisms living today in fact are *bacteria*, which are composed by a single cell. In fact, most of the discussion about functional modules in biology (for example, Hartwell et al. 1999) is referred to the bacterial world. In the bacterial domain functional modules can be transferred across lineages, the distinctive process called Horizontal gene Transfer, or HGT. This importance of this process is much more limited among multicellular organisms (eukaryotes) where vertical inheritance is dominant.
18. As noted earlier, this account doesn't include self-organization.

REFERENCES

Adner, R. and Levinthal, D.A. 2002. The emergence of emerging technologies. *California Management Review*, 45(1), pp. 50–66.

Andriani, P. and Cohen, J. 2013. From exaptation to radical niche construction in biological and technological complex systems. *Complexity*, 18(5), pp. 7–14.

Andriani, P. and Carignani, G. 2014. Modular exaptation: a missing link in the synthesis of artificial forms. *Research Policy*, 43(8), pp. 1608–1620.

Arthur, W.B. 2009. *The Nature of Technology*. New York: Free Press.

Barve, A. and Wagner, A. 2013. A latent capacity for evolutionary innovation through exaptation in metabolic systems. *Nature*, 500, pp. 203–206 (8 August).

Basalla, G. 1988. *The Evolution of Technology*. Cambridge: Cambridge University Press.

Bell, G., 2008. *Selection: The Mechanism of Evolution*. Oxford: Oxford University Press.

Bickhard, M.H. and Campbell, D.T. 2003. Variations in variation and selection: the ubiquity of the variation-and-selective-retention ratchet in emergent organizational complexity. *Foundations of Science*, 8, pp. 215–282.

Brown, M.J. and Feldman, M.W. 2009. Sociocultural epistasis and cultural exaptation in footbinding, marriage form, and religious practices in early 20th-century Taiwan. *Proceedings of the National Academy of Sciences of the United States of America*, 106(52), pp. 22139–22144.

Buss, D.M. et al. 1998. Adaptations, exaptations, and spandrels. *American Psychologist*, 53(5), pp. 533–548.

Carignani, G. 2016. On the origin of technology: the invention and evolution of the bow and arrow. In F. Panebianco and E. Serrelli (eds) *Understanding Cultural Traits. A Multidisciplinary Perspective on Cultural Diversity*. Berlin: Springer, pp. 315–340.

Cattani, G. 2005. Preadaptation, firm heterogeneity, and technological performance: a study on the evolution of fiber optics, 1970–1995. *Organization Science*, 16(6), pp. 563–580.

Cattani, G. 2006. Technological pre-adaptation, speciation, and emergence of new technologies: how Corning invented and developed fiber optics. *Industrial and Corporate Change*, 15(2), pp. 285–318.

Cuénot, L. 1914. Théorie de la préadaptation. *Scientia*, 16, pp. 60–73.

Darwin, C. 1859. *On the Origin of Species*. London: Murray.

Dew, N. 2009. Serendipity in entrepreneurship. *Organization Studies*, 30(7), pp. 735–753.

Dew, N., Sarasvathy, S.D. and Venkataraman, S. 2004. The economic implications of exaptation. *Journal of Evolutionary Economics*, 14(1), pp. 69–84.

Dew, N. et al. 2008. Outlines of a behavioral theory of the entrepreneurial firm. *Journal of Economic Behavior & Organization*, 66(1), pp. 37–59.

Doolittle, W.F. and Bapteste, E. 2007. Pattern pluralism and the Tree of Life hypothesis. *Proceedings of the National Academy of Sciences of the United States of America*, 104, pp. 2043–2049.

Dosi, G. 1997. Opportunities, incentives and the collective patterns of technological change. *The Economic Journal*, 107(444), pp. 1530–1547.

Fagerberg, J. 2003. Schumpeter and the revival of evolutionary economics: an appraisal of the literature. *Journal of Evolutionary Economics*, 13(2), pp. 125–160.

Fitch, W.T. 2005. The evolution of language: a comparative review. *Biology and Philosophy*, 20, pp. 193–230.

Ganfornina, M.D. and Sanchez, D. 1999. Generation of evolutionary novelty by functional shift. *BioEssays*, 21(5), pp. 432–439.

Garud, R. and Nayyar, P.R. 1993. Transformative capacity: continual structuring by intertemporal technology transfer. *Strategic Management Journal*, 15, pp. 365–385.

Gavetti, G., Levinthal, D.A. and Rivkin, J.W. 2005. Strategy making in novel and complex worlds: the power of analogy. *Strategic Management Journal*, 26(8), pp. 691–712.

Gentner, D., 1983. Structure-mapping: a theoretical framework for analogy. *Cognitive Science*, 7(2), pp. 155–170.

Gentner, D. and Jeziorski, M. 1993. The shift from metaphor to analogy in Western science. In A. Ortony (ed.) *Metaphor and Thought* (2nd edition). Cambridge UK: Cambridge University Press, pp. 447–480.

Gentner, D. and Markman, A. 1997. Structure mapping in analogy and similarity. *American Psychologist*, 52(1), pp. 45–56.

Gero, J. 1990. Design prototypes: a knowledge representation schema for design. *AI Magazine*, 11(4), pp. 26–36.

Gibson, J.J. 1950. *The Perception of the Visual World*. Boston: Houghton Mifflin.

Giffard, H.S. 2016. *Making Jet Engines*. Chicago, IL: The University of Chicago Press.

Gould, J.S. 1985. Not necessarily a wing. *Natural History*, 94, pp. 12–25.

Gould, S.J. 1991. Exaptation: a crucial tool for evolutionary psychology. *Journal of Social Issues*, 47(3), pp. 43–65.

Gould, S. and Vrba, E. 1982. Exaptation – a missing term in the science of form. *Paleobiology*, 8, pp. 4–15.

Grandori, A. 2007. Discovery in natural selection and knowledge processes: a commentary on an agent-based model of exaptive processes. *European Management Review*, 4, pp. 153–156.

Hargadon, A. 2003. *How Breakthroughs Happen*. Cambridge: MA: Harvard Business Press.

Hartwell, L.H. et al. 1999. From molecular to modular cell biology. *Nature*, 402(6761 Suppl), pp. C47–C52.

Heyde, H. 1978. On the early history of valves and valve instruments in Germany (1814–1833), *Brass Bulletin*, 24, pp. 9–33.

Hilpinen, R., Artifact E., and Zalta, N. (eds) 2011. *The Stanford Encyclopedia of Philosophy*. https://plato.stanford.edu/archives/win2011/entries/artifact.

Hofstadter, D.R.. 2001. Analogy as the core of cognition. In D. Gentner, K.J. Holyoak and B.N. Kokinov (eds) Boston MA: MIT Press, pp. 499–538.

Jacobs, 1969. *The Economies of Cities*. New York: Random House.

Jacobs, J. 1985. *Cities and the Wealth of Nations*. New York: Random House.

Jacobs, J. 2000. *The Nature of Economies*. New York: Random House.

Jeffreys, D. 2008. *Aspirin: The Remarkable Story of a Wonder Drug*. New York: Bloomsbury Publishing USA.

Kauffman, S. 1995. *At Home in the Universe*. Oxford: Oxford University Press.

Kauffman, S. 2000. *Investigations*. Oxford: Oxford University Press.

Knight, F.H. 1921. *Risk, Uncertainty and Profit*. Boston. MA: Hart, Schaffner & Marx; Houghton Mifflin Co.

Krohs, U. and Kroes, P. 2009. Philosophical perspectives on organismic and artifactual functions. In U. Krohs and P. Kroes (eds) *Functions in Biological and Artificial Worlds: Comparative Philosophical Perspectives*. Cambridge: MA: MIT Press, pp. 233–245.

Lane, D.A. 2011. Complexity and innovation dynamics. In C. Antonelli (ed.) *Handbook on the Economic Complexity of Technological Change*. Cheltenham, UK and Northampton, MA, USA: Edward Elgar Publishing, pp. 63–80.

Lass, R. 1990. How to do things with junk: exaptation in language evolution. *Journal of Linguistics*, 261, pp. 79–102.

Levinthal, D. 1998. The slow pace of rapid technological change: gradualism and punctuation in technological change. *Industrial and Corporate Change*, 7(2), pp. 217–247.

Linde-Medina, M. 2011. Adaptation or exaptation? The case of the human hand. *Journal of Biosciences*, 36(4), pp. 575–585.

Lombard, M. and Phillipson, L. 2010. Indications of bow and stone-tipped arrow use 64,000 years ago in KwaZulu-Natal, South Africa. *Antiquity*, 84(325), pp. 635–648.

Lorenz, K.Z. 1974. Analogy as a source of knowledge. *Science*, 185(147), pp. 229–234.

March, J. 1976. The technology of foolishness. In J.P. March and J.G. Olsen (eds) *Ambiguity and Choice in Organizations*. Oxford: Oxford University Press. pp. 69–81.

Mayr, E. 2001. *What Evolution Is*. New York: Basic Books.

Merton, R.K. 2004. *The Travels and Adventures of Serendipity: a Study in Sociological Semantics and the Sociology of Science*. Princeton: NJ: Princeton University Press.

Millikan, R.G. 1999. Wings, spoons, pills, and quills: a pluralist theory of function. *Journal of Philosophy*, 96, pp. 192–206.

Mivart, G. 1871. *On the Genesis of Species*. London and New York: MacMillian & Co.

Mokyr, J. 1990. *The Lever of Riches. Technological Creativity and Economic Progress*. New York: Oxford University Press.

Mokyr, J. 2000. Knowledge, technology, and economic growth during the Industrial Revolution. In B. van Ark, S.K. Kuipers and G.H. Kuper (eds) *Productivity, Technology, and Economic Growth*. Boston: MA: Kluwer Academic Press, pp. 253–292.

Mokyr, J. 2002. *The Gifts of Athena: Historical Origins of the Knowledge Economy*. Princeton: NJ: Princeton University Press.

Neander, K. 1991. Function as selected effects: the conceptual analyst's defense. *Philosphy of Science*, 58, pp. 168–184.

Nelson, R.R. and Winter, S.G. 1982. *An Evolutionary Theory of Economic Change*. Cambridge: MA: Harvard University Press.

Osepchuk, J.M. 1984. A history of microwave heating applications. *Microwave Theory and Techniques, IEEE Transactions*, 32(9), pp. 1200–1224.

Penrose, E. 1952. Biological analogies in the theory of the firm. *The American Economic Review*, pp. 804–819.

Preston, 2009. Biological and cultural proper functions in comparative perspective. In U. Krohs and P. Kroes (eds) *Functions in Biological and Artificial Worlds: Comparative Philosophical Perspectives*. Cambridge: MA: MIT Press, pp. 37–50.

Prigogine, I. and Stengers, I. 1997. *The End of Certainty : Time, Chaos, and the New Laws of Nature*. New York: Free Press.

Read, S. et al. 2010. *Effectual Entrepreneurship*. Oxford: Taylor & Francis.

Schumpeter, J.A. 1934. *The Theory of Economic Development*. Cambridge: MA: Harvard University Press.

Simon, H. 1962. The architecture of complexity. *Proceeding of the American Philosophical Society*, 106, pp. 467–482.

Tappi, D. 2005. Clusters, adaptation and extroversion a cognitive and entrepreneurial analysis of the Marche Music Cluster. *European Urban and Regional Studies*, 12, pp. 289–307.

Tuomi, I. 2002. *Networks of Innovation: Change and Meaning in the Age of the Internet*. Oxford: Oxford University Press.

Ulrich, K. 1995. The role of product architecture in the manufacturing firm. *Research Policy*, 24(3): 419–440.

Utterback, J.M. 1996. *Mastering the Dynamics of Innovation*. Cambridge: MA: Harvard Business School Press.

Vermaas, P.E. and Houkes, W. 2003. Ascribing functions to technical artefacts: A challenge to etiological accounts of functions. *The British Journal for the Philosophy of Science*, 54(2), pp. 261–289.

Vermaas, P.E. and Houkes, W. 2006. Technical functions: a drawbridge between the intentional and structural natures of technical artefacts. *Studies In History and Philosophy of Science Part A*, 37(1), pp. 5–18.

Vincenti, W.G. 2000. Real-world variation-selection in the evolution of technological form: historical examples. In J. Ziman (ed.) *Technological Innovation as an Evolutionary Process*. Cambridge: Cambridge University Press, pp. 174–189.

Wagner, A. 2011. *The Origins of Evolutionary Innovations: A Theory of Transformative Change in Living Systems.* Oxford: Oxford University Press.

Wagner, A. and Rosen, W. 2014. Spaces of the possible: universal Darwinism and the wall between technological and biological innovation. *Journal of the Royal Society, Interface/the Royal Society*, 11(97), 2013.1190.

De Winter, J. 2010. *A Pragmatic Account of Functions*, Ghent University.http://logica.ugent.be/centrum/pre prints/De_Winter_-_Functional_Explanation__DRAFT_24_.pdf (accessed 12 April 2014).

Zhu, L., Zhu, Y., Zhang, Y. and Li, Y. 2012. Engineering the robustness of industrial microbes through synthetic biology. *Trends in Microbiology*, 20, 94–101.

Ziman, J. 2000. *Evolutionary Models for Technological Change*. Cambridge: Cambridge University Press.

23. Emergence and radical novelty: from theory to methods

Professor Jeffrey A. Goldstein, Adelphi University

> We are suffering today from the lack of suitable images of the phenomena that are currently
> receiving our most ardent scientific attention. . . (Suzanne Langer)

23.1 INTRODUCTION

Over the past several decades, emergence and emergent phenomena have become cornerstones in the science of complex systems. Despite the host of attention devoted to them, however, their conceptualization has retained not a small measure of ambiguity and elusiveness. This problematic status cannot but have an adverse effect on the methods and tools devised for studying them. This pertains to one of the generally accepted working principles of research which, according to the late well-regarded philosopher of social science Abraham Kaplan (1964, Chapter 18), has it that

> [T]he experimental method is inseparable from the development and application of hypotheses and theories. For these are needed to tell us when we can generalize from the experimental situation, what controls must be imposed, what corrections are afterwards to be made. Without such conceptual guides the [the methods and tools] remain haphazard or wholly problematic in significance ('This proves something, but I don't know what!').

In other words, right and appropriate methods to conduct research necessitate right and appropriate 'conceptual guides' in the form of correct theories that do not mislead or misconstrue what they are theories about. Likewise, flawed methods will come from flawed theories and these flawed methods will in turn feed back onto the flawed theory rendering it that much more questionable.

This kind of vicious circle can be found in those approaches to emergence involving misleading conceptual models that spawn misleading methods which in turn reinforce the problematic biases of the models (please note we are using the terms 'approaches', 'models', 'perspectives', 'bases', 'frameworks', and 'perspectives' interchangeably for the express purpose of this chapter, whereas a much more careful usage of terminology would obviously be required for a more philosophical chapter).

23.2 MODELS OF EMERGENCE

In complexity science, the predominant 'conceptual guide' to emergence is self-organization as it has been interpreted according to three interrelated bases (the term 'self-organization' has become so intertwined with the term 'emergence' that today they are just about synonymous). The first basis comes from an amalgam of constructs and methods developed in both the far-from-equilibrium platform established by the Nobel Laureate Ilya Prigogine (Glansdorf and Prigogine 1971) and in the closely related framework developed in the research program of synergetics established by the preeminent German physicist Hermann Haken (2011). The second basis comes out of the cybernetics approach going back to around the time of the Second World War. This basis of self-organization has incorporated elements of information theory, control and guided systems, inquiry into homeostasis, and related themes. The third basis is that heritage of the idea of self-organization understood as circularly causal, adaptive self-regulation in biological organisms, a framework going back two centuries to the idealist, Romanticist orientations of Immanuel Kant, Friedrich Schelling, and other *naturphilosophischen*. Most of complexity research and theorizing concerning emergence relies on some mixture of these three bases.

The aim of this chapter is to examine in what ways the self-organization model (and related models) of emergence is misleading as to an accurate account of emergence as well as how it has led to misleading methods of study. To do so, we need to first draw-out the assumptions underlying the self-organization models as well as the methods that stem from it. The second is to provide correctives to these presuppositions.

Of course, there are other models of emergence besides self-organization including, for example, the nonlinear dynamical systems (*nds*) approach which seeks to mathematically understand processes of emergence along the lines of the unfolding of nonlinear dynamical systems. The interactions among the factors/variables are usually represented by nonlinear differential or difference equations operated on by control parameters that can lead to bifurcation through different regimes of organization known as 'attractors'. To the extent that self-organizing processes are accessible and then amenable to a dynamical systems approach, such processes can be a source of mathematical insight into emergence.

Another model of emergence is the quite recent appropriation of emergence on the part of particle physicists and cosmologists in their utilization of such expressions as *emergent* spacetime, *emergent* space, *emergent* time, *emergent* gravity, *emergent* continuity, *emergent* discreteness and so on (for example, revealed by a search of 'emergent' in the physics preprint database *ArXiv*). The reason for adding the prefix 'emergent' is a significant step in a strategy aiming at a grand unification of all forces or interactions in an originary, extremely high energy condition that is so beyond our normal (as well as classical) notions of space and time that the latter are relegated to the 'derived', 'secondary', even 'epiphenomenal' status of being 'emergent'. However, this model is quite vague as to what it is that is actually going on that justifies the label 'emergent' and the consequence that this 'emergent' something emerges from some ultimate unification. Moreover, this approach to emergent phenomena comes up with a scheme of disvalorization of emergence mostly opposite to the customary models of emergence.

Whatever the merits of such labeling with "emergent", what emergence comes down to in this approach can be considered a type of *brute emergence* which, as the philosopher Galen Strawson (Strawson et al. 2006) who coined this expression indicates, emergence is really no emergence at all but instead is 'a miracle every time it happens' since it is assumed to just happen spontaneously. By pointing out the shortcomings of this trend in particle physics, we will also be indirectly critiquing all the other 'logics' of emergence which veer near to the claim of spontaneity, including that frequently associated with self-organization.

There are other models, for example, self-organized criticality, evolutionary and adaptive models, computational models and so forth. They each add interesting twists and insights about emergence but self-organization has been around longer, is much more ubiquitous, includes aspects of these other models, possesses attributions and scope more coincident with emergence than the others. Hence, we will take self-organization as our 'test case' because of its predominance and because the others somehow or other allude to it, either in an outright and explicit sense as in the nonlinear dynamical systems approach or in simply assuming some of the key ingredients of the self-organization model such as: a differentiation into two levels: a 'lower' level of parts interacting with one another and another level, the 'higher' level of the macro-or global patterns plus a sense that one level is more basic and antecedent and the other derived or secondary.

Explicit comparisons/contrasts of self-organization with emergence have been entertained previously (for example, see Adiscott 2011; de Wolf and Holvoet 2004; Halley and Winkler 2008). These comparisons, however, lean toward vague generalizations and thus have not proved themselves of much worth regarding the project of this chapter.

Getting back to the case of the self-organization model, all it takes is a little rumination to recognize that this model contains a quandary about the novelty of the higher level resulting from self-organizational processes, namely, is it sufficient for the *radical* novelty demanded for genuine emergent phenomena? For the fact is that the core of emergence is the explanatory gap between the lower and higher levels and the core of this explanatory gap is radical novelty, and the core of radical novelty is the uncomputability of the higher from knowledge of the lower level alone. 'Radical novelty' captures in one phrase all of the defining claims about emergent phenomena, that is, that they are unpredictable, nondeductible, irreducible ('non-traceable' in earlier accounts), and most recently uncomputable (see Goldstein 2014).

To be sure, the idea of self-organization has benefited complex systems theory in many ways, the most valuable being its revisionary idea that system change (its reorganization) can be inner-driven (the 'self') and take place by means of internal resources (the claim of spontaneity) and not just through hierarchical command and control or external imposition. Yet, nowhere in the model of self-organizations' interactions of lower level components does there appear to be a power to generate radically novel outcomes.

23.3   THE EXPLANATORY GAP AND RADICAL NOVELTY

An initial definition of emergence can help in coming to grips with the kind of novelty that defines emergence as well as the generating potency of emergence processes that are needed to generate this kind of novelty. Such a definition would need to head what no less an authority in emergentist circles as the complexity pioneer John Holland (1998, p. 3) once stated, "It is unlikely that a topic as complicated as emergence will submit to a concise definition". Moreover, to avoid the prejudgments underlying self-organization and other models, it is necessary to base a definition on observation and neutral description as much as possible. Below is a first stab at a brief definition of emergence, after which its basis in observation and neutral description will be laid-out:

Processes in complex systems which contain a capacity for the self-transcending construction of higher/macro level, uncomputably novel, and coherent phenomena out of more particulate, lower/ micro level- substrates and their dynamics.

The first thing to notice is that emergence is something that happens in *complex* systems. Complex systems are nonlinear, contain a large amount of lower level components, the components are diverse in nature, are subject to scaling/power laws, and so forth. And, one of the defining signs of a complex system lies precisely in its capacity for emergence.

Second, there is an emphasis on a 'capacity' and not a proclivity or impetus. There is no contention that emergence results from some teleological drive no matter how sophisticated the discourse which might mask such a contention. Certainly, a capacity is something that can be said to pre-exist as a potentiality in the emergent system in the same sense that all literate human beings have a capacity to learn Portuguese but none possess a drive or final causative goal to learn it. Capacity is a potential but not in the sense of a potential seeking actualization as in Aristotle's famous notion of change. That is, the capacity for emergence doesn't possess an inclination or impetus for emergence as in the proposal for a drive towards greater evolvability (as entertained by Kauffman 1995). Nor does the capacity for emergence signify a push for some sort of growth spurt in 'complexity'. Since there is no other term in complexity theory more prone to vagueness than 'complexity', we shouldn't expect that a drive towards an increase in complexity is behind the capacity for emergence. Hence, capacity as such doesn't connote anything more than the possibility of something happening when the conditions are appropriate. This also means that research into emergence need not concern itself with methods trying to ascertain when the conditions are right for self-organization which is what research into self-organization would include. As long as the system is complex (nonlinear, and so on), the capacity for emergence is present. Therefore, time is more appropriately spent devising methods that probe the radical novelty of emergent phenomena; methods we will return to later in the chapter.

As products of a capacity existing in complex systems, emergent phenomena are neither rare nor everywhere but appear when internal and external conditions are right. This capacity of complex systems for emergence has existed in countless complex systems comprised by countless different kinds of *media*. 'Media' is a term expressing the constitutive 'material' in which emergence takes place. Thus, there is the media of biochemical reactions in which biochemical emergence takes place, or the excitable media in which scroll-ring and similar emergent phenomena emerge, or the computational media of computational emergence, and so on. Recognizing the type of media doesn't explain emergence yet awareness of the media involved avoids having to make an exclusive commitment to physicalism, non-reductive physicalism, materialism, and so forth. Trying to explain emergent phenomena by assuming there is only one absolute media, like physical matter, cannot do justice to emergent situations where physicality is just not relevant, for example, social emergence. Cognizance of the media in which emergence is occurring helps dissolve the aura of mysteriousness.

Regarding the definition above, next we find the phrase 'self-transcending construction' (or '*stc*') which is decidedly not introduced as some sort of highfalutin abstract or even esoteric process, a connotation which the word 'transcend' may bring to mind. Rather, etymologically, 'transcend' simply means to surpass, outdo, surmount, or move beyond. Here, 'self-transcending' refers merely to the observed shift from micro-component interaction to macro-level patterns or structure. Hence, the word 'self-' in *stc* is merely a reference to the system's substrates or earlier conditions before emergence ensues. This implies that 'self' in this context has nothing to do with a supposed locus of an agency for organizing or self-regulating a system as the model of self-organization would have it, nor does it entail any kind of putative inner-driven impetus which self-organization connote. With respect to emergence, 'self-transcending construction' is just a more accurate descriptive phrase for what is actually observed: novel patterns/entities/dynamics arise out of and at the same time transcend the system's antecedent status. What is observed in emergence is simply the noticeable and 'radical' change from micro to macro levels and thus any concept of a locus for a 'mechanism' referred to by self-organization one is not relevant.

The 'construction' of 'self-transcending construction' refers simply to the building-up of new order, structures, patterns, and so forth, all of these novelties bringing with them novel properties. In addition by speaking in terms of 'construction' the idea of an *stc* does not carry with it the sense of spontaneity associated with the self-organizational model of emergence. To be sure, the construction of emergent phenomena can be sudden or gradual, the only way to tell being observation of specific cases.

It must also be noted that *stc* also does not signify the bringing about of some kind of momentous event which the term 'transcending' might suggest. Early emergentists did in fact tend to only consider momentous events as emergent phenomena, for example, the origin of life, or mind, or speciation, and so on. However, such events are rare phenomena and as was stated above emergence is neither rare nor everywhere, sometimes momentous but mostly everyday occurrences, much of which take place unnoticed.

Self-transcending constructions bring about radical novelty (for an insightful exposition on novelty in general see North 2013), the term 'radical' meant to qualify the unique kind of novelty which emergent phenomena possess and *not* the kind of emergence as in the phrase 'radical emergence' (Silberstein and McGeever 1999). In the account of emergence being developed in this chapter, there is no radical or unradical type of emergence, there is only emergence or there is no emergence. The same holds for claims of 'weak' versus 'strong' emergence, or 'epistemological' versus 'ontological' emergence (see, for example, Stephan 1999). This chapter takes the position that emergence is a genuine phenomenon which expresses both epistemological and ontological features for if it was one or the other, it would not satisfy the criterion of being real.

To be sure, some commentators on novelty (for example, North 2013) talk about novelty as being relative or absolute. The notion of *relative* novelty can be understood mathematically through the notion of an equivalence class. For example, in the operation of congruency, one number is defined as equivalent to another according to whether it has the same remainder after being divided by some arbitrary number (the modulo operation). Thus, if the operation is defined as division of an integer by 4, then an equivalence class would consist of those multiples of 4 that when divided by 4 leave a remainder of 0, for example, the class made-up of $\{\ldots -8, -4, 0, 4, 8 \ldots\}$. Clearly, 4 and -8 are novel with respect to each other, yet in this specific example of congruence they are equivalent relative to the operation being performed. This implies that relative novelty is a feature that depends on previously established schemes demarcating what is to count as novel.

Another way to comprehend relative novelty can be found in the analysis of emergent novelty offered some eighty years ago by the distinguished British philosopher W.T. Stace (for an account of Stace's ideas on emergent novelty, see Goldstein 2006). Stace presented a compelling argument that the only meaning of novelty which emergentists could hold to be intelligible was *unexpectedness*: that is, because surprise accompanies all novelty, it might be thought that anything surprising had to be, at the same time, novel. Stace, however, pointed out that the sense of unexpectedness was an arbitrary assignment dependent on the subjective interests of the recognizers. What is unexpected is novel. But this is a rather spare kind of novelty since even a random event would possess the novelty of the unexpected. What is needed is some way to get beyond the attribute of being unexpected.

There currently exist mathematical formalisms that can be used to represent/define how radical novelty differs from ordinary novelty. One, developed by Richard Boyle (2016) in his study of what he calls 'natural novelty' in emergence, is founded on a kind of functional analysis. One of the advantages of Boyle's work for our purposes is that his formalism intimates how substrates as particular functions can be seen as transforming each other's specific functional formalism during processes of emergence (later on we will go over the critical importance of a specific meaning of transformation for radical novelty generation). However, at its nascent stage, Boyle's formalism is rather clunky and there is no transparent way to directly apply it to instances of emergent novelty.

Another formalism uses the mathematical field of category theory to represent the type of deep rooted change needed for a viable theory of emergence. This is the remarkably insightful approach of Ehresmann and Vanderbreesch (2007) which will be discussed shortly. Another is the mathematico-logical formalism of Goldstein (2002, 2014) which borrows from the work of Cantor, Gödel, Turing, and others in their specific lineage. This formalism is termed 'self-transcending constructions' (hence, the use of this phrase in the definition of emergence given above) for what is essentially a rather prosaic but observationally more accurate perspective despite the arcane connotations of the terminology. This formalism will be described below in greater detail since it is important in coming to an understanding of emergence and the requirement of radical novelty and its relation to certain facets of creative process in engendering radically novel outcomes.

To better understand 'radical' novelty we also need to appreciate the role of the *explanatory gap* as a defining characteristic of emergence since its inception. The explanatory gap is the breach in the course of explanatory connectivity linking substrate to outcomes. Whereas in most other cases of inquiry, explanation consists of the narrowing of any gap of understanding as much as possible, emergence is only emergence when the explanatory gap between substrate and emergent phenomena is not completely bridgeable. This is not a warrant for holding there is a causal rupture in the case of emergence. Rather, it is to claim that any operations, processes or constraints generating emergent phenomena will possess enough potency for *sui generis* novelty to be generated.

Because all identifications of novelty are relative in this sense, how can the radical novelty supposedly characterizing emergent phenomena be defined so as to insure it supports the explanatory gap necessary for defining emergence? If 'absolute' novelty is to have any sense at all, it must have something to do with an 'objective' determination of the degree of novelty attributed to an emergent phenomenon. Such a determination has come along in the form of the construct of uncomputability to which we now turn.

23.4   RADICAL NOVELTY AS UNCOMPUTABILITY

Historically, the radical novelty of emergent phenomena which supports the contention of an explanatory gap was a catch-all term for such properties of emergent phenomena as unpredictability, nondeducibility, irreducibility, and ostensiveness when viewed from the perspective of the lower level substrates. In fact, there does not seem to be any account of emergent phenomena that doesn't touch on one of these attributes in defining emergence. The most recent and most rigorous way of defining radical novelty is that of uncomputability, an idea derived from mathematical logic and computer science which was brought out and made prominent by studies of computational emergence (like artificial life). Interestingly, it was John Stuart Mill (see Blitz 1992) who initiated the idea of emergence in the mid-19th century way before the advent of computers, who offered the first intimation that an inability to compute the outcome of his term for emergence, heteropathic causality, was the way to think about the novelty of emergent phenomena. Accordingly, it was not possible to deduce all the combination of causes since new laws take the place of the old laws. For Mill, heteropathic processes needed to transgress (read as 'transcend') the rule-following of linking cause to effect at the bottom of more traditionally understood causation.

In research into computational emergence (for example, cellular automata, artificial life and similar simulations), the computational modeler Vince Darley (1994) measured the amount of computation needed to predict emergent outcomes using the metric of Kolmogorov/Chaitin/Solomonoff or *KCS* algorithmic information. The KCS of a bit string of data equals the most-compressed program capable of outputting the string. This means the KCS of a random string is as long as the string itself since randomness cannot be compressed, for example, the KCS of a series of values taken by random coin tosses would be as long as the number of separate coin tosses. Darley found that the KCS measure of emergent phenomena observed in computational simulations transcend the measure of information contained in the substrates such as those found in rule tables, initial conditions, and so forth. Darley argues that since emergent systems are those for which even perfect knowledge allows for little predictive information, such systems are uncomputable which translates into the conclusion that watching the system evolve is the *optimal means of prediction*.

Uncomputability is due to two factors according to Darley: an *accumulation of interactions*; and the *undecidability* residing in the universality characterizing such emergent simulations. By the *accumulation of interactions*, Darley means something like the *reductionist nightmare* which the biologist Jack Cohen and the mathematician Ian Stewart (Cohen and Stewart 1995) propose happens when complex system behavior is explained from the bottom-up. The branches represent possible causal pathways which rapidly accumulate upwards as reductionist explanations uncover more and more micro-units. This becomes an explanatory nightmare since the accumulation of causal chains soon occludes which causal pathways or branches are most promising.

Cohen and Stewart's (1995) Existence Theorem for Emergence becomes relevant here (p.438): "This kind of uncomputability occurs because the chain of logic that leads from a given initial configuration to its future state becomes longer and longer the further you look into the future, *and* there are no short cuts". Cohen and Stewart furthered observed, "[emergent phenomena] are not outside the low-level laws of nature; they follow from them in such a complicated manner that we can't see how" (pp.438–439). This in turn implies that, "[emergence] *transcends* its internal details, and there's a kind of scale *transcendence*" (pp.440–441; my emphasis). Notice that here 'transcendence' is a neutral term expressing how higher level emergent outcomes/phenomena consistently exceed whatever knowledge of lower level interactions is attained.

Regarding the second source of the uncomputability of emergence, Darley brings up the closely associated property of *undecidability* which has been demonstrated in the *universality* characterizing computational emergence. The inimitable British mathematician and code breaker, Alan Turing had shown in his famous theorem of 1936 on uncomputable numbers, that undecidability is equivalent to uncomputability (Turing 1937), a fact intimated by Gödel's famous theorems a few years earlier (see Hintikka 2000; Fitzpatrick 1966; Petzold 2009). Without needing to get bogged down in the intricate logic of their proofs, the fact is that both Gödel and Turing relied on a mathematical method going back to the great German mathematician Georg Cantor (Dauben 1979) in the last quarter of the 19th century. Cantor's method was devised in order to produce a radically new outcome with a radically novel property although it was Gödel/Turing who saw in it a way to generate undecidability and uncomputability. It is this method which was called a 'self-transcending construction' (Kaufmann 1978) that is the foundation of the logic for emergence we are developing here and to which we now turn (see Goldstein 2014).

A self-transcending construction, "consists in it being in principle unclosed: [whereas] a determinate constructional principle, however widely conceived will never lead to the goal, but it is only in the course of constructional activity itself that new instructions for continuing the procedure emerge" (the German mathematician Oskar Becker quoted in Kaufmann 1978, p.135). The self-transcending aspect of an *stc* works because "only in the course of constructional activity itself that new instruction . . . emerge". Moreover, an *stc* can continually transcend the self/substrates only because its resources are not already established and determinate. If an *stc* didn't work in a manner that is *in contrast to* an already fixed series, whatever resources it might contain for bringing about novelty would have to be introduced as an axiom, that is, something either discovered as already there but now explicit or simply postulated. This inclusion of 'in contrast to' alludes to the how during the operation of an *stc* previous rules are negated which is essential to novelty since for something to be new, the past from which newness emerges must somehow be negated.

It was this notion of a self-transcending construction that was borrowed from the mathematical context of Cantor since it seemed a particularly accurate description of what occurred in emergence, namely, a constructional process which operates on substrates, transforming them (including negating them), and resulting in something not explicitly contained in the antecedent lower level.

The *stc* that Turing used to demonstrate uncomputable numbers and that Darley then appealed to in his finding about the excess KCS complexity in emergent phenomena works by listing redundant patterns and then methodically negating each redundancy. This kind of logic of following (identifying, listing, recognize redundancy) and then negating these same redundant patterns is what ensures that the outcome will be uncomputable since the negation eliminates what is being followed. This perhaps dense statement can be explicated by appreciating a very similar way of considering emergence's irreducibility that has been expounded by the philosopher Paul Humphries (2008). He has argued that emergent phenomena cannot be reduced to lower level substrates (a crucial component of radical novelty and the explanatory gap) precisely because the lower level substrates no longer exist (have been negated) in that they have been taken-up and subsumed into the resulting emergent phenomena.

23.5   THE LESS-THAN RADICAL NOVELTY OF SELF-ORGANIZATION

The very close even synonymous relation between emergence and self-organization is largely based on their respective claims of a lower and higher level with the higher level being novel with respect to the lower one. Thus, in the self-organization model, dissipative structures (from the Prigogine school) or the partially ordered structures (from the Haken approach) arising out of interactions at lower, component levels have become prototypical for the novel outcomes of processes of self-organization. The novelty said to characterize the outcomes of self-organizational process hinges on some modicum of unpredictability resulting from the incorporating of randomness made possible from the far-from-equilibrium conditions to which the system is subjected. Thus, the physicist Gregoire Nicolis (1992; a protégé of Prigogine) points to unpredictability of the direction of currents in the Benard convection cells, a direction due to an amplification of random fluctuations. This amplification is made possible by the far-from-equilibrium conditions since at equilibrium such spontaneous randomness washes out and does not reach the macro-pattern level. Hence, the unpredictability of the random is what supplies the novelty of self-organizational products.

But there is another source of novelty as well. In this perspective, 'spontaneous fluctuations' are intensified and reinforced though the means of feedback loops, stabilized by 'coordinating mechanisms' (see Chiles, Meyer and Hench 2004). In addition, though, the interactions of the lower level parts lead to recombinations of the lower level parts resulting in novel arrangements. Hence, we can understand self-organizational novelty as amplified randomness plus some measure of recombination.

One needs only a moment's reflection, however, to recognize that this is a rather thin kind of novelty. Yet, for enthusiasts of self-organization claims are made that it is a much greater degree of novelty. But the 'stories' of explanation given for this greater novelty don't tell a convincing story. The reason lies in two presumptions of self-organization that we will go over below. One is its close association with self-regulation and the other is that recombination as such may only result in a rearrangement of the parts which does not suffice for the kind of radical novelty demanded by emergence.

The concept of self-organization goes back at least as far as Kant, Schelling, and other *naturphilosophen* (see Keller's excellent and insightful history, 2008a, 2008b). Self-organization was how Kant understood organic beings in terms of a circularly causal influence of parts on other parts and parts on wholes as well as wholes on parts. In those early days, self-organization could be considered as having its goal being a type of self-regulation in that the 'self' referred to the whole biological unit, and 'organization' was how this whole biological unit regulated its functioning. Self-regulation then was about maintaining what much later was termed 'homeostasis' and consequently was not at all about innovation or the generation of novel processes or outcomes. It needs to be emphasized that although self-organization at that time was mostly a philosophical pursuit, any methods that were developed to research biological organisms as sites of self-organization were limited to ways of determining the "holistic" functions operating in living forms. Perhaps the most egregious example of this research trend was Goethe's alternative qualitative methods which aimed at uncovering tendencies towards the generation of wholes (see Bortoft 1996). To be sure, Goethe's work was intentionally known as an alternative to the dominant scientific themes and methods of the time and afterwards, nevertheless one can see holistic, self-regulative ideas shaping mainstream biology (see Harrington 1999).

The association between self-organization and self-regulation received a major boost around the time of the Second World War as cybernetics and related systems theories came along with their focus on *homeostasis* (Pickering 2011) which is a strong version of self-regulation since any departures from some set norm is lessened by 'mechanisms' to restore equilibrium (any novelty would count as a departure from the norm and therefore be subject to being diminished, that is, negative feedback).

In fact, cybernetics' research methods were generally founded on various ways of inquiring into how self-organization as homeostasis worked (see the construction of Ross Ashby's notorious 'Homeostat' (see Pickering 2011). Ashby, certainly one of the brightest among the bright, in fact, rejected the very possibility of self-organization as being contradictory, yet his own approach did allude to self-organizing functionalities. One of the main methods was to build homeostatic, self-organizing machines, an exemplification of the paradoxical-sounding use of mechanisms to understanding self-organization in a putatively non-mechanical manner. Other cyberneticians also built supposedly self-organizing machines seeking to see how far they could go in replicating the self-regulating effects found in both intelligent machines and in living creatures.

Remember, that the 'self' of 'self-organization' is principally a marker for the *locus* of a system's organizing activity, that is, the organizing of lower level dynamics into a macro-order (such as the hexagonal shape of the Benard cells) is driven and provided for from *internal* resources and is not imposed from outside. Synonyms for 'self' in this case are 'innate', 'inherent', 'automatic', 'spontaneous', 'unplanned', and 'natural' plus the closely related 'self-causing', 'self-generating', 'self-modifying'. Many methods have been developed from this prototype of emergence in diverse kinds of systems including chemical, electronic, geological, urban growth, and social. Such methods include:

1. Identifying the micro-components and their dynamics on the 'lower' level;
2. Identifying the macro-patterns and order and their dynamics on the 'higher' level;
3. Probing the interactions of the micro-components with each other;
4. Recognizing the 'non- or far-from equilibrium constraints';
5. Observing instances in which occurs the dismantling of hierarchical command and control as effectuating the turn of the system towards its internal resources.

Notice that there are several crucial things that these methods disregard. The most critical is a neglect of changes in the micro-components themselves because of the presumptions that the micro-components are invariant during the process of self-organization and thus the only real change is the relation of the micro-components to each other. Another thing neglected, related to the first, is the nature of the novelty characterizing the new emergent phenomena. Certainly, the self-organization theory of emergence consists of claims of unpredictability like Nicolis' recounted above regarding randomization plus the above-mentioned spare novelty of ordinary recombination. But for unpredictability and the incorporation of some element of randomization to be the whole story of self-organization's novelty means that this particular model of emergence does not rise up to being able to produce the radical novelty needed for emergence's explanatory gap.

23.6 THE NEED FOR TRANSFORMATION

To better grasp why processes of self-organization are not up to the task of producing radical novelty, let's consider an analogy: shaking a bunch of fishhooks in a pail. First we place 30 to 40 differently sized and shaped fishhooks (these are the lower level parts that will interact) in an appropriately sized pail with a lid. Fasten the lid and shake the pail vigorously (the process of interaction). Next, take the lid off, let the sides of the pail fall away and observe the resulting fishhook structures. Take a photo of the resulting structures. Next, disentangle the fishhooks. Repeat the process, each time observing the new structure and comparing the photos of each trial. Repeat this process ten or 15 times.

Again, we are letting the fishhooks be the lower level components and the shaking of the pail is the self-organizational interactional processes. The 'mechanism' which leads to the sticking of fishhooks together (in couples, triads, quartets and so on) into a type of coherence on the higher level has to do with the interlocking of the hooks whose curved and sharp ends tend to clutch onto each other when they are in contact, and remain that way because of their geometrical structure. This does indeed produce novelty in the resulting fishhook structures, a type of novelty we can label a 'new relatedness' following the suggestion of the early emergentist philosopher and animal behaviorist C.L. Morgan (see Blitz 1992) in his depiction of emergent novelty. But how novel is each 'new relatedness'?

This new relatedness can only go so far as a representative for radical novelty since each lower level component, that is, each fishhook, remains invariant throughout the interactional processes. What's required for a *sui generis* novelty of emergent phenomena is a *transformation* of substrates and not just their recombination via interaction. The necessity for a transformation of the substrates themselves in order to fulfill the category of *radical* novelty is sorely lacking in methods devised for studying emergence in whatever systems it appears.

The necessity of a transformation of substrates in emergence has been brilliantly expounded by the philosopher and Indologist Jonardan Ganeri (2011, 2012) in his linkage of the Indian philosophical school of the Carvakas to contemporary issues surrounding emergence. In this regard Ganeri offers a presciently pertinent quote from the great Greek physician and philosopher Galen of Pergamon:

> For anything constituted out of many things will be the same sort of thing the constituents happen to be, should they continue to be such throughout; it will not acquire a novel characteristic from outside, one that did not also belong to the constituents. But if the constituents were altered, transformed, and changed in manifold ways, something of a different type could belong to the composite that did not belong to the elements . . . Consequently, something heterogenous cannot come from elements that do not change their qualities. But it is possible from ones that do. . . .. (Galen quoted in Ganeri 2011, p. 687)

A radically novel emergent whole must be constituted by radically transformed parts since a whole is precisely that which is constituted by the parts in a congruity. As H. Bortoft has put this point, "The whole is nowhere to be encountered by stepping back to take an overview, for it is not over and above the parts, as if it were some superior, all-encompassing entity. The whole is to be encountered by stepping right into the parts" (see Bortoft 1996, p.12). It follows that if there is to be a radically novel whole, there must be transformation of its parts. It is clear from the example of the fishhooks in the shaken pail, the interactions prompted by the shaking of the pail leave the fishhooks intact as they were before the operation. This is the case in general for self-organization and this is is what keeps self-organization from containing a capacity for producing radically novel, *sui generis*, and uncomputable outcomes. Genuinely novel emergent phenomena demand resources of creative potency that can result in the transformation of substrates or micro-level components. It does not seem at all likely that self-organizational processes contain this potency – certainly they have not demonstrated this is the past.

What such transformation entails can be better understood by another analogy. Imagine a tangram-like puzzle composed of 30 small plastic geometrical pieces like triangles (all three types), squares, rectangles, trapezoids, circles, and so on. Each time these parts are combined in a different way, the yield will be as Morgan held (recounted above): a novel relatedness.

But for a decidedly richer and more significant novel outcome, the tangram pieces themselves would need to change. To be sure, it is critical rule in the puzzle game that tangram pieces cannot be broken apart, nor glued together to form new shapes. Yet decomposition and the creation of new parts through some process of gluing, is what is needed for transformation to occur. That is, the previous rules for the tangram puzzle must be transcended. We can imagine the processes involved in transformation by noticing at which system 'level' emergence is thought to be happening. The novel structures of the resulting emergent phenomena take place at level l, and the variously shaped invariant pieces are found at $l-1$. Thus we can restate the rule against cutting and gluing as: only combinations of pieces at level are $l-1$ are permitted.

These simple rules, though, provide a hint as to what a shift beyond new relatedness or novel aggregation to emergent transformational integration must include. First, the rules indicate that any process aiming at transformation of the substrates must operate at a level lower than $l-1$ since that is where any decomposition necessary for transformation of the micro-components must lie. An example is the kind of transformation seen in chemical reactions which require operations on substrates at levels lower than $l-1$ such as at $l-2$ or $l-3$ ... that is, molecules can be considered at l, ions and atoms at $l-1$, subatomic particles at $l-2$ or $l-3$. In chemistry, explanations of the reactions typically involve the action of ions at levels lower than that of molecules (see the analogue of this in Cantor's *stc*, in Goldstein 2014).

The need for decomposition into lower levels and the subsequent transformation thereby made possible is highlighted in the mathematical theory of categories which the aforementioned mathematicians Ehresmann and Vanderbreesch (2007) have applied to emergence. A 'complexification' process of emergence is described in terms of how substrates exist together 'around' a binding constraint. These constraints need to change so that novel types of binding can occur leading to emergent novelty of components. Since the components of each level represent a binding of the components at the next lower levels, it is only when the bindings shift that novel components can emerge, these novel components being the result of novel bindings.

Ehresmann and Vanderbreesch point out that components must have the property of multifoldedness, or decomposability, of each bound emergent unified components. Accordingly, substrates at their own level can be decomposed into a multitude of sub-substrates at levels lower like what needs to happen to the original tangram pieces or tangram wholes. These newly shaped multi-objects (or sub-substrates) at lower levels may function to carryover sub-substrates from the past and sub-strates anticipations of the future (which is why Ehresmann and Vanderbreesch employ multifoldedness in their theories of memory and anticipation of novel futures).

Transformation leading to the kind of novelty required for emergence depends on the decomposition of bound units but this is not permissible by the normal 'rules' of the system. Consequently, there must be a transcendence of normal rules for emergent phenomena to be novel in the sense needed for genuine emergence.

An example of this kind of transformational operation in emergence can be found in the phenomenon of superconductivity which contemporary solid states physicists (spearheaded by such eminent physicists as the Nobel Prize Laureates Philip Anderson and Robert Laughlin) view as a kind of prototype of emergence. The BCS theory of superconductivity (for John Bardeem, Leon Cooper, and Robert Schieffer who won the Nobel Prize for their work) demonstrates such transformation as the key to the emergence of the macro-level, coherent, resistance-free 'quantum wave' out of micro-level quantum mechanical substrates which undergo a radical transformation (all of the material on superconductivity in this section comes from Buckel and Kleiner 2004; and Schmüser 2003).

The transformation discovered in the BCS theory was unprecedented, remarkable fermionic electrons becoming bosons. For a fermion to act like a boson is like the crazy glue used to mend the broken pieces of a ceramic bowl to suddenly act instead like the pieces of the mug and the pieces of the mug acting like the glue! As fermions, electrons repel each other and but as boson they can co-exist in a collective coherence. We see here that, as in the analogy of the trangram, transformation relies on decomposition into sub-components on lower levels, these decomposed subcomponents then joining together in a manner not allowable on higher levels.

The multifoldedness of a low temperature superconductive system consequently allows for decomposition and then reintegration into an emergent outcome not adequately understood as new relatedness. Yes, there is a new relationship among the lower level parts but this novel relation is much more than self-organization can bring about since self-organization does not contain decomposition and reintegration, that is, only new arrangements of components that are invariant. Radical novelty requires decomposition and reintegration for otherwise any new relatedness could be fairly easily traced back to the antecedent lower level dynamics. That is why the BCS theory requires appeal to the quantum mechanical level.

The most startling discovery had to do with 'Cooper pairing' between electrons forming an integrated unity comprised by entire pools of such bonded pairs. The Cooper pairs exhibit a much greater overall length when paired, and they overlap each other so that in the space occupied by a Cooper pair there are about a million other Cooper pairs. Cooper pairing has been found to be responsible for other types of emergence as well, for example, Helium 3 and superfluidity. This is a sign of the universality of emergence in solid state systems where similar even identical outcomes can come from very different substrates, an indication that higher level organizing principles must be at work (Laughlin and Pines 2000).

The BCS theory of superconductivity also shows that the notion of spontaneity pushed by advocates for self-organization (the apotheosis of this take on self-organization being Stuart Kauffman's 'order for free') is not the mainstay of emergence. For that matter, spontaneity is a misleading way of seeing self-organization, a fact which is demonstrated by the great many manipulations a system must go through in order to generate lasers which Haken holds as emergent phenomena *par excellence* for its extremely tight coherence. Although emergence may, of course, include spontaneous processes, the heart of emergence includes much else besides spontaneity.

23.7   CONCLUSION: EMERGENCE, RADICAL NOVELTY, AND RESEARCH

If self-organization falls short through its inability to produce the radical novelty and explanatory gap demanded for a viable model of emergence, then an alternative model of emergence needs to be developed. This model will require both a rethinking of what emergence is if not self-organization as well as a revising of research methods with a capability of probing deeply into means for generating radically novel outcomes. This rethinking takes off from the definition of emergence stated above, particularly through the emphasis on the transformational efficacy implicated in the self-transcending construction of higher/macro level, uncomputably novel, and coherent phenomena out of more particulate, lower/ micro level- substrates and their dynamics. This definition and its exposition also shows the way toward a revised set of methods for researching emergence.

Regarding revised methods, the first thing mandated are ones that focus, not just on processes of interaction which methods based on self-organization concentrate on, but instead on outcomes that demonstrate a radical degree of novelty. Innovation in complex systems is one of the areas that lends itself to such an inquiry. To be sure, the nature of innovation and how it is best brought about are already rich arenas of study in organizational and social system research. What needs to happen in specifically examining innovation from the perspective of the new model of emergence suggested in this chapter is the integration of methods capable of appraising the radically novel aspect of innovation.

This would include the development of novel methods for measuring the degree of novelty, that is, how the novelty of innovation expresses a self-transcending type of novelty characterized as uncomputable. Of course, this doesn't mean having to rely on the mathematical logical foundations of uncomputability. What it does mean is a recognition that the novelty must be radical enough that the players in the social system undergoing innovation will be experiencing it first hand and will at the same time be realizing how the very nature of the social system they are in is being transformed in a radical fashion. This might, for example, include ways of probing into experiential content plus some sort of measure of how the organization of the system is being transmuted according to the requirement of the particular type of transformation of substrates that was described above. This might also consist of metrics of how change is experienced such as it being gradual or sudden or some mix of the two.

An apt example of research methods which follow through on the above requirements for a revised set of ways for probing social innovation along the line of the radically novel nature of emergence can be found in a path-breaking paper by the emergence-oriented organizational researchers and theorists Benyamin Lichtenstein, Kevin Dooley, and G.T. Lumpkin (2006; see also the truly excellent exploration of emergence in the world of organizations containing many telling insights relating emergence to innovation in Lichtenstein 2014).

In this study, several schemes for measuring emergence were employed in order to recognize the degree of novelty found in a firm undergoing deep-seated innovation. More specifically, the researchers investigated innovation that involved three crucial aspects of organizational undergoing emergence, namely, corporate vision organizing, strategic organizing, and tactical organizing. The inclusion of all three was a wise move on the part of the authors for its recognition that if innovation is to really stick it must involve at least three aspects. This move also indicates one of the vital strengths of an emergence-oriented approach to innovation research: all the major facets of an organization or social system need to 'feel' and be impacted by the transformation going on. Moreover, the chief players will experience innovation both emotionally and cognitively and therefore the research modalities need to have a way to measure both. This study linked emergent novelty to innovation in how the firm was actually organized according to the participants' personal experience of the accompanying reorganization. We can call this studying emergence *from within* since it concentrated on how the novelty of what was happening effected the experiential component of the staff.

The strengths of this method for investigating emergence also consisted in the creative use of varied metrics whose findings could act as correctives to each other. Moreover, the researchers focused on the actual language used by participants through textual analysis and related methods and not simply by their observations of what was occurring (although, of course, the latter has its important role). For example, they found evidence supporting the emergence of a new organizing vision occurring in a punctuated manner, and even a nearly instantaneous change in strategic organizing. We can appreciate in this study how the sciences of complex systems have proven they can shed light on organizational and social system dynamics which previously were occluded from view.

Another line of approach for methods to use in order to research emergent novelty consists of examining how social systems can utilize the plethora of novelty generation techniques taken from studies into creativity, art, and related areas. Such techniques are intentionally devised in order produce original outcomes. Certainly, social system innovation can be classified as an arena where novel outcomes require creative processes, that is, ways that aim at creating new possibilities of effective functioning.

In fact, it is rather surprising that creativity techniques have not received nearly the attention they deserve for showing how radically novel outcomes can be generated. Creativity, after all, aims at the production of original, *sui generis* results, whether in the arts, engineering, product design, organizational functioning, and so forth. After all, radically novel outcome production lies precisely at the intersection of emergence and creative process.

The explicit association of emergence with creativity methods is found, for instance, in work by the systems thinker Peter Cariani (2008; Cariani's work is among the most innovative in emergentist circles) he makes a distinction between processes that combine all existing elements and those that, which he terms 'creative' that bring about new elements.

It is only the 'creative' systems that possess the capacity for 'open-ended' transformation that transcends previously enclosed, rule based, and norm obeying systems. The latter, the 'creative' is related to the above section on transformation in which it was stated that as transformation, emergence calls for fundamental change of substrates and not just a new relatedness of invariant substrates.

Furthermore, that the result of the 'creative' systems is radically novel is underscored by Cariani's description of the difference between the combinatorics of unchanging elements which are defined and bound by rule-following versus the creation of new primitives as ill-defined and open-ended. It can be said that from the perspective of the antecedent, lower level substrates the emergent outcomes need to be ill-defined and open-ended (this is what 'transcend' implies) in order to be classified as radically novel ('ill-defined' and 'open-ended' connote unpredictability and nondeducibility).

Cariani's distinction between these two kinds of novelty can be incorporated into methods that aim at uncovering this distinction in the varied created products of organizations and social systems. Again, some kind of metric can be devised to identify each kind of novelty and then research can back-into whatever system processes are responsible for the two types of created products. The latter strategy is like reverse engineering in that one takes an already completed product and takes it apart in order to ascertain how it was made in the first place.

Another approach to creativity, one devoted to exploring the originality of art students' created works, has been developed by the psychiatrist and creativity researcher Albert Rothenberg (1989) and his team. In his study, Rothenberg varied the stimuli given to art students and then had art experts rate the originality of the ensuing created product. The stimuli consisted of slides of different things projected on a screen either next to other slides on the screen or right on top of other slides projected on the screen (these experiments were conducted before the invention of PowerPoint). One set of stimuli were objects that could be considered harmonious with each other or had a figure ground relationship, for example, one slide would show a mermaid and another slide a beach. At one time the slides were shown side by side while at another time there were laminated on top of each other. A different set of slides consisted of sharply contrasting images, for example, a flower in one slide and a gun in another. The raters rated the most original created products to be those stimulated by contrasting images one on top of the other.

Rothenberg concluded that art was at its most creative, defined as productive of most original outcomes, when the stimuli stimulating the art work was constituted by strongly contrasted images existing in the same space, what he called 'homo-spatial' stimuli.

The originality of a created product can be considered one of the criteria for radical novelty in the sense that original implies it is new, unprecedented, *sui generis*. Developing methods of social research into the emergence of radically novel outcomes could utilize a search for sharp contrasts that stimulate novel ways of thinking, creating products, organizing work and so on. Such methods could consist of questionnaires and interviewing that elicit strong contrasts in varied aspects of social system functioning, whether in organizations or communities. An example would be inquiry into the techniques of 'difference questioning' and 'purpose contrasting' to facilitate organization change as proposed by Goldstein (1994).

Another example of research into emergent radical novelty looks at yet other creative methods for producing highly innovative outcomes as found in Kerne et al. (2008) which describes radically novel 'information discovery' in different organizational settings. Like Lichtenstein, Dooley and Lumpkin, Kerne et al. came up with innovative methods for measuring the emergence observed in novelty generation. This points to the wide-open field that beckons researchers regarding the development of new metrics for measuring emergence. One source for such measurement schemes could be the study of emergent cognition (see, for example, Finke, Ward and Smith 1992) which involves how cognitive processes use creative techniques that foster the emergence of creative responses not already present in various stimuli.

There is literally an untapped reservoir of methods for producing original outcomes, that is, *sui generis* ('one of a kind') novelty. All that it would take to make these applicable to social system dynamics is to enlarge the scope of what the substrates and emergence processes would be in a social setting.

Finally, there is the remarkable social and organizational innovations that have been prompted by the Networked Age in which we live. An example is the vast new world of social entrepreneurship for which complexity models are being proposed (see, for example, Goldstein, Hazy and Silberstang 2009). As it is stated in the latter book, social entrepreneurship has "complexity written all over it" (p. 13).

REFERENCES

Adiscott, T. (2011), 'Emergence or self-organization? Look to the soil population', *Communicative & Integrative Biology* 4 (4), 469–470.

Blitz, D. (1992), *Emergent Evolution: Qualitative Novelty and The Levels Of Reality*. Dordrecht: Kluwer Academic Publishers.

Bortoft, H. (1996), *The Wholeness of Nature: Goethe's Way Toward a Science of Conscious Participation in Nature*. Lindisfarne, UK: Lindisfarne Press.

Boyle, R. (2016), *Natural Novelty: The Newness Manifest in Existence*. Lanham, MD: University Press of America.

Buckel, W. and Kleiner, R. (2004), *Superconductivity: Fundamentals and Applications*. Hoboken, NJ: Wiley.

Cariani, P. (2008), 'Design strategies for open-ended evolution', in S. Bullock, J. Noble, R.A. Watson, and M.A. Bedau (eds), *Proceedings of the Eleventh International Conference on Artificial Life*. Cambridge, MA: MIT Press, pp. 94–101.

Chiles, T., Meyer, A. and Hench, T. (2004), 'Organizational emergence: The origin and transformation of Branson, Missouri's musical theater', *Organization Science*, 15, 499–519.

Cohen, J. and Stewart, I. (1995), *The Collapse of Chaos: Discovering Simplicity in a Complex World*. New York: Penguin Press.

Darley, V. (1994), 'Emergent phenomena and complexity', in R.A. Brooks and P. Maes (eds), *Artificial Life IV: Proceedings of the Fourth International Workshop on the Synthesis and Simulation of Living Systems*. Cambridge MA; NY: Bradford Books, pp. 411–416.

Dauben, J. (1979), *Georg Cantor: His Mathematics and Philosophy of the Infinite*. Princeton, NJ: Princeton University Press.

de Wolf, T. and Holvoet, T. (2004), 'Emergence versus self-organisation: Different concepts but promising when combined', Lecture Notes in Computer Science, 3464, 1–15.

Ehresmann, A.C. and Vanderbreesch, J.P. (2007), *Memory Evolutive Systems: Hierarchy, Emergence*. Amsterdam: Elsevier.

Finke, R.A., Ward, T.B., and Smith, S.M. (1992), *Creative Cognition: Theory, Research, and Applications*. Cambridge, MA: MIT Press.

Fitzpatrick, P.J. (1966), 'To Gödel via Babel', *Mind*, 75 (299), 322–350.

Ganeri, J. (2011),'Emergentisms, ancient and modern', *Mind*, 120, 671–703.

Ganeri, J. (2012), *The Self: Naturalism, Consciousness, and the First-person Stance*. New York: Oxford University Press.

Glansdorf, P. and Prigogine, I. (1971), *Thermodynamic Theory of Structure, Stability and Fluctuations*. New York: John Wiley and Sons.

Goldstein, J. (1994), *The Unshackled Organization*. Portland, OR: Productivity Press.

Goldstein, J. (2002), 'The singular nature of emergent levels: Suggestions for a theory of emergence', *Nonlinear Dynamics, Psychology, and Life Sciences*, 6 (4), 293–309.

Goldstein, J. (2006), 'Novelty, indeterminism, and emergence', *Emergence: Complexity and Organization*, 8 (2), 77–95.

Goldstein, J. (2014), 'Reimagining emergence, Part 3: Uncomputability, transformation, and self-transcending constructions', *Emergence: Complexity and Organization*, 16 (2), 116–176.

Goldstein, J., Hazy, J. and Silberstang, J. (2009), Complexity Science and Social Entrepreneurship: Adding Social Value through Systems Thinking. Boston, MA: ISCE Publications.

Haken, H. (2011), 'Self-organization', *Scholarpedia*, accessed on April 3, 2015 at http://www.scholarpedia.org/article/Self-organization.

Halley, J.D. and Winkler, D. (2008), 'Classification of emergence and its relation to self-organization', *Complexity*, 13 (5), 10–15.

Harrington, A. (1999), *Reenchanted Science: Holism in German Culture from Wilhelm II to Hitler*. Princeton, NJ: Princeton University Press.

Hintikka, J. (2000), *On Gödel*. Boston: Cengage.

Holland, J. (1998). *Emergence: From Chaos to Order*. San Francisco: Addison-Wesley.

Humphries, P. (2008), 'How properties emerge', in M. Bedau and P. Humphries (eds), *Emergence: Contemporary Readings in Philosophy*. Cambridge, MA: MIT Press, pp. 111–126.

Kaplan, A. (1964), *The Conduct of Inquiry: Methodology for Behavioral Science*. New York: Chandler Publishing.

Kaufmann, F. (1978), *The Infinite in Mathematics: Logico-Mathematical Writing*. New York: Springer.

Kauffman, S. (1995), *At Home in the Universe: the Search for the Laws of Self-Organization and Complexity*. New York: Oxford University Press.

Keller, E.F. (2008a), 'Organisms, machines, and thunderstorms: A history of self-organization, Part One', *Historical Studies in the Natural Sciences*, 38 (1), 45–75.

Keller, E.F. (2008b), Organisms, machines, and thunderstorms: A history of self-organization, Part Two, *Historical Studies in the Natural Sciences*, 39 (1), 1–31.

Kerne, A. et al. (2008), 'Measuring emergence in information discovery', *International Journal of Human-Computer Interaction*, 24 (5), 460–477.

Laughlin, R. and Pines, D. (2000), 'The theory of everything', *Proceedings of the National Academy of Sciences*, 97 (1), 28–31.

Lichtenstein, B. (2014), *Generative Emergence. A New Discipline of Organizational, Entrepreneurial, and Social innovation*. New York: Oxford University Press.

Lichtenstein, B., Dooley, K. and Lumpkin, G. (2006). 'Measuring emergence in the dynamics of new venture creation', *Journal of Business Venturing*, 21, 153–175.

Nicolis, G. (1992), Physics of far-from-equilibrium systems and self-organisation', in P. Davies (ed.), *The New Physics*. Cambridge: Cambridge University Press, pp. 314–347.

North, M. (2013), *Novelty: A History of the New*. Chicago, IL: University of Chicago Press.

Petzold, C. (2009), *The Annotated Turing: A Guided Tour through Alan Turing's Historic Paper on Computability and the Turing Machine*. New York: Wiley.

Pickering, A. (2011), The Cybernetic Brain: Sketches of Another Future. Chicago, IL: University Chicago Press.

Rothenberg, A. (1989), The Emerging Goddess: The Creative Process in Art, Science, and other Fields. Chicago, IL: University of Chicago Press.

Schmüser, P. (2003), *Superconductivity*. Hamburg, Germany: Institut für Experimentalphysik der Universität Hamburg.

Silberstein, M. and McGeever, J. (1999), 'The search for ontological emergence', *The Philosophical Quarterly*, 49, 182–200.

Stephan, A. (1999), 'Varieties of emergentism', *Evolution and Cognition*, 5, 49–59.

Strawson, G. et al. (2006), *Consciousness and its Place in Nature: Does Physicalism Entail Panpsychism?* Exeter: Imprint Academic.

Turing, A. (1937), 'On computable numbers, with an application to the Entscheidungsproblem', *Proceedings London Mathematical Society* (series 2), 42 (1), 230–265.

24. Applying the 15 complexity sciences: methods for studying emergence in organizations
Associate Professor Benyamin Lichtenstein, University of Massachussetts

24.1 INTRODUCTION – THE 15 COMPLEXITY SCIENCES

Since the formal introduction of complexity science into management (Anderson et al. 1999), research using these methodologies has progressed, with hundreds of publications that explain patterns emerging in complex systems. There has also been growth in the number of unique methodologies – individual sciences of complexity (Cohen 1999) – that can analyze the nonlinear dynamics of social systems. From the early applications of Kauffman's (1993) NK Landscape model (Levinthal 1991, 1997), complexity science has expanded to include a whole series of methods or disciplines in the sciences of complexity.

We have learned a good deal about these disciplines, as exemplified in key reviews by Goldstein (1999), Sorenson (2002), Maguire et al. (2006), Davis, Eisenhardt and Bingham (2007), and the recent *Sage Handbook for Complexity Science in Management* (Allen, Maguire and McKelvey 2011). Over that time, 15 separate complexity sciences have been identified in the literature (Lichtenstein 2011), each with a distinct methodology or operationalization for research in social systems. Understanding how these 15 sciences work can open up research in the field, as researchers learn about other methods and how to combine them in unique ways.

Recognizing the full range of methods can solve two key problems in the field. First, the original promise of complexity science – to provide a vibrant theoretical and methodological pallet for understanding complex dynamics in social systems – has not yet been fulfilled. Although a trickle of complexity papers are still published, the initial enthusiasm has long since waned. In its place are a few strong research groups that have mastered one particular methodology, applying it in systematic ways. Rather than building up a broad, integrated understanding of complex systems, complexity science in management has become focused on separate phenomena, each one unrelated to the next. Without some reframing, complexity science may simply dry up.

Second, the methodologies that tend to get published are computational models – agent-based, computer-generated simulations that illustrate how agents, following simple rules, can produce new structures in complex systems. NK Landscapes has been exemplary in this regard, being the basis for multiple top-tier research articles explaining a variety of phenomenon, including networks of knowledge creation (Sorenson, Rivkin and Fleming 2006), social structure in new markets and innovation (Ganco and Agarwal 2009), new product development (Oyama, Learmonth and Chao 2015), and search speed and effectiveness (Gavetti and Levinthal 2000; Fleming and Sorenson 2001; Billinger, Stieglitz and Schumacher 2013). Some suggest that complexity science *means* computational modeling.

If so, researchers who want to examine nonlinear phenomenon in *real* organizations and social systems would appear to have limited choices.

This chapter makes two suggestions for solving these problems and creating new avenues for research using complexity science. First, rather than a competition between methods, the entire project of complexity science could usefully be reframed as the study of *emergence* – how order and structure are created and expressed in dynamic systems (Goldstein 1999; Chiles et al. 2004; McKelvey 2001; McKelvey 2004). That frame promotes the use of *all* the sciences of complexity, each with a different perspective and approach. Table 24.1 introduces that idea, by suggesting how each complexity science uniquely examines emergence phenomena, and one or more research questions each is designed to explore. The table also reviews the data requirements for each method, an important consideration for social science researchers.

Second, a close examination of complexity research in human and social systems shows that these methods are rarely used in their pure form. Many interesting and fruitful complexity analyses use combinations of these sciences, integrating longitudinal and qualitative analysis into emergent findings. In this light, the second part of the chapter will provide examples of such combinations, and consider some future options and directions of this work.

24.2   COMPLEXITY SCIENCE METHODOLOGIES

The underlying methodologies of each of the 15 sciences is presented with more detail in Lichtenstein, 2014 (Chapter 3). These can be organized on a micro–macro scale, from the most mathematical to the most social-ecological. Thus, the first six disciplines reviewed here use mathematical techniques to study emergence. One of these is **deterministic chaos theory,** which can mathematically identify 'attractors' in longitudinal time series data – dimensions of the system within which all data will fall. Applied to organizations, an attractor is a kind of configuration (Meyer et al., 1993), a dynamic state (Levie and Lichtenstein 2010) or business model (Osterwalder and Pigneur 2010) that is mainly stable for long periods of time. An exemplary study used this analogy to examine a shift from one attractor to another in long-term innovation teams (Cheng and Van de Ven 1996). With more than 50 data points for each time series they analyzed, Cheng and Van de Ven revealed two statistically different attractors – distinct 'phases' of innovation activity in the MIRP innovation data (Van de Ven et al. 1989). This led to insights into the drivers of those different attractors (for example, learning, randomness), and how they help explain the dynamics in the 'innovation journey' (Van de Ven et al. 1999).

Attractive though this method is, it does have a statistical requirement of metrics over 50 time periods, which is the minimum needed for statistical confidence. Although few longitudinal studies can gather that depth of information about an organization, striving to do so opens up untapped areas of research, for example examining when systems or schema move into a new state. A main outlet for articles using this and other mathematical complexity methods is *Nonlinear Dynamics, Psychology and Life Sciences*, which some years ago sponsored a special issue on Nonlinear Organizational Dynamics (Dooley et al. 2013).

A related complexity science is **catastrophe theory,** which provides seven typologies for

Table 24.1   Research studies of emergence across the 15 complexity sciences

Complexity Science	Examines the Emergence of...	General Research Questions	Data Needed for Analysis
Deterministic chaos	New attractors: stable configurations of activity	When does a project move from one phase to the next? At what point does one schema shift into another?	50 data points that are contiguous or longitudinal
Catastrophe theory	System change that is nonlinear or discontinuous	Is there a nonlinear equation (typology) that explains a social phenomenon better than linear regression?	Same as regression. The method allows for higher-dimension modeling
Fractals	Similar patterns across scales	Are there patterns of behavior that are occurring at multiple levels, e.g. individual, team, unit, organization?	Highly accurate numerical data, at more than one unit of analysis
Positive feedback, Increasing returns	Reinforcing loops and amplifying cycles of activity	Why do some outcomes become locked in at the expense of others? What is the origin of reinforcing cycles of activity?	For a positive feedback cycle, data must be in a continuous time series
Power laws; self-organized criticality	Phenomena that follow a 80/20 rule, where 20% of the instances cause 80% of the outcomes	If most variables – including outcomes – are Pareto distributed (i.e. non-Gaussian), how do we understand the influences that cause them?	Full set of quantitative data including presumed 'outliers.' For analytic methods see Crawford et al. (2015)
System dynamics	Interdependencies across a system that yield unexpected outcomes	How do delayed responses and unseen interdependencies alter (reduce) the effectiveness of a process? Can these system breaks be understood and changed to improve managerial decisions?	Describing the activity (system) fully in terms of stocks and flows. Data are the amounts and rates of each. Analysis needs specific software
Complex adaptive systems	Collective behavior, in (computational) systems of interacting agents	What small number of simple rules, shared by agents with evolving schema, can explain stable collective behavior in a system?	Access to the schema of agents in the system, plus a description of agent interactions
Cellular automata	Stable aggregations of agents in a computational matrix	In a system of simple (computational) agents, how do increases in rule complexity and heterogeneity influence the stability of the final configuration?	An array of computational agent-based data that includes traits and rules, with some heterogeneity. CA models can be tested using human/sociological data

Table 24.1   (continued)

Complexity Science	Examines the Emergence of. . .	General Research Questions	Data Needed for Analysis
NK landscapes	Structure internal to an agent-based matrix. The structure is a 'landscape' of adaptation	What is the minimum amount of interdependence (K) across agents that still leads to differentiated adaptation? How does agent heterogeneity affect that structuring?	Computational data on traits of agents, interdependence, and changes across agents. Outcomes are levels of adaptation over time
Genetic algorithms	Evolving (improving) agents, due to their learning and sharing it with others	How can (computational) agents learn and adapt most quickly? How do re-combinations of rules (learning) influence all agents in a system?	Specifications of the rules/ rule strings of computational agents, with elements that can combine in unique ways
Agent-based + Multi-agent simulations	Large-scale (macro) effects and adaptation arising from micro-level interactions	At what point do aggregations of agents supervene on (influence) individuals within the group, and those entering the group?	Computational agents with traits, schema, roles and the ability to learn. Also needed are the algorithms that allow agents to form hierarchical layers of decision-making
Autogenesis, Autopoiesis, Autocatalysis	Mutually dependent systems or cycles that share positive-feedback loops	When will independent/ interdependent systems/ processes arise? What resources allows them to self-organize and be sustained?	Time series or historical data on inputs, nodes, outputs and interactions of (at least) two connected systems
Dissipative structures	New structures and sustained layers of order/organizing in social systems	How do new levels of order/ structure come into being in social systems? To what degree does this new order increase the capacity of the system?	Process data on elements and the system-as-a-whole, on e.g. disequilibrium, experiments, thresholds, re-combinations, capacity
Ecosystem resilience	Ecosystems that are sustained amidst changing conditions	How far from its normal regime can an ecosystem be pushed before it loses its resilience? What is required for organizations to gain resilience?	Time series data on each component of an ecosystem, and on external conditions that affect it
Ascendency, Directionality in evolution	Order writ-large, and how it arises in a progressive way (i.e. in a direction)	What quality or characteristic can explain the progressive development of organizing and organizations?	Long-time sequential data on an evolutionary or organizational form

nonlinear analysis of data in dynamic systems. Each typology is a kind of statistical algorithm that reveals alternative causes and drives. For example, the 'cusp catastrophe' is an S-type response surface that reveals hysteresis, that is, a shift from one state to another in which the same path cannot be used to return to the original state. These response surfaces – seven have been demonstrated consistently – allow for much more sensitive understanding of social dynamics. An exemplary study, by the leader in this method, is Guastello's (1998) explanation of leadership emergence. Using the 'swallowtail' model, he was able to explain three times more variance of emergence in a leadership situation, compared to an analysis of the same data using normal, linear regression. These models allow for a deeper understanding of social phenomena through these nonlinear frames.

Fractals is a method for identifying 'self-similarity' in a system, that is, patterns in data across levels of analysis. Early fractal analyses, for example, revealed the Mandelbrot Set, which became the icon for the early introduction of complexity science (Gleick 1987). Applications to organization are rare, but a recent exemplar is the study by Thomas et al. (2012) who did an analysis of urban areas in Europe. They found that urban areas are often more similar *across regions* then they are across a single city. By looking for similar patterns across levels, even using metaphors, researchers might gain insight into what appear to be disconnected phenomenon in and across organizations. Likewise, this algorithm might be useful when combined with others in large-scale studies.

Positive feedback processes are central to the growth of most biological systems and social ecologies, for these structures generate the continuous flow of resources from which order is produced. In their simplest form, positive feedback loops occur when the outputs of a process reinforce the inputs, leading to accelerating change, reinforcing cycles, and snowball effects (May 1976; Sterman and Wittenberg 1999). Mathematically introduced as cybernetics by Ashby (1956), this idea was expanded by Maruyama (1963) as 'the second cybernetics,' which in contrast to traditional cybernetic analysis of negative feedback loops, focused on positive increases that accelerate dynamic systems. Some scholars still work with cybernetic concepts, including Beer's work on the Viable System Model (DiRenzo and Greenhaus 2011; Espinosa et al. 2004; Brewis et al. 2011). Perhaps the most important application was developed by Arthur (1989, 1990), who was able to show how small differences between competing products can lead to a sharp rise of one product over another. In industries with a dominant design, this effect will lead to 'lock-in' that limits all future innovation, even if the winning design is not the best one (Arthur 1994). Finding the origin of this lock-in would provide significant advantage in every management area.

Power laws refer to systems that follow an 80/20 rule, where 80 percent of the outcomes are produced by only 20 percent of the sample. In such systems, the vast majority of agents have virtually no impact on the overall outcomes of the system, but a small number have a disproportionately large effect. More recently, scholars are finding that this pattern is repeated by a very large percentage of social phenomena, which are distributed in a non-normal (non-Gaussian) way. A string of studies have shown that the underlying distribution of most social systems is Paretian – an 80/20 curve (Andriani and McKelvey 2009; O'Boyle and Aguinis 2012; Aguinis and O'Boyle 2014; Crawford et al. 2015). The challenge this poses is that virtually all statistical analyses in the social sciences – from ANOVA to regression to Structural Equation modeling – are viable only for normally distributed data. In contrast, this research shows that for most social outcomes, including the size of companies, the wealth of individuals, or entrepreneurial performance, data are

Paretian not Gaussian; thus, all previous findings based on normal statistical analysis may be incorrect. This insight is leading to speculation on alternative forms of causality and drivers of outcomes (Crawford et al. 2015), and rethinking the structure of social systems.

System dynamics is a method for modeling the interdependencies of an organizational process thorough the 'stocks' and 'flows' of its resources and outputs. Using programming languages that model organizational variables and metrics, researchers have been able to identify key leverage points that have a very high impact in a system. For example, Hall (1976) found that the demise of the *Saturday Evening Post* magazine was caused by a simple but unreflective decision about the ratio of content to advertising, which inadvertently caused a negative spiral of fewer subscriptions and ultimately, bankruptcy. More recently, Rudolph and Repenning (2002) found how stress can build up on a system until a threshold will completely shift the quality of decision-making. A broad range of applications have examined issues around management decision-making (Repenning and Sterman 2001; Sterman 2002) and organizational learning (Senge 2006). Complexity science researchers could extend this literature by examining the organizational effects of system delays, and thresholds for interdependency effects.

The next five complexity sciences are computational methodologies – computer-based simulations in which heterogeneous agents interact with each other, creating order over time. The basic insight shown by these sciences is that simulated agents, following simple rules and at a moderate degree of interdependence, will create structural order in the system, as a normal result of their interactions (Sorenson 2002).

Complex adaptive systems (CAS) is the reference point for these complexity sciences (Holland 1975; Holland 1995; Holland 1998), and many applications of complexity are framed in terms of CAS, including Uhl-Bien and Marion (2008), Brown and Eisenhardt (1997), Fuller and Moran (2001), and Nair et al. (2009). In most of these applications, agents are connected through nested systems into a hierarchy of sorts, and agents adapt by changing their schema or their decision-rules. A common premise of computational methods is that each agent's traits and decision-rules are central to the structure that emerges. The core research questions explore how a few simple rules followed by (micro) agents can lead to collective (macro) order in the system.

Is there a technical methodology associated with CAS studies of organizing? In computational language, four such methods are most common: cellular automata, NK landscapes, genetic algorithms, and agent-based models (Lichtenstein and McKelvey 2011).

Cellular automata use a two-dimensional matrix such that each agent interacts with their nearest neighbors on either side. This is the basic paradigm of all agent-based computational models. Depending on the decision rule of each cell (agent) with its immediate neighbors, it changes in some way. Through a stream of interactions order emerges. CA models can vary in complexity, depending on the number of neighboring agents (2, 4, 8. . .) and the combinations of rules and possible outcomes. In one of the most insightful applications of this method, Schelling (1978) showed that individual agents, living randomly but with a slight preference to live near people who are similar to them, will always generate segregation in their region, an outcome that turns out to be very hard to shift once the structure emerges. Further applications have been developed to explain the emergence of political alliances (Axelrod and Bennett 1993), the organization of supply chains (Nair et al. 2009), and elements of economic geography (Krugman 1996; Lomi and Larsen 1996; Lomi and Larson 1997). In general, these research questions revolve around how changes in the number and complexity of rules influences the time it takes to reach a stabile configuration of order in the system.

NK landscape models extend the matrix to three dimensions, allowing the programmer to tune the agents to higher or lower degrees of interdependence, through their level of interaction. In these models, each agent (cell in the matrix) has N traits or attributes with a certain K level of interdependence. Across iterations of the simulation, a structure will emerge across the surface (landscape), based on the combinations of attributes that are most adaptive. It turns out that the order in the landscape is primarily a function of the degree of interdependence, rather than the number or quality of traits in the agents. The optimal overall outcomes occur when agents are only moderately interdependent. When interdependence gets too high, a 'complexity catastrophe' ensues, eradicating all adaptive change (McKelvey 1999).

Three exemplars give a view of how applications from NK Landscapes may be fruitful. Fleming and Sorenson (2001), studying technology innovation, use the NK model to explain the usefulness of an invention, measured as six-year citation count. They find that regardless of the number of components an innovation has, its usefulness is maximized when there is a moderate interdependence across components, rather than high or low. Similar findings of moderate interdependence have been found in the context of production processes (Ethiraj and Levinthal 2004), strategy-making (Baum 1999), competitive advantage (Porter and Siggelkow 2008), and in start-ups (Ganco and Agarwal 2009). Further research could explore the minimum amount of interdependence necessary for effective structuring, and the role of increased heterogeneity on the final outcome.

Genetic algorithms describe a simulation in which agents can literally change their traits, by deciding to give away or receive new traits from other agents – this allows agents to extend their capabilities over time (Holland 1995, 1998). In contrast to CA or NK models, agents in GA models can accumulate multiple rules, and even change rule strings (Macy and Skvoretz 1998). By modeling the evolution of learning in these agents, insights can be applied to organizational learning and action. For example, Paul et al. (1996) identified the characteristics of financial trading firms that had the most impact on their performance; others find implications in strategic leadership. Since this method nicely corresponds to human learning, researchers can extend current findings by examining the minimal thresholds and optimal combinations of traits that lead to viable learning in different contexts.

Agent-based models, by far the most complex of these approaches, allow agents to "exhibit behavioral adaptation, goal-directed learning . . . the formation of new networks . . . [and they] can self-activate and self-determine their actions on the basis of internal data" (Tesfatsion 2011, p.36). This complexity allows for an even closer analogy to human learning and action. Some of the most sophisticated agent-based models have been developed by Carley and her colleagues, which combine elements of CA, GA, and neural networks (Carley 1991; Carley and Hill 2001) into their CONSTRUCT-O model. Another powerful application is their ORGAHEAD program (Carley 1990, 1999; Carley and Lee 1998), which captures multiple levels of action across agents – an executive team, making decisions about strategy and design, to organizational agents within a changing environment. As is true for the others, the theoretical insights from these simulations can make important contributions to understanding organizations (Davis et al. 2007). Organizational researchers could use this method to uncover how aggregations of agents supervene on (influence) individual agents within the group, and how this process operates for new agents to the group.

Computational models are more well-known and easier to publish than rich, longitudinal and qualitative work, or the earlier mathematical approaches. But complexity science is not limited to computation. The following sciences are drawn from the biological sciences of biochemistry, thermodynamics, and ecology. When carefully operationalized, they provide very interesting opportunities to understand emergence in human organizations in real time.

Autocatalysis (Eigen 1971; Eigen and Schuster 1979) is related to autogenesis (Drazin and Sandelands 1992; Csanyi and Kampis 1985), autopoiesis (Maturana and Varela 1980), and the replicative model (Pantzar and Csanyi 1991; Rosen 1971, 1973). At the heart of these approaches is a unique dynamic in which a pair of reactions that need each other's outputs in order to start, are spontaneously activated by the catalysts that each reaction generates. In such cases, both reactions emerge at the same time – this is one definition of 'self-organization.' This shared creation plays an important role in biochemical evolution (Padgett 2011), the emergence of organizations (Drazin and Sandelands 1992), and the development of institutions (Padgett and Ansell 1993; Padgett et al. 2003; Padgett and Powell 2011); these approaches have also been applied to social systems more generally (Kickert 1993; Luhmann 1986). Asking when such interdependent systems will arise, and examining the micro-processes that lead to their maintenance, are valuable next steps for this methodology.

One prominent complexity science is **dissipative structures theory**, which has been used to help explain the emergence of group dynamics (Smith 1986; Smith and Comer 1994), innovation (Nonaka 1994; Saviotti and Mani 1998), entrepreneurship (Lichtenstein 2000; Foster 2011), radical organizational change (MacIntosh and MacLean 1999; Plowman et al. 2007), and collaboratives and regional clusters (Chiles et al. 2004; Browning et al. 1995). The underlying science was developed through empirical studies of thermodynamic order-creation processes (Bénard 1901; Prigogine and Stengers 1984), summarized as follows (Lichtenstein 2014, p.89):

> According to the theory, increasing energy flows through a system (e.g. in the form of heat or energy) can cause the system to move . . . into far-from-equilibrium dynamics. At a critical threshold, there will emerge a new 'level of order' – literally, macrostructures will appear spontaneously in the system, increase[ing] the capacity of the system to dissipate heat energy . . . by several orders of magnitude.

This process has been operationalized by multiple scholars into five phases or conditions that lead to emergence in organizations, through a sequence that reveals the underlying logic of emergence (Lichtenstein 2014). These phases have been seen in multiple inductive and longitudinal studies of emergence (Nonaka 1988; Browning et al. 1995; Chiles et al. 2004; Plowman et al. 2007; Lichtenstein 2014); more work can be done to validate them in other contexts as well. Further, researchers can use this method to measure increases in the capacity of a system that has undergone this type of order creation.

Two other complexity sciences are less easy to operationalize in organizations. **Resilience** draws from the science of biological and social ecosystems (Gunderson and Holling 2001), which has examined how far away from its normal activity regime can an ecosystem be pushed and still 'bounce back' to optimal functioning. Researchers have identified the properties that affect an 'adaptive cycle' in natural ecosystems (Walker et al. 2004; Folke et al. 2010); more work is needed to apply these insights to social or economic ecologies. Note that this is parallel but different from the interesting work by Positive Organization Science on building organizational resilience and resonance (Sutcliffe and Vogus 2013; Caza and Milton 2011). In both cases, researchers can inquire into how far a (social) ecosystem can be pushed from its norm before losing its resilience, or regaining a new degree of resilience.

Finally, **ecological ascendency** refers to a complex systems analysis of physical and social evolution which shows an undeniable 'trajectory' of ecosystem development; explanations include the 'law of maximum energy' (Lotka 1922), maximum power output (Odum and Pinkerton 1955), or 'ascendency' (Ulanowicz 1980, 1987). Evolutionary thinkers have made similar claims that extend neo-Darwinian models (Chaisson 2001; Wicken 1979, 1980; Morowitz 2002), and Hildago (2015) has recently updated those arguments to link evolution with the growth of economies. Organizational scholars have not yet worked on applications of this approach, but would be encouraged to ask what characteristic(s) can explain the progressive development in individual companies and the economy as a whole.

In sum, each of the 15 complexity sciences is based on a certain methodology or framework, which significantly influences the kinds of complex patterns that can be identified, and highlights specific dynamics that drive order-creation in different contexts. Although there's no meaning to the number – other scholars might find 16 or 13 – this is my best formulation after 30 years of active involvement with the field (Lichtenstein 1988, 1991). On the surface, each of these is a coherent stream of research, which contributes to the phenomenon of interest and to complexity science as a whole.

At the same time, there are complexity studies that don't directly fit into one of these 15 categories. Mainly these are multi-method studies, which incorporate two or more complexity studies; in other cases these draw on a complexity science which is applied more metaphorically in organizations. Other combinations are possible as well, as is shown next.

24.3   FROM COMPLEXITY METHODS TO ORGANIZATIONAL RESEARCH ON EMERGENCE

According to good social science practice, the choice of which complexity science to use should be driven by one's research question, not by one's familiarity with the approach (Edmondson and McManus 2007). This is easier if one knows all of the available options for theory and method, which is one reason for presenting all 15 complexity research methods. In addition, young scholars may be unaware of research designs that complexity scholars have adopted in the past. Thus, this second part of the chapter reviews several broad avenues for pursuing complexity research on emergence – computational modeling, analogies and metaphors, and narrative and multi-method studies.

24.3.1   Computational and Mathematical Models

A relatively large percentage of current complexity-styled work is being done by computational modelers and others using quantitative methods, using, for example, CA or agent-based models, NK landscapes, or analytic tools like deterministic chaos or power laws. Each of these sciences has been used in artful ways, based on quantitative data, in some cases time series data. The analysis has high validity, but the range of findings is dependent on the types of questions the underlying methodology can answer. For example, NK landscapes are (only) valid in contexts where interdependence is measurable and relatively consistent; as another example, deterministic chaos theory can examine a change of configurations (attractor change), but not the dynamics that lead to it.

Like other quantitative methods in organization science, making good use of computational and mathematical methods requires an investment of time and learning, to understand the algorithms and the software, and to become facile at making the appropriate links between the programming language and organizational reality. It can take many years to reach this degree of sophistication, and thus to make publishable contributions from one's findings. An unintended outcome, however, is that once trained in one method, a researcher may become focused on (limited by) that one, mainly because using a new method means essentially starting from scratch.

Another caveat of computational models is their use in theory-development, rather than theory-testing or application (Davis et al. 2007; McKelvey 2002; Sorenson 2002). Although a number of studies aim to test a particular relationship, arguments from philosophy of science strongly suggest that the knowledge gained from agent-based modeling is not the same as can be gained through (human) empirical examinations. Instead, this category of methods is best suited to find patterns across computational agents, which may have parallels in human organizations. When researchers go beyond those boundaries, their findings may suffer from a lack of validity and generalizability (McKelvey et al. 2013).

At the same time, the statistical robustness of computational approaches makes them a powerful tool for organization science. Many of the top programs are allowing for more flexibility and potential in agent-based models. Equally important for young scholars, these models have a higher likelihood of being published in top journals than qualitative methods, due to the bias around quantitative analysis that is present in the academic field. One could say that over the past 10 years, almost all the complexity science publishing has been done using computational models, most of them being NK studies. This does not mean that NK studies are most apt to be correct – see the important critiques by McKelvey et al. (2013) and Sawyer (2001). However computational sciences are tractable and yield defined results, even across multiple dimensions (Johnson 2009; Prietula 2011; Mitchell 2009).

24.3.2   Natural Sciences and Ideographic Analogies

A second, very different complexity approach, is based on insights gained from natural science experiments in complex systems. This work reflects the *European school* of complexity (McKelvey 2004) which emphasizes the processes of order creation in social systems, through analogical or metaphorical extensions of a complexity science into organizations (Maguire et al. 2006). In contrast to the *American school's* emphasis on computational agents, these other studies examine the complex behaviors of individuals in social situations. This approach is more applicable from an empirical standpoint, and it offers a view of emergence which may go beyond the epistemological constraints of computational models.

These approaches can be arrayed in a kind of continuum. At the more formal end, a good number of complexity researchers pursue ideographic analogies (Tsoukas 1989) between a particular complexity science and how it is applied in organizations. In this approach, the validity of an application depends on the rigor (integrity) of the mapping between the natural system and the target behavior in social systems (Tsoukas 1991). Most complexity researchers spend time describing the underlying science; two in particular go to great lengths to make the rigorous ideographic mapping – Padgett and Powell (2011, Chapters 3 and 4), and Lichtenstein (2014, Chapters 6, 7 and 8); Earlier studies are also notable (Penrose 1952; von Foerster 1960; Meadows et al. 1972; Rosen 1971). This approach retains the logic of action from the science because it entrains the researcher to start with the core driver (generative mechanism) of emergence as a natural system. The researchers then work backwards to insure that the generative mechanisms at play are fully comparable to the mechanisms experienced by agents in the application (that is, people). Many successful studies have used this approach.

On the less formal side of the continuum, some of these analogies are more like metaphors, which describe an insight from the underlying science in purely common language. One example is the core idea from CAS, namely that agents interacting according to 'simple rules' can create emergent order, like Canadian Geese flying in a 'V'. Many people use this notion of interaction according to 'simple rules' as the basis for their findings, to good effect. At the same time, this simplifying approach can leave a number of things underspecified. Simple rules can, for example, refer to a schema (cognitive frame), decision-making criteria (rules of thumb), or formal directions that guide routine organizational behavior. This range means that the source of an outcome could be mis-classified.

At the same time it gives researchers a lot of latitude in how the analogy is made, in some cases yielding useful innovations for organizations. An exemplary is Brown and Eisenhardt's (1997) appropriation of the term 'edge of chaos,' as a metaphor to explain semi-structuring in innovation teams. Even though the 'edge of chaos' is a disputed finding (a number of scholars consider the finding to be weak at best, especially as it is based on one interpretation of a single outcome in a computational study), its application to product development teams is helpful, especially in highlighting a dynamic mode of organizing that incorporates logics of structure and as well as flexibility/innovation.

Whether using rigorous ideographic mapping or informal metaphors, there is room to expand on this approach. For example, studies have identified the 'nonlinearity' of most organizational situations (Boisot and McKelvey 2010; Andriani and McKelvey 2009). What does this 'feel like' in organizations? How does this show up to individuals, groups, ventures, collaboratives, clusters, and industries? Likewise, in the context of autocatalytic processes (Padgett and Powell 2011), how do these generative interactions get started and flourish in fast-paced entrepreneurial environments? Drawing on catastrophe theory we can ask, what are the thresholds of organizational action (Guastello 1995)? How do the factors for ecological resilience (Holling 1973; Gunderson and Holling 2001) compare with POS insights into organizational resilience (Sutcliffe and Vogus 2013)? And so on.

A caution for researchers is that metaphors must be made carefully. As Maguire and McKelvey (1999) show, it was inaccurate metaphors that turned the initial round of complexity science into a managerial fad, with practitioners claiming all kinds of things that were not validated in organizations and which ended up doing more harm than good. In the context of this chapter, simply note that it is important for researchers to use care – have high integrity – when making analogies from science to social science, and especially from computational studies to real-time human activity.

24.3.3   Narrative and Multi-method Studies

A third broad category of research designs for studies using complexity science, is qualitative, narrative and multi-method designs. These approaches collect rich, in-depth interviews and other qualitative data, and apply inductive or other rigorous analytic techniques to reveal instances of emergence and its perceived qualities. In some cases, these data/analyses are supported by quantitative metrics and even statistical tools for tracking outcomes of emergent change over time. Although there is cross-over between these designs and the analogical designs mentioned above, these are not necessarily grounded in a particular complexity science, but can draw on a range of theories and insights around complex systems.

The narrative approach is perhaps best exemplified in the work by Garud and his colleagues (Garud et al. 2014; Garud and Guiliani 2013; Garud et al. 2015). These studies use the stories and images of interviewees to construct a narrative about a phenomenon which highlights the co-creative and shared emergent processes within it. By emphasizing the embodied experience of participants and the ways their stories help create organizational reality, this approach contextualizes emergence in a meaningful way. These recent studies, which examine the constitutive nature of organizing and innovation, offer a new understanding of the dynamics and complexity of innovation and change.

Other studies combine an in-depth qualitative understanding with a rigorous quantitative analysis of metrics, to show how systemic patterns and processes yield complex outcomes. The progenitors of this category are the MIRP studies by Van de Ven and his students – Venkataraman, Garud, and Polley; these are classics of combined qualitative-quantitative research (Van de Ven et al. 1989; Van de Ven et al. 1999). In brief, the team tracked several major innovations (internal venturing projects), collecting multiple interviews and measures every six weeks in each project across many years. Data on each innovation were qualitatively analyzed, and also coded into two or three quantitative variables, e.g. the degree of volatility, and the amount of new activity in the endeavor. These variables were then aggregated into time series, which could be graphically charted or statistically analyzed. An exemplar is the Cheng and Van de Ven (1996) study mentioned earlier, which analyzed these time series using deterministic chaos theory, and found a singular shift in these innovation projects, from one 'attractor' to another, during the six+ years they were under investigation.

As second exemplar of mixed method complexity research is Chiles' dissertation, published in Chiles et al. (2004), which reports on an in-depth longitudinal analysis of the emergence of Branson, Missouri – the second most popular tourist site in America (after the Grand Canyon). The data include dozens of interviews and hundreds of documents from a wide range of organizations. Their inductive analysis of the data revealed five key dynamics of self-organization across four 'cycles of emergence' in the region. Their claims were also rigorously tested using quantitative data – six time series of 100+ years each, analyzed through Poisson time series regressions. Like many successful multi-method papers, the quantitative analysis supports the insights from the qualitative and inductive work.

In parallel to the design question of how integrated (qualitative + quantitative) a study will be, a second core design feature of these studies refers to the temporality of the data, that is, whether data is to be collected in real time as the system emerges, or whether researchers collect retrospective accounts of an emergence process from those who have experienced it. This choice can be described as a continuum, with some research being more real-time based, and others being more retrospective. It turns out that most qualitative emergence studies use retrospective interviews – whether the data are analyzed through a complexity lens (for example, Padgett and Ansell 1993; Plowman et al. 2007) or not (for example, Garud et al. 2006; Purdy and Gray 2009; Sonenshein 2009; Garud et al. 2010). This approach can capture the 'entire' process of emergence, and gains the benefit of knowing there are outcomes which have occurred.

At the other end of the scale, some complexity studies have collected ongoing, real-time observations by multiple informants over many months or years. These studies strive to capture the actual week-by-week experience of emergence as it unfolds (Lichtenstein 2000; Lichtenstein et al. 2006). While more risky – for one doesn't know *ex ante* whether an organizing effort will actually yield emergent effects – this approach is able to identify *sequences* of behavior and complex systems change. Such a sequence can reveal a causal logic for emergence (Lichtenstein 2014).

A third design issue in narrative and mixed-method studies, involves the strategies for analyzing qualitative data. Overall, the majority of these studies rely on inductive qualitative analysis, in which interviews are examined for themes, which are coded and aggregated into meaningful concepts and insights. Exemplars for this approach including Plowman et al. (2007), Purdy and Gray (2009), Sonenshein (2009) and others; this is a well-known approach in organization science more generally.

An alternative method is to use constructs from the complexity science as the basis for coding interviews, as Van de Ven and his group did for innovations. Here the researchers code the data by looking for the presence of each construct in each interview (Lichtenstein 2000, 2014). The presence of each of these codes in the data can be counted, aggregated and graphed, the result is a tangible expression of the influence of each construct over time. Here, a type of panel data is necessary, that is, repeated observations of a few constructs that vary over time. Data can be organized by data collection periods (DCPs) in which the period is relatively brief – one-week, two-weeks, perhaps a month. Each DCP includes the interviews from that week or period, as well as any organizational memos, documents, financials and other secondary data. As such, each DCP is a snapshot, a frame, within the emergence process. An exemplar of this panel study approach is Amabile and colleagues' research in team innovation and learning (Amabile et al. 1996). Analytically, the value of organizing a study by DCPs is in developing a real-time visualization of emergence and change across each of its key dynamics (constructs), through a longitudinal metric of each attribute. These can be examined individually and in combination, to provide a visual analysis (Monge 1990) of the emergence process.

24.4   CONCLUSION

In summary, I've attempted to do a few things in this chapter. First, I have claimed that complexity is an amalgam of 15 sciences, each with a distinct theory, ontology, and methodology for identifying emergence and order creation in dynamic systems. These 15 are in some ways parallel to the list that Middleton-Kelly suggested (Mitleton-Kelly, 2003) as the core disciplines of complexity science, and they have been shown to be the origina-tors of the field itself (Goldstein 1999, p.55), also (McKelvey 2004). My main goal is to help complexity researchers take an expansive view of the field as a means of strength-ening it, making it more vibrant and effective. By holding the most open boundaries of complexity science we can incorporate the breadth of its research, findings, and practice. In the ideal, working across these frameworks (theories, methods) would generate syner-gies and lead to some important extensions of complexity science into management.

In addition, by presenting each science on its own, my hope is to offer the broadest palate of theoretical explanations for complexity and emergence. Each of the 15 has a different view about the generative mechanisms of emergence. As researchers we often get caught in one view or another, without knowing how to integrate alternative views that might be more interesting or relevant. For example, questions about how social structures emerge can be answered in many different ways: Autopoiesis (Luhmann 1986; Archer 1982) and autocatalysis (Padgett and Powell 2011) would posit one origin of social structures, whereas NK Landscapes would give a very different view, as would catastrophe theory (Guastello 1987, 1998), and so on. By understanding the differences between these three theories, a more thoughtful integration can be achieved to explain how and why new order emerges. Of course, these don't include the more mainstream explanations, includ-ing institutional logics (Thornton et al. 2012) among other theories.

My second main goal for the chapter is to introduce a range of research designs for applying complexity science to organizations. Especially for scholars who are considering how to pursue a research study, it is useful to see the range of options one has, the exem-plars of complexity studies that have had positive impact in the field. Likewise, before making a commitment to one or another of the methodologies, one may find a neighbor-ing complexity science is better suited as a theoretical and empirical base.

Similarly, these exemplars – drawn from mathematical, computational, and analogical approaches – allow us as scholars to appreciate the range of options that are available for answering any given question. In cases like Cheng and Van de Ven (1996) the effort was in re-analyzing existing longitudinal data into a nonlinear computer analysis; in Fleming and Sorenson (2001) it was in coding every citation of every patent for 10 years into a computational simulation; in Lichtenstein (2000) it was coding 750 interviews into a visual time series analysis of emergence. Some complexity research has been theory-based hypothesis-testing (Lichtenstein et al. 2007), others have relied on induc-tive qualitative analyses (Brown and Eisenhardt 1997; Chiles et al. 2004; Plowman et al. 2007). Some believe that computational simulations *are* what complexity means (Johnson 2009; Mitchell 2009), others have developed entire journals focusing on other complexity approaches, for example, *Nonlinear Dynamics in Psychology and the Life Sciences*, and *Emergence: Complexity and Organization*.

Perhaps most important is the advice from Edmondson and McManus (2007), who remind us that a tool is only as good as the question being asked of it. Scholars asking questions about emergence and complexity must be flexible with our tools, to know enough about the toolbox to figure out which tool (science) would be best to answer the particular question at hand, and which design will collect and analyze the ideal data. As scholars we should be aware of the degree to which our designs are based on metaphors versus ideographic mapping, especially so that when we present our findings to practitioners we maintain high integrity in our descriptions (c.f. Maguire and McKelvey 1999).

Last, I would suggest that the leading edge of this work is not in 'using complexity science' per se, but using complexity science tools to *better understand emergence* – the coming-into-being of new structures and social entities. Emergence studies are becoming increasingly common in the top journals; recent examples include Cardon et al. (2017), Howard-Grenville et al. (2017), Wiedner et al. (2017) Tatarynowicz et al. (2016), Fulmer and Ostroff (2015), Selden and Fletcher (2014), Bingham and Kahl (2013), and Kozlowski et al. (2013). Complexity science is uniquely able to explore emergence, and provide unique explanations for how it occurs and what it brings. Thus, complexity researchers could make powerful contributions to this phenomenon. I hope my chapter has offered some direction in that regard.

REFERENCES

Aguinis, H. and O'Boyle, E. 2014. Star performers in twenty-first century organizations. *Personnel Psychology*, 67, 313–350.

Allen, P., Maguire, S. and McKelvey, B. (eds) 2011. *The Sage Handbook of Complexity and Management*. Thousand Oaks, CA: Sage Publications.

Amabile, T., Conti, R., Coon, H., Lazenby, J. and Herron, M. 1996. Assessing the work environment for creativity. *Academy of Management Journal*, 39.

Anderson, P., Meyer, A., Eisenhardt, K., Carley, K. and Pettigrew, A. 1999. Introduction to the special issue: Application of complexity theory to organization science. *Organization Science*, 10, 233–236.

Andriani, P. and McKelvey, B. 2009. From Gaussian to Paretian thinking: Causes and implications of power laws in organizations. *Organization Science*, 20, 1053–1071.

Archer, M. 1982. Morphogenesis versus structuration: On combining structure and action. *British Journal of Sociology*, 33.

Arthur, B. 1989. Competing technologies, increasing returns, and lock-in by historcial events. *Economic Journal*, 99, 116–131.

Arthur, B. 1990. Positive feedbacks in the economy. *Scientific American*, February, 92–99.

Arthur, B. 1994. *Increasing Returns and Path Dependence in the Economy*. Ann Arbor, MI, University of Michigan Press.

Ashby, R. 1956. *An Introduction to Cybernetics*. New York, John Wiley & Sons.

Axelrod, R. and Bennett, D.S. 1993. A landscape theory of aggregation. *British Journal of Political Science*, 23, 211–233.

Baum, J. 1999. Whole–part coevolutionary competition in organizations. In: Baum, J. and McKelvey, B. (eds) *Variations in Organization Science*. Thousand Oaks, CA: Sage Publications, 113–136.

Bénard, H. 1901. Les tourbillons cellulaires dans une nappe liquide transportant de la chaleur par convection en régime permanent. *Annales de Chimie et de Physique*, 23, 62–114.

Billinger, S., Stieglitz, N. and Schumacher, T. 2013. Search on rugged landscapes: An experimental study. *Organization Science*, 25(1), 93–108

Bingham, C. and Kahl, S. 2013. The process of schema emergence: Assimilation, deconstruction, unitization and the plurality of analogies. *Academy of Management Journal*, 56, 14–34.

Boisot, M. and McKelvey, B. 2010. Integrating modernist and postmodernist perspectives on organizations: A complexity science bridge. *Academy of Management Review*, 35, 415–434.

Brewis, S.J., Papamichail, K.N. and Rajaram, V. 2011. Decision-making practices in commercial enterprises: A cybernetic intervention into a business model. *Journal of Organizational Transformation and Social Change*, 8, 35–49.

Brown, S. and Eisenhardt, K. 1997. The art of continuous change: Linking complexity theory and time-based evolution in relentlessly shifting organizations. *Administrative Science Quarterly*, 42, 1–34.

Browning, L., Beyer, J. and Shetler, J. 1995. Building cooperation in a competitive industry: Sematech and the semiconductor industry. *Academy of Management Journal*, 38, 113–151.

Cardon, M., Post, C. and Forster, W. 2017. Team entrepreneurial passion: Its emergence and influence in new venture teams. *Academy of Management Review*, 42, 283–305.

Carley, K. 1990. Group stability: A socio-cognitive approach. In: Thye, S., Lawler, E.J., Macy, M.W., and Walker, H.A. (eds) *Advances in Group Processes*. Stamford, CT: JAI Press, 1–44.

Carley, K. 1991. A theory of group stability. *American Journal of Sociology*, 56, 331–354.

Carley, K. 1999. On the evolution of social and organizational networks. In: Andrews, S. and Knoke, D. (eds) *Research in the Sociology of Organizations*. Stamford, CT: JAI Press, 3–30.

Carley, K. and Hill, V. 2001. Structural change and learning within organizations. In: Lomi, A. and Larson, E. (eds) *Dynamics of Organizations: Computational Modeling and Organizational Theories*. Cambridge, MA: MIT Press/AAAI Press, 63–92.

Carley, K. and Lee, J.-S. 1998. Dynamic organizations: Organizational adaptation in a changing environment. *Advances in Strategic Management*, 15, 269–297.

Caza, B.B. and Milton, L. 2011. Resilience at work: Building capability in the face of adversity. In: Cameron, R. and Spreitzer, G. (eds) *Oxford Handbook of Positive Organizations*. New York: Oxford University Press, 895–908.

Chaisson, E. 2001. *Cosmic Evolution: The Rise of Complexity in Nature*. Cambridge, MA, Harvard University Press.

Cheng, Y. and Van de Ven, A. 1996. The innovation journey: Order out of chaos? *Organization Science*, 6, 593–614.

Chiles, T., Meyer, A. and Hench, T. 2004. Organizational emergence: The origin and transformation of Branson, Missouri's Musical Theaters. *Organization Science*, 15, 499–520.

Cohen, M. 1999. Commentary on the *Organization Science* Special Issue on Complexity. *Organization Science*, 10, 373–376.

Crawford, G.C., Aguinis, H., Lichtenstein, B., Davidsson, P. and McKelvey, B. 2015. Power law distribution in entrepreneurship: Implications for theory and practice. *Journal of Business Venturing*, 30, 696–713.

Csanyi, V. and Kampis, G. 1985. Autogenesis: Evolution of replicative systems. *Journal of Theoretical Biology*, 114, 303–321.

Davis, J., Eisenhardt, K. and Bingham, C. 2007. Developing theory through simulation methods. *Academy of Management Review*, 32, 480–499.

DiRenzo, M. and Greenhaus, J. 2011. Job search and voluntary turnover in a boundaryless world: A control theory perspective. *Academy of Management Review*, 36, 567–589.

Dooley, K., Kiel, D. and Dietz, A.S. 2013. Introduction to the special issue on nonlinear organizational dynamics. *Nonlinear Dynamics, Psychology, and Life Sciences*, 17, 1–2.

Drazin, R. and Sandelands, L. 1992. Autogenesis: A perspective on the process of organizing. *Organization Science*, 3, 230–249.

Edmondson, A. and McManus, S. 2007. Methodological fit in management field research. *Academy of Management Review*, 32, 1155–1179.

Eigen, M. 1971. Molecular self-organization and the early stages of evolution. *Quarterly Reviews of Biophysics*, 4, 149–212.

Eigen, M. and Schuster, P. 1979. *The Hypercycle: A Principle of Natural Self-Organizing; In Three Parts*. New York: Springer.

Espinosa, A., Harnden, R. and Walker, J. 2004. Cybernetics and participation: From theory to practice. *Systemic Practice and Action Research*, 17, 573–589.

Ethiraj, S.K. and Levinthal, D. 2004. Modularity and innovation in complex systems. *Management Science*, 50, 159–173.

Fleming, L. and Sorenson, O. 2001. Technology as a complex adaptive system. *Research Policy*, 30, 1019–1039.

Folke, C., Carpenter, S., Walker, B., Scheffer, M., Chapin, T. and Rockstrom, J. 2010. Resilience thinking: Integrating resilience, adaptability and transformability. *Ecology and Society*, 15, 20–28.

Foster, J. 2011. Energy, aesthetics and knowledge in complex economic systems. *Journal of Economic Behavior and Organization*, 80, 88–100.

Fuller, T. and Moran, P. 2001. Small enterprises as complex adaptive systems: A methdological question? *Entrepreneurship and Regional Development*, 13, 47–63.

Fulmer, C.A. and Ostroff, C. 2015. Convergence and emergence in organizations: An integrative framework and review. *Journal of Organizational Behavior*, doi: 10.1002/job.1987.

Ganco, M. and Agarwal, R. 2009. Performance differentials between diversifying entrants and entrepreneurial start-ups: A complexity approach. *Academy of Management Review*, 34, 228–253.

Garud, R. and Guiliani, A.P. 2013. A narrative perspective on entrepreneurial opportunities. *Academy of Management Review*, 38, 157–160.

Garud, R., Gehman, J. and Guiliani, A.P. 2014. Contextualizing entrepreneurial innovation: A narrative perspective. *Research Policy*, 43.

Garud, R., Kumaraswamy, A. and Karnøe, P. 2010. Path dependence or path creation? *Journal of Management Studies*, 47, 760–774.

Garud, R., Kumaraswamy, A. and Sambamurthy, V. 2006. Emergent by design: Performance and transformation at Infosys Technologies. *Organization Science*, 17, 277–286.

Garud, R., Simpson, B., Langley, A. and Tsoukas, H. 2015. How does novelty emerge? In: Garud, R., Simpson, B., Langley, A. and Tsoukas, H. (eds) *The Emergence of Novelty in Organizations*. Oxford: Oxford University Press, 1–26.

Gavetti, G. and Levinthal, D. 2000. Looking forward and looking backward: Cognitive and experiential search. *Administrative Science Quarterly*, 45, 113–137.

Gleick, J. 1987. *Chaos: Making a New Science*. New York: Penguin.

Goldstein, J. 1999. Emergence as a construct: History and issues. *Emergence*, 1, 49–72.

Guastello, S. 1987. A butterfuly catastrophe model of motivation in organizations. *Journal of Applied Psychology*, 72, 165–182.

Guastello, S. 1995. *Chaos, Catastrophe, and Human Affairs: Applications of Nonlinear Dynamics to Work, Organizations, and Social Evolution*. Mahway, NJ: Erlbaum.

Guastello, S. 1998. Self-organization and leadership emergence. *Nonlinear Dynamics, Psychology, and Life Sciences*, 2, 301–315.

Gunderson, L. and Holling, C.S. (eds) 2001. *Panarchy: Understanding Transformations in Human and Natural Systems*. Washington, DC: Island Press.

Hall, D. 1976. *Careers in Organizations*. Glenview, IL: Scott, Foresman and Company.

Hildago, C. 2015. *Why Information Grows: The Evolution of Order from Atoms to Economies*. New York: Basic Books.

Holland, J. 1975. *Adaptation in Natural and Artificial Systems*. Ann Arbor, MI: University of Michigan Press.

Holland, J. 1995. *Hidden Order: How Adaptation Builds Complexity*. New York: Addison-Wesley.

Holland, J. 1998. *Emergence: From Chaos to Order*. Cambridge, MA, Perseus Books.

Holling, C.S. 1973. Resilience and stability of ecological systems. *Annual Review of Ecology and Systematics*, 4, 1–23.

Howard-Grenville, J., Nelson, A., Earle, A.G., Haack, J. and Young, D. 2017. If chemists don't do it, who is going to? Peer-driven occupational change and the emergence of green chemistry. *Administrative Science Quarterly*, 62, 524–560.

Johnson, N. 2009. *Simply Complexity: A Clear Guide to Complexity Theory*. Oxford: Oneworld Publications.

Kauffman, S. 1993. *The Origins of Order*. New York: Oxford University Press.

Kickert, W. 1993. Autopoiesis and the science of (public) administration: Essence, sense and nonsense. *Organization Studies*, 14, 261–278.

Kozlowski, S.W.J., Chao, G.T., Grand, M.T.B. and Kuljanin, G. 2013. Advancing multilevel research design: Capturing the dynamics of emergence. *Organizational Research Methods*, 16, 581–615.

Krugman, P. 1996. *The Self-Organizing Economy*. Cambridge, MA: Bradford Press.

Levie, J. and Lichtenstein, B. 2010. A terminal assessment of stages theory: Introducing a dynamic states approach to entrepreneurship. *Entrepreneurship Theory and Practice*, 34, 314–354.

Levinthal, D. 1991. Organizational adaptation and environmental selection: Interrelated processes of change. *Organization Science*, 2, 140–144.

Levinthal, D. 1997. Adaptation on rugged landscapes. *Management Science*, 43, 934–950.

Lichtenstein, B. 1988. The New Science Lectures. *Public lecture series*. Santa Cruz, CA: University of Santa Cruz.

Lichtenstein, B. 1991. A difference that makes a differance: Cybernetic inquiry and post-modern philosophy. In: Geyer, F. (ed.) *The Cybernetics of Complex Systems*. Salinas, CA: Intersystems Publications, 13–19.

Lichtenstein, B. 2000. Self-organized transitions: A pattern amid the "chaos" of transformative change. *Academy of Management Executive*, 14, 128–141.

Lichtenstein, B. 2011. Complexity science contributions to the field of entrepreneurship. In: Allen, P., Maguire, S. and McKelvey, B. (eds) *The Sage Handbook of Complexity and Management*. Thousand Oaks, CA: Sage Publications, 471–493.

Lichtenstein, B. 2014. *Generative Emergence: A New Discipline of Organizational, Entrepreneurial, and Social Innovation*. New York City, Oxford University Press.

Lichtenstein, B. and McKelvey, B. 2011. Four types of emergence: A typology of complexity and its implications for a science of management. *International Journal of Complexity in Leadership and Management*, 1, 339–378.

Lichtenstein, B., Dooley, K. and Lumpkin, T. 2006. Measuring emergence in the dynamics of new venture creation. *Journal of Business Venturing*, 21, 153–175.

Lichtenstein, B., Carter, N., Dooley, K. and Gartner, W. 2007. Complexity dynamics of nascent entrepreneurship. *Journal of Business Venturing*, 22, 236–261.

Lomi, A. and Larsen, E. 1996. Interacting locally and evolving globally: A computational approach to the dynamics of organizational populations. *Academy of Management Journal*, 39, 1287–1321.

Lomi, A. and Larson, E. 1997. A computational approach to the evolution of competitive strategy. *Journal of Mathematical Sociology*, 22, 151–176.

Lotka, A. 1922. Contribution to the energetics of evolution. *Proceedings of the National Academy of Sciences*, 6, 147–151.

Luhmann, N. 1986. The autopoiesis of social systems. In: Geyer, F. and Van Der Zouwen, J. (eds) *Sociocybernetic Paradoxes: Observation, Control and Evolution of Self-Steering Systems*. Thousand Oaks, CA: Sage Publications, 172–192.

MacIntosh, R. and MacLean, D. 1999. Conditioned emergence: A dissipative structures approach to transformation. *Strategic Management Journal*, 20, 297–316.

Macy, M. and Skvoretz, J. 1998. The evolution of trust and cooperration between strangers: A computational model. *American Sociological Review*, 63, 638–660.

Maguire, S. and McKelvey, B. 1999. Complexity and management: Moving from fad to firm foundations. *Emergence*, 1, 19–61.

Maguire, S., McKelvey, B., Mirabeau, L. and Oztas, N. 2006. Complexity science and organization studies. In: Clegg, S., Hardy, C., Nord, W. and Lawrence, T. (eds) *Handbook of Organization Studies* (2nd edition). London: Sage Publications, 165–214.

Maruyama, M. 1963. The second cybernetics. *American Scientist*, 51, 164–179.

Maturana, H.R. and Varela, F. 1980. *Autopoiesis and Cognition*. Dordrecht: Reidel Publishing.

May, R. 1976. Simple mathematical models with very complicated dynamics. *Nature*, 26, 455–467.

McKelvey, B. 1999. Avoiding complexity catastrophe in coevolutionary pockets: Strategies for rugged landscapes. *Organization Science*, 10, 294–321.

McKelvey, B. 2001. What is complexity science? It is really order creation science. *Emergence*, 3, 137–157.

McKelvey, B. 2002. Model-centered organization science epistemology. In: Baum, J. (ed.) *Companion to Organizations*. Thousand Oaks, CA: Sage Publications, 752–780.

McKelvey, B. 2004. Toward a complexity science of entrepreneurship. *Journal of Business Venturing*, 19, 313–342.

McKelvey, B., Li, M., Xu, H. and Vidgen, R. 2013. Re-thinking Kauffman's NK fitness landscape: From artifact and groupthink to weak-tie effects. *Human Systems Management*, 32, 17–42.

Meadows, D., Meadows, D., Randers, J. and Behrens, W. 1972. *The Limits to Growth: A Report to the Club of Rome*. New York: Universe Press.

Meyer, A., Tsui, A. and Hinings, C.R. 1993. Configurational approaches to organizational analysis. *Academy of Management Journal*, 36, 1175–1195.

Mitchell, M. 2009. *Complexity: A Guided Tour*. New York, Oxford University Press.

Mitleton-Kelly, E. 2003. Ten principles of complexity and enabling infrastuctures. In: Mitleton-Kelly, E. (ed.) *Complex Systems and Evolutionary Perspectives of Organizations: The Application of Complexity Theory to Organizations*. Oxford: Elsevier, 23–50.

Monge, P. 1990. Theoretical and analytical issues in studying organizational processes. *Organization Science*, 1, 406–430.

Morowitz, H. 2002. *The Emergence of Everything*. New York: Oxford University Press.

Nair, A., Narasimhan, R. and Choi, T. 2009. Supply networks as a complex adaptive system: Toward simulation-based theory building on evolutionary decision making. *Decision Sciences*, 40, 783–815.

Nonaka, I. 1988. Creating organizational order out of chaos: Self-renewal in Japanese firms. *California Management Review*, 30, 57–73.

Nonaka, I. 1994. A dynamic theory of organizational knowledge creation. *Organization Science*, 5, 14–37.

O'Boyle, E. and Aguinis, H. 2012. The best and the rest: Revisiting the norm of normality of individual performance. *Personnel Psychology*, 65, 79–119.

Odum, H. and Pinkerton, R. 1955. Time's speed regulator: The optimum efficiency for maximum power output in physical and biological systems. *American Scientist*, 43, 331–343.

Osterwalder, A. and Pigneur, Y. 2010. *Business Model Generation: A Handbook for Visionaries, Game Changers, and Challengers*. New York: Wiley & Sons.

Oyama, K., Learmonth, G. and Chao, R. 2015. Applying complexity science to new product development: Modeling considerations, extensions and implications. *Journal of Engineering and Technology Management*, 35(1), 1–24.

Padgett, J. 2011. Autocatalysis in chemistry and the origin of life. *The Emergence of Organizations and Markets*. Princeton, NJ: Princeton University Press.

Padgett, J. and Ansell, C. 1993. Robust action and the rise of the medici, 1400–1434. *American Journal of Sociology*, 98, 1259–1319.

Padgett, J. and Powell, W. 2011. *The Emergence of Organizations and Markets*. Princeton, NJ: Princeton University Press.

Padgett, J., Lee, D. and Collier, N. 2003. Economic production as chemistry. *Industrial and Corporate Change*, 12, 843–877.

Pantzar, M. and Csanyi, V. 1991. The replicative model of the evolution of the business organization. *Journal of Social and Biological Structures*, 14, 149–163.

Paul, D.L., Butler, J.C., Pearlson, K.E. and Whinston, A.B. 1996. Computationally modeling organizational learning and adaptability as resource allocation. *Computational and Mathematical Organization Theory*, 2, 301–324.

Penrose, E. 1952. Biological analogies in the theory of the firm. *American Economic Review*, October.

Plowman, D.A., Baker, L., Beck, T., Kulkarni, M., Solansky, S. and Travis, D. 2007. Radical change accidentally: The emergence and amplification of small change. *Academy of Management Journal*, 50, 515–543.

Porter, M. and Siggelkow, N. 2008. Contextual interactions within activity systems and sustainability of competitive advantage. *Academy of Management Perspectives*, 22, 34–56.

Prietula, M. 2011. Thoughts on complexity and computational models. In: Allen, P., Maguire, S. and McKelvey, B. (eds) *The Sage Handbook of Complexity and Management*. Thousand Oaks, CA: Sage Publications, 93–109.

Prigogine, I. and Stengers, I. 1984. *Order out of Chaos*. New York: Bantam Books.

Purdy, J. and Gray, B. 2009. Conflicting logics, mechanisms of diffusion, and multi-level dynamics of emergence in institutional fields. *Academy of Management Journal*, 52, 355–380.

Repenning, N. and Sterman, J. 2001. Nobody ever gets credit for fixing problems that never happened: Creating and sustaining process improvement. *California Management Review*, 43, 64–88.

Rosen, R. 1971. Some realizations of (M,R)-systems and their interpretation. *Bulletin of Mathematical Biophysics*, 33, 309–319.

Rosen, R. 1973. On the dynamical realization of (M,R)-systems. *Bulletin of Mathematical Biology*, 35, 1–9.

Rudolph, J. and Repenning, N. 2002. Disaster dynamics: Understanding the role of quantity in organizational collapse. *Administrative Science Quarterly*, 47, 1–30.

Saviotti, P.P. and Mani, G.S. 1998. Technological evolution, self-organization, and knowledge. *Journal of High Technology Management Research*, 9, 255–270.

Sawyer, K. 2001. Simulating emergence and downward causation in small groups. In: Moss, S. and Davidsson, P. (eds) *Multi-agent-based Simulation*. Berlin: Springer, 49–67.

Schelling, T. 1978. *Micromotives and Macrobehavior*. New York, W.W. Norton.

Selden, P. and Fletcher, D. 2014. The entrepreneurial journey as an emergent hierarchical system of artifact-creating processes. *Journal of Business Venturing*, 30, doi:10.1016/j.jbusvent.2014.09.002.

Senge, P.M. 2006. *The Fifth Discipline (Revised Edition)*. New York: Doubleday.

Smith, C. 1986. Transformation and regeneration in social systems: A dissipative structure perspective. *Systems Research*, 3, 203–213.

Smith, C. and Comer, D. 1994. Change in the small group: A dissipative structure perspective. *Human Relations*, 47, 553–581.

Sonenshein, S. 2009. Emergence of ethical issues during strategic change implementation. *Organization Science*, 20, 223–239.

Sorenson, O. 2002. Interorganizational complexity and computation. In: Baum, J. (ed.) *Companion to Organizations*. Malden, MA: Blackwell Publishers, 664–685.

Sorenson, O., Rivkin, J. and Fleming, L. 2006. Complexity, networks and knowledge flow. *Research Policy*, 35, 994–1017.

Sterman, J. 2002. All models are wrong: Reflections on becoming a systems scientist. *System Dynamics Review*, 18, 501–532.

Sterman, J. and Wittenberg, J. 1999. Path dependence, competition, and succession in the dynamics of scientific revolutions. *Organization Science*, 10, 322–341.

Sutcliffe, K. and Vogus, T. 2013. Organizing for resilience. In: Cameron, K., Dutton, J. and Quinn, R. (eds) *Positive Organizational Scholarship*, San Francisco, CA: Berrett-Koehler, 95–110.

Tatarynowicz, A., Sytch, M. and Gulati, R. 2016. Environmental demands and the emergence of social structure. *Administrative Science Quarterly*, 61, 52–86.

Tesfatsion, L. 2011. Agent-based modeling and institutional design. *Eastern Economic Journal*, 37, 13–19.

Thomas, I., Frankhauser, P. and Badariotti, D. 2012. Comparing the fractality of European urban neighborhoods: Do national contexts matter? *Journal of Geographic Systems*, 14, 189–208.

Thornton, P., Ocasio, W. and Lounsbury, M. 2012. *The Institutional Logics Perspective: A New Approach to Culture, Structure, and Process*. Oxford: Oxford University Press.

Tsoukas, H. 1989. The validity of idiographic research explanations. *Academy of Management Review*, 14, 551–561.

Tsoukas, H. 1991. The missing link: A transformational view of metaphors in organizational science. *Academy of Management Review*, 16, 566–585.

Uhl-Bien, M. and Marion, R. (eds) 2008. *Complexity Leadership. Part 1: Conceptual Foundations*. Charlotte, NC: Information Age Publishing.

Ulanowicz, R. 1980. An hypothesis on the development of natural communities. *Journal of Theoretical Biology*, 85, 225–245.

Ulanowicz, R. 1987. Growth and development: Variational principles reconsidered. *European Journal of Operational Research*, 30, 173–178.

Van de Ven, A., Angel, H. and Poole, M.S. 1989. *Research on the Management of Innovation*. New York: Ballanger Books.

Van de Ven, A., Pooley, D., Garud, R. and Venkataramen, S. 1999. *The Innovation Journey*. New York: Oxford University Press.

von Foerster, H. 1960. On self-organizing systems and their environments. In: Yovitz, M. and Cameron, S. (eds) *Self-organizing Systems*. New York: Pergamon Press, 31–50.

Walker, B., Holling, C.S., Carpenter, S. and Kinzig, A. 2004. Resilience, adaptability and transformability in social-ecological systems. *Ecology and Society*, 9, 5.

Wicken, J. 1979. The generation of complexity in evolution: A thermodynamic and information-theoretical discussion. *Journal of Theoretical Biology*, 77, 349–365.

Wicken, J. 1980. A thermodynamic theory of evolution. *Journal of Theoretical Biology*, 87, 9–23.

Wiedner, R., Barrett, M. and Osborn, E. 2017. The emergence of change in unexpected places: Resourcing across organizational practices in strategic change. *Academy of Management Journal*, 60, 823–845.

25. Mixed methods research: a method for complex systems

Professor Liz Varga, Cranfield University

> Whether or not we agree about the source of our knowledge and ideas (e.g. empiricist – sensed
> or rationalist – innate/deduced) the problem is about content of our knowledge and
> how the constant conjunction of events differs from our extant explanations.
> (Markie, 2013, attributed to Plato)

25.1 INTRODUCTION

Mixed methods research generally refers to a research strategy (or approach) which uses more than one research method in a single research project, but more precisely it is the use of different types of *data*, most likely both qualitative and quantitative data, in a research design to solve a research problem. Different types of data can lead to the use of different research methods, and when more than one method is used it is critical to understand the role of each method, its sequence within the research project as a whole, and the transformation of knowledge from one stage in the research design to the next until the research problem is addressed.

In order to understand mixed methods, specific knowledge of a broad range of research methods is not required. Neither is a knowledge of social systems research methods. Awareness is needed that researchers hold philosophical and paradigmatic preferences and that there exist choices for the researcher regarding research strategy (for example, mixed methods), research design (for example, triangulation), research methods (for example, action research), and the specific techniques used for data collection (for example, interview) and analysis (for example, modeling). A review of the domain is provided, identifying different methods but without any detailed explanations of each since these can be found elsewhere. The findings are intended for the researcher investigating real world problems however the applicability of the findings is broader than social research since mixed methods are also recognized in traditionally scientific disciplines, such as physics and biology.

Complex systems arise in all global systems from the atmosphere, hydrosphere, biosphere and lithosphere, to the anthroposphere. So regardless of a researcher's 'home' discipline the methods selected will need to address a problem in the real world. The real world is one in which complex systems co-evolve, not just within global systems but between them and at multiple scales, giving rise to unexpected and emergent phenomena.

All real-world problems involve the use of different types of data although traditional disciplinary related research tends to prefer its own largely mono types of data and related methods, for example, financial markets' use of quantitative pricing data for methods to track stock markets, or marketing research using qualitative questionnaire data for methods to analyse consumer preferences. Greene (2008, p.7) notes the practical demands upon practitioners, from the contexts in which they work, to identify "defensible patterns of recurring regularity as well as insight into variation and difference. And they called for results that conveyed magnitude and dimensionality as well as results that portrayed contextual stories about lived experiences. And they called for dispassionate neutrality as well as engaged advocacy for such democratic ideals as equity and justice." So it is desirable that appropriate methods are used to solve, understand, explain, and shed light on real-world problems which cross disciplinary boundaries. The opportunity of using multiple methods, across multiple disciplines, to gain cross-cutting insight into real world problems is driving a mixed methods strategy. Demands for multi-disciplinary, interdisciplinary and trans-disciplinary knowledge need research strategies which integrate and blend data and methods, to generate new fields of knowledge, to be closer to practice and to be more pertinent to decision-making.

But research methods and knowledge are co-evolving. Methods and techniques mature over time as they are better understood. The researcher also learns about the data and the methods during the research project and so when the methods are next used, they may be applied slightly differently. The quality, range, context, and interpretation of data also varies across research projects, leading to different results. In addition, the phenomena or subjects of research will also evolve over time and so repeating the same research project will almost inevitably create new findings. Even the use of the same research design for two concurrent research projects, are likely to differ somewhat in findings, as a consequence of location, different data, different choices of classification, categorization, validation, interpretation of results and so on.

In addition, the environment, and thus the research context, is always different. "You cannot step into the same river twice" (Johnston 2013, attributed to Heraclitus 500BC). Even mechanical systems, such as production lines, involve people, who influence system outcomes and performance based on their own goals and ambitions, which influence others, such as suppliers, in real-world systems who can learn and respond differently on future occasions. Interactions occur between the research team, the subjects or objects of the research, and the real world in which it is embedded. Importantly these interactions and the feedback they generate are nonlinear and many, and as the loops interact with each other, prediction in real world systems using standard analytical methods is "problematic (at best) for most intents and purposes" (Richardson and Tait 2010, p.91).

The introductory quote from Plato observes the need to recognize the anomalies between extant knowledge and experience. It is always the case that human knowledge is incomplete and that certainty in scientific enquiry is futile (Maxwell 1990). This is exacerbated by the pace of change in the real world. The juxtaposition of knowledge and experience will increasingly identify new research problems. Experiences are idiosyncratic, and knowledge is not omnipresent, so individuals and society reflect on learning through the process of naming and re-defining 'things', such as, phenomena, objects, units of analysis, and so on, updating repertoires of knowledge. And of course, the pace of change in the real world is not constant because evolution occurs at different rates in different places. The best way to understand the co-evolving real world is by using a mixed methods strategy which can respond to the various types of data constituting a real world problem. The demand for mixed methods research will increase as society develops higher performing computing, more knowledge, more access to data, and to different forms of data (for example, tweets, images), and are asking ever more ambitious research questions for an ever increasing variety of research problems.

Mixed methods is an appropriate, not just pragmatic, strategy for complex systems research just as mixed methods are underpinned by complexity science. They straddle disciplines and philosophical perspectives, to detect emergent phenomena, and to integrate across scales and boundaries.

25.2   INTEGRATION OF METHODS, VALIDITY AND RELIABILITY

When the *Journal for Mixed Methods Research* was established, the first issue's editorial recognized the evolving nature of mixed methods research, and set out the breadth of studies in mixed methods strategies by use of two types of questions, procedures, data types, analyses, and conclusions (see Box 25.1).

Fundamentally, mixed methods research concerns the integration of qualitative and quantitative data. The question of whether or not there exists a serious integration of findings from the use of multiple methods distinguishes mixed studies from quasi-mixed studies (Teddlie and Tashakkori 2006). Integration of findings occurs not always in the final stage of a research project, but may occur between stages and so between quantitative and qualitative methods or vice versa. When quantitative and qualitative data analyses are conducted separately, neither type of analysis builds on or interacts with the other during the data analysis stage and the findings from each type of analysis are neither compared nor consolidated until both sets of data analyses have been completed. This final stage strategy is referred to as non-cross-over mixed analyses by Onwuegbuzie et al. (2009). The mixed method process is located at the end of the research project, and the research strategy (also referred to as research paradigm by some authors) is quasi-mixed method. These are distinct from cross-over mixed analyses which require the integration of data and insights during the research project, and include techniques for integrated data reduction, data correlation, data consolidation, and so on (Onwuegbuzie et al. 2009, p.119).

The means of integration is at the heart of mixed methods challenges. Methods for integration have been reviewed by Greene (2008) who finds a variety of alternative strategies

---

BOX 25.1   SCOPE OF MIXED METHODS RESEARCH

- two types of research questions (with qualitative and quantitative approaches),
- the manner in which the research questions are developed (participatory vs. pre-planned),
- two types of sampling procedures (for example, probability and purposive),
- two types of data collection procedures (for example, focus groups and surveys),
- two types of data (for example, numerical and textual),
- two types of data analysis (statistical and thematic), and
- two types of conclusions (emic and etic representations, 'objective' and 'subjective,' etc.).

Source: Tashakkori and Creswell (2007) p.4.

Analysis Phase	Mixed Methods Analysis Strategy
Data transformation	Data transformation, one form to another (Teddlie & Tashakkori, 2003)
	Data consolidation or merging, multiple data sets into one (Louis, 1982)
Data comparison, looking for patterns	Data importation–using interim results of analyses of one data set to inform the analysis of another data set (e.g., extreme case analysis) (Li, Marquart, & Zercher, 2000)
	Integrated data display–presenting data from multiple sources in one display, thereby enabling cross-method comparisons and analyses (Lee & Greene, 2007)
Major analyses for inferences and conclusions	Warranted assertion analysis–iteratively reviewing all data for purposes of directly generating inferences (Smith, 1997)
	Pattern matching (Marquart, 1990)
	Results synthesis (McConney, Rudd, & Ayres, 2002)

Note: Mixed methods analysis strategies take place *after* data have been cleaned and descriptively analyzed. References cited characteristically refer to an example of the strategy noted.

Source: Greene (2008 p.15).

Figure 25.1   Integrated mixed methods analysis strategies

for data transformation, data comparison, and analyses for inferences and conclusions. See Figure 25.1.

It is generally agreed that certainty in scientific enquiry is regarded as futile (Maxwell 1990) so there is a focus on validity (Whittemore, Chase and Mandle 2001). This underlines the challenges to the attainment of complete knowledge in any scientific enquiry. Even where technical measures are used, different environmental and contextual conditions may lead to different insights. So knowledge is at best contingent. But in the real world it is not only the behaviour of technical components but also the behaviour of people which influences knowledge. As a result of this, categoric certainty cannot be assured. Rather than attempt to find certainty, the focus changes towards assuring the validity of the process, for example, repeatability, transparency, and so on. However validity is focused towards different aspects in qualitative and quantitative research. Rizzo, Corsaro and Bates (1992) highlighted the most important methodological differences being the relative emphasis placed on internal validity, that is, the accuracy of interpretations and reliability, for example, the transparency of coding schemes. For qualitative research, internal validity is paramount, whereas for quantitative research, it is reliability. Qualitative research also attends to reliability, as does, quantitative research to internal validity, but just not in the same measure.

The question of concept validity was addressed in one of the earliest articles on mixed methods which proposed a practical research strategy. Campbell and Fiske (1959) created the multitrait-multimethod matrix and introduced convergent and discriminant as new sub-categories of construct validity describing the degree to which concepts that should (or should not) be theoretically related are interrelated (or not) in reality. Despite the development of qualitative and quantitative methods, their integration and conjoint validity and reliability remain areas in need of further research.

25.3   MIXED METHODS DESIGNS

Several mixed methods research designs have been devised to describe and explain appropriate ways of arranging quantitative and qualitative methods in a research project. Greene, Caracelli and Graham (1989) describe the logics and purpose for the integration of methods whilst Teddlie and Tashakkori (2003) provide clarity on the understanding of the integration of methods. Nearly all research projects have multiple stages, each of which is driven by a different logic. Logic elaborates the reasons for sequencing the stages, so that ordering of methods has explicit relevance and influences the questions and/or focus of the next stage (Brannen 2005).

Mixed methods research designs provide a rational way to determine three possible dimensions of integration of methods: the location of the mixing, the priority of the method and the timing of the method. The location of mixing the methods may occur throughout or during the research design or at the end of the research design which reflects whether the methods are implemented independently or interactively (Crump and Logan 2010). The priority states the dominance of the method, if any, such that methods have parity or one can dominate. And the timing refers to whether methods are implemented concurrently or sequentially, and if the latter, the chronological order of the methods. Location, priority and timing of mixed methods research designs are shown in Figure 25.2. Key mixed methods research designs include: complementary, triangulation, embedded, explanatory, exploratory, initiation, contradiction, and expansion.

A Complementary Design treats qualitative and quantitative results differently but juxtaposes them for greater overall insight (Brannen 2005) or uses the strengths of one method to enhance the other in an additive way not a combinatorial way (Sale, Lohfeld and Brazil 2002). A complementarity design seeks elaboration, enhancement, illustration and clarification of the results from one method with the results from the other method (Creswell and Plano Clark 2010). Quantitative and qualitative results are used to measure overlapping but different phenomena.

A Triangulation Design brings together two different datasets which have been collected to answer the same question (Creswell and Plano Clark 2010). Alternatively, the same phenomenon is researched in order to seek convergence and corroboration and to eliminate the inherent biases from using only one method (Denzin and Lincoln 1994). This design is not without difficulties. Hammersley (2008) notes that triangulation may involve using different qualitative sources of data or quantitative methods rather than mixing either. Furthermore there is a philosophical issue with triangulation concerning the notion of the existence of a single reality which can be validated using alternative data sources and methods. This may fit a post-positivist philosophical stance, but not a constructivist or postmodernist one which allows multiple realities, and in which triangulation may at best only draw attention to their incommensurability.

An Embedded Design is one in which one dataset provides a supportive, secondary role in a study based primarily on the other (Creswell and Plano Clark 2010) or provides additional sources of information not provided by the first (Creswell 2013). Elaboration or Expansion involves data analysis of one method to add to the understanding being gained by another (Brannen 2005).

An Explanatory Design is a two-phase mixed methods design in which quantitative data is explained by qualitative investigation in the second phase (Creswell and Plano

Research Design	Location	Priority	Timing	Logic	Reference
Complementary	End		Concurrent	Use one method to add to another and measure overlapping but different phenomena	Brannen 2005 Sale et al. 2002
Complementarity					Creswell and Plano Clark 2010
Triangulation	End			Interpretation, validation or contradiction via comparison or juxtaposition	Caracelli and Greene 1993 Creswell and Plano Clark 2010
Convergent Parallel		Equal	Concurrent		Creswell 2013
Embedded	During	Unequal; usually Qn		The dominant method is augmented by the other	Creswell and Plano Clark 2010 Creswell 2013
Elaboration (or Expansion)					Brannen 2005
Explanatory	End	Qn	Sequential Qn then Ql	Quantitative data is explained by qualitative investigation	Creswell and Plano Clark 2010
Explanatory Sequential					Creswell 2013
Exploratory	During	Usually Ql	Sequential Ql then Ql or Qn	Qualitative study guides choice of subsequent method	Creswell and Plano Clark 2010
Exploratory Sequential		Ql	Sequential Ql then Qn	Quantitative study explains relationships from Ql study	Creswell 2013
Initiation	During		Sequential Ql then Qn Or Qn then Ql	Hypotheses or research questions generated from one method triggers research using a different method	Brannen 2005
Development					Creswell and Plano Clark 2010
Contradiction	End			Discount one method favouring validity of the other	Brannen 2005
			Sequential	Looks for paradox, contradiction and new perspectives for explanations	Creswell and Plano Clark 2010
Expansion				Extend the range of enquiry	Creswell and Plano Clark 2010
Transformative			Usually: sequential Qn then Ql	Address a social issue to bring about change	Creswell 2013
Multi-phase				Interrelates other design for common research objective	Creswell 2013

Note: Empty cells indicate that choices are not prescribed; Ql means qualitative, Qn means quantitative.

Figure 25.2   Examples of mixed methods research designs

Clark 2010). Explanatory Sequential describes a design in which quantitative results are explained and refined by qualitative research in a second stage (Creswell 2013).

An Exploratory Design carries out qualitative research in the first phase that guides the study of further qualitative and/or quantitative research (Creswell and Plano Clark 2010). Exploratory Sequential describes a design in which qualitative data and methods are followed by quantitative research to explain the relationships in qualitative data (Creswell 2013). Both are similar to Initiation which uses new hypotheses or research questions generated from one method to trigger research using a different method (Brannen 2005).

Contradiction discounts one method over another where results contradict each other, favouring validity or reliability (Brannen 2005).

Transformative and Multi-phase designs (Creswell 2013) are meta-designs, and so employ other designs but recognize that the purpose of the study is significant to its design. For example, the purpose may be to address a social issue of a marginalized or under-represented population in order to bring about change, or to align the different phases of a large-scale research project to its research objective.

25.4   PHILOSOPHICAL AND PARADIGMATIC INTERESTS

Complexity science is firmly located in a pluralistic view of the world permitting multi-paradigmatic views of the world, from the post-positivist to the postmodernist. This pluralism is able to inform learning and adaptation through its capability to embrace multi-ontological and multi-epistemological perspectives which represent real world systems, and employ mixed methods (quantitative and qualitative) to explore, explain and understand real world phenomena (Varga 2014).

But the question of methodological validity has concerned mixed methods scholars since its early days especially in respect of philosophical or paradigmatic reconciliation. Kuhn's (1970) notion of paradigm incommensurability reaffirms the thinking that qualitative and quantitative research is intrinsically different being underpinned by different philosophical assumptions (Brannen 2005). This leads to qualitative researchers typically employing an interpretivist or constructivist paradigm (a belief that reality depends on the beholder), although researchers may hold realist assumptions. Quantitative researchers by contrast may employ a post-positivist or critical realist paradigm (that there exists some shared reality independent of the beholder) although they can agree that social reality is constructed. This is a generalized version of researchers' metaphysical positions based on preference of data type and many variations exist. The reconciliation of philosophical perspectives in mixed methods research has been considered by reference to the degree of difference. Using this approach, it is possible to suggest stances which act as solutions to resolve differences. Indeed Greene (2008) has recently discovered some of the 'signal accomplishments' of mixed methods, including a number of alternative paradigmatic stances that have emerged to address the potential conflict in mixed methods research. See Figure 25.3.

At the strictest degree, the paradigmatic difference might be understood by a research project team as incommensurable. From this perspective, a purist stance emerges for which mixed methods has no response. If the research position is that paradigms upon which the methods are based have a different view of reality, thus a different view of the phenomenon, then mixed methods should be used for only complementary purposes, not

Degree of paradigmatic difference	Philosophical/Paradigmatic stance	Use Mixed Methods?	Suited mixed methods designs
Incommensurable	PURIST STANCE (Lincoln and Guba 1985)	No	None
Different but not fundamentally incompatible	COMPLEMENTARY STRENGTHS STANCE (Brewer and Hunter 1989; Morse 2003, Sale , Lohfeld and Brazil 2002)	Yes, but use methods discretely and compare and contrast findings	Complementary
Different but not sacrosanct; can usedifferences to achieve insights	Critical Theory DIALECTIC STANCE (Greene and Caracelli, 1997; Maxwell and Loomis 2003)	Yes but recognize the paradigm in which method contributes	Triangulation Contradiction
Converging to new paradigms, such as pragmatism, scientific realism, or transformation–emancipation	ALTERNATIVE PARADIGM STANCE (Howe 2003; Johnson and Onwuegbuzie 2004; Mertens 2003; Teddlie and Tashakkori 2003; others)	Yes	Embedded Explanatory Exploratory
Irrelevant because its assumptions are logically independent	A-PARADIGMATIC STANCE (Patton 2002; Reichardt and Cook 1979) Participatory paradigm (Heron and Reason 1997)	Yes	All
Irrelevant because it is embedded in or intertwined with practice which takes priority	SUBSTANTIVE THEORY STANCE GROUNDED THEORY (Glaser and Strauss 1967)	Yes	All
Ignored by uncritical new generation of researchers	'absent' STANCE (Sale, Lohfeld and Brazil 2002)	Yes but with caution	All

Figure 25.3   Paradigms and mixed methods designs (extending Greene 2008, Table 1, p.12)

for cross-validation or triangulation (Sale et al. 2002). A dialectic stance enables the possibility of considering the same phenomenon from different philosophical perspectives, allowing mixed methods designs such as Triangulation and Contradiction.

The most relaxed paradigmatic stance occurs when the underlying assumptions behind the qualitative-quantitative debate are perceived as merely technical and so can be ignored (Sale et al. 2002). The studies carried out under this 'absent' stance may be valid but should be used with caution as the philosophical considerations will not have been examined. For research to allow integrated, combined methods, requires either that an alternative paradigmatic stance is taken such as pragmatism, or that it is treated as technically irrelevant by using an A-paradigmatic stance or by using a Substantive Theory stance.

25.5   CONSISTENCY AND CHOICE IN METHODOLOGICAL DESIGN

The configuration of research methods and selection of a research design is not simply a technical choice. The methods to be used in the research design should be aligned with

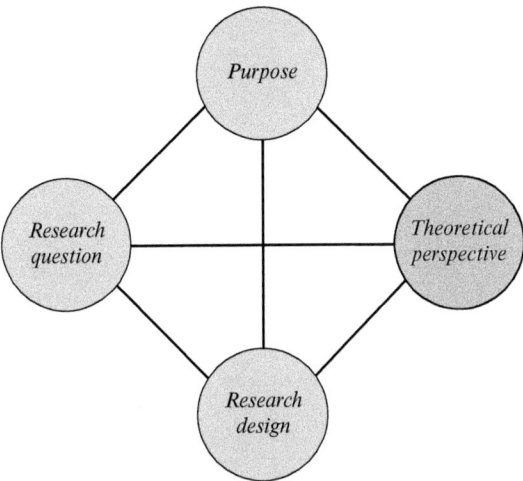

Source: Partington (2002).

Figure 25.4   Components of a research project

philosophical and theoretical assumptions; ideological, ethical, and political assumptions driving the purpose of the research; and the research questions which are to be addressed by the research. See Figure 25.4.

Brannen (2005) defines a 3Ps framework – paradigms, pragmatics and politics. Pragmatics covers the 'Research Question' theme from Partington's framework. Brannen (2005) notes that the framing of research questions may be underpinned by both philosophical and pragmatic issues. And indeed the research purpose from Partington's framework is closely aligned to the politics of Brannen's framework, and concerns the motivation for research, highlighting the value of mixed methods for use in critical studies or transformative research, such as those raising the profile of the marginalized or where social injustice occurs.

Creswell (2013) highlights the need for alignment between philosophical worldview, research approaches (strategies), designs and research methods. See Figure 25.5 which shows that 'Research Methods' provides a grouping of the various processes undertaken during research, for example, data analysis and validation.

25.6   INFLUENCES ON RESEARCH DESIGN

The key components of a mixed methods research project are: the philosophical stance and/or the theoretical stance; the research purpose and research questions; and the research design including all methods, their location in the design, their priority and timing, including methods for data integration and analysis between them. However other influences on research design need to be understood. These other influences are often practical drivers suggested by researchers (see Acknowledgements) despite being inappropriate under most circumstances for independent research. The purpose here is

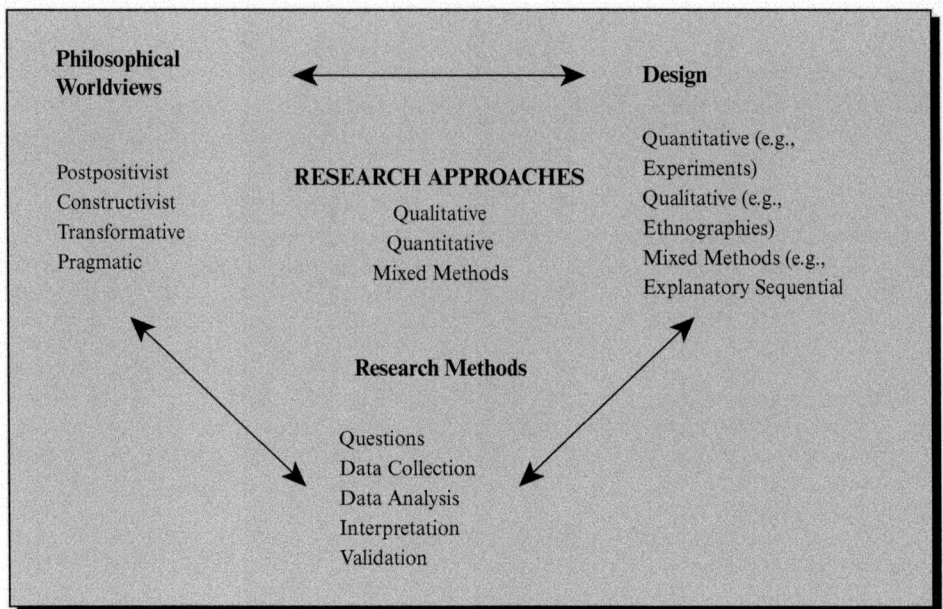

Figure 25.5   *A framework for research – the interconnection of worldviews, design and research methods*

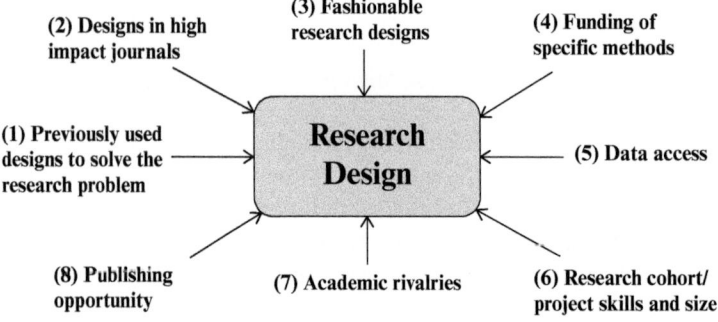

Figure 25.6   *Other influences on research design*

to recognize some of the pressures and challenges of research design selection and not to suggest ways to determine research design. See Figure 25.6.

A major influence on research design selection is (1) previously published research designs addressing the same or similar research questions but perhaps in another context or at another scale. Full details of the research design may not have been reported with complete transparency and sometimes the author may be able to provide clarity together with reflective comments about its utility. However, apparent success in one research project is not an indicator of future success. Similar research problems may be addressed successfully by using alternative methods, and articles using the most up-to-date research designs may not yet have been published. In fact, recommended examples and previous case studies are deliberately avoided here since they can constrain experimentation in research by suggesting past choices are the successful ones to be used in new research. For the interested reader, several are available (see for example: Tomos et al. 2013; Lopez-Fernandez and Molina-Azorin 2011; Wheatley 2012). And in any event, every research project is unique so previous examples can only be a proximate guideline at best. Complex systems research on the real world requires experimentation of alternative methods and techniques which do not reinforce existing choices and which encourages innovative thinking.

(2) Selection of a research design used in a peer-reviewed journal with the highest impact factor or with the greatest influence in the field of the research may occur with the hope of being able to publish in the same journal. However journal rankings change quite regularly and new journals are instantiated without impact factors but with high potential for the research to have impact.

(3) Theoretical perspectives are connected to philosophical domains, and both are evolutionary, even co-evolutionary, rising and waning in popularity. Extant perspectives and stances may be fashionable, trendy, and desirable, but not all will be long-lived, robust and valid. A fashionable research design may be relevant only for the social context in which it fits. Mixed methods itself is sometimes seen as a fad, or a universalistic discourse that believes mixed methods research produces better outcomes than single method research (Bryman 2007, quoted in Brannen 2005). The endorsement for interdisciplinary research programmes (RCUK 2015) may also been seen as fashionable and is also a driver of mixed methods designs.

(4) The availability of funding sources may also drive research design. Some funding sources demand the use of specific methods, for example, interdisciplinary methods to solve nexus-type (overlapping sector) problems such as Beddington's (2010) water, energy, food 'Perfect Storm'. Funders increasingly agree that "complex and pluralistic social contexts demand analysis that is informed by multiple and diverse perspectives" (Sammons et al. 2005, p.221) and the need for pluralism (Bohmann 1996) whereby "linear, scientistic conceptions of innovation are giving ground to more plural, socially situated understandings" (Stirling 2008, p.262).

(5) Access to data, either primary or secondary, may influence the research design. A particular research design may be preferred but it may not be practical, for example, due to difficulties with access to confidential or sensitive data. Convenience sampling which relies on conveniently available data or respondents, may influence choice of research design, as may availability of public data or data obtained via networks or relationships. The issue of data access is related to the feasibility of particular methods (Brannen 2005).

(6) The skills and time (or capability and capacity) allocated to the research project to collect, code, analyse and report on data, that is, the resources available to the research project (Brannen 1992), will create a constraint on the research design. The demand for resources is perceived as being greater to implement both quantitative and qualitative aspects in a mixed methods project (Niglas 2004). Many research projects are multi-institutional and have the potential to use mixed methods whereby partners bring

knowledge of different methods which together may provide insights into intractable, real world problems. However, selection of a mixed methods strategy requires the Principle Investigator to show leadership and innovation, to overcome practical limitations of truly integrating data and methods when researchers are not co-located and not of the same theoretical or philosophical disposition.

(7) Researchers with established track-records may become synonymous with specific methods. This may create rivalries between professionals who persist in advocating the methods which have established their careers. Research centres may attract lucrative funding for specific academics due to their experience and expertise in specific methods, reducing the incentive to try new methods.

(8) Conferences, invitations to contribute to special issues of journals, and other opportunities which provide attractive outlets for research, may influence research design choices. Preferred methods used by the conference or journal, lead time to conference/publication, access to data which is relevant for the conference or journal, availability of resources and skills to collect and process data, may lead to the selection of a research design which is not ideal.

Other influences on research design are possible and Figure 25.6 is not intended to be exhaustive.

25.7   RESEARCH TECHNIQUES – DATA COLLECTION

Strauss and Corbin's (1998, p.3) definition of research method "A set of procedures and techniques for gathering and analyzing data" serves to exemplify the need to consider data collection and analysis as discrete activities. Both activities must be mindful of data type: quantitative and qualitative.

Quantitative data is data which can be measured using some numerical scale, for example, length, frequency, rate, whilst qualitative data represents words or narrative which can be divided into different groups, and so is categorical (grouped by convenience or purpose) or is arbitrary (determined by chance). For rigorous research, all data must be collected systematically using data collection techniques which align with philosophical and theoretical stances, generally under the auspices of a particular research method. For example case study research in a Hermeneutics tradition could use questionnaire data collection techniques.

It is possible that specific techniques may be employed by different methods. Figure 25.7 presents recognized techniques of data collections that have the potential to be pertinent to a variety of research methods. Complex systems are real world systems which are messy with open boundaries, and the ability to experiment with different techniques using alternative methods, is possible. The figure aligns data collection techniques in the first column with research methods in the second column. These are the traditional ways of collecting data and then processing the data to gain insight. The reason for the 'ladder' in the middle is to show that these data collection techniques can be distanced from traditional methods to gain new insights. So whilst the choice of method, for example, ethnography might indicate collecting data through lived experiences, the technique should not be regarded as exclusive and other techniques such as interview could and are used.

Data collection research technique	Research method	Philosophical or Theoretical Stance[1]
Replicating data in electronic, paper, sound, or multi-media formats[2]	Critical emancipatory psychology Reflective phenomenology Collaborative research Dialogical analysis	Critical Theory Critical Realism
Interview - Open ended - Semi-structured - Structured	Action research Population simulation Longitudinal studies	Pragmatism
Survey instrument; Census instrument	Survey Census	
Questionnaire	Case study Life history study Focus groups	Hermeneutics
Observation; Obtain reflections	Heuristic inquiry Narrative research	Heuristics Phenomenography
Diary/journal; Continuous recording	Phenomenology Naturalistic / interpretive inquiry Symbolic interactionism	Grounded Theory
Experiment - Quasi - Field	Computational methods Artificial Intelligence methods	Post-positivism
Randomized Control Trial		
Collect data through lived experiences	Ethnography Ethnoscience Qualitative Evaluation	Social Constructionist

Notes:
1. The figure attempts to classify in a static way extant philosophies and theories which the reader is asked to examine from other sources; these philosophies and theories have typical methods shown in the second column however these are constantly shifting and dynamic.
2. Replication is a data collection technique itself and does not involve novel empirical data collection. Researchers often refer to the use of secondary data sources and replication is the technique they use prior to analysis.

Figure 25.7   Alignment of techniques for data collection to research method and philosophical and theoretical stance; building on Niglas (2004)

25.8   DATA COLLECTION CHALLENGES

Mixed methods research designs are likely to involve more than one technique for data collection, for example, the use of both experiments and interviews. The additional data collection requirements (Bryman 2012; Tashakkori and Teddlie 2010) for mixed methods designs will create challenges and will require strategies to achieve successful research outcomes. These additional data collection challenges include (see Figure 25.8): (a) Multiple forms of data access requirements to the same organization (or to the community which is the subject of research); (b) Greater knowledge of a particular research problem will be gleaned raising the potential for ethical considerations; (c) The same respondent may become fatigued if asked to provide the same information in different ways; (d) The time required to collect multiple datasets and keep good records of the data can be time-consuming for the researchers; (e) The absolute and relative sizes of different samples from different sources may need to be comparable so attention will need to be given to sampling sufficiency; (f) Sample selection criteria and bias will need to be considered across datasets to avoid marginalization or exclusion of respondents; (g) The division of data collection work across multiple researchers needs consideration in respect of the skills and development opportunities for researchers. Strategies are needed to solve the additional burdens of mixed methods data collection challenges and suggestions are shown in Figure 25.8.

Mixed Methods Requirement	Mixed Methods challenge	Suggested Strategies
a) Multiple access needs within one organization	Avoid excessive demands upon one organization, leading to withdrawal	• Good communication • Manage expectations • Providing interim updates • Use multiple organizations
b) Ethical concerns	Uncover insights which may harm individuals or organizations	• Use data anonymization • Avoid small area statistics, e.g. by data merge
c) Participant selection	Avoid lack lustre or conflicting responses due to participant fatigue	• Feedback previous results • Ask for datasets concurrently • Ask different people in the organization
d) Data collection time	Employ a mixed methods design that is parsimonious	• Recognize data saturation and avoid unnecessary duplication
e) Different sample sizes	Statistical procedures needed for quants methods may require more data than is possible to collect qualitatively	• Preparation and planning for datasets • Realistic estimation of response rates
f) Sample selection criteria and bias	Avoid marginalization and exclusion due to selection choices	• Expose inherent biases in different data collection choices, e.g. use paper and electronic capture
g) Multi-researcher	The quality of data collection may be compromised if skills are inadequate	• Planning and training for skills acquisition • Use of mentoring by skilled researchers

Figure 25.8   Strategies to address challenges created by more demanding data collection requirements of mixed methods designs

Techniques	Qualitative strategy	Mixed methods strategy	Quantitative strategy
Descriptive techniques describe data by categorization or interpretation	• Word count • Cognitive mapping • Thick description • Content analysis • Theoretical Coding • Grounded Coding • Taxonomic analysis	• Integrated data display	• Frequency count • Correlation • Cluster analysis • Measures of central tendency and dispersion • Principal components analysis
Comparative techniques compare two or more datasets	• Multi repertory grids • Analytic induction • Inter-rater analysis • Concordancing	• Data transformation • Cross-Over analysis • Data consolidation • Results synthesis • Pattern Matching	• Mann-Whitney 'U' test • t-tests • ANOVA • ANCOVA (co-variance)
Predictive techniques uncover rules and patterns in data in an attempt to predict future states	• Induction • Theory building • Abductive inference • Framework development, e.g. BCG Matrix • Qualitative models, e.g. Porter's 5 forces		• Regression • Path analysis • Genetic algorithms • Modeling • Simulation • Network Analysis • Data mining

Note: Shading shows typical priority.

Figure 25.9   *Examples of descriptive, comparative, and predictive data analysis techniques*

25.9   RESEARCH TECHNIQUES – DATA ANALYSIS

Data analysis is the process of carrying out consolidation, transformation, categorization, comparison, or more generally processing of datasets, but excluding data collection or gathering. The reason for choosing a particular data analysis technique is its capability to describe, compare or predict from available qualitative or quantitative data. Data analysis techniques can be categorized as descriptive, comparative or predictive. Alternative data analysis techniques are employed for different types of data: qualitative, quantitative and mixed. Some examples are shown in Figure 25.9.

Predictive techniques aim to identify rules and patterns of relationships which could predict the likely performance of the whole population. The data analysis techniques used for prediction embody the explanations or extant theories as to why the prediction is possible or likely. Descriptive techniques may find very many relationships between data and may turn up insights which can be developed using comparative and predictive techniques. Denzin and Lincoln (1994, p.2) argue that qualitative research is "multimethod in focus, involving an interpretive, naturalistic approach to its subject matter". Due to the use of both qualitative and quantitative data, comparative techniques are critical for mixed methods research and so it is not a surprise more development has gone into this area. There are clear gaps in descriptive and predictive techniques for mixed methods research, for example how can quantitative and qualitative data be described 'together'. Arguably some forms of modelling, such as agent-based models (ABMs), may be classified as a mixed methods predictive technique since ABMs require and integrate both qualitative and quantitative data.

25.10   DISCUSSION

Benefits of mixed methods research have been expanded elsewhere (see for example Hurmerinta-Peltomäki and Nummela 2006; Brannen 2005; Johnson and Onwuegbuzie 2004), and include: (1) data variety and richness of study; (2) reduced bias; (3) more meaningful research; (4) appeal to a wider audience (academics, practitioners, policymakers and so on); (5) better risk mitigation by using the strengths of one method to help cover weaknesses of another; (6) more practical and relevant to action; (7) more illustrative of real-life phenomena; (8) more generalizable; (9) surprising and contradictory findings create insight and opportunities for new knowledge, theories, research directions and so on; (10) self-validation via the research design itself; (11) adds value by the combination of methods and their timing; (12) more representative of (or reflects more completely) the business complexity; (13) more representative of increased blurring between disciplines; (14) enables thinking outside the domain of existing theories specific to particular disciplines; (15) aids skills enhancement and broadening of the researcher's methodological repertoire.

Generalizability, one of the anticipated benefits of mixed methods, is regarded as an "ideal a goal to be achieved, rather than an accurate depiction of what transpires in real-world research" (Polit and Tatano Beck 2010, p.1452), yet quantitative ideals of statistical generalization persist as do the ideals of analytic generalization, most often used in qualitative research which strives to generalize from particulars to broad constructs or theory (Polit and Tatano Beck 2010). Case-to-case generalization, called *transferability* generalization, is most often a collaborative enterprise and it is the readers and users of the research who transfer the results. The use of rich and complementary data sources together with well-grounded meta-inferences (Teddlie and Tashakkori 2009), claim to enhance analytic generalization (Polit and Tatano Beck 2010).

Overall, Onwuegbuzie and Leech (2005) suggest there are more similarities between the research strategies than most people have considered. For example: using safeguards to minimize bias, attempting to explain complex relationships, using data reduction and data verification techniques, and selecting analytical techniques to gain maximum meaning. However, whilst these similarities help to improve quality of a mixed methods project there exist also many differences.

The key differences between qualitative, quantitative and mixed methods research strategies (see for example, Johnson and Harris 2003; Denzin and Lincoln 2000) are tabulated in Figure 25.10 and include: research aim, typical timing of method, ability to plan research design, means of data collection, nature of data, data detail, unit of analysis, pre-knowledge, contribution, assessment of truisms, researcher detachment, ethical implications and scale. Some of which have already been discussed.

Dimension	Qualitative (Ql) data and methods	Quantitative (Qn) data and methods	Mixed Methods (Ql & Qn)
Research aim	The aim is a complete, detailed description	The aim is to classify features, count them, and construct statistical models in an attempt to explain what is observed	The aim is to answer the research question in a real world setting recognizing the need for both description and explanation
Typical timing of method	Recommended during earlier phases of research projects	Recommended during latter phases of research projects	Both can be accommodated in a Mixed Methods Research design according to the purpose of the research
Ability to plan research design	The design emerges as the study unfolds	All aspects of the study are carefully designed before data is collected	A balance of research planning and research responsiveness can be achieved through smaller projects
Means of data collection	The researcher is the data gathering instrument	The researcher uses tools, such as questionnaires or equipment to collect numerical data	Data collection may become onerous, and the resources needed to collect the various types of data need estimating
Nature of data	Data is in the form of words, pictures or objects	Data is in the form of numbers and statistics	Cross-over mixed methods analysis provide a way to integrate data and findings
Data detail	Qualitative data is more 'rich', time consuming and has great variety of format	Quantitative data is more efficient and lends itself to test hypotheses	The level of detail required from each method to contribute to answering the overall research purpose will need to be planned
Unit of analysis	May be narrower or more specific	May be broader or more cross-cutting	The units of analysis may change from one stage of the design to the next, provided there is a logic for this
Pre-knowledge	Researcher knows roughly in advance what he/she is looking for	Researcher knows clearly in advance what he/she is looking for	Given the fast pace of change in real world systems, pre-knowledge may be less assured in the future
Contribution (inductive/ deductive)	Theoretical or moderatum generalization through inductive means	Axiomatic or probabilistic generalization through deductive means and limited by sample	Analytic generalizability may be enhanced via meta-inferences and complementary data
Assessment of truisms	Subjective – individuals' interpretat ion of events is important e.g. uses participant observation, in-depth interviews, ...	Objective – seeks precise measurement & analysis of target concepts, e.g. uses surveys, questionnaires, ...	The assessment of research truisms for real world problems is embraced directly by using both subjective and objective methods
Researcher detachment	Researcher tends to become subjectively immersed in the subject matter	Researcher tends to remain objectively separated from the subject matter	Researchers in a mixed methods project have the opportunity to become more or less separated from the subject through meta-integration and shared roles
Ethical implications	Greater due to proximity of research focus of research on 'why and how' to understand participants' perspectives	Less due to distance from specific cases and focus of research on 'what'	Ethical implications may become greater via diverse access to the research subject; allows mixed methods to answer 'why', 'how' and 'what'
Scale	Human participants are studied in natural settings and conditionst	Artificially constructed populations and context independence or global interests may be studied	Both local and global scales can be embraced through integrated cross-over methods

Note: Ql means qualitative, Qn means quantitative.

Figure 25.10 Dimensions of research strategies

The research aim of mixed methods is to include both rich descriptions as well as explanations, allowing the research to answer questions from a real world setting, whereas independently qualitative and quantitative research aims at complete description and explanation through numerical analysis respectively.

The ability to plan by selecting a research design at outset is a challenge especially for mixed methods research since the integration outcomes will not be known at outset. Experimentation with different designs is recommended. For qualitative and quantitative research, plans emerge and are fixed at outset respectively.

For qualitative research, the researcher is often the data gathering instrument whereas for quantitative research, tools are used. For mixed methods, the need to gather both types of data may become onerous and need careful resourcing.

The unit of analysis may be narrow for qualitative research, due to the depth of research, whilst broader for quantitative research. The challenge for mixed methods research is to reconcile differences across the stages of the research design. The incorporation of different scales for mixed methods research is an opportunity for unique insights whilst qualitative research may limit scale to local settings, and quantitative research may use artificial and non-real world scales.

Many of the challenges of mixed methods research have also been identified, and it is pertinent to start by considering research bias evident in scholars educated within disciplinary silos rather than with contemporary mixed methods lenses. The extremes can be found in two quotes from Miles and Huberman (1994, p.40): the quantitative researcher Fred Kerlinger says "There is no such thing as qualitative data. Everything is either 1 or 0" and at the other end of the spectrum Donald T. Campbell states "All research ultimately has a qualitative grounding". Research bias toward particular data types, or using specific data collection or analysis techniques may be tempered in (a) larger mixed methods research projects as collaboration and integration of interdisciplinary research teams are essential, and in (b) single person mixed methods studies through the use of data and methods from non-home disciplines.

Conflicting paradigms and the *incompatibility thesis* may deter some orthodox researchers although even as long ago as 1988, Howe argued that for abstract (non-real-world) paradigms to determine research methods (in a one-way fashion) was untenable. And since then several paradigmatic stances have emerged as shown in Figure 25.3. Regardless of this real world motivation, scholars have found existing methods of social inquiry to have insufficient explanatory power leading to theoretical and epistemological concerns (Greene 2008). Even so, some researchers will want reconciliation of epistemological and philosophical perspectives and may find that some methods cannot be combined, in a single research project, to a shared paradigm.

Despite the growing catalogue of mixed methods studies and advice from mixed methods scholars and researchers, there still exists uncertainty about which mixed methods design to select. Uncertainty of the order (concurrent use of methods, quantitative methods first, then qualitative methods, vice-versa, or more elaborate combinations) may arise as might uncertainty about how to connect mixed methods research designs to a particular research question. Examples of existing applications of research designs can be detrimental to the creation and experimentation of novel and possibly more fruitful research designs for new research. Equally challenging is that there may be no precedent of successful mixed methods research projects in the researchers' field of study which would suggest the need for greater convincing of funders, industrial partners and/or editors of journals. The issue of research design selection could arise due to fashion of mixed methods and the belief that it will solve many of the mono-methods limitations. Thomas (2003, p.7) declares that "each research method is suited to answering certain types of questions but not appropriate to answering other types. Furthermore the best answer frequently results from using a combination of qualitative and quantitative methods" and continues to propose ways of blending different methods for research theses. Connecting research designs to research questions is critical for the success of mixed methods project.

The validity of mixed methods, that is, the additional validity required by both the research strategy and the specific methods integration processes, over and above the validity of the specific qualitative and/or quantitative methods used, remains a barrier to mixed methods research. This concerns the logic of integration which may not be pre-plannable. Questions may arise such as: is the scope of a qualitative method sufficient for the following quantitative method? What is a fair trade-off between planning and emergence, especially when outcomes and insights will emerge from each method (and stage of the research design) and so are unknowable? Does planning stifle creativity and innovation by preventing emergent designs? Furthermore, "some of the advantages of mixed method research may not emerge until the *end* of the research process" (Brannen 2005, p.9, italics in original).

Concerns about anomalous or irreconcilable results may provide a challenge to researchers. Strategies for integrated data analysis were called for in 1989 (Greene, Caracelli and Graham 1989), and methods such as data transformation were devised (see Figure 25.1 for others), for example, cross-over mixed analysis methods for data integration (Onwuegbuzie et al. 2009). More methods and techniques are needed for: data cleansing; processing of outliers; the need to create comparable sample size; aligning knowledge from different methods; and generally ways to explain, expose, identify, and rationalize differences. The overall volume of data and data management processes will come increasingly under pressure as more research projects grapple with big data and attempt to solve real-time data problems, such as for example, congestion, major incidents, and surges in demand. Other concerns include time lags between the uses of two different methods. For example, in a fast moving industry such as clothing or food, insights may become dated before they can be acted upon. Also, a particular population sample may be identified as leading to interesting insights in one stage, but access is not possible in a later stage.

The contrary position to the benefit of skills acquisition through mixed methods research is the fear of incompetence in using methods and manipulating data in ways for which the researcher has not been prepared, either philosophically or technically. How much familiarization with other disciplines is needed? How much training in methods that are not the 'home' discipline(s) of the researcher is required before sufficient capability is created for the research project needs? Also, how much training is needed to develop skills required in mixed methods techniques themselves? Receptivity and acknowledgement of the value of multiple perspectives will help to foster a new generation of researchers who are able to flip between methods, learn and respect the opinions of scholars from potentially very distant disciplines or functional areas, and use the integrated insights to further knowledge of the real world. Greene (2008, p.20) says that mixed methods "engages with the differences that matter in today's troubled world, seeking not so much convergence and consensus as opportunities for respectful listening and understanding".

25.11   CONCLUSIONS, REFLECTIONS AND FURTHER RESEARCH

Mixed methods research has achieved critical mass. There are numerous examples of studies, two international journals devoted to the subject: *Journal for Mixed Methods Research*, and the *International Journal of Multiple Research Approaches*, plus many special issues in peer-reviewed journals such as the *Strategic Management Journal*.

The field is an evolving one, in which extant challenges can be considered and novel solutions developed. Some of the challenges may not disappear but they will need consideration in the planning of mixed methods research projects. It would be useful to have a searchable catalogue developed using a systematic literature review (Tranfield, Denyer and Smart 2003) of mixed methods research articles that could expose the specific processes used to integrate data and findings, during and at the end of a research project; the disciplines which were mixed; the stances (paradigms, philosophies and theories) taken; and the research problems they addressed. This would identify gaps and therefore opportunities for further development of the field, as well as mature designs. It would also shed light on why most of the citations here appear to indicate that social sciences research is leading the way to incorporating quantitative methods. Can using mixed methods to address hard sciences' research problems be as radical and ground-breaking as the use of mixed methods in the social sciences?

This contribution aims to elaborate the burgeoning field of mixed methods to complexity researchers. It also attempts to show that mixed methods is not only relevant for complexity research but that it is underpinned by complexity thinking. Furthermore, several opportunities in mixed methods research are highlighted to which complexity researchers could respond. A contribution is made to mixed methods research through various categorizations and comparisons, highlighting gaps such as those in descriptive and predictive mixed methods. Hopefully this contribution continues to open up the field of mixed methods to complexity researchers and leads to new discoveries and value for society.

ACKNOWLEDGEMENTS

Helpful insights and feedback is acknowledged from doctoral students at the School of Management, Cranfield University, where mixed methods classes have run since 2011.

REFERENCES

Beddington, J. (2010), *Food, Energy, Water and the Climate: A Perfect Storm of Global Events*. UK Government Office for Science.
Bohmann, J. (1996), *Public Deliberation: Pluralism, Complexity, and Democracy*. Cambridge, MA: MIT Press.
Brannen, J. (1992), *Mixing Methods: Qualitative and Quantitative Research*. Aldershot: Ashgate Publishing.
Brannen, J. (2005), *Mixed Methods Research: A Discussion Paper*, ESRC National Centre for Research Methods, NCRM Methods Review Papers, http://eprints.ncrm.ac.uk/89/1/MethodsReviewPaperNCRM-005.pdf.
Brewer, J., and Hunter, A. (1989), *Multimethod Research*. Thousand Oaks, CA: Sage Publications.
Bryman, A. (2007), 'The research question in social research: what is its role?', *International Journal of Social Research Methodology*, 10, pp. 5–20.
Bryman, A. (2012), *Social Research Methods*. Oxford: Oxford University Press.

Campbell, D.T. and Fiske, D.W. (1959), Convergent and discriminant validation by the multitrait-multimethod matrix. *Psychological Bulletin*, 56, pp. 81–105.

Caracelli, V.J. and Greene, J.C. (1993), Data analysis strategies for mixed-method evaluation designs. *Educational Evaluation and Policy Analysis*, 15(2), pp. 195-207.

Creswell, J. W. (2013), *Research Design: Qualitative, Quantitative, and Mixed Methods Approaches*. Hardcover. Thousand Oaks, CA: Sage Publications.

Creswell, J.W. and Plano Clark, V.L. (2010), *Designing and Conducting Mixed Methods Research*. Thousand Oaks, CA: Sage Publications.

Crump, B. and Logan, K. (2010), *A Framework for Mixed Stakeholders and Mixed Methods*, at www.ejbrm.com (accessed 12 September 2017).

Denzin, N.K. and Lincoln, Y.S. (1994), *Handbook of Qualitative Research*. Thousand Oaks, CA: Sage Publications.

Denzin, N.K. and Lincoln, Y.S. (2000), Introduction: The discipline and practice of qualitative research. In: N.K. Denzin and Lincoln, Y.S. (eds), *Handbook of Qualitative Research* (2nd edition). Thousand Oaks, CA: Sage Publications, pp. 1–28.

Glaser, B.G. and Strauss, A.L. (1967), *The Discovery of Grounded Theory: Strategies for Qualitative Research*. Chicago, IL: Aldine Publishing Company.

Greene, J.C. (2008), Is mixed methods social inquiry a distinctive methodology? *Journal of Mixed Methods Research*, 2(1), pp. 7–22.

Greene, J.C. and Caracelli, V.J. (eds) (1997), *Advances in Mixed-Method Evaluation: The Challenges and Benefits of Integrating Diverse Paradigms* (New Directions for Evaluation No. 74). San Francisco: Jossey-Bass.

Greene, J.C., Caracelli, V.J. and Graham, W.F. (1989), Toward a conceptual framework for mixed-method evaluation designs. *Educational Evaluation and Policy Analysis*, 11(3) pp. 255–274.

Hammersley, M. (2008), Troubles with triangulation. In: M.M. Bergman (ed.), *Advances in Mixed Methods Research*. London: Sage Publications, pp. 22–36.

Heron, J. and Reason, P. (1997), A participatory inquiry paradigm. *Qualitative Inquiry*, 3, pp. 274–294.

Howe, K. R. (1988), Against the quantitative–qualitative incompatibility thesis or dogmas die hard. *Educational Researcher*, 17(8), pp. 10–16.

Howe, K.R. (2003). *Closing Methodological Divides*. Boston: Kluwer Academic.

Hurmerinta-Peltomäki, L. and Nummela, N. (2006), Mixed methods in international business research. *Management International Review*, 46(4), pp. 439–459.

Johnson, P. and Harris, D. (2003) Qualitative and quantitative issues in research design. In: D. Partington (ed.), *Essential Skills for Management Research*. London: Sage Publications, pp. 99–116.

Johnson, R.B. and Onwuegbuzie, A.J. (2004), Mixed methods research: A research paradigm whose time has come. *Educational Researcher*, 33(7), pp. 14–26.

Johnston, W.M. (2013), *Encyclopedia of Monasticism*. London: Routledge, p. 196.

Kuhn, T. (1970), *The Structure of Scientific Revolutions* (2nd edition). Chicago, IL: University of Chicago Press.

Lincoln, Y.S. and Guba, E.G. (1985), *Naturalistic Inquiry*. Thousand Oaks, CA: Sage Publications.

Lopez-Fernandez, O. and Molina-Azorin, J.F. (2011), The use of mixed methods research in the field of behavioural sciences. *Quality and Quantity*, 45(6), 1459–1472.

Markie, P. (2013), Rationalism vs. empiricism. In: E.N. Zalta (ed.), *The Stanford Encyclopedia of Philosophy* http://plato.stanford.edu/entries/rationalism-empiricism/#2 (accessed August 17, 2017).

Maxwell, J.A. (1990), Up from positivism. *Harvard Educational Review*, 60, pp. 497–501.

Maxwell, J.A. and Loomis, D.M. (2003), Mixed methods design: An alternative approach. In: A. Tashakkori and C. Teddlie (eds), *Handbook of Mixed Methods in Social and Behavioral Research*. Thousand Oaks, CA: Sage Publications, pp. 241–271.

Mertens, D.M. (2003), Mixed methods and the politics of human research: The transformative-emancipatory perspective. In: A. Tashakkori and C. Teddlie (eds), *Handbook of Mixed Methods in Social and Behavioral Research*. Thousand Oaks, CA: Sage Publications, pp. 135–164.

Miles, M.B. and Huberman, A.M. (1994), *Qualitative Data Analysis*. Thousand Oaks, CA: Sage Publications.

Morse, J.M. (2003), Principles of mixed methods and multimethod research design. In A. Tashakkori and C. Teddlie (eds), *Handbook of Mixed Methods in Social and Behavioral Research*. Thousand Oaks, CA: Sage Publications, pp. 189–208.

Niglas, K. (2004), *The Combined Use of Qualitative and Quantitative Methods in Educational Research*. Tallinn: Tallinn Pedagogical University Press.

Onwuegbuzie, A.J., Burke Johnson, R., and Kathleen Mt Collins, K. (2009), Call for mixed analysis: A philosophical framework for combining qualitative and quantitative approaches. *International Journal of Multiple Research Approaches*, 3(2), 114–139.

Onwuegbuzie, A.J. and Leech, N.L. (2005), On becoming a pragmatic researcher: The importance of combining quantitative and qualitative research methodologies. *International Journal of Social Research Methodology*, 8(5), pp. 375–387.

Partington D. (2002), Grounded theory. In: D. Partington (ed.), *Essential Skills for Management Research*. Thousand Oaks, CA: Sage Publications, pp. 136–158.

Patton, M.Q. (2002), *Qualitative Research and Evaluation Methods* (3rd edition). Thousand Oaks, CA: Sage Publications.

Polit, D.F. and Tatano Beck, C. (2010), Generalization in quantitative and qualitative research: Myths and strategies. *International Journal of Nursing Studies*, 47(11), pp. 1451–1458.

RCUK (2015), RCUK Framework, Research Councils UK, accessed 24 July 2015 http://www.rcuk.ac.uk/Publications/policy/framework/.

Reichardt, C.S. and Cook, T.D. (1979), Beyond qualitative versus quantitative methods. In: T.D. Cook and C.S. Reichardt (eds), *Qualitative and Quantitative Methods in Evaluation Research*. Thousand Oaks, CA: Sage Publications, pp. 7–32.

Richardson, K.A. and Tait, A. (2010), The Death of the Expert?, *E:CO*, 12(2), pp. 87–97.

Rizzo, T.A., Corsaro, W.A. and Bates, J.E. (1992), Ethnographic methods and interpretive analysis: Expanding the methodological options of psychologists. *Development Review*, 12, pp. 101–123.

Sale, J.E.M., Lohfeld, L.H. and Brazil, K. (2002), Revisiting the quantitative-qualitative debate: Implications for mixed-methods research. *Quality & Quantity*, 36, pp. 43–53.

Sammons, P., Siraj-Blatchford, I., Sylva, K., Melhuish, E., Taggard, B. and Elliot, K. (2005), Investigating the effects of pre-school provision: Using mixed methods in the EPPE research. *International Journal of Social Research Methods*, 8(3), pp. 207–224.

Stirling, A. (2008), 'Opening up' and 'closing down' power, participation, and pluralism in the social appraisal of technology. *Science, Technology & Human Values*, 33(2), pp. 262–294.

Strauss, A. and Corbin, J. (1998), *Basics of Qualitative Research: Techniques and Procedures for Developing Grounded Theory* (2nd edition). Thousand Oaks, CA: Sage Publications.

Tashakkori, A. and Creswell, J.W. (2007), The new era of mixed methods. *Journal of Mixed Methods Research*, 1(1), pp. 3–7.

Tashakkori, A. and Teddlie, C. (eds) (2010), *Sage Handbook of Mixed Methods in Social and Behavioral Research*. Thousand Oaks, CA: Sage Publications.

Teddlie, C. and Tashakkori, A. (2003), Major issues and controversies in the use of mixed methods in the social and behavioral sciences. In: A. Tashakkori and C. Teddlie (eds), *Handbook of Mixed Methods in Social & Behavioral Research*. Thousand Oaks, CA: Sage Publications, pp. 3–50.

Teddlie, C. and Tashakkori, A. (2006), A general typology of research designs featuring mixed methods. *Research in the Schools*, 13(1), pp. 12–28.

Teddlie, C. and Tashakkori, A. (2009), *Foundations of Mixed Methods Research*. Thousand Oaks, CA: Sage Publications.

Thomas, R.M. (2003), *Blending Qualitative and Quantitative Research Methods in Theses and Dissertations*. UK: Corwin.

Tomos, F., Djebarni, R., Rogers, A., Thomas, A., Clark, A. and Balan, C. (2013), *Mixed Research Methods: Former and new Trends in Women Entrepreneurship Research*, 374–XIII. Kidmore End: Academic Conferences Intern.

Tranfield, D., Denyer, D. and Smart, P. (2003), Towards a methodology for developing evidence-informed management knowledge by means of systematic review. *British Journal of Management*, 14, pp. 207–222.

Varga, L. (2014), *Complexity science: The integrator in The Social Face of Complexity Science: A Festschrift for Professor Peter M. Allen*, Strathern, M. and McGlade, J. (eds). New York: ISCE Publishing.

Wheatley, D. (2012), Work–life balance, travel-to-work, and the dual career household. *Personnel Review*, 41(6), 813–831.

Whittemore, R., Chase, S. and Mandle, C L. (2001), Validity in qualitative research. *Qualitative Health Research*, 11(4), pp. 522–537.

26. Dynamical systems therapy (DST): complex adaptive systems in psychiatry and psychotherapy

Professor Yakov Shapiro, University of Alberta and Associate Professor J. Rowan Scott, University of Alberta

> The brain is wider than the sky –
> For – put them side by side –
> The one the other will contain
> With ease – and you – beside. (Dickinson, in Edelman 2004)

26.1 INTRODUCTION AND BACKGROUND

The study of complexity relates to the detection and mapping of systemic algorithms found across many levels of the emergent natural hierarchy in physical, chemical, biological, psychological, and social processes. The human brain/mind is a mid-scale phenomenon of vast complexity that behaves as a *complex adaptive system* (CAS; Gell-Mann 1994) characterized by nested hierarchies of multi-level causal chains and an intrinsic potential for adaptive self-organization in response to changing environmental demands. In clinical contexts, psychiatry and psychotherapy deal directly with the evolved complexity of the brain/mind, conscious and unconscious intentionality, and multiple layers of individual and social meaning. A nonlinear dynamical systems approach can be used to model emergent properties arising at each level of organization, from neural network dynamics to subjective experience, and on to interpersonal, family, group and cultural processes.

The language of information inherent in complexity theory (Von Bayer 2003) can serve to bridge the persistent brain–mind divide that has plagued psychology and psychiatry ever since Rene Descartes' famous dictum: "Cogito ergo sum" – "I think, therefore I am." Re-casting the language of psychopathology in informational terms has the potential to re-unify the third-person 'objective' perspective of neuroscience with the subjective meaning of the patient's experience and his or her intersubjective patterns of relating to others. In addition, the adaptive dimension of the CAS model allows us to capture the *systemic relational process* of the patient in their environment, including the treatment milieu. The nonlinear relational perspective suggests the need for a paradigm shift in psychiatry, from an emphasis on a disembodied reductionist framework of disease classifications and towards incorporating subjective participant-observers into a comprehensive picture of psychobiological reality. A similar shift was instrumental in incorporating conscious participant-observers into the psychophysical picture of the physical reality in the quantum micro-world.

The emerging Dynamical Systems Therapy approach (DST; Shapiro 2015) stands as a trans-theoretical model with the explanatory power to integrate physiological systems of synaptic networks with psychological systems of meaning. It powerfully argues for shifting the focus of psychiatric intervention from the patient's presenting symptoms as problems to be fixed, to understanding symptoms as breakdowns of adaptation, the patient's imperfect solutions to his or her constitutional or developmental adversity. Instead of looking at our patients as passive victims of their mental illness, we begin to see them as active *intentional agents* who continuously shape their subjective and interpersonal reality based on the interplay between their inborn adaptive algorithms and specific developmental templates.

The emphasis of the DST model is on the emergent patterns of psychological complexity in individual health and psychopathology, integrating the hardwired '*adaptive algorithms*' acquired in the course of brain/mind phylogeny with the individual templates developed in the course of ontogeny. From the DST perspective, the cornerstone of psychological health is the individual capacity for emotional homeostasis and self-system coherence coupled with flexible adaptation to changing environmental demands. In contrast to linear models of psychopathology, which rely on identifying symptom-based categorical disorders requiring generic manualized treatment, the nonlinear approach focuses on the multi-level, emergent patterns of systemic self-organization in the individual's inherited and developmental milieu. Of particular importance are developmental misadaptations to adverse and traumatic experiences that may result in dysfunctional ways of experiencing oneself in relation to others and to the world at large. These recurrent patterns of thinking, feeling, and relating can be analyzed by using modified fitness diagrams (adaptive or *A-landscapes*), which integrate objective, subjective and intersubjective clinical data. The A-landscape model enables psychiatric practitioners to chart the patient's unique life trajectory through its malleable *attractor/repellor states* in health and psychopathology, constructing a virtual 'map of the mind' that guides both psychotherapeutic and psychopharmacological interventions.

The DST model also conceptualizes the treatment provider (psychotherapist and/or psychopharmacologist) as an intricate part of the emergent patient-therapist system formed by the partial interpenetration of the therapist's and the patient's A-landscapes. This position stands in contrast to the traditional view of the treatment provider as an 'objective expert-observer' who prescribes evidence-based treatment for the patient's psychopathology. The doctor–patient relationship in all medical settings, including the prescribing process, becomes an indispensable tool in re-shaping the topography of the patient's A-landscape and re-establishing the self-organizing process.

26.2   THE FRACTAL LANDSCAPE OF PSYCHOBIOLOGY

26.2.1   The CAS Model as a Meta-framework for Psychiatry

The nonlinear dynamical systems approach to psychiatric nosology and treatment integrates advances in understanding individual and group processes as complex adaptive systems characterized by the emergent properties of subjective and cultural experience. Living organisms are open, negentropic systems that operate in the intermediate zone between the rigid thermodynamic equilibria of inorganic structures and chaotic dynamics. They build quasi-stable networks of emergent complexity fueled by energy and information from their environment (Gell-Mann 1994).

Living systems are characterized by a nested hierarchy of processes which organize themselves into emergent levels of complexity in response to changing environmental demands. We cannot calculate the trajectory of a foraging ant in part because the emergent systemic behavior of an organism or a community is not fully reducible to the forces of molecular bonding or atomic interactions. Living systems, however simple, behave as *minimal autonomous agents* (Kaufman and Clayton 2006), where the systemic whole becomes more than the sum of its parts. Each level of organization builds upon causal chains from the lower levels but adds new layers of interactions that affect lower-level processes in a top-down fashion, a process described as *reciprocal causation* in systems biology (Laland et al. 2015). For instance, molecular interactions at the individual neuron level are modulated by the wider network dynamics, such as higher cortical regions having an inhibitory effect on lower-level processes, the mechanism critical in emotional regulation and behavioral response. The emergence of subjective experience introduces another level of causal interactions, such as an anorexic's choice to forego food intake because of his or her distorted body image. Group-level processes result in yet another level of reciprocal causality, such as cyber-bullying triggering suicidal ideation. Cultural norms, in turn, mold both individual and group processes, such as in body weight stereotypes and normative emotional responses. Each level of organization introduces qualitatively new causal dynamics and a level of discourse not reducible to lower-level phenomena: it would be meaningless to limit the analysis of cultural phenomena to biochemical interactions even though cultural experience as we know it would be impossible without biochemistry.

The CAS model is eminently suited for psychiatry because it captures the dimension of adaptive, nonlinear, multi-level processes that must be considered to describe systemic organization in real time – from synaptic network dynamics, to subjective emotional/cognitive experience, to individual behavior or group processes – all without artificial separations into biological versus psychological domains. It brings to the forefront three criteria vital to understanding the evolution of psychopathology:

1. *We are complex systems.* Both synaptic networks and individual/group behavior are dissipative systems that operate in the 'edge of chaos' zone and follow nonlinear dynamical trajectories characterized by sensitivity to initial conditions and attractor/bifurcation dynamics.
2. *We adapt.* Neither synaptic networks nor human/group behavior can be understood outside of their ongoing process of adaptation to the external environment and self-organizing processes within their internal milieu. Adaptive processes show self-similarity on evolutionary (phylogenetic) and individual (ontogenetic) scales.
3. *We operate with multiple levels of emergent complexity.* Both synaptic networks and individual/group dynamics are nested hierarchies of processes with multiple levels of emergent complexity that interact along horizontal (within-level) and vertical (between-level) bottom-up and top-down causal loops.

From the CAS perspective, the therapist–patient dyad can be described as an interaction of three nested systems involving two human agents (Stolorow 1997). Both the psychiatrist/psychotherapist (CAS 1) and the patient (CAS 2) bring in their own unique ways of organizing experience and developmental adaptations. In coming together for the purpose of the treatment (whether psychopharmacological or psychotherapeutic), they create a shared intersubjective system (CAS 3) in the environment of the consulting room. What happens in the subjective space of each participant (CAS 1 and CAS 2) can be stunningly complicated, engaging conscious and unconscious meanings, metabolic pathways, behavioral and emotional templates, past and recent association networks, and self-reflective awareness. The intersubjective space shared by the therapeutic dyad (CAS 3) is just as complex, each participant bringing in complicated patterns of expectation and relatedness, constantly shifting in the process of the therapeutic engagement (Marks-Tarlow 2011).

Both the psychotherapeutic dyad as a whole and its individual participants are intentional agents that display nonlinear dynamics such as extreme context sensitivity, making the prediction of the future therapeutic trajectory indeterminable on principle (Mitchell 2009). Context sensitivity arises from the fact that systemic sub-components (such as the hierarchically organized neurons in the brain or the conscious agents in the room) continually interact with each other, the potential outcome trajectories diverging very rapidly depending on the implicit and explicit meaning of the interaction (the so-called 'butterfly effect'). The treatment trajectory is further influenced by qualitatively new emergent physiological, behavioral, emotional and cognitive configurations that are mutually organized and co-regulated, becoming a part of the evolving insight that contributes to the therapeutic outcome. Whether we introduce a psychopharmacological agent, a psychodynamic interpretation, or cognitive-behavioral intervention, we impact on complex objective, subjective and intersubjective cycles of nonlinear, multi-level interactions that change the patient, the therapist and the therapeutic relationship.

The CAS framework as described above helps to capture the adaptive dimension of psychopathology in the context of the patient's lived experience, including the treatment setting. It strongly argues for shifting the emphasis of psychiatric interventions from generic algorithms for symptom-based DSM or ICD categorical syndromes, toward targeting specific, process-sensitive interventions for each patient (Roth and Fonagy 2006). In addition, the CAS construct and its language of information helps to bridge the persistent Cartesian gap between 'material brain' and 'immaterial mind' that underlies the conceptual divide between biological and psychosocial interventions. In his seminal paper on psychobiology, "A new intellectual framework for psychiatry" (1998) Eric Kandel, a Nobel laureate and leading pioneer in integrating neuroscience and psychodynamics, states: "Insofar as psychotherapy or counseling is effective and produces long-term changes in behavior, it presumably does so through learning, by producing changes in gene expression that alter the strength of synaptic connections" (Kandel 1998, p. 460). We begin to treat brain/mind as a *psychobiological system* that possesses a fundamental objective/subjective complementarity; psychotherapy becomes a form of biological intervention with the power to rewire the patient's synaptic networks, and biological treatments carry a wealth of subjective meaning that can facilitate or undermine their efficacy in the context of the doctor–patient relationship.

26.2.2   Shifting Paradigms: From Reductionism to Systemic Psychobiology

The work of psychiatry and psychotherapy has a paradoxical relationship with basic assumptions held within many branches of reductive natural science. The rigorously applied logic of reductive thought, our *only* formulated scientific paradigm, sweeps consciousness and intentionality into a bin of anomalous problems. These are manifested by the 'Heisenberg cut' associated with the limit on quantum measurement (Atmanspacher, Weidennman and Atmann 1995); the 'Cartesian cut' connected with resolving Descartes' dualism of brain/mind; and the 'explanatory gap' related to the limit on understanding subjective experience such as 'qualia' by means of functional neuroscience (Chalmers 1995). These issues lie at the core of scientific inquiry and neuropsychological science but stubbornly resist materialist neo-Darwinian solutions (Nagel 2012).

The paradox of consciousness is often sidestepped through the use of contextual adaptations of the reductionist worldview that implicitly or explicitly abandon the logic of the reductive approach. One such adaptation treats the 'explanatory gap' as if it had been filled in, functionally equating brain and mind. Another one denies the existence of mind, consciousness or intentionality altogether (Libet 1999). Conversely, a third approach makes consciousness as 'fundamental' as the fundamental phenomena of quantum physics, implicitly abandoning the materialist position for a dualistic worldview (Seager 2012). A fourth approach applies reduction within discrete levels of evolved complexity but stops short of complete reduction to quantum science; it leaves the evolved complexity of the emergent, higher order level of organization intact but effectively avoids an argument about the limits of reductive thought (Dennett 1995). This is a variation frequently used in biology, where the existence of higher-level causal interactions is accepted implicitly without dealing with the consequent abandonment of the reductionist model that this entails (Brigandt and Love 2015). No fundamental solution has arisen that can effectively address this Gordian knot of problems surrounding consciousness, emergence and intentionality in natural science.

Many quantum and emergent phenomena including consciousness invoke the use of the ambiguous term 'holism.' In spite of the fact that there is no defined holistic scientific paradigm, the holistic approach aims to study emergent complexity from a top-down perspective while reductionism addresses it from the bottom up. A similar contraposition exists in psychiatric treatment, where psychotherapeutic interventions are seen to utilize top-down functions while psychopharmacological treatments primarily operate through bottom-up mechanisms (Mayberg 2003). In spite of the fact that emergence is a critical aspect of evolutionary complexity, conservative reducible emergence that does not possess any causal efficacy is the only form allowed by reductive logic. Within the reductionist framework, causally efficacious conscious intention as a form of radical nonreducible emergence becomes an extremely elusive phenomenon (Seager 2012).

The unsolved problems associated with consciousness and intentionality point directly at the limits of the reductionist paradigm. The conscious experience of agency or mindful intention is seen as a subjective illusion (Libet 1999), with no causal influence in the world beyond that assigned by quantum physics. Yet, despite these reductive arguments, the world, natural evolution and the experience of consciousness are full of irreducible examples of radical emergence and downward causation (Noble 2006). Reductive epiphenomenalism of consciousness stands absurdly in direct contradiction with the day-to-day experience of any complexity scientist or psychiatrist who chooses to study their discipline or intervene in their patient's psychopathology. It also collides with the empirically validated role of participatory, causally impactful observers in quantum experimental contexts (Stapp 2009). This situation is analogous to the contradictory picture of reality found at the intersection of quantum physics and relativity theory, where relativity emphasizes continuity, locality and determinism on the macro scales, while quantum mechanics emphasizes the fundamental nature of discontinuity, non-locality and indeterminism on the micro scales. Coupled with the fact that both frameworks are among the most internally consistent and experimentally validated theories in the history of science, this paradox suggests the need for a new overarching paradigm in the physical sciences (Pylkkanen 2007). The conundrums surrounding consciousness similarly point to the fact that the reductive model is fundamentally limited in modeling natural evolution, and a qualitative paradigm shift may be necessary in biological and psychological domains (Kuhn [1962]1996).

The clinical work of psychiatrists and psychotherapists implicitly assumes a functional solution to the problems surrounding consciousness and reductive natural science. It incorporates most of the previously defined but individually unsuccessful adaptations of the reductive paradigm. This patchwork solution is employed in various settings, often without fully integrating the psychobiological model of care provision:

a.   We approach clinical work as if reductive science had already filled the 'explanatory gap' while resisting the reductionist push towards denying the existence of subjective experience and conscious intentionality. Instead, we incorporate a functionally more encompassing biopsychological paradigm that treats conscious phenomena as an emergent product of brain/mind complexity.

b.   We implicitly embrace the causal efficacy of higher-order emergent complexity, such as focusing on our patients' life choices. Top-down and bottom-up causation can operate in parallel and violate no physical laws, even under the most rigorous application of reductionism.

c.   We treat consciousness and group phenomena as if they were irreducible to lower-level mechanisms. However, we resist assuming a dualistic position in favor of the perspective of emergent evolutionary complexity. Attempting to reduce higher-order interactions, such as subjective awareness or social dynamics, to lower-order effects loses the level-specific informational complexity that defines the very nature of these phenomena.

d.   We implicitly modify the metaphysics of science; in choosing between consciousness and intentionality versus the preservation of the reductive method – we utilize the third-person reductive methods in the service of attending to the patient's subjective experience.

Consequently, the practice of psychiatry and psychotherapy assumes a *dualistic view of natural science*, rather than a dualistic view of consciousness. This view assumes the 'reductive epiphenomenalism' debate can be resolved by embracing a dual definition of emergence that integrates reductive 'bottom-up' with holistic 'top-down' evolutionary relationships. The isolated bottom-up perspective leads to the 'mindless brain' model common in conservative neuroscience, which loses the higher-level meaning of the experience (equivalent to an attempt to understand a lovers' kiss in terms of elementary particle interactions). By contrast, the isolated top-down perspective results in the 'brainless mind' of conservative psychology, which misses the vital component interactions that underlie neural network dynamics and psychopathology. Instead of contradicting each other, these two perspectives stand in fundamental complementarity; the residue of Descartes' dualism is transformed into a reductive/holistic synthesis requiring an interdisciplinary scientific approach that models the evolutionary process on both macro- and micro-scales in parallel holistic and reductive modes. Psychopathology cannot be understood as either an aberrant brain or a deviant mind in isolation from each other without losing the informational complexity of level-specific interactions, since reductive analysis to either higher or lower level processes ignores distinct cause–effect relationships on other levels and reciprocal interaction between them.

Insights arising from nonlinear science and the evolutionary emergence of complexity point to a new psychobiological paradigm conceptualized as *systemic psychobiology* – an emergent CAS of brain/mind nested mid-way within a hierarchy extending from quantum to atomic, molecular, cellular and neural network dynamics on the lower scales, and individual, group, cultural, ecological, and technological processes on the higher scales. Systemic psychobiology conceptualizes brain/mind as a dynamic, nonlinear system with multi-level emergent properties that evolve in a continuous diathesis with their internal and external environments. It arises at the intersection of objective, subjective, and inter-subjective science (Velmans 2009) as a synthesis of nonlinear complexity and evolution-ary frameworks. The systemic psychobiology model allows us to transcend the brain/mind dichotomy by utilizing informational language to describe parallel fractal dynamics between three emergent levels of complexity:

1. An objective level of observable neural network processes;
2. The emergent level of subjectivity with associated patterns of perception, feeling, thinking, and intentionality (conscious and unconscious); and
3. The emergent level of intersubjectivity with associated verbal, non-verbal and behav-ioral patterns of relational interactions.

A functioning human brain is seen as a complex, hierarchical system incorporating both an 'objective' level of nonlinear synaptic network dynamics and an emergent 'subjective' level of self-awareness and intentionality. These aspects of brain/mind reality are distinct but irreducible to one another; therefore, asking whether mental illness is 'really' a neuro-chemical imbalance in the brain or a subjective experience in the mind is as meaningless as asking whether a photon is 'really' a particle or a wave. We endow our biology with subjective meaning, while shifts of meaning alter the very synaptic networks that give rise to our capacity for self-awareness (Beauregard 2007). In psychiatry, a functional system that achieves the level of complexity we describe as 'a living person' is inseparably psy-chobiological; we can only address psychopathology in a meaningful way by attending to *both* the objective symptomatic presentation and the patient's subjective experience of his or her biology.

26.3   DYNAMICAL SYSTEMS THERAPY (DST)

26.3.1   The Systemic Psychobiology View of Psychopathology: Adaptive Landscapes

The nervous system in general and the human brain in particular are first and foremost pattern analyzers and prediction devices (Lowenstein 2013). Our brain constantly scans for patterns of events in our external environment; patterns of response in our social sur-roundings; and patterns of homeostasis in our internal milieu. Primary neural networks monitor changes in an organism's environment, with emergent qualities of perception and basic intentionality, such as an amoeba moving away from hypertonic solution. Secondary neural networks monitor changes in primary networks, with emergent properties of core consciousness and reflex action (pain when touching a hot stove). Tertiary networks monitor changes in secondary networks, with emergent properties of self-awareness and personal identity extended in time, which allow for *hindsight* of the past, *insight* into the present, and *foresight* of the future (Edelman 2004).

Predictability is key to an organism's survival and its ability to adapt – to maintain physiological and psychological homeostasis in the face of changing internal and environ-mental demands (Tooby and Cosmides 1990). The adaptive function is a running thread across both the phylogenetic history of our species and the developmental history of the individual. Our evolutionary imperative is short: we adapt or we die. The systemic psycho-biology view of psychopathology makes the process of *multi-level reciprocal adaptation* the cornerstone for understanding our evolution as a social species and our 'coming into being' as individuals endowed with reflective awareness; we develop our distinct sense of 'self' out of the early intersubjective matrix with our caregivers (Stern 1985), and shape both our inner and external environments in doing so. Conversely, deficits in adaptability or developmental *misadaptations* are seen as a major factor contributing to the develop-ment of mental illness (Slavin and Kriegman 1992).

The evolutionary process has optimized our brain to adapt to recurrent ancestral dilem-mas; for instance, children in many cultures have an instinctive aversion to snakes, the his-torical predator of our species, and feel fascinated by cars, even though deaths and injuries due to MVAs in the modern world far outweigh those caused by reptiles. These hardwired instinctual strategies, or *Darwinian algorithms*, represent evolved patterns of perceiving and organizing information relevant for our survival, and responding to it (Nesse 1998). Consciousness itself may represent an adaptive strategy for predicting and sharing in the immense complexity of communal group social interactions (Rochat 2009), its evolution driven by ever-increasing synaptic network complexity required for nonlinear behavioral pattern analysis in the tribal group and beyond. The historical imperatives of inseparable self- and other-oriented dispositions go hand in hand in the course of *Homo sapiens* phy-logeny since no individual, however well adapted, would have been able to survive alone.

An organism's success in adapting and reproducing is referred to as its Darwinian fitness, which can be plotted on a fitness landscape with the peaks representing higher reproductive success (Arnold, Pfrender and Jones 2001). The process of evolution, which involves a constant struggle to maintain or increase reproductive fitness, has thus been likened to climbing local fitness peaks. However, in natural environments, adaptation demands do not remain static; the landscape perpetually shifts its configurations, some-times gradually and sometimes abruptly (the model described as punctuated equilibria – Gould and Eldredge 1993). Rather than 'climbing fitness hills,' a more apt metaphor for evolutionary adaptation would be trying to get to high ground during an earthquake in order to avoid the coming tsunami. In addition, multiple simultaneous fitness peaks are possible depending on the organism's genetic alleles and their phenotypic expressions in a given environment. The overall picture resembles a changing terrain described as a *rugged landscape* (Figure 26.1). In CAS terms, rugged landscapes allow for multi-level

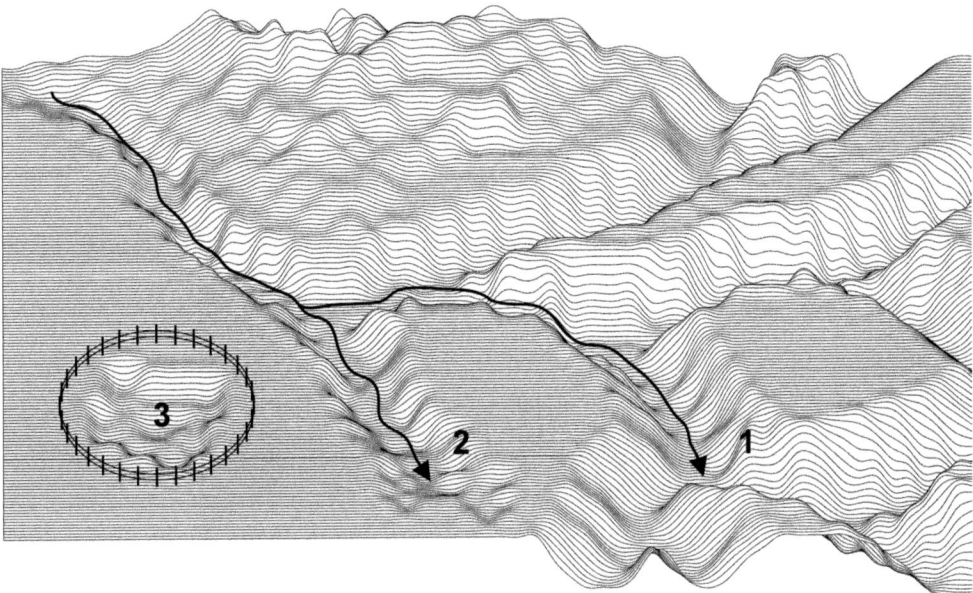

Figure 26.1   A rugged A-landscape

individual system solutions while *dancing landscapes* involve the interaction of multiple adaptive agents (Page 2011), either interacting subsystems within a system under study (such as a patient's conflictual motivations), or multiple co-evolving systems such as a patient-therapist dyad.

If we want to describe individual behavior from the dynamical systems perspective, we can plot the parameters of interest (such as the individual's dispositions, emotional responses, thinking patterns, or relational cycles) in a psychological phase-space, monitoring changes in given parameters over time. The diagram would show our most likely choices clustering in *attractor basins*, and the choices we are least likely to make on *repellor peaks* (Figure 26.2a). In CAS terms, attractor and repellor coordinates represent the most likely versus the least likely states of the system respectively. Just as a heavy object depresses the surface of a trampoline, our habitual preferences 'gravitate' toward attractor basins; our life trajectory follows the 'lowest energy' paths, often outside our conscious awareness. Of course, there is no gravity in psychological space; what defines the attractors is our prevalent *affective valence* (a feeling of interest, contentment, or relief in making a particular choice of action). In terms of proximate (ontogenetic) causation, attractors can represent both one's action tendencies (for example, preference for solitary activities in an introvert), and patterns of organizing experience that shape our perceptual, cognitive and relational dynamics (for example, internal working models, such as experiencing the world and others as hostile based on one's developmental experiences – Bowlby 1984). In terms of distal (evolutionary) causation, attractors reflect a specific set of hardwired Darwinian algorithms optimized for the Environments of Evolutionary Adaptedness (EEA).

Individual behavior of any CAS, such as living organisms, is inherently unpredictable from the outset (think of a particular ant's or bee's trajectory from moment to moment).

A. Pathological attractors/repellor with an unstable equilibrium

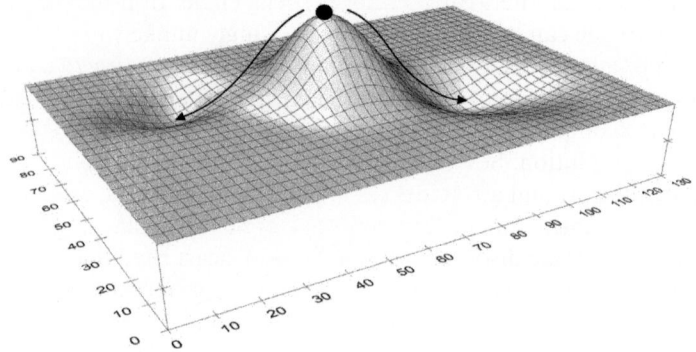

B. Re-shaping the landscape by creating a functional intimacy template

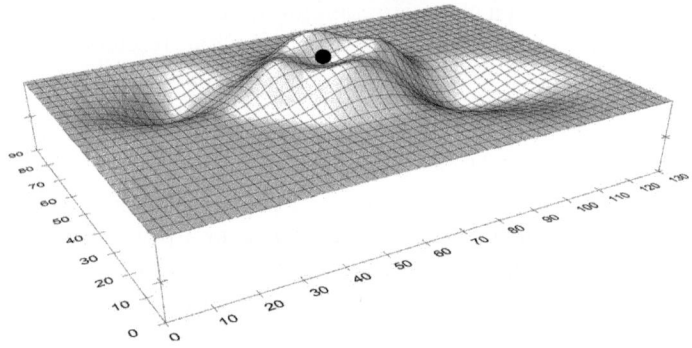

C. Reversing the polarity of pathological attractors/repellors

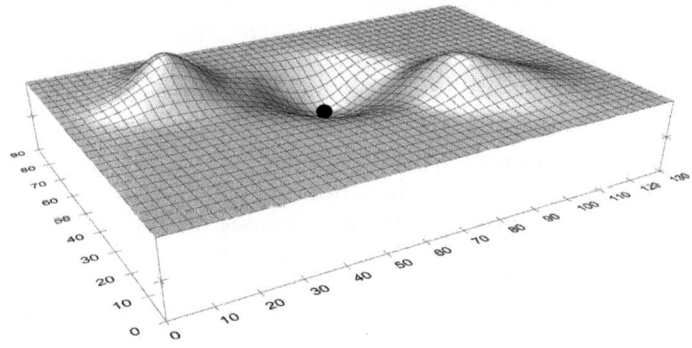

Figure 26.2   Reshaping the topology of the patient's A-landscape

However, *patterns of behavior* such as an ant's role in the colony or a bee flying from one flower to another, remain stable over time even though their precise trajectories will never be repeated. In spite of the fact that as human beings we are free to choose from a multitude of actions, each of us tends to follow a habitual routine, such as sleeping in a bed (rather than on a pavement), eating our preferred foods, engaging in habitual occupational tasks, and interacting with our social circle. In principle, we could choose to eat from a garbage can but we would be exceedingly unlikely to do so. These patterns of feeling, thinking and relating are the hallmark of *strange attractors* initially described by Edward Lorenz in trying to analyze the behavior of weather systems (Lorenz 1993). Attractors map a complete set of all possible states that a complex system may assume in the course of its evolution. Several other attractor types have applications to biological systems, such as fixed-point attractors (as in unipolar depression, obsessions or addictive disorders); periodic attractors (as in sleep-wake cycle or bipolar disorders); and chaotic attractors (as in psychotic disorders or variations of heart rhythmicity).

The entire set of an individual's attractor/repellor configurations comprises an *adaptive* or *A-landscape*, which can be visualized as a virtual set of all attractor basins and repellor peaks unique to a given person (Shapiro 2015). Similar to phylogenetic fitness landscapes, the distinguishing characteristic of individual A-landscapes is that they do not represent a static cross-section of the patient's symptoms or traits (as in categorical psychiatric diagnoses) but a changing configuration of their adaptive patterns over time. The A-landscape dynamics of attractors evolving in time resembles an intricate canyon or river delta, the basins transforming into a multitude of interconnected channels where our life trajectories flow under most circumstances (Figure 26.1). Similarly, individual peaks extend into ridges, which we are normally reluctant or unable to climb. The attractor valleys may grow deeper or shallower over time, such as a marriage developing out of a casual dating relationship (Figure 26.1-1), or a potential career interest fading off (Figure 26.1-2). This fractal branching pattern is fundamentally psychobiological, involving *both* synaptic network configurations and subjective meaning of our experiences. The progression is epigenetically driven, where differentiating synaptic networks both guide the individual's subjective choices and are molded by them, in an ongoing cycle of reciprocal causation described as *neural Darwinism* (Edelman 2004).

Subjectively, we experience attractors as *patterns of feeling, thinking, and relating* that we habitually engage in, either healthy or pathological. Healthy attractors relate to *emotional needs*, whether physiological or psychological, while unhealthy repetition-compulsion scenarios (such as addictions, compulsions, or maladaptive patterns of relating) are indicative of pathological attractor states. Pathological attractors can be manifested both behaviorally, such as compulsive behaviors or abusive relational cycles, and intrapsychically, such as recurrent obsessive thoughts or dysphoric emotional states. The DST approach to mapping and modifying such dysfunctional attractor-bound patterns starts with bringing them to conscious awareness by focusing on *why we find the cycle attractive* and continue to engage in it.

In contrast to conventional fitness landscapes in population genetics, where local peaks represent higher levels of reproductive success for which the organisms compete, the repellor peaks in A-landscapes represent ways of feeling, thinking and relating that the individual tends to avoid. These avoidance patterns can be healthy, such as being mindful of objective dangers in our environment, or they can represent psychopathology, such as in phobic, anxiety and trauma-based disorders, where patients find themselves unable to engage in otherwise adaptive behaviors and suffer progressive lifestyle restriction and functional impairment.

Our unique *A-landscape* molds our conscious and unconscious choices and preferences in a wide range of settings, from behavioral reactions to personal tastes; from temperamental traits to personality characteristics; from relational and sexual preferences to internal patterns of self-regulation. It represents a psychobiological 'map of the mind,' without artificial separations into physical versus mental, or physiological versus psychological. Both 'objective' behavioral and 'subjective' meaning domains can be mapped equally well in informational terms. Healthy A-landscapes are able to adapt flexibly to environmental demands while preserving the individual's physiological and psychological coherence and homeostasis. They occupy a far-from-equilibrium zone between rigid equilibrium and chaotic dynamics, allowing for the buildup of systemic complexity (Siegel 2006). Pathological A-landscapes, on the other hand, are shifted towards either rigid or chaotic ways of organizing experience and relating to others.

At the one extreme, pathological landscapes are shifted towards the rigid boundary (*PL-type I*), such as traumatic experiences metaphorically scarring the psychobiological terrain. In this scenario, survival is achieved at the cost of adaptability. Such landscapes are characterized by rigid and maladaptive patterns, the patient feeling like a perpetual misfit in his or her current surroundings, and re-enacting dysfunctional cycles from their past. Such presentations are common in depressive, trauma-based and high-functioning borderline disorders. The DST approach to PL-type I treatment focuses on de-stabilizing the pathological landscape and processing traumatic affects in a healthier relational setting, where more functional attractor systems can be incorporated and re-assembled in a more adaptive configuration.

At the other extreme, pathological landscapes are shifted towards chaotic dynamics (*PL-type II*), where survival is achieved at the cost of internal coherence, resulting in a perpetually shifting and chaotic psychobiological terrain. This presentation is common in dissociative, psychotic, and severe borderline disorders, where the lack of coherent self-structure prevents stable functioning in most areas of the patient's life. A fitting metaphor for this presentation is an attempt to build a complex architectural design in an earthquake-prone zone. The DST approach to these patients focuses on stabilization and building basic trust, with eventual therapeutic exploration of the underlying traumatic templates and creation of more stable self-structure and functional adaptations in the present.

It is important to note that in both cases (PL-type I and II), pathological templates are seen as multi-level psychobiological constructs that incorporate *both* synaptic brain connectivity and the emergent meaning of the patient's experience. They ultimately arise as an adaptive diathesis between the individual's self-organizing capacities and his or her environmental milieu. In this light, the definition of a 'traumatic experience' ceases to be external to the person and represents a mismatch between the hardwired psychobiological make-up versus demands placed on it by the patient's developmental environment. For example, autistic patients are severely compromised in their ability to engage in social interactions or adapt to even minute changes in their environment; such normative experiences as staying in a hotel or meeting an unfamiliar person may be traumatizing for them. On the other hand, an extreme environmental stressor, such as a concentration camp experience, would make it virtually impossible for anyone to preserve their psychological integrity. The centrality of the psychobiological adaptive fitness perspective means that in the DST definition, *psychopathology is fundamentally tied to the meaning of the person's experience.*

The fact that higher-level psychological and social processes may rest in part on fully deterministic CAS functions does not detract from the possibility of indeterministic behavior on the higher emergent levels of organization; such transitions are common in physical hierarchies, such as fully deterministic laws of statistical mechanics arising from fundamentally indeterministic quantum processes (Musser 2015). Conversely, relying on quantum processes as a source of free will (Hameroff and Penrose 1996) commits a category error; even if the postulated quantum interactions do underpin conscious choice, such a process would only free the quantum-sensitive structures in one's brain to behave indeterministically, rather than allowing for free intentional human agents. By contrast, the DST model forces us to consider an emergent hierarchy of processes contributing to human behavior; we begin to see our patients not as passive victims of their psychopathology but as active intentional agents with the power to choose their future life trajectory and re-mold the topography of their A-landscapes. The primary question in therapy becomes: *Why do we do what we do* (consciously or unconsciously)?

The property of self-organization in CAS systems (Kaufman 2006) brings to the forefront the fact that both evolutionary and individual adaptations are driven by the interplay between environmental factors external to the CAS in question and the system's internal dynamics. From the DST perspective, we have to shift away from linear models of patient–provider interaction, where the psychiatrist's function is to analyze and 'fix' the patient's objective pathology with an armamentarium of biological or psychotherapeutic tools. The therapeutic change in DST is re-defined as the development of *healthier adaptations*, which do not come from the psychiatrist, the medications they prescribe, or the specific theories or techniques they utilize. Instead, we engage with the patient's intra-psychic and relational dynamics by bringing our own A-landscape to bear on theirs, the new psychobiological environment acting as a supervenient dyadic system that catalyzes changes in the patient's A-landscape. In effect, it is the patient who shapes the therapeutic process and re-adapts to the new dyadic environment, generalizing the new adaptive algorithms to their outside interactions. In keeping with the interpersonal neurobiology model (Siegel 2006), the DST approach reinforces the fact that we do not fix our patients, we only provide the optimal environment for their growth.

There are two categories of change in Complex Adaptive Systems: incremental *first-order adaptations* are characterized by gradual, quantitative changes, the fine-tuning of a generally successful adaptive strategy. This process occurs in a linear mode, which parallels the classical Darwinian view of gradual evolutionary adaptations. On the other hand, *second-order adaptations* represent nonlinear changes described as *bifurcations*, where the entire system abruptly transitions into a different attractor/repellor configuration, sometimes with a minimal change in one of its parameters.

Bifurcations become more frequent in far-from-equilibrium conditions; in fact, chaotic dynamics is precipitated by increased frequency of bifurcations, which destabilize the build-up of systemic complexity (Mitchell 2009). Even though the power laws describing chaotic dynamics are fully deterministic, the timing and direction of bifurcations are unpredictable in principle and parallel the punctuated equilibrium evolutionary model (Gould and Eldredge 1993), where periods of linear adaptations alternate with periods of abrupt evolutionary transitions. Many qualitative transitions in human knowledge, such as inventions and flashes of creativity can be linked to bifurcation dynamics (Skar 2004). Bifurcation points also have important implications for the treatment process; successful psychiatric outcome depends on building more adaptive patterns by re-organizing A-landscape topography in far-from-equilibrium conditions, when the patient's adaptive capacity is by definition inadequate to his or her current internal/environmental demands. Working in the far-from-equilibrium zone means that we have to incorporate not only gradual type I changes but also abrupt system bifurcations, such as 'lightbulb moments' in psychotherapy that may lead to qualitative change in the patient's self-awareness and relational dynamics.

26.3.2   From Categorical Diagnoses and Generic Treatment Algorithms to Modeling the Meaning of the Patient's Experience

The dynamical systems approach outlined above strongly suggests the inadequacy of linear reductionistic models, whether biological or psychological, in dealing with the evolving complexity of the brain/mind system. In addition, static diagnostic categories and treatment algorithms fail to capture the subjective meaning of individual psychopathology, from its ontological misadaptations to the complex dynamics of the here-and-now therapeutic interaction. We do not treat disembodied DSM or ICD diagnoses but the people who have them. It is a well-known clinical fact that the same patient may present with different symptoms at different phases of the treatment, and respond differently to different treatment providers. Each patient–therapist pair has unique dynamics that may highlight different aspects of the patient's presentation at different times.

The DST model forces clinicians to move from fixed symptom-based syndromes to conceptualizing psychiatric formulation as a *process of knowing* our patients that grows out of the deepening therapeutic relationship. The objective third-person perspective of psychopathology is only part of the clinical picture, which also includes the patient's and the treatment provider's subjective systems of meaning (the first-person perspective), and their intersubjective interaction in the treatment setting (the second-person perspective). In this view, psychopathology and its treatment occur at the intersection of objective, subjective and intersubjective science, and have to be addressed as a *component-process function*. Maladaptive symptoms, thoughts and behaviors emerge out of the ongoing interaction of the patient's internal dynamics (whether synaptic networks or habitual ways of organizing experience) with his or her external environment. Consequently, symptoms do not remain constant over time but change their configurations in the course of the treatment, not unlike melting icebergs in the ocean current. It is the implicit 'underwater topography' that directs the ocean currents; in addition, continents and islands, just as symptom clusters, are not fixed and separate structures as they may appear to us. Rather, they are visible projections of a unified whole – mountains and plateaus on the Earth crust projecting above the sea level. Just as the continents gradually shift over geological times forming new geographical configurations, so the A-landscape topography gradually re-organizes over the course of one's life and psychiatric treatment.

In DST terms, psychiatric treatment is defined by the interaction between the person of the treatment provider (CAS-1) and the person of the patient (CAS-2), their relational dynamics defining a supervenient dyadic system (CAS-3) that catalyzes therapeutic change. We can visualize two A-landscapes coming into contact and molding each other in the process (Figure 26.3). The therapist-system (CAS-1, left) is seen to be more flexible and coherent, performing a stabilizing and re-organizing function for the patient's system (CAS-2, right), which is seen as more rigid or disorganized. The shared intersubjective

Figure 26.3   A graphic representation of interpenetrating therapist–patient A-landscapes

system (CAS-3, middle) defines the dynamics of the therapeutic alliance, which is unique for each patient–therapist pair and acts as a catalyst for therapeutic change. In effect, the therapist's and the patient's systems interpenetrate and co-evolve together in the course of the treatment within the emergent dancing landscape of the intersubjective space. This process re-shapes the topography of both individual A-landscapes, affecting the patient and the therapist asymmetrically, in keeping with many experienced therapists' wisdom that we learn from our patients just as they learn from us (Karasu 1992). In these terms, any form of psychiatric intervention, including the process of prescribing, is fundamentally a psychobiological transaction tied both to the meaning of the patient's experiences and to their synaptic network dynamics.

26.3.3   The Fractal Dynamics of the Treatment Process: Changing A-landscape Topography to Re-establish Adaptive Capacity

In contrast to categorical diagnostic classifications and symptom-reduction treatment algorithms, the DST model focuses on emergent adaptive function, both at the level of neural network activity and the patient's life experience (Post and Weiss 1997). Symptoms, in this view, are manifestations of underlying attractor dynamics; they reflect pathological attractor-bound trajectories in the same way that turbulent river rapids are influenced by rock formations below the surface. The turbulence model is a particularly apt metaphor for psychiatry because it is a nonlinear process that cannot be reduced to static environmental components or water molecule interactions; it has to incorporate the fractal dynamics of water flow over the underlying landscape.

Table 26.1 DST case formulation algorithm

Part I.	**Identifying attractor/repellor configurations**
	• Problems or symptoms as part of a pattern that proposed treatment will attempt to address
Part II.	**Identifying attractor progression (adaptive psychobiology)**
	• Nature factors: biological vulnerabilities, temperamental characteristics, family history
	• Nurture factors: early dyadic interactions, family functioning, extended relational experiences, developmental trauma/neglect, cultural considerations
	• Core misadaptations: mapping A-landscape (how did the patient end up where (s)he is?)
Part III.	**Identifying bifurcation points (core pattern re-enactment)**
	• Core dysfunctional attractor/repellor – its origin and manifestations (relate to presenting problems)
	• The patient's choices that serve to perpetuate the cycle (what does the patient contribute to the impasse, consciously or unconsciously?)
	• What drives the choices (unresolved feelings/needs)?
	• What meaning does the patient attribute to her choices (why we do what we do?)
Part IV.	**Proposed ways of modifying A-landscape topology**
	• What treatment does the patient need?
	• Psychotherapy – what kind?
	• Psychopharmacology – what kind?
	• Behavior modification/skills training – what areas?
	• Key strengths and weaknesses: insight, ego-strength, motivation, quality of object relatedness, defensive organization
	• What is the patient willing to change at this time?
	• Anticipated problems in treatment: how is the postulated attractor pattern expected to affect here-and-now therapeutic interactions?

The equivalent of turbulence in psychopathology is the patient's ego-dystonic (distressing) patterns of feeling, thinking and relating to others, which are often re-enacted in the treatment setting. Patients present for treatment because they are dissatisfied with the way they feel about themselves or their ability to relate to the outside world. These patterns represent a stochastic dynamical function triggered by diverse events and sensitive to initial conditions, both in the course of individual development and in therapy interactions. The trajectory of such a therapeutic system is unpredictable from the onset; however, it also follows a fractal pattern of self-similarity, where the patient actively re-enacts developmentally significant scenarios from his or her past, gravitating towards maladaptive ways of relating or experiencing here-and-now interactions in the light of affectively-laden experiences from the past, the process referred to as therapeutic *transference*. The DST-informed formulation (Table 26.1) focuses on identifying these recurrent maladaptive patterns (attractor configurations); charting their course across the patient's life span; and bringing them to conscious awareness, which allows for different choices in the present. The primary question in treatment becomes: "What can we do differently – here, now, together?"

CASE VIGNETTE 1: MAPPING A-LANDSCAPE TOPOGRAPHY

Thirty-two-year-old amateur musician with a long-standing pattern of brief and unsatisfying relationships, occupational instability, and frequent moves around the country. Comes from an unstable and traumatic family background. Referred for intensive group-based day hospital treatment. Initially presents as psychologically minded and motivated to engage in therapy; however, re-enacts a crisis in her second week, feeling criticized by her primary therapist and ostracized by the group members. Acknowledges thoughts of leaving the program. Group discussion ensues.

T: *"Is this a familiar place for you?"*
Pt: *[Silent momentarily] "Now that you mention it, I feel this way quite often. I have a sense I've been given a score to play, and I keep repeating the same notes..." [Breaks down crying] "I try to vary the tempo or play them differently sometimes, but it always ends the same way."*
T: *"Do you think we can play it differently here? Can stay in the group and look at what is driving the cycle for you?"*
Pt: *[Making eye contact with tearful eyes] "I think I can."*

Reconceptualizing psychopathology as an *emergent process* rather than static diagnostic categories has a major impact on the treatment process, whether psychopharmacological or psychotherapeutic. The patient's problems are not seen to be 'out there' (whether in the outside world or in her head) but in the ongoing diathesis between established misadaptations and reactions to the here-and-now therapy environment. The patient–therapist interaction represents a fractal stochastic system sensitive to initial conditions: we know that a similar pattern is likely to be re-enacted in the treatment setting but the exact triggers, trajectory and the final outcome cannot be predicted from the treatment outset. The stochastic function incorporates the patient's conscious choices in changing the established trajectory of her life, bringing to the forefront *intentionality* as an agent of therapeutic change (Owen 2009).

In introducing a new psychobiological environment (both relational and psychopharmacological), we begin to explore and stabilize the patient's A-landscape, bringing into awareness her maladaptive attractors (pre-emptive 'reject before you get rejected') and repellors ('intimacy is a threat'). In doing so, we enable a choice to face her traumatic feelings and allow meaningful connections with others, gradually modifying the pre-treatment A-landscape topography to make dysfunctional attractors (running away) less of a pull and repellors (healthy intimacy) more accessible. In the words of Amini et al. (1996, p.234), "The therapist's job is to allow the duet to begin and to take up his/her place in the melody, so that the piece can gradually be directed to a different ending."

CASE VIGNETTE 2: CHANGING RIGID A-LANDSCAPE TOPOGRAPHY: PATHOLOGICAL LANDSCAPES PL-TYPE I

Thirty-six-year-old mother of two who presents with a history of ongoing physical, sexual and emotional abuse in her 17-year marriage. A successful professional, she finds herself progressively more depressed and socially phobic to the point of being unable to function outside of her home. Combined psychopharmacological, individual and group psychotherapy treatment is recommended. The patient presents as a motivated and psychologically minded individual who works well in the individual sessions, however consistently stays silent in the group setting. When the therapist interprets the pattern in one of the groups, she panics and runs out of the room. The situation is explored in the next individual session – the patient compares her experience to "being put on a hot seat."

T: *"I understand that speaking in the group feels difficult for you. But do you see the self-fulfilling prophecy you recreate?"*

Pt: *[Looks up with surprise] "What do you mean?"*

T: *"Well, if I leave you sitting silently and focus on others in the group, you are left to struggle on your own and your needs are left unmet – just as they were unmet in your family growing up. So, I become a neglectful parent, like your mother. But if I try to bring attention to you, you experience it as being punished, 'on a hot seat' as you said, so you make me into an abusive figure like your father or husband. It feels like a no-win scenario."*

Pt: *[Breaking down crying] "Why do I do that?"*

T: *[Smiling] "It's an excellent question: can we look for the answer together?"*

In this case, the patient manifests two discrete attractor states, re-enacting both neglectful and abusive experiences in her interpersonal interactions. Even though these states are experienced as ego-dystonic, they are unconsciously recreated in most of her significant relationships, from family of origin dynamics, to her marriage, occupational, social, and therapy settings. By contrast, genuine or caring interactions represent a repellor hill (Figure 26.2a), which does not allow for stable intimacy; even if she is able to climb it momentarily (the ball on the top), she enters a state of an unstable equilibrium that will inevitably decompensate back into one of the attractor basins.

One way of conceptualizing therapeutic process is building a new template of 'being self-with-other' (Lyons-Ruth and BCPSG 2001) in here-and-now individual/group interactions, which allows for a more stable intimacy equilibrium and gradually flattens the repellor hill (Figure 26.2b). With repeated interactions, such as working through traumatic feelings and looking for the meaning of the patient's reactions in the treatment process, successful therapy can potentially reverse the attractor/repellor polarity, making intimate interactions feel more rewarding and nurturing than abusive/neglectful ones (Figure 26.2c). The new attractor of *earned secure attachment* (Roisman et al. 2002) can then be generalized from the therapy setting to other relationships in her outside life.

The qualitative shifts in attractor/repellor shape and polarity represent a cardinal difference between in-depth therapies as opposed to either symptomatic pharmacological management or brief solution-based treatment, where a dysfunctional attractor state is simply 'fenced off' (such as in using an antidepressant to stabilize the patient's depressive presentation without more in-depth trauma work, or telling her to leave her abusive husband), while leaving the underlying A-landscape topography intact (Figure 26.1-3). This difference may be reflective of the findings that patients successfully completing psychodynamic psychotherapy continue to improve following treatment discontinuation, which is not the case for psychopharmacological or shorter-term cognitive interventions (Maina, Rosso and Bogetto 2009).

The DST perspective would argue against the model of the therapist as a detached 'objective observer' but rather a genuine partner in the process of the patient's self-discovery. We co-determine the therapeutic trajectory by participating in therapeutic alliance punctuated with micro-ruptures and repairs. In a conceptual similarity to quantum systems, alternate paths and possibilities co-exist and can be explored within the intersubjective space, each patient–therapist interaction serving as a potential bifurcation that may serve as a tipping point altering the configuration of the patient's A-landscape (Bak, 1996).

CASE VIGNETTE 3: UTILIZING BIFURCATIONS

Fifty-five-year-old male engineer with a history of physical and sexual abuse, and emotional neglect throughout his childhood and adolescence. He has internalized, and lives as his own identity, the sum of these assaults, blaming, shaming, and degrading stories told by those around him, translating them into his internal object relationships. Similar conflicts have been re-enacted in his professional career in spite of his success as a head of his architectural firm, and in his marriage. The patient is in individual psychotherapy for chronic panic attacks and possible underlying post-traumatic stress disorder, describes a recent crisis that led to an exacerbation of his suicidal ideation.

Pt: "When I realized how hopeless my situation is and how little control of anything I have, I crashed into the most profound depressed and panicked emotional state."

T: "Oh my, what do you think triggered it?"

Pt: "I wrote it out as an equation last night. The sum of all expectations is less than or equal to 0: $\sigma e \leq 0$."

T: "You really sink into a black place, don't you?" [Short silence follows in which patient and thera- pist sit quietly with the shared feel] "It's odd what just came to my mind. I'm not quite sure what to do with it. The sum of all expectations is less than or equal to 0 feels so negative and empty, a very tragic place, but what came to my mind is how close it is to a mindfulness exercise or a Buddhist teaching on letting go of expectations so you can live in the moment."

The shift in perspective evident in this particular 'alternate path' resembles a well-known 'two faces versus candlestick' illusion; we cannot perceive both aspects of the picture simultaneously but we can hold them together in awareness and learn to shift back and forth between them. In this case, the alternative perceptions and associated action tendencies hold vastly different meanings, which spell the shift between suicidal despair and agency towards change, utilizing the present moment of authentic connection as an opportunity to change the trauma-bound trajectory of the patient's life. The therapeutic push is towards a more adaptive dancing landscape, where therapist and patient influence each other, introduce new links and break old ones, explore a diversity of associations and emotional states, and enhance more functional adaptations and learning.

CASE VIGNETTE 4: THE DANCING LANDSCAPE

Nine-year-old boy with a history of incapacitating obsessive-compulsive disorder, the extensive rituals preventing him from attending school or participating in any social activities. When a family member would try to intervene, the patient would fly into a rage, almost autistic in intensity. Yet, the patient is a bright boy, in sufficient distress from his symptoms to come for individual treatment.

T: "You know I was reading about people's hearts in a medical journal. . . Can you feel your heart beating in your chest? Is it really a regular beat?

Pt: "Yes, I can feel it."

T: "Apparently a healthy heart is quite regular if you feel the pulse but if you look even closer at what it is doing, it varies a lot within that regular beat. It varies and changes with every breath, thought, action change in your body or the room around you. The variation in that regular beat is what is really healthy."

Pt: "I suppose you are now going to tell me that has something to do with my mind and my obses- sions and compulsions?"

T: "Yup, I'm predictable aren't I? Let's do a little game together. Let's walk around the room like this, in a circle and keep going round and round making sure we use exactly the same path .... Now, let's do the same thing, staying in the room, but now let's make sure we never use the same path if we can help it." [The patient following along] "Can you see you can be anxious about using the same path all the time, or you can be anxious about trying to never use the same path? That's weird isn't it?"

Pt: "Yes, the anxiety kind of moves around with what we are doing . . . it's kind of fun!"

In the course of the therapeutic interaction, the patient and therapist not only utilize the existing A-landscape topography but create new attractor channels where none had existed before. This patient's rigid, OCD-bound landscape, unaffected by conventional cognitive-behavioral treatment, forces the therapist to improvise a new mode of interaction, qualitatively different from either the patient's or the therapist's past experiences. In introducing the *play space* (Winnicott 1971; Marks-Tarlow 2011), the boy's compulsive anxiety ceases to be the unconquerable 'disease' controlling his behavior, and becomes a quirky subjective experience that can be shifted at will like the faces–candlestick illusion, putting the patient firmly in control of his choices. In doing so, the therapist has to leave the safety of his chair and engage with the patient at his own level, 'dancing' into the waiting room in full view of his staff and other patients. The two landscapes co-evolve together, leaving the mutual rigidity of the established attractor channels and re-organizing in a more adaptive configuration.

CASE VIGNETTE 5: RE-ESTABLISHING THE SELF-ORGANIZING CAPACITY

Fifty-one-year-old professional chef coming from a background of emotional neglect in the family of origin. He experienced severe emotional trauma at the age of 18, when he hit another vehicle and an 8-year-old girl died, for which he has blamed himself ever since. Involved with drugs intermittently since his late teens, he has been locked in an emotionally abusive marriage for 26 years. The patient attended a weekly psychodynamic group for three years followed by a 20-session individual contract. Able to make major life changes in the course of the treatment such as setting healthier limits with his wife; forming a closer relationship with his aging father; engaging in more rewarding work interactions; and letting go of the guilt and self-punishing behaviors about the MVA. Presents for his last psychotherapy session.

T: *[smiling] "Well, how does it feel to be here for the last time?"*
Pt: *"You know, I thought I'd be panicking when I can't come here anymore. But I dealt with that [problem employee] at work and it went OK. And I've been talking to [his estranged wife] and it doesn't feel like I have to please her now. . . You know, there are all these plants at home, and plants were always her thing, and I'd forget to water them all the time or put too much water in. And now I just walk into the house, and I can see right away how much water they need . . . It feels a bit sad to leave, but I'm also excited – I know things won't always be great, but I think I can deal with that now . . ."*

The goals of psychiatric treatment go beyond symptom reduction or even remission. Two very complex co-evolving human psyches join in an intersubjective exploration, traversing along an uncertain and only partly definable terrain. The mutually engaged participants explore their shared contexts in search of new emotional and associative meanings in order to facilitate a state of wellness rather than mere absence of disease (Cloninger 2004). In DST terms, wellness is evidenced by re-establishing a flexible A-landscape to achieve a more adaptive homeostasis and inner coherence in the patient's life. The therapist's greatest reward is to be a part of such transformation process.

26.4   CONCLUSION

The conceptual framework incorporating complexity, emergence, nonlinear science and evolutionary dynamics is poised to revolutionize the process of psychiatric diagnosis and treatment. It serves as a trans-theoretical model which allows us to capture the real-time trajectory of the treatment process, tailoring it to specific patient needs. At the same time, the nonlinear perspective forces us to abandon the seeming certainty of categorical diagnostic classifications and generic treatment algorithms; the deceptive simplicity of linear statistical predictions in treatment response; and the false safety of the disengaged observer perspective. Instead, we enter the patient's experiential raft to explore the currents of his or her psychobiology, and the underlying A-landscape topography that guides them. There are no ready-made prescriptions for emergent reality: any answers emerge in the present from the patterns of relational improvisation and the participants' inner dynamics rather than external theory-driven solutions. We have to tolerate the uncertainty of nonlinear trajectories and rely on our knowledge and experience to strengthen the raft and guide it through turbulent waters while re-molding the patient's pathological attractor/repellor configurations.

The DST model extends the traditional biopsychosocial orientation in several important respects. First, the DST psychobiological perspective allows us to bridge the persistent conceptual gap between biological and psychosocial interventions by mapping both objective neurophysiological/behavioral or subjective meaning data on the A-landscape, which is equally applicable to pharmacotherapeutic, psychotherapeutic, and behavior modification interventions. It provides a visually accessible map that can guide individualized treatment interventions rather than focus on disembodied clusters of symptoms often defined without reference to the patient's unique experience.

Second, the DST-informed formulation integrates the intersubjective dimension into the treatment process, which makes the psychiatrist or psychotherapist an integral part of the dyadic treatment system rather than an illusory 'impartial expert-observer' of the patient's psychopathology. It captures the *process of treatment*, which has been largely ignored in the conventional reductive medical models and outcome research.

Third, the DST approach puts the emphasis of psychiatric intervention on the patient's adaptive capacity within his or her environment, rather than on their external symptomatic presentation. It treats patients as active intentional agents driving the process of therapeutic change, who consciously or unconsciously choose their future life trajectory rather than passively respond to the treatments administered to them.

Finally, the 'map of the mind' that the DST paradigm allows us to construct serves as an invaluable training tool to teach integrated, individualized treatment to trainees of all orientations, from psychiatric residents to psychotherapists, psychopharmacologists, clinical psychologists and social workers.

Perhaps the most far-reaching implication of the systemic psychobiology framework as outlined above is that it may help to clarify the limits of the reductive scientific method itself, and assist in constructing a complementary reductive-holistic paradigm, which can pave the way to integrating consciousness and intentionality into a comprehensive view of natural science and evolution of complexity in the Universe.

BIBLIOGRAPHY

Amini, F., Lewis, T. et.al. (1996). Affect, attachment, memory: contributions toward psychobiologic integration. *Psychiatry, 59*, 213–239.

Arnold, S.J., Pfrender, M.E., and Jones, A.G. (2001). The adaptive landscape as a conceptual bridge between micro-and macroevolution. *Genetica, 112*(1), 9–32.

Atmanspacher, H., Weidennman, G., and Atmann, A. (1995). Descartes revisited. *Complexity, 1*(3), 15–21.

Bak, P. (1996). *How the Nature Works: The Science of Self-organized Criticality.* New York: Springer Verlag.

Beauregard, M. (2007). Mind does really matter: evidence from neuroimaging studies of emotional self-regulation, psychotherapy, and placebo effect. *Progress in Neurobiology, 81*, 218–236.

Bowlby, J. (1984). Psychoanalysis as a natural science. *Psychoanalytic Psychology, 1*(1), 7–21.

Brigandt, I. and Love, A. (2015). Reductionism in biology. *The Stanford Encyclopedia of Philosophy*, Edward N. Zalta (ed.), http://plato.stanford.edu/archives/fall2015/entries/reduction-biology/ (accessed August 17, 2017).

Chalmers, D. (1995). Facing up to the hard problem of consciousness. *Journal of Consciousness Studies, 2*, 200–219.

Cloninger, C.R. (2004). *Feeling Good: The Science of Well-being.* New York: Oxford University Press.

Dennett, Daniel C. (1995). *Darwin's Dangerous Idea: Evolution and the Meanings of Life.* New York, London, and Toronto: Simon & Schuster.

Edelman, G. (2004). *Wider than the Sky: The Phenomenal Gift of Consciousness.* New Haven and London: Yale University Press.

Gabbard, G. (1994). Mind and brain in psychiatric treatment. *Bulletin of the Menninger Clinic, 58*(4), 427–446.

Gell-Mann, M. (1994). Complex adaptive systems. In: *Complexity: Metaphors, Models, and Reality*, Cowan, G. et al. (eds), SFI studies in the sciences of complexity, XIX, 17–45.

Gould, S.J. and Eldredge, N. (1993). Punctuated equilibrium comes of age. *Nature, 366*, 223–227.

Hameroff, S. and Penrose, R. (1996). Orchestrated reduction of quantum coherence in brain microtubules: a model for consciousness. *Mathematics and Computers in Simulation, 40*(3), 453–480.

Kandel, E. (1998). A new intellectual framework for psychiatry. *American Journal of Psychiatry, 155*, 457–469.

Karasu, T.B. (1992). *Wisdom in the Practice of Psychotherapy.* New York: Basic Books

Kauffman, S. and Clayton, P. (2006). On emergence, agency, and organization. *Biology and Philosophy, 21*, 501–521.

Kuhn, T. ([1962]1996). *The Structure of Scientific Revolutions* (3rd edition). Chicago and London: University of Chicago Press.

Laland, K.N., Uller, T., Feldman, M.W., Sterelny, K., Müller, G.B., Moczek, A., Jablonka, E, and Odling-Smee, J. (2015). The extended evolutionary synthesis: its structure, assumptions and predictions. *Proceeding of the Royal Society of London B, 282* (August) (1813), 1–14.

Libet, B. (1999). Do we have free will? *Journal of Consciousness Studies, 6*(8–9), 47–57.

Lorenz, E. (1993). *The Essence of Chaos.* Seattle: University of Washington Press.

Lowenstein, W.R. (2013). *Physics in Mind: A Quantum View of the Brain.* New York: Basic Books.

Lyons-Ruth, K, and BCPSG. (2001). The emergence of new experiences: relational improvisation, recognition process, and non-linear change in psychoanalytic theory. *Psychologist-Psychoanalyst, 21*, 13–17.

Maina, G., Rosso, G., and Bogetto, F. (2009). Brief dynamic therapy combined with pharmacotherapy in the treatment of major depressive disorder: long-term results. *Journal of Affective Disorders, 114*(1), 200–207.

Marks-Tarlow, T. (2011). Merging and emerging: a nonlinear portrait of intersubjectivity during psychotherapy. *Psychoanalytic Dialogues, 21*(1), 110–127.

Mayberg, H.S. (2003). Modulating dysfunctional limbic-cortical circuits in depression: towards development of brain-based algorithms for diagnosis and optimised treatment. *British Medical Bulletin, 65*, 193–207.

Mitchell, M. (2009). *Complexity: A Guided Tour.* Oxford, New York: Oxford University Press.

Musser, G. (2015). Is the Cosmos random? *Scientific American, 313*(3), 88–93.

Nagel, T. (2012). *Mind and Cosmos: Why the Materialist Neo-Darwinian Conception of Nature is Almost Certainly False.* New York: Oxford University Press.

Nesse, R. (1998). Emotional disorders in evolutionary perspective. *British Journal of Medical Psychology, 71*, 397–415.

Noble, D. (2006). *The Music of Life.* Oxford, New York: Oxford University Press.

Owen, I.R. (2009). *Talk, Action and Belief: How the Intentionality Model Combines Attachment-Oriented Psychodynamic Therapy and Cognitive Behavioral Therapy.* New York, Bloomington: iUniverse.

Page, S.E. (2011). *Diversity and Complexity.* Princeton and Oxford: Princeton University Press.

Post, R. and Weiss, S. (1997). Emergent properties of neural systems: how focal molecular neurobiological alterations can affect behavior. *Development Psychopathology, 9*, 1–23.

Pylkkanen, P. (2007). *Mind, Matter and the Implicate Order.* Heidelberg: Springer.

Rochat, P. (2009). *Others in Mind: Social Origins of Self-consciousness.* New York: Cambridge University Press.

Roisman, G.I., Padrón, E., Sroufe, L.A., and Egeland, B. (2002). Earned-secure attachment status in retrospect and prospect. *Child Development, 73*(4), 1204–1219.

Roth, A. and Fonagy, P. (2006). *What Works for Whom? A Critical Review Of Psychotherapy Research* (2nd edition). New York and London: The Guilford Press.

Seager, W. (2012). *Natural Fabrications: Science, Emergence and Consciousness.* Heidelberg and New York: Springer Verlag.

Shapiro, Y. (2015). Dynamical Systems Therapy (DST): Theory and practical applications. *Psychoanalytic Dialogues, 25*(1), 83–107.

Siegel, D. (2006). An interpersonal neurobiology approach to psychotherapy: awareness, mirror neurons, and neural plasticity in the development of well-being. *Psychiatric Annals, 36*(4), 248–256.

Skar, P. (2004). Chaos and self-organization: emergent patterns at critical life transitions. *Journal of Analytical Psychology, 49*, 243–262.

Slavin, M. and Kriegman, D. (1992). *The Adaptive Design of the Human Psyche: Psychoanalysis, Evolutionary Biology, and the Therapeutic Process*. New York/London: The Guilford Press.

Stapp, H.P. (2009). *Mind, Matter and Quantum Mechanics* (3rd edition). Heidelberg: Springer.

Stern, D. (1985). *The Interpersonal World of the Infant: A View From Psychoanalysis and Developmental Psychology*. New York: Basic Books.

Stolorow, R. (1997). Dynamic, dyadic, intersubjective systems: an evolving paradigm for psychoanalysis. *Psychoanalytic Psychology, 14*(3), 337–346.

Tooby, J. and Cosmides, L. (1990). The past explains the present: emotional adaptations and the structure of ancestral environments. *Ethology Sociobiology, 11*, 375–424.

Velmans, M. (2009). *Understanding Consciousness* (2nd edition). London and New York: Routledge.

Von Bayer, H.C. (2003). *Information: The New Language of Science*. Cambridge, MA: Harvard University Press.

Winnicott, D.W. (1971). *Play and Reality*. London: Tavistock.

Index